CANADIAN BULLETIN OF FISHERIES AND AQUATIC SCIENCES 215

Canadian Aquatic Resources

EDITED BY

M. C. HEALEY

Department of Fisheries and Ocean
Fisheries Research Branch
Pacific Biological Station
Nanaimo, B.C. V9R 5K6

AND R. R. WALLACE

Dominion Ecological Consulting Ltd.,
No. 200, 601-17th Avenue S.W.
Calgary, Alta. T2S 0B3

THE RAWSON ACADEMY OF AQUATIC SCIENCE

L'ACADÉMIE RAWSON DES SCIENCES DE L'EAU

DEPARTMENT OF FISHERIES AND OCEANS
Ottawa 1987

The *Canadian Bulletins of Fisheries and Aquatic Sciences* are designed to interpret current knowledge in scientific fields pertinent to Canadian fisheries and aquatic environments.

The *Canadian Journal of Fisheries and Aquatic Sciences* is published in annual volumes of monthly issues. *Canadian Special Publications of Fisheries and Aquatic Sciences* are issued periodically. These series are available from authorized bookstore agents and other bookstores, or you may send your prepaid order to the Canadian Government Publishing Centre, Supply and Services Canada, Ottawa, Ont. K1A 0S9. Make cheques or money orders payable in Canadian funds to the Receiver General for Canada.

Information and Publications Branch

Johanna M. Reinhart, M.Sc.   Director and Editor
Peter M. Burke   Editorial and Publishing Services

Editorial Office: Department of Fisheries and Oceans
Communications Directorate
Information and Publications Branch
200 Kent Street
Ottawa, Ontario, Canada K1A 0E6

Typesetter: Paul Aubut & Associates Ltd., Ottawa, Ontario
Printer: Kromar Printing Ltd., Winnipeg, Manitoba
Cover Design: Diane Dufour

Correct citation for this publication:

HEALEY, M. C., AND R. R. WALLACE [ED.]. 1987. Canadian aquatic resources. Can. Bull. Fish. Aquat. Sci. 215: 533 p.

Contents

Contents (concluded)

Preface

This book represents the final stage of a three-part project on the status of Canadian aquatic resources undertaken by the Rawson Academy of Aquatic Science. The project was stimulated in 1983 by two significant events. First, Canadians were in the process of reevaluating the established machinery of economic prosperity through the MacDonald Royal Commission on the Economic Union and Development Prospects for Canada. Since aquatic resources make a significant contribution to the dollar economy of Canada, the findings of the MacDonald Commission were likely to have important consequences for aquatic resource management. Second, Environment Canada was about to launch an independent inquiry, headed by Dr. Peter Pearse, into Federal water policy and the institutions of aquatic resource management. Over the past decade a number of large scale aquatic resource issues had emerged (such as acid rain, toxic contamination of the Niagara and St. Clair rivers, salinization of prairie farm land) that were intractable and exceeded the scope of existing institutions. Canadians were having to make hard choices about environmental, economic, and public health trade-offs that were poorly understood. The time seemed ripe for a review of the status of Canadian aquatic resources and the multifaceted problem of their management.

Consequently, in 1984 the Rawson Academy of Aquatic Science initiated an investigation of the status and management of Canadian aquatic resources. Financial support for this investigation was obtained from the Donner Canadian Foundation, Suncor Inc., The Royal Canadian Geographical Society, Esso Resources Canada Ltd., and Procter and Gamble Inc. We gratefully acknowledge the support and encouragement of these organizations.

The Academy's project had three main thrusts. The first was to commission reviews by Canadian experts on a wide range of issues related to aquatic resource management that would form the basis for a book on Canadian aquatic resources. This volume is the culmination of those reviews. The second was to host a national symposium on Canadian aquatic resources that would bring together a broad mix of professionals concerned with aquatic resource management and stimulate exchange of information among their professions. The symposium was held in Toronto, Ontario, in May 1985. Papers selected from those given at the national symposium have been published by the Canadian Water Resources Association in Volume 11, Nos. 1 and 2, of the *Canadian Water Resources Journal* (1986). The third was to stage a series of touring lectures on important issues affecting Canadian aquatic resources. The lectures, which dealt with issues ranging from acid rain to the legal and constitutional basis of aquatic resource jurisdiction, were delivered at numerous Canadian centres ranging from St. John's, Newfoundland, to Victoria, British Columbia, and from Calgary, Alberta, to Whitehorse in the Yukon.

The Rawson Academy of Aquatic Science is a nonprofit Canadian organization named in honour of the late Donald S. Rawson (1905-1961). Rawson was a pioneer in Canadian aquatic science and in the ecosystemic and comparative ecological approaches to aquatic resource management. The objectives of the Rawson Academy include the study of scientific, social, economic, and legal matters that relate to aquatic resource management and the dissemination of information about Canadian aquatic resources to policy makers and to public and private concerns interested in aquatic resources. An important assumption underlying the work of the Academy is that no single profession has a monopoly on the information needed for wise management of human uses of aquatic resources. Consequently, the Academy's approach to aquatic resource issues is both multidisciplinary and transdisciplinary. This approach is reflected in the broad range of topics dealt with in this book. Indeed, it is our hope that this work will stimulate and enhance what we feel is an essential requirement for effective management of aquatic resources in Canada — the working together of many disciplinary professionals to solve complex problems of aquatic resource management.

The book is aimed at senior students of aquatic resources management and policy makers. Although many of the chapters deal with highly technical issues, we have taken care to minimize specialized jargon and to ensure that the authors wrote in a way that would be understandable to professionals in other fields and to informed laypersons. Each chapter is an overview of the topic. Detailed bibliographies appended to each chapter, however, will provide the serious student with source material for more thorough study of any topic, and policy makers with access to a burgeoning literature.

Numerous individuals contributed time and energy to the development of this book. The board of directors of the Rawson Academy acted as an overall steering committee for the project. An editorial board provided valuable assistance in the early stages of planning and review of various outlines for the book's contents. Two members of the editorial board in particular, Fred Ward and Angelo Grima, contributed long hours to the review of preliminary chapter outlines and of some chapter manuscripts. Mrs. Angela Lynch provided valuable service in her role as coordinator of the national symposium. We appreciate the assistance provided by these dedicated individuals.

All of the chapters were refereed by two or more of the authors' peers. These professionals provided us with an essential check on the accuracy, completeness, and objectivity of the authors. This was a role that could not be reliably performed by ourselves as editors because so many of the chapters explored topics that were outside our own areas of expertise. With gratitude we append the names and affiliations of the referees to this preface.

The Department of Fisheries and Oceans through the Information and Publications Branch (IPB), Communications Directorate, have assisted in many ways with the publication of this volume. We are particularly indebted to the production staff of IPB for their assistance with copy editing and with all the details of production essential to a volume such as this. Having the Department of Fisheries and Oceans as publisher also makes it possible for this book to be offered at a modest cost. The Academy's desire that information of this sort should be readily available to every interested person is, thereby, fulfilled.

The Rawson Academy endorses the general contents of this book as being consistent with its objective of encouraging open and informed discussion of aquatic resource issues. Individual authors and their supporting agencies are, however, responsible for the specific contents of their chapters. Inclusion of a chapter by an author associated with a particular agency does not imply that that agency endorses any other part of the contents of the book. Nor should any of the granting agencies be held responsible for any part of the publication.

M. C. HEALEY AND R. R. WALLACE

Referees

R. BARTLETT, Professor of Law, University of Saskatchewan, Saskatoon, Sask.

R. M. BAXTER, Environment Canada, Canada Centre for Inland Waters, Burlington, Ont.

J. U. BAYLY, Barrister and Solicitor, Yellowknife, N.W.T.

N. G. BENSON, 1330 Northface Ct., Colorado Springs, CO, USA

F. BERKES, Associate Professor of Urban and Environmental Studies Brock University, St. Catherines, Ont.

R. B. B. DICKISON, Professor of Forestry, University of New Brunswick, Fredericton, N.B.

D. ELTON, Canada West Foundation, Calgary, Alta.

I. D. FOX, Professor Emeritus, University of British Columbia, Vancouver, B.C.

G. GEEN, Dean of Science, Simon Fraser University, Burnaby, B.C.

D. L. GOLDING, Associate Professor of Forestry, University of British Columbia, Vancouver, B.C.

D. M. GRAY, Professor of Agricultural Engineering, University of Saskatchewan, Saskatoon, Sask.

K. HALL, Westwater Research Centre, University of British Columbia, Vancouver, B.C.

R. A. HALLIDAY, Environment Canada, Inland Waters Directorate, Regina, Sask.

D. HARVEY, Environment Canada, Inland Waters Directorate, Ottawa, Ont.

G. HODGSON, Arctic Institute of North America, Calgary, Alta.

R. JANOWICZ, Indian and Northern Affairs Canada, Whitehorse, Y.T.

R. KALLIO, Environment Canada, Inland Waters Directorate, Ottawa, Ont.

S. KERR, Department of Fisheries and Oceans, Bedford Institute of Oceanography, Dartmouth, N.S.

G. D. KOSHINSKY, Department of Fisheries and Oceans, Freshwater Institute, Winnipeg, Man.

G. K. MCCULLOUGH, Department of Fisheries an Oceans, Freshwater Institute, Winnipeg, Man.

R. A. MCGINN, Assistant Professor of Geography, Brandon University, Brandon, Man.

G. A. MCKAY, 122 Brooke St., Thornhill, Ont.

D. MARSHALL, Federal Environmental Assessment and Review, Vancouver, B.C.

H. MUNDIE, Department of Fisheries and Oceans, Nanaimo, B.C.

W. NICHOLAICHUK, Environment Canada, National Institute of Hydrology, Saskatoon, Sask.

D. R. PERCY, Professor of Law, University of Alberta, Edmonton, Alta.

F. QUINN, Environment Canada, Inland Waters Directorate, Ottawa, Ont.

H. A. REGIER, Professor of Zoology, University of Toronto, Toronto, Ont.

R. REID, Bobolink Enterprises, Washago, Ont.

J. E. ROBINSON, Assistant Professor of Man-Environment Studies, University of Waterloo, Waterloo, Ont.

H. ROGERS, Department of Fisheries and Oceans, Fisheries Research Branch, Vancouver, B.C.

P. G. SLY, Glenora Fisheries Station, Picton, Ont.

D. W. STEWART, Agriculture Canada, Land Resources Institute, Ottawa, Ont.

R. H. SWANSON, Northern Forest Research Centre, Edmonton, Alta.

G. B. WARNER, NWT Water Board, Yellowknife, N.W.T.

J. ROSS WIGHT, U.S. Department of Agriculture, Boise, ID, USA.

J. BRAIN WILSON, NWT Water Board, Yellowknife, N.W.T.

MING-KO WOO, Professor of Geography, McMaster University, Hamilton, Ont.

S. C. ZOLTAI, Environment Canada, Canadian Forestry Service, Edmonton, Alta.

Abstract

HEALEY, M. C., AND R. R. WALLACE [ED.]. 1987. Canadian aquatic resources. Can. Bull. Fish. Aquat. Sci. 215: 533 p.

Canada ranks first among nations in terms of the amount of surface water per capita of population and fifth in terms of total surface water discharge. These and other statistics of water supply, together with relatively low total demand for water in Canada, make this nation appear extremely water rich. The appearance of superabundance has led to high per capita demand for water and the treatment of water as a virtually "free" commodity. Yet shortages of supply and inadequacies of quality are emerging in particular locations, and at the same time, pervasive problems, such as acid rain and leaky waste dumps, endanger a significant fraction of Canada's aquatic resources. Many of the emerging problems of supply, demand, water quality, and the integrity of aquatic ecosystems are beyond the capacity of existing institutions and legislation to solve. Issues of aquatic resource management are both multidisciplinary and transdisciplinary so that traditional, single-discipline, compartmental approaches to management are inadequate. New approaches are needed that take better account of the interrelatedness of the components of aquatic ecosystems, that recognize that water is an essential part of terrestrial as well as aquatic ecosystems, and that acknowledge that man is a part of the systems he presumes to manage, not outside them. This book explores issues of Canadian resource management such as these and illustrates the, almost, inexorable way in which problems develop when we take a piecemeal approach to management.

Résumé

HEALEY, M. C. ET R. R. WALLACE [ED.]. 1987. Canadian aquatic resources. Can. Bull. Fish. Aquat. Sci. 215: 533 p.

Le Canada se classe premier au monde sur le plan de la quantité d'eau de surface par habitant et cinquième sur le plan du débit total d'eau de surface. Ces statistiques et d'autres sur l'approvisionnement en eau, jointes à une demande totale relativement faible d'eau au Canada, font paraître ce pays extrêmement riche en eau. La surabondance apparente a conduit à une demande élevée en eau par habitant et au fait que le traitement de l'eau est considéré comme un service pratiquement gratuit. Cependant, des problèmes de pénurie d'approvisionnement et de qualité d'eau inadéquate commencent à se faire sentir à certains endroits et, en même temps, des problèmes envahissants, comme les pluies acides et les dépôts de déchets non étanches compromettent une partie importante des ressources aquatiques du Canada. Une bonne partie des problèmes d'approvisionnement, de demande, de qualité de l'eau qui se font jour et l'intégrité des écosystèmes aquatiques dépassent la capacité qu'ont les institutions et les lois existantes de les solutionner. Les questions touchant la gestion des ressources aquatiques sont multidisciplinaires et transdisciplinaires de sorte que les approches traditionnelles de gestion, compartimentées et mettant en cause une seule discipline, sont inadéquates. De nouvelles approches sont nécessaires qui tiennent mieux compte de l'interrelation qui existe entre les constituants des écosystèmes aquatiques, qui reconnaissent que l'eau constitue une partie essentielle des écosystèmes aussi bien terrestres qu'aquatiques et qui admettent que l'homme fait partie des systèmes qu'il se permet de gérer, plutôt qu'en être exclu. Ce livre examine les questions touchant la gestion des ressources canadiennes comme celles susmentionnées et montre la façon presque inexorable avec laquelle les problèmes surgissent quand nous adoptons une approche de gestion fragmentée.

CHAPTER 1

Canadian Aquatic Resources: An Introduction

M. C. Healey

Department of Fisheries and Oceans, Fisheries Research Branch, Pacific Biological Station, Nanaimo, B.C. V9R 5K6

and R. R. Wallace

Dominion Ecological Consulting Ltd., No. 200, 601-17ᵗʰ Avenue S.W., Calgary Alta. T2S 0B3

Canada possesses abundant aquatic resources, both living and inanimate. This simple statement, which every schoolchild knows, implies a great deal.

First, it implies that by some measures, Canada is richly endowed with water. Canada ranks third among nations in terms of river flow (after Brazil and the USSR), for example, and first in available water per individual in the population (128 000 m^3, twice as great as any other country). Our great rivers, which are among the largest in the world, debouch into three of the world's oceans and were the access routes for and the lifelines of Canada's pioneers. Seven of the world's great lakes are partly or wholly within Canada's borders. These lakes, and countless smaller ones, constitute a great storehouse of accessible freshwater as well as a recreational and aesthetic resource of inestimable value.

Second, it implies an organic richness. An image of wheeling flocks of waterfowl, or of trout dancing on the end of a fisherman's line, is evoked, the kind of abundance that was central to the lives of our hunter-gather ancestors and of pastoral philosophers such as Thoreau (Torrey and Allen 1962).

Third, it implies a measure of responsibility through ownership, for what the country possesses surely belongs individually or collectively to the Canadian people. The key word is "possesses". One implication of this word is that aquatic resources are like market commodities — that rights to them may be bought and sold. But who can truly own the waters and is the concept of property appropriately applied to aquatic resources? Canadians are struggling with this question as the recent Inquiry on Federal Water Policy has shown (Pearse et al. 1985). Political jurisdictions respecting aquatic resources are fragmented and confused. Water moves through both space and time and carries with it the residues of mankind's activities. It thus defies the traditional European concept of property.

Fourth, the statement underscores a smugness and complacency that Canadians have displayed in the past about aquatic resources and still display in regions and communities that have not yet been touched by the emerging problems of water supply and water quality.

These and other images that we believe are implicit in the statement that "Canada possesses abundant aquatic resources" are the subject of this book. Twenty-six authors explore a variety of technical, social, and philosophical issues in the management of human uses of aquatic resources. The list of topics addressed is not comprehensive. Rather it is intended to reflect the scope of aquatic resource management, to emphasize the fact that no single discipline has a monopoly on the information or skills needed for wise management, and to reveal some of the significant challenges that policy makers and managers must confront.

Despite its vital importance to every facet of Canadian society, water is so pervasive a substance in Canada, and the technology of supply is so well developed, that most

Canadians take it for granted. Our right to "pure water" is often assumed to be part of our heritage. In the few places where water has been too scarce or too abundant, from the narrow perspective of certain uses to which we would like to put the land, we have acted to correct the perverseness of nature by augmenting the supply or diverting the flood. In fact, Canada ranks first among nations in terms of the volume of water that has been redirected from its natural course to serve a perceived human need elsewhere. Yet, all of this remodelling of nature's waterways has been done without any certain information of the quantity available, and when one asks "How much water does Canada have?", it turns out that the answer is not a simple one. Translating the perception of an abundant supply into a precise measure requires technical information on such things as stream flow, precipitation, dew fall, lake basin shape, soil water content, aquifer volume, and many other factors that simply is not available for large areas of Canada. Arleigh Laycock addresses this question in Chapter 2. He shows that there has been considereable evolution in estimates of Canada's surface water supply since Cass-Beggs (1961) first estimated the quantity, and suggests that measures of surface water supply will probably continue to evolve for some years to come. Without such knowledge, and equivalent knowledge of subsurface storage and recharge, it is difficult, if not impossible, to make appropriate decisions about allocation of supply. As augmentation of supply becomes less and less attractive for both economic and environmental reasons, the challenge to managers will be not only to muster and allocate local surpluses but also to begin to redefine land use patterns in terms of available supplies.

The difficulty in precisely estimating either the total reserves or the annual supply of water in Canada arises in part because water is not only a pervasive but also a very dynamic substance. At any particular moment there may be water in the ground, lying or running on the surface of the land, contained within the vegetation and fauna, and swirling overhead in the atmosphere. Each of these great reservoirs of water is in intimate connection with the others and with the world's oceans. The continual exchange of water among these reservoirs constitutes what is called the hydrological cycle (Fig. 1). Through this cycle, water vapour in the atmosphere condenses and falls to the earth in various forms as precipitation, and thereby contributes to the surface and soil reservoirs. Water on the surface of the land evaporates back into the atmospheric reservoir, percolates down into the soil reservoir, or runs away to the sea. Water from the soil reservoir evaporates, or is transpired by plants, back into the atmosphere, or discharges into rivers or lakes, or to the ocean. The oceans are by far the greatest reservoir of water, containing more than 97% of the world's supply. The dynamic association between ocean and atmosphere has much to do with our climate, which in turn has much to do with our freshwater supply and its distribution. Since each of the great water reservoirs is in a constant state of flux, one can never be certain of the exact amount of water in any one. Nor can we categorically claim that all of this water is truly Canadian, since, with a few exceptions, we possess it for only a brief period during which it is on its way to somewhere else, either around or, literally, through us. As was pointed out in the final report of the Inquiry on Federal Water Policy (Pearse et al. 1985), the challenge for policy makers and managers in the future is to find ways to satisfy reasonable and legitimate human uses of aquatic resources while maintaining the integrity of the hydrological cycle.

Perhaps more important than the total reserves, or even the total annual supply of water in Canada, is the amount available for various human uses now and in the future, bearing in mind the need, as Pearse et al. (1985) acknowledged, to protect the integrity of the Canadian ecosystem. Canadians make prodigious demands upon their water resources. We are, by and large, wasteful users of water, taking much more than we need, despoiling it, and throwing it back for nature to repair. Already this prodigal approach to water use has led to scarcity in some locations. These are a glimpse of what could be a much more general shortage of supply in the future if current trends

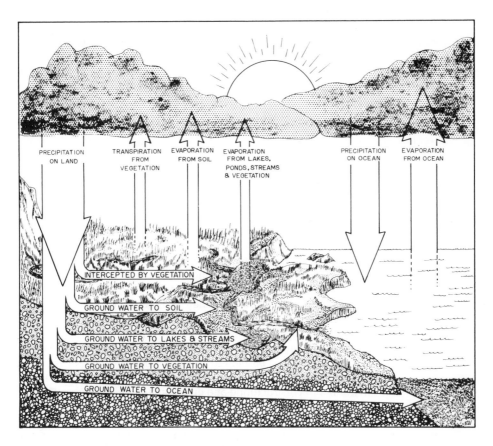

FIG. 1. The hydrological cycle: the movement of water between the atmospheric, oceanic, surface, and underground reservoirs.

in demand for water continue. Don Tate describes the demand for water in broad regions of the country by different sectors of the Canadian economy. He goes on to describe scenarios of water demand for the future based on different assumptions about economic growth, technological advance, and our attitudes toward water conservation. Some of these scenarios will be disquieting to those who subscribe to the myth that Canada possesses inexhaustable water resources. A clear message from Tate's chapter is that we must, sooner or later, begin to manage demand for water in Canada instead of rushing to augment supply every time withdrawals exceed supply. An early implementation of demand management policies would constitute an important first step toward resolving emerging problems of supply while preserving the integrity of the hydrological system.

In spite of the fact that many users already compete for a limited supply of water in some parts of Canada, and that with increasing population and decreasing water quality such competition must ultimately intensify, there seems to be no shortage of grand schemes to export Canadian water to the United States (e.g. Kierans 1984; Bourassa 1985). Arguments pro and con over the export of water are often more emotional than analytic. After a careful review of the evidence, Richard Bocking concludes that shortages of water in the United States are either mythical or a consequence of wastefulness no less profligate than our own. No serious experts on water supply in the United States are calling for imports from Canada. In fact, the issue appears to be generated in Canada largely by individuals or corporations that stand to gain politically

3

or economically from promoting export of water. Furthermore, the cost of such exported water would be so great as to render economically unviable the industry and agriculture it was imported to sustain, that is, provided the users were required to pay the real cost of the water. Undoubtedly the issue of water export will remain a topic of political debate in Canada for many years to come. We note, with circumspection, that water is possibly Canada's only valuable mineral resource that is not exported to be consumed or processed into a higher valued good elsewhere. We are uncertain whether this inconsistency reflects economics or simply a lack of opportunity. The current negotiations on free trade with the United States may provide the opportunity. Bocking cautions that we should not negotiate away our future options or risk the integrity of a large part of the Canadian ecosystem for a short-term gain.

Even without considering water export, Canada has an abundance of impoundments and diversions that are impressive in scale. The "supply management" approach to water scarcity in Canada is one root cause of our massive river diversions. In addition, the suitability of our rivers for hydroelectric power generation, coupled with the availability of markets in the United States for our surplus energy, has led to Canada becoming one of the world's largest producers of hydroelectric power. For decades we have languished in the belief that such engineering works were ecologically benign. A reservoir was just a lake, and a river redirected was still a river. David Rosenberg and his coauthors effectively debunk that myth. They analyse the evolution of environmental impact assessment associated with major impoundments in Canada and the evolution of our attitude toward such megaprojects as the Churchill–Nelson diversion and the James Bay hydroelectric project. It is clear from their analysis that we still have much to learn and some philosophical barriers to surmount before we can expect to have a mature ecological and a sociological ethic with respect to the consequences of such large-scale environmental restructuring and engineering works. The challenge, politically and institutionally, is to be more sensitive to the diffuse costs and the harm that can result from such projects and less captivated by the concentrated benefits.

An important question for water managers, for which none of our authors was able to offer a very satisfying answer, is how much of the water in our lakes and rivers (and of what quality) is essential to their continued functioning as desirable biological communities? There is no unequivocal answer to this question. The answer depends entirely on how much we are willing to risk or alter the ecological integrity of Canada. Obvious catastrophes, like extreme eutrophication, mercury contamination, or acid precipitation, can, to some extent, be dealt with, since they arouse general public concern and force the political machinery to action. More subtle and smaller scale degradation either goes unnoticed or is ignored. Yet ultimately, the cumulative effects of the small-scale changes may be the more devastating and all the more difficult to correct because their causes are dispersed and unregulable. Nor is it simply a matter of the colour, odour, taste, and toxicity of the water itself or the living resources inhabiting our myriad of lakes and rivers. Our tendency to regard water as simply another commodity, the value of which can be measured by its use in a market economy, has blinded us to the fact that water is integral to the functioning of all ecosystems. Water is the forests and the grasslands as much as it is the lakes and rivers. Water is also a part of our expanding urban regions. Five chapters explore various facets of the intimate relationship between water and the living landscape of Canada.

Three of these chapters deal with water in relation to the great vegetation regions of Canada, the forests, grasslands, and tundra. In his chapter on forests, Eugene Hetherington shows how trees can act as important mediators in the transfer of water among the atmospheric, soil, and surface reservoirs and how man's harvest of the forests alters this relationship. Among the many images evoked by his chapter is that of a tree as a living fountain drawing moisture from the soil and surface reservoirs and exhaling it in great clouds into the atmospheric reservoir. The romance of such images is tempered, however, by a pragmatic recognition of the intimate relationship between the water, the forests, and man the forest harvester. This relationship is of vital importance to Canadians, since the economy of the forests is a very large part of the dollar economy of the country. Yet the needs of the forest seem hardly to be considered in our decisions about water management. We must begin to incorporate such vital relationships into our policies and into our management.

Eeltje de Jong and his coauthor describe the relationship between water and the grasslands region of Canada, most of which is now given over to agriculture. They stress the need, and the oppportunity, for moisture conservation in this naturally arid region. If we are to continue to enjoy the agricultural production of Canada's breadbasket, we need to take heed of their works. Once again, they bring the intimate relationship between the land, water, and man into focus by, for example, their analysis of the contribution of cultivation techniques to soil salinization. Nowhere is the need for creative, yet pragmatic water conservation more pressing. Once again the challenge is to implement policies and practices that maintain the integrity of the natural system rather than to focus narrowly on yield maximization or cost minimization.

In the rolling tundra prairies of northern Canada the problem is not lack of soil moisture but rather the fact that the subsoil is permanently frozen. The presence of permafrost alters the hydrological cycle and severely restricts the species of vegetation and the natural productivity of a large area of Canada. The vegetation, soil, and ice in regions of permafrost exist in a delicate balance. The effects on the land of disturbing that balance can be devasting. Robert van Everdingen discusses the unique hydrological features of this vast region of Canada and some of the technical difficulties associated with human activity in the area. As always, we have been ingenious in our technical solutions to the difficulties permafrost creates for the expansion of human domestic and economic activity in the north. We have been less sensitive to the undesirable side effects of our invasion of the north.

Two chapters deal specifically with living aquatic resources. Henry Regier discusses freshwater fisheries, but not in the strict sense of the economy of a renewable resource. Through Regier's eyes we begin to see fishes as a component of the freshwater ecosystem, and that ecosystem in turn as a component of the human ecosystem. The interlinking of these ecosystems reveals that, when we assign water for industrial or domestic uses, or consign our wastes to rivers or lakes, we are, in fact, allocating or trading-off some of the useful biological productivity of those waters. We continually fail to take such trade-offs fully into account in decisions about water allocation. The legacy of this failure is evident in altered fish species composition and inedible fishes in the Great Lakes, in dying sugar maples in Ontario and Quebec and salmon in the Maritimes, and in decimation of fish and wildlife populations dependent on the Peace-Athabasca delta, to mention only three well-publicized examples.

Tom Whillans introduces us to the intricacies of wetlands, those "boggy bits and pasture sloughs" that people have always been so eager to dry up. Yet, on closer inspection we discover that these are not wet wastelands, but are important contributors to the water economy of cultivated soils and to the production of economically valuable waterfowl, fishes, and mammals. Certain types of wetland may even provide safe sites for toxic waste disposal. As has so often happened in the past, what was once regarded as a nuisance and a disadvantage is now coming to be regarded as an asset. Wetlands conservation is still very much an issue, however, and the disbenefits of wetland drainage are often not well understood. We need more and better technical information and policies that recognize the value of wetlands and their role in the hydrology and productivity of the landscape.

Most of us do not dispute the charge that human activity alters the natural landscape and the aquatic resources, even when we disagree about how great an alteration is acceptable. We are, however, less likely to think of the urban landscape as a dynamic ecosystem involving water. Yet cities are complex ecosytems with their own hydrological cycle. Frank MacKenzie discusses how urbanization affects the exchange of water among

5

are blowing for the institutions that have controlled aquatic resource management in the past. First, Canada's native indians, like peoples everywhere in the world who consider themselves to be disenfranchised, are calling for a new order and a new political reality that will recognize their aspirations. Aquatic resources will be a part of that new reality. Barry Barton reviews the legal position of natives vis à vis water under the present, paternalistic order and the implications for aquatic resource sharing through land claims agreements. Second, and intertwined with the emerging political strength of the aboriginal peoples, is the growing demand for political independence by our northern Canadian territories. In spite of our tendency as a people to congregate in the uncertain warmth along our country's southern border, it may be that our destiny, like much of the nation, lies in the north (Stephansson 1922). Terry Fenge presents a vigorous argument for a rapid devolution of authority for northern aquatic resource management from centralization in Ottawa to those who live in the north. Certainly the legislative base exists there. What is lacking is the political will to allow northern peoples to take charge of their affairs. The challenge for policy makers and managers in both these areas is to find a way to satisfy the legitimate aspirations of natives and of northerners within the context of a national approach to aquatic resource management.

Whatever the inadequacies and injustices of the existing system, invariably the rules society is prepared to live by, and the institutions that administer them, are resistant to change. Aquatic resource managment in Canada will be governed, for many years to come, by legislation and by institutions created in the past. Important aspects of the Canadian system are the separation of jurisdiction between Provincial and Federal heads of power in the *British North America Act* and the rights and responsibilities of individuals owning land adjacent to bodies of water. Andrew Thompson explores the tangled web of legislation, jurisdiction, responsibility, and individual rights that is inherent in any significant dispute over the utilization of aquatic resources. Such disputes are likely to become more frequent and intense as competition for useable water resources increases. Existing institutions are likely to become even more strained and their inadequacies even more apparent as disputes over water escalate.

Is it possible to draw any conclusions about the state of Canadian aquatic resources and our treatment of them from this sampling of issues and ideas? We believe that it is and that, notwithstanding some aches and pains as evidenced by significant local problems, the resource is, in the main, healthy. The prognosis for the future, however, is less certain. Distressing symptoms, which could be the early warning signs of serious disease and of the failure of our management system, are present. Some of these, such as acid rain, contamination of the St. Clair and Niagara rivers, dioxins, and contaminated fishes in the Great Lakes, are well publicized and familiar to everyone. Others, such as groundwater pollution and the waste of water resources due to unrealistic pricing and access policies, are less well known but no less integral to any developing problems. The presence of these symptoms indicates that a reevaluation of our attitudes toward aquatic resources, and toward institutions that regulate our use of aquatic resources, is in order. We have already identified a number of significant challenges facing policy makers and managers that emerge from the essays in this book. How Canadians will cope with issues of aquatic resource management in the future is uncertain. The Inquiry on Federal Water Policy and the MacDonald Commission on the Economy have set the stage for a restructuring of the institutions and a redefining of the objectives of natural resource management in Canada. In our final chapter, Jack Vallentyne and Andy Hamilton explore some of the implications of these national inquiries for the future of Canada's waters. They conclude that neither inquiry went far enough nor was bold enough in its recommendations for change, even though submissions by ordinary Canadians to both inquiries were creative and suggested a need and a desire for significant change. Some fundamental changes in the way we approach the management of aquatic resources are required if we are to meet the challenges outlined above.

We have argued that our waters and the ecosystems that depend on them are integral to the Canadian pattern of life. Historically, however, we have used our aquatic resources as though they were simply another set of renewable resources, the value of which could only be realized through exploitation. Yet aquatic resources have a value that transcends the narrow perspective of the market economist (Farnworth et al. 1981; Talhelm 1983). We have tried to illustrate this value through our choice of topics for this book and through our emphasis on the interconnections between water and all facets of Canadian social and economic life. To date, Canadians have not been forced to come to grips with the intangible values of water. We have so much water that serious consequences of our use and abuse of aquatic resources are only now becoming apparent. We have arrived, however, at the time when difficult choices have to be made, particularly with respect to chemical pollution and future large-scale engineering works involving water. How we deal with those choices will depend greatly on our perception of our place in the natural ecosystem.

As we pointed out earlier, western man's historic view of his relationship to the natural world has been an anthropocentric one. Under this view, mankind held domination over the natural world, and science and technology provided the power to transcend any obstacle. This being the case, mankind was logically exempted from the checks and balances that constrained the natural world. Although man's anthropocentric view was rooted in Christian dogma, our postindustrial technological society has tended to reinforce the idea that man, the toolmaker, can invent his way out of any crisis. This view, which sees man as separate from and ultimately in control of the natural world, has been labelled "The Human Exemptionalism Paradigm." Humphrey and Buttel (1982) listed four major assumptions underlying this view of the world: (1) humans have a cultural heritage in addition to, and distinct from, their genetic inheritance and thus are unlike all other animal species; (2) social and cultural factors (including technology) are the major determinants of human affairs; (3) social and cultural environments are the crucial context for human affairs, and the biophysical environment is largely irrelevant; (4) culture is cumulative and thus, technological and social progress can continue indefinitely, making all social problems ultimately soluable.

These are the assumptions that have underlain the western industrial philosophy of progress through unfettered growth. They have led to previously unimagined material prosperity. But this prosperity was gained in part through the deferral of significant ecosystemic consequences, consequences that are now coming to the fore. The kind of fundamental change in our approach to aquatic resource management that we believe is necessary requires that we reevaluate the assumptions of the Human Exemptionalism Paradigm and, perhaps, discard some of them in favour of others that are more holistic.

During the past few decades an alternative assessment of mankind's position in the world has been gaining strength, stimulated by the growing frequency of ecosystemic problems. This is the view that mankind is integral with, and ultimately totally dependent on, the limited resources of the natural world. This view, which has been labelled "The New Ecological Paradigm," has been espoused by numerous natural philosophers in the past and is a component of some eastern religions. With few exceptions, however, it was without significant influence in the west prior to the environmental revolution of the 1960s and 1970s (Humphrey and Buttel 1982). During those two decades the new philosophy gained a strong beachhead in western industrialized societies. It now numbers among its adherents Nobel laureates and influential administrators from all continents (Pauling et al. 1982). Four assumptions underlying The New Ecological Paradigm include the following (adapted from Humphrey and Buttel 1982).

(1) Even though human beings have exceptional characteristics, they are but one among many species within the global ecosystem.

(2) Human affairs are influenced not only be social and cultural factors but also by intricate linkages of cause and effect in the web of nature. Thus, purposive human actions have many unintended consequences.

7

the atmospheric, soil, and surface reservoirs. Water dynamics are considerably altered in an urban area compared with adjacent natural areas, and this creates problems of management no less complex and compelling than those associated with forestry or agriculture. Cities are also major producers of aquatic contaminants; controlling and containing a city's undesirable waste products is a major and unresolved problem for city planners and aquatic resource managers.

A cliche of the 1950s and 1960s was that consumerism had created a throw-away society. In the past decade a cliche has arisen that science has created a chemical society. These cliches are no less true of Canada, with its primary resource-based economy, than they are of more highly industrialized nations. Over 1000 new chemicals enter into the industrial society of North America each year (Environment Canada 1984). Most of these will eventually find their way into our waters where they will have undefined and incalculable impacts on living aquatic resources and on human health. Unfortunately, our technology of waste control has not kept pace with our capacity for waste creation. With a few exceptions, our approach to waste management is still very similar to that of our nomadic, stone-age ancestors; we dump waste on the ground or into the water and let nature "take care of it." Even where we have implemented apparently economically and ecologically acceptable waste management, our past practices can be a legacy of deadly surprises. Contamination of our waters has been an issue among scientists for a long time. In North America, however, contamination of our waters did not really become an issue of strong public concern until Rachel Carson (1962) brought it, compellingly, to everyone's attention. In the 25 yr since Carson's book appeared, the variety of man-created compounds accumulating in our waters has, more than ever, outstripped our capacity to monitor and to evaluate them in terms of their consequences for human and ecosystem health. Bob Hamilton and his coauthors provide a mind-boggling introduction to surface water contamination and what we are not doing, and are not able to do, about it. Just as with major engineering works, we need to develop policies with respect to toxic substances that take better account of the diffuse costs of employing such substances and less account of the focused benefits.

Even more disquieting, because it is even less well understood, is the contamination of water beneath the land surface. The vast reservoirs of water in underground aquifers, which feed thousands of rural Canadian wells, are uniquely vulnerable to certain kinds of contamination. The water in many aquifers is replaced only very slowly. Contaminants that sink into the ground may lie in pools hundreds of metres below the surface where they slowly enter into the groundwater. Consequently, once the underground water is contaminated, it can remain so for hundreds or even thousands of years. This contamination will wander as the aquifer wanders, discharging toxic substances into wells and creeks as it goes. John Cherry reviews the unique characteristics of groundwater contamination and the difficulties of predicting, monitoring, or controlling the underground dispersal of contaminants. In our opinion, the problems of groundwater contamination will soon greatly eclispe those of surface water contamination. This will be the Canadian aquatic resource issue of twenty-first century.

Each author, at our request, took a peek into the future. Some were bolder and more ambitious prophets than others. This is not surprising, since forecasting is a risky business. One author who's topic demanded that he be a forecaster was Earl Ripley who wrote on the effects of climate change on Canadian hydrology. The effects of anthropogenic increases in atmospheric carbon dioxide and other substances on world climate have been hotly debated in recent years. It is widely believed that the increases in the concentration of these gasses in the atmosphere will create a "greenhouse effect" whereby short wavelength energy from the sun will pass through the atmosphere to warm the earth but the longer wavelength energy will be less able to escape back throught the atmosphere. Ripley attempts to come to grips with how the greenhouse effect will influence regional Canadian climate and water supply. The effects of significant climatic warming are potentially traumatic for established societies and for the balance of power in the world. What will be the overall effect of anthropogenic climate alteration upon Canadian society is very uncertain. Predictions of the effects of climate change on elements of the hydrological cycle, such as precipitation, evaporation, and water yield, are perhaps les tenuous. Certainly Canada is likely to be warmer in the future, and wetter in some areas but dryer in others. For example, cereal crops may be cultivated further north on the prairies but it may become too dry for such crops in the southern prairies. Whether the net benefits of these changes will be positive or negative will depend on how we plan for change and are able to adapt to it. Such planning needs to be initiated well in advance of any perceptible climatic change.

The ways in which we use, and abuse, our aquatic resources reflect a complex set of value judgements and trade-offs among conflicting societal objectives. It is our view that these judgements ought to be based on the best possible technical information about the issues and with as full an understanding as possible of the consequences of alternative actions. We have, therefore, weighted this book toward the technical issues of aquatic resources management. This does not mean that we believe environmental and resource allocation problems to be ultimately soluble through technical means alone. We acknowledge, unreservedly, that judgements about pollution and aquatic resource use will reflect the national philosophy of the day and the needs of the human economy. The way in which technical information is used to make decisions about aquatic resource management will, therefore, evolve as our society evolves.

Just how deep are the cultural roots of our society's behaviour was brought out in a lecture by the political philosopher, Leo Strauss (1967):

> All the hopes that we entertain in the midst of the confusions and dangers of the present are founded positively or negatively, directly or indirectly on the experiences of the past. Of these experiences the broadest and deepest, as far as we Western men are concerned are indicated by the names of the two cities Jerusalem and Athens. Western man became what he is and is what he is through the coming together of biblical faith and Greek thought. In order to understand ourselves and to illuminate our trackless way into the future, we must understand Jerusalem and Athens.

Passmore (1974) explored a similar theme within the contex of resource management and pointed out that the Judeo-Christian view of God as transendential, and of man created in his image, effectively separated man and nature. This view of man as separate from nature, when combined with the developing Greek traditions of critical scientific thought, and ultimately with the mechanistic philosophy of Descartes, legitimized the idea that man held dominion over nature. Freed from any moral responsibility for the preservation of natural systems, western man embarked upon the industrial, technological, and scientific revolutions that have moulded our society for the past two centuries. In reality, however much we may believe otherwise, mankind is a part of nature and is subject to nature's limitations. We may manouvre around, or outrun, the unforseen natural consequences of our actions for a while, but eventually we have to face up to those consequences. The technical chapters of this book identify some of the present and emerging consequences of our use and abuse of aquatic resources. The sociological and philosophical chapters explore our past and present approaches to dealing with the consequences of our actions and suggest some needed changes for the future.

Canadians have experienced a rather dramatic evolution of approach and philosophy toward aquatic resource management throughout the country's brief history as a nation (Quinn 1985). Tony Dorcey reviews the rise and fall of expectations in aquatic resource management over the past few decades and offers some insightfull suggestions for improvements to the links between science and management that will help us achieve our goal of informed decision-making. He argues that the scientific community needs to develop more effective tools for communicating its ideas and for barganing to ensure that science plays a role in decision-making. On two other fronts the winds of change

(3) Human beings live in and are dependent on a finite biophysical environment that imposes potent physical and biological restraints on human affairs.

(4) However much the inventiveness of human beings and the powers derived therefrom may seem to transcend for a while the carrying capacity of the biosphere ecological laws cannot be repealed. Mankind is subject to checks and balances and feedback mechanisms similar to those that regulate the behaviour of other organisms.

The New Ecological Paradigm embodies the spirit of the changes in attitude we would prescribe. Furthermore, it is a viewpoint that has been gaining credibility in Canada. In underlies the growing emphasis on managing ecosystems as wholes rather than as a set of semi-independent components. The ecosystem approach, revolutionary at the time, was the guiding philosophy of the Canada-U.S. Great Lakes Water Quality Agreement of 1978 (International Joint Commission 1982). The New Ecological Paradigm is hinted at, if not totally embraced, in some recommendations of the Pearse Inquiry on Federal Water Policy. Pearse et al. (1985) recommended for example, that the natural and most appropriate management unit for water is the watershed and that water management must treat surface water, groundwater, and atmospheric water as a total system, the integrity of which should be preserved. The paradigm was very clearly the guiding philosophy in Environment Canada's (1984) submission to the MacDonald Royal Commission on the Economy. It remains to be seen, however, whether the new philosophy has been sufficiently assimilated to bring a halt to the degradation of our aquatic ecosystems by piecemeal and seemingly insignificant alterations. The opportunity exists to pursue integrated, ecosystem-oriented conservation and aquatic resource management, management that will benefit not just our generation but future generations as well. It is in the spirit of The New Ecological Paradigm, and in the belief that ecosystem health and economic prosperity for Canada are inextricably intertwined, that we offer these essays on Canada's aquatic resources.

References

BOURASSA R. 1985. Power from the north. Prentice-Hall Canada, Scarborough, Ont.

CARSON, R. 1962. Silent spring. Houghton-Mifflin, Boston, MA.

CASS-BEGGS, D. 1961. Water as a basic resource, p. 173-189. In Background papers, Vol. 2. Resources for Tomorrow Conference. Queen's Printer, Ottawa, Ont.

ENVIRONMENT CANADA. 1984. Sustainable development, p. 1-18. In A submission to the Royal Commission on economic union and development prospects for Canada. Environment Canada, Ottawa, Ont.

FARNWORTH, E. G., T. H. TIDRICK, C. F. JORDAN, AND M. W. SMATHER JR. 1981. The value of natural ecosystems: an economic and ecological framework. Environ. Conserv. 8: 275-284.

HUMPHREY, C. R., AND F. R. BUTTEL. 1982. Environment, energy, and society. Wadsworth Publishing Co., Belmont, CA.

INTERNATIONAL JOINT COMMISSION. 1982. First biennial report under the Great Lakes Water Quality agreement of 1978. International Joint Commission, Ottawa, Ont., and Washington, DC.

KIERANS, T. W. 1984. The Grand Recycling and Northern Development (GRAND) Canal, p. 124-143. In Futures in water. Proceedings of the Ontario Water Resources Conference. Sponsored by the Government of Ontario, Toronto, Ont.

PASSMORE, J. 1974. Man's responsibility for nature. Gerald Duckworth and Co., London, England.

PAULING, L., F. BENAVIDES, F. T. WAHLEN, M. LASSAS, B. B. VOHRA, AND G. A. KNOX. 1982. Open letter: to all who should be concerned. Environ. Conserv. 9: 89-90.

PEARSE, P. H., F. BERTRAND, AND J. W. MACLAREN. 1985. Currents of change. Final report, Inquiry of Federal Water Policy. Environment Canada, Ottawa, Ont. 222 p.

QUINN, F. 1985. The evolution of Federal water policy. Can. Water Resour. J. 10: 21-33.

STEPHANSSON, V. 1922. The northward course of empire. George G. Harrap Ltd., London, England.

STRAUSS, L. 1967. The beginning of the bible and its Greek counterparts. The Frank Cohen Public Lectures in Judaic Affairs, March 13 and 15, 1967. The City College, The City University of New York, New York, NY.

TALHELM, D. R. 1983. Unrevealed extra market values: values outside the normal range of consumer choices, p. 275-286. *In* R. D. Rowe and L. G. Chestnut [ed.] Managing air quality and scenic resources at national parks and wilderness areas. Westview Press, Boulder, CO.

TORREY, B. AND F. H. ALLEN [ED.] 1962. The Journal of Henry D. Thoreau (1837-1861). 2 vol. Dover, New York, NY.

CHAPTER 2

The Amount of Canadian Water and Its Distribution

Arleigh H. Laycock

Department of Geography, The University of Alberta, Edmonton, Alta. T6G 2H4

Introduction

My first reaction, when asked to prepare this chapter, was to want to provide a revised and updated version of the *Hydrological Atlas of Canada* (Fisheries and Environment Canada 1978). This is not possible because of the scale and space limitations of this volume and the enormity of the task of revising and adding maps of national patterns. The 1978 atlas is a very good first attempt at mapping and describing the water resources of Canada and it should be used by readers as a major background for this discussion.

Much of the more conventional treatment of data from weather and stream gauging stations is included in the atlas. The contributions of agencies such as the Atmospheric Environment Service and the Inland Waters Directorate of Environment Canada were not fully related to each other, however, and there are gaps and inconsistencies. I have, therefore, decided to employ an integrating water balance approach in describing and discussing the freshwater resources of Canada, as this approach will best complement the atlas material.

The first section is on world patterns and the place of Canadian resources within them. With these perspectives we may gain a better insight into what water resources we have in comparative abundance and recognize some of the opportunities and limitations we have in the use of these resources.

The main section is on the amount and some of the characteristics of Canada's freshwater supply. The water balance approach is used in explaining some of the distribution patterns in space and time and in an attempt to resolve some of the inconsistencies and fill is some of the gaps in the atlas and other conventional references. This may stimulate some discussion (encouraged by the editors) but my intent is that our understanding of Canadian patterns be improved by such cross-examination.

The final section is on water supply in the future. Some of the most striking changes in Canada's water supply will take place as a result of "greenhouse warming" and we will not be able to project future patterns from past records. We must be aware of the trends in supply, however, and their impacts upon future water resources management. If we have a better understanding of what we have, we will better appreciate the effects of the changes that we anticipate.

World Water Balance

The Hydrological Cycle

Approximately 577 000 km^3 of water is evaporated and transpired into the atmosphere from the surface of the globe every year (Korzun 1974). Most of this (505 000 km^3) is from the world oceans and a smaller part (72 000 km^3) is from land areas. This moisture falls back to the surface as precipitation upon ocean (458 000 km^3) and land (119 000 km^3) areas. The precipitation upon land areas is partly evaporated and transpired (72 000 km^3) and the balance runs into the oceans as streamflow (45 000 km^3) and ice and groundwater flow (approximately 2000 km^3).

This cycle, driven by solar radiation, has relatively small variation from year to year on a world scale but the regional variations can be large. In reality, the world hydrological cycle is a composite of many movements of water through many systems with greatly different and varying rates of flow and amounts of storage (Miller 1977). Misconceptions are common. It is sometimes suggested, for example, that the greater part of the precipitation on land areas is derived from evaporation and transpiration from the continents, and from the net amounts quoted above, this might appear to be so. In fact, most precipitation upon land areas is derived from moist marine air masses, and much of the evaporation and transpiration from land areas is into drier continental and modified marine air masses which later pass out over ocean areas to pick up more moisture. The consequences of this fact are significant. For example, the potential for us to increase precipitation over land areas by increasing the supply of water available for evaporation (e.g from expanded irrigation and forest areas and augmented lakes and wetlands) is small. Despite this, local recycling does occur and it is of some significance, especially in interior areas.

World Water Supply

Data on the amount of water present in various forms at any point in time should be accompanied by data on the replacement period required in the hydrologic cycle. In this way, both total reserves and the rate at which quantities withdrawn for industrial or other uses can be determined. In Table 1, both are shown.[1]

TABLE 1. World water supply. Long-term average volume and replacement rate in each category of the hydrological cycle. Based upon Korzun (1974) p. 43 and 109). Other sources include L'vovich (1974), Nace (1969), Kalinen and Bykov (1969), and U.S.S.R. Committee for the International Hydrological Decade (1974).

Category	Total volume (km³)	% of total	% of fresh	Replacement period	Annual volume recycled (km³)
World ocean	1 338 000 000	96.5	—	2 650 yr	505 000
Groundwater (to 2000 m)	23 400 000	1.7		1 400 yr	16 700
Predominantly fresh groundwater	10 530 000	0.76	30.1	—	—
Soil moisture	16 500	0.001	0.05	1 yr	16 500
Glaciers and Permanent snow	24 064 100	1.74	68.7	—	—
Antarctica	21 600 000	1.56	61.7	—	—
Greenland	2 340 000	0.17	6.68	9 700 yr	2 477
Arctic Islands	83 500	0.006	0.24	—	—
Mountain areas	40 600	0.003	0.12	1 600 yr	25
Ground ice (permafrost)	300 000	0.022	0.86	10 000 yr	30
Lakes	176 400	0.013		17 yr	10 376
Freshwater	91 000	0.007	0.26	—	—
Salt water	85 400	0.006		—	—
Marshes	11 470	0.0008	0.03	5 yr	2 294
Rivers	2 120	0.0002	0.006	16 d	48 400
Biological water	1 120	0.0001	0.003	—	—
Atmospheric water	12 900	0.001	0.04	8 d	600 000
Total water	1 385 984 610	100[a]			
Freshwater	35 029 210	2.53[a]	100[a]		

[a]Some duplication in subcategories and categories.

---

[1]Korzun (1974) and his associates showed the volume and average replacement period data but did not project these to show the amount of recycling and replacement per year. The river and atmospheric water volumes are only approximately equal to those noted earlier. Despite this, a better impression of supply available through time is indicated.

Ocean

Most of the world water reserve is contained in the world oceans. Oceanic waters are a major resource for navigation, fishing, recreation (especially on the coasts), and many other uses, are an important determinant of world climate patterns, and are the major source of atmospheric moisture for precipitation and streamflow on land. Although Canada has a longer coastline than any other country and the role of oceans in Canadian development has been of major importance, the topic of oceans is beyond the scope of this chapter in which freshwater supplies are stressed.

Groundwater

The groundwater reserves of the world are large but the natural replacement period is long so that the supply per year is much smaller than that which is available in rivers or which falls as precipitation. Much of the groundwater reserve (to 2000 m depth) is saline and much of that which is fresh is very costly to extract. Despite these and other limitations, groundwater is relatively ubiquitous in its distribution, is more abundant than streamflow in many areas, and is an excellent reservoir that is relatively free from evaporation losses. A substantial part of surface streamflow during the drier seasons and winter comes from groundwater, if recharge in the wetter seasons and wetter years has been sufficient. The flow in a groundwater-fed stream has a much more even seasonal distribution than that in a stream dominated by surface runoff. Conjunctive use of surface and groundwater is often possible because of their complementary features, surface water being used during wet seasons and groundwater during dry.

The Canadian groundwater supplies available for human uses are much smaller than might be indicated by our area (6.6% of the world land area). This is a consequence of a number of factors. Recharge is limited in the north by permafrost (see van Everdingen, this volume) and in the dry Interior Plains (where much of the groundwater reserve is brackish) by low precipitation. The massive igneous and metamorphic rocks in Precambrian Shield, Cordilleran, and Appalachian regions have low porosity and hence low reserves.

Use of groundwater in Canada is low because of these limitations and because we have an abundant surface supply in rivers and lakes in many areas where water demands are high. Nevertheless, groundwater is an important source of water for a wide range of domestic, urban, industrial, and other uses in the more populated southern parts of Canada.

Soil Moisture

The world reserve of moisture in the soil is much smaller than that in groundwater or ice but the replacement period for soil moisture is short so that the amount recycled annually is large. In the warmer and more humid areas of the world the replacement period is well under a year (L'vovich 1974; Miller 1977).

The Canadian share of the world soil moisture reserve is low relative to our area because soils in northern areas are shallow and poorly developed and large, glaciated areas have little parent material over the bedrock. Despite this, our vegetative cover is highly dependent upon soil moisture stored between rains and following snowmelt recharge.

Glaciers

Glaciers and permanent snows occur predominantly in Antarctica and Greenland and these remote reservoirs of water appear to be of little use to mankind. Perhaps their greatest value lies in the fact that they are stored where they are. If these masses were to melt, sea level could rise as much as 70-80 m inundating extensive low lying coastal areas (Environmental Protection Agency 1983b). Canada has a substantial part of the remaining glacier ice in our western and northern mountain areas. The replacement periods for these water reserves are long and the volumes present are approximately in balance. The flow regimen of many western and northern rivers are affected by glacier melt and there is some net storage from wet to dry years, but it would be incorrect to attribute more than a very small part of the streamflow of most regions to this source.

In addition to permanent and semipermanent glaciers and icefields, most of Canada is snow covered in winter. The volume of water precipitated as snow is large, as much as 2000 km^3/winter. Canada has probably over 20% of the snow detention storage of the world. The duration of snow cover ranges from not much more than the duration of the parent storms in lowland Vancouver Island to almost all of the year in some high-level and far-northern areas where snow drifts may be transitional to glaciers. Variations in the amount, duration, and timing of snowfall and snow melting contribute to major variations in streamflow volumes and regimen.

Permafrost

Over half of Canada has permafrost (perhaps one third of the world permafrost area). The effect of permafrost upon drainage, lake, marsh, river, and soil moisture systems is pronounced. The "active layer" above the permafrost is wet during most of the summer thaw period and wetland habitats are widespread. Surface runoff predominates and the abundant lakes and marshes detain flow well into summer and fall. Permafrost creates many problems for construction, drainage, and water supply but technological solutions have been found for most of these. Only 1% of the population of Canada resides in regions with continuous and discontinuous permafrost, however, and many northern communities now have insulated or heated water and sewage lines and draw water from lake and river supplies below freezing depth.

Lakes

The lakes and reservoirs of the world supply human demands for water to a much greater degree than is indicated by the volumes involved. This is because these reserves are much better located than other water sources relative to mankind's diverse demands for withdrawal, consumption, and other uses, e.g. navigation, recreation, fishing, aesthetic, water storage, etc. Lakes have a levelling effect upon flow regimen and can improve water quality by sedimentation of particulate matter, oxidation, and diffusion.

Canadian lakes are relatively shallow (compared with Baikal and the African rift lakes) and we may have no more than 15% of the freshwater lake volume of the world. However, Canada has close to half of the lake surface area of the world and approximately 20% of the reservoir area.[2] The Great Lakes are a dominant geographic feature and are important because of their proximity to centres of population. Yet the Canadian portion of the Great Lakes constitutes only 11.7% of the lake area of Canada (Statistics Canada 1981). By contrast, the American portion of the Great Lakes makes up 76% of the natural lake area of the conterminous United States.

Canada has only a small area of brackish and salt lakes, mostly in the Prairies, and the Canadian portion of world salt lake volume is very small. Most of the world saltwater lake volume indicated in Table 1 is in the Caspian Sea. Many of our smaller prairie lakes are eutrophic, but in areas of higher precipitation, expecially in the Precambrian

---

[2]Although Korzun (1974), based in part upon L'vovick (1974), is a better reference source than any other for the world as a whole, the information on Canada is surprisingly poor. For example, the lake area indicated for Canada is less than one third of that reported in Canada Yearbooks a decade earlier, mainland mountain glacier areas in Canada are ignored or included with those of Alaska, and wetland areas are estimated to be similar to those of the United States. This is partly due to Korzun's failure to use available data but the data base provided by Canada in the past has also been weak.

Shield where nutrient inputs are low, most are oligotrophic. Most Canadian lakes freeze over in Winter and are cool to cold in summer — but many are suitable for swimming and other recreational activities for part of the summer. Most provide diverse habitat for wildlife and fishes and much of the scenic variety in the Canadian landscape can be attributed to the fact that we have so many lakes.

Marshes

Marshes and other wetlands were estimated by Korzun (1974) to occupy about 2% of world land areas and only 0.9% of North American land areas. Canadian extimates are that wetlands comprise about 18% of the nation (Zoltai 1979). Additional areas are wet in some seasons. Wetlands are no longer considered to be wastelands (see Whillans, this volume). Their importance as wildlife habitat, in water storage, and filtration, and in nutrient use is coming to be recognized. Drainage of wetlands for agricultural, forest, and other land development is likely to continue for some time to come but wetland maintenance and preservation is of growing significance as these resources come to be properly evaluated for alternative and sometimes complementary uses.

Rivers

Rivers are the major source of water for most withdrawal uses (e.g. domestic, urban, industrial, irrigation, mining) and are very important for a wide range of instream, onstream, and beside-stream uses (e.g. fishing, navigation, hydro power, wildlife habitat, recreation, viewing). They will be stressed in this chapter. The amount of water in rivers at any point in time is relatively small, but with a world average replacement period of 16 d, the amount available for use through time is very large. In Canada, the average replacement period for river water is much longer than 16 d because of storage delays in the many lakes and reservoirs along most rivers and because of ice in winter.

Biological Water

Biological water is that which is consumed by plants and animals (including aquatic) (L'vovich 1974). The physiological consumption by man is about 3 L/d and by cattle about 40 L/d. Spring wheat might use 150-300 mm/season in transpiration and evaporation. Biological water is small in quantity at any point in time but the replacement period is short and the amounts cycled annually are very large. The seasonal variations are large in the higher latitudes and the totals for Canada are well below that indicated by our area (relative to world area) because of our relatively short and cool growing seasons.

Atmospheric Water

The atmospheric water reserve is small but this small reserve is recycled extremely rapidly so that total annual supply of atmospheric water is very large (Table 1). Atmospheric water is the most important source of moisture for most terrestrial plant life, either directly or after a period of soil moisture storage. The atmospheric water reserve over Canada is well below the world average because temperatures are cool to cold for so much of the year. Canada is a higher latitude country and atmospheric moisture content is closely related to latitude. At 70° latitude, the average water vapour content of the atmosphere is 0.2%, at 50° is 0.9%, and at the equator is 2.6% (Korzun 1974). Canadian precipitation is roughly the same as that of conterminous United States (48 states) but evaporation and transpiration are less than one half as great, so that much more of the precipitation is available for streamflow.

Water Balance, by Country

World streamflow is approximately 45 000 km^3/yr and this is supplemented by approximately 2000 km^3 of ice and groundwater flow directly to the oceans (Korzun 1974). L'vovich (1974) provided estimates of total streamflow by nation (Table 2). These

TABLE 2. Average annual water balance (km³) by country. Based on L'vovich (1974) and U.S.S.R. Committee for the International Hydrological Decade (1974). The estimate for Canada does not include the Arctic Islands (possibly 500-km³ streamflow). On an annual basis, precipitation must be balanced by evapotranspiration plus streamflow plus or minus storage change. On the average, storage change may be assumed to be zero and may be omitted. The evapotranspiration is potential evapotranspiration minus deficit, both calculated using the Budyko (1956) procedures (U.S.S.R.) which are similar to those of Thornthwaite (1948) and Thornthwaite and Mather (1955, 1957) in North America.

Country	Ppt.	=	(P.E.	−	D)	+ streamflow	%[a]	Per capita, 1971 (000 m³)
Brazil	15 800	=	(11 912	−	1 700)	+ 5 668	13.8	59.5
U.S.S.R.	10 960	=	(9 810	−	3 200)	+ 4 350	10.6	17.8
China	7 488	=	(6 908	−	2 300)	+ 2 880	7.0	3.8
Canada*	4 930	=	(2 690	−	500)	+ 2 740	6.7	128.0
United States	6 398	=	(6 753	−	2 700)	+ 2 345	5.7	11.4
India	3 662	=	(3 676	−	1 600)	+ 1 586	3.9	2.9
Indonesia	3 579	=	(2 669	−	600)	+ 1 510	3.7	13.0
Columbia	3 079	=	(2 467	−	500)	+ 1 112	2.7	52.7
Zaire	3 744	=	(3 636	−	900)	+ 1 019	2.5	58.4
France	528	=	(376	−	80)	+ 232	0.6	4.6
Sweden	299	=	(135	−	30)	+ 194	0.5	24.1
Great Britain	259	=	(127	−	20)	+ 152	0.4	2.7
Egypt	20	=	(1 316	−	1 300)	+ 4(+91)[b]		0.12
Netherlands	28	=	(21	−	3)	+ 10(+79)[b]		0.78
Australia	1 686	=	(8 804	−	7 500)	+ 382	0.9	30.0
World land areas	113 500	=	(122 500	−	40 000)	+ 41 000	100	10.9

[a]Percentages of world streamflow.
[b]The streamflows originating in Egypt and the Netherlands are small but the inflow from upstream areas are relatively large.

estimates were based upon earlier data so that the total for all nations is smaller than the world estimate in Korzun (1974). National totals are, however, not available in Korzun and other more recent sources.

L'vovich (1974) estimated the streamflows of all countries but only those of the top nine countries with over 1000 km³ of annual flow, and selected others, are shown in Table 2. Brazil, with most of the Amazon flow plus other major rivers, has the largest annual flow. The U.S.S.R. is second and Canada is third if we add the flow of the Arctic Islands which was not included in the L'vovich estimate. Canada is fifth among nations in precipitation but evapotranspiration is much smaller than in China or the United States. Consequently, the residual streamflow is larger. Canada has a relatively small population so that the per capita streamflow is very large (128 000 m³/yr, again excluding the streamflow of the Arctic Islands from the total).

The data shown in Table 2 are for streamflows originating within each country. Inflow to some countries from upstream areas can be very important, e.g. for Egypt and the Netherlands. This aspect of water supply will be noted for Canadian provinces in a later section.

Water Balance in Canada

Canadian Streamflow Estimation

Streamflow measurements in Canada were confined largely to hydroelectric power, navigation, and irrigation needs and to international streams at or near the United States borders (for water apportionment purposes) until well after World War II. Many people recognized that northern rivers were large but most estimates of provincial and national flow were very conservative. Precipitation measurements for northern areas were usually low and unrepresentative because they were taken in or near settlements that were in the warmer and drier lowland sites. The assumption that streamflow must be small because measured precipitation was small was widely held. As recently as 1961, Cass-Beggs, using available streamflow data for southern Canada and precipitation and other data for the North, estimated that Canadian streamflow might be as large as that of the conterminous (48) United States, or about 1532 km^3 (Table 3).

Streamflow gauging stations were extended into northern areas before and during the International Hydrological Decade of 1965-74. By 1967 it was becoming apparent that northern rivers were much larger than had been previously estimated. At the same time, regional estimation based on water balance techniques was being used to supplement streamflow data. Laycock (1968) estimated that Canadian streamflow was at least 2 billion acre feet (2 467 000 000 dam^3 or 2467 km^3)/yr and Sanderson and

TABLE 3. Evolution of streamflow estimates for Canada. Many other estimates have been made but the trends are indicated. Based on Cass-Beggs (1961), Laycock (1967, 1985), Sanderson and Phillips (1967), Prince (1969), Canadian National Committee for the International Hydrological Decade (1969), Fisheries and Environment Canada (1978), L'vovich (1974), Pearse et al. (1985), and United States Water Resources Council (1968, 1978).

	Year	Annual flow estimate (km^3)	% increase from 1961 (km^3)	Note
Cass-Beggs	1961	1532	0	Assumed equivalent to conterminous U.S. (48 states)
Laycock	1967	2467	61	
Sanderson and Phillips	1967	2537	66	Does not include Arctic Islands
Prince	1969	Up to 4000	161	Speculative estimate
Preliminary maps, I.H.D.	1969	2963	93	
L'vovich	1974	2740	79	Does not include Arctic Islands
Hydrological Atlas of Canada	1978	3098	102	
Laycock	1983	3368.5	120	
Laycock	1985	3750	145	Estimate made for this chapter
Pearse Inquiry	1985	3318	117	Estimate made by Water Survey of Canada
Alaska[a]	1968	801		
	1978	1250		56% increase

[a]For comparison.

Phillips (1967) estimated that Canadian flow, not including that of the Arctic Islands, was 2 057 000 000 acre feet (2537 km^3). The upward revisions of Canadian steamflow estimates based upon new northern data were such that Prince suggested in 1969 that Canadian totals could be as high as 4000 km^3/yr. The Canadian National Committee for the International Hydrological Decade was much more conservative in its 1969 estimates and it made only minor upward revisions for the *Hydrological Atlas of Canada*. L'vovich, expanding a bit upon the Sanderson and Phillips estimate, was still conservative in 1974 if one allows for up to 500 km^3 for Arctic Islands flow.

In 1983, I estimated total streamflow originating within Canada to be 3368.5 km^3 (Laycock 1985). In making this estimate, I recognized apparent underestimates in the *Hydrological Atlas of Canada*, particularly for the Northwest and Yukon Territories and for the coastal one sixth of British Columbia. Using the most recent data available in 1985, plus projections of water balance patterns to areas without streamflow measurements, I have revised this total for a submission to the Pearse Inquiry and for this chapter. I believe that an estimate of 3750 km^3 is now reasonable if we include ice flow, interflow (lateral flow above the water table), and groundwater flow across boundaries and directly to the oceans and the flow of streams in internal drainage basins. These could total at least 75 km^3 or 2% of the total. These flows are included with streamflow in the following section.

In 1985, the Inland Waters Directorate, Environment Canada, provided an estimate of stream flow originating within Canada for the Pearse Inquiry (Pearse et al. 1985) that was 105 135 m^3/s or approximately 3318 km^3/yr.

I believe it is likely that additional upward revisions of total Canadian streamflow will be made in the future, especially for northern and coastal areas. It is probably not coincidental that similar revisions upward have been made for Alaskan streamflow in recent years, e.g. in the estimates used in the First and Second National Assessments of the United States Water Resources Council (1968, 1978).

Water Balance and Water Yield

A number of water balance procedures have been used to estimate water yield from climatic data. One of the best for use in Canada is that of C. W. Thornthwaite (Thornthwaite 1948; Thornthwaite and Mather 1955, 1957). It has been used by Sanderson (1948), Sanderson and Phillips (1967), Laycock (1967), and many others, to estimate Canadian water supplies, sometimes with modifications for better local application. The procedure is described well in recent volumes by Mather (1978, 1984).

In this chapter, I shall use the Thornthwaite procedure to illustrate some regional patterns of water surplus and deficiency and as a basis for projecting yield patterns from gauged basins to ungauged areas.

The major factors in determining water surplus or yield are identified in a water balance equation: Precipitation = (potential evapotranspiration − deficit) + surplus ± storage change.

Precipitation (Ppt.) is measured at many locations throughout Canada and precipitation patterns have been mapped, but these data are often biased by undercatch in precipitation gauges,[3] especially for snowfall. Furthermore, the precipitation measured

---

[3]Meteorological records are of recorded measurement data and there is usually no attempt to "correct" these data to provide better values for water balance study purposes. Precipitation maps show the recorded values and these, even for station areas, may be below actual precipitation by 10-50%. There is little attempt to correct the map patterns for the greater precipitation of adjoining upland areas. Topographically corrected isohyetal mapping is still rare in Canada.

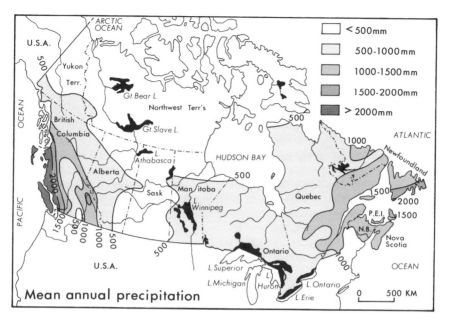

FIG. 1. Mean annual precipitation, Canada. Source: adapted from Fisheries and Environment Canada (1978). From Pearse et al. (1985, plates 3 and 85).

at gauging stations is often unrepresentative of that received in more elevated areas nearby. Moisture received by condensation on surfaces is not measured and the relocation of snow by drifting can result in greater yields in some areas and smaller yields or compensating increases in deficits in others that go unrecorded. Usually, snow drifting creates a net increase in yield because compensating deficits occur in areas swept bare of snow. In the drier regions of the country, especially in the drier years, much of the limited streamflow is derived from the surpluses of snow drift areas.

The map of mean annual precipitation for Canada (Fig. 1) is taken from the Pearse Inquiry Report (Pearse et al. 1985) and this map was based upon plates 3 and 25 of the *Hydrological Atlas of Canada*. A cautious but commendable attempt by the Atmospheric Environment Service was made in plate 25 to adjust for elevation, snow to water ratio, and undercatch. The scale of the map and the limited range of precipitation categories result in very generalized patterns of precipitation. The low values shown for the St. Elias Range, Baffin Island, and some other mountain areas seem overly cautious. However, this is still a better map for illustrating water balance patterns than more traditional precipitation maps (such as plate 3 in the atlas). More topographically adjusted isohyetal mapping on a larger scale is needed (e.g. as in Laycock 1978) and much of this mapping might be based in part upon water balance calculations working back from yield data. For hydrological purposes, precipitation should be redefined to include major condensation (including fog drip) additions to the water balance.

Potential evapotranspiration, the amount of water that might be evaporated and transpired by the energy supply available if moisture supplies were not limiting, is calculated from temperature data with an allowance for day length (latitude). The Thornthwaite procedure for doing this is empirical and is based upon observations in numerous environments throughout the world. Mather (1984), in commenting on this procedure, stated the following (p. 64): "Thornthwaite, fitting data of evaporation from watersheds and irrigation plots to air temperature, obtained the following expression for unadjusted potential evapotranspiration (in cm/month)

$$e = 1.6\left[\frac{10t}{I}\right]^{a}$$

where t is monthly temperature (°C), I is an annual heat index (determined from the sum of the 12 monthly heat index values, $I = \Sigma i$, where $i = (t/5)^{1.514}$) and t is mean monthly temperature; a is a nonlinear function of the heat index equal to

$$a = 6.75 \times 10^{-7}\, I^{3} - 7.71 \times 10^{-5}\, I^{2} + 1.79 \times 10^{-2}\, I + 0.49.$$

These expressions are complicated and mathematically inelegant. They can only be evaluated readily with the use of tables and nomograms (Thornthwaite and Mather 1957) or computer programs (Willmott 1977). Unadjusted potential evapotranspiration is the loss for a 30-day month with each day 12 hours long. This value is adjusted by a factor which expresses how the actual day and month length differ from these values."

The importance of expressing climatic water demand (P.E.) in the same units as supply (Ppt.) is evident. A representative soil moisture storage capacity (relating to soil texture and vegetative cover) can be selected and the bookkeeping calculations of balancing supply (Ppt.) and demand (P.E.), with allowances for soil moisture recharge and withdrawals, can be performed. If Ppt. exceeds P.E. and soil moisture storage, surpluses result, whereas if P.E. exceeds Ppt. and soil moisture is depleted, deficits result. Many refinements in procedure can be introduced as needed.

In most areas in Canada there are moisture shortages in some months, because P.E. exceeds Ppt. and a deficit (D) is experienced after soil moisture is depleted. The actual evapotranspiration is P.E. − D. The deficit is one of the best indices of drought intensity available and it has been widely used in climatic studies.

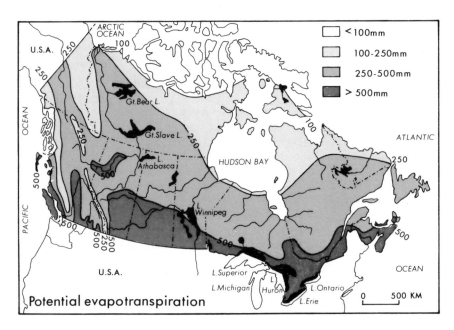

FIG. 2. Mean annual potential evapotranspiration, Canada (as defined by Thornthwaite 1948; calculation procedures of Thornthwaite and Mather 1957). Source: based on Sanderson and Phillips (1967, palte 25), Laycock (1967), Rowe (1972), and calculations for this chapter.

The map of potential evapotranspiration for Canada (Fig. 2) is based upon plate 25 of the *Hydrological Atlas of Canada* (for those areas with zero or minor deficits because actual evapotranspiration is shown in the plate rather than potential evapotranspiration), as well as on Sanderson and Phillips (1967), Laycock (1967), Rowe (1972), and calculations for this chapter (including information shown in Table 5). Annual precipitation is greater than potential evapotranspiration in most of Canada with the exception of the drier prairie and western mountain valley areas. Much larger areas have seasonal deficits (Fig. 3), however, because many areas that have surpluses in winter, spring, or early summer have greater P.E. than Ppt. in summer and fall and deficits are experienced after soil moisture has been depleted. A single deficit map based upon calculations for 100-mm storage capacity can be used to display general patterns but deficit maps for other storage capacities can be selected to revise more specific local paterns. For example, a map based upon 250-mm storage capacity would be more appropriate for forest areas in coastal British Columbia and the deficits shown in it would be much smaller than those in the map based upon 100-mm storage capacity (Fig. 3). Maps based upon 13- or 50-mm storage capacity would be more appropriate for many bare rock, shallow soil, and paved areas and the deficits shown in these maps would be larger than those shown in Fig. 3.

The calculated water surplus is the residual of Ppt. − P.E. after soil moisture has been recharged to the capacity selected. Different capacities relate to different soil types and rooting depths, e.g. 13 mm for a bare sand or bare rock or a largely paved urban commercial district, 50 mm for fallow, 100 mm for cereal grains and moderately grazed pasture, 150 mm for forage crops and better pastures, and 250 mm for mature woodland (adapted from Mather (1978) and a number of regional studies in the University of Alberta, Department of Geography (1964-84)).

The map of mean annual runoff for Canada (Fig. 4) was also taken from the Pearse Inquiry Report (Pearse et al. 1985) and was adapted from plate 24 in the *Hydrological*

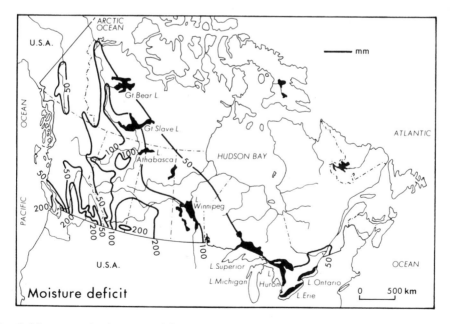

FIG. 3. Mean annual soil moisture deficits, Canada (as defined by Thornthwaite 1948; calculation procedures of Thornthwaite and Mather 1957). Source: based on Laycock (1967) and calculations for this chapter.

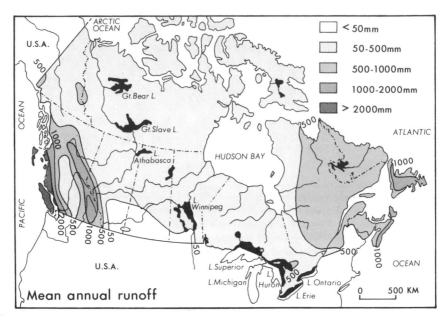

FIG. 4. Mean annual runoff, Canada. Source: Pearse et al. (1985).

Atlas of Canada and other reports. The patterns for southern Canada appear to be reasonable for most areas where streamflow gauging stations are abundant. Some changes to western patterns that would show higher values in the St. Elias Range and some other mountain areas would improve the accuracy of the map. The 50- to 500-mm category covers too broad a geographic area and it should be subdivided. Controversy concerning runoff in northern areas, where gauging stations are few, has been avoided by the use of such a broad runoff category.

More detailed maps of local runoff patterns would be useful but not many are available. Some have been prepared for Newfoundland (Shawinigan Engineering Co. Ltd. 1970), the eastern slopes of the Rockies (Laycock 1957), and parts of Quebec, etc., for particular purposes. Some provincial atlases have improved general maps reflecting topographic patterns (e.g. *Atlas of British Columbia* (Chapman et al. 1956; Farley 1979) and the atlas on *Water Quantity Resources of Ontario* (Ontario Ministry of Natural Resources 1984)).

Local variations in water balance under different assumed soil moisture storage capacities are indicated in Table 4 for Saskatoon and Toronto. In Saskatoon, surpluses are zero in most years when higher storage capacities are assumed because precipitation is rarely sufficient to recharge soils to these levels. Only when lower storage capacities are assumed are there substantial surpluses, varying greatly from year to year. The effect upon surpluses of a change in land use from natural grassland or cropland to urban uses can be dramatic (from 4 or 14 mm of surplus to as much as 91 mm of surplus). Similarly, the larger surpluses deriving from fallow (49 mm) than from crop areas (14 mm), with increased problems of salinization from groundwater movement, can be illustrated.

In Toronto, during the period 1971–80, there were surpluses in all years for all assumed storage capacities but yields would have differed substantially with different land uses. Similar variations in seasonal deficit can be shown to be dependent on land use patterns.

TABLE 4. Average water balance (1971-80) for a range of storage capacities in the Saskatoon and Toronto areas. Based upon Environment Canada (1971-80) and Thornthwaite and Mather (1957).

Storage capacity (mm)	Ppt. = (P.E. – D) + surplus					Surplus range
Saskatoon						
13	349 =	(548 –	288) +	91		52 to 176
50	349 =	(548 –	246) +	49		1 to 138
100	349 =	(548 –	211) +	14		7 × 0 to 88
150	349 =	(548 –	201) +	4		9 × 0 to 38
250	349 =	(548 –	197) +	0		10 × 0
Toronto						
13	815 =	(610 –	177) +	391		255 to 479
50	815 =	(610 –	137) +	344		214 to 440
100	815 =	(610 –	88) +	283		214 to 388
150	815 =	(610 –	46) +	252		164 to 388
250	815 =	(610 –	1) +	198		65 to 298

TABLE 5. Water balance at selected Canadian stations, averaged for the period 1971-80. Soil storage capacity assumed to be 100 mm (Thornthwaite and Mather 1957). Based upon Environment Canada (1971-80) and Thornthwaite and Mather (1957).

	Ppt. = (P.E. – D) + surplus			
Whitehorse	282 =	(392 –	132) +	17
Yellowknife	275 =	(413 –	171) +	32
Inuvik	252 =	(284 –	73) +	41
Resolute	135 =	(54 –	0) +	78
Prince Rupert	2679 =	(542 –	1) +	2136
Vancouver	1161 =	(643 –	143) +	660
Oliver	274 =	(649 –	395) +	23
Glacier	1606 =	(366 –	5) +	1249
Fort Vermilion	402 =	(508 –	140) +	41
Grande Prairie	448 =	(512 –	110) +	43
Medicine Hat	328 =	(593 –	273) +	7
Saskatoon	349 =	(548 –	211) +	14
Estevan	439 =	(536 –	171) +	85
Winnipeg	464 =	(564 –	116) +	30
The Pas	445 =	(490 –	106) +	59
Churchill	386 =	(252 –	5) +	132
Thunder Bay	707 =	(518 –	64) +	250
Sudbury	874 =	(553 –	59) +	392
Windsor	882 =	(624 –	74) +	325
Toronto	815 =	(610 –	88) +	282
Ottawa	957 =	(607 –	74) +	417
Montreal	934 =	(586 –	64) +	405
Quebec	1295 =	(549 –	1) +	742
Chibougamau	1021 =	(456 –	0) +	571
Sept Îles	1167 =	(421 –	4) +	726
Charlottetown	1165 =	(542 –	37) +	667
Fredericton	1182 =	(567 –	30) +	652
Halifax	1536 =	(544 –	16) +	1013
St. John's	1483 =	(468 –	1) +	1032

An average soil moisture storage capacity of 100 mm is often selected for regional comparisons (as in table 5). The capacities selected should, however, be appropriate for the study area soil and vegetative cover (depth of rooting, etc.) patterns.

The surplus water may take different routes to streams, as surface runoff, interflow, or groundwater flow. Although the routing pattern is important to the streamflow regime (e.g. groundwater flow has a flatter seasonal curve than surface runoff) and quality (e.g. surface runoff is usually more turbid than interflow and groundwater flow is higher in dissolved solids), the annual yield will not be greatly affected.

The storage change, in both soil moisture and snow detention, may vary greatly seasonally and from year to year. However, if we calculate water balance over a 10-yr period (as in Tables 4 and 5) or more, we might anticipate that the average change will close to zero and that it can be ignored despite an imperfect balance.

Water balances are usually calculated using monthly average data (as in Tables 4 and 5), but surpluses and deficits may be hidden by this bookkeeping procedure, especially for areas with lower storage capacities. The use of daily data can result in more accurate estimates because brief storms in which precipitation exceeds soil storage capacity are often lost in a composite monthly P.E. − Ppt. balance. Urban storm runoff, for example, may be much greater than that indicated by monthly averages of P.E. − Ppt. (see chapter by MacKenzie, this volume).

Northern centres such as Whitehorse, Yellowknife, Inuvik, and Resolute, as was mentioned previously, are in the warmer and drier sites of their regions (Table 5). The calculated water surpluses, although large relative to the precipitations listed, are small compared with those in southern Canada. Streamflow measurements made near these northern centres in recent years indicate much larger yields (two to five times larger) and it is apparent that the cooler, wetter uplands nearby have larger surpluses. Precipitation undercatch and higher yields from bare rock and shallow soil areas with storage capacities well under 100 mm are other reasons why northern yields are probably much larger than those shown in Table 5. If water balance calculations for these areas are corrected on the basis of recent streamflow measurements and other considerations, and these corrected values are projected to other northern areas, one must conclude that streamflow in the north is much greater than has been previously estimated.

The Cordilleran regions of the west have many areas with very high yields (e.g. Prince Rupert, Vancouver, and Glacier in Table 5) and interior valleys that are very dry (e.g. Oliver in Table 5). Some areas that have large yields from winter precipitation have droughts in summer. In other areas the forest cover can draw upon more than 100 mm of soil moisture storage and thus suffer a smaller deficiency than the one indicated for Vancouver (Table 5). Elsewhere there are bare rock surfaces that provide large yields, in part because most of the precipitation runs off quickly, leaving dry surfaces and larger deficits than those listed.

The western interior plains have relatively small yields because the precipitation which occurs largely in summer is usually exceeded by potential evapotranspiration (Fort Vermilion, Grande Prairie, Medicine Hat, Saskatoon, Estevan, Winnipeg, and The Pas in Table 5). Soil moisture storage capacities are large and much of the rain and snowmelt of other seasons is stored for summer evaporation and transpiration. Surpluses are derived largely from snowmelt and occasional heavy rains and they vary greatly from year to year. There is major variation in the drainage area of many basins from one year to the next because local drainage into depressions (a heritage of glaciation) is often not large enough for overflow into streams that will eventually reach the sea (Stichling and Blackwell 1957). Evaporation from the sloughs and lakes and transpiration from vegetation on their margins account for much of the local surface and groundwater flow. In the muskeg areas of the northern plains the summer evapotranspiration is usually in excess of precipitation and much of the water detained on land surfaces is lost so that yields are often very low. Phreatophyte (plants that require more water than is available from precipitation and obtain it largely from interception of groundwater flow) growth in groundwater discharge areas is evident in many areas of the Prairies (Toth 1962, 1968; Meyboom 1963; Freeze 1969). Despite this, groundwater recharge in most plains areas is too small, and the amount of phreatic growth is too meagre, for me to suggest more than a limited role for groundwater in plains water balances. A large part of plains streamflow is from surface runoff, and in the wetter years, this can be large and flashy with widespread flooding despite the low average yields.

The Canadian Shield has many bare rock and shallow soil areas and water yields are larger than those indicated (Churchill, Thunder Bay, Sudbury, Chibougamau, and Sept Îles in Table 5). In the drier western Shield, evaporation from lakes and marshes often exceeds precipitation upon them and outflows are exceeded by inflows in these years. The lakes and marshes in the Shield provide detention storage for the runoff from adjoining areas and the flow regimen of rivers are, thereby, greatly improved for hydro power and other uses. Reservoir storage can be added relatively cheaply. The greater precipitation in the east (e.g. Chibougamau and Sept Îles) and lower P.E. in the North are reflected in greater yields.

The Great Lakes–St. Lawrence lowlands and the Appalachian regions in southeastern Canada have moderate to large yields reflecting a relatively high precipitation, especially in the east (e.g. Halifax and St. John's in Table 5). P.E. is also relatively large, however, and deficits are experienced in summer, especially in the southwest (e.g. Windsor, Toronto, and Ottawa in Table 5). The greatest yields are from snowmelt but summer rains are often great enough to produce a secondary summer peak in yield. Flooding can be attributed to both snowmelt and heavy rains.

As our understanding of water balance patterns and local variations improves, we shall be able to more accurately estimate water surpluses for gauged and ungauged rivers and streams. By projecting regionally corrected values to areas in which the data base is poorer, we shall obtain better estimates of total water yield in Canada. Regime and natural quality patterns are closely related. More detailed studies of water balance relationships to complement stream gauging programs are needed.

Provincial Distribution of Streamflow

It is useful to have streamflow data for the provinces and territories of Canada as well as for the individual drainage basins. The administration of natural resources, including water, is primarily a provincial responsibility, with the federal government having a growing role in interprovincial and international concerns and in certain sectors such as navigation, fisheries, and agriculture and a cooperative role in data gathering, research, and many other sectors (Pearse et al. 1985; Thompson, this volume). The water supply is not evenly distributed between or within provinces and the per capita availability varies greatly (Table 6).

The total of 3750 km^3 of "streamflow" originating within Canada includes ice flow, interflow, and groundwater flow across boundaries and directly to the oceans plus the flow of streams in internal drainage basins (see earlier section). The auxillary flows noted above, conservatively estimated at 75 km^3, are in need of closer study.

The estimate of 160 km^3 of streamflow for the Yukon Territory includes a large ice and meltwater flow into Alaska and British Columbia. The yields of the high-snowfall St. Elias Range are undoubtedly large and can only be guessed at, as there does not appear to have been a close study of the volumes involved on the Canadian side of the border.

In the Northwest Territories, the flows of most of the smaller basins are unrecorded and too few representative streams are gauged (especially in the Arctic Islands) to permit more than a very rough estimate of total streamflow. In many areas, the water yields of the streams that are gauged are greater than the recorded precipitation (e.g. near Resolute). Thus, we know that the precipitation station data are unrepresentative for their areas, that the precipitation measurements are characterized by severe undercatch,

TABLE 6. Streamflow (km^3) originating within and passing through the provinces and territories of Canada. Based upon Laycock (1985).

	Originating in province or territory	% of Canadian streamflow	Passing through province or territory	Available per capita (census 1981)
Yukon Territory	160	4.2	25	7 980
Northwest Territories	900	24.0	190	23 696
British Columbia	900	24.0	70	353
Alberta	72	1.9	68	63
Saskatchewan	58	1.5	17	77
Manitoba	98	2.6	78	172
Ontario	335	8.9	175	59
Quebec	810	21.6	280	170
New Brunswick	47	1.3	21	98
Nova Scotia	46.5	1.2	—	55
Prince Edward Island	3.5	0.1	—	28
Newfoundland	321	8.6	—	565
	3750.0	100	255	165

or that condensation upon surfaces is a large and unrecorded component of moisture receipt. Summer evapotranspiration losses are significant in interior mainland areas but they are much lower in coastal areas and on most islands because of the cold summers. In my view, the Arctic Islands probably have a yield of over 500 km^3 but water balance projections have not been widely used to evaluate streamflow in this area. Environment Canada estimates are still around 300 km^3 (Kallio 1985). The groundwater and interflow components are small because of permafrost but glacier flow to the sea is significant on eastern and northern islands.

The estimate in Table 6 for British Columbia is 100 km^3 greater than my earlier estimate (Laycock 1985) because it has become apparent that the yields calculated for central and northern coastal and coastal mountain areas were much too low. Glacier and meltwater flows in these areas are also large and the very long coastline must have substantial interflow and groundwater flow directly to the Pacific. The flows of additional representative coastal streams should be recorded.

The estimates for the remaining provinces have undergone little change from those which I made earlier (Laycock 1985). I have made some additional allowance for internal drainage basin streamflow in the Prairies but the amounts are small. Larger allowances may be justified because many local depressions have little or no outflow in most years and, to be consistent, we should include streamflow, interflow, and groundwater flow into these depressions. As noted by Stichling and Blackwell (1957), Mayboom (1963), and others, these volumes can be substantial — perhaps as much as 2 or 3 km^3/yr for the Prairies.

Larger allowances than those which I used can probably be made for interflow and groundwater yields directly to the Atlantic. The very long coastlines, permeable glacial drift deposits, and lack of permafrost are factors in such estimation. Condensation upon surfaces (fog drip, etc.) and the cool summer temperatures (lower P.E.) in coastal areas result in greater yields than are present even short distances inland.

Yields from the Great Lakes basin are not proportional to the areas (lake and land) in the United States and Canada. The Canadian side is cooler in summer with lower evapotranspiration and there are larger areas of bare surface and shallow soils in the Shield that have relatively high yields. The expanding urban, paved highway, and other

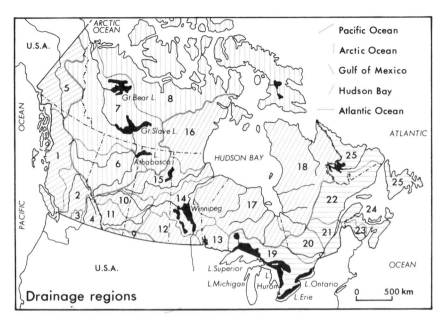

FIG. 5. Drainage regions, Canada. Source: Pearse et al. (1985).

Ocean basin region	River basin region	Area (10^3 km^2)	Population in 1981 (10^3)
Pacific	1. Pacific coastal	352	616
	2. Fraser - lower mainland	234	1 722
	3. Okanagan-Similkameen[a]	14	189
	4. Columbia[a]	90	161
	5. Yukon[a]	328	23
Arctic	6. Peace-Athabasca	487	286
	7. Lower Mackenzie	1300	43
	8. Arctic Coast - Islands	2025	13
Gulf of Mexico	9. Missouri[a]	26	14
Hudson Bay	10. North Saskatchewan	146	1 084
	11. South Saskatchewan[a]	170	1 282
	12. Assiniboine-Red[a]	190	1 300
	13. Winnipeg[a]	107	77
	14. Lower Saskatchewan - Nelson	363	224
	15. Churchill	298	68
	16. Keewatin	689	5
	17. Northern Ontario	694	157
	18. Northern Quebec	950	109
Atlantic	19. Great Lakes[a]	319	7 579
	20. Ottawa	146	1 270
	21. St. Lawrence[a]	116	5 193
	22. North Shore - Gaspé	403	653
	23. St. John - St. Croix[a]	37	393
	24. Maritime coastal	114	1 314
	25. Newfoundland-Labrador	376	568
Canada		9974	24 343

[a]Canadian portion only; area and population on American side of international basin regions are excluded from totals.

impermeable surfaces on especially the American side, however, are resulting in increased flows characterized by degraded quality and flashy flow regimen.

Much of the streamflow passing through most provinces orginates in other provinces or the United States. In Manitoba, Ontario, and Quebec this flow of external origin is available largely in the south. In Alberta, it is largely in the north but some of this could be available to the south, at considerable cost, by diversion. Southern Saskatchewan has moderate inflow but demands for it are already competitive.

Per capita supplies in Canada are large (Tables 2 and 6) but there are regional shortages, only partly indicated in Table 6 (the demand patterns are discussed by Tate in this volume). For example, Alberta accounts for close to half of the consumptive use of water in Canada (two thirds of the irrigation and large mining use in the recovery of oil and gas including from oilsands). The irrigation demand is concentrated in the South Saskatchewan Basin, which has about 6% of the Alberta water supply (half of which is committed to downstream use in Saskatchewan and Manitoba). The area under irrigation could be doubled with better storage from wet to dry seasons and from wet to dry years, and with more efficient use of the available supply. There is growing competition for water from other users, however, and the pressures for interbasin transfer from the north are becoming stronger (see Bocking, this volume). In the more humid parts of Canada the demands for consumptive use are relatively small and the potential for reuse is greater. In addition, there are smaller variations in flow from season to season and year to year, and storage in lake and groundwater supplies is much greater. Thus, the shortages of the drier years are much less acute than those of the drier regions. For a better perspective on regional supply/demand relationships, we must have more information on basin supply patterns within provinces, seasonal and annual variations in flow and storage, and the consumptive, withdrawal, onstream, instream, and beside-stream demand patterns.

Drainage Basin Streamflow

The major river basins of Canada are outlined in Fig. 5 and the areas and populations of these basins are shown in the table in the caption to Fig. 5. The streamflows of these basins are listed in Table 7 and summaries of the above tables with percentages are listed in Table 8.

The very great diversity, and the lack of correspondence in the distribution of areas, populations, and streamflow in Canada, are illustrated in these tables. The pacific drainage basins have just over 10% of the area and population and about 25% of the water yield. Within this region, close ot half of the population lives in or near Vancouver, and other cities and towns in the southwestern corner of British Columbia contain most of the remainder. Water yields of over 2500 mm occur in many coastal mountain and some interior mountain basins, yet rain-shadow valleys, such as the Okanagan, have local yields approaching zero.

The Arctic and Hudson Bay drainage basins have close to 75% of the area of Canada and less than 20% of the population. They have almost 50% of the streamflow despite the relatively low precipitation over much of these regions. This occurs because evapotranspiration is low, especially in the north (see earlier water balance discussion). Averages for such a large basin are deceiving with respect to supply/demand imbalances, however. The southern portions of the Prairies, for example, including the South Saskatchewan, Assiniboine-Red, and Missouri basins (regions 11, 12, and 9 in Fig. 5 and Table 7) have 3.9% of the area of Canada, 10.7% of the population, but only 0.3% of the streamflow. Water demands, largely for consumptive use, are approaching supply levels in this region, and shortages in the drier years are increasingly widespread.

The Great Lakes, Ottawa, and St. Lawrence basins (regions 19, 20 and 21 in Fig. 5 and Table 7) have well over half of the population of Canada, living in only 5.8% of the area, and with only 6-7% of the streamflow originating within Canada. This region has inflow from the United States equal to less than 3% of Canadian streamflow,

TABLE 7. Canadian annual river flow rates. Based upon Pearse et al. (1985, p. 28).

| Ocean basin region | River basin region | Annual flow rates (m^3/s)[a] | | |
		Reliable[b] (low)	Mean	High[c]
Pacific	1. Pacific coastal	12 570	16 390	20 200
	2. Fraser-lower mainland[d]	3 044	3 972	4 900
	3. Okanagan-Similkameen[e]	31	74	116
	4. Columbia[e]	1 644	2 009	2 373
	5. Yukon[e]	1 806	2 506	3 206
Arctic	6. Peace-Athabasca	1 862	2 903	3 946
	7. Lower Mackenzie[f]	6 114	7 337	8 561
	8. Arctic Coast - Islands	5 920	10 251	14 582
Gulf of Mexico	9. Missouri[e]	3	12	41
Hudson Bay	10. North Saskatchewan	160	234	373
	11. South Saskatchewan	147	239	418
	12. Assiniboine-Red[e]	16	50	188
	13. Winnipeg[d,e]	382	758	1 137
	14. Lower Saskatchewan-Nelson[e,f]	1 108	1 911	2 714
	15. Churchill[d]	323	701	1 070
	16. Keewatin	2 945	3 876	4 806
	17. Northern Ontario[d]	3 733	5 995	8 258
	18. Northern Quebec[d]	12 820	16 830	20 830
Atlantic	19. Great Lakes	2 403	3 067	3 733
	20. Ottawa	1 390	1 990	2 590
	21. St. Lawrence[e,f]	1 504	2 140	2 777
	22. North Shore-Gaspé	6 437	8 706	10 980
	23. St. John-St. Croix[e]	507	779	1 050
	24. Maritime coastal	2 079	3 081	4 085
	25. Newfoundland-Labrador	6 908	9 324	11 739
Canada		75 856	105 135	134 674

[a]From recorded flows except in Prairie basins where natural flows have been estimated.
[b]Flow equalled or exceeded in 19 out of 20 years.
[c]Flow equalled or exceeded in 1 out of 20 years.
[d]Excludes flow transferred into neighbouring basin region; because this flow is recorded in importing basin, transfers have little effect on national total.
[e]Excludes inflow from United States portion of basin region.
[f]Excludes inflow from upper basin region.

a mixed blessing because of water quality problems. The water balance patterns of the region result in relatively small deficits, and supplementary irrigation is used in relatively small areas for some crops, Consumptive uses of water are relatively small and supplies, aside from quality problems, are abundant.

The Atlantic drainage basins east of the St. Lawrence River have 9.3% of the area, 12.0% of the population, and close to 20% of the streamflow. The water use patterns involve relatively small consumptive use, and since water deficits are small or absent, there are large, dependable yields for nonconsumptive, instream, onstream, and beside-stream uses. Hydroelectric power, recreational fishing, and urban and industrial uses are among those most favoured.

TABLE 8. Summary, Canadian basin areas, streamflows, and populations. Source: Pearse et al. (1985).

Basin	Area (10³ km³)	%	Mean steamflow (m³/s)	%	Population in 1981 (10³)	%
Pacific	1018	10.2	24 951	23.7	2 711	11.1
Arctic	3812	38.2	20 491	19.5	342	1.4
Gulf of Mexico	26	0.3	12	0.01	14	0.1
Hudson Bay	3607	36.2	30 594	29.1	4 306	17.7
Great Lakes - St. Lawrence[a]	581	5.8	7 197	6.8	14 042	57.7
Atlantic[b]	930	9.3	21 890	20.8	2 928	12.0
Canada	9974	100	105 135[c]	100	24 343	100

[a]Regions 19, 20, and 21 in Fig. 5.
[b]Regions 22–25 in Fig. 5.
[c]The estimate by Laycock for this chapter is 118 830 m³/s with most of the addition being in the Arctic and Pacific basins. The percentages would then be Pacific 25.3%, Arctic 23.1%, Hudson Bay 26.3%, Great Lakes - St. Lawrence 6.1%, and Atlantic 19.2%.

The largest river basins, in area, in Canada are the Mackenzie, Nelson, St. Lawrence, Chirchill, and Yukon. The largest rivers, in annual flow, are the St. Lawrence (rising partly in the United States), Mackenzie, Fraser, Columbia, and Nelson. There is a problem of defining where in a delta or estuary the flow becomes sufficiently brackish that it is no longer "fresh." The St. Lawrence River above the mouth of the St. Maurice and the Mackenzie in its delta are approximately equal in annual flow, changing in leadership from high- to low-flow years (the Mackenzie is more variable) and with different periods of record.

The *Hydrological Atlas of Canada* has a useful table (accompanying plate 22) showing annual large river flow in which the drainage areas and mean discharges of 40 rivers are listed. The full basins are represented rather than those portions above specific gauges thus some estimation was needed. Such estimation can best be done using water balance projections.

The Water Survey of Canada collects streamflow data at more than 3000 gauging stations. The *Historical Streamflow Summaries* and *Surface Water Data for the Provinces* are the basic data sources (Environment Canada 1970–85). The data collection program is excellent but if the interpretation program was oriented more toward assessments of water balance and representative stations, it is probable that a more efficient distribution of stations could be selected and better conclusions might be reached. New stations might be used to extend and correct or confirm interpreted patterns rather than just add to gauged area totals. Program rationalization, always in progress, might be better based.

Flow Regime Patterns and Flow Variation

Plate 23 in the *Hydrological Atlas of Canada* is a set of 36 graphs, over a map of Canada, showing maximum and minimum monthly mean flow and mean monthly river flow at selected gauging stations in Canada. The periods of record and the drainage areas are listed. Reference to these graphs is recommended for those who wish to review seasonal flow patterns in Canada. Similar maps and graphs may be found in regional atlases (e.g. Drinkwater et al. 1969) and professional reports and journal articles for most parts of Canada. The stress tends to be upon the larger river patterns, and the flashier flow characteristics of the smaller tributaries may be lost in lake and reservoir storage and in basin lag and averaging patterns in the larger basins. The spring and summer maximum flow is evident in most parts of Canada, and since precipitation in most areas east of the Rocky Mountains also has a spring or summer maximum, it is widely assumed that seasonal streamflow is in direct response to seasonal precipitation. In most areas, however, it is not. Seasonal water balance patterns reveal that winter snowfall is subject to relatively small losses due to evapotranspiration whereas much (in some instances, most) of the rainfall in summer is lost to evapotranspiration, usually after a brief period of soil moisture storage. There are regional exceptions — e.g. where soil moisture storage is limited, as in bare rock, shallow soil, and paved areas, or where evapotranspiration is very low because of low temperatures, or where precipitation is very high or concentrated in exceptional storms — but in most areas and most years, the yield from snowmelt is greater than that from rainfall.

A very good, but little known, study of seasonal flow regimes in Canadian rivers is that by Seifried (1972). She has analyzed most of the Canadian streamflow data available on computer tapes to 1970 according to procedures used by the International Hydrologic Program Committee of the International Geographical Union. She concluded that, in most parts of Canada, the maximum monthly flow is determined more by thermal conditioning for melting snow than by direct precipitation.

The major exceptions to a snowmelt maximum in the flow regime are in low level areas on the Pacific Coast where winter rains contribute directly to runoff (at higher elevations, snowfall is again significant) and in some high-elevation and high-latitude locations where meltwater flows from glaciers and icefields create a later summer maximum (this latter pattern is also a consequence of temperature conditions, however).

Latitude and elevation will obviously affect the timing of snowmelt and maximum flow. February and March are maximum flow months in middle-level coastal and low-level southern interior British Columbia, the southwesternmost parts of the Prairies, and parts of southern Ontario. April is the month of maximum runoff in much of British Columbia, the southern interior plains, much of southern Ontario, the St. Lawrence lowlands, and the more populated parts of the Atlantic Provinces. Runoff maxima in May and June are most common in the higher and northern areas of the western Cordillera, the northern interior plains, the Precambrian Shield, and the higher areas of the Atlantic Provinces. Most of the rivers that peak in July or later because of snowmelt are at high elevation or high latitude. In addition, some rivers with major basin lag, and/or storage in lakes, have their peak discharges a month or more after local runoff takes place. Secondary maxima are often experienced in October–December in the more humid parts of Canada, after the main period of evapotranspiration and before freeze-up.

A minimum in flow occurs in late winter in much of Canada, especially in the North, and is caused by seasonally low temperatures. In the drier parts of the Prairies, local flow is frequently zero in mid- to late summer and sometimes in autumn when potential evapotranspiration greatly exceeds precipitation and soil moisture storage (recharged in spring) has been fully depleted.

The lag time between snowmelt or precipitation and flow response and the spreading out of flow is greater in Canada than in most countries because of our extensive natural lakes and wetlands and our growing artificial storage in reservoirs. The Great Bear River flowing out of Great Bear Lake has a coefficient of discharge[4] that ranges from 1.12 to 0.89 from the highest to lowest monthly flow (Seifried 1972). The coefficients for the Niagara and St. Lawrence rivers range from 1.08 to 0.90 but some artificial storage and diversions are present (Seifried 1972). The contrast in regimen between the Red

---

[4]If annual flow is evenly distributed, all months would have a coefficient of 1.

River and Winnipeg River in southern Manitoba is striking: the coefficients for the Red River range from 4.34 to 0.17 and those for the Winnipeg River range from 1.3 to 0.89 (Seifried 1972). The natural contrast would not have been as great as this because the Winnipeg River has both storage and diversion for hydro power whereas many of the marshes and small lakes in the Red River Basin have been drained (uncontrolled) for agricultural land development so that the extremes in discharge in the Red River have been accentuated.

The role of groundwater in determining streamflow regimen is relatively small in Canada. In permafrost areas, groundwater movement is limited and most runoff is surficial. In the drier plains, groundwater recharge is very small and much of the limited streamflow is the product of snowmelt (including snow drift melt even in the drier years). Surface runoff from intense storms occurs with little groundwater or even interflow[5] routing. The massive igneous and metamorphic rock of much of the Precambrian Shield and parts of the western Cordillera and Appalachian regions have limited groundwater flow. Much of what appears to be a groundwater recession curve in streamflow hydrographs is actually basin and lake storage lag, deferred meltwater flow from higher elevations, and small storm yield from low soil moisture storage areas. In the more humid areas of southern Canada, however, the role of groundwater in flow regime modification is significant.

Duration curves of monthly flow, such as the 60 graphs published in the preliminary maps of the *Hydrological Atlas of Canada* (Canadian National Committee for the International Hydrological Decade 1969) are very useful in showing the proportions of time a stream has high, medium, and low flows. These were not included in the atlas but some are shown in Fig. 6, in part because the preliminary maps are no longer widely available. A very flat curve — e.g. St. Lawrence River at Cornwall, Winnipeg River at Slave Falls, and Churchill River at Granville Falls — is usually indicative of major lake storage, which in many cases is augmented by artificial storage. A curve with flooding and low flow extremes — e.g. Matane River near Matane, Grand River at Galt (Cambridge), and Assiniboine River at Headingly — is indicative of limited marsh and lake storage, little, if any, artificial storage, and short-duration snowmelt or heavy rain periods. In general, the larger basins will have more basin lag and will be less affected by individual storms. The smaller basins will have less basin lag and may be greatly affected by individual storms.

The variations in flow from year to year for major rivers in Canada are partly indicated in Table 7 in which mean flow data are compared with the low flows that might be equalled or exceeded in the driest 5% of the years and the high flows that might be equalled or exceeded in the wettest 5% of the years. Most of these river basins contain tributary basins with a wide range of water balance characteristics, e.g. the North and South Saskatchewan basins encompass very humid mountains, humid foothills, and humid and subhumid plains. The mountains provide large and relatively dependable yields and the great variations in the flow from drier plains may not be evident in total flow. In general, the reliable low flow[6] tends to be over 75% of the mean flow in the larger, more humid basins but it is less than 25% of the mean flow in the smaller, drier basins (Table 7). With artificial storage and interbasin transfers, some variations in seasonal or interannual flow can be reduced but it is very costly to store much water for a succession of dry years. In some regions, the use of natural storage in lakes and groundwater supplies may help to resolve the shortages of dry years.

Artificial storage may accentuate rather than reduce flow variation, e.g. storage for irrigation may result in greater flow in summer upstream of the irrigation districts but

---

[5]Interflow is lateral transfer above the water table yet below the ground surface.
[6]That which can be expected in 19 out of 20 years.

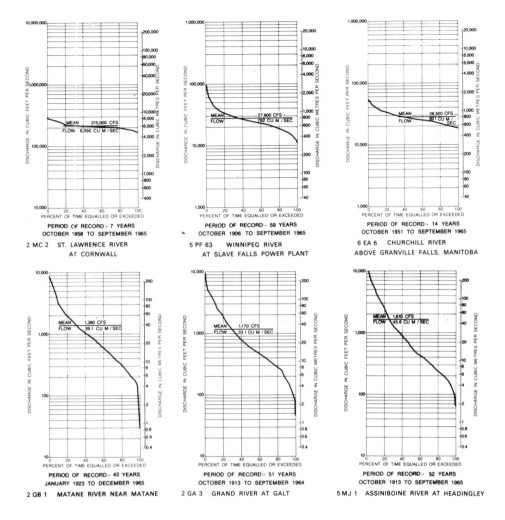

FIG. 6. Duration curves of monthly streamflow at selected streamflow gauging stations, Canada. The duration curves are of monthly flow only. The shorter term extremes of flooding and low flow in the periods of each graph lie beyond the range indicated. Source: Canadian National Committee for the International Hydrological Decade (1969).

withdrawals for irrigation, especially in the drier years, may greatly deplete summer flow downstream. Storage for hydro power will usually result in a less variable flow. In those areas with thermal power, however, the flow for hydro power may be largely in peak demand periods and significant flow variations (diurnal, seasonal) may result, e.g. on the Bow River and some of its tributaries upstream from Calgary.

Many other types of graphs, tables, and maps may be used to illustrate streamflow patterns but most are more amenable to local detailed analyses (e.g. of flood events), and summary discussions of national patterns would be difficult to provide even if events in most parts of the country had been analyzed.

Various means of overcoming supply deficiencies have been outlined in the literature. In most cases if the alternative means studied include improvements in the efficiency of use and ways of conserving water to reduce demand, it will be found that massive engineering programs to increase supply will not be needed, or may be deferred well into the future. Despite this likely conclusion for many deficiency studies (especially in periods of economic recession), the widespread regional imbalances of supply and demand noted earlier are a continuing basis for proposals for interbasin transfers of water from water surplus to water deficiency areas.

Interbasin Tranfers

Canada already has a larger volume of interbasin transfer than the United States and the U.S.S.R. combined (Quinn 1981). Most transfers are for hydro power production, and irrigation, municipal, and other demands are still very subordinate in actual development. The proportion of total streamflow that had been diverted was still small in 1980, about 4400 m^3/s (approximately 140 km^3/yr) of 3.7% of total streamflow in Canada. Nevertheless, this was 43% greater than the flow of the St. Lawrence below Lake Ontario. The proportions of these water diversions by Province were Quebec 42%, Manitoba 18%, Newfoundland 15%, Ontario 13%, British Columbia 8%, Alberta 1.5%, and other 1.1% (Day 1985). Transfers specifically for irrigation amounted to only 1% of the total and these were largely in Alberta and Saskatchewan.

Many proposals for major interbasin transfer involve Canadian water export to the United States. It should be apparent that the physical potential for transfer is large. However, the costs of transfer including construction, operation, and maintenance costs, opportunity costs, environmental costs (of keeping environmental damage to a minimum and providing for at least compensating environmental enhancement), carrying charges, and a profit (to cover much of the risk involved) would greatly exceed the direct capacity of potential users to pay (National Water Commission 1973). The political, legal, administrative, and other costs must also be considered and there are enough problems of development involved that large-scale transfers are unlikely for many years to come (see Bocking, this volume). Despite the unattractive economics of transfers to the United States, promoters of various schemes continue to be enthusiastic and political support is always possible. Most of the major proposals are irrigation oriented but it is possible that political support may be greater in the long run for transfer into the Great Lakes from Hudson Bay drainage basins with multiple-purpose objectives in mind. These could include greater transfer southward to carry more urban effluent away from the Great Lakes (as is happening at Chicago now) and to provide better recreational opportunities, greater hydro power, better urban and industrial supplies, and higher levels in low-level years for improved navigation, etc. Products pipelines, with the water medium for sale, as well as the products carried, may be significant in the future but the volumes transferred would be relatively small. Planning on a major scale must be long range but assurances of long-range availability of extra water and extra construction can be valuable stimuli for the economies of the areas involved long before any transfers take place. We must be extremely careful not to make inflexible commitments, e.g. in exchange for short-term advantages in trade in other sectors.

Interbasin transfers within Canada, and especially within provinces, are still expanding where users (e.g. hydro) can cover the engineering costs involved and where the opportunity, environmental, and social (e.g. native claims, expropriation) costs are accorded low values. Hydro power projects involving transfer have been phenomenally profitable (for crown and private corporations) with high energy prices in the late 1970's and early 1980's. Many proposals for interbasin transfer of water, especially in western Canada, involve transfer for irrigation rather than hydro power. In most of these, the costs will greatly exceed the capacity of irrigators to pay. Agricultural prices have risen slowly or remained static while engineering costs, energy costs, and the cost of financing have grown rapidly. Environmental issues have delayed the initiation of projects and increased costs. Despite these factors, political pressures for government-subsidized transfers are growing and massive engineering programs to divert northward flowing waters have been outlined. There are many cheaper alternatives available, however, including smaller scale sequential transfer. The attractiveness of the large, "unused" northern water resources and the presence of proglacial spillway channels, leading from supply to demand areas and excellent for storage and transfer, will be continuing bases for new proposals, promotions, and political promises.

Natural Water Quality

In a country as large as Canada there is great range in natural water quality, ranging from the colourful brown, organic carbon laden outflows of northern muskegs and the white, chloride precipitate laden playa lake waters of parts of the drier prairies to the virtually pure, sediment-free runoff of some bare rock areas in the western Cordillera, Precambrian Shield, and Atlantic Appalachian regions. Generalizations cannot be applied everywhere, yet some are valid because of the nature of our climate and water resource patterns.

Canada's widespread lakes and wetlands have modified water quality in many ways. These water bodies serve as sediment traps and many of our rivers are relatively free of mineral sediments through much of their courses. Turbidity takes other forms, however: organic matter, plankton, and microscopic organisms abound in shallow surface waters and summer flows may be clogged by this growth. In part because permafrost underlies half of the country, and rock is widely exposed in Precambrian Shield, Cordilleran, and Appalachian areas, surface runoff is a larger component of total runoff in Canada than in most countries. The converse of this is that groundwater flow is a smaller component of total runoff and our rivers and streams have a relatively low total dissolved solids content because of this. There are many regional exceptions to these generalizations. The water of many rivers in the Prairies and parts of southern Ontario is high in dissolved solids (and hardness) and streams rising in glaciers and recent glacial deposits are often high in mineral sediments.

Natural or geologic erosion varies greatly from place to place within Canada. There is relatively little erosion with low flow in winter, but this results in more concentrated erosion with high snowmelt runoff in spring. Stream gradients vary greatly, largely because of Pleistocene glaciation. There are many stretches of quiet water with lakes and marshes and aggradation alternating with stretches with waterfalls and rapids and rapid degradation. Freshwater deltas (e.g. Peace-Athabasca, Slave, Saskatchewan), floodplains, and alluvial fans are widespread but only some of our rivers carry heavy sediment loads to the oceans (e.g. MacKenzie, Fraser). The large surface runoff is not as erosive as might be anticipated because widespread forest and other cover protects surfaces from rapid erosion. In areas of cultivation, burning, and forest cutting there are problems of accelerated erosion and large sediment yields are not uncommon (see chapters by Hetherington and de Jong and Katchanoski, this volume).

Although much of what we read about pollution is related to the activities of man (see chapters by Hamilton et al. and Cherry, this volume), every stream is naturally polluted to some degree. Streams and lakes differ in turbidity, alkalinity, dissolved oxygen, temperature, dissolved solids, mineral sediment loads, and in many trace elements. Many of these, even in naturally occurring concentrations, can place severe limits upon the biota that can survive, and upon human uses. Fish and other aquatic life require that dissolved oxygen be present in moderate amounts, while industrial users know that the smaller the amount of dissolved oxygen the smaller the costs from the corrosion of equipment. A completely clear, distilled water which would be ideal for indoor swimming would not support desirable stream fauna (White 1971). Standards of cleanliness appropriate to one use need not be appropriate for others. Such standards should be defined in terms of intended uses, natural and existing qualities, and the costs of management for each water body. National standards should be modified to some degree according to basin needs, or the costs of strict enforcement would be prohibitive and the priorities accorded to fish and navigation would be misplaced relative to other uses. A greater use of the *Canada Water Act* and a lesser dependence upon the *Fisheries Act*, etc., is recommended (see Thompson, this volume).

Water Supply in the Future

Many projections of supply and demand patterns are based upon the assumption that future water balance and supply patterns will be similar to those of the past. There is growing evidence that they will not be. Increases in atmospheric levels of carbon dioxide and other greenhouse gases (methane, nitrous oxide, chloro- and fluorocarbons, low-level ozone) will increase global temperatures and alter precipitation patterns by trapping infrared radiation. The combined concentrations of these greenhouse gases will be eqivalent to a doubling of carbon dioxide from preindustrial levels (now up 20%) by 2080 to 2100 (Environmental Protection Agency 1983a, 1983b, 1984) or as early as the 2030's (Environment Canada 1986; see also Ripley, this volume). The timing will depend in part upon what we may do to slow down the process but there is little doubt about the trend and its probable effects.

With a doubling of the atmospheric carbon dioxide concentration, or equivalent, mean global temperatures will rise between 1.5 and 4.5°C. Numerous general circulation models of global climate have been employed in projections and that of the Goddard Institute for Space Studies (GISS), indicating a 4.16°C temperature rise is used in the Environmental Protection Agency (1984) analysis. North American patterns of change, centering upon conterminous United States, have been mapped for 23 grid units of 8° latitude by 10° longitude from 23° to 55°N latitude and from 55° to 135°W longitude (fewer in the south). Temperature increases of 4.2°C in the eastern United States and 4.9°C in central and western areas were indicated, with winter increases 40% greater than those of summer. Greater increases would be present in Canada, partly because albedos of surfaces would be lower for more of the year with reduced ice and snow cover, and increases of 5 to 8°C in winter and 3 to 5°C in summer may be present in much of our sub-Arctic. Growing season increases of 25-65 d (21-57%) are forecast for southern Canada (south of 55°N latitude) and proportional increases will be greater in the North. The season at Fort Vermilion, Alberta, near the present northern limits of cultivated crops, may be doubled.

The temperature changes could result, for example, in both a greater use of traditional recreational lakes and an expanded use of northern lakes in Canada by southern Canadians and Americans interested in escaping the hotter southern summers. July average daily and monthly maximum temperatures would be 5-7°F higher at 85 and 101°F for Portland, 89 and 97°F for Los Angeles, 113 and 120°F for Phoenix, 102 and 110°F for Oklahoma City, 90 and 103°F for Bismarck, 90 and 99°F for Chicago, 96 and 103°F for St. Louis, 88 and 102°F for New York, 84 and 101°F for Boston, and 93 and 103°F for Washington, DC (Environmental Protection Agency 1984).[7] In winter, with increases of 7-10°F in American centres, the southern ski resorts would suffer and those in the North would flourish, despite their greater relative temperature increases. Many other changes would be present — e.g. Winnipeg would have a climate like present-day Minneapolis and Calgary or Edmonton like present-day Denver; the northward shift of corn, winter wheat, and soybeans would add to the diversity of crops in the Prairies; northern boreal forest growth would be more rapid and the present tree line would shift far to the north.

Global precipitation will increase an average of 11% in the GISS model and most Canadian areas will experience increases of more than this (especially interior British Columbia - western Alberta up 23%). Precipitation in the Great Lakes - St. Lawrence

---

[7]Similar data are not yet available for Canadian cities.

area, however, will be down by 3-5%. Global evaporation will also increase by an average of 11%, with greater increases in most areas in Canada and a decrease of 8% in southern Ontario. Runoff will increase 41% in coastal British Columbia, 34% in interior British Columbia - western Alberta, and 14% in western Saskatchewan - eastern Alberta and Newfoundland. Runoff will be down 14-15% in southern Ontario to New Brunswick, however, and 0-22% in eastern Saskatchewan to central Quebec. The significance of changes in runoff for hydro power generation and other uses of water is great. In the United States, precipitation will increase by 58% in Oregon and decline by 26% in Kansas to Texas (Environmental Protection Agency 1984). Drought indices are negative for southern Ontario to New Brunswick and positive for western Saskatchewan to British Columbia.

The seasonal patterns are as interesting. Coastal and interior British Columbia and western Alberta will have greater precipitation in all seasons (up over 25% in winter) but summer increases will be balanced by increased evaporation. Western Saskatchewan - eastern Alberta will have greater winter and especially greater spring precipitation with little change in summer and fall. The warmer summers and falls will be drier but a shift to winter wheat and greater irrigation will result in at least as great production. Eastern Saskatchewan and Manitoba will have moderate increases in precipitation in all seasons but increased summer evaporation will make this season drier. Northwestern Ontario will have greater precipitation, especially in the fall, but annual runoff will be less because of increased evapotranspiration. James Bay to Newfoundland will have precipitation increases in winter (Newfoundland especially) and summer (James Bay) and less in fall with decreased annual runoff in central Quebec. Southern Ontario to New Brunswick will experience little change in winter, spring, and summer precipitation but pronounced decreases in fall. With greater evaporation, annual runoff will decline 14-15%. Northern Canada, not illustrated in the grid analyses, will probably have substantially greater precipitation, possible more than enough to compensate for evaporation increases in most areas, and runoff increases might be expected. The present data base is weak in the North and any projection of percentage changes from it must be unreliable.

Supplemental irrigation in subhumid areas will probably grow more rapidly than dry land irrigation and some of this will be in southern Ontario and Quebec which will be drier in summer and fall in the future. The demands for drainage of wetlands for agriculture will grow in northern areas from the Clay Belt of northern Ontario and Quebec to the northern interior plains well into the Northwest Territories. The wetland losses and erosion damage may be high in some areas but reasonable compromises in development can probably be reached. There will be many regional changes in supply patterns but Canada will probably have greater water supplies as a result of the climatic changes outlined. The reduced supplies in the southeastern parts of Canada (southern Ontario - New Brunswick) will still be abundant by most standards.

The patterns suggested above are rough but the trends seem probable. A doubling of carbon dioxide is not an ultimate level and further climatic change might be anticipated. Secondary changes resulting from the Great Lakes (and later Hudson Bay) no longer having ice cover in winter, and from albedo changes on land and other water surfaces, can be pronounced. Summer use of Arctic waterways will be commonplace and Canadian development will probably have greater latitudinal depth. Sea level rises of 100-140 cm in the next century and much more later with major melting of Antarctica and Greenland ice sheets can be devastating for most ports, estuaries, and low-lying coastlands. We can slow down the process by reducing world consumption of mineral and wood fuels, etc., but it is unlikely that the trends will be flattened much (see Ripley, this volume). We might gain more by planning for change and adapting to it.

The changes in water quality due to acid rain, toxic substance dumping and escape, nuclear disaster, and other man-made disruption may be problems of even greater urgency. The trends are not encouraging.

Technological advances in weather modification and desalination may be more significant in the future but these alternatives hold only limited promise, largely for the wetter years, and the cost will be high. Engineering solutions to problems of supply are becoming increasingly costly with decreasing return as the nearer and better sites are developed. Massive transfers seem to be increasingly unlikely but much of that now present was unanticipated three or four decades ago. Perhaps some transfer to dry regions will be undertaken, in part to reduce the rate of sea level rise and to modify regional climates by adding recycled precipitation. The limited apparent effect on climates resulting from large-scale irrigation by groundwater mining of the Ogallala aquifer in the United States would appear to make this rationale a weak one but it is partly plausible. Pipeline technology is improving and it may be a means of transfer, especially if products pipelines are more widely used, but the volumes will probably be small.

Our changing priorities in water demand, with greater stress upon wildlife, recreation, and a clean environment, may be significant. Irrigation and drainage for increased agricultural (and forest) production will be important but only part of it will be in new areas of production. Much of it will be supplemental to increase (or maintain) productivity in areas that are now producing at less than their maximum capacity. Industrial demands for new water may decline with increased reuse and recycling. Municipal demands will grow but greater efficiencies in use might be expected, as in other sectors. A conservation ethic will become stronger and more widespread. We will have substantial increases in water use but our stress in development will be less upon engineering structures and conquering nature and more upon making more efficient and better use of available high-quality water supplies and retaining attractive natural environments.

White (1971) ably identified changes taking place in man's attitude toward nature and his concomitant role in water management: "The view of man the transformer or man the conqueror that permeates so much of the single and multiple-purpose construction and that shows itself in the chart of the future as a contest between rising human demands for water and bounded natural supplies is replaced by another. In the view of man as the co-operator, man the harmonizer, construction is only one means of coming to terms with an environment he never fully explores and that is constantly changing under his hand. With the adoption of this view, the means and instruments of handling water become increasingly complex, the concern with tracing environmental impacts more acute, the adjustments to human preferences increasingly sensitive, and the demand for citizen participation heavier. The emphasis shifts from construction to scientific probing, and from long-term commitment to short-term flexibility."

References

BUDYKO, M. I. 1956. Heat balance of the earth's surface. Gidrometeoizdat, Leningrad. 255 p. (In Russian)

CANADIAN NATIONAL COMMITTEE FOR THE INTERNATIONAL HYDROLOGICAL DECADE. 1969. Preliminary maps, Hydrological atlas of Canada. Ottawa, Ont.

CASS-BEGGS, D. 1961. Water as a basic resource, p. 178. In Background papers. Resources for Tomorrow Conferences, Montreal, Que.

CHAPMAN, J. D., D. B. TURNER, A. L. FARLEY, AND R. I. RUGGLES [ED.]. 1956. British Columbia atlas of resources. British Columbia Natural Resources Conference, Victoria, B.C. 48 plates.

DAY, J. C. 1985. Canadian interbasin diversions. Inquiry on Federal Water Policy Research Paper No. 6. Environment Canada, Ottawa, Ont. 111 p.

DRINKWATER, T. A., E. S. HUESTIS, D. I. ISTVANFFY, J. J. KLAWE, A. H. LAYCOCK, AND W. C. WONDERS [ED.]. 1969. Atlas of Alberta. Government of Alberta and The University of Alberta, Edmonton, Alta. 162 p.

ENVIRONMENT CANADA. 1971-80. Monthly record. Atmospheric Environment Services, Ottawa, Ont.

          1970-85. Historical streamflow summaries and surface water data for the provinces of Canada. Inland Waters Directorate, Ottawa, Ont.

1986. CO$_2$ climate report. Climate Program Office, Atmospheric Environment Service, Issue 86-1, 10 p.

ENVIRONMENTAL PROTECTION AGENCY. 1983a. Can we delay a greenhouse warming? United States Government Printing Office, Washington DC. 203 p.

1983b. Projecting future sea level rise. United States Government Printing Office, Washington DC. 127 p.

1984. Potential climatic impacts of increasing atmospheric CO$_2$ with emphasis on water availability and hydrology in the United States. United States Government Printing Office, Washington DC. 108 p.

FARLEY, A. L. 1979. Atlas of British Columbia. University of British Columbia Press, Vancouver, B.C. 136 p.

FISHERIES AND ENVIRONMENT CANADA. 1978. Hydrological atlas of Canada. Ottawa, Ont. 34 plates and text.

FREEZE, R. A. 1969. Regional groundwater flow — Old Wives Lake drainage basin, Saskatchewan. Science Series No. 5, Inland Waters Branch, Department of Energy, Mines and Resources, Ottawa, Ont.

KALININ, G. P., AND V. D. BYKOV. 1969. The world's water resources, present and future. *From* Impact of science on society. Vol. XIX, No. 2, 1969, UNESCO. *In* R. L. Smith [ed.]. The ecology of man: an ecosystem approach. Harper & Row, New York, NY. p. 335-346.

KALLIO, R. 1985. Personal communication with Engineering and Development Division, Water Planning and Management Branch, Inland Waters Directorate, Environment Canada, Ottawa, Ont.

KORZUN, V. I. [ED.]. 1974. World water balance and water resources of the earth. (Translated from Russian by UNESCO, 1978, Paris, France. 664 p.)

LAYCOCK, A. H. 1957. Precipitation and streamflow in the mountain and foothill region of the Saskatchewan River basin. Prairie Provinces Water Board Report No. 6, Regina, Sask. 47 p.

1967. Water deficiency and surplus patterns in the Prairie Provinces. Prairie Provinces Water Board Report No. 13, Regina, Sask. 176 p.

1968. Water, chapter 5, p. 112-136. *In* John Warkentin [ed.] Canada, a geographical interpretation. Methuen, Toronto, Ont. 608 p.

1978. Precipitation mapping in Alberta, p. 133-149. *In* K. D. Hage and E. R. Reinelt [ed.] Essays on meteorology and climatology in honour of Richmond W. Longley. Department of Geography, The University of Alberta, Edmonton, Alta. 427 p.

1985. Water, p. 1921-1922. *In* The Canadian Encyclopedia. Vol. III. Hurtig, Edmonton, Alta.

L'VOVICH, M. I. 1974. World water resources and their future. (*In* Russian; English edition edited by R. L. Nace, 1979, American Geophysical Union, Washington, DC. 415 p.)

MATHER, J. R. 1978. The climatic water budget in environmental analysis. Lexington Books, Lexington, MA. 239 p.

1984. Water resources, distribution, use, and management. Wiley, New York, NY. 439 p.

MEYBOOM, P. 1963. Patterns of groundwater flow in the prairie profile, p. 5-20. *In* Groundwater, Proceedings Hydrology Symposium No. 3. National Research Council of Canada, Ottawa, Ont.

MILLER, H. 1977. Water at the surface of the earth. Academic Press, New York, NY. 557 p.

NACE, R. L. 1969. World water inventory and control, p. 31-42, chap. II. *In* R. S. Chorley [ed.]. Water, earth and man. Methuen, London. 588 p.

NATIONAL WATER COMMISSION. 1973. Water policies for the future. Washington, DC. 579 p.

ONTARIO MINISTRY OF NATURAL RESOURCES. 1984. Water quantity resources of Ontario. Toronto, Ont. 71 p.

PEARSE, P. H., F. BERTRAND, AND J. W. MACLAREN. 1985. Currents of change. Final Report, Inquiry on Federal Water Policy. Environment Canada, Ottawa, Ont. 222 p.

PRINCE, A. T. 1969. Comments to the National Advisory Committee on Water Resources Research by the Director, Policy and Planning, Inland Waters Directorate, Ottawa, Ont. (Unpubl.)

QUINN, F. J. 1981. Water transfers — Canadian Style. Can. Water Resour. Jl. 6(1): 64-76.

ROWE, J. S. 1972. Forest regions of Canada. department of the Environment, Canadian Forestry Service, Ottawa, Ont. 172 p.

SANDERSON, M. E. 1948. The climates of Canada according to the new Thornthwaite classification. Sci. Agric. 28: 501-517.

SANDERSON, M. E., AND D. W. PHILLiPS. 1967. Average annual water surplus in Canada. Climatological Studies No. 9, Meteorological Branch, Department of Transport, Toronto, Ont. 76 p.

SEIFRIED, A. 1972. Canadian river regimen, p. 119-158. In R. Keller [ed.] Runoff regimen and water balance II. 2nd Report of the I.G.U. Commission of the Internarational Hydrological Decade. geographical Institute, Albert-Ludwigs Univeristy, Freiburg, Germany.

SHAWINIGAN ENGINEERING CO. LTD. 1970. Water resources of Newfoundland. Atlantic Development Board.

STATISTICS CANADA. 1981. Canada Year Book 1980-81. Department of Supply and Services, Ottawa, Ont. 1004 p.

STICHLING, W., AND S. R. BLACKWELL. 1957. Drainage area as a hydrologic factor on the glaciated Canadian Prairies. Int. Assoc. Sci. Hydrol. Gen. Assem. Toronto Proc. 3: 365-376.

THORNTHWAITE, C. W. 1948. An approach toward a rational classification of climate. Geogr. Rev. 38(1): 55-94.

THORNTHWAITE, C. W., AND J. R. MATHER. 1955. The water balance. Publications in climatology. Vol. 8. No. 1. Laboratory of Climatology, Centerton, NJ. 104 p.

1957. Instructions and tables for computing potential evapotranspiration and the water balance. Publications in climatology. Vol. 10. No. 3. Laboratory of Climatology, Centerton, NJ. p. 186-311.

TOTH, J. 1962. A theory of groundwater motion in small drainage basins in central Alberta, Canada. J. Geophys. Res. 67: 4375-4381.

1968. A hydrological study of the Three Hills area, Alberta. Bulletin 24, Research Council of Alberta, Edmonton, Alta.

UNIVERSITY OF ALBERTA, DEPARTMENT OF GEOGRAPHY. 1964-84. Theses employing water balance procedures by MacIver, Kakela, Spence, Erxleben, Wong, Witter, Hallock, Wiche, Woodburn, Park, and Gregg. University of Alberta, Edmonton, Alta.

UNITED STATES WATER RESOURCES COUNCIL. 1968. The nation's water resources. First National Assessment. Washington, DC 421 p.

1978. The nation's water resources 1975-2000. Second National Water Assessment. Vol. 2. Washington, DC 536 p.

U.S.S.R. COMMITTEE FOR THE INTERNATIONAL HYDROLOGICAL DECADE. 1974. Atlas of world water balance. The UNESCO Press, Paris. 65 plates. (Translated from Russian edition 1974 Hydrometeorological Publishing House, Moscow-Leningrad. 65 plates.)

WHITE, G. F. 1971. Strategies of American water management. The University of Michigan Press, Ann Arbor, MI. 155 p.

WILLMOTT, C. J. 1977. Watbur: a Fortran IV algorithm for calculating the climatic water budget. Publications in climatology. Vol. 30. No. 2. Laboratory of Climatology, Centerton, NJ. p. 1-55.

ZOLTAI. S. C. 1979. An outline of the wetland regions of Canada, Proceedings. Workshop on Canadian Wetlands. Lands Directorate, Environment Canada. Ecol. Land Class. Ser. 12: 1-8.

CHAPTER 3

Current and Projected Water Uses in Canada, 1981-2011

Donald Tate

Environment Canada, Water Planning and Management Branch, Inland Waters Directorate, Ottawa, Ont. K1A 0E7

Introduction

In any society, the use of water is one of the most fundamental interactions between man and nature. Water enters into almost all human activities, from the very basic acts of drinking and sanitation to support for recreation to water use in industrial and energy production. Water is essential for all of these activities, and, in turn, the activities themselves may have profound impacts on the water resource. Not only does water pervade almost all socioeconomic activities, it also has considerable economic value. An unpublished analysis completed for Environment Canada in 1979[1] showed that the aggregate estimated value of water to Canada lay between $10 billion and $20 billion annually. Muller (1985), using updated numbers, recently confirmed that this range was essentially correct.

This chapter examines the uses which Canadians make of their abundant water resources. Its specific purpose is to present estimates of current water use by various major sectors of the economy throughout the country, to project these uses to the year 2011, and to discuss some of the implications of these forecasts.[2]
While it is not possible to predict precisely how much water Canadians will use at the end of the 30-yr projection period, the paper presents projections for a series of plausible scenarios based upon a variety of assumptions. This procedure, hopefully, will create "a fence around the ballpark" of future water use.

In addition, this analysis serves several more general purposes. First, in all areas of resource management, periodic inventories both of supplies and uses are necessary. This chapter provides the water use side of such an inventory. Second, while Canada is generally a water-rich country, there are regions where water is often in short supply (e.g. the western Canadian interior). Comparisons of supplies and uses, such as the one made here, identify these areas and describe them quantitatively. Third, knowledge of emerging water use trends is vital to the planning process. Massive publicly funded water development projects must compete for scarce capital and satisfy environmental impact requirements. Increasing constraints upon public funding and more stringent environmental requirements mean that water demand management options will likely become more attractive in the future and that water use studies will likewise assume more importance. This chapter describes the impact of certain demand management concepts on future water use.

---

[1]Water and the Canadian Economy. Available from the Inland Waters Directorate, Environment Canada.
[2]In this chapter, the terms "forecast" and "projection" and their derivatives are used synonymously.

Scope of the Study

Recent research carried out for the Inquiry on Federal Water Policy (Tate 1985) forms the basis for this chapter, augmented to include a more substantial review of past studies and a preliminary examination of nonwithdrawal water uses. The reader should refer to the Inquiry research paper for further details on the methodology and results contained in this chapter.

Time Horizon

The water use forecasts described in this chapter cover a 30-yr time horizon, from 1981 through 2011. The year 1981 was the most recent one for which a complete set of water use data was available, making it the logical starting point for the study. Due to uncertainties involved in long-term forecasting, water use projections become more guesswork than analysis beyond a 30-yr period. Also, over the next 30-yr period, we need not be too concerned about the impacts of climatic change on water use. Only in the longer term future is climatic change likely to affect water supply, and hence water use patterns (see chapter by Ripley, this volume).

Uncertainty also occurs during the forecasting period, and to cope with this factor, the research employs an "alternative futures" framework (U.S. National Water Commission 1976) to compile a range of future water uses. Accordingly, this chapter presents a number of projections of water use reflecting varying assumptions about the future. The alternative futures approach is more effective than producing just a single projection, since it examines the effects of different economic, social, and policy assumptions on evolving water use patterns. Hence, the focus of the chapter rests in defining a *range* of water uses for the future, and not in determining a "most likely projection."

Approaches to Forecasting

The "futures" community today is a large one, which has spawned a wide variety of approaches to forecasting. These break down into essentially two distinct approaches to "futures" study: analytic and futuristic. The analytic approach relies on variables which are, by and large, quantifiable, and on the use of mathematical models. The futuristic approach is much more holistic in nature, and involves speculating not only about quantitative variables, but also about qualitative ones, such as changes in lifestyles and future social ethics and philosophies.

This chapter employs the analytic approach. It develops a model encompassing four variables thought to be closely related to water use: economic activity levels, production technology, water price, and water use practices. Precise definitions of these variables are given in a later section. In brief, I have assumed that unquantifiable factors affecting water use (e.g. lifestyles) will have no net effect on the overall levels of use.

Water Use Terminology and Measurement

Analysts generally divide water use into withdrawal and nonwithdrawal categories. Withdrawal uses remove water from an ambient source such as a river or a lake, use it in some kind of distinct water circulation system, and discharge some or all of it back in aqueous form, usually quite close to the withdrawal point. Typical withdrawal uses include municipal water supplies, intake by a self-supplied industrial operation, and irrigation. Nonwithdrawal uses employ characteristics of water in its natural state, examples being navigation, recreation, and fish and wildlife production. Hydroelectric power generation here constitutes a nonwithdrawal use, although it may, depending upon site conditions, have some withdrawal characteristics.

With regard to the withdrawal uses, the following definitions apply throughout the chapter: *intake* (or *withdrawal*) — the amount of water withdrawn from an ambient water source to satisfy a given need; *gross water use* — the total amount of water required to satisfy a given need; *recirculation* — the arithmetic difference between gross water use and intake; *discharge* — the amount of water returned to the watercourse following use; *consumption* — the arithmetic difference between intake and discharge; *use rate* — a measure of the degree of recirculation; the ratio of gross water use divided by intake; *consumption rate* — a measure of the degree of consumptive use; the ratio of (intake − discharge) divided by intake.

Comparisons of water use with available water supplies made here use both intake[3] and consumption (Wolman and Bonem 1971) measures. Neither measure is completely satisfactory. Water intake denotes the total instantaneous withdrawal from an ambient water source by a given set of demands. Most, or in some cases, all of this water returns to the watercourse from which it was withdrawn, and therefore does not constitute a "loss" in water quantity from a watershed. The chapter acknowledges but does not address alterations which may occur as the result of withdrawal use, such as loss of water quality.

Water consumption, in contrast, is a mesure of water loss such as evaporative losses, water incorporated into products, water removed from the plant site to landfill sites, and other such losses. Water consumption thus measures the water loss at a particular location, water which is generally unavailable for further use. But this measure does not necessarily indicate losses to the water resource system as a whole, even locally. Evaporative losses can fall in the withdrawal area as precipitation; product use often returns some water, which has been considered consumed, to watercourses near the point of withdrawal; water contained in wastes deposited at landfill sites eventually reenters the local water system as part of the groundwater system. Thus, water consumption is not a true indicator of water loss to local or regional water supplies, despite being the best indicator we have of water not available for reuse within a basin. As used here, the concept of water consumption refers to the *specific withdrawal use being considered* and not to the hydrologic cycle.

Considerable difficulty arises in addressing nonwithdrawal water uses because they are more diverse than the withdrawal uses and accordingly have a wider range of measurement parameters. Also, water use analysts have devoted much less study to them than to water withdrawal. Hence, there are few established methodologies for estimating water demands by nonwithdrawal users (Albery 1968). The chapter employs a straight forward and somewhat arbitrary measure of nonwithdrawal use in major drainage areas in order to estimate complete water supply use balances. This measure considers nonwithdrawal uses in a drainage basin as a fixed percentage of reliable flow, and will be described in a later section.

Spatial Focus

Producing national and regional forecasts of water uses is a principal aim of this study. The regions in the analysis include the five major Canadian political-economic regions: British Columbia, the Prairies (Manitoba, Saskatchewan, and Alberta), Ontario, Quebec, and the Atlantic region (Newfoundland, Prince Edward Island, Nova Scotia, and New Brunswick). The two northern territories are not included because they account for only a negligible amount of water withdrawal and because water use:supply balances are generally not a problem in the two territories. The national and regional work provides a uniform set of forecasts for the major regions of the country. The model used here broke the regional forecasts down further into major drainage areas in order to make preliminary supply:use comparisons.

Economic Sectors

The forecasting model used here to project water use consists of 30 distinct economic sectors (Table 1), 5 primary sectors (e.g. agriculture, mineral extraction), 19 secondary

_____

[3]National Water Needs Study. Unpublished working paper available from Inland Waters Directorate, Environment Canada.

45

TABLE 1. Industrial sectors used in the input-output model.

A. Primary industries

1	Agriculture
2	Forestry, fishing, hunting, and trapping
3	Metallic minerals
4	Mineral fuels
5	Nonmetallic minerals

B. Secondary industries

6	Food and beverages
7	Tobacco products
8	Rubber and plastics
9	Leather products
10	Textiles, knitting mills, and clothing
11	Wood products
12	Furniture and fixtures
13	Paper and allied products
14	Printing and publishing
15	Primary metals, excluding iron and steel
16	Iron and steel
17	Metal fabricating
18	Machinery
19	Transportation equipment
20	Electrical products
21	Nonmetallic mineral products
22	Petroleum and coal products
23	Chemical and chemical products
24	Miscellaneous manufacturing

C. Tertiary industries

25	Construction
26	Transportation, communication, and storage
27	Electric power
28	Gas and other utilities
29	Wholesale and retail trade
30	Other (i.e. finance, insurance, real estate, etc.)

sectors (manufacturing industries, such as food and beverages, textiles, etc.), and 6 tertiary sectors (e.g. electric power production), which include 2 sectors related to municipal water use. Three of the tertiary economic sectors (i.e. construction; transportation, communication, and storage; and "other") did not use water directly and are included merely to preserve the comprehensiveness of the model.

Rural domestic water use, although accounted for in the 1981 water use totals, was not forecasted. Rural domestic usage accounted for less than 1% of total Canadian water withdrawals in 1981 and was thus a minor water use. It did not correspond, for forecasting purposes, to any of the economic sectors of Table 1. It could not form part of the agriculture sector, since agricultural water use is increasing, whereas farm population, and thus rural domestic water use, is falling. Neither did it correspond to any of the population-related sectors used here (i.e. the utilities and the trade sectors).

For ease of reporting, the water use forecasts for individual sectors are aggregated into five composite sectors reflecting major withdrawal water use activities: agriculture, mineral extraction, manufacturing, power generation, and municipal uses.[4]

---

[4]Detailed data on each of the 30 sectors are available from the author upon request.

A Review of Forecasting Methods with Canadian Examples

In the past, Canadian water managers have concentrated on manipulating the country's massive water supplies to solve water resource problems. As a consequence of this supply management orientation, they have seldom made in-depth examinations of the characteristics of water demand or the possibilities for demand management (Mitchell 1984; Tate 1984b). Few water demand studies have been carried out on any but the most local of levels. This section reviews a number of studies of water demand done in the past, concentrating on Canada, but drawing from other areas where necessary to illustrate the broad range of approaches to the forecasting problem. Whittington (1978), Kindler and Russell (1984), and Tate (1978, 1984a) covered these methodologies in more detail.

Cass-Beggs (1961) completed one of the few Canadian studies of emerging water uses on the national level. He used the *fixed direct requirements*, or "coefficients," approach in projecting water uses, as have most water use forecasting studies in the past. The fixed direct requirements approach assumes a constant relationship between water use and one (independent) variable, such as time, employment, population, or level or value of industrial output. The amount of water used per unit of the independent variable is calculated or assumed from other sources. This amount (the "coefficient") is assumed to be constant over the forecasting time horizon. Forecasts of the independent variable are multiplied by the coefficient to produce water use forecasts. Such forecasting is cheap to implement in both time and cost. In the absence of any theoretical basis for linking water use to one independent variable and for projecting this relationship into the future, however, such forecasts are normally unreliable.

Cass-Beggs' study employed various American-based water use coefficients multiplied by forecasts of economic activity or population level to produce water use forecasts. For example, with respect to industry (manufacturing only), he assumed that "the water used in manufacturing would bear a fixed ratio to the value added in manufacturing" (Cass-Beggs 1961, p. 182). He made analogous assumptions to develop water use coefficients for other economic sectors. A later section of this chapter summarizes the results of this study for comparison with the forecasts developed here. Most water use studies in Canada have been of this nature. Nearly all consulting reports on individual municipal water systems project population and multiply this by a water use coefficient per capita to produce water use forecasts for the municipality under study. Many studies of regional or national water use have also used the fixed direct requirements approach. Examples include the "National Water Needs Study",[5] Lane and Sykes (1982) study of water use in western Canada, and my own study entitled "Water and the Canadian Economy".[6]

Dissatisfaction with the fixed direct requirements approach has led researchers such as Grima (1972), Rees (1969), do Rooy (1970), and others to employ a *regression analysis approach* to analyzing water use. Studies of this nature begin by hypothesizing statistical relationships between water use and a number of independent variables, such as the output volume of industrial output, water recirculation rates, water price, etc. The hypothesized relationships form the basis of data collection and subsequently the calibration of multiple regression equations relating water use to the independent variables. Future values of the independent variables, projected by some independent forecasting method, are then used in conjunction with the multiple regression equation to project water use. This methodology marks a substantial improvement over the fixed direct requirements approach because it is based upon testable hypotheses and has a firm statistical foundation. Its application in studies such as the present one is limited, however, because it must be tailored to fairly homogeneous groupings of industrial or municipal water uses. This limitation is of little consequence if the required regression

---

[5]op. cit.
[6]op. cit.

TABLE 2. Comparison of water use forecasting methodologies. *Criteria descriptors*: data availability: ease with which data can be collected (high is best); alternative futures: ability to consider a wide variety of alternative futures (high is best); technological change: ability to incorporate technological change (high is best); economic interrelationships: incorporation in the model of relationships between economic sectors (high is best); spatial resolution: ability to consider river basin and subbasin detail (high is best); overall cost: total required expenditure in terms of money and person-years (low is best); theoretical basis: degree to which the methodology is based in theoretical principles (high is best).

Methodology	Criteria for comparison						
	Data availability	Alternative futures	Technological change	Economic interrelationships	Spatial resolution	Overall cost	Theoretical basis
Coefficients	High	Medium	Low	Low	High	Low	Low
Regression	Low	Medium	Low	Low	Low	High	Medium
Demand management	Low	Medium	Low	Low	Low	High	High
Structural modeling	High	High	High	High	Low	Low	Medium
Simulation modeling	Medium	High	High	High	High	High	Medium

models of water use are available for each sector under study. However, it is an insurmountable limitation in terms of the resources required to generate the necessary data in a relatively short time if they do not, as is the situation in Canada. These and other reasons summarized in Table 2 led to rejecting the regression approach to water use forecasting on a national scale for Canada.

Building upon the regression analysis framwork, Sewell and Rouche (1974), Grima (1972), and Kitchen (1975) provided Canadian examples of the *demand management approach* to water use analysis. This methodology assumes that the price of water to users is an explicit variable in determining water demand. Its behavioral basis is that consumers are responsive to the price of water and that the quantity of water demanded is an inverse function of its price. The demand management approach implies that water managers should meet supply shortages through correct water pricing practices (to rationalize demands) and should use supply augmentation only where it is the cheapest means of meeting shortages. In Canada, few studies in the past employed this approach, and the statistical basis required for its use is unavailable. Thus, the present research could not use a full demand management approach although its underlying philosophy formed the basis for one of the scenarios for water use forecasting (see below).

In contrast with methodologies which rely upon formal statistical methods, the *input-output modeling approach* extends the fixed direct requirements approach into one which takes a multi sectoral view of the economy. Several Canadian water resource studies have used this approach because it is comprehensive in its view of water use in an economy and is relatively straight forward to implement. Input-output methods underlie the forecasts presented in this chapter, and are described more fully in the following section.

One of the first Canadian water resource studies employing input-output modeling was an analysis of the economy of the Okanagan River basin of British Columbia (Anonymous 1972). This work did not focus on future water demands *per se,* but did include some elements of a demand forecasting study. A study of current and projected consumptive water use in the Canadian section of the Great Lakes basin employed input-output methods (Tate 1978). Earmme (1979) used input-output to study water balances in the North Saskatchewan River basin. Finally, my recent research for the Inquiry on Federal Water Policy (Tate 1985) is based largely upon input-output modeling.

Simulation modeling constitutes the final approach to water use forecasting outlined here. Under this approach, the variables underlying water use form the basis of a logical and consistent computerized model, which will replicate current water use. The analyst then makes projections of the independent variables, which are then combined in the model to produce water use forecasts. In using such a procedure, the simulation approach is similar to other approaches outlined above. However, simulation modeling is more flexible in that it can be applied to any sort of region, thus being useful in modeling at the river basin level. Simulation models often include water supply records, thereby allowing the user to compare supplies and demands and to test the impacts of various demand scenarios on water supply conditions. Also, simulation models can be extended to consider water quality issues (Victor 1972).

Current federal government studies of water use employ simulation models to compare water supplies and uses in various Canadian drainage basins (Acres International Ltd. 1983, 1984, 1985). These simulation models recreate both water use and water supply conditions in the basins under study. The models project water uses based upon assumptions about economic and population growth, water use practices, water prices, and other forecasting parameters of concern and then compare the water use projections with data on historical streamflow conditions. The output of the analysis consists of a listing of periods when critical water supply conditions would be exceeded by projected water withdrawals or consumptive uses. As mentioned, development of the simulation modeling approach is currently underway, but was not sufficiently advanced for use in this study.

Methodology Underlying the Forecasts of This Chapter

Description of the Water Use Forecasting Model

An augmented input-output model provided the basis for the water use forecasts presented in this chapter. Input-output analysis is an econometric technique that examines the flows of goods and services in an economy, both between the industrial production sectors themselves and from those sectors to the points of final consumption, or final demand.

The input-output approach proceeds from an input-output transactions table (Fig. 1) that reflects the flow of products from the producing sectors of the economy (the x_{ij} of Fig. 1) to final demand sectors such as households, government purchases, exports, etc. (The C, I, G, and E of Fig. 1). The table also contains entries for input from primary sources external to the production system, in this case labor and other value added, respectively (The L and V in Fig. 1). The rows of the transactions table demonstrate the product distribution of each industry to itself, to other producers, and to the various

To		Purchasing sectors		Final demand sectors				
From		$1 \ldots\ldots i \ldots\ldots n$	Households	Private investment	Government	Exports	Total gross outlays
P	1	$X_{11} \ldots X_{1f} \ldots X_{1n}$	C_1	I_1	G_1	E_1	X_1
r							
o	•	• • • •	•	•	•	•	•
d							
u	•	• • • •	•	•	•	•	•
c							
i	•	• • • •	•	•	•	•	•
n							
g	i	$X_{i1} \ldots X_{ij} \ldots X_{in}$	C_i	I_i	G_i	E_i	X_i
s	•	• • • •	•	•	•	•	•
e							
c	•	• • • •	•	•	•	•	•
t							
o	•	• • • •	•	•	•	•	•
r							
s	n	$X_{n1} \ldots X_{nj} \ldots X_{nn}$	C_n	I_n	G_n	E_n	X_n
Labour		$L_1 \ldots L_j \ldots L_n$	L_C	L_I	L_G	L_E	L
Other value added		$V_1 \ldots V_j \ldots V_n$	V_c	V_I	V_G	V_E	V
Total gross outlay		$X_1 \ldots X_j \ldots X_n$	C	I	G	E	X

FIG. 1. A simplified input-output transactions table (from Richardson 1972, p. 19).

final demand sectors. Conversely, the columns of the table show the source of each industry's inputs. As portrayed in the bottom row and right-most columns, total inputs to each industry balance total outlays.

Input-output tables can be used to formulate accounts comparable with the system of Keynesian national accounts used traditionally in government descriptions of production in the economy (e.g. gross national product). However, they also serve another analytical purpose by forming the basis for a model to link changes in the final demand for industrial products to changes in industrial output and to examine the processes of structural and technological change as they affect water use. Miernyk (1965) and Richardson (1972) described in detail the derivation of input-output models, and I have described elsewhere the particular model used to produce the projections for this chapter (Tate 1985).

The model states that the set of total industrial outputs (i.e. by sector) is the product of an "inverse" matrix, derived from the input-output table, and the set of final demands for industrial products. In a study of the California water industry, Lofting and McGaughey (1963) augmented the basic input-output model to include water use considerations. I used their augmented model to generate the national and regional water use forecasts discussed in this chapter. In mathematical terms, the model is

$$w = (I - A)^{-1} \cdot W \cdot y$$

where w = a vector of total industrial water use (i.e. for any water use parameter), containing entries for each industry in the economy, y = a vector of the values of final demands for industrial products, $(I - A)^{-1}$ = the inverse matrix derived from an input-output table, and W = a matrix of water use coefficients measured in volumes per million dollars of output. The coefficients form the principal diagonal of the matrix, and all off-diagonal elements are zeroes.

The model contains explicit terms for the three factors used here to project water use in the various sectors of the Canadian economy. The final demand factor (y) captures the effects of economic activity and its growth on water use. The coefficients of the inverse matrix are proxies for the state of industrial production technology. Finally, the water use coefficients incorporated by Lofting and McGauhey (1963) describe water use practices (e.g. trends in recirculation). Modifications to the water use coefficients provided the means for examining the impacts on water use of rising water prices and other conservation measures.

Production Levels and Economic Growth

Manipulation of the final demand variable permits the modeling of production level and growth effects on water use. The analysis uses regional input-output models from 1974,[7] the latest year for which complete input-output data were available. Final demand data for 1974 were projected forward to the 1981 base year of this study by means of economic growth rates experienced between 1974 and 1981. The model then multiplies the 1974 regional inverse matrices by the 1981 final demands to produce an estimate of 1981 regional gross outputs in each of the 30 industries under consideration (Table 1). Final demand values for industry 28 (gas and other utilities, including water utilities) and industry 29 (wholesale and retail trade) have close ties to population levels. For these industries, therefore, population rather than economic growth rates provided the means of calculating 1981 final demand and output levels.

Forecasting future water use requires that future growth rates in industry and the population also be forecast. Producing such growth rates is a complex undertaking done normally by specialist agencies using sophisticated econometric models. These agencies

---

[7]The input-output material was provided by staff of the Input-Output Division, Statistics Canada.

51

rarely attempt forecasts for a 30-yr time period. After reviewing most available sources, I decided to use the industrial growth projections recently produced by Informetrica, Ltd. (1984) as the basis of the water use forecasts and population forecasts by Statistics Canada (1982). I extrapolated the resultant time series to the year 2011 for use here. These same sources of economic output and population projections also provided the basis for selecting alternative high, medium, and low growth rates.

Production Technology

The modeling of trends in technological change presents difficulty, largely because there is much controversy about the variables which lead industries to make modifications to their production processes. Trends are the product of many individual decisions, and few summary statistics are available about the outcome of these decisions.

To estimate trends in production technology, I assumed that the coefficients of the inverse matrices used for this chapter reflected the state of production technology for the time period covered by the input-output tables. Accordingly, changes in such coefficients between time periods constitute a production technology time trend. Provided that input-output tables are available for a sufficient period of time, linear regression techniques allow the calculation of trends for each coefficient, making appropriate adjustments for autocorrelation. The statistically significant trends can then be projected, while those which are not significant can remain constant. This method of analyzing production technology time trends assumes linear time trends in the coefficients of the input-output inverse matrices. This assumption allows analysis of the impacts of production technology on water use, but should be subject to further research.

All regressions to estimate the production technology time trends were calculated at the national level, for which a 20-yr time series of annual input-output tables was available. Time series of regional tables would have formed a better basis for this analysis, but these were unavailable. As a "second-best" alternative, national trends for the 1961-80 period were superimposed on the regional industrial sectors.

Water Use Practices

As used in this paper, water use practices refer to the amount of intake, recirculation, gross water use, consumption, and discharge that occurred in the industrial sectors covered by this study. Trends in water use practices, likewise, refer to trends observed or anticipated in these five parameters. I quantified water use practices for use in the forecasting model by calculating coefficients of water use (i.e. by parameter) per million dollars of 1981 total output for each sector of the model. Survey results (Tate and Scharf 1985) and several secondary data sources provided the water use data for this chapter (Tate 1985, p. 28-29).

Modifying the set of water use coefficients provided the means for analyzing both the potential impacts of raising the price of water in various activities and the effects of other conservation measures. This water conservation analysis is theoretical, and results are indicative in nature rather than certain projections of what would happen under a water management philosophy which tries to influence water use by pricing and other conservation techniques.

Several studies (Grima 1972; de Rooy 1970; Kindler and Russell 1984) have shown that water demand, like demand for the vast majority of goods and services in our society, behaves in a "conventional" economic fashion such that when its price rises, the quantity demanded falls. In Canada, although country-wide water pricing data fail to provide an adequate basis for constructing regional water demand curves, the surveys of industrial water use referred to earlier did collect data on water costs. In these surveys, the water cost at a plant was defined as the sum of intake cost, intake treatment cost, recirculation cost, and effluent treatment cost. By summing these costs and dividing by the amount of intake, the average cost of water for each industry can be calculated. This amount, which has been used in the past as a proxy for water price (de Rooy

1970), can be taken as the price of water to industries. In other words, industries face a zero or minimal commodity price for their water. Water costs estimated by this means are downwardly biased, since no allowance is made for the capital cost of water conveyancing facilities in plants.

Another important piece of information required for the pricing analysis is a set of price elasticities of demand for water in the various industries. Price elasticity of demand refers to the percentage change in the quantity of water demanded for a given percentage change in water price. I assumed average price elasticities for each industry and held these constant through the range of price for water in the forecasts presented here. Estimates of average price elasticity of water demand were unavailable in many cases for Canada and were taken from various secondary sources (Hanke 1978); Grima 1972; Leone 1975; Boland et al. 1984). Where no data were available from any source, plausible estimates were made by reference to related industrial groups for which data were available. The water intake coefficients calculated from survey data as described earlier show how much water is used by industry per unit of output in the absence of a commodity price for water. The pricing analysis indicates how the coefficients will alter as an increasing commodity price is charged for water. In other words, by applying the price elasticity data to the water use coefficients, the model estimated how water intake by the various groups would respond to given percentage increases in the average cost (i.e. price) of water. In the analysis, I assumed also that water prices would rise 10% by 1991, 25% by 2001, and 35% by 2011.

There are many possible water conservation measures apart from price increases, such as time-of-day metering, flow control (e.g. water restricting shower heads), and leak detection. To allow for these additional impacts, the effects of the hypothesized water pricing arrangements were increased by an arbitrary 50%. This analysis is hypothetical in nature, but I believe that it indicates the potential impacts of manipulating water prices to control demand.

Nonwithdrawal Uses

For the purposes of this chapter, the amount of water required for all nonwithdrawal uses in a drainage basin consists of one-half of the reliable annual flow available at the mouth of the basin. Reliable annual flow refers to that amount of flow available during 19 out of 20 yr (i.e. the 95% level), as computed by Pearse et al. (1985, p. 46). At this level of annual flow, many rivers, particularly in drier areas, would pass very little water during dry months. While this nonwithdrawal flow requirement is arbitrary, it is not without precedent. The 50% flow criterion was used for fish maintenance requirements in certain Alberta streams (Anonymous 1977). Durrant (1983) suggested another possible minimum flow constraint, viz. 50% of the flow experienced in two consecutive drought years. This criterion was rejected because appropriate data were not available country-wide. The reliable flow criterion used here ignores such factors as location, stream reach, season, and streambed morphometry, which are important determinants of environmental suitability for many nonwithdrawal uses. However, until the research necessary to refine measures of these important water uses is conducted, analysts must employ rough approximations of nonwithdrawal use in regional water use studies. In fact, this is the first Canadian study to attempt to assess nonwithdrawal requirement on a nation-wide basis. Because the flow criterion method is largely hypothetical at this stage, *the nonwithdrawal uses of water were not projected*.

Alternative Futures

Five scenarios for the future (Table 3) formed the basis for making alternative water use projections. I used three alternative projections of population in the scenarios, based upon the latest Statistics Canada (1982) work. With regard to economic production, the scenarios incorporate various future mixes of primary, secondary, and tertiary industrial activity that result from using high, medium, and low alternative growth rates.

TABLE 3. Conditions assumed in each future scenario for the major variables of the forecasting model.

Scenario	Principal components of the forecasting model		
	Economic growth	Production technology	Water use practices
1. Reference	Medium	Constant	Constant
2. Conservation	Medium	Constant	Altered
3. Technological change	Medium	Altered	Constant
4. Recession	Low	Constant	Constant
5. High growth	High	Constant	Constant

Energy use assumptions are implicit in these projections of industrial growth. Technological change was allowed in one of the scenarios according to the methodology outlined earlier. Finally, environmental policy as it affects water use was modified in one scenario to simulate the effects of a conservationist policy stance.

In any forecasting exercise, only a few factors can be treated as variables. The remainder must be held constant to establish common socioeconomic background for the scenarios. In general, the more time available for the analysis, the fewer the number of factors which must be held constant. In this analysis, since time limitations were quite severe, even some of the analytically tractible variables were assumed to be common to all scenarios. Some of the more important assumptions made include the following. Lifestyles will not change radically with regard to their water use. This means that lifestyle changes may occur, but without radically altering current water demand patterns. Current institutional and administrative arrangements will continue throughout the time period forecast. The exception to this is an assumed rise in water price and a concerted water conservation effort in scenario 2. Urbanization trends, implicit in the population projections, were not considered explicitly in the forecasts. Finally, no new industrial classifications will develop during the forecasting period.

Five Views of the Future

Three sets of population and economic growth rates (low, medium, and high), two types of technological conditions (stable and changing), and two sets of water use coefficients (stable and price-altered) underlie the forecasts produced for this project. Thus, 12 scenarios were possible by altering just the major variables. Within each major variable, moreover, assumptions could have varied by industrial sector, thereby increasing enormously the number of alternative future outlooks.

In selecting scenarios, the analysis strikes a compromise between designing scenarios which could be described in detail, and presented as feasible views of the future, and those which would describe the sensitivity of the model to broad changes in assumptions about each major variable. The analysis emphasizes the latter task, that of sensitivity testing. The reader is free, then, to recombine the variables in other combinations, and thus to examine other futures. None of the scenarios described here is designated "most likely."

Scenario 1: A reference case

A reference scenario holds all factors constant except the economic and population growth, which increase at their medium rates of growth (Table 3). This scenario is therefore essentially an extrapolation of past trends. Production technology remains constant, as do the water use coefficients. The scenario assumes the absence of severe energy shocks such as those experienced during the 1970's. This scenario is a "business as usual" view of the future, with no significant shocks to the socioeconomic system.

Water management policy remains one of supply management, with major capital works occurring as required.

Scenario 2: A conservationist scenario

In the water management field, Pearse et al. (1985) have suggested that water conservation and the principles of water demand management (Mitchell 1984; Tate 1984b) should form a major part of a renewed federal water policy. One major instrument offered for demand management is the use of the price system and increases in water price to allocate water supplies, accompanied by a major incentive system to implement other non-price-related water conservation measures. To analyze some aspects of a conservationist policy, scenario 2 (Table 3) assumes medium rates of population and economic growth, constant production technology, and the set of price-altered water use coefficients specified earlier.

Scenario 3: The effects of technological change

This scenario demonstrates the effects of production technology changes on national and regional water use (Table 3). It is difficult to describe the exact characteristics of this scenario, for changes in economic growth and water use practices would likely accompany changes in production technology, but the exact relationship among these factors is uncertain. Operationally, this scenario uses medium population and economic growth rates, constant 1981 water use coefficients, and the technology-altered set of production coefficients (i.e. inverse matrices).

Scenario 4: A recession scenario

Scenario 4 assumes a prolonged period of slow economic growth throughout the country (Table 3). Under such an assumption, recessionary conditions would ensue, and accordingly the scenario uses low growth rates throughout the economy. Both the state of production technology and water use practices remain constant through the forecasting period due to the scarcity of capital.

Under this scenario, unemployment would climb. No attention at all would be afforded to environmental programs, and few efforts at water conservation efforts would be made. Increasing forced leisure time would generate the need for more nonconsumptive recreational resources.

Scenario 5: High economic growth

Scenario five emphasizes maximizing economic efficiency and achieving a high rate of economic growth (Table 3). Accordingly, public policy and private development would focus on exploiting to the full the raw material wealth of the country and developing its primary industry base. Rapid development would also occur in the manufacturing sectors, based upon increased rates of growth in the primary sectors of the economy (i.e. industries 1 to 5 in the model). The tertiary sectors also grow rapidly in support of growth in other areas of the economy. Thus, all sectors achieve high rates of growth in this scenario. Technological change and water use practices remain constant in order to isolate the effects of the high growth rate.

Under this scenario, economic growth is the primary public objective. Environmental programs on water conservation receive little or no attention. Water use practices remain at the 1981 levels.

Results of the Study

Water Use in 1981

Water Withdrawal

In 1981, the base year for this study, Canadians withdrew just over 37 500 million cubic metres (mcm) from all sources. Ontario accounted for 56% of the total, mostly from the Great Lakes. In Ontario, thermal cooling at power plants accounted for about 70% of the 21 230 mcm of total water withdrawal. Thermal power plants in Ontario employed once-through, or nonrecycling, water systems. Lakeshore locations obviate the need for water recycling. Also, cheap water policies have provided disincentives for water conservation and reuse. The Prairie region accounted for the next highest regional water withdrawal (5363 mcm), of which a large amount supports irrigated agriculture. The semiarid climate, high public subsidies, and the desire to promote regional development have all contributed to high irrigation water use in this region. Quebec stood third in regional water use in 1981. The region had a diversified industrial base and a large population, but its reliance on hydro power meant that thermal generation requirements were small in comparison with Ontario's. The same set of factors, but a somewhat smaller industrial base, gave British Columbia its fourth place with respect to regional water withdrawal (3789 mcm). Finally, Atlantic Canada, with its small industrial base and population, accounted for the smallest regional water use (2884 mcm).

Gross water use followed essentially the same pattern of regional distribution as that outlined for water withdrawal, and totalled 53 436 mcm in 1981. In other words, recirculation practices "stretched" the available water supplies some 1.4 times. In all, about 16 000 mcm of water was recirculated throughout Canada in 1981, most of it by manufacturers. The aggregate use rate (i.e. gross water use divided by water intake) of 1.4 varied regionally, being lowest in Ontario (1.2) and highest in the Prairies (1.9). Manufacturing was responsible for most water recycling, and the use rates show the Prairie region had water use practices somewhat more efficient than in other parts of the country. With the region's semiarid climate, this higher water use efficiency is not surprising.

Water consumption totalled 3906 mcm in 1981, just over 10% of total intake. Agricultural use (mainly irrigation) in the Prairie region, with 1892 mcm, used the lion's share of this consumption. The consumption rate (i.e. water consumed divided by water intake) in this region was 0.4, very high in comparison with the other four regions. The lowest consumption rate (0.03) was in Ontario, where none of the intake water at thermal power plants was recirculated and very little water consumed.

Water discharged from Canadian industries and municipalities totalled 33 612 mcm

TABLE 4. Volumes (mcm) of significant water use parameters by region and major economic sector, 1981. All water use figures in this and subsequent tables, except where specified, were derived from Tate (1985).

	Water intake	Consumption	Gross water use	Recirculation[a]	Discharge[b]
Region					
Atlantic	2 884	139	3 849	965	2 475
Quebec	4 252	435	7 346	3 094	3 817
Ontario	21 230	589	25 352	4 122	20 641
Prairies	5 363	2 256	10 038	4 675	3 107
British Columbia	3 789	487	6 851	3 062	3 302
Canada	37 518	3 906	53 436	15 918	33 612
Industry					
Agriculture	3 125	2 412	3 125	0	713
Mineral extraction	648	179	3 440	2 792	469
Manufacturing	10 201	507	21 459	11 258	9 694
Power generation	19 281	168	21 149	1 868	19 113
Municipal	4 263	640	4 263	0	3 623
All industries	37 518	3 906	53 436	15 918	33 612

[a]Imputed figure: recirculation = gross water use − water intake.
[b]Imputed figure: discharge = water intake − consumption.

in 1981. The discharge pattern differed regionally from that of intake because of the large consumptive water use in the Prairies. Consequently, Quebec assumed second place with regard to the regional pattern of discharge, but still remained far behind Ontario, with its discharge of cooling water from thermal power stations. The waterborne wastes contained in industrial discharges contribute to Canada's water quality problems.

Withdrawal, recirculation, and consumption may also be viewed from the perspective of the major economic sectors included in the study (Table 4). Thermal power plants accounted for 51% of the total Canadian water withdrawal. Ontario contained most of these plants, but they were also important in the Prairie and Atlantic regions. The mineral extraction and manufacturing sectors accounted for most of the water recycling practiced in Canada, with aggregate use rates of 5.3 and 2.1, respectively. A small amount of water recycling also occurred in the thermal power sector, at two plants in Alberta. Recirculation occurred in neither the agriculture nor the municipal sectors.

The agriculture sector accounted for the greatest proportion of consumptive use, primarily as a result of irrigation in the Prairie region. High evaporative and conveyance losses in irrigation led to a consumption rate in agriculture of 0.77, greatly above the Canadian average for all sectors of 0.12. The mineral extraction sector, where significant quantities of water are used in deep-well injection for enhanced petroleum recovery, also showed relatively high consumptive usage. The consumption rate was lowest in the thermal power sector, at 0.008, and was 0.15 in the municipal sector.

Water intake in the rural domestic sector, which was included in neither Table 5 nor the subsequent forecasts, totalled 348 mcm for Canada in 1981. The sector did not recycle water, and all withdrawals are counted as being returned to source. Water withdrawal in this sector was distributed as follows:

Region	% of total intake
Atlantic	17
Quebec	24
Ontario	28
Prairies	20
British Columbia	11

TABLE 5. Projected volumes (mcm) of total water intake and total consumption by region and major economic sector, 1981 and 2011.

	1981		2011	
	Intake	Consumption	Intake	Consumption
Region				
Atlantic	2 884	139	5 584	244
Quebec	4 252	435	7 629	690
Ontario	21 230	589	42 861	1 093
Prairies	5 363	2 256	11 172	4 443
British Columbia	3 789	487	7 085	893
Canada	37 518	3 906	74 331	7 363
Industry				
Agriculture	3 125	2 412	5 897	4 567
Mineral extraction	648	179	1 733	433
Manufacturing	10 201	507	20 274	1 038
Power generation	19 281	168	39 558	349
Municipal	4 263	640	6 869	975
Total	37 518	3 906	74 331	7 363

Nonwithdrawal Use

Nonwithdrawal water uses in this chapter consisted on half the reliable annual flow as defined earlier (Table 6). The estimated nonwithdrawal requirements cover all instream uses: fishing, wildlife support, environmental maintenance, navigation, and power generation. Nonwithdrawal uses are sensitive to many conditions not modeled here, such as location, stream reach, season, streambed morphometry, and other factors. They are particularly sensitive to *seasonal* water supply. Thus, the use here of an annual flow criterion underestimates requirements to maintain wildlife or to provide power in low-flow years.

Drainage Region Water Use

Regional water use estimates, when broken into major drainage areas, and compared with estimates of reliable annual flow (Table 6), allow the estimation of basin water supply:demand balances. Two major qualifications apply to making such com-

TABLE 6. Volumes (mcm) or reliable annual flow compard with 1981 and 2011 water uses. Adapted from Pearse et al. (1985) and Tate (1985).

Drainage region	Reliable annual flows[a]	Nonwithdrawal use[b]	Withdrawal use[f]			
			1981		2011 reference case	
			Intake	Consumption	Intake	Consumption
Pacific coastal	396 400	198 200	2 134	69	3 922	132
Fraser-lower mainland	96 000[c]	48 000	975	219	1 780	381
Okanagan	971[d]	486	312	146	578	256
Columbia	51 850[d]	25 925	305	33	577	60
Yukon	56 950[d]	28 475	16	1	47	1
Peace-Athabaska	58 720	29 360	251	155	548	374
Lower Mackenzie	192 800[e]	96 400	26	2	63	5
Arctic coastal	186 670	93 335	—	—	—	—
Missouri	105[d]	53	156	38	324	74
North Saskatchewan	5 046	2 523	1 405	154	2 941	321
South Saskatchewan	4 636[cd]	2 318	2 578	1 680	5 031	3 263
Assiniboine-Red	497[d]	249	1 012	207	2 041	403
Winnipeg	12 040[cd]	6 020	143	3	275	7
Lower Saskatchewan-Nelson	34 937[de]	17 468	122	25	281	51
Churchill	10 190[c]	5 095	1	—	—	—
Keewatin	92 870	46 435	1	—	—	—
Northern Ontario	117 700[c]	588 50	128	4	236	4
Great Lakes	75 780[d]	37 890	20 850	567	41 506	1 006
Ottawa	43 840	21 920	638	77	1 139	134
Northern Quebec	404 300[c]	202 150	121	9	241	16
St. Lawrence	47 430[de]	23 715	3 320	343	5 750	527
North Shore and Gaspe	203 000	101 500	443	39	782	66
Saint John-St. Croix	15 990[d]	7 995	892	54	1 687	98
Maritime Coastal	65 564	32 782	1 736	68	3 344	118
Newfoundland-Labrador	217 900	108 950	299	14	524	24
Total	2 392 186	1 196 094	37 864	3 907	73 617	7 321

[a]Flow equalled or exceeded, on average in 19 of 20 yr.
[b]50% of reliable annual flow.
[c]Excludes flow transferred into neighboring basin region.
[d]Excludes inflow from United States portion of basin.
[e]Excludes flow from upper basin region.
[f]For municipal, rural residential, manufacturing, mineral extraction, agricultural, and thermoelectric power uses. Rural residential uses are excluded from the 2011 forecasts.

parisons at a broad spatial level and on an annual basis. First, data compiled at the drainage basin level may mask local variations in supply:use balance. In other words, the data pertain to basin-wide supply and demand conditions, and comparisons between the two parameters are for locations at the mouths of each drainage area. Thus, the table does not identify local variations in water balances. Second, the data pertain only to annual balances and thus cannot identify seasonal variations. In many cases, seasonality is a vital consideration in measuring water supply:demand balances. These qualifications notwithstanding, there is usually a close relationship between annual water supply:demand problems and seasonal or local problems. Thus, the annual balances implicit in Table 6 are indicators of potential trouble spots. Further research can build upon these indicators to detail local and seasonal problems.

By definition, all basins require at least 50% of reliable annual flow to satisfy non-withdrawal uses. In most cases, these volumes dwarf total consumptive use by the five major withdrawal sectors (Table 6). On a national basis, the total water consumed by all the withdrawal uses (3907 mcm) accounted for just 0.03% of the remaining reliable annual flow, confirming the fact that water availability presented little problem nationally on an annual basis. However, on a drainage area basis, three basin regions showed net water use (i.e. consumptive use plus nonwithdrawal requirements) approaching reliable annual flow. These were the Missouri (Milk), South Saskatchewan, and Assiniboine-Red drainage regions. The Okanagan basin also showed a relatively high net water use relative to reliable flow. These four drainage regions, all located in the "rain shadow" of the Cordilleran mountain chain, constitute Canada's regions of relative water shortage, keeping in mind the caveats of the preceding paragraph.

Veeman (1985) pointed out that considerable care is necessary in drawing conclusions from such findings, particularly with respect to "absolute" water shortages. He argued that most "water scarcity" arguments treat water as a free good, when in fact we know that water use is price sensitive, and that significant opportunities exist for demand management. In other words, as indicated previously in this chapter, there are many demand management options which can be brought into action in times of water shortages, such as rationing, recycling, production rescheduling, and the like. Veeman also addressed issues and opportunities in each of the basins previously deemed water short by Foster and Sewell (1981). He showed that in each area, "water scarcity is more complicated than a mere physical comparison of consumptive use with short-run available supplies" (Veeman 1985, p. 19). Thus, while the drainage regions identified in the last paragraph are ones of relative water shortage, this does not necessarily imply a crisis or a situation of impending doom. It merely means that these are the areas where significant efforts in water demand management might result in large cost savings (vis-à-vis expanding the water supply network.

Projected Water Use, 1981-2011

Withdrawal Uses

Under the assumptions of the reference case scenario, the average annual growth rate for total Canadian water intake from 1981 to 2011 was about 2.3% per annum and for consumptive use was about 2.1% (Table 5). The difference in growth rate between the two components of water use occurs because some of the smaller industrial groups incorporated in the model have a water intake but no consumptive use. The Ontario and Prairie regions, under the reference case scenario, will experience growth rates slightly above average at about 2.4% while Quebec, at about 2.0%, will be somewhat below average. The higher than average rate of increase on the Prairies is a response to agricultural growth slightly higher than the national average, as the region continues to dominate the country's agricultural sector. Ontario's relatively high rate of increase is a consequence of thermal power production expanding faster than the national average. Neither of these two major water use sectors is important in Quebec,

TABLE 7. Volumes (mcm) of regional water intake and consumption, 1981 and 2011, with scenario impacts (%) on the reference case. 1, reference scenario; 2, conservation scenario; 3, technological change scenario; 4, recession scenario; 5, high growth scenario.

	1981	Volume in 2011 for scenario:				
		1	2	3	4	5
Water intake						
British Columbia	3 789	7 085	3 726 (−47)	8 177 (15)	5 481 (−23)	8 057 (14)
Prairies	5 363	11 172	6 687 (−40)	13 865 (24)	8 786 (−21)	12 895 (15)
Ontario	21 230	42 861	29 235 (−32)	53 646 (25)	33 847 (−21)	48 258 (13)
Quebec	4 252	7 629	4 327 (−43)	8 441 (11)	6 029 (−21)	8 901 (17)
Atlantic	2 884	5 584	3 764 (−33)	6 749 (21)	4 579 (−18)	5 929 (6)
Canada	37 518	74 331	47 738 (−36)	90 879 (22)	58 719 (−21)	84 039 (18)
Water consumption						
British Columbia	487	893	464 (−48)	991 (11)	691 (−23)	1 046 (17)
Prairies	2 256	4 443	2 262 (−49)	4 784 (7)	3 418 (−23)	5 318 (20)
Ontario	589	1 093	625 (−43)	1 235 (13)	842 (−23)	1 291 (18)
Quebec	435	690	380 (−45)	782 (13)	566 (−18)	804 (17)
Atlantic	139	244	151 (−47)	281 (10)	203 (−17)	263 (8)
Canada	3 906	7 363	3 882 (−47)	8 074 (10)	5 721 (−22)	9 025 (18)

which will continue to rely on hydro electric power (not included in the forecasts) and its traditional and relatively old economic base of manufacturing. The growth rate of water use in British Columbia will be slightly below the national average, while that in the Atlantic region will approximate the average.

Scenario 3, the technological change scenario, shows the highest rate of growth in total water intake over the 30-yr forecast period, with an average annual rate of growth of 3.0%, while the lowest rate of growth occurs in conjunction with scenario 2, the conservation scenario (0.8%) (Table 7). The large increase in total water intake in the technological change future was somewhat surprising. Previous work had shown that production technology trends, taken by themselves during the 1966-76 period, tended to *lower* industrial water use during the decade (Tate 1984a). Thus, with the growth rate held constant, I would have expected the technological change scenario to track *below* the reference case. More research is required before the effects of technological change on water use are properly understood.

With regard to scenario 2, the impact of the emphasis on conservation mechanisms to lower water use is clear. Water intake under this scenario will be only 65% of that predicted using the reference case assumptions by 2011. This is an important finding, for it shows that substantial water savings could accrue by adopting demand management principles as an integral part of a renewed water policy.

National trends in consumptive water use for the five scenarios (Table 7) indicate patterns essentially the same as those described for water intake, although they showed slightly slower growth rates, for the reason previously mentioned. The water consumption projections for the different scenarios relative to the reference case were essentially the same as those for water intake.

Among industry groups, the thermal power and agricultural sectors will continue to dominate the water intake and consumptive use projections, respectively. The average annual growth rates in water use by the various sectors cluster between 2 and 2.5%, with the lowest rate being 1.6% for muncipal intake and the highest being 3.3% for the mineral extraction sector.

Projected Water Use in the Drainage Regions

Projected drainage area breakdowns of projected water uses permit comparisons with water availability and nonwithdrawal uses (Table 6 shows the reference case). The distribution of regional water use amongst the basins remains constant throughout the period, making no allowance for possible shifts in socioeconomic activity. Data and techniques for building a fully dynamic model of water use distribution were unavailable when this research was carried out. For comparative purposes, both supply and nonwithdrawal use also remain unchanged over the forecasting period.

In most areas, net water use will remain just above 50% of reliable annual flow, indicating little probability of water availability problems for Canada as a whole. However, consumptive use in the Missouri, South Saskatchewan, and Assiniboine-Red basins will exceed 50% of the reliable annual flow in 2011. This indicates that net water use will exceed reliable annual flow, thereby creating possible problems of water shortage and requiring substantial water demand management efforts. In the Okanagan basin, net water use in 2011 will exceed 75% of reliable annual flows, indicating more severe water shortages in this area than are experienced currently. The use of conservation measures such as water pricing would avert some of these problems, as indicated in Table 6, although net water use in these water-short basins would still exceed 75% of reliable annual flow. Comments similar to those made earlier with respect to doctrines of water scarcity apply equally to future water supply:demand balances.

Comparison of the Current Forecasts with Two Earlier Studies

Cass-Beggs (1961) carried out the only study in the past to produce national estimates of withdrawal water use. A comparison of Cass-Beggs' projection for 1981 with the actual 1981 water use estimates contained in this chapter shows that the 1961 forecast

TABLE 8. Comparison of two water use forecasts (mcm) for the 1980–81 period with current study results.

	Cass-Beggs (1961) study	Water and the Canadian economy (op. cit.)	Current study
	1981	1980	1981
		Water intake	
By sector			
Agriculture	3 946	3 028	3 125
Mineral extraction	—	1 620	648
Manufacturing	39 456	14 208	10 201
Power	—	20 837	19 281
Municipal	3 822	4 128	4 611
Total	47 224	43 821	37 866
		Consumptive use	
Agriculture	—	1 609	2 412
Mineral extraction	—	826	179
Manufacturing	—	564	507
Power	—	156	168
Municipal	—	726	640
Total	—	3 881	3 906
		Water intake	
By region			
British Columbia	—	3 612	3 789
Prairies	—	9 315	5 363
Ontario	—	19 528	21 230
Quebec	—	5 877	4 252
Atlantic	—	5 491	2 884
Total	—	43 823	37 518
		Consumptive use	
British Columbia	—	517	487
Prairies	—	2 020	2 256
Ontario	—	751	589
Quebec	—	415	435
Atlantic	—	180	139
Total	—	3 883	3 906

was too high in total (47 223 mcm as compared with 37 518 mcm in the current study) (Table 8). Osborne et al. (1984) observed similar overestimation of water use in the United States. These two instances illustrate the dangers inherent in any projection of water use, but particularly for a single scenario only. The sectoral projections made by Cass-Beggs were also inaccurate, most severely in the industrial sectors. Cass-Beggs included neither a mineral extraction sector nor a thermal power generation sector in his projection, and his estimate for the manufacturing sector is four times greater than that measured in 1981. The overestimation of the manufacturing sector was caused by the assumption of industrial growth rates much higher than those actually experienced. By contrast, estimates of water use for the agricultural and municipal sectors made in 1961 were quite close to those that actually occurred.

My study of the role of water resources in the Canadian economy[8] also contained forecasts of water use for 1980. Foster and Sewell (1981) and Veeman (1985) both used data from this study, thus making it important to review the forecasts briefly. Projected water use for 1980 totalled 43 821 mcm, broken down into sectors and regions as shown in Table 8. Major inaccuracies occurred in the mineral extraction and manufacturing sectors. With respect to mineral extraction, "*Water and the Canadian Economy*" included all the water used for enhanced recovery of petroleum resources, principally in Alberta. It later became apparent that much of this water was saline ground-water and not fresh surface water and thereby constitutes no loss to the water supplies effectively available for use. Thus, in preparing the data base for the current study, only the volumes of water considered by Alberta officials as used in enhanced petroleum recovery were included. With respect to manufacturing, the overprojection of water use in my earlier study resulted from using the higher economic growth rates commonly used in the pre-1980's era. Other discrepancies arose because of the better data base available when the current study was carried out.

Implications of the Results

Detailed studies and results can often lead to an understanding of the "trees" but not the "forest." Certainly, water use in any nation comprises a vast and complex field, and it is easy for one to get lost in detail and to ignore the wider implications of findings. To conclude this chapter, I will address some of these wider issues briefly, hoping to provide a perspective on the results of this research and to identify some important research directions.

Water Use and Growth

Since we first began to keep systematic water use records, total water withdrawals have grown much faster than the general population (Fig. 2). For the most part, growth of thermal power generation facilities, especially in Ontario, and expanded levels of Prairies irrigation led this growth pattern. Moreover, the other industrial sectors have grown at rates faster than the population. The forecasts in this chapter indicate that this growth may continue over the next 30 yr.

Several implications flow from this finding. Even Canada, with its abundant water supplies, cannot continue to use more and more water without thought for the consequences. Over half of the nation's total renewable water supplies in any event flows northward through sparsely populated areas, while most of the socioeconomic activity is in the south. Some drainage areas in western Canada are already relatively water-short, and developments threaten to make them even more so. The expansion of irrigation in western Canada at rates suggested by some public agencies is very serious in its implications for consumptive water use, public funding, water transfers, and environmental disruption. For example, the recently published South Saskatchewan River Basin study (Anonymous 1984) projects an additional 98 000 ha of irrigated land in the Alberta portion of the basin alone by the year 2000. This additional area could rise to as much as 364 000 ha. The additional consumptive use for irrigation, given a constant water supply, implied by these projected increases may create water shortages in other sectors such as hydro and thermal power generation, municipal water supply, and fish and wildlife habitat maintenance. Demands for public funds for irrigation system expansion mean that other, possibly higher valued uses (e.g. waste treatment, transportation system construction, etc.) may experience capital shortages. In the event

---

[8]Water and the Canadian Economy, op. cit.

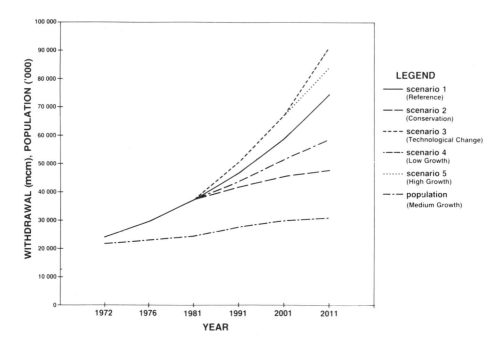

FIG. 2. Total water withdrawal and population projections for Canada, 1972-2011.

of interbasin water transfers, substantial environmental damages may occur (Howe and Easter 1971, p. 106-107; see also chapter by Rosenberg et al., this volume). Increasing water supplies to support industrial and municipal development, without first examining the effects of other options, likewise will lead to inefficient use of both water supplies and investment capital.

Water Uses, Public Policy, and Water Conservation

In Canada, the federal and provincial levels of government share jurisdiction over water resources, and thus over matters of water use (Pearse et al. 1985, p. 63-66; see also chapter by Thompson, this volume). Except in the north, however, where the federal government has quasi-provincial powers, and in federal lands such as national parks, the provinces exercise most direct control over water use. Water use policies across the country are relatively consistent. All provinces license the withdrawal of water. For manufacturing industries, licences generally permit a maximum rate or volume of withdrawal, with fees set to cover administrative but not commodity charges. Water rentals for hydropower generation are flat rates based on station capacities or production. As shown above, municipalities often charge flat rates for supplying water to homes and industries and, if they employ metering, normally set declining block rate pricing schedules. In general, declining block rates mean that the more water used, the less paid per unit. Neither flat rates nor declining block rates encourage water conservation and the rationalization of water supply systems.

The analysis in this chapter shows that significant conservation can be achieved given a certain set of policies. Traditionally, water has been treated and managed as a "free" good, and, as a consequence, wastage of water supplies, public overexpenditure both for capital, and operating and maintenance charges as well as water pollution have resulted. Indeed, this failure to employ pricing concepts, the backbone of western economic systems, to water management may impede efforts for effective water management (Bower 1968; Mitchell 1984). Examples of overuse of water resulting from a failure to price water correctly are legion in Canada. For example, in the Prairies, the City of Calgary, which, for the most part, has unmetered water supplies and flat rate charges, uses twice as much water per capita as Edmonton, which has metered supplies (Edmonton Journal 1983). In many other areas throughout the country, the same observation holds true. Even with metered water supplies, promotional rate structures (e.g. flat rates) discourage conservation of municipal water supplies and overbuilding of water supply systems. In irrigation, large public subsidies to promote community and regional development have led to enhanced irrigation use, since users pay an artificially low price for the water and associated infrastructure. In manufacturing, water recycling in Canada is used less frequently relative to the United States, partially due to the failure to impose economically based water prices. This situation also contributes substantially to the water pollution problem by encouraging industries to deposit their wastes untreated into surface or ground waters.

As stated by Pearse et al. (1985, p. 98–99), suitable water pricing arrangements would

> . . . create incentives to avoid waste and use water efficiently thus contributing to water conservation. By reducing the water needed, and the waste disposal capacity, it will reduce infrastructure costs. By demonstrating users' willingness to pay for water, prices help allocate supplies among the users so that the highest value is generated from limited resources. Pricing will generate revenue to cover the cost of water supply and waste disposal systems. Suitable pricing can ensure that the cost of water services is equitably born by the beneficiaries according to the benefits they receive.

Therefore, there is little doubt that water use can be substantially reduced, thereby alleviating the water problems inherent in the process of an expanding economy.

The phrase "water conservation" as used in this paper does not refer wholly to the reduction of water use. Rather, following Baumann et al. (1980), the term is defined as "any beneficial reduction in water use or water losses." Here, beneficial is used in a social welfare sense, occurring in those situations where the aggregated benefits of the water management measure exceed the aggregated adverse effects of implementing the measure. Thus, as in supply management, the net benefits of proposed water conservation measures should be positive and greater than those of other alternatives if they are to be implemented.

Synergisms, Water Use, and Water Quality

While individual water use sectors have major problems of their own, such as inflated demands in irrigation, and water pollution by industry, these problems become much more serious if we look at interactions between uses. As Canada grew, the uses of water multiplied, and users once separated from one another now must share water courses. Today, conflicts between water users occur frequently, spilling through the media into public agendas. Industrial pollution may damage adjoining recreational areas and wildlife habitat and threaten public water supplies. Increases in irrigation water use may curtail power production downstream. Increased municipal demands place burdens on public treasuries which could be avoided by demand management. An exhaustive list of potential conflicts is beyond the scope of this chapter, but conflicts are increasing in number and complexity, and solving them constitutes a major water management task for the future.

This chapter has not examined water quality issues in detail, but has noted that they are an integral part of Canada's total water use picture (see chapter by Hamilton et al., this volume). In fact, water quality problems are much broader geographically and potentially much more serious in their effects on Canadians than are water quantity problems. As Pearse et al. (1985, p. 48) stated, ". . . with few exceptions, Canada's water problems are not related to inadequate supply at all, but to degraded water quality and to disrupted flow regimes." This is not to say that water shortages are not serious where they occur, but to emphasize that water quality considerations are also essential in looking at future water use.

Future Research Needs

The research undertaken for this chapter is quite comprehensive in its examination of future Canadian water use. Nevertheless, it has significant limitations in both scope and technique, which suggest areas for future research.

Integration of Water Use, Water Supply, and Water Quality

Some areas of Canada can without question be termed areas of water shortage. These areas are located in western Canada. Also, although not discussed in detail, areas throughout the country have substantial water quality problems which threaten effects even more serious than those of water quantity.

The real value of water demand forecasting will be realized when demand, supply, and quality are integrated at levels effective in identifying and quantifying water imbalance and degradation problems. To achieve such an integration at the drainage basin level is a long-term and very complex undertaking. Simulation modeling, along lines outlined earlier, holds promise as an effective methodology for achieving such an integration. Such a methodology should be refined and applied to major basins in Canada, beginning with those basins where water supply:demand problems are current or threatening, and with those areas which have significant water quality problems. Further, the development and improvement of river basin simulation modeling should be emphasized as the basis for future water use and demand studies.

Continued Development of Regional Water Use Modeling

This chapter concentrates on projecting water use in the major regions of Canada. A number of assessments, such as watching briefs on overall national and regional water use, studies of future consumptive water use, and policy evaluations, will continue to be based on such projections and modeling activities. Development and application work on these aggregate models should be continued. Further, the regional models should be augmented to include a waterborne waste component similar to the one suggested by Victor (1972).

Water Demand in Specific Sectors

The chapter shows that very little research has been done in Canada on detailed aspects of water demand in various economic sectors. Particularly in industry and agriculture, although we have aggregate statistics on water uses and the responses to questionnaires of water users, we have none on the process modeling results, such as those carried out by Russell (1973), Russell and Vaughn (1976), and those outlined by Kindler and Russell (1984). There is a real need to conduct this type of research for major water users such as irrigation, thermal power, pulp and paper, iron and steel, and petroleum refining.

Further Investigation of Conservation Measures

Scenario 2 demonstrated that significant reductions in water demand relative to other growth paths may be achievable through improved water pricing practices and other conservation measures. The analysis was quite hypothetical, however, and relied upon a number of assumptions and secondary data sources. The impacts of various conservation measures need to be investigated fully, not only on future water use but also on the requirements for new water supply development initiatives.

Water Use Data

The forecasting model used in this research relied heavily on the availability of reliable Canadian data on various aspects of water use. These data have been collected regularly over the past 10 yr. The collection of such data should continue, both to assess the accuracy of the forecasts made here and to provide data for improved forecasting. Since improved forecasting is likely to derive from river basin level simulation models, the data need to be collated on a river basin basis as well as on the basis of political regions.

Nonwithdrawal Water Uses

The forecasts presented here relate only to the major withdrawal uses of water. Preliminary indications, however, show that the nonwithdrawal water uses (e.g. recreation, wildlife, hydroelectric power, etc.) are larger in quantitative terms, and probably just as important, albeit unquantified, in economic terms (Muller 1985). There is a real need for more research in the area of nonwithdrawal uses to identify the most efficient method of incorporating these considerations into investigations of future water demand.

Acknowledgements

The project used as the basis for this chapter was made possible only with the assistance of several persons. Ms. P. Dossett served as research assistant throughout the project and performed most of the mathematical computations and graphical presentations contained in the report. Dr. A. Kassem provided material on current water use simulation work. The above persons also provided useful day-to-day discussions and advice and reviewed the draft chapter. D. Scharf, D. Lacelle, and P. Hess provided many data pertaining to current water use on which the study is based. Ms. C. Lefebvre provided clerical assistance throughout the project. The Advisory Committee established by the Inquiry on Federal Water Policy, for the project on which the chapter is based, provided many useful ideas and suggestions. This committee consisted of D. Baumann, P. Harrison, F. Quinn, C. Simmonds, C. Sonnen, B. Stokes, D. Vallery, and T. Veeman. Finally, the research done for the Inquiry was supported by the Inland Waters Directorate, Environment Canada. However, the research and discussion reflect the views of the author alone.

References

ACRES INTERNATIONAL LTD. 1983. Water supply constraints to energy development — phase II summary report. 135 p. (Available from Environment Canada, Inland Waters Directorate, Ottawa-Hull).

　　　1984. Water supply constraints to energy development — phase III summary report. 207 p. (Available from Environment Canada, Inland Waters Directorate, Ottawa-Hull).

　　　1985. Water supply constraints to energy development — phase IV summary report. 150 p. (Available from Environment Canada, Inland Waters Directorate, Ottawa-Hull).

ALBERY, A. C. R. 1968. In-stream or flow water demands, p. 204-233. *In* W.R.D. Sewell, and B.T. Bower [ed.] Forecasting the demands for water. Department of Energy, Mines and Resources, Ottawa-Hull.

ANONYMOUS. 1972. Final report of the Okanagan River Basin Board. Okanagan River Basin Board, Penticton, B.C. 365 p.

　　　1977. A review of instream flow needs: methodologies for fish and recommendations for instituting protective measures in Alberta. Alberta Environment, Planning Division, Edmonton, Alta. 12 p.

　　　1984. South Saskatchewan River Basin Planning Program Scenario Report. Alberta Environment Planning Division, Edmonton, Alta. 322 p.

BAUMANN, D. D., J. J. BOLAND, AND J. H. SIMS. 1980. The problem of defining water conservation. Cornett Papers, University of Victoria, Victoria, B.C. p. 127-133.

BOLAND, J. J., B. DZIEGIELEWSKI, D. D. BAUMANN, AND E. M. OPITZ. 1984. Influence of price and rate structures on municipal and industrial water use. U.S. Army Corps of Engineers, Fort Belvoir, VA. 117 p.

BOWER, B. T. 1968. Industrial water demands, p. 204-233. *In* W. R. D. Sewell and B. T. Bower [ed.] Forecasting the demands for water. Department of Energy, Mines and Resources, Ottawa-Hull.

CASS-BEGGS, D. 1961. Water as a basic resource, p. 173-189. *In* Background papers. Vol. 2. Resources for Tomorrow Conference. Queens Printer, Ottawa-Hull.

DE ROOY, J. 1970. The industrial demand for water resources: an econometric approach. University Microfilms, Ann Arbor, MI. 300 p.

DURRANT, E. F. 1983. An appraisal of future water needs and resources. Futurescan, Regina, Sask. 55 p.

EARMME, S. Y. 1979. A water use projection model for the North Saskatchewan River basin, Alberta: an input-output approach., Dissertation, University of Alberta, Edmonton, Alta. 226 p.

EDMONTON JOURNAL. 1983. Water on the brain. May 4 issue.

FOSTER, H. D., AND W. R. D. SEWELL. 1981. Water: the emerging crisis in Canada. Lorimer and Co., Toronto, Ont. 117 p.

GRIMA, A. P. 1972. Residential water demand: alternative choices for management. University of Toronto Press, Downsview, Ont. 211 p.

HANKE, S. H. 1978. A method for integrating engineering and economic planning. Am. Water Works Assoc. 70 (9): 487-492.

HOWE, C. W., AND K. W. EASTER. 1971. Interbasin transfers of water: economic issues and impacts. Johns Hopkins Press, Baltimore, MD. 196 p.

INFORMETRICA, LTD. 1984. The Canadian provincial economies to 2005: provincial assumptions and summary tables, I-84 Forecast. Ottawa, Ont. 325 p.

KINDLER, J., AND C. S. RUSSELL. 1984. Water demand modelling. Academic Press, Toronto, Ont. 248 p.

KITCHEN, H. M. 1975. Statistical estimation of a demand function for residential water. Environment Canada, Inland Waters Directorate, Soc. Sci. Ser. No. 11: 9 p.

LANE, R. K., AND G. N. SYKES. 1982. Nature's lifeline: prairie and northern waters. Canada West Foundation, Calgary, Alta. 467 p.

LEONE, R. A. 1975. Changing water use in industry, p. 339-349. *In* J. E. Crews and J. Tang [ed.] 1981. Selected works in water supply, water conservation and water quality planning. U.S. Army Corps of Engineers, Fort Belvoir, VA. IWR Research Report 81-R10.

LOFTING, E. M., AND P. H. MCGAUGHEY. 1963. An interindustry analysis of the California water economy. University of California, Water Resources Center, Berkley, Contribution No. 116: 83 p.

MIERNYK, W. H. 1965. The elements of input-output analysis. Random House, New York, NY. 156 p.

MITCHELL, B. 1984. The value of water as a commodity. Can. Water Resour. J. 9 (2): 30-37.

MULLER, R. A. 1985. The socio-economic value of water in Canada. Inquiry on Federal Water Policy, Research Paper 5, Ottawa-Hull. 104 p.

OSBORNE, C. T., J. E. SCHEFTER, AND L. SHABMAN. 1984. Forecasting water use: past predictions and reality. Unpublished paper given at AWRA Meeting, Virginia Tech., Blacksburg, VA. 15 p.

PEARSE, P. H., F. BERTRAND, AND J. W. MACLAREN. 1985. Currents of change: final report of the Inquiry on Federal Water Policy. 222 p. (Available from Environment Canada, Inland Waters Directorate, Ottawa-Hull.)

PRAIRIE PROVINCES WATER BOARD. 1982. Historical and current water uses in the Saskatchewan-Nelson Basin. Prairie Provinces Water Board, Regina, Sask. 147 p. plus appendices.

REES, J. A. 1969. Industrial demand for water: a study of South East England. Weidenfield and Nicolson, London. 150 p.

RICHARDSON, H. W. 1972. Input-output and regional economics. Weidenfield and Nicolson, London. 172 p.

RUSSELL, C. S. 1973. Residuals management in industry: a case study for petroleum refining. Johns Hopkins Press, Baltimore, MD. 193 p.

RUSSELL, C. S., AND J. VAUGHN. 1976. Steel production: processes, products and residuals. Johns Hopkins Press, Baltimore, MD. 193 p.

SEWELL, W. R. D., AND L. ROUCHE. 1974. The potential impact of peak load pricing on urban water demands in Victoria, British Columbia, a case study, p. 141-161 *In* F. Leversege [ed.] Priorities in water management. Victoria, West. Geogr. Ser. 8.

STATISTICS CANADA. 1982. Population projections for Canada and the provinces, Catalogue No. 91-520. Ottawa-Hull. 472 p.

TATE, D. M. 1978. Water demand forecasting in Canada: a review. International Institute for Applied Systems Analysis, Laxenberg, Austria. Research Memorandum 78-16. 53 p.

1984a. Industrial water use and structural change in Canada and its regions: 1966-1976. Dissertation, University of Ottawa, Ottawa, Ont. 256 p.

1984b. Canadian water management: a one-armed giant. Can. Water Resour. J. 9 (3): 1-7.

1985. Alternative forecasts of Canadian water use, 1981-2011. Inquiry on Federal Water Policy. Research Paper 17 Ottawa-Hull. 383 p.

TATE, D. M., AND D. N. SCHARF. 1985. Manufacturing water use in Canada 1981: a summary of results. Environment Canada, Inland Waters Directorate, Soc. Sci. Ser. 19. Ottawa-Hull. 52 p.

U.S. NATIONAL WATER COMMISSION. 1976. Forecasts and the role of alternative futures. Staff pap. J. Water Resour. Plann. Manage. Div. ASCE Nov.: 365-383.

VEEMAN, T. S. 1985. Water and economic growth in western Canada. Economic Council of Canada Discussion Paper 117, Ottawa. 95 p.

VICTOR, P. A. 1972. Pollution: economy and environment. George Allen and Unwin, London. 247 p.

WHITTINGTON, D. 1978. Forecasting industrial water use. International Institute for Applied Systems Analysis, Research Memorandum 78-71. Laxenberg, Austria. 50 p.

WOLMAN, N., AND C. W. BONEM. 1971. The outlook for water. Johns Hopkins Press, Baltimore, MD. 286 p.

CHAPTER 4

The Environmental Assessment of Hydroelectric Impoundments and Diversions in Canada

D. M. Rosenberg, R. A. Bodaly, R. E. Hecky, and R. W. Newbury

*Department of Fisheries and Oceans, Freshwater Institute,
501 University Crescent, Winnipeg, Man. R3T 2N6*

Introduction

The rivers of Canada transport more than $100\ 000$ m³ of water per second (Environment Canada 1975a). As the water descends from elevated surfaces to sea level, the energy of flow maintains the stream and river channels that combine to form the major drainage basins of this country. In many of these drainage basins, a portion of the energy of flow is captured in reservoirs and transformed into hydroelectric power.

Canada has an estimated 95 000 MW of economically feasible hydropower potential (Efford 1975), of which from one third to one half already has been developed. "By 1972, there were some 365 hydroelectric projects in Canada, affecting over 200 rivers and streams" (Efford 1975, p. 201). Many of these projects include diversions (Efford 1975; Day 1985), and several schemes have been proposed that would transfer major portions of northern rivers southward for irrigation and other water uses in Canada and the United States (see chapter by Bocking, this volume).

Proponents of large-scale hydroelectric developments and water redistribution systems commonly espouse two mistaken perceptions concerning the environment (e.g. Bourassa 1985). In the first, they claim that hydroelectric development has minimal environmental consequences; that is, renewable hydropower is "clean." In the second, they consider that water allowed to flow freely into the ocean is a "wasted" resource. Nothing is further from reality than these two notions.

Hydroelectric developments usually are accompanied by deleterious environmental effects. The alteration of river flows and capture of their hydraulic energies profoundly affect the aquatic ecosystems and landscape forms involved. Because each drainage basin has evolved its present form over thousands of years, a major alteration of flow in one branch removes part of the energy that maintains river form and forces the landscape to evolve into a new configuration. In turn, biological communities must adjust to the evolving landscape. We will describe typical environmental disruptions that result.

Water flowing unimpeded to the sea is far from being wasted. In the case of north-temperate rivers, natural seasonal runoff patterns heavily influence downstream deltaic, estuarine and coastal environments, and modification of this runoff by interbasin water diversion and storage for power production can have severe impacts (e.g. Townsend 1975; Berkes 1981; Neu 1982a, 1982b). Thus, the deleterious consequences of extensive hydroelectric development in a watershed also can extend to areas thousands of kilometres away.

Nevertheless, these incorrect perceptions often emanate from otherwise respectable sources (e.g. Abelson 1985), so it is important that *correct* information regarding the environmental consequences of hydroelectric development be disseminated widely. Such information can originate only from the proper scientific study of such developments.

Thus, the objective of this chapter is to answer the following important questions: Is the best framework of scientific and management techniques and procedures being applied to the prediction and assessment of changes caused by major hydroelectric and water-diversion projects? As a scientific community, do we have adequate techniques and knowledge to assess and predict the effects of such changes? We shall critically assess selected environmental impact studies of major Canadian hydroelectric developments (reservoirs and/or water diversions) with a view to identifying scientific and management needs for future impact assessments of water developments.

Methods

Hydroelectric Developments Chosen for Critical Analysis

We chose the following five Canadian hydroelectric developments (or areas directly affected) for critical analysis: Kemano (B.C.), downstream effects of the Williston Reservoir (i.e. on the Peace-Athabasca Delta, Alberta), Churchill Falls (Labrador), Churchill-Nelson (Manitoba), and James Bay (Quebec) (Fig. 1). Characteristics of these projects are summarized in Table 1. We chose these projects because of their wide geographical distribution and their occurrence in different physiographic regions, because of the relatively large amount of documentation available on them, and because of the time span covered (from the 1950's to the present). In addition, we selected one major United States hydroelectric development, impoundments along the mainstem Missouri River, for comparison because it provided a midcontinental location and expanded our perspective.

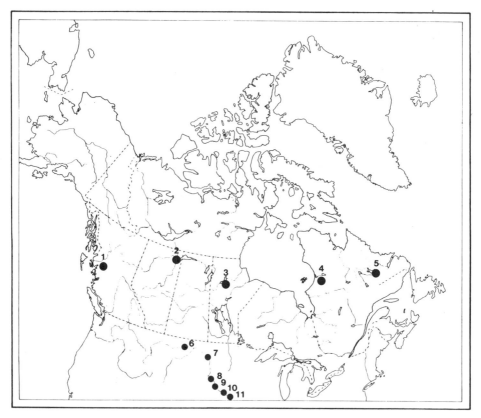

FIG. 1. Hydroelectric developments (or areas directly affected) chosen for analysis. 1, Kemano Phase I; 2, Peace-Athabasca Delta; 3, Churchill-Nelson; 4, James Bay; 5, Churchill Falls; 6-11, Missouri mainstem reservoirs (6, Lake Fort Peck; 7, Lake Sakakawea; 8, Lake Oahe; 9, Lake Sharpe; 10, Lake Francis Case; 11, Lewis and Clark Lake).

Elements of bias may exist in the selection of large-scale, completed projects for review. First, in order to do the analysis, we found it necessary to choose *completed* projects; projects that were proposed, examined, and subsequently cancelled are not included. However, abandonment of projects after completion of the predevelopment tunnel phase is rarely attributable solely to environmental reasons. Factors such as international disputes or more economical alternatives are more typically responsible. There is no reason, therefore, to believe that the projects chosen for analysis here had unusually weak environmental assessments. Second, projects having extensive postdevelopment assessments, such as those considered here, may be considered atypical because post-development assessment of hydroelectric projects is rare (Geen 1974). Third, large-scale projects may not reflect the norm because they are usually contentious, attract considerable attention, and lead to comprehensive documentation. (Nevertheless, the availability of such documentation is basic to the analysis presented here.) Fourth, emphasis on large-scale projects neglects the cumulative impacts of small developments. Although this effect is inadvertent, large-scale projects deserve rigorous appraisal because of their potentially enormous influence on vast areas of landscape. Moreover, the appearance of general trends from the analysis would indicate that the large-scale projects examined are representative not only of small-scale hydroelectric developments but, perhaps, of other types of developments as well.

Selection of Criteria Used for Evaluation of Hydroelectric Developments

Hecky et al. (1984) combined information from the environmental impact assessment literature in general with the environmental impact assessment experience for the Churchill–Nelson development (mainly Southern Indian Lake) in particular to identify an optimal framework for assessing hydroelectric developments. The optimal framework requires (1) a study design that involves a predevelopment predictive phase, and a postdevelopment monitoring and evaluation phase, (2) the use of quantitative predictions, and the integration of these predictions among trophic levels, and (3) institution, before development, of adequate compensation schemes for resource users, and mitigation schemes for resource protection. These elements of the framework were elaborated into a checklist of questions that we used in analyzing the hydroelectric and diversion projects selected above (see Table 3).

Description of Reservoir and Diversion Projects Reviewed

Kemano River

In December 1949, the government of British Columbia granted a water license to the Aluminum Company of Canada (Alcan) to store and divert the headwaters of the Nechako and Nanika Rivers in central British Columbia. The water would supply a high-head power generating station on the Kemano River.

Phase I of the project involved diverting almost half of the available water (Table 1) out of the Nechako–Fraser basin and into the Kemano basin through a 16-km tunnel that drops 792 m to a powerhouse on the Kemano River, near sea level. The station produces an average of 635 MW, and this power is transmitted 75 km to the Kitimat aluminum smelter. The other half of the available water was retained in the Nechako–Fraser basin, but was rerouted over a spillway at Skins Lake into the adjacent Cheslatta River channel.

Nechako River waters were stored in a series of seven lakes arranged in a circular pattern at the base of the Quanchus Mountain Range. These lakes were inundated when the Kenney Dam was closed at the head of the "Grand Canyon" reach of the upper Nechako Valley. This valley leads away from the Quanchus Mountain Range into the upper Fraser River drainage to the east. The reservoir created by flooding the seven lakes is 890 km^2, has a total flooded area of 334 km^2, and a tributary drainage area of 13 900 km^2.

TABLE 1. Characteristics of hydroelectric projects selected for assessment here. NA = not applicable.

Project	Location	Period of construction	Total generating capacity (MW)	Reservoirs				Diversions		Comments
				No. involved	Total surface area of impounded water (km²)	Range of mean depths (m)	Range of drawdown depths (m)	No. involved	Total annual mean flow (m³/s)	
Kemano Phase I	West-central British Columbia	1950-54	896	1	890	>30ᵃ	2	2	(1) 115 (2) 130	(1) Diversion of Nechako River into Kemano River for power generation. (2) Diversion of Nechako River into Cheslatta River for spillway. For details see Envirocon Ltd. (1981)
Peace-Athabasca Delta	Northeastern Alberta	Construction of Bennett Dam 1962-67; drying of Delta noticed 1968-71 (coincident with filling of Williston Reservoir, Peace River, B.C.)	NA	NA	NA	NA	NA	NA	NA	Downstream effects of construction of Bennett Dam and filling of Williston Reservoir on the Peace River in British Columbia of interest here. Descriptions of the upstream hydroelectric development and the Peace-Athabasca Delta can be found in Peace-Athabasca Delta Project Group (1972, 1973) and Townsend (1975)
Churchill Falls	Labrador, Newfoundland	1967-75	5 230	3 (including the west forebay, i.e. Jacopie Lake)	6 705	Not given	3.5-4	3	Not given: total drainage area of Upper Churchill was increased from 56 000 to 69 000 km² by diversions	Primary purpose of this development is hydroelectric power generation. For details, see Duthie and Ostrofsky (1975)
Churchill-Nelson development	Northern Manitoba	1968-76	2 378	3	3 299	7-21.3	0.7-2.4	1	850 (licensed maximum capacity of Churchill River diversion through Southern Indian Lake)	Kelsey generting station excluded. Only three reservoirs (Southern Indian Lake, Notigi, Stephens Lake) involved in development. Lake Winnipeg is regulated within its natural range of elevation, and all other forebays, although flooded, are of limited sizes. See Lake Winnipeg, Churchill and Nelson Rivers Study Board (1975) for description of Churchill-Nelson development and Newbury et al. (1984) for description of impoundment of and river diversion through Southern Indian Lake

TABLE 1. (Concluded)

Project	Location	Period of construction	Total generating capacity (MW)	Reservoirs				Diversions		Comments
				No. involved	Total surface area of impounded water (km²)	Range of mean depths (m)	Range of drawdown depths (m)	No. involved	Total annual mean flow (m³/s)	
James Bay (La Grande complex)	Northern Quebec	1975-84	10 300	5	11 369	8-29	4-12.9	3	50 km³	Information here refers to system as built through 1984. Studies currently in progress to add more generating capacity at LG2 and another reservoir and powerhouse at LG1 on the La Grande River. All information except drawdown depths is from Berkes (1985). Drawdown estimates are from Marsan and Coupal (1981)
Missouri mainstem reservoirs	Montana, North Dakota, South Dakota, Nebraska	1935-65	2 098 (see comments)	6	6 260	5-24	0.6-11	None	NA	Missouri reservoirs serve flood control, navigation, hydropower, irrigation, water supply, and other functions. For details see Benson and Cowell (1967)

[a]Depth of flooding alone; natural lake levels not measured.

The spillway configuration for the Nechako Reservoir was, environmentally, a major disruptive component of the project. The decision to create a spillway rather than build a control structure at Kenney Dam was made because of cost. The Nechako River canyon below Kenney Dam has been abandoned, and the spillway is located 68 km west of the dam at Skins Lake on the Cheslatta River. This river was formerly a small tributary of the Nechako which entered the main channel at Cheslatta Falls, a few kilometres downstream from Kenney Dam. Reservoir releases into the Cheslatta River Valley exceed its natural flow by two orders of magnitude (maximum inflows: 500 m^3/s; average flows since 1956: 130 m^3/s) and have caused the river channel to erode as much as 20 m below the former valley floor. Sediment eroded from the valley is deposited in Cheslatta Lake, 25 km below the spillway, where it has formed a large gravel delta. The level of Cheslatta Lake is controlled by the Murray Dam, below the lake outlet, to provide for riparian flows downstream. Several small lakes along the course of the original Cheslatta have drained due to erosion of their outlet channels, and concentrations of suspended sediments exceeding 200 mg/L are carried through the system into the Nechako River below Cheslatta Falls (Envirocon Ltd. 1981).

Land for the project was purchased from the Province of British Columbia or directly from local landowners. Water rentals and taxes based on energy consumption and aluminum prices are charged for the remaining Crown lands affected by the development. Alcan was required to clear navigation channels in the Nechako Reservoir, between lakes, at a cost not exceeding $250 000. In recent years, an additional $500 000/yr has been spent on underwater clearing, using specially designed barges.

Approximately 10 ranches were purchased in the area that became the Nechako Reservoir, and 10 reserves belonging to the Cheslatta Indian Band were affected along the spillway route. The best reserve lands (e.g. natural hay meadows) were flooded (Day 1985). The water license to flood Cheslatta Lake was issued 4 mo after the flooding started. Seven months later, the Federal Minister of Indian Affairs approved the sale of Cheslatta Band lands to Alcan for $125 000. Assistance in moving was provided for nine Indian families, and two monuments were erected above the flood line to mark the location of two Indian graveyards. "More than 30 years after these events, the displaced families are considering legal action to regain the losses…" (Day 1985, p. 41; see also Telkwa Foundation 1984).

No water was released from the Nechako Reservoir during its filling from October 1952 to June 1955, and water was released only intermittently until 1957 (Day 1985). The exposure of spawning beds and the effects of sedimentation and increased temperatures severely reduced stocks of the Nechako River salmon. These stocks have recovered to only 40% of their predevelopment levels since 1957 when regular releases from the reservoir were started (Department of Fisheries and Oceans 1984, p. 1). In 1980, the Government of Canada obtained a mandatory injunction in the Supreme Court of British Columbia ordering Alcan to provide a suitable release schedule in the Cheslatta River to maintain flows for the protection of fish habitat and migration. On the other hand, the Kemano River fishery has been enhanced by diversion waters released from the generating station (Department of Fisheries and Oceans 1984, p. 41–46).

In 1978, Alcan began studies of the existing Nechako impoundment and diversion and of the uncompleted Nanika River diversion that was part of the original water license (see above). The proposed "completion" project would increase the total diversion flow into the Kemano generating station by 84% from 115 to 202 m^3/s. An additional 798 MW would be generated and would supply two new aluminum smelters.

Diversion of the upper Nanika River would be accomplished by damming it at the outlet of Kidprice Lake, 25 km north of the western arm of the Nechako Reservoir, and digging a tunnel between the new 58-km^2 reservoir in the Nanika Valley and the Nechako Reservoir. The average diversion flow would be 12 m^3/s or approximately 62% of the Nanika flow at Kidprice Lake. The diversion would transfer 890 km^2 from the headwaters of the Skeena River basin to the Fraser and Kemano River basins.

Another tunnel, twinning the one from the Nechako Reservoir to the Kemano River, also would be built.

In contrast with the single-purpose view taken for the first phase of the Kemano project, Phase II is based on "...the most thorough and open prediversion assessment undertaken in Canada to date..." (Day 1985, p. 43). Phase II studies have considered the effects of Phase I and how they might be mitigated by altering the project configuration. For example, one of the proposed mitigation schemes would reduce flows in the Cheslatta Valley by constructing release facilities at the Kenney Dam. This would allow partial restoration of flows to the upper canyon of the Nechako and better temperature control in the upper Fraser River.

To date, over $20 000 000 has been spent on historical assessment of Phase I and redesign of the project. Of this, approximately 15% went to public consultation. However, plans for Phase II have been suspended due to poor world markets for aluminum (Day 1985).

Peace-Athabasca Delta

The first large hydroelectric project built in the Mackenzie Basin was the W.A.C. Bennett Dam on the upper Peace River in British Columbia (Mackenzie River Basin Committee 1981). Effects of the filling of Williston Reservoir and flow regulation of the Peace River on the downstream Peace-Athabasca Delta have become a classic example of the kind of major ecological disturbance that can result when hydroelectric development is inadequately planned.

The Peace-Athabasca Delta covers 3800 km^2 and comprises the deltas of three rivers: the Athabasca which flows into the western end of Lake Athabasca from the south, the Birch to the west, and the Peace to the north. The main outlet of Lake Athabasca is the Rivière des Rochers. It is joined by the Peace River north of the lake to form the Slave River which flows into Great Slave Lake. Other important outlets include the Chenal des Quatre Fourches and the Revillon Coupé.

Before construction of the Bennett Dam, high, early summer flows in the Peace River impeded regular flow out of Lake Athabasca, and this caused Lake Athabasca waters to back up and flood the Delta. Later, when water levels on the Peace River decreased, the outlets of Lake Athabasca were no longer blocked, waters from the lake resumed their northward flow, and flood waters receded from the Delta. The seasonal flooding of the Delta maintained the vegetation in an early successional stage of high productivity and this, in turn, led to high productivity of wildlife.

The Bennett Dam was closed in December 1967, and Williston Reservoir was filled (61.574 × 10^6 m^3) from 1968 to 1971. Although the Dam is located 1100 km upstream from the Delta, it controls approximately half the flow of the Peace River at Lake Athabasca. Normal peak flows of 4000-9000 m^3/s on the Peace River were reduced to 280 m^3/s during filling; flood flows in the river adjacent to the Delta were reduced by as much as 5663 m^3/s. Water levels in the river dropped 3.0-3.5 m below normal, and Lake Athabasca waters flowed out of the Delta without causing flooding (Townsend 1975).

In the summers of 1968-71, the Delta landscape began to change appreciably. Perched basins which depended on flooding for refilling and larger lakes connected to Lake Athabasca or to river channels in the Delta began to dry out. Vegetational succession continued unchecked, creating new meadow and willow communities.

In January 1971, a symposium was held at the University of Alberta to address concerns regarding drying of the Delta (Reinelt et al. 1971). As a result of a recommendation from this conference, the Peace-Athabasca Delta Project Group was established (PADPG 1972, 1973). This was a cooperative study involving the Canadian government and the governments of Alberta and Saskatchewan. The government of British Columbia was not a task-force member despite the fact that a British Columbia crown corporation caused the problem.

The deleterious downstream effects of filling Williston Lake included a nearly 40% reduction in shorelines and water-surface areas of perched basins, exposure of 50 000 ha of mud flats of larger lakes connected to Lake Athabasca and the river channels, and severe reductions in the muskrat population (a furbearer important to local trappers) (Townsend 1975). Long-term effects of operation of the Bennett Dam were predicted using hydrological and wildlife computer simulation models. The hydrologic model predicted a marked departure from past flow patterns of the Peace River and long-term reductions in summer and peak flows. The wildlife model used results from the hydrologic model, translating effects of changes in the water regime into habitat and wildlife changes. Fish populations were not included in the model because of a lack of quantitative information. Nevertheless, predictions for fish populations were made on the basis of lower water levels (Townsend 1975). The deleterious effects predicted to result from continued operation of the Bennett Dam included a reduction of perched-basin shorelines of 50-55%, a decrease in duck production of 20-25%, a reduction of fall populations of muskrat of 40-60%, and reduced spawning success of walleye. Reductions in muskrat and walleye populations would exacerbate already serious unemployment and low-income problems in Fort Chipewyan, a major Indian and Metis settlement in the area.

As a remedial measure, PADPG recommended that a fixed-crest weir be built on the Rivière des Rochers to compensate for the decline in water level of Lake Athabasca due to the changed hydrological regime. This weir, and another on the Revillon Coupé, were built and are currently operational.

Postdevelopment resource monitoring was recommended (PADPG 1973; Alberta Environment Conservation Authority (AECA) 1973) to fill gaps in ecological knowledge of the Delta and to evaluate effectiveness of the weirs. Hydrological monitoring has been continued through the regular network of Alberta Environment stations. Some ecological monitoring was done between 1973 (completion of the PADPG studies) and 1976 (construction of remedial weirs), after which monitoring gradually diminished. Parks Canada has sponsored vegetation and bison surveys, and numerous federally and provincially sponsored fisheries studies have been done (see review in Stanley Associates Engineering Ltd. 1980). There is considerable controversy over the effectiveness of the weirs. According to Townsend (1982), they have raised the minimum (winter) levels of Lake Athabasca without raising the summer levels, although the objective was to do the latter. However, subsequent hydrological monitoring efforts by the Inland Waters Directorate of Environment Canada suggest that the weirs have been effective in raising summer lake levels (Peace-Athabasca Delta Implementation Committee 1983). Currently, there is a revival of interest in the efficacy of the weirs because of a possible hydroelectric development on the Slave River.

Although several forms of compensation for resource users were recommended by PADPG (1973) and AECA (1973), it appears that nothing resulted. At one time, the town of Fort Chipewyan launched a lawsuit against B.C. Hydro, but the case never came to court.

Churchill Falls

The Churchill Falls hydroelectric development (Table 1) is situated on the Labrador plateau in a remote region of western Labrador. Prior to hydroelectric development, much of the region was uninhabited. The Labrador plateau is like a shallow saucer with notches in its rim (Duthie and Ostrofsky 1975). Water drains off the plateau through these notches, the main one being the Churchill River. A complex series of diversions now retains the water on the plateau and channels it to turbines above the river and east of the original rapids and falls. The point where the water is channelled to the turbines is only 2 km from the river, but it is 300 m higher.

The two main generating stations in the Churchill Falls development are at Twin Falls and Churchill Falls. Twin Falls was constructed in three stages from 1959 to 1968 to supply mining developments at Wabush and Labrador City. Its construction on the Unknown River formed the Ossokmanuan Reservoir in the southern part of the plateau. Eventually, water from this reservoir was diverted into the Smallwood Reservoir.

The Churchill Falls development was built after Twin Falls. Its main reservoir, and the main reservoir of the Churchill Falls development, the Smallwood, was filled from 1971 to 1974. The Smallwood Reservoir was formed behind a series of low dikes 80 km upstream from the power plant. Flow to the forebay and intake is regulated by a series of control structures (see below). The Smallwood Reservoir receives natural drainage from the Ashuanipi River basin and diverted drainage from three other basins: the Atikonak-Ossokmanuan (the Twin Falls development), the Naskaupi-Michikamau (most of the eastern part of the reservoir formed by diverting the natural outlet of Michikamau Lake into the Naskaupi River westward into the Churchill River basin), and a small part of the Kanairiktok (the extreme northeastern lobe of the reservoir formed by diverting a small portion of the headwaters of the Kanairiktok River).

The Lobstick control structure regulates flows out of Smallwood Reservoir into the west forebay (Jacopie Lake). The Jacopie control structure, on the west forebay, diverts water that originally flowed down Churchill Falls through the Whitefish control structure into the east forebay and the intakes for the generating station. The Gabbro control structure now links the Smallwood and Ossokmanuan reservoirs, and the Twin Falls generating station no longer operates.

No environmental assessment was done for the Twin Falls portion of the project. The Churchill Falls development was studied, although concurrently with its construction. The main environmental study was commissioned by the Churchill Falls (Labrador) Corporation, a subsidiary of British Newfoundland Corporation (BRINCO), and done by private and university consultants. Completed "predevelopment" studies were concerned mainly with water quality and lower trophic levels (e.g. Sheppard T. Powell Associates (Canada) Ltd. (STP) 1970-72; Duthie and Ostrofsky 1974, 1975) and the effects of flooding on vegetation (Bruneau and Bajzak 1973). Unfortunately, some of the predevelopment studies were ended prematurely. The STP study was intended to provide baseline data against which future changes in the limnology of Smallwood Reservoir could be evaluated, but it also predicted changes. It relied on the "reservoir paradigm" to predict a trophic upsurge during early phases of reservoir formation. "The reservoir paradigm makes the general prediction of an initial trophic upsurge during which all biotic communities have higher standing crops and productivities" (Hecky et al. 1984, p. 725). It is attributable to Baranov (1961), although earlier observations of a trophic upsurge due to impoundment are available (e.g. Ellis 1936). Predictions made by the STP study were generally not quantitative, and typically stated that a particular physical, chemical, or biological variable probably would increase, decrease, or remain unchanged. An experimental approach involving nutrient additions to 100-L plastic bags containing lake water was utilized to predict effects of increased nutrient supply on primary production, but the results of these experiments were not related to possible increases in nutrients that might be expected in the new reservoir.

Numerous studies, mainly concerned with limnology and fish populations, were done after the Smallwood Reservoir was created (e.g. Bruce 1974, 1975, 1979; Bajzak 1975, 1977; Ostrofsky and Duthie 1975, 1980; Davis 1978; Bruce and Spencer 1979; Duthie 1979; Barnes et al. 1984; and numerous unpublished reports). Unfortunately, these studies were done in an uncoordinated fashion by a variety of groups. Limnological conditions were followed effectively over the immediate postimpoundment period and generally were predicted correctly except for an initial decline in phytoplankton biomass and primary production during the filling phase (Duthie 1979). Unfortunately, predictions of shoreline erosion neither were made before impoundment of the Smallwood Reservoir nor quantified after impoundment was completed. Fish populations were studied mainly from the point of view of the potential for commercial exploitation (e.g. Bruce 1984), but the veracity of predictions concerning the fate of fish populations was not verifiable because of a lack of preimpoundment baseline data. High mercury levels in fish were noted after impoundment, and although there were no pre-impoundment data for comparison, the cause was attributed to reservoir formation (Bruce and Spencer 1979).

Environmental mitigation and compensation to resource users were never a consideration because the area was largely uninhabited. Neither commercial nor domestic fisheries existed in the area, although there has been attention to the development of postimpoundment fisheries.

Thus, the Churchill Falls environmental assessment, which actually preceded the current formal requirements for this type of analysis, was mainly a series of unconnected studies, done by a variety of groups (e.g. consultant, government, university, and utility) both before and after development occurred. The postdevelopment studies were especially disjointed. Many of these were oriented toward a single component of the affected ecosystem, and few relied on preimpoundment data for comparison because such data were not always available.

Churchill-Nelson Development

The hydroelectric potential of the Churchill River in northern Manitoba was recognized in the early 1900's (Hecky et al. 1984). Feasibility studies for development of this potential began in the 1940's, and by the mid-1950's the possibility of diverting the Churchill into the Nelson was identified. The first generating plant on the Nelson, Kelsey (224 MW), was completed in 1961; it supplied power to the International Nickel Co. mine and refinery at Thompson. In 1964, federal-provincial studies examined the feasibility of further hydroelectric development on the Churchill and Nelson rivers to provide power for southern markets. The decision to proceed with Nelson development was announced by the government of Manitoba in 1966, and several diversion schemes were proposed that would combine the Churchill and Nelson systems to increase flows on the lower Nelson River.

In 1976, most of the flow of the Churchill River was diverted into the Nelson basin. A control dam was placed at the outlet of Southern Indian Lake (Missi Falls), a large (1977 km^2) multibasin system through which the Churchill River flowed, and the lake level was raised 3 m above its long-term mean. Water was diverted through a channel constructed between the southeast part of the lake (South Bay) and the headwaters of the Rat River in the Nelson basin. A control structure at Notigi Lake was built to regulate flows down the Rat River Valley. The control structure was closed in 1974 and retained local runoff until water levels in the Rat River Valley and Southern Indian Lake were similar. This created the 584-km^2 Notigi Reservoir which is contiguous with the 2391-km^2 Southern Indian Lake Reservoir. The diversion channel was opened in 1976, and the diversion was operating at licensed capacity by late 1977. The diverted flow joins the Burntwood River at Threepoint Lake, below the Notigi control structure, and enters the Nelson River at Split Lake (Bodaly et al. 1984c).

The diversion is licensed to operate at a maximum capacity of 850 m^3/s, although it has been operating at a mean discharge of 760 m^3/s, or 75% of the Churchill flow. Minimum releases to the lower Churchill currently are set at 14.2 m^3/s during the open-water period and 42.5 m^3/s during the period of ice cover (Newbury et al. 1984).

Hydroelectric power from the project is produced in the lower Nelson where the combined annual mean flow of the Churchill and Nelson amounts to 3500 m^3/s (Table 1). Power production involves the regulation of Lake Winnipeg (23 750 km^2), within natural lake levels, for winter storage. The level of Lake Winnipeg is controlled by the Jenpeg power dam (126 MW). Flows from Lake Winnipeg into Jenpeg are regulated by a complex series of artificial channels, control dams, and dikes. Jenpeg was constructed at the main inflow to Cross Lake in 1974, and the flow entering Cross Lake, and the lake's water level, are controlled by discharges through Jenpeg. In addition to Kelsey, Kettle (1272 MW) and Long Spruce (980 MW) generating stations, which were completed by 1978, are below Jenpeg. Only the forebay of the Kettle station (Stephens Lake) is of reasonable extent. Construction of another power dam at Limestone Rapids (1200 MW) was suspended in 1979 due to insufficient power demand, but plans to complete it were announced recently by the Manitoba government and construction has resumed. To date, only one quarter to one third of the 8400-MW potential of the Nelson has been developed.

A major environmental impact assessment of the Churchill–Nelson development was conducted by the Lake Winnipeg, Churchill and Nelson Rivers Study Board (LWCNR 1975). This was a cooperative study between the governments of Canada and Manitoba, and was initiated in 1971. The objectives were "…to determine the effects that regulation of Lake Winnipeg, diversion from the Churchill River and development of hydroelectric potential of the Churchill River diversion route are likely to have on other water and related resource uses and to make recommendations for enhancing the overall benefits with due consideration for the protection of the environment" (LWCNR 1975, p. 4). Unfortunately, the study began after Manitoba Hydro had fixed the configuration, operating regime, and timing of construction for the Churchill River diversion, and construction of the control works and the diversion channel proceeded at the same time as the study (Hecky et al. 1984). The best-studied parts of the Churchill–Nelson development are Southern Indian and Cross lakes. Because of the relatively large amount of information available for these two reservoirs, much of the rest of the discussion will deal with environmental effects on them.

Cross Lake, located on the Nelson River upstream from the confluence of the diverted Churchill River water with the Nelson, had a mean open-water depth of 2.4 m and a surface area of 460 km^2. As a result of regulation by Jenpeg, minimum flows result in a drawdown of 1.7 m below the historic mean open-water elevation, a decrease in volume of 53%, a decrease in area of 26%, and a decrease in mean depth to 1.5 m (Bodaly et al. 1984c). Submerged vegetation proliferated after regulation, and widespread oxygen depletion has occurred during the winter. Maximum surface temperatures during the open-water period have increased from <20°C before regulation to 21–26°C after. The shallowness of the lake has resulted in high turbidity due to wind resuspension of lake sediments. The regulation of Cross Lake, especially increased winter drawdown, has adversely affected the year-class strengths and abundance of shallow-water, fall-spawning whitefish and cisco (Bodaly et al. 1984c; Gaboury and Patalas 1984). The relative abundance of whitefish declined 65% after regulation, and catch per unit effort declined from 35 fish in 1977 to 9 in 1980 and 4 in 1981.

Southern Indian Lake was also severely altered by impoundment and diversion (see Hecky et al. 1984 for summary). After impoundment, the lake became deeper and colder. Substantial erosion occurred because the postimpoundment water surface intersected glacial and organic deposits along most of the shoreline (Fig. 2). Large volumes of eroded materials fell into the lake, most of which were deposited in the nearshore zone. Eroded material also went into suspension, significantly increasing offshore suspended sediment concentrations and reducing light penetration.

At one time, Southern Indian Lake supported the largest commercial whitefish fishery in northern Manitoba. However, the catch per unit effort on the traditional whitefish grounds declined after impoundment, apparently due to a redistribution of stocks caused by reduced water transparency, and blockage by the Missi Falls control structure of natural migration between Southern Indian Lake and lakes on the lower Churchill River. This resulted in a redistribution of commercial fishing effort and exploitation of lower-quality stocks that were formerly avoided. Total whitefish catch was maintained for 5 yr after flooding by increasing fishing effort, but then effort dropped and total catch declined. Mercury concentrations in muscle increased in all commercial fish species (whitefish, northern pike, walleye). Concentrations in pike and walleye exceeded Canadian marketing limits, and sometimes export limits as well, and threatened an important domestic food source. Predictions made for the commercial fishery were concerned with potential problems of production and overlooked problems with quality.

FIG. 2. Shoreline in South Bay of Southern Indian Lake (A) before and (B) after impoundment.

Thus, the decline in whitefish market quality contributed to the economic decline of the commercial fishery on the lake (Wagner 1984), and mercury contamination threatened marketability of walleye and pike.

Southern Indian Lake received more study than other parts of the development. Underwood-McLellan and Associates Ltd. (U-M 1970) assessed the impact of various diversion configurations on natural resource utilization before development. U-M (1970) based its predictions on experience with other reservoirs, as derived from the scientific literature (the reservoir paradigm, cf. Hecky et al. 1984). U-M (1970) was required, by its terms of reference, to quantify impacts, so it scaled impacts against depth of flooding (Hecky et al. 1984). The "scalar" approach allowed quantitative predictions to be made, but they were not derived from confirmed relationships in the reservoir literature because such relationships did not exist at the time. This approach led to erroneous conclusions. For example, U-M (1970) "...underestimated the effect that shoreline erosion would have on the whole lake at low levels of flooding. Consequently, the study erred in accepting an apparently minor change in lake level as insignificant for fisheries because impacts on SIL did not increase in a linear manner with depth of flooding. Rather, the ecosystem endured a discrete change with the first water-level increase over the natural range at which shorelines had become stable over several thousand years" (Hecky et al. 1984, p. 728).

LWCNR (1975) examined the environmental effects of a low-level impoundment of Southern Indian Lake concurrently with construction activities. LWCNR (1975) refined and extended the U-M study predictions, but abandoned the reservoir paradigm as not applicable to Southern Indian Lake. "The existing reservoir paradigm was unsatisfactory for Southern Indian Lake because many expectations from the paradigm were qualitatively incorrect or, if correct, they were not quantifiable" (Hecky et al. 1984, p. 729). LWNCR (1975) did not need quantitative predictions of resource impact to evaluate project alternatives because construction of the diversion was completed at about the same time as the study. The only quantitative prediction made concerned the long-term effects of river diversion on the productivity of northern regions of Southern Indian Lake. Nutrient-loading theory was used to predict a 30% decline in primary productivity and fish production due to diversion of riverborne phosphorus.

Recognition of the lack of quantitative theory relating flooding and diversion to water quality and biology prompted continued research by the Freshwater Institute (Department of Fisheries and Oceans, Winnipeg, Man.) on the lake after the impoundment and diversion were completed (see Can. J. Fish. Aquat. Sci. 41: 548–732). The *Northern Flood Agreement*, signed in 1977 by the governments of Canada and Manitoba, Manitoba Hydro, and the Northern Flood Committee (representing five Indian communities affected by the project), provided for further ecological monitoring, but this monitoring is only now being initiated.

As a remedial environmental action, limited shoreline protection was instituted at selected locations such as cemeteries and road crossings. Limited, and largely ineffective, timber clearing also was done before flooding for aesthetic, hydraulic, and fisheries purposes (Hecky et al. 1984; Rosenberg et al. 1985).

There were no predevelopment payments or provisions for compensation to resource users in the Churchill–Nelson system. The main vehicle for compensation to resource users is the *Northern Flood Agreement*, which was signed *after* major construction activities were completed. The community of South Indian Lake was excluded from the Agreement, but the commercial fishermen negotiated a series of 1-yr compensation agreements (1978–82) with Manitoba Hydro and a final settlement in 1983. According to Wagner (1984), the postimpoundment commercial fishery was not viable without compensation payments, and fishermen were undercompensated because the extra effort required to fish Southern Indian Lake was not recognized as an increased cost. Institution of compensation has been retrospective and contentious and ultimately has penalized the resource user more than the resource developer (Hecky et al. 1984; Wagner 1984).

The government of Manitoba is moving ahead with development of the Limestone generating station (for power sales to the United States) and is considering developing the Conawapa generating station (1272 MW) in the near future. There is some doubt that the Manitoba government has learned any ecological or social lessons from past studies of the Churchill-Nelson development.

James Bay

This hydroelectric project can certainly be described as Canada's most ambitious (Table 1). The project was announced in 1971 after two independent groups of consulting engineers advised Quebec Hydro that 12 000 to 15 000 MW could be generated by developing the hydroelectric potential of the major Quebec rivers flowing into James Bay and southern Hudson Bay (Penn 1975). Two different development sequences were proposed: (1) first developing the more southerly option which would divert flows from the Nottaway and Broadback rivers into the Rupert in which a series of power dams would be built or (2) first developing the more northerly option by diverting the Eastmain, the headwaters of the Caniapiscau (which flows into Ungava Bay), and the Great Whale rivers into the La Grande River in which a series of power dams would be built.

Eventually, a modified version of the second option was constructed (Berkes 1985; Day 1985). This first phase of the La Grande River development did not include diversion of the Great Whale River, but did include two other major diversions into the La Grande: (1) waters of the Caniapiscau River which are delivered via the Caniapiscau Reservoir and the Laforge Diversion to the upper part of the La Grande River and (2) waters from the Eastmain and Opinaca rivers which are delivered via the Opinaca Reservoir and Sakami Lake to the lower part of the La Grande. A third, smaller diversion transfers water from Frigate Lake on the Sakami River into the middle reaches of the La Grande rather than letting it flow downstream into Sakami Lake. Three powerhouses (LG2, LG3, and LG4) are currently operating, and five more remain to be constructed in the second phase of La Grande development.

The Quebec government established the Société de développement de la Baie James (SDBJ) in 1971 to oversee economic development in the region. In turn, the SDBJ established the Société d'énergie de la Baie James (SEBJ) to deal specifically with hydroelectric development.

The decision to develop the hydroelectric potential of the James Bay region was made without prior assessment of environmental impact. The first formal impact assessment was published less than a year later, however, by a federal-provincial working group specially convened for this task (Groupe de Travail Fédéral-Provincial 1971). Although the group had no input into a go/no go decision, it was specifically asked to recommend which of the two proposed developments would be least environmentally objectionable, and also to recommend a long-term program of study to address environmental concerns. The working group had only 5 mo to report and was hampered by the nearly total lack of information (e.g. topographical, ecological) for the area (cf. Groupe de Travail Fédéral-Provincial 1971, p. 27). Consequently, their report identified in a general way the problems involved with hydroelectric development and discussed these in relation to the James Bay proposals.

Despite their misgivings about rendering advice on such a limited data base, the working group concluded that the La Grande development would be less environmentally objectionable than the Nottaway-Broadback-Rupert (NBR) scheme because of the lower fishery and wildlife productivity of the watersheds involved in the La Grande development. This lower productivity was reflected in Indian settlement patterns. No resettlement was foreseen in the La Grande development, whereas resettlement would be required for the NBR development. The working group also recommended that the Caniapiscau diversion be delayed until further on-site analysis could be done, and it outlined immediate and long-term studies needed to provide relevant information for future development in the region.

The first environmental impact studies based on new research were done by scientists cooperating with native groups that had funding from the federal government (Berkes 1985). These on-site studies substantiated that native land use was essential to their economic well-being, a fact not appreciated in the earlier working group report. Data from these studies also formed a major part of the evidence in a court action launched by the natives to halt construction until outstanding issues such as property rights and compensation for damage to traditional land and water use were settled (Penn 1975). De facto, Penn's (1975) review of the use of environmental information in court provides the first comprehensive assessment of the La Grande development before construction began!

Although the court action was ultimately unsuccessful in halting construction, it did stimulate settlement of land claims. The *James Bay and Northern Quebec Agreement* of 1975 (JBNQA) was signed by the Crees and Inuit of northern Quebec, the governments of Quebec and Canada, and the three crown corporations involved in the development. The JBNQA contains several sections specifically related to environmental assessment, environmental mitigation, and compensation for native land claims (Berkes 1981). It also established organizations such as La Société des travaux correcteurs du Complexe La Grande (SOTRAC) to study, recommend, and implement mitigation works and programs, with the participation of local users (Soucy 1978).

The SDBJ is charged with the (somewhat contradictory) task of protecting the natural environment and developing the territory (Penn 1975), so almost all environmental studies subsequent to the JBNQA have been conducted by SDBJ or SEBJ. The federal government also has been involved, however, and has conducted a series of studies on possible effects of hydroelectric development on the estuarine fish populations, waterfowl, and water balance and circulation of James and Hudson Bays (e.g. Prinsenberg 1980, 1982).

An impressive array of environmental studies covers various aspects of the La Grande development. Some highlights are as follows.

(1) A comprehensive program of baseline data collection instituted before impoundments and diversions were operational (Roy 1982b).

(2) Systems models developed and tested on an experimental reservoir prior to completion of major reservoirs (Thérien 1981). This effort was a significant improvement in prediction-making and testing over the simple qualitative application of the reservoir paradigm used by the Group de Travail Fédéral-Provincial (1971). The developed models did not, however, include fish or wildlife.

(3) Effects on fish stocks of saltwater encroachment at the mouth of the La Grande as a result of flow reduction caused by filling of the LG2 reservoir (Roy 1982a, 1982b). This was an impressively successful quantitative prediction of fish survival based on a physical model of estuarine circulation that involved saltwater–freshwater balance under different conditions of ice cover and on knowledge of fish biology.

In general, many predictions have been reasonably accurate (D. Roy, pers. comm.), as indicated by examples 2 and 3 above. However, sometimes the accuracy of environmental assessment predictions improved as more details became available, and some predictions were obviously wrong. For example, Berkes (1982) chronicled the evolution of environmental impact assessment for a particular fishery in the La Grande estuary. Biological predictions changed over time as more detailed information and analysis became available. Original predictions of catastrophic effects on the fishery (Groupe de Travail Fédéral-Provincial 1971) did not materialize, at least in the short term. However, substantial changes in the utilization of the fishery resulted from socioeconomic alterations caused by the development, and these were not predicted by any of the environmental assessments done. (A parallel can be found in the changed fishing patterns in Southern Indian Lake, cf. Bodaly et al. 1984b.)

Unpredicted consequences of the La Grande development included (1) higher-than-expected mercury concentrations in fish after impoundment (Boucher and Schetagne 1983; cf. Bodaly et al. 1984a for a similar occurrence in the Southern Indian Lake impoundment) and (2) the need to relocate Fort George, the largest settlement in the area, partly because of erosion by higher flows on the La Grande (cf. reasons given above for choosing the La Grande development over the NBR!).

The general approach taken in the La Grande development emphasizes mitigation and compensation rather than substantial alteration of design or operational regime. Thus, SOTRAC undertakes studies of resource dislocation and the implementation of remedial actions. It is managed by native and SEBJ representatives and has SEBJ funding. As an example of the mitigation and compensation approach, Groupe de Travail Fédéral-Provincial (1971) and Environment Canada (1975b) recommended that minimum acceptable flows be established for dewatered river channels. Instead, weirs were built to maintain water levels in river channels (e.g. on the Eastmain-Opinaca), exposed river channels were replanted, and intensive monitoring of dewatered reaches was instituted (Société d'énergie Baie James 1984). In addition, Soucy (1978, p. 21) outlined an expensive program of corrective management, and some of the mitigation programs concerned with clearing and reservoir management for fish production are innovative (e.g. Faubert 1982; Therrien 1982).

However, Berkes (1985) has criticized mitigation as being too late if not too little. The main problem has been the lack of background environmental information upon which mitigation can be planned. SOTRAC may also be less effective than desired because it is a reactive body and does not have advisory powers to require prevention (Berkes 1981).

With regard to compensation programs for resource users, the JBNQA contains a provision for an income security program which provides monetary support to hunters, trappers, and fishermen to insure their livelihoods. This provision is funded directly by the province of Quebec and was developed to strengthen the traditional hunting economy in northern Quebec. Implementation of the program was precipitated as a result of hydroelectric development. However, the success of the program in maintaining native lifestyle also depends on fulfillment of other provisions of the JBNQA funded by hydroelectric development (e.g. SOTRAC) (Feit 1982). The program seems to be successful in stablizing the numbers of Cree utilizing natural resources in the traditional manner.

Missouri River Reservoirs

Six reservoirs are dotted along the Missouri River mainstem in Montana, North and South Dakota, and Nebraska (Tables 1, 2). These reservoirs were constructed over a 30-yr period ($\cong$ 1935-65) by the U.S. Army Corps of Engineers (USACE). All except Lake Fort Peck were constructed under the *Flood Control Act* of 1944, which authorized construction of dams on the mainstem by the USACE. It also authorized provision of water for irrigation from the mainstem reservoirs and the construction of dams on tributaries of the Missouri River by the U.S. Bureau of Reclamation.

Reservoirs of the Missouri mainstem are truly multipurpose. All serve flood-control, navigation, irrigation, recreation, and power-generation functions. Their hydroelectric capacities are large (2098 MW, Table 1), and further capacity is being considered.

The reservoirs vary greatly in size (113-3060 km^2) and extent of annual drawdown ($\cong$ 1-10 m) (Table 2). The three upstream reservoirs, Lakes Oahe, Sakakawea, and Fort Peck, are the largest and are used for water storage during dry periods. Their water levels vary depending on storage and runoff. The three downstream reservoirs, Lewis and Clark Lake and Lakes Francis Case and Sharpe, are smaller and have little storage capacity. The water levels of these reservoirs are managed quite uniformly from year to year.

Predevelopment assessments were done for five of the six projects under the authority of the *Fish and Wildlife Coordination Act* of 1934. Only Lake Fort Peck was ignored

TABLE 2. Characteristics of reservoirs on the Missouri River mainstem (data from Benson and Cowell 1967).

Reservoir	Dam	Location	Date of dam closure	Surface area (km^2)	Maximum depth (m)	Mean depth (m)	Average annual drawdown (m)
Lake Fort Peck	Fort Peck	Montana	1937	991	66	24	2-3
Lake Sakakawea	Garrison	North Dakota	1953	3060	57	17	3
Lake Francis Case	Fort Randall	South Dakota	1953	420	41	15	7-11
Lewis and Clark Lake	Gavins Point	Nebraska	1955	113	17	5	<1.5
Lake Oahe	Oahe	North and South Dakota	1958	1450	62	19	3
Lake Sharpe	Big Bend	South Dakota	1963	226	24	9	<1.5

because its dam was closed in 1937, and an amendment to the Act requiring predevelopment inventory studies was not passed until 1946. The Act was further amended in 1958 to include more stringent requirements such as mitigation of potential fish and wildlife losses. The 1958 amendment required the project proponent to consult the U.S. Fish and Wildlife Service to prevent losses of natural resources as a result of the project and to develop and improve natural resources in the area of the project.

Predevelopment reports prepared for Lakes Sakakawea and Francis Case and Lewis and Clark Lake under the 1946 amendment (U.S. Fish and Wildlife Service 1946, 1948, 1950) all assumed that the projects would proceed. Predictions of effects were based on limited original data and field work. They were centred on fish and were based mainly on knowledge of species biology. Other trophic levels were not considered. These reports noted the relatively minor utilization of the river reaches for commercial and sport fishing prior to development and predicted an increase in the value of the fisheries as a result of development. Some recommendations for environmental mitigation were made (e.g. minimum discharge for protection of downstream areas, maintenance of suitable water levels for fish biology, construction of hatcheries, creation of stable-water-level subimpoundments).

Predevelopment reports prepared for Lakes Oahe and Sharpe under the 1958 amendment (U.S. Fish and Wildlife Service 1960, 1962) were similar to the ones cited above, except there was some reliance on experience gained from work on the other Missouri reservoirs. However, the Lake Sharpe report was based partly on field work to determine the species composition of fish, and it briefly considered other trophic levels.

Postdevelopment studies on the Missouri reservoirs were conducted mainly by the North Central Reservoir Investigations Group of the U.S. Fish and Wildlife Service, and some university groups and the USACE also were involved. In contrast with the predevelopment studies, postdevelopment studies were based broadly and included physical aspects and all trophic levels (e.g. Benson 1968, 1980; June 1974). However, the main emphasis remained on fish, and especially on the modification of reservoir operation to improve the production of valuable species (e.g. Nelson 1978). These studies have provided valuable information on the spawning habits and reproductive requirements of valuable fish species in relation to water levels, temperature, and shoreline characteristics of reservoirs. A number of recommendations concerning the regulation of water levels resulted from these studies, but few of the recommendations were practical with respect to other reservoir uses, and so few changes were implemented.

Compensation for resource users was not directly considered for any of the Missouri reservoirs. As noted above, many of the impounded river reaches had little or no commercial or recreational fishing, and predevelopment reports generally predicted that impoundment would improve fishing. However, Indian residents on reserves adjacent to sections of impounded river were severely affected. Land belonging to seven Indian reservations was flooded by four of the projects. Domestic and sport fishing were disrupted, and the reservations lost their most productive river bottom-land. The bottomlands had provided timber, wildlife, and grazing and overwintering areas for cattle. Relocation of much of the population of some of the reservations to less productive upland areas was necessary.

The U.S. Congress, the only body that can authorize the taking of Indian lands, passed statutes authorizing the USACE to negotiate with most of the Indian tribes to be affected by impoundment. These statutes proposed certain tentative compensation amounts. However, most negotiations between the USACE and the Indians failed (e.g. Macgregor 1948). Congress then imposed compensation by a further series of statutes (Shanks 1974). The compensation was mainly for terrestrial parts of the ecosystem (e.g. losses of lands, forests, wildlife, roads, buildings), but not for fisheries. A number of studies were done by the Missouri River Basin Investigations Project of the U.S. Bureau of Indian Affairs to evaluate the efficacy of compensation paid to the affected Indian groups.

Results

Answers to the questions in the checklist relating to the pre- and postdevelopment phases of the projects examined here are given in Table 3. It can be interpreted vertically, by comparing pre- and postdevelopment phases, and horizontally, by comparing assessments over time. Reservoirs of the Missouri River mainstem can be used as a comparison for trends observed in the Canadian reservoirs. The projects examined are arranged in approximately chronological order. Construction of the Bennett Dam, which led to drying in the Peace-Athabasca Delta, predated construction of the Churchill Falls project, so the Peace-Athabasca Delta study precedes Churchill Falls in Tables 1 and 3. The Churchill Falls and Churchill-Nelson developments were approximately contemporaneous, but diversion activities causing major environmental perturbations in the Churchill-Nelson development did not occur until the mid-1970's. Hence, Churchill Falls precedes the Churchill-Nelson in the tables.

By comparison with the other projects examined, the amount of work done in the predevelopment phase of the Kemano project was very small and narrowly defined. Environmental studies commenced after the water license was granted and construction had started (Telkwa Foundation 1984; Day 1985). These studies were of limited duration and scope, mainly concerning fisheries. Reports produced were arcane and appeared to be mainly internal documents (e.g. Lyons and Larkin 1952). Although some of these deficiencies also occurred in other projects examined here, only in Kemano did they occur simultaneously. Therefore, predevelopment studies of Kemano Phase I were not considered to represent a viable predevelopment analysis for the purpose of this review.

Vertical Comparisons in Table 3: Predevelopment vs. Postdevelopment Phases

An increase in the performance of predevelopment studies is obvious over time (Table 3). Predevelopment assessments were considered not to have been done for Kemano Phase I, were not done for the Peace-Athabasca Delta, but were done for the remaining Canadian projects. Among the three projects for which these assessments were done, there was a marked improvement in their completeness with time. Whereas Churchill Falls polled only two *yes* and three *yes (limited)* answers out of a possible 13, Churchill-Nelson had seven *yes* and one *yes (limited)*, and James Bay had 11 *yes* and one *yes (limited)*. Thus, by the mid-1970's, fairly complete predevelopment studies were being done for major hydroelectric projects.

For the Missouri mainstem, no predevelopment studies were done for Lake Fort Peck (mid-1930's). Predevelopment studies were done for the two later groups of reservoirs (Tables 1, 2), but there was little improvement from the Lake Sakakawea (mid- to late 1940's) to the Lake Oahe (early 1960's) groups. The Lake Sakakawea group was credited with three *yes* and three *yes (limited)* answers, and the Lake Oahe group was credited with four *yes* and three *yes (limited)* answers.

Thus, there were no predevelopment studies of the earliest Missouri mainstem reservoir (Lake Fort Peck) and none acceptable for the earliest Canadian reservoir examined (Kemano). Predevelopment studies on the second group of Missouri mainstem reservoirs (the Lake Sakakawea group), done in the mid-1940's, were similar in scope to those done on Churchill Falls in the late 1960's to early 1970's. Predevelopment studies on the Lake Oahe group of the Missouri system done in the early 1960's were also similar to Churchill Falls, indicating that significant progress had not been made over a 20-yr period on the Missouri studies. However, predevelopment studies of the Peace-Athabasca Delta, which was approximately contemporaneous with Churchill Falls, were not done at all, indicating variability among individual projects.

The postdevelopment phase always was done for the Canadian reservoirs examined. The Kemano study, however, only was done recently (Envirocon Ltd. 1981), as part

TABLE 3. Checklist for evaluation of environmental impact assessments of selected hydroelectric projects. "Limited" applies to one or two components, whereas "yes" means more broadly based.

Checklist questions	Kemano Phase I	Peace-Athabasca Delta	Churchill Falls	Churchill-Nelson	James Bay	Lake Fort Peck	Missouri[a] Lake Sakakawea Lake Francis Case Lewis and Clark Lake	Lake Oahe Lake Sharpe
PREDEVELOPMENT								
I. Were any predevelopment environmental studies/reports done/prepared?	No	No[b]	Yes	Yes	Yes	No[c]	Yes	Yes
A. Were studies intended to influence go/no go decisions?	No	No	No	No	No	No	No	No
B. Were studies intended to influence alternative designs?	No	No	No	Yes (limited)	Yes	No	No	No
C. Were studies intended to provide baseline information for future monitoring?	No	No	Yes (limited)	Yes	Yes	No	Yes (limited)	Yes (limited)
D. Were studies ecosystem oriented?	No	No	No	Yes	Yes	No	No	No
II. Were predictions made?	No	No	Yes (limited)	Yes	Yes	No	Yes (limited)	Yes (limited)
A. Were predictions/models quantitative?	No	No	No	No	Yes (limited)	No	No	No
B. Were systems models used?	No	No	No	No	Yes	No	No	No
C. Was the reservoir paradigm (sensu Hecky et al. 1984) used?	No	No	Yes	Yes	Yes	No	No	Yes (limited)
D. Was knowledge of species biology applied?	No	No	No	Yes	Yes	No	Yes	Yes
E. Were predictions/models tested experimentally?	No	No	Yes (limited)	No	Yes	No	No	No
III. Were environmentally based mitigation measures recommended?	No	No	No	Yes	Yes	No	Yes (limited)	Yes
IV. Where economic compensation schemes emplaced?	No	No	No	No	Yes	No	Yes	Yes
POSTDEVELOPMENT								
V. Were any postdevelopment environmental studies/reports done/prepared?	Yes[d]	Yes	Yes	Yes	Yes	Yes	Yes	Yes
A. Were studies done to address predicted consequences?	No[e]	No[f]	Yes	Yes	Yes	No[e]	Yes	Yes

TABLE 3. (Concluded)

Checklist questions	Kemano Phase I	Peace–Athabasca Delta	Churchill Falls	Churchill-Nelson	James Bay	Lake Fort Peck	Missouri[a] Lake Sakakawea Lake Francis Case Lewis and Clark Lake	Lake Oahe Lake Sharpe
B. Were studies done to address unpredicted consequences?	No	No	Yes	Yes	Yes	No	Yes	Yes
C. Was a comprehensive monitoring program emplaced?	No	Yes (limited)	Yes (limited)	Yes (limited)	Yes	No	Yes	Yes
D. Were studies sufficiently broad to explain effects observed?	No	Yes	Yes (limited)	Yes	Yes	No	Yes (limited)	Yes
E. Were experimental studies conducted to test postdevelopment hypotheses?	No	No	No	Yes	No	No	No	No
VI. Were environmentally based mitigation measures emplaced?	Yes	Yes	No	Yes (limited)	Yes	No	No	No
A. Was efficacy of mitigation measures studied?	Yes	Yes	No	Yes (limited)	Yes	No	No	No
VII. Was compensation for damage to natural resources paid to users of these natural resources?	Yes	No	No[g]	Yes	Yes	No	Yes	Yes
A. Were socioeconomic studies done to evaluate efficacy of compensation?	No	No	No	Yes (limited)	Yes	No	Yes	Yes

[a] Development of Missouri River reservoirs followed three stages. Dam closures were as follows: Lake Fort Peck, 1937; Lakes Sakakawea and Francis Case, and Lewis and Clark Lake, 1953–55; Lakes Oahe and Sharpe, 1958 and 1963, respectively.

[b] Peace–Athabasca Delta Project formed in 1971 to investigate drying of Delta due to filling of Williston Reservoir upstream on Peace River in British Columbia (1968–71).

[c] Predevelopment environmental impact assessments not required at the time, under authority of Fish and Wildlife Coordination Act of 1934.

[d] For proposed Phase II completion.

[e] Predevelopment stage lacking, so no predictions made.

[f] Predevelopment stage lacking so no predictions made, although monitoring studies examined some predictions made by Peace–Athabasca Delta Project Group (1973) and efficacy of weirs (e.g. Townsend 1982) (see Stanley Associates Engineering Ltd. 1980 for review).

[g] No resource users in area before reservoir construction.

of the Kemano completion project, so it should be regarded as occurring between Churchill-Nelson and James Bay, temporally. Real improvement in the scope of postdevelopment assessment appears with Churchill-Nelson and James Bay. Out of a possible 10 *yes* answers, the former had six *yes* and four *yes (limited)*, and James Bay had nine *yes*. Peace-Athabasca Delta had four *yes* and one *yes (limited)*, and Churchill Falls had three *yes* and three *yes (limited)*. Although Kemano was done approximately between Churchill-Nelson and James Bay, it polled only four *yes* answers. At this point, two observations can be made: (1) there seems to be improvement over time in the scope of postdevelopment assessments and (2) project variability is possible in light of the first observation, as evidenced by Kemano.

Although the Missouri mainstem reservoirs were constructed over a 30-yr period, the bulk of postdevelopment assessments for them were done at approximately the same time (late 1960's to early 1970's). When compared with contemporaneous Canadian postdevelopment studies, Peace-Athabasca Delta and Churchill Falls, it can be seen that the Lake Sakakawea group (six *yes* and one *yes (limited)* answers) and the Lake Oahe group (seven *yes*) are similar, but Lake Fort Peck (one *yes*) is not. Lake Fort Peck is located in Montana, quite far from the main group of people that conducted postdevelopment studies of the Missouri mainstem reservoirs (see section above on Missouri), and this may explain why its postdevelopment assessment is so different from the other reservoirs in the system.

Horizontal Comparisons in Table 3

All Answers NO

"IA: Were studies intended to influence go/no go decision?" This situation has not changed over time for either Canadian projects or the Missouri reservoirs. "Tokenism" is a common failure of environmental impact assessments (Rosenberg and Resh et al. 1981, p. 609). Obviously, environmental information should be one of the factors used in decision-making. Just as obviously, from this survey, it is seldom used in this way.

Several questions received virtually all *no* answers. The only *yes* for the questions "IIA: Were predictions/models quantitative?, IIB: Were systems models used?, and IV: Were economic compensation schemes emplaced?" was for James Bay. For IIA and IIB, this result may indicate that the application of modelling to environmental impact assessment is a relatively recent phenomenon. This conclusion is supported by the *no* answers for all the Missouri studies, the latest of which were done in the early 1960's.

For IV, the special nature of the arrangements made in the James Bay project for land claims settlements (see section above on James Bay) indicates that this answer is not time related. No compensation schemes were emplaced in the Lake Fort Peck development (probably because no Indian reserves were directly affected) but they were in the latter two Missouri reservoir groups. It is to be hoped that, once begun in a particular hydroelectric or water-diversion project, compensation schemes remain a feature of any further development and that this element will become a routine part of the initial planning for all such projects (Hecky et al. 1984).

The only *yes* answer for "VE: Were experimental studies conducted to test postdevelopment hypotheses?" concerned the Churchill-Nelson. This is related to the long-term involvement of the Freshwater Institute in studying this development and to the relatively long postdevelopment period of ecological monitoring (see section above on Churchill-Nelson).

All Answers YES

"V: Were any postdevelopment studies/reports done/prepared?" It is noteworthy that postdevelopment studies were done for all the hydroelectric projects/water diversions examined (although the one done for Kemano resulted from a desire by Alcan to complete Phase II of the project!). This may be an artifact of involving only large, extant projects in this review, although several such projects are known that have not had systematic postdevelopment assessments. However, it more likely reflects the fact that development produces serious unexpected consequences requiring study and mitigation.

Several questions received virtually all *yes* answers. For "VC: Was a comprehensive monitoring program emplaced? and VD: Were studies sufficiently broad to explain effects observed?", the only *no* is for Kemano, although a number of *yes (limited)* answers occurred (especially for VC). Given that the postdevelopment assessment of Kemano was fairly recent, these answers indicate that the scope of the Kemano study may have been too narrow. Answers to these questions in the Missouri mainstem reservoirs were *yes* for all but Lake Fort Peck. Since Missouri postdevelopment studies were done at approximately the same time, the reason for the lack of a comprehensive monitoring program and studies explaining effects for Lake Fort Peck perhaps can be ascribed to its relative remoteness. It may also be that attitudes to a long-standing reservoir are different from those for relatively recent ones.

The questions "VI: Were environmentally based mitigation measures emplaced? and VIA: Was efficacy of mitigation measures studied?" are related and so the answers to them were identical. The only *no* answers were for Churchill Falls. The area affected by the Churchill Falls development was virtually uninhabited (see section above on Churchill Falls), which explains these results. However, attempts at environmental mitigation were very limited for the Missouri reservoirs, and this represents a major divergence between the Missouri and the Canadian projects examined.

Mixed Responses Showing Patterns

Several questions received replies virtually equally split between *yes* and *no*. The questions "IB: Were studies intended to influence alternative designs?, ID: Were studies ecosystem oriented?, IID: Was knowledge of species biology applied?, III: Were environmentally based mitigation measures recommended?, and VIIA: Were socioeconomic studies done to evaluate efficacy of compensation?" received three *no* and two *yes* replies. Kemano, Peace-Athabasca Delta, and Churchill Falls consistently had *no* answers to these questions, whereas Churchill-Nelson and James Bay consistently had *yes* or *yes (limited)* answers. Once again, this indicates increasing assessment sophistication with the later developments, and this appears to be true for scientific (ID, IID), sociological (VIIA), and administrative (IB, III) matters.

The Missouri system yields a less straightforward picture by comparison. Although two of the questions (IB, ID) were all *no* answers, and the remaining three were *no* for Lake Fort Peck and *yes* for the two other Missouri groups, no pattern emerged when the questions were classified as scientific, sociologic, or administrative. Therefore, *no* answers to the influence on alternative designs (IB) and the ecosystem orientation of studies (ID) agree with Canadian studies of a similar time period, whereas *yes* answers to use of species biology (IID), the recommendation of environmental mitigation (III), and performance of socioeconomic studies (VIIA) in the Lakes Sakakawea and Oahe groups indicate that the Missouri system was advanced in comparison with Canadian projects of a similar time period (Kemano, Peace-Athabasca Delta, and Churchill Falls). Once again, environmental mitigation was not a concern of the Churchill Falls development because the area was virtually uninhabited.

The questions "I: Were any predevelopment environmental studies/reports done/prepared?, IC: Were studies intended to provide baseline information for future monitoring?, II: Were predictions made?, IIC: Was the reservoir paradigm (sensu Hecky et al. 1984) used?, VA: Were studies done to address predicted consequences?, and VB: Were studies done to address unpredicted consequences?" received three *yes* and two *no* replies. Churchill Falls, Churchill-Nelson and James Bay consistently had *yes* or *yes (limited)* answers to these questions, whereas Kemano and Peace-Athabasca Delta consistently had *no* answers. The *no* answers for Kemano and Peace-Athabasca Delta relate to the lack of (or lack of a suitable) predevelopment phase of study for these two projects (I, IC, II, IIC). Although questions VA and VB are in the postdevelopment phase, the presence of a predevelopment phase is an initial requirement for a positive reply.

The Missouri mainstem reservoirs agree well with results for the Canadian projects for these questions. No predevelopment assessment was done for Lake Fort Peck, and all of the answers to the questions are *no*. All of the answers for the Lakes Sakakawea and Oahe groups are *yes* or *yes (limited)* except for IIC for the Lake Sakakawea group. Predevelopment assessments of the Lake Sakakawea group (mid- to late 1940's) may have predated general use of the reservoir paradigm, which was only used in a limited way in the Lake Oahe group (early 1960's) and not until Churchill Falls (late 1960's to early 1970's) in the Canadian projects examined.

No Discernible Pattern

Truly variable responses were obtained for questions "IIE: Were predictions/models tested experimentally? and VII: Was compensation for damage to natural resources paid to users of these natural resources?", and no trends were obvious in the results. However, it appears that very few studies of hydroelectric or water-diversion projects include extensive experimental testing of predictions/models in the predevelopment phase of the assessment. James Bay was the only one to do so unequivocally in our survey (including the Missouri projects). Inclusion of such testing probably depends on the orientation of the individuals doing the work as well as on the terms of reference of the study.

Results from our checklist indicate that compensation payments usually are made to users of natural resources damaged by hydroelectric or water-diversion projects (VII). Discounting Churchill Falls because of the absence of resource users in the area, the only Canadian project that did not do so was the Peace–Athabasca Delta. This was the only example in which resource users in a different province were affected by resource development. In addition, the exact damage suffered by an already depressed economy such as existed in Fort Chipewyan in the Peace–Athabasca Delta (see section above on Peace–Athabasca Delta) may have been difficult to identify. Compensation was a feature of the two latest groups of Missouri reservoirs, but not Lake Fort Peck, probably because Indian reservations were not affected by formation of this reservoir.

Discussion

The abundance of Canadian inland waters makes it logical that hydroelectricity has been, and remains, the preferred option for production of electrical energy in this country. Demand for such energy is concentrated along Canada's southern border and in the even-larger energy market of the northern United States, to which Canadian electrical energy is exported.

Exploitation of Canadian hydroelectric potential began in the last century and has advanced northward like a wave that is now cresting in the midnorthern latitudes. New hydroelectric projects have become larger, have become increasingly more remote from their consumers, have come into increasing conflict with people who are still dependent on natural, renewable resources, and have entered into terrain formerly untouched by major industrial development. Hydroelectric development in the north poses problems which may differ in scope and scale from earlier, southern developments.

Increased environmental awareness, in the late 1960's, and regulations requiring environmental assessment have revealed hydroelectric developments to be major, and often contentious, modifiers of northern environments and land-use patterns. Therefore, it is essential that sufficient knowledge and experience exist to predict correctly the consequences of such developments in order to ensure rational development of Canada's northern hydroelectric potential.

North-South Spatial and Cultural Isolation

The projects we chose for analysis (Table 1) are remote from the energy markets they serve as well as the population and political centres where development decisions are made. Nearly all remaining potential hydroelectric development in Canada has the same characteristics. Most Canadians are unfamiliar with the northern environments affected and the use that northern native people make of them. Values and attitudes of the southern consumer and decision-maker differ significantly from those of the northerner who occupies the lands to be flooded and uses the waters that will be altered. Hydroelectric development brings these different cultures into conflict.

Southerners tend to view the north as a vast, under- or even unutilized wilderness awaiting appropriate development to release its wealth. They often regard northern native people as being colourful but poor, and as pursuing anachronistic lifestyles that should have been abandoned long ago. They perceive economic development as something that will improve the well-being of these people. The natives, who have successfully occupied the land for thousands of years, do not always share this view. The southern view encourages a disregard for the hopes and fears of northern native people. Such disregard is reinforced by the physical isolation of the two cultures.

Attitudes toward land tenure best exemplify differences between southerners and northern natives. In the south, all land of value is owned, and competition for its use has led to a system of societal checks and balances on its permissable uses. A proposed hydroelectric development in a populated area of southern Canada would raise concerns about whose farm, business, or house would be flooded, and such a proposal could be tied up in litigation for years while settlements were reached with the people affected. In contrast, individual land ownership is a foreign concept to the indigenous peoples of Canada, and they are still more concerned about the natural resources upon which they depend for their livelihoods (and about their access to these resources) than they are about land ownership. Shanks (1974) described a similar situation of physical and cultural isolation between the consumers who benefited from development of the Missouri mainstem reservoirs and the local inhabitants, mainly native, who suffered the impacts resulting from these projects.

A common southern misconception is that native Canadians are only entitled to northern, reserve lands established by treaty. However, native land use traditionally has extended well beyond reserve boundaries to the land as a whole (cf. Penn 1975). Ignorance of this use was evident, for example, in the recent James Bay development where the announcement of the intent to build the project, and initial construction, preceded settlement of land claims. Such an event would be unthinkable in the south.

Evolution of Environmental Impact Assessment

A number of negative and positive features characterize environmental assessments of Canadian hydroelectric and water-diversion projects. With few exceptions, the Canadian experience is similar to reservoirs of the Missouri River mainstem.

The analysis presented in Table 3 demonstrates that there has been an evolution in concern about environmental issues pertaining to hydroelectric development. The earliest projects examined in both the reservoirs of the Missouri mainstem and in Canadian projects lacked predevelopment consideration of environmental consequences. Sequentially later developments showed an increasing depth of analysis used earlier in the planning stages. However, none of the projects examined was subjected to a predevelopment analysis which affected the decision to proceed.

It is disconcerting to see that, at least until the mid-1970's, environmental assessments of major hydroelectric and water-diversion projects in Canada still did not play a role in the fundamental decision of whether or not to proceed with development. Furthermore, the positive influence of studies on alternative designs is only a relatively recent phenomenon, is usually very limited in extent, and is often tokenistic.

95

The sequence of *predevelopment impact assessment-public hearings-decision*, which was followed for the proposed development of the Churchill River in Saskatchewan (Churchill River Study (Missinipe Probe) 1976), should be a model for future hydro-electric developments. The relative importance of (1) the predevelopment environmental assessment, (2) the public opposition that emerged at the hearings, or (3) the availability of cheaper alternative sources of power to the subsequent decision not to proceed with the development is unknown. However, the sequence which was followed offers an oppurtunity for environmental information to influence the go/no go decision.

Predevelopment environmental assessments serve two other important purposes, after a decision has been made to proceed with development. First, good assessments will offer a comprehensive catalogue of predicted impacts, along with clear statements on the likelihood of their occurrences (Penn 1975). These predictions should be quantitative whenever possible, so that benefits and costs can be calculated. Such predictions are guides to planning and are only the beginning of the process of impact analysis (Rosenberg and Resh et al. 1981; Beanlands and Duinker 1983; Hecky et al. 1984). In particular, these predictions should guide the development of programs for environmental mitigation and social compensation, especially because such programs usually eventually result (Table 3). Although compensation usually is paid to users of natural resources damaged by development, it is significant that *predevelopment* compensation schemes almost never have been included in Canadian hydroelectric and water-diversion projects. Calculating the costs of such programs before development begins would make their institution more timely and less contentious. A number of studies (e.g. Berkes 1981; Wagner 1981, 1984) have shown that timeliness of such programs is extremely important because delaying their implementation shifts costs to the group affected by development.

The second important function of predevelopment impact studies is the establishment of natural baselines against which postdevelopment conditions can be assessed. This remains the most important and most utilized function of such studies. Whether or not predevelopment studies are done, postdevelopment studies are always required, usually to address unpredicted consequences (Table 3). Such studies are generally broad in scope and attempt to establish the causes of impacts. The value of these expensive postdevelopment studies, which appear to be inescapable, would be greatly enhanced by establishing natural conditions and initiating baseline monitoring programs before construction started. In the projects considered here, only the James Bay development made a commitment to such an approach.

It is gratifying to see that comprehensive monitoring programs, environmentally based mitigation measures, and study of the efficacy of these measures are usually emplaced in assessments of Canadian hydroelectric and water-diversion projects. However, this is only a qualitative judgement, and the extent and success of such programs are what really matter.

The fact that most postdevelopment studies are sufficiently broad to explain effects observed is a positive feature of Canadian studies. Most of these explanations are hypotheses that should be tested by experimental studies or in future hydroelectric developments. There appears to be a need for more use of the experimental testing of predictions or models in the predevelopment phase of project assessments.

A number of features of Canadian projects are characterized by improvement over time. First, there is an increase in performance of predevelopment studies and their completeness. These is also real improvement in the completeness of postdevelopment studies. Second, the ecosystem orientation of studies and the application of knowledge of species biology, the performance of socioeconomic studies to evaluate efficacy of compensation, and the recommendation of environmentally based mitigation measures have also improved with time. These represent scientific, social, and administrative matters, respectively. Third, although two other scientific aspects, the use of quantitative predictions or models and the use of systems models, were part of only the last project examined (James Bay), this may indicate the start of widespread application of biological as well as physical modelling to environmental assessment of water-development projects.

The apparent relationship between the effectiveness of environmental assessment and time may be due to two factors: (1) emerging public interest in the environment, which has led to statutory and nonstatutory environmental assessment requirements, and (2) a steady improvement in scientific capability to do such studies. However, despite this improved capability, important technical difficulties still remain (see below).

Assessments such as for the Churchill–Nelson and James Bay developments show that existing frameworks and tools are adequate. The desire of decision-making agencies to do a "good job" is really the crux of the matter. A critical review, similar to this one, of developments constructed in the 1980's will reveal whether real progress has been made in assessing Canadian hydroelectric and water-diversion projects.

Technical Deficiencies in Environmental Impact Assessment

Although the environmental impact assessment of hydroelectric developments is maturing, substantial technical deficiencies remain even in the most recent ones. The lact of quantitative prediction is a fundamental weakness in nearly all studies. Hecky et al. (1984) noted the general lack of empirical and especially systems models for new reservoir ecosystems and the absence of models that have been rigorously tested by application to a variety of reservoir situations. This is surprising given the existence of the widely used, but unfortunately qualitative, reservoir paradigm. Ultimately, systems models predicting qualitative and quantitative effects of impoundment and/or diversion on commercially important fish stocks are required. In contrast, presently available models usually deal with interactions between hydrological variables and lower trophic levels (e.g. the relationship between phosphorus inputs after flooding and primary productivity; see Hecky et al. 1984 for references). Systems modelling efforts in the James Bay study (Thérien 1981) were the most advanced among the projects examined, but fish populations were not included. Hopefully, as data are accumulated on new reservoirs by the James Bay monitoring program, the models developed will be assessed critically and eventually combined to form true systems models. Data accumulated over time by pre- and postdevelopment monitoring will be invaluable for testing models, but this will occur only if model development is given priority in research accompanying environmental impact assessment.

Experimental approaches are rarely used in the assessment of new hydroelectric developments. Experimental studies, especially using mesocosms (e.g. Grice and Reeve 1982; Rudd and Turner 1983), can circumvent more costly and time-consuming monitoring programs as a way of testing hypotheses and models. Modelling and model testing through experimentation are essential if such assessments are to improve rapidly.

Existing knowledge of species ecology, and especially environmental physiology, of even widespread and commercially important species is inadequate. However, a common assumption in environmental assessment is that such knowledge is available. Often, the impact of a development on a species of interest is predicted by starting with a qualitative statement about environmental change (e.g. turbidity will increase) and inferring what effect this will have based on knowledge about the biology of the species. This approach would have failed to predict the effects of flooding Southern Indian Lake on the lake whitefish, *Coregonus clupeaformis*, even if conditions in the physical environment following impoundment and diversion had been predicted *quantitatively* (Hecky et al. 1984). Furthermore, even the modelling efforts of Thérien (1981) and his associates for the James Bay project did not extend to species of fish. More research is required on the basic ecology and physiology of important or target species before predictability will improve substantially. Because such species are usually large, and have complex habitat requirements and life histories, they are more difficult to work with experimentally while maintaining reasonably natural conditions. Innovative approaches in field research methods will be required in order for progress to be made on this problem.

Predevelopment, baseline conditions must be measured over a period of years sufficient to reveal the degree of natural variability. By this criterion, the lengths of predevelopment studies in the projects analysed were naively short. In fact, a relatively long period of time is required to define annual variability of ecosystems (Likens 1983; Schindler 1987), and a short record for the predevelopment period will contain unknown bias no matter what the length of the postdevelopment study done to assess the impact. Alternatively, statistical models must be developed to describe and assess natural variability in aquatic ecosystems.

Despite advances in scientific capability to predict the environmental effects of hydroelectric developments, a great deal of uncertainty still surrounds this activity (cf. Penn 1975). Indeed, even some major impacts resulting from hydroelectric development are still being identified. For example, discovery in the last decade of contamination of fish by mercury in new reservoirs (e.g. Bodaly et al. 1984a) challenges the sanguine view that all significant impacts associated with reservoir formation in temperate regions are known (Baxter 1977).

The problem of predicting effects of cumulative impacts is particularly vexing when we are still discovering new phenomena within reservoirs. "Destruction by insignificant increments" (Gamble 1979, p. 1) may be the most important cumulative impact in northern developments (Berkes 1981) and can involve scales of magnitude ranging from overhunting in a small area because of new access by roads to changing the circulation of Hudson Bay (Prinsenberg 1980). For example, altered patterns of resource utilization imposed by increased access to and within the affected area was one of the major impacts resulting from the James Bay development (Berkes 1981, 1982). Altered patterns of resource utilization also had major effects on the Southern Indian Lake fishery (Bodaly et al. 1984b). In both projects, these effects were unpredicted.

Historically, the impact zone has been viewed too narrowly. The Peace–Athabasca Delta study was required because the downstream effects of constructing the Bennett Dam on the Peace River in British Columbia and filling Williston Reservoir were not foreseen. The altered natural flow regime of Quebec rivers has changed circulation in the Gulf of St. Lawrence and possibly reduced fisheries productivity as far away as the Grand Banks (Neu 1982a, 1982b). Thus, cumulative impacts are often transboundary in nature and can extend over thousands of kilometres. Much better efforts must be expended in predicting such impacts, including the development of new regulatory arrangements to ensure their adequate consideration.

Allowing for Uncertainty

After the decision to develop has been made, the primary use of the environmental impact assessment is to predict and assess impact. However, by themselves, environmental impact assessments protect neither the natural resource nor the natural resource user (Hecky et al. 1984).

Assessment, including identification of the causes of significant impacts, is within our present capability. Such assessment, however, requires broadly based monitoring studies that include coverage of the entire zone of impact, begin early enough, and are followed by experimental hypothesis testing. Realistically, the experience required to satisfy all the qualifiers in the previous sentence is still being acquired. Predictive capability for environmental matters apparently evolved only slowly over the time period of the projects analyzed here. Environmental impact assessment is still struggling to develop quantitative prediction within a single reservoir, let alone coping with prediction for cumulative impacts resulting from several reservoirs within one large project, for many large projects, or for large-scale international water diversions.

Our ability to predict the consequences of hydroelectric development will improve only if a scientific approach is taken to impact assessment (Rosenberg and Resh et al. 1981). Certainly, the literature of environmental impact assessment must become more rigorous and subject to peer review (Schindler 1976) if the topic is to attract serious scientific inquiry. The apocrypha of limited-circulation impact literature associated with the "well-known" hydroelectric developments analyzed here attests to this situation.

The conclusion is inescapable that a great deal of uncertainty must be associated with any prediction about the state of environment following a major development. Until environmental science develops panaceas, is there a recourse for the resource user?

Citizens of industrialized countries commonly acquire insurance to protect against major disasters as a way of dealing with uncertainty in daily life. Similarly, in developing hydroelectric plants, a significant portion of the capital costs is set aside in the form of a contingency fund to pay for liability insurance, extra costs that may arise from innovative design schemes, the risk of undiscovered construction problems, or the uncertainties of future interest rates. One of the components not included in such a contingency fund is the potential cost of unanticipated environmental damage. Resource users in a region about to be affected by a development usually must assume this risk. In effect, they need insurance or access to some portion of the contingency fund to guarantee that their livelihoods and lifestyles will survive the development.

The concept of an environmental contingency fund to protect resource users is fundamentally different from existing Canadian methods of determining and instituting compensation with regard to hydroelectric developments. Compensation agreements have been either proactive (e.g. the JBNQA of the James Bay development) or retroactive (e.g. the *Northern Flood Agreement* of the Churchill–Nelson development). Agreements made before development have two disadvantages: (1) the amount of payments may bear no relationship to the damage that subsequently occurs and (2) the funds may be used in a manner that does not compensate resource users directly, either as a group or as individuals. Retroactive agreements also have disadvantages: (1) Payment of compensation can be extremely contentious because the agencies responsible for it have not set aside appropriate funds to service claims. Arbitration then becomes a complex and hotly contested legal and bureaucratic process. (2) Retroactive schemes require effective systems of damage documentation and assessment as a basis for establishing the extent of compensation, but such systems rarely are emplaced. The existence of an environmental contingency fund would overcome problems with existing compensation schemes.

The cost of an environmental contingency fund should be borne by the developer, and if the developer cannot afford it then the project should not proceed. In addition, the resource user should not be required to waive rights to have a say in the development as a result of acquiring such insurance. It seems appropriate to set a fixed fraction of gross revenue earned by a development to establish and maintain the fund in order to cover claims for mitigation, compensation, or new investment initiated by resource users. The royalty system currently applied to oil extraction may be a suitable model. In fact, in most provinces, hydroelectric companies pay a water rental to the provincial government, but the money usually enters general revenues, and appropriate amounts may or may not be assigned to impacted areas and resource users. A suitable role for government to play in this scheme would be to set the amount of the fund and payments made into it, to define the region eligible to receive compensation, and to establish a management framework composed of local representation to allocate monies.

Some such solution will be required until sufficient knowledge is acquired to make adequate predictions about impacts and to devise adequate mitigation measures. If the system of compensation adopted leaves the occupants of an affected region better off than before development occurred, then few will challenge that the development was successful.

Looking Ahead

Canadian hydroelectric developments are increasing in scale and complexity. Development of the first phase of the La Grande River, described above, is a good example, as is the plan described by Bourassa (1985) to develop other major rivers in northern Quebec. However, schemes to redistribute water on a massive scale may pose the most serious threat to the environment in the future. For example, Canadian water for export to the United States may become a bargaining item in free-trade negotiations between the two countries (Hall 1985; Reisman 1985).

Our ability to predict and assess the environmental effects of single impoundments is progressing. This is less so for complex schemes involving several impoundments, but the knowledge and experience to deal with major water redistribution schemes is almost nonexistent. Predicting and assessing the environmental impact of such water-diversion schemes on freshwater, estuarine, and coastal marine ecosystems thousands of kilometres away poses our biggest challenge for the future.

Acknowledgments

We thank the following individuals for providing information concerning the hydroelectric developments reviewed here: M. Atton, F. Berkes, C. Day, R. Delk, H. Duthie, A. Hostyk, A. Keast, B. Kemper, D. Kiell, P. Larkin, G. Mick, D. Roy, G. Townsend, and W. Williams. Reviews and critical comments on the manuscript by R. Baxter, N. Benson, F. Berkes, P. Campbell, H. Duthie, R. Gorton, G. Koshinsky, K. Patalas, D. Roy, G. Townsend, and D. Schindler are appreciated. A. Wiens helped in various aspects of producing the manuscript. The many hours of word processing were done by D. Laroque.

References

ABLESON, P. H. 1985. Electric power from the north. Science (Wash., DC) 228: 1487.

ALBERTA ENVIRONMENT CONSERVATION AUTHORITY. 1973. The restoration of water levels in the Peace-Athabasca Delta. Report and recommendations. Alberta Environment Conservation Authority, Edmonton, Alta. 136 p.

BAJZAK, D. 1975. Interpretation of flooding damage to vegetation in the Smallwood Reservoir Churchill Falls, Labrador, p. 405-412. In G. E. Thompson [ed.] Third Canadian Symposium on Remote Sensing, September 22-24, 1975, Edmonton, Alta.

    1977. Physical and environmental changes in the reservoirs of the Churchill Falls Hydro Electric Power Plant. Interim Report, 1977. Faculty of Engineering and Applied Science, Memorial University of Newfoundland, St. John's, Nfld. 53 p.

BARANOV, I. V. 1961. Biohydrochemical classification of the reservoirs in the European U.S.S.R., p. 139-193. In P. V. Tyurin [ed.] The storage lakes of the U.S.S.R. and their importance for fishery. Israel Program for Scientific Translations, Jerusalem, Israel.

BARNES, M. A., G. POWER, AND R. G. H. DOWNER. 1984. Stress-related changes in lake whitefish (Coregonus clupeaformis) associated with a hydroelectric control structure. Can. J. Fish. Aquat. Sci. 41: 1528-1533.

BAXTER, R. M. 1977. Environmental effects of dams and impoundments. Annu. Rev. Ecol. Syst. 8: 255-283.

BEANLANDS, G. E., AND P. N. DUINKER. 1983. An ecological framework for environmental impact assessment in Canada. Institute for Resource and Environmental Studies, Dalhousie University, Halifax, N.S., and Federal Environmental Assessment Review Office, Hull, P.Q. 132 p.

BENSON, N. G. 1968. Review of fisheries studies on Missouri River main stem reservoirs. U.S. Fish Wildl. Serv. Res. Rep. 71: 61 p.

    1980. Effects of post-impoundment shore modifications on fish populations in Missouri River reservoirs. U.S. Fish Wildl. Serv. Res. Rep. 80: 32 p.

BENSON, N. G. AND B. C. COWELL. 1967. The environment and plankton density in Missouri River reservoirs, p. 358-373. In Reservoir Fishery Resources Symposium, April 5-7, 1967, University of Georgia, Athens, GA. American Fisheries Society, Washington, DC.

BERKES, F. 1981. Some environmental and social impacts of the James Bay hydroelectric project, Canada. J. Environ. Manage. 12: 157-172.

    1982. Preliminary impacts of the James Bay hydroelectric project, Quebec, on estuarine fish and fisheries. Arctic 35: 524-530.

1985. Environmental impacts of major hydro developments in Quebec: the James Bay project. Bull. Can. Soc. Environ. Biol. 42(2): 43-50.

BODALY, R. A., R. E. HECKY, AND R. J. P. FUDGE. 1984a. Increases in fish mercury levels in lakes flooded by the Churchill River diversion, northern Manitoba. Can. J. Fish. Aquat. Sci. 41: 682-691.

BODALY, R. A., T. W. D. JOHNSON, R. J. P. FUDGE, AND J. W. CLAYTON. 1984b. Collapse of the lake whitefish (*Coregonus clupeaformis*) fishery in Southern Indian Lake, Manitoba, following lake impoundment and river diversion. Can. J. Fish. Aquat. Sci. 41: 692-700.

BODALY, R. A., D. M. ROSENBERG, M. N. GABOURY, R. E. HECKY, R. W. NEWBURY, AND K. PATALAS. 1984c. Ecological effects of hydroelectric development in northern Manitoba, Canada: the Churchill-Nelson River diversion, p. 273-309. In P. J. Sheehan, D. R. Miller, G. C. Butler, and Ph. Bourdeau [ed.] Effects of pollutants at the ecosystem level. SCOPE 22. John Wiley & Sons Ltd., New York, NY.

BOUCHER, R., AND R. SCHETAGNE. 1983. Répercussions de la mise en eau des réservoirs de La Grande 2 et Opinaca sur la concentration de mercure dans les poissons. Société de l'énergie de la Baie James, Montreal, P.Q. 38 p.

BOURASSA, R. 1985. Power from the north. Prentice-Hall Canada, Scarborough, Ont. 182 p.

BRUCE, W. J. 1974. The limnology and fish populations of Jacopie Lake, west forebay, Smallwood Reservoir, Labrador. Fish. Mar. Serv. Tech. Rep. Ser. No. NEW/T-74-2: 74 p.

1975. Experimental gill net fishing at Lobstick and Sandgirt Lakes, Smallwood Reservoir, western Labrador, 1974. Fish. Mar. Serv. Intern. Rep. Ser. No. NEW/1-75-4: 35 p.

1979. Age and growth of brook trout (*Salvelinus fontinalis*) in the Churchill River watershed, Labrador. Fish. Mar. Serv. Tech. Rep. 907: 10 p.

1984. Potential fisheries yield from Smallwood Reservoir, western Labrador, with special emphasis on lake whitefish. N. Am. J. Fish. Manage. 4: 48-66.

BRUCE, W. J., AND K. D. SPENCER. 1979. Mercury levels in Labrador fish, 1977-78. Can. Ind. Rep. Fish. Aquat. Sci. No. 111: 12 p.

BRUNEAU, A. A., AND D. BAJZAK. 1973. Effect of flooding on vegetation in the main reservoir, Churchill Falls, Labrador. Project Establishment Report. Faculty of Engineering and Applied Science, Memorial University of Newfoundland, St. John's, Nfld. 27 p.

CHURCHILL RIVER STUDY (MISSINIPE PROBE). 1976. Summary report, Churchill River Study, Saskatchewan Department of Environment, Regina, Sask. 52 p.

DAVIS, C. C. 1978. Notes on zooplankton from some Labrador lakes. Int. Ver. Theor. Angew. Limnol. Verh. 18: 233-239.

DAY, J. C. 1985. Canadian interbasin diversions. Inquiry on Federal Water Policy Research Paper No. 6. Natural Resources Management Program, Simon Fraser University, Burnaby, B.C. 111 p.

DEPARTMENT OF FISHERIES AND OCEANS. 1984. Toward a fish habitat decision on the Kemano Completion Project. A discussion paper. Government of Canada, Department of Fisheries and Oceans, Vancouver, B.C. 79 p.

DUTHIE, H. C. 1979. Limnology of subarctic Canadian lakes and some effects of impoundment. Arct. Alp. Res. 11: 145-158.

DUTHIE, H. C., AND M. L. OSTROFSKY. 1974. Plankton, chemistry, and physics of lakes in the Churchill Falls region of Labrador. J. Fish. Res. Board Can. 31: 1105-1117.

1975. Environmental impact of the Churchill Falls (Labrador) hydroelectric project: a preliminary assessment. J. Fish. Res. Board Can. 32: 117-125.

EFFORD, I. E. 1975. Assessment of the impact of hydro-dams. J. Fish. Res. Board Can. 32: 196-209.

ELLIS, M. M. 1936. Some fishery problems in impounded waters. Trans. Am. Fish. Soc. 66: 63-75.

ENVIROCON LTD. 1981. Kemano completion hydroelectric development baseline environmental studies. Vol. 2, Sect. B, p. 175-201. Envirocon Ltd., Vancouver, B.C.

ENVIRONMENT CANADA. 1975a. Hydrological atlas of Canada. Information Canada, Ottawa, Ont.

1975b. James Bay hydro-electric project: a statement of environmental concerns and recommendations for protection and enhancement measures. Environment Canada, Environmental Monitoring Service, Hull. P.Q. 45 p. and attachment.

FAUBERT, N. 1982. Travaux d'aménagement piscicole au réservoir de LG 2. Can. Water Resour. J. 7(2): 172-180.

FEIT, H. A. 1982. The income security program for Cree hunters in Quebec. An experiment in increasing the autonomy of hunters in a developed nation state. Can. J. Anthropol.

3: 57-70.

GABOURY, M. N., AND J. W. PATALAS. 1984. Influence of water level drawdown on the fish populations of Cross Lake, Manitoba. Can. J. Fish. Aquat. Sci. 41: 118-125.

GAMBLE, D. J. 1979. Destruction by insignificant increments. Arctic offshore developments: the circumpolar challenge. North. Perspect. 7(6): 1-4.

GEEN, G. H. 1974. Effects of hydroelectric development in western Canada on aquatic ecosystems. J. Fish. Res. Board Can. 31: 913-927.

GRICE, G. D., AND M. R. REEVE [ED.]. 1982. Marine mesocosms: biological and chemical research in experimental ecosystems. Springer-Verlag, New York, NY. 430 p.

GROUPE DE TRAVAIL FÉDÉRAL-PROVINCIAL. 1971. Étude préliminaire des impacts écologiques du projet de développement de la Baie James, Province de Québec. Rapport du Groupe de Travail Fédéral-Provincial. 75 p. and attachments.

HALL, T. 1985. Huge water deal with U.S. could imperil Canada's freedom. The Citizen, Ottawa, Ont. November 28, 1985.

HECKY, R. E., R. W. NEWBURY, R. A. BODALY, K. PATALAS, AND D. M. ROSENBERG. 1984. Environmental impact prediction and assessment: the Southern Indian Lake experience. Can. J. Fish. Aquat. Sci. 41: 720-732.

JUNE, F. C. 1974. Ecological changes during the transitional years of final filling and full impoundment (1966-70) of Lake Oahe, an upper Missouri River storage reservoir. U.S. Fish Wildl. Serv. Tech. Pap. 71: 57 p.

LAKE WINNIPEG, CHURCHILL AND NELSON RIVERS STUDY BOARD. 1975. Summary report. Canada-Manitoba Lake Winnipeg, Churchill and Nelson Rivers Study, Winnipeg, Man. 64 p.

LIKENS, G. E. 1983. A priority for ecological research. Bull. Ecol. Soc. Am. 64: 234-243.

LYONS, J. C., AND P. A. LARKIN. 1952. The effects on sports fisheries of the Aluminum Company of Canada Limited development in the Nechako Drainage. Book 1 Text. British Columbia Game Department, Vancouver, B.C. 57 p.

MACGREGOR, G. 1948. Attitudes of the Fort Berthold Indians regarding removal from the Garrison Reservoir site and future administration of their reservation. U.S. Department of the Interior, Bureau of Indian Affairs, Missouri River Basin Investigations, Billings, MT. 25 p.

MACKENZIE RIVER BASIN COMMITTEE. 1981. Mackenzie River Basin study report. A report under the 1978-81 Federal-Provincial Study Agreement respecting the water and related resources of the Mackenzie River Basin. Environment Canada, Inland Waters Directorate, Regina, Sask. ISBN 0-919425-08-9. 231 p.

MARSAN, A., AND B. COUPAL. 1981. The role of mathematical models in assessing environmental impacts, p. 7-16. In N. Thérien [ed.] Simulating the environmental impact of a large hydroelectric project. Simulation Proceedings Series 9(2). The Society for Computer Simulation (Simulation Councils Inc.), La Jolla, CA.

NELSON, W. R. 1978. Implications of water management in Lake Oahe for the spawning success of coolwater fishes. Am. Fish. Soc. Spec. Publ. 11: 154-158.

NEU, H. J. A. 1982a. Man-made storage of water resources — A liability to the ocean environment? Part I. Mar. Pollut. Bull. 13: 7-12.

     1982b. Man-made storage of water resources — A liability to the ocean environment? Part II. Mar. Pollut. Bull. 13: 44-47.

NEWBURY, R. W., G. K. McCULLOUGH, AND R. E. HECKY. 1984. The Southern Indian Lake impoundment and Churchill River diversion. Can. J. Fish. Aquat. Sci. 41: 548-557.

OSTROFSKY, M. L., AND H. C. DUTHIE. 1975. Primary productivity, phytoplankton and limiting nutrient factors in Labrador lakes. Int. Rev. Gesamten Hydrobiol. 60: 145-158.

     1980. Trophic upsurge and the relationship between phytoplankton biomass and productivity in Smallwood Reservoir, Canada. Can. J. Bot. 58: 1174-1180.

PEACE-ATHABASCA DELTA IMPLEMENTATION COMMITTEE. 1983. Status report for the period 1974-1983. A report to the Ministers. Peace-Athabasca Delta Implementation Committee, Canada, Alberta, Saskatchewan. 44 p.

PEACE-ATHABASCA DELTA PROJECT GROUP. 1972. Peace-Athabasca Delta. A Canadian resource. Summary Report, 1972. A report on low water levels in Lake Athabasca and their effects on the Peace-Athabasca Delta. Environmental Ministers of Canada, Alberta, and Saskatchewan, Edmonton, Alta. 144 p.

     1973. The Peace-Athabasca Delta Project. Technical Report. A report on low water levels in Lake Athabasca and their effects on the Peace-Athabasca Delta. Environmental Ministers of Canada, Alberta, and Saskatchewan, Edmonton, Alta. 176 p.

PENN, A. F. 1975. Development of James Bay: the role of environmental impact assessment in determining the legal right to an interlocutory injunction. J. Fish. Res. Board Can. 32: 136-160.

PRINSENBERG, S. J. 1980. Man-made changes in the freshwater input rates of Hudson and James bays. Can. J. Fish. Aquat. Sci. 37: 1101-1110.

    1982. Present and future circulation and salinity in James Bay. Nat. Can. (Rev. Ecol. Syst.) 109: 827-841.

REINELT, E. R., R. KELLERHALS, M. A. MOLOT, W. M. SCHULTZ, AND W. E. STEVENS [ED.]. 1971. Proceedings of the Peace-Athabasca Delta Symposium, Edmonton, January 14 and 15, 1971. The University of Alberta, Water Resources Centre, Edmonton, Alta. 359 p.

REISMAN, S. 1985. Canada-United States trade at the crossroads: options for growth. Article based on paper presented at Ontario Economic Council conference "Canadian trade at a crossroads: options for new international agreements", Toronto, Ont., April 16 and 17, 1985.

ROSENBERG, D. M., R. A. BODALY, R. E. HECKY, R. W. NEWBURY, AND K. PATALAS. 1985. Hydroelectric development in northern Manitoba: the Churchill-Nelson River diversion and flooding of Southern Indian Lake. Bull. Can. Soc. Environ. Biol. 42(2): 31-42.

ROSENBERG, D. M., AND V. H. RESH ET AL. 1981. Recent trends in environmental impact assessment. Can. J. Fish. Aquat. Sci. 38: 591-624.

ROY, D. 1982a. Répercussions de la coupure de la Grande Rivière a l'aval de LG 2. Nat. Can. (Rev. Ecol. Syst.) 109: 883-891.

    1982b. Le réseau de surveillance d'écologie aquatique de la Société d'énergie de la Baie James. Can. Water Resour. J. 7(1): 229-250.

RUDD, J. W. M., AND M. A. TURNER. 1983. Suppression of mercury and selenium bioaccumulation by suspended and bottom sediments. Can. J. Fish. Aquat. Sci. 40: 2218-2227.

SCHINDLER, D. W. 1976. The impact statement boondoggle. Science (Wash., DC) 192: 509.

    1987. Detecting ecosystem responses to stress. Can. J. Fish. Aquat. Sci. 44. (In press)

SHANKS, B. D. 1974. The American Indian and Missouri River water developments. Water Resour. Bull. 10: 573-579.

SHEPPARD T. POWELL ASSOCIATES (CANADA) LTD. 1970-72. Churchill Falls Power Project. Water quality and biological study. Vol. 1A, 1970-1971, p. 1-132; Vol. 1B, 1970-1971, p. 133-214; Vol. 2, 1971, 91 p.; Vol. 3, 1972, 106 p.

SOCIÉTÉ D'ÉNERGIE DE LA BAIE JAMES. 1984. Étude des effets du détournement des rivières East-main et Opinaca en aval des ouvrages de dérivation. Rapport d'étape 1983. Société d'énergie de la Baie James, Montréal, P.Q. 58 p. and attachments.

SOUCY, A. 1978. L'environnement du Territoire de la Baie James et les principales répercussions du projet. L'ingénieur 325: 13-24.

STANLEY ASSOCIATES ENGINEERING LTD. 1980. Slave River hydro feasibility study. Task Area 4B. Interim Report. Stanley Associates Engineering Ltd., Edmonton, Alta. 350 p. + references + Appendix A.

TELKWA FOUNDATION. 1984. DFO Kemano meetings. Found. Newsl. 7(1): 1-20.

THÉRIEN, N. [ED.]. 1981. Simulating the environmental impact of a large hydroelectric project. Simulation Proceedings Series 9(2). The Society for Computer Simulation (Simulation Councils Inc.), La Jolla, CA. 118 p.

THERRIEN, D. 1982. Les aménagements correctifs d'environnement au complexe La Grande. Can. Water Resour. J. 7(2): 147-162.

TOWNSEND, G. H. 1975. Impact of the Bennett Dam on the Peace-Athabasca Delta. J. Fish. Res. Board Can. 32: 171-176.

    1982. An evaluation of the effectiveness of the Rochers Weir in restoring water levels in the Peace-Athabasca Delta. Canadian Wildlife Service, Edmonton, Alta. 47 p.

UNDERWOOD-MCLELLAN AND ASSOCIATES LTD. 1970. Manitoba Hydro Churchill River diversion: study of alternative diversions. Underwood-McLellan and Associates Ltd., Winnipeg, Man. 124 p.

U.S. FISH AND WILDLIFE SERVICE. 1946. A preliminary report on fish and wildlife problems in relation to the water development plan for the proposed Missouri Basin Garrison Dam Project. U.S. Department of the Interior, Fish and Wildlife Service, Chicago, IL. 21 p. and attachments.

    1948. A preliminary evaluation of the effect of the Fort Randall Dam and Reservoir on fish and wildlife resources. Missouri River — main stem. Missouri River Basin. U.S. Department of the Interior, Fish and Wildlife Service, Missouri River Basin Studies, Billings, MT. 19 p. and appendices.

1950. A preliminary evaluation report on fish and wildlife resources in relation to the water development plan for the Gavins Point Reservoir. Missouri River — main stem. Missouri River Basin. U.S. Department of the Interior, Fish and Wildlife Service, Missouri River Basin Studies, Billings, MT. 28 p. and attachment.

1960. A plan for fish and wildlife resources of the Oahe Reservoir. North Dakota and South Dakota. U.S. Department of the Interior, Fish and Wildlife Service, Bureau of Sport Fisheries and Wildlife, Missouri River Basin Studies, Billings, MT. 95 p. and attachments.

1962. Fish and wildlife resources. Big Bend Reservoir. South Dakota. Missouri River Basin Project. U.S. Department of the Interior, Fish and Wildlife Service, Bureau of Sport Fisheries and Wildlife, Minneapolis, MN. 23 p. and attachment.

WAGNER, M. W. 1981. Economic performance of the summer commercial fishery of Southern Indian Lake, Manitoba. Master of Natural Resources Management thesis, Natural Resources Institute, University of Manitoba, Winnipeg, Man. 163 p.

1984. Postimpoundment change in financial performance of the Southern Indian Lake commercial fishery. Can. J. Fish. Aquat. Sci. 41: 715-719.

CHAPTER 5

Canadian Water: A Commodity for Export?

Richard C. Bocking

Richard C. Bocking Productions Ltd., 4274 West Hill Avenue, Montreal, Que. H4B 2S7

Water Export: Birth of a Controversy

Since 1939, the Colorado River Aqueduct has carried 1 million acre feet of water each year across desert sands and over mountain ranges to supply the cities of southern California. In 1963, after years of bitter wrangling, a U.S. Supreme Court decision required the state to give up half of that water. It was instead to be delivered to the dry lands of central Arizona.

The Colorado was already so dammed and diverted that no water had reached its mouth for years. The law, in fact, required it to deliver more water than flowed between its banks. Nature had clearly erred in the southwestern United States. To correct this situation, engineers in the United States and Canada sprang to their drawing boards to create imaginative plans for rearranging the flow of water in North America. Large-scale water divisions became the stuff of headlines.

In these plans, the rivers of the continent were portrayed as parts of a gigantic plumbing system. And that system simply needed some dams, pumps, and reservoirs added in order to pour massive quantities of fresh water, some of it Canadian, onto the desert lands of the American southwest. The projects had fine, rolling names — The Great Lakes Pacific Waterways Plan, the Central North America Water Project, the Western States Water Augmentation Concept; and there were many more (Quinn 1973).

The grandiose schemes had the enthusiastic backing of many politicians from the drier states of the southwest. Among them were Senator Frank Moss of Utah and Congressman Jim Wright of Texas. They wrote books and lobbied in Congress in support of their favourite plans (Moss 1967). Jack Williams, governor of Arizona, regarded Canadian water as a valuable export commodity flowing wasted into the sea. It would be, he stated, "a good neighbour gesture" on Canada's part to make this water available to Arizona. Joseph Jensen, Chairman of the Metropolitan Water District of southern California, expressed the dream of many of his fellow citizens when he envisioned the desert valleys of the southwest inhabited by millions more people. "That's what we're going to have, and we're going to get the water from somewhere and bring it down here" (Bocking 1972).

In Canada there were those who supported water export as a source of great wealth to Canadians and others who, disturbed by media stories of drought in the southwest, felt that Canada could not stand idly by while its good neighbour suffered a deepening thrist.

Early in the 1970s, the issue faded from the headlines. In Canada, the country's rivers and lakes began to be recognized as vital elements of national identity, a viewpoint discouraging to those who saw diversion of Canadian water south in purely economic terms. In both countries, deepening concern over environmental issues focused attention on the hazards of large-scale water development. Promoters of schemes to export water to the United States saw their cause undermined when a 1966 study by the U.S. National Research Council and National Academy of Sciences stated

> This report is not prompted by a national water shortage, for there is no nationwide shortage and no imminent danger of one.

105

Through the last few years of the 1970s the issue of Canadian water exports to the United States appeared to lie dormant, at least as far as the popular press was concerned. But the headlines returned in 1981. *Newsweek* proclaimed "The browning of American" (Feb. 23, 1981); *U.S. News and World Report* asked "Water — Will We Have Enough to Go Around?" (June 29, 1981). *Life* magazine lamented "America the Dry" (July 1981). In Canada in the same year, *Today* magazine displayed "The Big Thirst" on the cover of an issue which featured an article entitled "The Browning of North America."

Water, the quantity and quality of it, who controls and who profits by it, was once again in the forefront of public attention.

Two projects from the halcyon days of continental water planning were dusted off and once more thrust forward. The North American Water and Power Alliance (NAWAPA) would reverse the flow of many great rivers and create giant reservoirs drowning thousands of hectares of western Canada so that more acres of desert could be irrigated in the southwestern United States (Fig. 1). The scheme woud divert south some 136 km^3/yr (110 million acre feet/yr). About 240 reservoirs would be created by enormous dams, some of them much larger than any built in the world to date. The largest would flood 500 km of the Rocky Mountain Trench, the magnificient valley

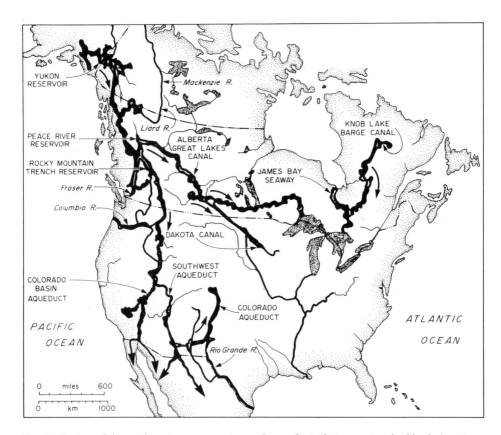

FIG. 1. System of dams, diversions, reservoirs, and aquaducts that comprise the North American Water and Power Alliance (NAWAPA), a scheme divised by the Ralph M. Parsons Co. of Pasadena, CA, to deliver Canadian and Alaskan water to serve the needs of consumers in the western, southwestern, and midwestern United States and the Prairie provinces.

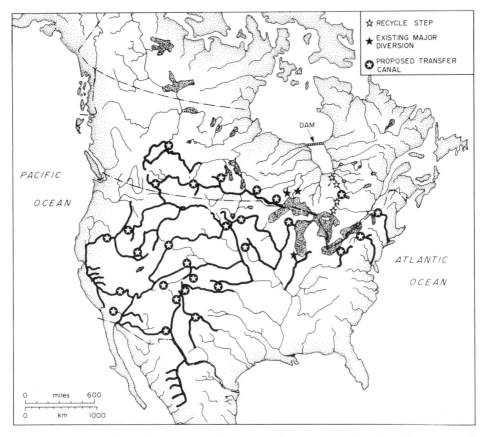

FiG. 2. System of dams, diversions, and reservoirs that comprise the Great Recycling and Northern Development (GRAND) scheme of Thomas Kierans to deliver northward flowing Canadian water to arid regions of southern Canada and the United States.

that extends from Monana to the Yukon in British Columbia (Sewell 1985). Once again, NAWAPA gained publicity for the Los Angeles engineering firm that originated it.[1]

The Great Recycling and Northern Development (GRAND) plan of Canadian engineer Thomas Kierans has also resurfaced with some significant political support (Fig. 2). The scheme envisions a dam 160 km long across James Bay. The fresh water that would accumulate behind it would be pumped southward to the Great Lakes at up to 2 million L/s for ultimate distribution within the United States and Canada (Kierans 1984).

Fifteen years ago, attention was primarily focused on western rivers as likely sources of water for the United States. Today, more attention is given to the Great Lakes as a possible source of water for export. Since the GRAND plan has been seriously promoted in recent years, it will be used in this chapter as an example when examining the implications of large-scale water development and export.

Like all other schemes of large-scale water transfer, the GRAND plan depends upon the perception that a crisis in water supply lies ahead int he United States. So that is the issue that must first be considered in any analysis of the water export question.

---

[1]For example, N. W. Snyder of the Ralph M. Parsons Company, the engineering firm that originated the NAWAPA scheme in 1963, highlighted the benefits of NAWAPA in a lecture entitled "Water from Alaska," presented at the California Conference on "A high technology policy of U.S. reindustrialization" in Los Angeles, California, October 15, 1980.

Is There a Crisis in Water Supply in the United States?

Every comprehensive study of water supply in the United States has concluded in much the same terms as that of the U.S. Water Resources Council in 1978 that

...there is no national water shortage problem now nor in the foreseeable future.

In 1981, economist Alan Kneese, a leading authority on American water resources, said

I don't see water, or the lack of it, as a source of impending disaster.

Gilbert White, of the University of Colorado, one of the world's most respected water experts, stated in the same year that

The people of the United States are in no danger of any early shortage of water — the facts just do not bear that out.

In 1983, Peter Rogers of Harvard University said that

Water is not limitless, but contrary to the views of alarmists, there is absolutely no danger that it will run out. The U.S. as a whole can count on at least fifty years without serious shortages, even at present wasteful rates of consumption.

Yet from time to time, the term "crisis" is applied to the water supply situation in various parts of the United States. Even in the well-watered east, cities like Boston and New York periodically face shortages when a normally bountiful nature delivers less than average rainfall. And with current use patterns in drier regions, the demands of population growth and irrigated agriculture may exceed the readily available supply. Are these really "crises"?

North Americans consume prodigious amounts of water, about 180 gal per person each day. The average per capita consumption in Britain and Sweden is less than one third of that; in Germany and France, it is less than a quarter as much (Rogers 1983). Much more water than is consumed must be fed into many North American systems, particularly in the American northeast, because 10-50% of it leaks out before reaching the consumer.

Polluting water reduces the quantity of it that is available for use. In communities across the continent, poisons accumulate in both surface and underground drinking water supplies. Little can be done to remove contaminants that have entered subterranean aquifers following decades of thoughtless waste disposal and heavy applications of chemicals on farmlands (Rogers 1973; see also chapter by Cherry, this volume).

The really big user of water in the United States is farming. Agriculture takes 80% of all water consumed. Most of that is used in the driest parts of the nation, where water obviously is in shortest supply. We must, then, turn to the farms of the west and southwestern United States to determine the adequacy of American water supplies.

There are three regions in the American southwest that are often claimed to be facing a water supply crisis. These are southern California, Arizona, and the High Plains region covering parts of six southwestern states.

The imagery is persuasive and familiar. The San Joaquin Valley of California is one of the world's richest agricultural areas. Excessive pumping of groundwater has allowed the land to settle and compact, and parts of the valley floor are 30 ft lower than they were 50 years ago.

In Arizona, near Phoenix, land settling because of the "mining" of groundwater through years of heavy pumping has created great gullies that scar the land. As groundwater levels drop, energy costs to pump the water rise. Some farmers growing lower value crops such as alfalfa or grains can no longer afford to do so. Tumbleweed rolls across lands that once were green, and they now have been reclaimed by the desert.

Rick Johnson of the Central Arizona Project said [2]

> ...We've used every drop of our surface water sources, we are over-draughting, mining
> our ground water resources, and we are just in potentially bad shape.

"Mining" is a term applied to the use of underground water at a rate faster than that at which it is being replaced. A classic case is found on the High Plains of Texas, New Mexico, Oklahoma, Nebraska, Kansas, and Colorado. This is a region of 225 000 mi^2 of flat, semiarid land beneath which lies the great Ogallala Aquifer. This may be the greatest underground freshwater reservoir in the world, containing an estimated 2465 km^3 (2 billion acre feet) of water. (An acre foot is the amount of water that will cover 1 acre of land to a depth of 1 ft.) But even a gift of nature on this grand scale has proven inadequate to meet the demands upon it.

Powerful pumps lift the water up out of the aquifer to nourish 6 million ha of farmland. That land produces 25% of the nation's cotton, and grain to fatten 40% of American cattle. More than 70 000 wells tap the Ogallala Aquifer; three decades ago there were only about 2000 (Szechenyi 1981).

The water "mined" from the Ogallala is equivalent to the annual flow of the Colorado River, about 17.3 km^3 (14 million acre feet). In the 17 western states, 27.6 km^3 (22.4 million acre feet) are "mined" in an average year (Frederick 1982). This is, of course, a practice that cannot be considered stable over the long term, and so farmers ar permitted a groundwater depletion allowance to reduce the taxes they pay. In this respect they are treated like oil companies who also are exhausting the resource they draw from beneath the surface of the earth.

National surveys stating that no significant shortage of water exists are not likely to be persuasive for a farmer dependent upon underground water which is being exhausted by excessive pumping. As the level of the underground water table drops lower and lower, the energy bills for running his pumps rise higher and higher, making his water more and more costly. He may well argue that even if there is no absolute shortage of water, there certainly is a shortage of *cheap* water.

That is why a "crisis" is often claimed to be inevitable in regions such as southern California, Arizona, and the High Plains. Scenarios of disaster are developed by projecting the water use patterns of today into a future of growing populations and economic growth, with little allowance for changing technology, institutions, or the capacity of human beings to adapt to new conditions.

Such a "crisis" projection is, then, the result of rather simplistic calculations. So it is not surprising to find that it can be misleading. The authoritative text *Water Supply — Economics, Technology and Policy* (Hirshleifer et al. 1960) explains "forecasts for water requirements for the year 2000, or 2020, or any other distant time based on extrapolations of recent aggregate trends are usually mistaken and dangerous. The forecasts are typically mistaken because they ignore the factor of water costs; the growing scarcity of water in relation to the needs and desires of a community will, we believe, bring into being in the not too distant future a shift toward techniques that are conserving rather than extravagant with respect to water inputs."

The term "crisis" can be useful if it causes people to look for imaginative solutions for problems. But the term can also be dangerous, particularly if a problem perceived as a crisis turns us to costly and potentially damaging megaprojects as solutions. Large dams and their associated canals and structures are simple in concept and easy to explain. They are profitable for their promoters and enable politicians to say that "something is being done" about a problem. But if there are more effective, less damaging, and less expensive ways of tackling water problems, we will recognize them only if we understand the origin and nature of what is so readily termed a "crisis."

---

[2] Rich Johnson. 1981. Interview for the C.B.C. televison documentary "After the Flood" (Richard Bocking, Producer).

The Origins of "Crises" in Water Supply

In the Grand Valley of Colorado, farmers use more than six times as much water as is required to grow their crops (Rogers 1983). Almost 30% of the water in the system, or 9.9 km^3 (8 million acre feet), is lost annually from unlined canals in the great federal irrigation projects of the western United States. In some of those projects, 60% of the water never reaches the crops it is supposed to nourish (Szechenyi 1981).

The waste that characterizes much of the use of irrigation water in the United States is encouraged by the laws under which water is allocated among various possible users. The key doctrine in water law in the west is that of Prior Appropriation. Basically, that means that the first person to get the water has legal right to it forever. It also implies, "if you don't use it, you lose it." While it played a useful role in the early development of the west, the doctrine now works against the introduction of water conservation policies appropriate to the needs of today.

Sometimes the principle seems to apply on a scale much larger than that of the individual farmer. John Carroll (1983) suggested that this is a principal reason for support by the state of North Dakota for the Garrison Diversion project. This scheme to divert Missouri River water onto farmlands of the state makes no economic sense at all. But state politicians may fear that if North Dakota does not establish its claim to the water of the river by diverting some of its flow into consumptive use, the water will be claimed by another state further downstream.

There is a more fundamental reason for the waste that characterizes water use in North America. To most consumers, whether urban or agricultural, water is so cheap that conserving it is not worth the bother. Abel Wolman of Johns Hopkins University, an authority of world renown in the field of water supply, said (*New York Times*, August 8, 1981)

> Water is cheaper than dirt. That means there is no orderly design as to when and where to use it. In a vast country such as ours we have never been able to organize a thoughtful, logical national plan, and I am very doubtful we ever will.

The High Plains region is always cited as a prime example of approaching crisis, and as a "raison d'être" for massive water transfer projects such as the GRAND scheme for the export of Canadian water. But how well has the region used the vast resources already available to it in the Ogallala Aquifer?

In five of the six High Plains states, there is some degree of control on the use of groundwater, by either state or local authorities. In Texas, there is none. Each person pumps all the water he wishes from beneath his land. A survey in the six states by Stephen White and David Kromm of Kansas State University showed that although depletion of the Ogallala Aquifer is most serious in Texas, farmers of that state are least willing to accept any kind of water management that would prolong the useful life of the aquifer (White and Kromm 1984).

On the other hand, a much higher percentage of Texans, 76% as opposed to 18% of Nebraskans, favoured the importation of water from outside their state to replenish diminishing groundwater supplies. Of course, their rugged "go it alone" preference in water use would not apply to such projects. Taxpayers would be warmly welcomed, indeed required, to participate!

Although the proportion varies from a small minority in some states to a majority in Texas, a considerable number of High Plains farmers actively promote the transfer into their region of water from the Missouri or Mississippi River, and ultimately from Canada. They are supported by regional real estate interests, bankers, and other businessmen of the region. Yet the costs of interbasin transfers from the Missouri or Mississippi would be at least 10 times greater than the maximum that farmers could afford to pay (Frederick 1982). And that figure includes no consideration of the cost of bringing water to those rivers from Canada. Kenneth Frederick emphasized that the recent Corps of Engineers study of the cost of water importation into the High Plains area "confirms the infeasibility of moving water long distances for agricultural uses."

The reason for support of massive water importation schemes entirely lacking in economic rationale is not hard to find. The history of western water development shows clearly that the beneficiaries will not be required to pay the costs of the operation. Water prices have traditionally been based on the ability of farmers to pay, not on the cost of delivering the water.

That is what is happening in Arizona, for example, where the Central Arizona Project (CAP) will soon be pumping water from the Colorado River into the Phoenix and Tucson areas of central Arizona. Arizona battled with California for many years to determine which state had prior right to the water. Following Arizona's victory in the costly court procedures, and the spending of $3 billion of federal tax money, water will be delivered to farmers and urban consumers at prices much higher than are presently being paid for water in central Arizona. It begins to seem like a strange sort of "victory" for Arizonans!

William Martin and Helen Ingram of the University of Arizona studied the reasons for support of the CAP by farmers even though project water will be so costly. The researchers found that, in their words, farmers were "ready to play, but not to pay." That is, they would support the project, "stay in the game," in spite of predictions of high cost water simply because they were absolutely sure they would never have to pay the real cost of it (Martin et al. 1981).

As CAP approached completion, it became clear that the farmers were right. Indeed, the capital costs being charged to farmers for CAP water, $2/acre feet (approximately 1233 m^3) will pay only a tiny part of 1% of the interest on money spent on the project. Even with this remarkable subsidy by the American taxpayer, the average central Arizona farmer will still have to pay operating charges of about $70/acre foot, considerably higher than present costs of pumped or surface water.

The cities of the region were to pay higher prices than the farmers. But when it became clear that this would raise their water cost far above current levels, they used political leverage to force their prices down to those paid by farmers.[3] So prices for CAP water are being set according to political necessity, and have nothing to do with the capital cost of bringing water from the Colorado River. Supporters of water importation to the High Plains are counting on similar treatment.

In the United States, many water experts are confident that CAP will be the last great water transfer project in the nation. In fact, there have been no new projects approved by the U.S. Government for the Bureau of Reclamation since 1978. It may be that the "frontier" approach, which views the search for "more" as the only real solution to water problems, has at last run its course. In American water literature, interbasin water transfer attracts little support and much critical comment. Importation of water from Canada is barely mentioned.[4]

There are still in the United States a few politicians, engineering firms, and other direct beneficiaries in local areas who promote large-scale water transfer. But most of the high-profile supporters of such projects have died or have been removed from office and have not been replaced. Serious examination of water problems now usually relates to better use of existing supplies, and to alternatives quite different from traditional massive structural ones. We will turn now to a consideration of some of these.

---

[3]William E. Martin, University of Arizona, Tucson, AZ, February 1985, personal communication.

[4]Arliegh Laycock, University of Alberta, Edmonton, Alta., 1985, personal communication.

Alternative Approaches to Water Problems

When University of Arizona economist William Martin suggested in the late 1960s that no water shortage existed in his state that would detrimentally effect its economy, he was denounced as a heretic and invited by a leading newspaper to leave the state.

But times have changed. The present governor of Arizona, Bruce Babbitt, replied recently to the question "Do you think lack of water is ever going to impede Arizona's growth?" by saying (USA Today, February 12, 1985)

> There is such a huge margin available in management conservation that it's not likely to happen in my lifetime; maybe fifty years out.

California, and particularly southern California, has for several decades been reaching further afield for new water supplies. Until recently, it was assumed that the growing population of arid southern California would require importation of vast quantities of water from the north. But even with the impending loss of 0.62 km^3 (0.5 million acre feet) of water a year to Arizona, the Director of California Department of Water Resources was able to state recently (Kennedy 1985)

> California in fact does have enough water to meet all its reasonable needs, including protection of its fisheries and environmental values. We must find the wisdom to develop and protect this resource to benefit the entire state.

Since waste is an important part of the problem, reducing waste is part of the solution. The obvious way to accomplish that is to let the price of water rise closer to its real cost (e.g. see chapter by Tate, this volume). As economist Kenneth Boulding (1978) explained:

> It is a very fundamental principle of the dynamics of the price system that, if we have something that is plentiful now but is going to be scarce later on, the sensible thing to do is make it expensive now, which will force us into improvements that will save and economize the expensive item, so that by the time it becomes really scarce we will have found means to use it more efficiently and we will also have found substitutes.

Certainly this has been demonstrated in the case of energy during the 1970s and the 1980s. Awareness of a problem in energy supply resulted in some reduction of waste. But higher prices encouraged most of the conservation and technological development which so dramatically changed the energy scene.

For households, the simple metering of consumption normally reduces water consumption by half. That is the case in Alberta, where Edmonton is metered and consumes half as much per capita as Calgary, which is less than one third metered. The same is true in Arizona. Tucson is metered and consumes half as much per capita as Phoenix, which is not metered.

The reason for the difference is not hard to find. After real needs have been met, additional or wasted water will have a definite cost for the metered consumer. Even if the unit cost is low, he will tend to reduce unnecessary consumption. In a nonmetered household, consumption of more water after normal needs have been satisfied will have a cost of zero (Gysi 1980). There will be no economic motivation to save, and waste will be the norm.

In communities where conservation has been promoted as an alternative to costly and damaging structural solutions to water problems, consumption has considerably decreased with no effect whatever on quality of living. This has been the case in the Kitchener-Waterloo region of Ontario, where community conservation programmes, at about one tenth the cost, have reduced the requirement for new facilities to supply more water and to process increased volumes of sewage.[5] Not only has the communi-

---

[5]James Robinson, University of Waterloo. Interview for the C.B.C. television documentary "After the Flood" (R. Bocking, Producer).

ty avoided the cost of building these structures, but a particularly beautiful part of the Grand River Valley was saved from the destruction that would have accompanied the construction of a proposed dam.

In the United States, water conserving devices in households and for lawn watering have reduced per capita use by 20-40% in some cities (White 1983). Wherever prices for water are raised toward a realistic reflection of the cost of providing it, the level of consumption drops. Waste is the consequence to be expected from the attitude common in much of North America that water is a "free and abundant resource."[6]

Marshall Gysi pointed out that the "promotional pricing" structures commonly used by water and electric utilities reward waste by reducing unit prices as consumption increases. He proposed instead "conservational pricing" for such essential commodities. Prices would be maintained at moderate levels for the meeting of normal requirements, but the per unit price would increase as consumption rises (Gysi 1980). Excessive consumption will obviously be discouraged.

Industry has shown that it can adapt to water scarcity. If water prices are raised sufficiently, new technologies will be developed and implemented that use less water. In Sweden, a new paper-pulping technique has been developed that uses half as much water while nearly doubling productivity (Rogers 1983). A shift from evaporation cooling towers to nonevaporative cooling towers for stream electric power plants can reduce the water consumed by thermal power plants almost to zero.

The second United States national water assessment projects a *decrease* of as much as 60% in industrial withdrawals in the United States after 1975, which was the baseline for the study (White 1983). Conservation, reuse, and recycling of water and other technological advances are reducing industrial use of water in many countries.

Better control of pollution is making water available for consumption which would otherwise be lost. Little water is actually consumed by cities. Cleaning it up after use makes it reusable, time and again. That is much cheaper than going further and further afield for new supplies.

Contamination of surface and underground water by chemical and other industrial waste is a serious and growing problem (see chapters by Hamilton et al. and Cherry, this volume). One sixth of the population of North America is threatened by dioxin and other substances of extreme toxicity which are seeping into the Great Lakes system, particularly in the Niagara area. Many aquifers in the eastern United States have had to be abandoned because toxic materials have contaminated them.

In some case it may be possible to purify water after it is brought to the surface. In others, it may be tempting to look elsewhere for new supplies of cleaner water. In fact, however, it is far safer and cheaper to prevent contamination from getting into water in the first place. Peter Rogers, professor of environmental engineering at Harvard University, insists that the technology does exist to avert water contamination. "There is no doubt that, at a price, the country can safely handle its wastes" (Rogers 1983).

In Agriculture, too, water consumption declines with increases in price. A U.S. Bureau of Reclamation study 10 years ago showed that when water was priced to farmers at $1-3/acre foot, 60% of it never reached the intended crop. An increase in price to $10/acre foot, which is still inexpensive, reduced water losses by 20% (Szechenyi 1981). With better practices, efficiency in the use of water in farming can reach much higher levels.

If water were priced at or closer to its real price, irrigation ditches would be lined, and there would be careful scheduling of irrigation according to crop needs. The growing of crops that need less water, or the use of techniques such as drip irrigation, could

---

[6]Dixon A. R. Thompson. 1981. Allocation policies for water use in Alberta. Paper presented to the Alberta Society of Professional Biologists Symposium on Biological and Social Issues in Water Management, Calgary, Alta.

reduce agricultural consumption by 20-50% (*New York Times*, August 9, 1981). Such methods would also reduce the rate at which lands are being lost to cultivation through salinization, a process which may be intensified by overwatering.

Higher energy costs are also responsible for greater efficiency of water use in areas dependent upon pumping from underground sources. In Arizona, farmers using unsubsidized power to pump water are installing drip irrigation systems, planting more water-efficient crops, and reducing flood irrigation. Where flooding still is advantageous, laser beam technology improves water use by making fields absolutely level. Where electricity is subsidized by the Bureau of Reclamation, or where surface water is supplied at cheap subsidized rates, improvements in efficiency are much less marked.[7]

Kenneth D. Frederick (1981) noted

> In some areas, the inevitable adjustment to declining groundwater supplies will not be pleasant. Nevertheless, the socially most expensive response would be to provide subsidies that either enable farmers to pump to greater depths or to import water. We have begun to recognize that water is becoming increasingly valuable; now we need to provide the incentives to assure that we treat it as such. The worst social costs associated with the changing water situation will arise if we attempt to keep water cheap when it is not.

Realistic water pricing will not threaten the "Sun Belt" of the American southwest, which has become home to millions of people seeking a warm climate. It will continue to be a favoured location for some kinds of industry, and it will continue to be a year-round source of fruit and vegetables for the rest of North America. Frederick (1982) stated

> Since irrigation is such a predominant user of western waters, a small improvement in the efficiency of irrigation might save enough water to satisfy nonagricultural water needs without significantly reducing irrigated acreage.

When realistic pricing is combined with a legal and market structure that enables water to be transferred between users and between water agencies, a bit of imagination can work miracles. Let us look again at the principal regions in the United States often perceived to be facing water shortages.

Alternatives in California

The Metropolitan Water District of Southern California (MWD) transports water from the Colorado River across the Mojave Desert to the Los Angeles area. It also brings water 240 mi from the Owens Valley and draws about 1.85 km^3 (1.5 million acre feet) from northern California by means of the State Water Project canal.

As noted earlier in this chapter, the MWD predicts that by the year 2000 it will be short of water to supply demand by about 173 million m^3 (140 000 acre feet). A principal reason for this is that as CAP begins operations in the late 1980s and pours Colorado River water into the Phoenix-Tucson area, the MWD's Colorado aqueduct will lose more than 600 million m^3 (0.5 million acre feet), about half of its current annual flow. And so the MWD is fighting hard for additional water from northern California to make up the difference and provide for growth.

But new structures will be the most costly way of avoiding shortages in southern California, according to U.S. Department of the Interior economists Richard Wahl and Robert Davis (1984). They showed that by charging consumers the real cost of water (land taxes subsidize water prices in the MWD area), consumption would decline by 300 000 to 500 000 acre feet/yr. That is more than the shortage estimated by the MWD for the year 2000.

---

[7]William E. Martin, 1985, personal communication.

Wahl and Davis (1984) pointed out other possibilities for alleviating potential shortages. If the MWD were to invest in improvements to the canals and other fixtures of the Imperial Irrigation District, about 500 million m³ (400 000 acre feet) could be made available to the MWD. The MWD has accepted this idea, originally promoted by the Environmental Defense Fund, and is trying to conclude agreements with the Irrigation District.

In the San Joaquin Valley, the federal Central Valley Project has about 1.2 km³ (1 million acre feet) surplus to its needs for irrigation. With appropriate state-federal arrangements, this is water which could be shipped to Los Angeles by means of the State Water Project canal. Water from both of these sources would be less expensive than that made available through new structures such as the Peripheral Canal or the storage dams proposed by the MWD. The MWD is following up on this proposal too, promoting in the U.S. Congress the legal measures that would make sure a transfer possible. The winds of change are indeed blowing across the American southwest.

Alternatives on the High Plains

In their study of the High Plains situation, agricultural economist Raymond Supalla and his associates at the University of Nebraska found no role at all for importation of water into the High Plains from outside the region (Supalla et al. 1984). They were convinced that the lowering of water in the Ogallala Aquifer is by no means the cata-strophic event that promoters of the "crisis" mentality suggest. In areas where a return to dry land farming is necessary, it will take place over decades, a rate gradual enough to preclude widespread economic stress and disruption. Improvement in dryland technology is likely to offset most of the losses due to gradual reductions in irrigated acreage.

Economist Allan Kneese participated in a study of the High Plains situation for the U.S. Department of Commerce and the High Plains Council of Austin, Texas. "Even a cursory look at the economics of imported water starkly reveals what an outlandish idea it is," he stated (Kneese 1984). He pointed out that the willingness to pay would be, at maximum, in the region of $75/acre foot. But the cost of delivered water, based on studies by the Army Corps of Engineers, would be around $1000/acre foot.

The study by the High Plains Associates (1982) found that a strategy of voluntary conservation practises encouraged by education, research, demonstration, and economic incentives would provide for growth in the economy of the High Plains of about 3.3% annually over the next 50 years. Some land would go out of production as water tables dropped, but this would be more than balanced by the development of new land in other parts of the High Plains. Since the economy based on irrigation from the Ogallala Aquifer can continue to grow for many years, "the adjustment to depleting supplies can be gradual and non-catastrophic" (Kneese 1984).

Kneese (1984) concluded after this exhaustive study that "there is no water crisis on the High Plains. And water importation is a thoroughly bad idea, even if just the economics are considered; imagine an environmental impact statement for such a project."

Alternatives in Arizona

There are no indications that realistic pricing of water would damage the economy of the American southwest. In Arizona, agriculture consumes about 90% of the water but contributes only 3% of the state's total income. Economist William Martin instisted that with rational use of its existing water, the economy of the state can flourish[8]:

---

[8]William E. Martin. 1981. Interview for the C.B.C. television documentary "After the Flood" (Richard Bocking, Producer).

> We can grow a long way if water is transferred slowly to other uses as it is needed. Eventually water will be tranferred from low valued farm use (Alfalfa, grains, cotton) to higher valued uses in agriculture (fruits and vegetables) and finally even to higher valued uses in industry and municipalities

This is already happening to some extent in Arizona. It was insisted upon by the federal government as a condition of financing CAP. Another condition was the adoption of regulations that ensure more rational use of underground water. The real answer to water problems in central Arizona is being found in sound use and conservation policies, not in the enormous expense of CAP.

To most of us, there is something sacrosanct about land that produces food. In a hungry world, allowing irrigated land to revert once again to desert may seem almost morally wrong. But in North America, a principal problem of the farm economy is that of gigantic surplusses. And no informed person suggests that hungry nations can be fed indefinitely by shipments of food from North America. Except in cases of short-term emergency, the only real solution is to help such nations develop their own agriculture, and to feed their own people. Public money spent in this way would be much more useful to hungry nations than the expenditure of similar sums to produce surplus products on the dry lands of the southwestern United States.

It is clear that many alternatives are available to resolve water problems, even in those regions of the American southwest usually considered prime prospects for a "crisis" in water supply. Such alternatives are generally cheaper than the dams and other structures that may be proposed within their own regions. And they are a tiny fraction of the economic, environmental, and social cost of interbasin transfers, either within the United States or from Canada.

Returning therefore to our question "*Is there a crisis in water supply in the United States?*," we find that the answer must be no, not at the present time. In future there could be, but only if water remains far cheaper than it is worth; only if present levels of waste continue; only if the quality of water in streams, lakes, and underground aquifers continues to be degraded; only if inadequate use is made of new technology for efficient water use; only if laws persist that ensure inefficient use of water; and only if the future is viewed as a simple projection of trends in the face of which North Americans are helpless and incapable of exercising choice or control.

If a crisis of supply develops for these reasons in a given region, it is difficult to see why another region should be expected to bail it out, either through massive water transfers or through taxation. This is particularly true between nations; if the United States should ever ask Canada to ship water to an American region, Canadians should understand that the "need" is man-made and political, not biologic or economic. And when Canadian interests promote export of water to the United states, it will be for reasons unrelated to genuine need in that country. Analysis of such proposals and the motivation for them can be more readily accomplished when this central fact is recognized.

Power, Profit, and the Public Purse

Who profits and who pays, who benefits, and who is disadvantaged when rivers are dammed and their waters spread across dry land to grow food and fibre?

In the United States

Early irrigation project were gravity-driven systems which brought water economically from reasonably close sources onto land that was often ideal for irrigation. In Arizona, the Salt River Project distributed water across the desert in canals which followed closely those of the Indian irrigators of preceding centuries. Dams such as the Roosevelt generated electricity while controlling the flow of the Salt River. It was a system upon

116

which the prosperity of the Phoenix area was built.

The Salt River Project was the first of a vast number of federally funded irrigation systems built under the *Reclamation Act* of 1902. The new Bureau of Reclamation was charged with the provision of irrigation water to homesteaders to promote the opening and settlement of the western United States. Each person was limited to 160 acres of irrigation, to ensure that public money was indeed used for the development of family farms.

In the ensuing years, something like 4.9 million ha (12 million acres) were irrigated under the provisions of the Act. At the same time, project reservoirs submerged about the same amount of land, some 5.2 million ha (12.9 million acres), much of if productive. For example, from 1967 to 1975, 30 000 ha (75 000 acres) of producing cropland were inundated each year (Atkinson 1981).

To operate on such a scale, the Bureau of Reclamation grew to a vast bureaucracy that required new projects to keep it busy long after the west was settled and the original pupose of the Act was no longer appropriate. And those who benefited were no longer the family farmers that Theodore Roosevelt considered the basis of American democracy and for whom he signed the *Reclamation Act*.

In the irrigated regions of California and Arizona, one must now travel far to find a family farm. Publicly financed water pours onto lands owned by companies like Exxon, Prudential Life, The Southern Pacific Railroad, or J. G. Boswell, the world's largest cotton grower. The social goals of the *Reclamation Act* ceased to be a factor long ago, but these agribusinesses, through their enormous political clout, have managed to retain the benefits that were supposed to be for farm families. U.S. Senator Gaylord Nelson said "Reclamation law was not intended to subsidize the factory farmer. It was not supposed to subsidize the wealthy investor. But it does and this is a fraud; this is a rip-off of the taxpayer of unbelievable magnitude" (Atkinson 1981).

These huge agribusinesses would be the principal beneficiaries of any Canadian water that found its way into the southwestern United States.

When water is applied, desert lands under the scorching southern sun can be extremely productive. The Central Valley and Imperial Valley projects in California, and the deserts of central Arizona, produce billions of dollars worth of agricultural produce per year. But only a small part of these crops are high-value fruits and vegetables. Most irrigated land produces small grains or cotton or alfalfa, crops which can easily be grown elsewhere in North America, and which in any case are usually produced in surplus quantities.

In Arizona, alfalfa grows on 81 000 ha (200 000 acres), consuming almost 1.2 km^3 (1 million acre feet) of water. Two thirds of the alfalfa is shipped out of the state, and most of the cotton is exported to Japan. Agriculture in Arizona produces about 3% of the state's income, but it uses 88% of the water. Examination of federal irrigation projects reveals other strange anomalies. For instance, from 1956 to 1981, the federal government paid $15.5 billion to farmers to take land out of production in order to reduce surpluses. During the same period, the same government authorized $18 billion for programs designed to bring land into production through irrigation programs (Atkinson 1981).

Western irrigation programs have had other effects upon agricultural markets. Cotton was traditionally grown in the deep south of the United States, but cheap subsidized water applied to the deserts of the west produced cotton more cheaply. Many southerners were forced out of the cotton business. Elimination of this labour-intensive cash crop deprived a great many farmworkers of jobs and contributed to waves of emmigration from the deep south to the cities of the north.[9]

---

[9]Charles W. Howe. 1981. Interview for the C.B.C. television documentary "After the Flood" (Richard Bocking, Producer).

In 1981, the Comptroller General of the United States reported to Congress that it found, in examining six current federal irrigation projects, that payments by farmers for irrigation covered less than 10% of the actual cost to the federal government. The study found that water produced by federal irrigation projects cost from $54 to $130/acre foot. In the United States an average of 3.3 acre feet will be used on each acre irrigated (10 000 m³/ha) each year. The value of the crops grown was less than the real cost of the water required to grow them. But farmers continued to use the water because they were charged prices far below the cost.

The price charged a farmer is based on his ability to pay, not on the cost of the water. The Comptroller General joined a growing chorus in the United States called for a reexamination of the basis on which water is supplied by the federal government for agriculture under the old *Reclamation Act* of 1902. Dry land farmers, for instance, or farmers who irrigate by pumping from aquifers with unsubsidized electricity consider it unfair to have to compete in the market place with farmers receiving highly subsidized water from federal projects.

Why does the construction of patently uneconomic water projects proceed? The basic reason is that benefits can be clearly seen by the few who will receive them, and they will fight hard to get them. The costs are dispersed over the taxpayers of an entire nation, so that no one individual has a sense of paying for it (Hirshleifer et al. 1960).

In the United States, a Congressman who can deliver such a project will earn the appreciation of beneficiaries in his constituency, while no significant group of voters will censure him. So over the years the process of "log-rolling" has developed in the politics of American water development. A congressman will vote for a colleague's water project in return for support for the dam or other works projects he wants to deliver to his own constituents. The builders of the projects and the recipients of cheap water win. The politician wins not only votes but perhaps even his name in bronze on the finished project. "The gravy train runs in the same direction as the glory trail" (Hirshleifer et al. 1960).

There are in the United States two large federal bureaucracies dedicated to the construction of water projects. It is in the career interests of people in the Bureau of Reclamation and the Army Corps of Engineers to design and build dams and canals, and so they are very supportive of the log-rolling activities of politicians.

The structures required for hydroelectric generation or water control and diversion are extremely capital intensive. So the "cost of money" can be a vital element in deciding the fate of a project. There the government dam-building agencies have an ace to play. For in the United States and Canada, discount rates, or the rate of interest charged against capital used in public water projects, are normally lower than the interest that private sector investment would pay.

The rate for water projects in the United States is calculated by an arcane formula that is dismissed by resource enconomists as inevitably producing a low discount rate. This can create the impression in a cost benefit statement that the project is viable economically, when it is not (Carroll et al. 1979).

Many economists insist that to allocate funds properly between private and public sectors of the economy, and between various sections of the public sector, the "opportunity cost" of capital must be used to determine discount rates (Carroll et al. 1979). Vital though this may be to the sound evaluation of water projects, it appeals less to the public imagination than the bands and banners at the ribbon-cutting ceremony for a new dam. It will require imaginative political leadership to introduce realistic water and energy cost evaluation and pricing structures.

In Canada

The great hydroelectric utilities are the principal dam builders in Canada. The only Canadian counterpart to the U.S. Bureau of Reclamation is the Ministry of the Environment in Alberta, which, in spite of its name, is devoted to building dams. On a provincial scale, the Alberta situation bears a remarkable resemblance to the continental one upon which export plans are based, and ideas similar in principle have been proposed for water diversion from northern Alberta to the south. About the only difference is that there is no international boundary involved.

In southern Alberta an extensive irrigation system has made a vital contribution to the economy of the region. There are pressures to increase the quantity of water beyond that presently available by means of a new generation of water projects: a dam on the Oldman River, pumps to lift the water to areas that could most profitably use it, and ultimately a system of river dams and diversions that would reach progressively further north in order to turn northward flowing water to the south.

A plan to accomplish this called PRIME was evolved in the 1960s, but it was officially abandoned when the present Conservative government came to power. More recently, however, the now-retired provincial premier, Peter Lougheed, spoke favourably of the possibility of interbasin transfer. The Alberta Water Commission, which is presently examining the use of the province's water, is headed by Henry Kroeger, a man long dedicated to the transfer of northern waters to southern Alberta.

In Alberta, water diversion programs such as PRIME and its successors are justified by the fact that only 15% of the province's water supply is in the southern half of the province, while 80% of the water demand is to be found there. But the south has prospered with its 15%. And through improvements and additions to local storage and conveyance facilities, irrigated acreage on the southern prairies can be doubled (Lane and Sykes 1982). Combining better use of water for agriculture with more efficient domestic and industrial water use will provide plenty of room for economic growth. Fifteen percent of a very generous total provincial water budget is still a lot of water. And it is certain that agriculture, which is the only possible user of large additional amounts of water, cannot pay for diverted supplies.

Alberta Government policy is to subsidize irrigation costs at up to 75 or even 85% of cost (Mitchell 1984). The question of fairness is an obvious one — is it right to use public funds to subsidize some farmers in the production of food products to be sold in the same markets as products grown by unsubsidized farmers? D. M. Tate (1984) considered the distortions that such subsidies could create in agriculture:

> In Agriculture, massive subsidies lead to over-expansion of acreage, the growing of crops which could be best supplied more cheaply in other areas or by other means, and ultimately to 'demands' for more water, reallocation of water from perhaps higher valued users, and finally to agitation for water transfers.

This chain of events can be observed in both southern Alberta and in the southwestern United States. Tate (1984) continued

> Irrigation is a tricky and potentially explosive topic to address in this manner in some parts of the country. "Regional Development" is frequently put forth to meet this criticism. Yet, we must recognize that the end product of agriculture is food, not irrigated land, and that perhaps we should do analyses aimed at the cheapest possible production, not at massively funded irrigation schemes.

Building water projects for reasons often described as "regional development," rather than for meeting specific needs at the most economical cost, has provided North America with many uneconomic water developments. As the authoritative text *Water Supply — Economics, Technology and Policy* (Hirshleifer et al. 1960) pointed out:

> If water is scarce in an area and this constitutes a barrier to development, subsidizing the water will attract precisely the wrong kind of industries in view of the real natural advantages and disadvantages of the region...'development' is not an intangible worthy of consideration as a goal in water-supply projects over and above the economic consideration of costs and benefits.

Alternative approaches to the problem of a perceived shortage of water on the southern Canadian prairies are similar to those appropriate in much drier regions of the American southwest: more efficient use of the available water (the southern Alberta system loses vast amounts of its water through leaks in the canals); prices for water that will reward good management rather than providing water at flat rates per acre; metering and charging for the amount used; recycling of waste and runoff water; giving preference to the irrigation of high value crops where markets are available; ensuring that new subsidized water projects are not providing water for crops that can be grown elsewhere in the province under natural rainfall.

The Inquiry on Federal Water Policy has recommended that an appropriate pricing system be a condition of federal participation in water projects in Canada (Pearse et al. 1985). Perhaps this will prove to be a significant step toward more effective use of water in those regions often perceived to be facing shortages. And it may provide a damper on the use of costly structural solutions to water problems while so many economic measures leading to greater efficiency are still unused, and while many technological improvements to existing systems remain to be applied.

Current Attitudes Toward Interbasin Transfer of Water

Attitudes concerning diversion of water from one region to another historically have been distinctly different at opposite ends of the flow. Citizens of the area which would receive the water generally approved and promoted such developments. They expected to benefit by receiving a vital resource for which they would have to pay only a small part of the cost. Those in the area of origin usually feared and opposed exports to other basins, expecting such projects to offer few benefits and many threats to their region.

Within the United States, areas of origin have fought bitterly to protect their water resources from diversion to, or consumption in, other regions. Frank Quinn (1973) has described the struggles of the states of the upper Colorado River after the first world war to maintain their right to part of the flow of the river in the face of increasing demands on it by lower, drier states.

The citizens of northern California long resisted the transport of water out of their region to satisfy the voracious thirst of the burgeoning south, centred on Los Angeles. They lost their battles, and the State Water Project was built. It is a guargantuan development, featuring an aqueduct 869 km long shipping northern water south over the Tehachapi Mountains to the Los Angeles basin. It cost $2.8 billion. Now, northern Californians are fighting off demands from the south for still more water via a "periferal canal" skirting the Sacramento-San Joaquin Delta.

The states of the Pacific northwest regard the Columbia River as their own. For 20 yr, they have been able to muster enough political leverage to prevent American federal water agencies even from studying the possibility of transferring water out of the Columbia to the southwest.

In Canada, more than 90% of those who addressed the water export question in submissions to the 1985 Inquiry on Federal Water Policy opposed large-scale export of water to the United States (Inquiry on Federal Water Policy 1985). The few who supported the idea of water export were mainly proponents of particular schemes for the selling of Canadian water. Engineer Thomas Kierans used an Inquiry submission to promote his GRAND canal scheme (Fig. 2) for massive export of Canadian water (Kierans 1984), and a British Columbia firm proposed the export of small amounts of water by tanker. Both suggested that water shortages in the United States or elsewhere represent an economic opportunity for Canada.

The report of the Inquiry distinguished clearly between these two types of export. Water volumes involved in taker export are minuscule, not necessarily permanent, and environmental problems associated with them are of a scale that can be rationally assessed (Pearse et al. 1985). Large terrestial transfers of water, on the other hand, involve enormous environmental, economic, social, and political questions, for many of which no definitive answers are possible. Such factors place them in a category totally different from tanker exports.

As submissions to the Inquiry demonstrated, many Canadians take seriously projections such as those of *Newsweek* and other publications mentioned above that predict a crisis in water supply in vast regions of the southwestern United States. They fear that, as a nation perceived to have vast quantities of water flowing unused to the sea, Canada may be seen as a natural source of water for an area of the United States whose political strength is growing along with its water problems. We have seen that there is no economic or biological need for imported water supplies in the region. Let us consider, then, the politics of water export to the United States.

Political Postures, North and South

The first principal of water policy...is that rational thinking doesn't apply...Water is a political, not an economic, commodity.

— David Durenberger, United States Senator (Minn.), 1984

There is not now, nor has there ever been, any request from a United States government that Canada consider the exportation of water to that country. In june 1984, U.S. Ambassador at the time, Paul H. Robinson, emphasized this:

The United States Government believes that serious discussion of large-scale diversions is premature. No adequate demand now exists for such diversions, nor any serious interest in supplying such demand, nor any source of funding for the immense costs of such ventures...Canada has nothing to fear from the United States on this issue of regional water sharing.

The official Canadian attitude toward water export as expressed by succeeding federal governments has been that no identifiable market exists in the United States for Canadian water, and that, in any case, negotiations could not be considered because Canadian water supplies and future requirements have not been adequately assessed. Although federal policy has seemed firmly opposed to export of water until the present time, no clear statement of such a policy has even been published.

The Inquiry on Federal Water Policy (Pearse et al. 1985) commissioned by the former Liberal government reported in September of 1985 to the Progressive Conservative administration. It emphasized the fact that no market exists for large-scale water exports now or in the foreseeable future, and that costs for such projects would be far greater than any possible benefits. It outlined the heavy social and environmental costs inevitable in such projects. But it stopped short of a recommendation for or against exports, throwing the ball back at the federal government for a political decision:

The cabinet should determine whether the government is prepared to consider large scale diversions of water to the United States and, if so the criteria that must be met for approval of proposals.

It is not clear why the Commission felt unable to take a position on the issue after its in-depth study. Nor is it possible to predict what position the Conservative government will take on the issue. However, Prime Minister Brian Mulroney has appointed Simon Reisman to be Canada's Ambassador in talks aimed at free trade with the United States. Mr. Reisman has been an active member of the GRAND canal consortium which has been promoting the massive water export scheme. He proposed the export of Canadian water as a bargaining lever in establishing a free trade agreement with the United States.

121

Kierans and his backers in the huge American engineering firm Bechtel and Canadian engineering companies such as Montreal's SNC Ltd. actively lobbied in Ottawa in 1985 and 1986 to obtain financing for feasibility studies for their GRAND canal project. Although they gained support within the National Research Council of Canada and among a small number of senior civil servants, the request was vigorously denounced by water experts within the department of the environment and by concerned officials in other departments of government.[10] When the issue became a public one, the Minister of the Environment stated that such a grant would not be considered at least until the government had decided what its policy toward export would be.

But the National Research Council had in fact already given Kierans a grant of $30 000 for feasibility studies. Although this was only a small part of the $500 000 he was seeking, the principle of government contribution to it angered a great many observers in and out of government. Environmental, citizen, and native groups demanded that the government rescind the grant, terming the project an "environmental Frankenstein."[11]

At the provincial level, a brief flurry of indignation followed the invitation by Premier William Bennett of British Columbia to California businessmen in 1981 in which he suggested that, although British Columbia water is not for sale today, they should "come and see me in 20 years." Since any plans for 20 yr hence by an elected official are in the realm of fantasy, Premier Bennett's declaration was not taken seriously. But his attitude to water export was established on the public record.

Most other provinces probably share the attitude of Alberta, whose official policy states

> Alberta will not be a party to any undertaking for the possible export of water beyond Canadian borders. The priority of water use and allocation is based firstly on provincial, secondly on interprovincial, and finally on national considerations, and will not be influenced by international considerations.

Despite this policy, some Albertans fear that proposed diversions of northward flowing water to southern Alberta, even though intended for irrigation purposes within the province, could lead to sales of water across the border when it became evident that the transfers within the province were excessively costly.

The province of Quebec has opposed large-scale water export. With the election of the Liberal government of Robert Bourassa in December of 1985, however, that policy must be considered to be in doubt. Mr. Bourassa devoted a chapter of his recent book *Power from the North* to a vigorous support of the Kierans plan for damming James Bay and shipping water to the United States via the Great Lakes (Bourassa 1985).

So, at present, two provincial premiers indicate approval in principal of water export. The federal government is eagerly pursuing free trade with the United States, employing a chief negotiator who is publicly committed to the idea of trading Canadian water for access to American markets. A request, or political or economic pressure, for Canadian water from the United States probably would receive a more favourable reception in Canadian government circles than at any time in our history.

There is, however, no suggestion of interest in importations on the part of the Americans at present; in fact, the subject is almost completely absent from the water literature in the United States. As Kenneth R. Farrell (1984) of Washington's Resources for the Future has pointed out:

> Today, western water policy emphasizes efficient management of this increasingly valuable resource rather than developing additional supplies.

---

[10]*Montreal Gazette* report "James Bay Study causing rift in Ottawa over funding," January 8, 1986.

[11]*Globe and Mail* report "Water diversion plan called environmental Frankenstein," February 11, 1986.

Nor are such statements limited to professionals in the field of water management and studies. A decade ago, one could not have imagined western governors making the sort of statements that are now routine. Utah's Governor Scott M. Matheson (1984) has observed

> It seems clear that the great dam building days are over.

And he pointed out that discussing large-scale international imports that

> Given the costs of such large construction projects, and the associated impacts that would result, western U.S. water planners do not view such proposed transfers as a current viable alternative to meeting water needs in the foreseeable future.

The question of water exportation to the United States is, then, a "made in Canada" issue. It is increasingly recognized south of the border that large-scale interbasin transfer even within American borders is far more expensive than users can pay for and constitutes an undue drain on federal funds. So attention has been concentrated in recent years on strategies that will make better use of water presently available and to policies that may provide less costly alternatives than traditional dams and diversions.

This has important implications for Canadians. For whether or not Americans ask Canada for water sometime in the future depends to a considerable extent upon the evolution of water policy within that country. American economist Charles Howe of the University of Colorado said[12]

> I think that we have to realize that domestic water policy within the United States has a great deal to do with the development of political pressures of new water projects, including among those water projects the possibility of transfers from Canada.... Obviously, an increase in the efficiency of water use is always a substitute for new projects, and so, the more inefficient policies are perpetuated in the United States, the more pressure there will be for new projects; domestic projects and where appropriate international projects such as transferring water from Canada.

This was among the concerns of participants at a conference called "Futures in Water" convened by the Government of Ontario in June 1984. An international panel of speakers, including politicians, senior civil servants, and experts in water resource development, expressed deep concern about possible diversions out of the Great Lakes. The extraordinary importance of the Great Lakes to the economy and life of the region was stressed, as were the disastrous effects implicit in manipulation of lake levels.

Former Michigan Governor William Milliken (1984) pointed out that every inch by which lake levels were lowered would cost tens of millions of dollars in lost hydropower, reduced shipping, damaged recreational assets, and wildlife habitats. Diversion, he suggested, means destroying the advantages the Great Lakes region offers because of its reliable quantities of fresh water. Export of water, he insisted, means export of jobs and economic opportunity.

The governors of states bordering the Great Lakes and the Premiers of Ontario and Quebec have jointly made clear their opposition to further diversion of water out of the Great Lakes. They signed an agreement in early 1985 "to guard the Great Lakes against possible raids by Southern and Southwestern states" (*New York Times*, February 12, 1985).

The Council of Great Lakes Governors is currently seeking ways to legally strengthen their position, since they feel that the Great Lakes are an inevitable target of any major water diversion scheme that would direct water toward the American southwest. Their fear is that the choice could be taken out of their hands as power in the United States

---

[12]Charles Howe, 1981. Interview for the C.B.C. documentary "After the Flood" (Richard Bocking, Producer).

Congress shifts toward the expanding populations of the southwest at the expense of the northeast.

All of the Great Lakes except Lake Michigan, which is totally in the United States, fall under the jurisdiction of the International Joint Commission. Since the signing of the Boundary Waters Treaty in 1909, the Commission has been responsible for arbitrating water problems along the borders of the two countries. Divisions out of any of the Great Lakes except Lake Michigan would require the approval of Canada, the affected provinces, and the International Joint Commission. This international aspect of the Great Lakes protects them to a considerable extent from political power plays in the United States.

But the "wild card," as Senator David Durenberger called it, is Lake Michigan (Durenberger 1984). A diversion already exists at Chicago for sluicing the sewage of that city down the Illinois River and into the Mississippi. The flow is limited by a decision of the U.S. Supreme Court to 91 m^3/s (3200 ft^3/s). But the canal can in fact accommodate more than double that rate.

Since Lake Michigan is considered an American lake, withdrawals from it cannot be referred to the International Joint Commission. Only the states bordering that lake have legal standing in any effort to prevent increased exports from the Great Lakes. But although American states have traditionally exercised control over the movement of water across their borders, recent decisions of the U.S. Supreme Court may have weakened the authority of individual states (Micklin 1983).

It now seems possible that powerful political clout from another region might be able to force the Great Lakes states to give up some of their water via increased flows through the Chicago Canal out of Lake Michigan. However, bills are pending in Congress that would prohibit the diversion of water from the Great Lakes without the express consent of each of the Great Lakes states and the International Joint Commission (Durenberger 1984).

The political implications of the export of Canadian water to the United States need little elaboration. The canals and associated works carrying Canadian water south would have to be operated largely according to the requirements of the receiving region. An American dependency on Canadian water would in effect designate this country's water resources as "continental" rather than "Canadian." The arrangement would have to be a permanent one, as many American politicians have pointed out.

Normal economic arrangements between nations can be changed or terminated, but the physical integration of the most important resource clearly cannot. As someone has said about water export, perhaps facetiously, "you can turn it on with a treaty, but if you want to turn it off, you'll have to deal with the marines."

The United States is a good neighbour with which Canadians feel much kinship. But engineering projects that would send more water flowing across the border would increase the already enormous American economic "stake" in Canada and could bind Canada even more tightly to the destiny and decisions of its southern neighbour. That destiny may be great, and the decisions wise, but they would not be Canadian.

The Cost of Interbasin Transfer

Since 80% of American water is used for irrigated agriculture, the only possible reason for large interbasin transfers of water from Canada to the United States would be to supply this industry. An important consideration in any analysis of such projects therefore must be the capacity of agriculture to pay for imported water. Kneese (1984) has observed that the maximum willingness to pay for water on the High Plains, for instance, would be $75/acre foot, a figure considerably higher than that paid by most water users today. Any cost higher than that for transfer of Canadian water to the High Plains means subsidization by government.

124

The NAWAPA scheme (Fig. 1), first proposed in 1964, was estimated by its authors in 1975 to have a price tag of U.S.$120 billion; American federal water agencies estimated that a canal from the Mississippi River to West Texas and New Mexico to supplement the falling water levels of the Ogallala Aquifer would cost U.S.$53.5 billion (1985 dollars); more recent proposals would move smaller amounts of water from the Missouri or Arkansas River to the High Plains area at costs ranging from U.S.$36 billion (1977 dollars) to $20 billion (Micklin 1983). According to the High Plains study, these figures mean costs of about $1000/acre foot to supply the farmer with water for which he can afford to pay no more than $75/acre foot (Kneese 1984). And that is just the beginning of the expense.

Depriving the Mississippi, Missouri, and Arkansas rivers of normal flows would have very serious economic and environmental repercussions. To replace that water by means of a diversion from Lake Superior to the Missouri would cost U.S.$20 billion (1982 dollars) (Micklin 1983). The states and provinces of the Great Lakes have made very clear the enormous cost to them of any reduction in normal water levels and flows in the Great Lakes. And so the Grand canal plan would envisage pumping water south from James Bay to replace that exported Great Lakes water. The GRAND scheme's promoter estimated in 1984 that his project would cost $100 billion, half of which would be needed to deliver water from James Bay to Lake Huron (Kierans 1984).

Where would all the money come from? Deficit reduction and reduced spending programs are priorities in public finance in both Canada and the United States these days. Money would have to be taken from existing social and other public programs, or borrowed. Studies have shown that the perceived lower cost of capital to public corporations has been extremely costly to the Canadian economy in the case of large hydroelectric and nuclear power projects. York University economist John Evans estimated these costs at 1% of the Gross National Products.[13] The much larger capital requirements of water export projects such as the GRAND scheme would severely limit the availability of capital for genuinely productive uses in Canada and would increase its cost to Canadian businessmen and consumers.

An added problem is that, after they are built, all these schemes must push water uphill. Pumping so much water will entail vast quantities of electricity. T. W. Kierans (1984) estimated that his GRAND plan would require 10 000 MW of energy, or an amount equivalent to the entire production of the huge James Bay power project. Pushing water from Lake Superior to the Missouri would consume 33.5 MW-h/yr, and forcing water uphill to the high plains of Texas and adjoining states would cost over U.S.$1 billion/yr, just for electricity (Micklin 1983).

There are in addition the contributions that are normally made to water projects without charge by governments in Canada. The rivers and the water within them are provided at no cost; unless privately owned, the land that is flooded by reservoirs costs the project nothing; the forest, mineral, and recreational development possibilities forgone are not normally added as an opportunity cost of the project; there is no charge for the wildlife habitat eliminated.

Canadian governments are only beginning to recognize the right to adequate compensation of river basin residents, usually natives, whose lives are affected, often drastically, by massive water development (see also chapter by Rosenberg et al., this volume). These payments can be of considerable size — in the hundreds of millions of dollars in each of the recent Manitoba and Quebec hydro developments, for instance. But they are not usually considered a charge to the water development project. These costs are paid by general taxation rather than included as part of the cost of electricity, thereby keeping the price charged for the power artificially flow. Although Canadians

---

[13]John Evans. 1984. Interview for the C.B.C. television documentary "Electricity, the cost of too much power" (Toronto, May 18).

pay such costs from another envelope, American purchasers of Canadian electricity or water are excused from paying this part of the cost of the product.

In the case of large-scale water export projects, such costs would be many times greater than those incurred to date in hydroelectric developments. Many of the major structures would be in the north where southern Canadians have traditionally carried out projects involving massive environmental disruptions with relative equanimity. But engineering works and the accompanying economic, social, and environmental impact would occur also in populated and developed regions of both Canada and the United States. Who would pay the vast cost of human dislocation and property damage? Would it be reflected in the price charged Americans for Canadian water, or would it be paid by the Canadian taxpayer? The precedent noted above in the case of hydroelectric development is instructive.

It is useful to remember, too, that initial cost estimates for water projects normally bear little resemblance to final figures. For example, the Dickson Dam in Alberta was projected in 1975 to cost $42.5 million; by 1981 the estimate was $132 million.[14] In the United States, the story is the same; final costs of 103 federal water projects were calculated by the Bureau of Reclamation to be 277% of original estimates (Atkinson 1981).

When such costs are added to the construction bill in order to arrive at a price for water delivered to the American southwest, it becomes even clearer that such water would cost many times more than locally available alternatives. The cost of moving water from Canada to Texas would be far greater than the value of the agricultural products it would help to produce.

Since there is no economic reason whatever for the movement of water these vast distances, public money which could never be repaid would be required to build such projects. Justification for this in the name of job creation can be expected, yet temporary employment created by large water projects is many times more costly than permanent jobs created in almost any other sector of society.

It is clear that water export projects could only be undertaken as political manipulations or "boondoggles' of such proportions as to overshadow even those which are already considered in the United States to be scandalous. Just the feasibility and engineering studies are projected at over $100 million in the case of the GRAND plan (Harrison 1985). So there is much money to be made by a fortunate few, even if the schemes never get past the planning stage. We can, therefore, expect their promotional efforts for such schemes to continue.

If the users of the water cannot possibly pay for it, who will? Canadians who suggest that Americans would pay a great deal of money for Canadian water must be counting on the American government to pay a substantial sum for the water itself, in addition to the enormous amounts involved in building the transmission works and compensating for the economic, social, and environmental damage they would cause in Canada and the United States.

Within the United States, the principle has always been observed that water from one basin is never paid for by the receiving basin or the federal government. That Canadian water should be paid for would probably be admitted but as Bruce Mitchell (1984) has demonstrated, the value of water is extremely difficult to establish. Its price would have to be worked out through negotiation. As a figure to add to the astronomical costs of major diversion structures, the price would obviously be kept as low as the weight of the American government in its negotiations with Canada could accomplish.

---

[14]Dixon A. R. Thompson. 1981. Allocation policies for water use in Alberta. Paper presented to the Alberta Society of Professional Biologists Symposium on Biological and Social Issues in Water Management, Calgary, Alta.

Since the export of water is so clearly unrelated to economic reality, negotiations of any such arrangements would be purely a political matter. "Linkage" of the transaction with other questions at issue between the two countries would be inevitable. Canada's chief trade negotiator with the United States, Simon Reisman, has in fact suggested that Canada's water might be used as a bargaining lever to gain freer access to American markets (Harrison 1985). And as American economist Alan Kneese noted, "It appears that negotiations about international rivers frequently involve considerations in quite other arenas, and these often are dominant. In other words, it is impossible to understand the outcome of such international negotiations simply by looking at its apparent focus."

In such an "elephant and mouse" negotiation, it is difficult to imagine a scenario in which Canadians could expect to be well paid for their water. Canada's negotiating position would be further weakened by benefits claimed for Canada by promoters of such schemes as the GRAND plan. Thomas Kierans (1984) claimed such advantages for Canada as hydroelectric generation, irrigation within Canada, regulation of the Great Lakes, and "flushing" or dilution of Great Lakes pollution. Illusory though these Canadian "benefits" turn out to be when examined, American negotiators would assuredly use them in pressing for greater Canadian investment and lower water prices.

The real issues in any export discussion would be further submerged under institutional complications of an extraordinary scale. These are far from negligible in purely domestic water management questions, as Canadian water experts agreed at a Banff workshop in 1983 concerning "Institutional Arrangements for Water Management in the MacKenzie River Basin" (Sadler 1984).

Diversions within Alberta, for instance, would begin in the Saskatchewan River basins but would cascade north to find replacement water from the MacKenzie system which carries 85% of the water leaving the province. Any development implicating that enormous basin involves three provinces, two territories, the federal government, and native rights. An international project such as water export adds another, stronger federal government, those states which would be involved, an array of interest groups both pro and con, and deep concerns of national pride and control on both sides of the border.

It is apparent that the institutional and political complications of water export negotiations would be such as to overwhelm any rational consideration of environmental impact, social disruption, and economic cost or benefit.

A Leap in the Dark

Institutional complications pale into insignificance when compared with our vast ignorance of the environmental and social consequences of manipulating complex systems like those of the MacKenzie River or the Great Lakes. Biologists have always known, however, that where land meets water, life is richest. Man has always preferred to live at the margin, too, for both practical and aesthetic reasons. Unfortunately, it is precisely at this point that the greatest damage from large water projects occurs.

Since World War II, Canadians have been pushing ahead with huge water developments, principally for hydroelectricity, with little understanding of their effects on the natural systems involved. And there has been little study after the fact designed to learn more about the impact of what we have done (see chapter by Rosenberg et al., this volume). Peter Larkin (1984) pointed out

> In British Columbia alone, there are several major reservoirs created by dams built in the last three decades that have yet to be studied limnologically.... There are literally dozens of lesser reservoirs and hundreds of small reservoirs for which there may have been a prediction of what might happen but virtually no records of what did happen.... Comprehensive, after-the-event assessments of major projects are rare.

But in the case of the diversion of the Churchill River into the Nelson in northern Manitoba, scientists of the federal Freshwater Institute in Winnipeg studied through more than a decade the changes brought to the region by the diversion. Their work contributed greatly to our limited knowledge of the impact of such developments on northern water systems (Newbury et al. 1984). In another chapter of this volume, Rosenberg et al. describe some of the impacts inherent in the manipulation of northern Canadian water systems.

This work, and the limited research that has been carried out into the impact of other hydroprojects on northern Canadian water systems, requires that we be deeply concerned about the consequences to be expected if we build the great structures essential to any large water export project (Peace-Athabasca Delta Project Group 1972; Thompson 1984). Much more study is needed, but in such complex systems, no amount of research could adequately predict the range and depth of impact. "Surprise" would dominate in the wake of large-scale interbasin transfers. Larkin (1984) noted

> ...it has to be acknowledged that natural systems being what they are, it is simply not possible to predict their behaviour except in fairly general terms...long term prior studies will not necessarily give insights into many long-term effects of projects.

Removing water from the Great Lakes system in any sort of export scheme would certainly be a leap into the unknown. It is clear, however, that small drops in water levels would profoundly affect the shallow areas of the Great Lakes that are by far the most productive areas for fish. The shallow wetlands on the margins of the lakes are essential habitat for a great many species of birds, animals, and fish, who use them for the rearing of young, for feeding, and for protection from enemies and the elements (Andrews 1984). The wetlands are important staging areas for migratory wildfowl. If numbers are needed to indicate the importance of all this economically, the Ontario Minister of Tourism and Recreation, T. H. Gibson (1984), estimated that

> ...wetlands related activities produced more than 50 million user-days of fishing, hunting, camping and bird watching. That contributed more than 800 million dollars to our economy.

No matter how vast the water system, it is the first removal, water taken "off the top," that will have the greatest perturbing effect upon it. A tiny drop in the MacKenzie River will reduce the flow over the shifting sandbars enough to end the important river navigation that supplies settlements and industry. And it is the height of the spring flood that fills the "hanging lakes" of the MacKenzie Delta that is essential in the life cycle of northern fish populations.

Similarly, a drop of 2.5 cm in the Great Lakes-St. Lawrence system will reduce the amount of cargo shipped through the Great Lakes by 1 million t/yr. A 15-cm drop in lake levels would cost Ontario Hydro alone $20 million in reduced power production. Quebec is very concerned about the impact of reduced river levels on its hydroelectric production, on water quality in the province, and on the quality of life for millions of Quebecers who will live along the mighty river.

If the first removals are by far the most deleterious ones, those water transfer proposals that speak in terms of withdrawing only small fractions of Canada's water supply, or of depriving a given system of only a limited percentage of its flow, are ignoring the most important aspects of a river's function as a living system and as a commercial and social asset in its present condition.

Structural solutions to the problems created by lower river or lake levels are regularly put forward. For instance, among the principle benefits claimed in support of Kieran's GRAND plan is the regulation of water levels in the Great Lakes. These have always been a source of concern, particularly when higher than average, as in the 1950s and 1980s, or when lower than usual, as in the 1960s. Some regulation of Lake Superior and Lake Ontario is possible, but studies over many years by the International Joint

Commission have shown that major efforts to control lake levels would cost more than any benefits that could be expected (Carroll 1983).

The role of nature in varying the levels of the Great Lakes is so much greater than that which any man-made control could impose, that such efforts would be counter-productive. Changes made to the lake levels can be felt in the system for 15 years, so water added to the system at a time of low water would still be affecting levels years later when normal or high levels had returned. Damage would inevitably accompany such artificially high water.

B. C. DeCooke (1984) of the U.S. Army Corps of Engineers concluded that since climatic conditions are not predictable for any extended period, "continual interjection of water (into the lakes) to offset a withdrawal (such as would occur in diversion schemes to the south) may result in damage to users of the Great Lakes system." Since trying to control lake levels would do more harm than good, according to responsible studies, the International Joint Commission considered shoreline planning to minimize damage from fluctuating levels to be a more useful approach (Carroll 1983).

The GRAND plan also predicts improvement of water quality in the Great Lakes through the sluicing effect of greater quantities of water in the system. Since only 1% of the volume of the water in the lake flows through the system in an average year, such "flushing" would be nonexistent. In any case, modern pollution control depends upon reduction of the offending elements, not in the treatment of lakes as if they were gigantic toilet bowls.

An image begins to form of some of the consequences to be expected of large-scale diversion projects like NAWAPA or the GRAND plan. The Harricana River flows to James Bay though Quebec and finally Ontario. The GRAND plan requires that we imagine a series of dams along the river creating a string of head-to-toe reservoirs. These would be linked by giant pumps pushing the water uphill from reservoir to reservoir until it could flow down the southern side of the divide. As we have noted, these pumps would consume as much electricity as is produced by Quebec's James Bay Project.

Eventually the water would reach the Ottawa River, Lake Nippissing, and Lake Huron via the French River. Since T. W. Kierans described the output of his system as fluc-tuating from zero to 2 million L/s, Lake Nippissing would presumably be the necessary control reservoir, rising and falling according to the desired flow. The historic French River would become a sluiceway, flooding beyond its banks as it carried vastly increased amounts of water to Lake Huron. Serious erosion of river banks and weakening of bridge, dam, and building supports along the French, Ottawa, and the other rivers involved are among the more obvious impacts.

Rivers are often said to "waste to the sea" if they are not devoted to direct human use. In supporting the NAWAPA scheme, U.S. Senator Frank Moss (1967) stated that "vast amounts of water are pouring unused into northern seas and are irretrievably lost." In promoting his GRAND plan, T. W. Kierans (1984) spoke of the water of the rivers flowing into James Bay as "totally lost to the salt water of Hudson Bay." North American water supplies are often described as being "in the wrong place" when what is meant is that a particular human activity in a particular place might profit if greater quantities of water were readily and cheaply available to it.

For most forms of life, including most people, the water making up the natural envi-ronment is in precisely the *right* place, since species, populations, habitat, and land forms have evolved over the millenia in concert with the rivers and streams, their flows, and their floods.

J. P. Bruce (1984) explained

Diversions are often characterized as making use of "wasted waters". From the perspec-tive of the ecosystems of northern regions, none of the water flowing to the northern seas is wasted. Every drop is used to sustain the life forms in these northern systems.

In addition, the fresh water flows affect the carefully balanced energy exchanges in this region which are of great importance in shaping the climate of the whole hemisphere.

Oceanographers are finding that the natural flood and flow of rivers to the sea is of much greater influence on marine productivity than had heretofore been thought. R. H. Loucks (1984) and K. F. Drinkwater (1984) reported impacts of spring freshets on ocean currents, temperatures, salinity, and fish production thousands of likometres from the source of the fresh water.

The rivers of James and Hudson bays carry more water to the sea than the St. Lawrence pours into the Gulf of St. Lawrence. The impact of the peak discharge of those rivers on salinity, nutrient production, and fish population is felt all the way down the coast of Labrador to the Grand Banks of Newfoundland.

As far back as 1973, W. H. Sutcliffe, Jr. speculated that changes in the seasonal flow of the St. Lawrence and its tributaries could affect the abundance of some species of fish and shellfish. Subsequent work has confirmed the importance of spring flooding in the St. Lawrence on the fishery of the Gulf of St. Lawrence, as well as along the coast of Nova Scotia and in the Gulf of Maine. Drinkwater (1984) concluded.

> ...when considering Canada's freshwater resources and their possible modifications, two points in particular must be kept clearly in focus: firstly, freshwater entering the sea alters the marine environment both in the immediate vicinity and subsequently over a period of months at "downstream" distances on a scale of 1000-2000 km; secondly, the striking correlations between river discharge and subsequent commercial fish catch on Canada's major east coast fishing banks.

It appears that the timing and the amount of spring runoff from rivers into oceans affects not only fish populations, but also coastal currents and related large-scale atmospheric and oceanic patterns. For example, a high spring freshet in the St. Lawrence will be followed by a warmer subsequent winter in the Gulf of Maine region. J. P. Bruce (1984) suggested another sort of climatic effect which could be triggered by large-scale diversions of northward-flowing rivers:

> Decreased freshwater inflows to Hudson Bay and the Arctic Ocean would result in increases in the salinity of arctic waters. This lowers the temperature at which ice forms and would add to the already enormous Arctic warming conditions predicted due to the 'greenhouse' effect.

Canadians are not alone in their capacity to influence the biological and climatic future of the polar regions. Our neighbours on the other side of the Arctic Ocean, the Russians, have recently decided to proceed with the first stage of an enormous diversion scheme that would divert northward flowing rivers south. If they follow through with subsequent stages, very important reductions will occur in the flow of fresh water to the Arctic. Loucks (1984) concluded

> We must recognize that a widespread and complex role is played by run-off water. Although the evidence is fragmentary and the large-scale patterns are only partially revealed to us, it would seem risky and arrogant to set out to bring such vast, self-organizing systems under "management". The connectedness of nature is more pervasive than has previously been believed, and therefore the context for water policy must be more comprehensive than has been the case. It follows that proposals to increase the flow of water diverted from Lake Michigan and the St. Lawrence to the Mississippi, to dam James Bay and ship this run-off south for export, and to dam more rivers for hydro-electric generation, should come under much closer scrutiny in future. There may be a large price to pay for further altering or eliminating river freshets into the sea; the public should be apprised of the risks.

As we learn more about the consequences of large-scale manipulation of water resources, it becomes evident that the unexpected and the unpredictable can never be eliminated. Even if we continue and expand the essential research into the nature of our water resources, surprise will always be a dominating factor when dams are built and in particular when rivers are diverted, reversed, flooded, or deprived of natural flows. There are also uncertainties external to the resource itself. For instance, as discussed in another chapter (Ripley, this volume), changes in climate are considered likely in the near future due to rising carbon dioxide levels in the atmosphere. The implications for water management are very great. Bruce (1984) suggested that drier conditions will mean less runoff into the Great Lakes and greater evaporation,combining to decrease the flow of the Great Lakes by 21%. The impact on shorelines and shipping would be important, and decreased generation in existing power stations could amount to U.S. $0.75 billion (1984 dollars).

Drier and warmer conditions may increase the demand for irrigation water in southern Ontario, northern Michigan, and New York. Obviously, massive water management schemes undertaken under one set of circumstances may be totally inappropriate when other conditions prevail. Bruce (1984) concluded that

> ...we should be extremely cautious about considering any further diversions of water in or out of the Great Lakes System. Attempts to restore lake levels by major diversions of northward flowing rivers may well have the effect of further increasing the climatic warming and could have profound effects on polar ecosystems.

Since uncertainty is a dominating factor in any examination of future water supply and use, it must be the central consideration in water planning. This requires flexible policies that enable us to adapt and adjust as required by changing conditions; it implies the retention of a wide range of options and alternatives so that as much resilience as possible is retained in the system. It necessitates above all the avoidance of massive structural commitments that are not absolutely essential. By causing extensive and irreversible changes in natural systems and in human society, and by committing future generations to pay their very high costs, such enterprises can lock a society into a future that may not be in accordance with its needs and preferences.

Any water export scheme would of necessity impose long-term institutional and structural rigidities upon both Canada and the United States. The flexibility that is the first requirement of modern water policy would be lost. For this reason alone, Canada should have no difficulty in refusing to consider large-scale water export proposals, whether they originate in Canada or the United States.

It is sometimes recommended that Canada assess its water resources to determine amounts that might be available for export. The establishment of a body that would consider export demands has also been suggested. Adoption of either course would signal that in principle Canada is prepared to consider the export of water. This would provide support to those Americans who would use the prospect of a future bail-out by Canada to excuse profligate water use today. And it would encourage both Canadians and Americans to continue lobbying for the vast engineering proposals that, through publicly financed studies, can be profitable to their promoters even if they are never built.

Some Americans have expressed the view that the most neighbourly thing Canada could do is indicate a firm "no export" policy now, a gesture that would support the present thrust in American water policy that emphasizes measures that will enable the United States to live within the limites of its own resources (Bocking 1972).

The Perils of Predictions

The International Joint Commission in 1981 estimated that consumption of Great Lakes water would increase over the next 50 years by 2.7% in the United States and 3.3% in Canada, due in considerable measure to increased nuclear and thermal power production. Such a projection means water consumption would increase seven times by the year 2035. Lake levels could drop by 13 in., and the flow of the St. Lawrence could be reduced by 12%. As discussed earlier, such a drop in water level would be catastrophic for all the states and provinces of the Great Lakes and the St. Lawrence River.

Projections such as those of the International Joint Commission can be useful. In describing an unacceptable future, they may prompt policies and practices designed to ensure that the projections are not realized. In the Great Lakes case, this could lead to increased efficiency of water use.

Such projections can also be dangerous if viewed as inevitable and if used as arguments for costly and damaging water development schemes. For the simple fact is that such projections are almost always wrong. They belong to that class of predictions that has been responsible for some of the largest-scale errors in decision-making in recent decades: air travel projections that led to Mirabel Airport and bankrupt airlines for instance; or energy projections that provided the world with many billions of dollars worth of un-needed hydro and nuclear plants; or market projections that blighted the British Columbia economy with its billion dollar northeast coal project. We are finding that our rate of success in predicting markets, or prices, or technological or social change, or environmental impact, is close to zero. We need only consider the record in prediction of oil supply, demand, and prices during the past dozen years.

The International Joint Commission projection was badly out of line with reality only 3 years into its 50-year forecast. As former chairman of the Canadian section of the International Joint Commission, J. B. Seaborn, stated in 1984, "the Commission now tends to question the feasibility of projections so far ahead as 2035, given the rapid variability of so many of the assumptions about economic growth and technological change."

On a global scale, "The World Environment" study for UNEP, coauthored by Gilbert White (1983), showed

> ...how irrelevant any broad generalizations about water supply and use over large areas may be to understanding what, in fact, is happening to water resources as they affect human beings.

There is a growing understanding that neither human affairs nor natural systems move smoothly in a single direction, and that discontinuity is the norm in the behaviour of systems (Holling 1978). As examples proliferate in many disciplines, a "science of surprise" is developing that is particularly useful in the consideration of water development generally and large-scale water transfer in particular.

From promotion to planning to construction, large-scale water projects can easily take 20-30 years or more. Even if rationally conceived, in an era of rapid social, economic, and technological change they risk being totally inappropriate long before they approach completion. With present export proposals such as the NAWAPA or GRAND schemes, we start with no conceivable market, unquantifiable but very great environmental risk, inevitable large-scale social disruption, and the opposition of the vast majority of those Canadians who have expressed views on the subject. The possibility of a favourable outcome to such an enterprise reaches the vanishing point.

There are massive gaps in our understanding of the complexities of aquatic ecosystems. We have learned how difficult it is to predict the impact upon them of human endeavours. But as Andrew Hamilton (1985) of the International Joint Commission has pointed out, the long-term conservation, protection, and wise use of water is fundamental if we are to have a sustainable and secure society. This requires excellence in research into relationships between land, air, water, and living resources, including man. And yet, said Hamilton, the quality and long-term relevance of ecosystem science in Canada is declining.

Focussed research aimed at deepening our understanding of our freshwater resources and careful analysis of the impact of past water developments are among the pressing needs in water research. Exploration of the wide range of social, economic, and technological alternatives to massive water developments is essential. Then perhaps, when faced with real problems, we will be able to choose wisely, allowing for the ignorance and uncertainty that will always be dominant.

If the best alternative involves manipulation of streams and rivers, perhaps with such a basis of understanding we will be capable of doing that with sensitivity and elegance. For the experience of recent decades demonstrates that conventional large-scale structural solutions to water problems, whether domestic or international, comprise a simplistic, expensive, and outdated approach to water resource management.

References

ANDREWS, W.A. 1984. Some possible environmental effects of large-scale water transfer schemes, p. 210-217. *In* Futures in water. Preceedings of the Ontario Water Resources Conference. Sponsored by the Government of Ontario, Toronto, Ont.

ATKINSON, R. 1981. National water policy marked by waste, confusion. Kansas City Times, May 1981.

BOCKING, R. C. 1972. Canada's water: for sale? James Lewis and Samuel, Toronto, Ont. 188 p.

BOULDING, K. E. 1978. Ecodynamics — a new theory of societal evolution. Sage Publications, Beverly Hills, CA. 384 p.

BOURASSA, R. 1985. Power from the North. Prentice-Hall Canada, Inc., Scarborough, Ont. 181 p.

BRUCE, J. P. 1984. The climate connection, p. 50-55. *In* Futures in water. Proceedings of the Ontario Water Resources Conference. Sponsored by the Government of Ontario, Toronto, Ont.

CARROLL, J. E. 1983. Environmental diplomacy: an examination and a prospective of Canadian-U.S. environmental relations. The University of Michigan Press, Ann Arbor, MI. 402 p.

CARROLL, J., C. HOWE, AND J. DUFFIELD. 1979. Benefits or costs 11. An analysis of the Water Resources Council's Manual of Procedures for Evaluation of Benefits and Costs. National Wildlife Federation, Washington, DC.

COMPTROLLER GENERAL OF THE UNITED STATES. 1981. Federal charges for irrigation projects reviewed do not cover costs. Report to the Congress, March 13, 1981, Washington, DC.

DeCOOKE, B. G. 1984. Impacts of the Great Lakes water supply modification, p. 104-123. *In* Futures in water. Proceedings of the Ontario Water Resources Conference. Sponsored by the Government of Ontario, Toronto, Ont.

DRINKWATER, K. F. 1984. Effects of river runoff on the marine environment of eastern Canada. A submission to the Inquiry on Federal Water Policy, 1984, (Available from Environment Canada, Ottawa, Ont.)

DURENBERGER, D. 1984. Water policy in the 1980s, p. 172-197. *In* Futures in water. Proceedings of the Ontario Water Resources Conference. Sponsored by the Government of Ontario, Toronto, Ont.

FARRELL, K. R. 1984. Twenty-first century agriculture — Critical choices for natural resources. Resources No. 75 (Winter 1984): 14-16.

FREDERICK, K. D. 1981. Costs up on western water resources. Resources No. 68 (October 1981): 8-9.

1982. Water for western agriculture. Research Paper Series, Resources for the Future, Inc., Washington, DC. 241 p.

GIBSON, T. H. 1984. Water — The magnet for tourism and outdoor recreation, p. 180-187. *In* Futures in water. Proceedings of the Ontario Water Resources Conference. Sponsored by the Government of Ontario, Toronto, Ont.

GYSI, M. 1980. A guide to conservational pricing: three for 50 cents or five for a dollar. Vantage Press, Inc., New York, NY. 114 p.

HAMILTON, A. L. 1985. Managing the uses and abuses of freshwater ecosystems in Canada: challenges and opportunities. Environmental Colloquium sponsored by the Economic Council of Canada, Toronto, Ontario. (Available from the International Joint Commission, Ottawa, Ont.)

HARRISON, F. 1985. Should we add water to our trade mix? The Financial Post, Toronto, May 25, 1985.

HIGH PLAINS ASSOCIATES. 1982. Six-state High Plains — Ogallala Aquifer Regional Resources Study. A report to the U.S. Department of Commerce and the High Plains Council, Austin, TX.

HIRSHLEIFER, J., J. C. DE HAVEN, AND J. W. MILLIMAN. 1960. Water supply — economics, technology and policy. The University of Chicago Press, Chicago, IL. 386 p.

HOLLING, C. S. [ED.]. 1978. Adaptive environmental assessment and management. International Institute for Applied Systems Analysis. John Wiley & Sons, Toronto, Ont. 377 p.

INQUIRY ON FEDERAL WATER POLICY. 1985. Hearing about water — A synthesis of public hearings of the Inquiry on Federal Water Policy, Environment Canada, Ottawa, Ont.

KENNEDY, D. N. 1985. California needs more wisdom, not more water. Focus on Water No. 5. Metropolitan Water District of southern California.

KIERANS, T. W. 1984. The Great Recycling and Northern Development (GRAND) Canal. Notes for a submission to the Inquiry on Federal Water Policy in Canada, St. John's, Newfoundland, September 24, 1984. (Available from Environment Canada, Ottawa, Ont.)

KNEESE, A. 1981. How to come to grips with our water problems. An interview in U.S. News and World Report, June 29, 1981.

_____ 1984. High plains, low water. Resources No. 78 (Fall 1984): 7-9.

LANE, R. K., AND G. N. SYKES. 1982. Nature's lifeline: Prairie and northern waters. A technical report sponsored by Canada West Foundation with support from Devonian Group of Charitable Foundations. Canada West Foundation, 245 West Palliser Square, Calgary, Alta.

LARKIN, P. A. 1984. A commentary on environmental impact assessment for large projects affecting lakes and streams. Can J. Fish. Aquat. Sci. 41: 1121-1127.

LOUCKS, R. H. 1984. A glimpse of the role of river runoff in the sea. A submission to the Inquiry on Federal Water Policy on behalf of the Scientific Committee of the Canadian Meteorological and Oceanographic Society, September 1984. (Available from Environment Canada, Ottawa, Ont.)

MARTIN, W. E., H. INGRAM, AND N. K. LANEY. 1981. A willingness to play: analysis of water resources development. Arizona Agricultural Experiment Station Paper No. 381. The University of Arizona, Tucson, AZ.

MATHESON, S. M. 1984. The sharing of water on a continental basis, p. 18-27. In Futures in water. Proceedings of the Ontario Water Resources Conference. Sponsored by the Government of Ontario, Toronto, Ont.

MICKLIN, P. P. 1983. Interbasin water transfers in the United States. Prepared for the Working Group on the Environmental Effects of Large-Scale Water Projects, United Nations Environmental Programme, Nairobi, Kenya.

MILLIKEN, W. 1984. Impacts of the Great Lakes on the regional economy, p. 12-17. In Futures in water. Proceedings of the Ontario Water Resources Conference. Sponsored by the Government of Ontario, Toronto, Ont.

MITCHELL, B. 1984. The value of water as a commodity, p. 90-103. In Futures in water. Proceedings of the Ontario Water Resources Conference. Sponsored by the Government of Ontario, Toronto, Ont.

MOSS, F. E. 1967. The water crisis. Praeger Publishers, New York. NY. 305 p.

NATIONAL ACADEMY OF SCIENCES-NATIONAL RESEARCH COUNCIL. 1966. Alternatives in water management. Publication 1408. Washington, DC.

NEWBURY, R. W., G. K. MCCULLOUGH, AND R. E. HECKY. 1984. The Southern Indian Lake impoundment and Churchill River diversion. Can. J. Fish. Aquat. Sci. 41: 548-557.

PEACE-ATHABASCA DELTA PROJECT GROUP. 1972. The Peace-Athabasca Delta — A Canadian resource. Summary report.

PEARSE, P. H., F. BERTRAND, AND J. W. MacLAREN. 1985. Currents of change. Final Report, Inquiry on Federal Water Policy, Environment Canada, Ottawa, Ont. 222 p.

QUINN, F. J. 1973. Area-of-origin protectionism in western waters. Social Science Series No. 6. Inland Waters Directorate, Water Planning and Management Branch, Environment Canada, Ottawa, Ont.

ROBINSON, P. H. 1984. Banquet address, p. 158-164. In Futures in water. Proceedings of the Ontario Water Resources Conference. Sponsored by the Government of Ontario, Toronto, Ontario.

ROGERS, P. 1983. The future of water. The Atlantic Monthly, July 1983, p. 91.

SADLER, B. [ED.]. 1984. Institutional arrangements for water management in the MacKenzie River Basin. The Banff Centre for Continuing Education, Banff, Alta.

SEABORN, J. B. 1984. The International Joint Commission and the water resources of the Great Lakes, p. 226-233. *In* Futures in water. Proceedings of the Ontario Water Resources Conference. Sponsored by the Government of Ontario, Toronto, Ont.

SEWELL, W. R. D. 1985. Inter-basin water diversions: Canadian experiences and perspectives. *In* G. N. Golubev and A. K. Biswas [ed.]. Large-scale water transfers: emerging environmental and social experiences. Tycooly Publishing Ltd, Oxford, England.

SUPALLA, R. J., F. C. LAMPHEAR, AND G. D. SCHAIBLE. 1984. Economic implications of diminishing groundwater supplies in the High Plains Regions. University of Nebraska, Lincoln, NB.

SZECHENYI, C. 1981. Thirsty plains rapidly drain Ogallala Aquifer. *In* The next American crisis. Kansas City Times, May 1981.

TATE, D. M. 1984. Water management in Canada: a one-armed giant. An unpublished paper. (Available from Environment Canada, Inland Waters Directorate, Ottawa, Ont.)

THOMPSON, D. 1984. Special features of the MacKenzie Basin, p. 17-36. *In* B. Sadler [ed.] Institutional arrangements for water management in the MacKenzie River Basin. The Banff Centre for Continuing Education, Banff, Alta.

WAHL, R. W., AND R. K. DAVIS. 1984. Satisfying Southern California's thirst for water: efficient alternatives. Office of Policy Analysis, U.S. Department of the Interior, Washington, DC.

WHITE, G. F. 1983. Water resource adequacy: illusion and reality. Natural Resources Forum Vol. 7, No. 1. United Nations, New York, NY.

WHITE, S. E., AND D. E. KROMM. 1984. Interstate groundwater management preference differences: the Ogallala Region. Presented at the 7th Annual Applied Geography Conference, Florida State University, Tallahassee, FL.

CHAPTER 6

Climatic Change and the Hydrological Regime

Earle A. Ripley

Department of Crop Science and Plant Ecology,
University of Saskatchewan, Saskatoon, Sask. S7N 0W0

Climates of the Past

There is ample evidence that there have been much warmer, and much cooler, periods in the earth's past and that periodic droughts and floods are features of even climatically "stable" periods. A series of climate-related disasters during the early 1970s helped to focus attention on the possibility that climate may undergo changes rapid enough to affect us within a human life span. This raised the spectre of climatic change operating on an already highly stressed world food system with, perhaps, devastating effects.

The climatological temperature record (Fig. 1) shows variations on time scales ranging from a few years to many thousands of years, with amplitudes of several degrees for the globe as a whole. Since the last major glaciation, which ended about 15 000 yr ago, average global temperature has varied less than 3°C. A warm period about 5000 yr B.P., called the Hypsithermal, was followed by gradually declining temperatures until the Little Ice Age of the sixteenth to nineteenth centuries. Since the beginning of the present century, the mean temperature of the Northern Hemisphere rose over 0.5°C until about 1940 and since then has declined about half that amount (Fig. 2a). According to Idso (1983), the recent decline was more accentuated in northern latitudes than in the tropics and not evident at all in the Southern Hemisphere. However, an analysis of sea-surface and marine-air temperatures over the entire globe (Folland et al. 1984) showed very close agreement with the Northern Hemisphere data during this period, although not before.

This same general trend in temperature has been experienced in Canada since the turn of the century (Fig. 2b), except for minor variations due to local anomalies. A slight downtrend since 1950 has been noted in all parts of the country (Thomas 1975).

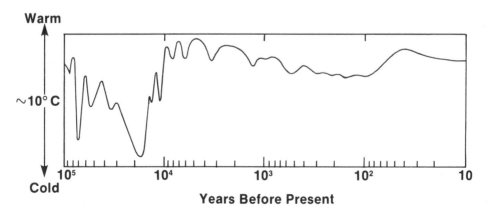

FIG. 1. Generalized trend in Northern Hemisphere surface temperature during the past 100 000 yr. (Adapted from numerous sources)

137

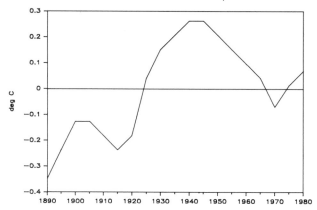

a. Northern Hemisphere

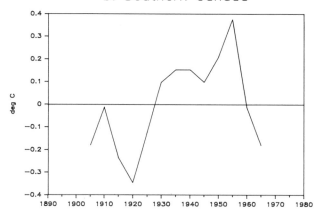

b. Southern Canada

FIG. 2. Recent temperature trends for (a) Northern Hemisphere and (b) southern Canada. (Adapted from Thomas 1975; Jones et al. 1982)

Trends in precipitation are more difficult to recognize, particularly for smaller regions and shorter time periods, due to its inherently greater variability (Landsberg and Jacobs 1951). Since 1940, there appears to have been a slight decrease in precipitation on the Prairies, little change in the central part of the country, and a slight increase along the east coast and in the high Arctic (Thomas 1975).

If we define "natural" as excluding the effects of human activities, then most of the variations displayed in Fig. 1 must be due to these "natural" causes, while those in Fig. 2 may be due to human activities as well. Consideration of the many components and interactions that comprise the terrestrial climate system (Fig. 3) has led to a better understanding of both the natural and human-related factors that are likely to affect the earth's climate. It is only through an understanding of these factors, and of the mechanisms through which they control the climate system, that we will be able to make useful forecasts of future states of the system.

The intention in this chapter is to look at climate, in relation to its natural variations, and to consider the effects that human activities may have on it during the next century. We will then choose a computer-modelling scenario for the climate 100 yr from now and use it as the basis for assessing the likely impacts on Canadian hydrology.

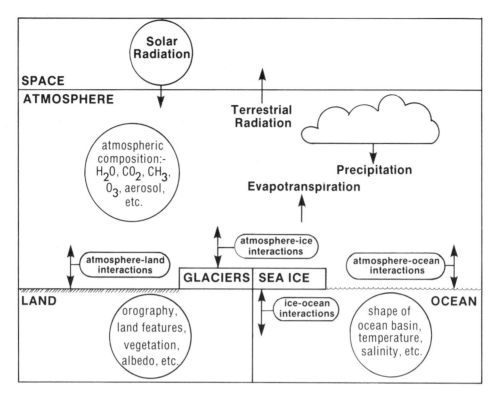

FIG. 3. The earth's climate system. Variations in the major "forcing functions" (encircled) may affect climate. (Adapted from Anonymous 1975)

Variations due to Natural Factors

The fundamental factors that determine the climate of the earth and its atmosphere (Fig. 3; Table 1) are the input of solar radiation and the properties of the atmosphere, the oceans, and the land masses. External influences such as the solar energy reaching the earth's atmosphere may be considered as "forcing functions" that drive the climate system. The characteristics of the system, such as the radiation-absorbing properties of the atmosphere, determine its response to the forcing functions. Over a long period of time, the system is more or less in equilibrium, so that the energy income from the sun (the main forcing function) is balanced by the loss of energy from the earth by terrestrial radiation.

The distribution of temperatures in the atmosphere and on the surface of the earth is a result of this balance and would presumably be constant if the forcing functions and the properties of the system were so. Unfortunately, a major characteristic of the climate system is its continual state of flux — not only does the solar input vary due to planetary motion and changes in earth–sun geometry, for example, but many properties of the system can vary markedly with time, and each component has a different response time. Thus, the system never reaches equilibrium, but is rather in a continuous state of transient adjustment. It is also possible that a given forcing function could produce more than a single resultant state of the system and that climate could alternate between different states independently of changes in the forcing functions or system properties (Lorenz 1976; Griffiths and Linden 1985).

Extraterrestrial Factors

Variations in the quantity and quality of electromagnetic energy emitted by the sun, its transmission through space, and its receipt by the earth constitute the main extrater-

TABLE 1. Major factors influencing climate (data from Degens et al. 1981; Gribbin 1978; Kondratyev 1980; Mitchell 1976; Anonymus 1975; Wollin et al. 1973).

	Time scale (approx.) (yr)	Likely importance during next century
Natural factors		
External to climate system		
Changes in solar emission	10^1-10^9	Minor
Changes in earth–sun geometry	10^4-10^6	Minor
Interstellar dust, etc.	10^8	Nil
Solar wind infuences	10^2-10^3	Minor
Solar and lunar tides	10^{-1}-10^2	Minor
Continental drift and mountain building	10^5-10^9	Nil
Changes in earth's magnetic field	10^2-10^6	Nil
Internal to climate system		
Atmospheric composition	10^{-1}-10^9	Moderate
Oceanic circulation, temperature, and salinity	10^{-1}-10^9	Major
Changes in earth's ice cover	10^{-1}-10^9	Moderate
Changes in vegetation	10^2-10^5	Moderate
Human-related factors		
Atmospheric composition	10^{-1}-10^2	Major
Surface changes	10^1-10^3	Local or regional
Thermal pollution	10^0-10^2	Local or regional

restrial forcing functions of the global climate system (Table 1). Of, perhaps, lesser importance are cosmic-ray fluxes, interactions of the solar wind with the earth's magnetic field, and solar and lunar tidal perturbations.

Variations in solar emission were once thought to be the underlying cause of the earth's ice ages. However, observations during past 100 yr indicate that the extraterrestrial solar flux has varied less than 1%, even during solar disturbances, although there may be significant fluctuations in some of the spectral components. There is model evidence that this magnitude of change, if sustained, would produce a global mean surface temperature change of about 1°C (Mason 1976).

The Milankovitch (1920) mechanism, proposed well over a half-century ago as a possible explanation of the ice ages, has recently gained considerable support (Hays et al. 1976; Mason 1976; Schneider and Thompson 1979). The mechanism is based on known cyclical changes in earth–sun geometry (Gribbin 1979), namely variations in the tilt angle (obliquity) of the earth's rotation axis, the precession of the earth's axis which changes the longitude of perihelion, and the eccentricity of the earth's orbit around the sun. While these variations result in little change in the total energy the planet receives from the sun each year, they do change its latitudinal and seasonal distribution and, according to many (e.g. Kukla 1975; Mason 1976; and Weertman 1976), provide a plausible explanation of the major climatic variations, including the ice ages, experienced by the earth during the past half-million years.

In comparison with the sun, the other planets of the solar system and the earth's moon are likely to have only minor effects on the earth's climate system. Nevertheless, some theories have been proposed, and some statistical evidence found, to relate these minor forcing functions to climatic variations (Gribbin 1973).

Terrestrial Factors

While extraterrestrial forcing functions are generally of a periodic nature, that is not the case for those originating on the earth (Degens et al. 1981). Geotectonics has probably been the major factor controlling terrestrial climate since the planet's creation.

Its time scale, measured in millions of years, is, however, much longer than most extra-terrestrial forcing functions. Even changes in land elevation and land to water ratio, associated with regional tectonic pulses, operate on time scales (10 000 to 100 000 yr) well outside the limits of the major known solar periodicities.

Continental topographical changes affect atmospheric circulation, cloud cover, precipitation, and surface temperature distribution, as well as have a strong influence on ice sheet growth. Changes in ocean-floor topography affect ocean currents, and both changes affect sea level and land to sea ratio (Gribbin 1978; Degens et al. 1981). Mountain building during the past several million years has led to the disappearance of many interior seas and has reduced the meridional heat transport by the oceans to its present value of about 50% of the atmospheric transport. Energy-balance considerations indicate that this has diminished the flow of heat from the tropics poleward and led to cooling at mid- and high latitudes.

It is almost certain that over the past few million years, i.e. since the appearance of man on the earth, the composition of the atmosphere has changed little. This is ignoring, for the moment, the very recent increase in CO_2 concentration. It is perhaps just as certain that the present atmospheric composition is strongly influenced by *biospheric processes*. According to Lovelock (1972), "if life on earth were to cease, the oxygen and the nitrogen would decline in concentration until they were both trace components in an atmosphere of water vapour, carbon dioxide and noble gases." This view considers the atmosphere as a "living entity," a component of the biosphere, the state (including the composition) of which is determined by exchanges of mass and energy within the system.

Thus, the atmosphere was very different before life appeared on the earth (about 3 Gyr B.P.) and started to evolve toward its present composition with the appearance of the first life forms. Except for the two components which have major transfers with the surface, CO_2 and water vapour, atmospheric composition is now relatively constant in both time and space.

The radiative regime of the atmosphere is largely controlled by the presence of water, CO_2, ozone, and atmospheric aerosols (Plass 1956; Kondratyev 1980). Although relatively transparent to the short wavelengths of solar radiation, the atmosphere absorbs strongly in the far-infrared region (at which the earth radiates) largely because of the above components.

Water plays a major role in many of the energy transformation processes in the biosphere. It exists in all three phases, and its concentration in the atmosphere varies markedly with geographic location, altitude, and time. The variation of the saturation vapour pressure of water with temperature (List 1966) has a major influence on the transfer of water within the biosphere.

CO_2 is exchanged with the surface through the processes of photosynthesis and respiration/decomposition, and hence varies in response to changes in the source/sink strengths of the processes. The amplitude of the seasonal variation (Fig. 4a) is greater over continents than oceans, and in the Northern rather than the Southern Hemisphere. The trend of increasing atmospheric CO_2 concentration (Fig. 4b) will be discussed in a later section.

Volcanic emissions have been quoted by Navarra (1979) as consisting of 58% water, 24% CO_2, 12% sulphur compounds, and 6% nitrogen compounds. According to Gilliland (1982), "sulphuric acid based particulates are much more important than dust particles in changing the radiative energy balance." While the output of volcanoes may not approach that of other sources on a long-term basis, major eruptions inject materials as high as the stratosphere that may reduce surface temperatures by several tenths of a degree, and persist for 2–3 yr (Kelly and Sear 1984). From 1945 to 1970 the annual number of volcanic eruptions increased from about 20 to 40, of which 5–10% were classed as being of "great" magnitude (Bryson and Goodman 1980).

As an indication of the difficulty of assessing the climatic effects of volcanic activity, an analysis by Stothers (1984) of one of the largest eruptions in recorded history failed to show any conclusive evidence of climate change. Although the vast amount of ejected material remained visible in the stratosphere for at least the following 2 yr, and caused 1816 to be known as "the year without a summer" in North America and Europe (Stommell and Stommel 1979), the mean annual temperature was only 0.7°C below normal in 1816, the same as it had been in 1814.

As land dwellers, we sometimes overlook the importance of the oceans in relation to the earth as a whole. The oceans cover about 72% of the earth and absorb well over half of the solar radiation reaching the surface. The main transfer process from ocean to atmosphere is the latent heat, or evaporative flux, which is closely related to the surface temperature and acts as the driving force for many atmospheric motions. Wind stress, or momentum flux, is the main transfer process from atmosphere to ocean and serves as the principal driving force for ocean currents. The greatest latent heat transfer from ocean to atmosphere is found over subtropical oceans, particularly in winter (Gribbin 1978). Peak values exceed $300 \text{ W} \cdot \text{m}^{-2}$ (equivalent to 1 cm of water per day) off the east coast of the southern United States and over the East China Sea. The key areas for momentum transfer are the midlatitudes during the hemispheric winter season.

Any persistent alterations in the above exchange processes, particularly those occurring in sensitive regions of the globe, could serve to alter global climate. Changes in major ocean currents, whether they result from long-term changes in atmospheric wind patterns or oceanic thermohaline structure, or from tectonic alterations in the sea floor, would affect sea-surface temperatures (SSTs) and the energy flow from ocean to atmosphere. This would, in turn, affect the atmospheric circulation and the wind stress on the oceans. Eustatic changes in sea level may be caused by diastrophism, growth, or decay of ice sheets and expansion or contraction of oceanic water due to temperature changes. A sea level change will alter the earth's land to water ratio and, because of the very different characteristics of the two surfaces, the fluxes between the surface and the atmosphere. Persistent changes in wind stress may affect, in a dramatic way, vertical oceanic currents (upwelling and downwelling), thereby producing changes in SSTs and in atmosphere–ocean energy exchange (Flöhn 1982). By bringing cooler deep-ocean water to the surface, upwelling reduces the transfer of both CO_2 and H_2O to the atmosphere, and since both are strong absorbers of infrared radiation, this change has greenhouse effect repercussions.

Most natural surfaces reflect between 5 and 30% of the solar radiation incident upon them (List 1966). Forests and water surfaces are at the low end of the range, and desert sands and short grass are at the high end. A covering of clean, dry snow, however, produces a dramatic change, raising the albedo to values approaching 90%. This change produces a striking difference in the amount of energy absorbed by the surface during the day and "constitutes by far the most important variable in the energy interactions of the earth and sun" (Gribbin 1978). As well as sharply reducing the energy absorbed by the surface, the snow cover provides an effective insulating blanket between the surface and the atmosphere, acts as an effective longwave radiator, and consumes a substantial amount of energy to melt or sublimate.

The distribution of snow and ice cover on land and sea has recently figured more prominently in theories of climatic change (Clark 1983; Walsh 1983). The mechanisms through which the cryosphere would exert its effect on climate range from those mentioned above to the ice-shelf theory of Wilson (1964). Wilson believes that the Antarctic ice sheet grows in thickness, with the weight of added snowfall, until the pressure at the underlying rock surface becomes so great that the ice reaches its pressure melting point. The ice then begins to flow out over adjacent waters and grows in size until it becomes large enough to increase the earth's albedo sufficiently to induce extensive glaciation in the rest of the world.

One of the most powerful means of studying past climates is the reconstruction of former biota and their growth rates, using such methods as pollen and tree-ring analysis, and lake and bog stratigraphy. These are based on the assumption that climate affects the presence, growth, and survival of organisms, and careful analysis of these indicators has provided invaluable knowledge concerning prehistorical climates.

The "Gaia" concept of Lovelock (1972), alluded to earlier, views plants and animals as not only responding to climate, but also playing an important role in shaping it. Indeed, the composition of the atmosphere itself is almost completely controlled by the earth's biota. Other effects include the composition of the oceans, the albedo and aerodynamic roughness of land surfaces, and evaporation from land surfaces covered by vegetation (Bolin 1980). Less obvious perhaps is the emission of natural halocarbons as a result of microbial decomposition on land and sea, which may have important roles in atmospheric chemical processes (Lovelock 1975; Harriss et al. 1985). Steudler and Peterson (1984) studied the global sulphur cycle and concluded that man was responsible for two thirds of all releases of sulphur compounds from the surface, the remaining one third being mainly biogenic. Though contributing only 1% of the total on a global basis, salt marshes were found to have the highest emissions on a unit area basis.

Freezing nuclei play a very important role in precipitation processes in the atmosphere (Lutgens and Tarbuck 1979). They are often in short supply and may limit precipitation in some situations. Some material of biological origin, such as decaying vegetation, marine plankton, and bacteria, is known to be more effective in initiating freezing of atmospheric water droplets than many nonbiological materials (Maki and Willoughby 1978). It is possible that variations in the availability of such materials, and in mechanisms to carry them aloft, may have played an important role in past climate variations.

Variations due to Human Activities

"Although human beings have to a considerable degree dominated the earth for a few millennia..., *the first really global impact that human beings have had upon the planet which they inhabit has been the increase in atmospheric CO_2 which has occurred since the start of the industrial era*" (Stewart 1978).

In the above statement, Stewart implies that even such monumental human achievements as the Great Wall of China, large-scale agricultural developments, and the eastern United States megalopolis are insignificant climatically in comparison with the increase in atmospheric CO_2 concentration attributable to human activities. Although this may be the most important, it is by no means the *only* anthropogenic perturbation of the climate system. In the next section (Atmospheric pollution) (see Table 1), in addition to CO_2, one may add numerous other gases, aerosols, and particulate matter, each of which has potential climatic effects (Wang et al. 1976; Kellogg 1977; Hameed et al. 1980; Kondratyev 1980). Perhaps as important, on local or regional scales, are effects due to large-scale modification of land surfaces and large-scale release of heat. Robock (1978) noted that most of these effects have been almost doubling every 20 yr.

There are many other minor activities which, while not climatically significant at the present, may become so in the future. These include cloud seeding, irrigation, and destruction of forests. In addition, such proposed activities as melting the Arctic ice cap by covering it with soot, or diverting river or Atlantic water to the Arctic Ocean, might have devastating climatic results (Kellogg and Schneider 1974).

Atmospheric Pollution

Human activities such as agriculture, industry, and slash-and-burn land clearing release a wide variety of substances into the atmosphere (Kellogg 1978). While increasing atmospheric CO_2 is now believed to have the most dramatic potential impact, at least on a global scale, there are many other substances, released in lesser quantities, that may turn out to be even more important (Cumberland et al. 1982). These include "trace gases" with potentially undesirable effects such as carbon monoxide, nitrogen and sulphur oxides, freon, and numerous hydrocarbons. Not only may these substances have direct climatic effects, but they also may interact with each other or with the natural constituents of the atmosphere to produce undesirable secondary effects. Aerosols and dust produced by industry and agriculture may also have impacts on climate, either directly by affecting solar radiation or indirectly by affecting precipitation processes and atmospheric chemistry. The transfer of both particles and gases between the atmosphere and the surface has been treated in a comprehensive review by Engelmann and Sehmel (1976).

The greatest effects of these materials are felt on local or regional scales, as most of them remain in the atmosphere less than a few days (Kellogg 1978). Brazel and Idso (1979) noted that the climatological effects of aerosols depend strongly on their vertical distribution in the atmosphere. The injection of dust high into the stratosphere by volcanic activity favours the reduction of surface temperatures, while increases in lower tropospheric dust concentrations usually produce surface warming. A modelling study of the regional effects of hygroscopic aerosols (Koenig 1974) found that cloudiness, and the temperature at the surface and in the lower atmosphere, increased. Precipitation was not significantly altered in the polluted region, but the simulation indicated a reduction in tropical precipitation, resulting from the midlatitude pollution. Another modelling exercise (Robock 1978) found that aerosols produced surface cooling on a global scale of the same magnitude as the warming predicted due to CO_2 increase. However, according to the author, "the magnitude, and even the sign of this effect, are open to much question due to our incomplete knowledge of the physical and chemical processes involved."

A wide range of gases are released to the atmosphere as a result of human activities (mainly industry, agriculture, and automobiles). These range from the relatively harmless (such as water vapour) to the highly toxic and from minute quantities to millions of tonnes per year (Stoker and Seager 1976). Most absorb infrared radiation, and hence contribute to the greenhouse effect. In terms of quantity produced, the most important emissions are those of carbon monoxide, nitrogen oxides, hydrocarbons, and sulphur oxides.

While many human activities may be inadvertently affecting the climate system, in the words of Kellogg (1977), "The atmospheric carbon dioxide increase emerges as by far the most dominant one." The "greenhouse" effect of CO_2 in the atmosphere was discussed over a century ago by Tyndall (1861). The eminent American meteorologist, W. J. Humphreys (1920), while dismissing carbon dioxide variation as the main factor in glaciation, used simple energy-balance calculations to estimate a 1.3°C global temperature rise for a doubling of CO_2 concentration in the atmosphere.

It has been less than 50 yr since it was recognized that the burning of fossil fuels was causing a general increase in atmospheric CO_2 content. In 1957, Revelle and Suess wrote: "human beings are now carrying out a large-scale geophysical experiment.... Within a few centuries, we are returning to the atmosphere and the oceans the concentrated organic carbon stored in the sedimentary rocks over hundreds of millions of years." It is hard to view this "experiment" objectively, since we are part of it. The interest of scientists and the general public has led to an increasing number of newspaper articles, scientific papers, and books on the subject.

There were very few direct measurements of atmospheric CO_2 concentration taken before the present century. Most of the early data have been summarized by Callendar (1958). In the same year as Callendar published his summary paper, a major monitoring station opened in Mauna Loa, Hawaii. Far from major sources and sinks of CO_2, the ideally located site has been the single most reliable source of evidence of global CO_2 concentration variation over the past quarter century (Keeling et al. 1976). Since the establishment of the Mauna Loa site, a number of other monitoring stations and programs have been initiated, including aircraft flights in Sweden, an ice-cap site at the South Pole, and a high-Arctic station at Alert, Canada (83°N, 62°W). The remarkable agreement between the data from the various sites (Kellogg 1978; Peterson et al. 1982; Hare 1981) attests to the efficiency of large-scale mixing processes in the troposphere. The historical record continues to be supplemented by newly "discovered" data (e.g. Preining 1983), but these are often of questionable value due to limitations of the methods used at the time. A recent analysis of air occluded in polar ice cores (Neftel et al. 1985) has extended the measured data record back to the mid-eighteenth century, at which time the concentration was about 279 vpm.

Since the beginning of the present century, the mean atmospheric concentration has increased from about 290 to 345 vpm at the present time (Fig. 4b). During the past decade, the rate of increase has averaged 1.3 vpm $\cdot$ yr^{-1}.

Apparently, both fossil-fuel combustion and decomposition of plant material following large-scale deforestation are releasing CO_2 to the atmosphere, about half of which remains there while the balance is transferred to the oceans (Hare 1981).

The way in which CO_2 affects climate is through the so-called "greenhouse effect" of differential radiation absorption (Schiff 1981). Many polyatomic gases (mainly H_2O and CO_2) have their main absorption bands in the longwave region of terrestrial radiation, while being virtually transparent at the shorter solar radiation wavelengths. Thus, they have little effect on radiation arriving at the earth from the sun, but tend to absorb radiation emitted by the earth. This means that the atmosphere retains some of the heat radiated by the earth's surface, keeping the earth–atmosphere system warmer that it would otherwise be. Higher concentrations of these absorbing gases increase the greenhouse effect, leading to a warmer earth. The effect is amplified because higher surface temperature leads to increased evaporation, and hence atmospheric humidity. The complexity of the climate system (Fig. 3), with its many interactions and feedbacks, makes quantification of the effect a difficult exercise, and there is considerable difference of opinion about its magnitude (Plass 1956; Robock 1978; Schlesinger 1984; Elsaesser 1984) and even its direction (Idso 1983).

Surface Modification

Human land use often produces considerable surface changes, which may affect the exchanges of energy and matter between the surface and the atmosphere (Fig. 3). These include, as well as the clearing of land for agriculture and building of cities, such things as irrigation, road and seismic-line construction, surface mining, and even oil slicks. Surface modification was probably the first mechanism through which early man impacted on the climate system, through burning of vegetation, land clearing, agriculture, and overgrazing, for example. Although some of the effects have been estimated, particularly on a local basis, there has never been a world-wide inventory which would permit a prediction of the overall impacts on the climate system (Kellogg 1977).

One of the main results of surface modification due to human activity is an increase (usually) in the surface reflection coefficient for solar radiation (or albedo). Burroughs (1981) indicated that about 5% of the earth's surface has had its albedo approximately doubled due to human modification of the surface. This higher reflectivity means that more of the incoming solar energy is lost to outer space, and thus less is available at the surface for other processes. Other surface changes brought about by human activity include changes in aerodynamic roughness, which affect the turbulent transfer between the surface and the atmosphere, changes in moisture availability, which affect the evaporation from the surface, and changes in the insulating blanket of vegetation or snow, which affect the heat flow in and out of the soil.

There is evidence that land-use changes may influence regional temperatures and precipitation, e.g. overgrazing in the Sahel (Charney et al. 1975). Although Fowler and Helvey (1974) found little effect of a large-scale irrigation project on actual climatic data in eastern Washington State, Rapp and Warshaw (1974) did find significant climatic effects of a simulated Saharan lake. While Shukla and Mintz (1982) concluded that large-scale tropical deforestation, in addition to reducing local rainfall, could have noticeable effects elsewhere, Henderson–Sellers and Gornitz (1984) found only local effects (Schneider 1984). There is no easy way to forecast such impacts, if for no other reason than the very great range of possible impacts and the wide variety of ecosystems to which they have been applied. Perhaps even more important is the notion that some areas, because of their geographical location and/or role in the climate system, may be far more sensitive, and susceptible to impact, than others.

Thermal Pollution

Release of heat from the surface, whether from fossil-fuel combustion or other energy sources, is a direct input of energy to the atmosphere. This is also true for release of water vapour, an input of latent energy, which is ultimately released when the moisture condenses out. Most fossil fuels, upon burning, release copious quantities of water, CO_2, and other substances along with the desired heat. Therefore, any prediction of climate change due to atmospheric changes resulting from fossil-fuel combustion should consider the combined effects of the inputs of CO_2, other gases, aerosols, and waste heat.

Release of heat from urban and industrial areas is, along with other factors, responsible for the so-called "urban heat island" (Oke and Maxwell 1975), with its characteristically higher temperatures and precipitation compared with the surrounding countryside (see also chapter by McKenzie, this volume). Thus the local effects of thermal pollution are well documented, although certain aspects are still controversial (Charton 1973), while regional effects are suspected but not yet proven. On a global scale, since the total heat release by human activities is only 0.01% of the solar input, it seems that such a small fraction is likely to have a negligible effect on world climate (Kellogg 1977). However, there is the possibility that the climate system may respond to even apparently negligible amounts if they are injected in "sensitive" geographical areas.

Main Factors in Climate Change

The preceding pages will surely have left the impression that there are a great many factors involved in the earth's climate system, and that they span a wide range of time scales. Each has at least some empirical evidence in support of it, and in many cases plausible mechanistic theories have been developed.

To prepare a climate scenario for 2085, it will be necessary to assess the relative importance of each of these factors and to quantify those deemed likely to be major "forcing functions" during the next century.

Table 1 and the preceding sections of this chapter indicate that the most important factors affecting global climate during the next century will be *anthropogenic* atmospheric pollutants. There is a considerable body of evidence (but, as yet, no *proof*) that increasing atmospheric CO_2 will cause global warming of several degrees Celsius by the end of the next century (Schlesinger 1984). Concurrent with the CO_2 increase will likely be increases in other "greenhouse" gases and in aerosols. Their effects could well be equivalent in magnitude to the CO_2 effect, but will likely be opposed (to each other) in direction (Robock 1978; Wang et al. 1976). Therefore, to a first approximation, we may assume that their effects will cancel one another, and omit them from consideration. The global effects of surface changes and heat input could well be important but, because they are less clear, will also be omitted.

The climatic variations attributable to *natural* factors are much less certain although the most important, in the short term, are likely to be sunspot activity, volcanic dust, and the Milankovitch mechanism (Gilliland 1982; Hays et al. 1976). The Milankovitch model predicts global cooling for at least the next few centuries, although the magnitude is expected to be less than 0.1°C per century (Mason 1976; Hengeveld 1984). With regard to the other factors, Schneider and Mass (1975) used a one-dimensional radiative–convective model to simulate the effects of sunspots and volcanic dust on global temperatures during the past 400 yr. In spite of good empirical agreement, the authors cautioned that the results only "demonstrate that these factors may have contributed to the shape of the record" and should not be "interpreted as indicating that

146

the external factors explain the observed record." Gilliland (1982), using an energy-balance model, was able to explain up to 93% of the variance of northern hemisphere mean annual temperature over the past 100 yr by simulating the possible effects of variation in solar radius, volcanic aerosols, and CO_2 increase. A more sophisticated energy-balance model was used by Gilliland and Schneider (1984) to test the response to CO_2, in the absence of solar and volcanic forcing. Their approach indicated global warming of 1.6°C, a rather lower value than that predicted by many general circulation models.

In the next section, we will assume that increasing concentrations of radiatively active gases in the atmosphere will be the main perturbing factor affecting global climate. For practical purposes, the overall effect will approximate a doubling of atmospheric CO_2 during the next century. Based on this assumption, an attempt will be made to predict its most likely effects on Canadian climate and hydrology.

Modelling the Climate System

A system as complex as the earth's climate (Fig. 3) taxes the capabilities of even the most powerful computers to model it. Even if all of the natural and anthropogenic forcing functions were well known, there would still be great difficulties because of the very wide range of relaxation times involved in the system and the intrinsically stochastic nature of many atmospheric and oceanic motions.

At first glance, it may seem strange that climate modellers attempt to simulate climatic response to human activities before being able to predict natural, or "background," variations. However, this is not as foolhardy an exercise as it may appear. Our ability to model *natural* variations depends on our understanding of not only the climate system itself, but of all of its major forcing functions as well. It is this latter knowledge that is lacking, although recent progress in rectifying this has been impressive. Simulating responses to *human activities*, on the other hand, requires only a good system model. We can define our own forcing functions and see how the model responds to them. In the words of Kellogg (1977) "It is important to distinguish between...the prediction of changes in the statistics of the atmosphere...due to the many interactions within the climate system itself, and...prediction of how these climate statistics will change as a result of an alteration in the external or boundary conditions of the system." Without this understanding of natural variations, however, it is impossible to verify if simulations of the effects of human activities are correct until their magnitude exceeds those of natural variations, i.e. until the "signal" emerges from the "noise".

The investigation of climate system variability in this chapter will be limited to the next century or so, thus excluding forcing functions which are important only over longer periods. Accordingly, the Milankovitch mechanism will be ignored because it is negligible over such a short time period, volcanic activity is not sufficiently predictable to be included, sunspot cycles could be accounted for, but will likely be unimportant compared with the effects considered, and trace gas and aerosol effects will be assumed to be of equal and opposite magnitude so that they cancel each other out.

Thus, we will focus on the CO_2 effect as the main climatic forcing function during the next century, and we will attempt to predict its impact on global, and in particular Canadian, climate in 2085. This will involve three stages: (1) an analysis of the global carbon cycle in order to determine the likely rate of increase of atmospheric CO_2 during the period, (2) use of a global climate model to determine the effects of increasing CO_2 on precipitation and surface temperature, and (3) an assessment of the impacts of the modelled changes in temperature and precipitation on Canadian hydrology.

The Global Carbon Cycle

The present concern about global greenhouse warming rests mainly on observations of atmospheric CO_2 concentration (Fig. 4) and their interpretation. To predict the likeli-

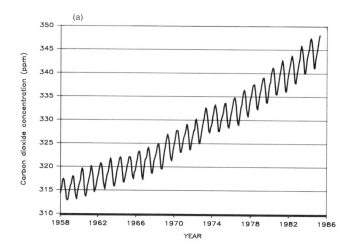

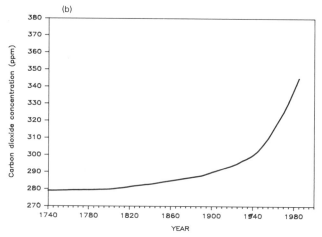

FIG. 4. Atmospheric CO_2 concentration (parts per million by volume). (a) Measurements from Mauna Loa, Hawaii, showing seasonal variation due to plants; (b) generalized Northern Hemispheric data, with seasonal variation removed. (Adapted from Bruce and Hengeveld 1985; Neftel et al. 1985; Sadler et al. 1982; Keeling et al. 1982; and unpublished data courtesy of C. D. Keeling)

hood and magnitude of future changes in atmospheric CO_2, it is necessary to examine the global carbon cycle, a simplified view of which is shown in Fig. 5.

The major global carbon reservoirs are relatively inaccessible to transfer processes so that the carbon they contain turns over at a very slow rate indeed. Most natural transfer occurs between the terrestrial biota, oceanic surface waters, and the atmosphere. Land plants, atmosphere, and surface waters contain roughly equivalent amounts of carbon, while dead organic matter, or the humus layer, containts about five times as much. Average annual exchanges between land and sea and the atmosphere amount to 10-20% of the reservoir capacities, so that mean residence times are 5-10 yr. These may be compared with time constants of about 1000 yr for the deep ocean (Woodwell and Pecan 1973; Schiff 1981) and considerably longer for ocean sediments.

In 1977, Kerr summarized our knowledge of the global carbon balance. The main points of his summary were (i) about half of the CO_2 produced by fossil-fuel combustion could be accounted for by the increase in atmospheric concentration, (ii) only part

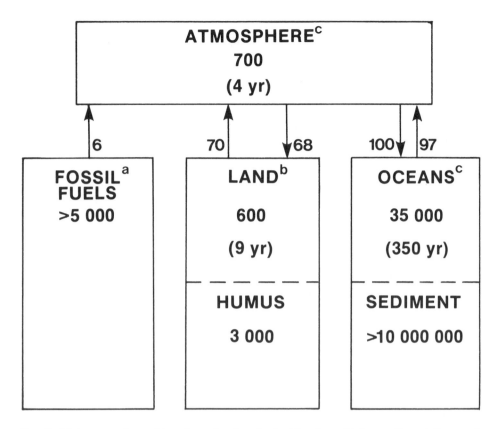

FIG. 5. Main reservoirs and transfers of carbon in the biosphere. Units are Gt and Gt • yr^{-1}, respectively. Parenthetical numbers are mean residence times. Notes: a, source strength depends on fuel consumption; b, source or sink depending on relative magnitude of land clearing and CO_2-enhanced growth; c, sinks. (Data from Goudriaan and Ketner 1984; Bolin et al. 1979; Hare 1981; Schiff 1981; Wong 1978; Woodwell and Pecan 1973)

of the remainder was believed to be entering the oceans, because of their slow rate of intake, and (iii) the terrestrial biosphere was suspected of being a source, rather than a sink, of CO_2, because of large-scale deforestation and other agricultural practices.

Assuming all of these hypotheses to be correct left a large surplus of CO_2 unaccounted for. Woodwell et al. (1978) reviewed the arguments in favour of the terrestrial biomass as a source of carbon, while Stuiver (1978) studied the carbon isotopic ratios of Douglas-fir cellulose and concluded that terrestrial biomass had decreased from the mid-nineteenth to the mid-twentieth century, but had changed little since then. Broecker et al. (1979) further questioned the land biota's role and concluded "there is no compelling evidence which establishes that the terrestrial biomass has decreased at a rate comparable to that of fossil fuel combustion over the past two decades, as has been recently claimed." This, and other evidence, led Schiff (1981) to suggest three possible answers to this problem: (1) in spite of deforestation, the terrestrial biomass is either not changing, or is actually accumulating carbon, (2) the oceanic uptake is more rapid than currently believed, or (3) there exists another, as yet unidentified, sink for atmospheric CO_2.

Killough and Emanuel (1981) carried out simulations using five oceanic models with different layering and concluded that "the role of the terrestrial biota...as source or sink for excess carbon may have shifted during the period 1860-1975, although the role of sink appears to have become dominant in the years immediately preceding 1975".

Thus the impasse of the late 1970s seems to have gradually resolved itself by allowing for an increase in the rate of oceanic CO_2 uptake, and by modifying the role of land biota from a strong source of CO_2 to a weaker variable source/sink.

The only forcing function in the carbon cycle (Fig. 5) is the human input resulting from the combustion of fossil fuels. The production rate of CO_2 from this combustion increased at an annual average rate of just over 4% from 1860 to 1975 (Broecker et al. 1979), except for the period 1915–45 when it dropped to just over 1%. It is expected that the growth rate of energy use will diminish in the future as supplies become depleted and prices continue to rise (Rotty 1980). Although Canada is responsible for only about 3% of global emissions, it ranks near the top on a per capita basis (Anonymous 1983a).

The International Institute for Applied Systems Analysis, in Austria, prepared two energy-use scenarios based on certain assumptions regarding future reliance on fossil fuels and other forms of energy (Häfele 1980). Their *low growth* scenario represents an annual increase of about 1% per year, while a *high growth* scenario is based on a 2% annual increase rate. The present release, of about 6 Gt C • yr^{-1}, would increase by 2085 to 18 and 40 Gt C • yr^{-1}, respectively, in the two cases. Corresponding atmospheric concentrations would be about 590 and 730 vpm in 2085 (Marland and Rotty 1980). Thus, predicted CO_2 doubling by 2085 would lie intermediate between the two scenarios.

The great difficulty in predicting future energy use, on which atmospheric CO_2 projections depend, should be kept in mind. If those who contend that society will adjust to a lower energy-use life-style are correct, however, the effect will be only to slow down the atmospheric increase of CO_2 and to delay the global warming.

Global Climate Models

If one assumes, on the basis of the foregoing, that atmospheric CO_2 doubling by 2085 is not unreasonable, then the next step is to determine what effects this will have on global climate. Although consideration of the "greenhouse effect" leads us to expect global warming, the situation becomes complicated by a number of secondary effects or "feedbacks" (Schiff 1981).

An increase in oceanic surface temperature would cause an increase in evaporation rate and hence air humidity, enhancing the greenhouse effect. This effect could lead to greater cloudiness, which in turn would reduce the radiation reaching the surface, and lead to cooling. Hunt (1981), using a radiative–convective model, found that a 10% increase in cloudiness (or even an increase in low cloud albedo from 0.69 to 0.74) could completely compensate for the warming produced by CO_2 doubling.

Rising surface temperatures would reduce the area and duration of snow and ice cover, thus lowering albedos. This would cause more of the incoming radiation to be absorbed, again enhancing the greenhouse effect.

Most models predict greater CO_2 warming in polar regions than near the equator, so that the temperature difference between equatorial and polar regions would decrease. This would tend to diminish the strength of the atmospheric circulation, which is driven by the temperature differential, and hence reduce wind stress on the oceans. The diminished ocean currents would be less effective in transporting equatorial heat poleward, leading to an increase in the equator-to-pole ocean temperature gradient.

The greater thermal capacity of the oceans, compared with the land, would result in global warming being delayed over the oceans, thus leading to a transient effect while the oceans responded more slowly to the CO_2 increase than the land. Bryan et al. (1982) found that the tropical oceans responded quite rapidly to the change, but that it took 25 yr for the effect to reach equilibrium at the poles.

The above examples illustrate that there are a number of secondary effects that may either enhance (positive feedback) or reduce (negative feedback) the primary effect of increasing temperature. Because many of these processes depend on one another, there is no way of dealing with them satisfactorily other than through the use of computer models.

The main types of computer models used to simulate the global climate system comprise energy-balance models (EBMs), radiative–convective models (RCMs), statistical-dynamical models (SDMs), and general circulation models (GCMs). These types differ considerably in complexity and in the way they treat the various components of the system, such as radiation, surface conditions, atmospheric motion, etc. For a description of the details of each, the reader is referred to Schiff (1981), Meehl (1984), and Schlesinger (1984). Only the fourth type will be described briefly here.

GCMs are the "heavy-weights" among global climate models, having the greatest complexity, requiring the most powerful computers, and producing the most detailed simulations (Meehl 1984; Schiff 1981; Schlesinger 1984). They solve the three-dimensional equations of motion, thermodynamics, and continuity of the atmosphere and simulate all of the major dynamic processes therein. Various GCMs differ mainly in the way they treat clouds, snow, soil moisture, sea ice, and oceanic processes. The most realistic, and the most complex, are coupled atmospheric and oceanic GCMs. Global warming scenarios for CO_2 doubling lie mostly in the range of 2-4°C for GCMs with interactive oceans (Smagorinsky 1982).

One important aspect of climate model assessment that was perhaps neglected until fairly recently is that of analysing the statistical significance of the simulation results. Strict attention to this might require that more of the total computational effort would have to go into the assessment of statistical significance than into the simulations themselves (Gribbin 1978).

The most direct test of a climate model, simulating CO_2 warming, would be a comparison with actual global, or hemispheric, temperature data for the past few decades. Unfortunately, the predicted increase by now of about 0.2-0.4°C is not very large compared with naturally occurring temperature variations, and the problem becomes one of recognizing the CO_2 warming "signal" in the background "noise" of natural climatic variations. As the atmospheric CO_2 concentration continues to climb, its "signal" should become stronger and stronger, until it eventually rises above the "noise" and becomes recognized.

Madden and Ramanathan (1980) investigated the problem and decided that the signal should be noticeable now. Since it was not, they concluded that either the model estimates were too high or there was an unconsidered lag in the system, such as ocean thermal inertia, that might delay the effect for 10-20 yr. Using a similar approach, but with different signal and noise data, Wigley and Jones (1981) concluded that natural warming during the early part of this century may have obscured the CO_2 effect, thus delaying detection until early in the next century. Epstein (1982) was more optimistic, suggesting that statistically convincing evidence should be available by 1986. In the meantime, we may have to look to changes in global mean sea level, polar sea-ice extent, global rainfall distribution, or the ratio of surface warming in high northern latitudes to general stratospheric cooling to find the evidence we seek (Etkins and Epstein 1982; Kellogg 1983).

Kellogg (1983) has also pointed out that when the signal-to-noise ratio reaches unity, we can have 85% confidence that the signal exists, and when the ratio reaches 2, the confidence value rises to 98%. How long can we afford to wait?

A Global Climate Scenario for CO_2 Doubling

GCM Consensus

A general circulation model simulation was chosen as the basis for assessing the effects of CO_2 warming on Canada's hydrological regime because only GCMs are capable of providing the required geographical resolution. The question then was reduced to a choice among the major existing models, namely those of the Geophysical Fluid

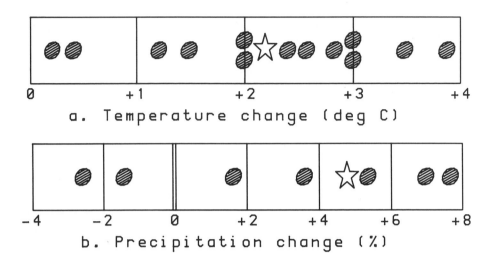

FIG. 6. Summary of GCM predictions for CO_2 doubling. (a) Global temperature change; (b) global precipitation change. The stars denote the predictions of the UKMO model used in the present study. (Data from Mitchell 1983; Schlesinger 1983)

Dynamics Laboratory (GFDL) (Manabe et al. 1981), Oregon State University (OSU) (Gates and Cook 1980), the National Centre for Atmospheric Research (NCAR) (Washington and Meehl 1983), the Goddard Space Flight Centre (GSFC) (Hansen et al. 1983), and the United Kingdom Meteorological Office (UKMO) (Mitchell 1983).

These various models have produced a range of global mean temperature predictions for CO_2 doubling ranging from 0.2 to 3.9°C (Fig. 6a). The lowest two values were for GCMs with noninteractive oceans, and the highest two were for the Goddard model of Hansen et al. (1978). Aside from these four extreme values, the remaining 10 GCM predictions span the range of 1.2–3.0°C, with an average of 2.25°C. It may be recalled that this value is also near the middle of the range of values predicted by EBMs and RCMs. Global precipitation changes, predicted by the GCMs, vary from a reduction of 2.5% to an increase of almost 8% (Fig. 6b). Closest to the middle are the NCAR and an OSU simulation and that of Mitchell (1983).

A recently published version of the UKMO GCM (Mitchell 1983) was chosen as the basis for the climate change scenario, partly because its predictions tend to be "middle-of-the-road" (Fig. 6) and partly because it is the most recently published version (and hopefully therefore the most advanced!). Although it does not include an oceanic circulation model, it does treat snow cover and soil moisture interactively, and the version used includes a realistic sea-surface temperature enhancement (2°C). It should perhaps be mentioned that this value for sea-surface temperature enhancement is based on the results of other GCMs. This wise choice virtually ensures good agreement between this model and the others in regard to global temperature rise. The model is based on the fundamental equations of meteorology in their undifferentiated form and uses a finite difference method, with a 10-min time step, for their solution. It divides the atmosphere into five equal-mass layers and has a horizontal grid length of 330 km.

The model predicts a mean global temperature increase of 2.2°C, the result of a 2.6°C increase over land and a 2.0°C increase over the sea. Global precipitation (P) and evaporation (E) are both increased about 90 mm • yr^{-1}. However, while the precipitation is slightly less than this over the land, the evaporation is much less, leading to an increase in (P − E) of about 55 mm annually over the continents. Most of the precipitation increase is in the high latitudes, particularly during the winter season. Temperature change is fairly constant with latitude, although slightly higher values are found in the midlatitudes, especially in winter.

Before proceeding with this scenario and elaborating its effects on Canada, and particularly Canadian hydrology, it might be appropriate to pause for a moment and consider some of the other very different scenarios proposed recently by a minority of scientists.

A multitude of global climate model predictions, made during the past few decades, have led to a virtual consensus that CO_2 doubling will raise global mean surface temperature $2-3°C$. Whether this consensus has been influenced by the "steamroller effect" or the feeling of "safety in numbers" is hard to say. However, there were few dissenting voices heard until the late 1970s, when Newell and Dopplick (1979) and Idso (1980) suggested that the modellers' predictions were 10 times too high. More recently, Idso (1983) has proposed that increasing CO_2 may *cool* rather than *warm* the planet, and Elsaesser (1984) has criticized current models for omitting, or treating inappropriately, a number of factors. Some of the dissenters' criticisms have been answered by Kandel (1981), who found no basic disagreement between the two groups.

Although the arguments put forward so far by the dissenters are not very convincing, they should not be dismissed. Consensus does not prove an hypothesis.

Current Trends

Climatologists have been very busy recently trying to find evidence of climate change in the observational records. This is not an easy task because of the scarcity of reliable, long-term weather stations and their uneven distribution on the earth. There has been considerable effort towards developing a dependable data set, which could be used to monitor climate change (Landsberg et al. 1978; Robock 1982).

Rising global temperatures between the beginning of the present century and the 1940s (Groveman and Landsberg 1979) lent credibility to the greenhouse warming theory. However, the switch to a slight cooling trend after that (Fig. 2) sent climatologists scurrying for a plausible explanation. The most successful of these involved sunspots and volcanic activity, in addition to the CO_2 effect (Agee 1980; Gilliland 1982). Idso (1983) noted that the post-1940 cooling trend was greatest in northern latitudes, much less nearer the equator, and absent in more southern latitudes. He suggested that this might be related to CO_2 production, which is greatest in the mid- to high latitudes of the Northern Hemisphere. This does, however, seem to be at variance with the noted uniformity of CO_2 concentration over the globe (Hare 1981), the apparent shift toward rising global temperatures since 1970 (Jones et al. 1982; Raper et al. 1984), the decreasing extent of Antarctic pack ice in summer (Kukla and Gavin 1981), and the "rapid rise in sea level over the past 40 years, and especially since 1970" (Etkins and Epstein 1982).

The temperature trends since the turn of the century in southern Canada (Thomas 1975) were very similar to those found by Kelly et al. (1982) for the entire Arctic region. Because of the low density of observing stations in Canada, apparent trends for particular regions are much more variable, although most show a downturn in temperature from 1950 to 1970 (Thomas 1975). There appears to have been a change to a global upswing since then, however (Jones et al. 1982; Kelly et al. 1982; Raper et al. 1984). Juday (1982) noted higher winter temperatures, and a trend toward longer growing seasons, during the late 1970s in the interior of Alaska.

Canadian Climate in 2085

Temperature

The present geographical distribution of mean annual temperature in Canada (Fig. 7a) is clearly dominated by latitude, moderated somewhat (particularly in the west) by the oceans. Only a thin strip of the country, along the United States border, has a mean temperature above the freezing mark. The global land surface temperature increase of 2.6°C, predicted by the Mitchell model, is amplified in the midlatitudes, particularly during the winter season. Although Mitchell (1983) presented separate simulation data for summer and winter seasons, these have been combined to yield annual values for use in the present work. Extreme winter-season temperature increases, predicted by the model, exceed 10°C in the European U.S.S.R. and 6°C in central Canada. Summer season increases over most of North America are much closer to the global mean. Thus, the model predicts for Canada a general 2-3°C warming, supplemented by several degrees more in the central part of the country during the winter. The distribution of average annual increase (Fig. 7b) exceeds 4°C in central Canada, dropping to about 2°C on the perimeter of the country. This is equivalent to an average annual increase of 2.8°C for Canada as a whole.

The present model's Canadian scenario has a general similarity to those of most of the other GCMs, except that it predicts the main winter temperature increase to be over central Canada, rather than the western Arctic. Some other GCMs predict up to 15°C increases in Arctic winter temperatures. Wigley et al. (1980) used an historical-analogue approach to produce geographical detail corresponding to a global prediction of greenhouse warming. They compared the five warmest and five coolest years in the period 1925-74 in terms of their distribution of differences in temperature and precipitation. After adjustment, their temperature scenario for CO_2 doubling shows a 6°C temperature increase over northwestern North America, dropping off to 3°C in southern British Columbia and southern Ontario and to 0°C over the Labrador Sea. In its general features, this scenario is not very different from that of Fig. 7b.

Precipitation

There is general recognition that the catch of most conventional precipitation gauges is reduced due to the effects of wind flow around the gauges (Rhoda 1971; Larsen and Peck 1974). Underestimations vary from about 10% for rainfall to as high as 45% in the case of snow (den Hartog and Ferguson 1978). For hydrological studies, then, it is necessary to correct the measured precipitation data to make them consistent with other components of the water balance. It is this "derived" precipitation field (den Hartog and Ferguson 1978) that is used in the present study (Fig. 8a) as the basis for simulation of the effects of climate change.

The present distribution of precipitation across Canada clearly shows the influence of the oceans and the western cordillera. Annual totals on both east and west coasts surpass 2000 mm in some areas, and 1000 mm in the Rocky Mountains. Amounts drop off inland and northwards, falling slightly below 400 mm in the Prairies and below 100 mm in the high Arctic. The national average is about 600 mm.

The geographical and seasonal distribution of the UKMO model's predicted 90-mm global precipitation increase shows large changes in many tropical and subtropical areas, but the main mid- and high-latitude change is a wintertime increase exceeding 2 mm $\cdot$ d^{-1} centered over British Columbia and the Gulf of Alaska (Mitchell 1983). From this peak, there is a gradient of winter-season increase across Canada dropping to zero on the east coast. Expected changes in summer-season precipitation average near zero across the country, with increases of about 0.5 mm $\cdot$ d^{-1} in the Arctic and southeastern Canada and corresponding decreases in British Columbia and northern Quebec. On an annual basis (Fig. 8b), the model predicts a national increase of almost 150 mm $\cdot$ yr^{-1}, varying from a slight reduction in northern Quebec to an increase in excess of 300 mm in northern British Columbia. The most dramatic change indicated by the model is in the high Arctic, where the 200-mm increase would result in a tripling of the present total.

There is much greater intermodel variation in the prediction of precipitation than there is for temperature. Most other GCMs indicate increases in Canadian precipitation, but of lesser magnitude and with a different geographical distribution than the UKMO model. For example, one GFDL GCM simulation predicts an average annual

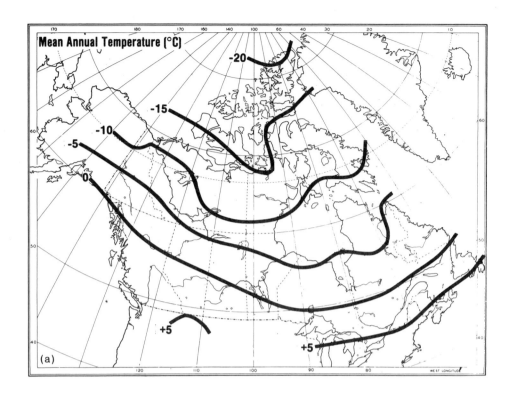

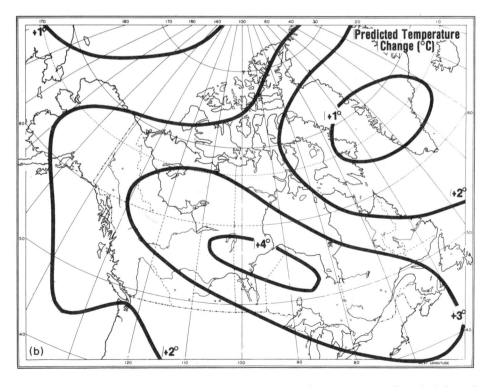

FIG. 7. (a) Present geographical distribution of mean annual temperature in Canada (adapted from Hare and Hay 1974); (b) model prediction of mean annual temperature change (2085–1985) for Canada (adapted from Mitchell 1983).

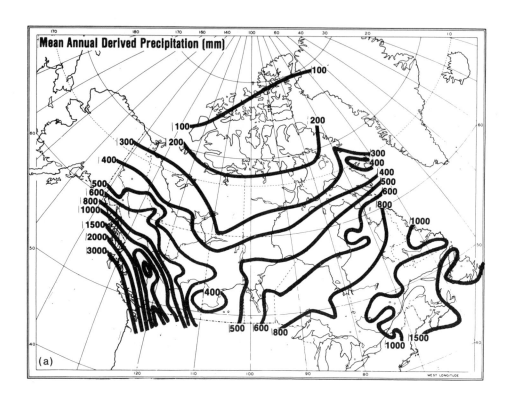

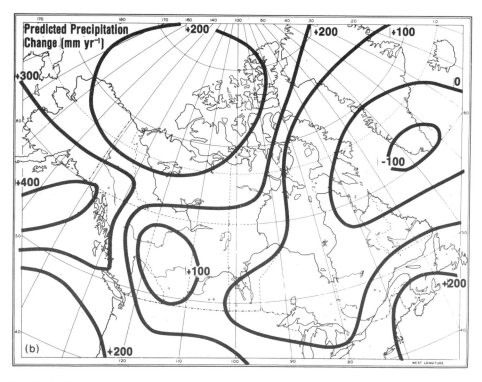

FIG. 8. (a) Present geographical distribution of mean annual derived precipitation in Canada (adapted from den Hartog and Ferguson 1978); (b) model prediction of mean annual precipitation change (2085-1985) for Canada (adapted from Mitchell 1983).

156

increase of about 100 mm nationally with up to 200 mm more on the Prairies. Referring to the historical-analogue study of Wigley et al. (1980), described in the previous section, provides little help, as it shows only a positive precipitation change over all of Canada, except southeastern British Columbia and the Atlantic Provinces, with no indication of the magnitude of the change.

Summary

It should be kept in mind that while climatic change may be described in terms of altered averages of temperature and precipitation, it will actually manifest itself as changes in day-to-day weather. Global warming will push the midlatitude belt of westerlies poleward, taking with them frontal systems and storm tracks. Not only will there be changes in the averages of the weather variables, but also in the frequency of occurrence of severe weather, such as hail and dust storms, droughts, and floods. This will make the climate of some regions more suitable for man, while making that of other regions less suitable. The model predictions of the effects of CO_2 doubling in Canada (Fig. 7b and 8b) certainly represent substantial changes from the present climate. If they occur, they would have dramatic impacts on many aspects of life in this country. It should not be overlooked that for the present at least, predictions of geographical distributions of climate change are at a very rudimentary level, and only the most conjectural use can be made of them. Nevertheless, the can at least make us aware of what possibilities exist and will hopefully be replaced in the not too distant future with geographical forecasts of acceptable accuracy. An assessment of the statistical significance of the present model's predictions has been given by Mitchell (1983).

Likely Effects on the Hydrological Regime

The most direct effects of changing climate on the hydrological regime will be through variations in precipitation amount, type, timing, and intensity and through changes in evaporation rate, which is closely related to temperature. Less direct effects will be on the times of arrival and disappearance of winter snowpacks and changes in the patterns of cloudiness, windspeed, vegetation characteristics, and soil properties. Even less direct will be the effects of changes in land use that occur in response to the changing climate.

The main hydrological effects of greenhouse warming, as summarized by Manabe (1983), are likely to be increases in the global mean rates of both precipitation and evaporation, an increase in the annual runoff rate at high latitudes, a decrease in coverage and thickness of Polar ice pack, and the earlier arrival of the snowmelt season.

Sea-level rise could have significant effects on the area of continental land, on global freshwater resources, and in particular on human populations through the flooding of low-altitude urban and agricultural land. The disappearance of even part of the Polar ice cap, for example the West Antarctic ice sheet, would raise sea level 5-7 m (Kellogg 1980). However, the response time of the cryosphere is so long (Fig. 9) that changes would be scarcely noticeable in even a 100 yr period. Recent studies of sea-level changes (Hansen 1985) attribute the current rise of about 30 cm • century[-1] to a combination of mountain glacier retreat, thermal expansion of the oceans, and residual postglacial crustal rebound effects. However, there is no conclusive evidence that any of the change is due to the melting of Polar ice caps. The present greenhouse warming scenario will likely produce a sea-level rise of 40-60 cm during the next century, mainly through the continued melting of mountain glaciers and oceanic thermal expansion.

Of the almost 1400 million km³ of water on the earth (Fig. 9), less than 0.003% of it is discharged from all of the rivers on our planet each year. Canadian rivers contribute about $1/12$ of this small fraction, as annual runoff, from a land area which is about $1/15$ of the earth's total. Future changes in climate will likely have a major impact on the patterns of precipitation and evapotranspiration, affecting water availability for

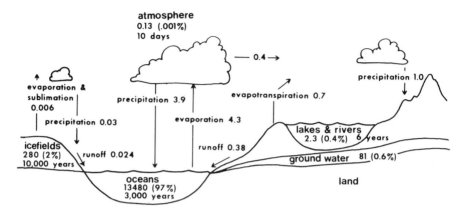

FIG. 9. Global hydrological cycle, showing the main reservoirs, transfers, and residence times (defined as reservoir size divided by transfer rate). Units are 10^{17} kg and 10^{17} kg • yr^{-1} for reservoir and transfer time, respectively. (Adapted from Anonymous 1972)

human consumption, irrigation, power generation, and direct use by natural and agricultural ecosystems.

Climatic Leverage Points

The term "climate" comprises much more than mere measurements of temperature and precipitation, and climate change must be measured in more ways than changes in temperature and precipitation fields (as in Fig. 7b and 8b). We can expect that a change in climate will involve changes in the frequency of occurrence of various types of weather conditions such as cloudiness, fog, heavy showers, thunderstorms, and tornadoes. Climate models of the GCM type do include many of the above factors and interpret correctly their influences on temperature, precipitation, and any other field one wishes to have plotted. However, details of the causative factors are lost when all are integrated into diagrams such as Fig. 7b and 8b. The best way to determine the effect of climate change on runoff, for instance, is to include a water balance section as part of the overall model and treat the water balance interactively and internally as part of that model (Klemes 1982).

Changes in precipitation will have the most obvious effect on the water balance, but in some situations, changes in evaporation rate may be just as important. Evaporation from a surface depends on a number of factors, including a supply of energy to vaporize the water and turbulent motion to carry the vapour up into the atmosphere. A simple relationship between evaporation and surface temperature is often adequate to describe its dependence on climatic factors. An expression used by Ferguson et al. (1970) shows evaporation increasing at about 4% per degree rise in temperature (at 5°C), with the rate decreasing somewhat as the surface temperature rises. Thus, global warming, by itself, will lead to higher evapotranspiration rates and may make climates more arid in spite of unchanged, or even slightly increased, precipitation income. Glantz and Ausubel (1984) warned that this effect in the U.S. Great Plains would result in a more rapid depletion of the already overused Ogallala aquifer that stretches from the Dakotas to Texas.

The timing of both precipitation and evaporation changes will also be important. Increased winter precipitation will be of value to agriculture only if it is not lost by direct evaporation or runoff. Higher wintertime temperatures in Polar regions will result in greater water loss due to snowmelt and runoff and will tend to lengthen the growing season, resulting in greater water use by vegetation. On the other hand, increasing atmospheric CO_2 concentration will raise the water-use efficiency of many plants (Rogers et al. 1983; Sionit et al. 1981) through partial stomatal closure. This may reduce water loss from vegetation, particularly in arid areas, sufficiently to cause significant increases in streamflow (Idso and Brazel 1984; Wigley and Jones 1985).

A factor that should not be overlooked is the potential feedback effect of large-scale management projects (such as river diversions, dams, and irrigation schemes) on climate (Vowinckel and Orvig 1982).

Global and Canadian Hydrology in 2085

Global Hydrology

The scenario chosen to illustrate the hydrological effects of CO_2 doubling is only one of a number of equally plausible scenarios currently available. The use of a different model might produce quite different results from those indicated here.

The UKMO model (Mitchell 1983) included water-balance terms, but the results were presented on a zonal, rather than geographical, basis. To provide a geographical representation, it was necessary to use the model's simulated temperature and precipitation changes and to calculate the effects on evaporation and runoff from these. While this procedure is not ideal, it is perhaps adequate in view of the low significance level of the model output and of the many other unknown factors involved.

The evaporation (E) changes predicted by the model are fairly evenly distributed with latitude, so the precipitation (P) changes dominate the runoff ($P - E$) field. The greatest increases in runoff, predicted by Mitchell, are in the mid- to high-latitude regions in the autumn through to spring seasons. Reductions in runoff are found in many equatorial and low-latitude regions and in much of the Southern Hemisphere. Model predictions of soil moisture changes are dominated by decreases, except for Northern Hemisphere mid- and high-latitudes in winter and early spring and Southern Hemisphere low-latitudes.

Canadian Water Budget

The *Hydrological Atlas of Canada* water balance map (den Hartog and Ferguson 1978) was used as the basis for the following analysis. As mentioned previously, the precipitation and evaporation isolines on this map had been "derived" from measured data by making adjustments bases on assumptions of consistency with other data, such as streamflow measurements. The authors have pointed out the provisional nature of their map and have indicated that improvements will be made to it as new data and improved methodology become available.

The present distribution of annual precipitation across Canada (Fig. 7a) was discussed previously. den Hartog and Ferguson's corresponding annual evapotranspiration map is shown in Fig. 10. Similar to the temperature map (Fig. 7a), the main geographical variation in evapotranspiration is with latitude, decreasing from 700 mm in extreme southern Ontario to less than 100 mm in the high Arctic, with a national average of near 250 mm. Water availability, of course, puts an upper limit on the evapotranspiration, which is especially noticeable in the southern Prairies and interior British Columbia.

In regions where P exceeds E, the excess of moisture is available for runoff. Therefore, the difference between Fig. 7a and 10 represents a map of annual runoff. To produce this, the data from the two figures were transferred to spreadsheets, using a 100 km square grid. The spreadsheets were subtracted from one another and the resulting runoff figures totalled for each of Canada's five major drainage areas (Fig. 11). These totals were then compared with the *Hydrological Atlas of Canada* drainage data (Greene and Godwin 1978) as a check on the accuracy of the operations that had been carried out. Discrepancies for the individual drainage areas ranged from 5 to 10%, and for the country as a whole amounted to less than 2%.

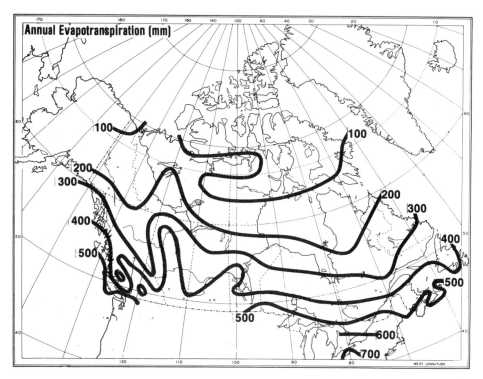

FIG. 10. Present geographical distribution of mean annual evapotranspiration in Canada. (Adapted from den Hartog and Ferguson 1978)

The next step was to apply the greenhouse warming predictions of changes in temperature (Fig. 7b) and precipitation (Fig. 8b) to the present fields in order to produce forecast fields of P and E for 2085. To do this, Fig. 7b and 8b were also encoded and transferred to spreadsheets. The predicted precipitation change was applied to the existing precipitation map to produce a forecast precipitation map for 2085. A slightly more complicated procedure had to be used for the evapotranspiration. The predicted temperature change map (Fig. 7) was converted to a change in evaporation rate by using the relationship derived from Ferguson et al. (1970), described above. This was then added to the present-day evapotranspiration map to yield a forecast map for 2085. The final step was to subtract the forecast maps for precipitation and evapotranspiration from one another to produce a forecast runoff map.

The various maps described above were summed by drainage area to yield summary totals which are presented in Table 2. For the country as a whole, the modelled increase in precipitation amounts to 142 mm, and in evapotranspiration to 32 mm. This results in a substantial increase in runoff (32%), most of which (52%) arises in the Arctic area, with more or less equal increases in the Hudson and Pacific areas (20% of total), and a lesser increase in the Atlantic area (8%). The simplified treatment of evaporation change may have resulted in an underestimation of evapotranspiration from this region.

A similar treatment, carried out for the United States by the Environmental Protection Agency (EPA) (Rind and Lebedeff 1984), extended over the southern part of Canada. Although their analysis used a very different grid size and geographical divisions, it is possible to make some comparisons. Their annual precipitation increases (up to 55°N) varied from 70 mm in Quebec to 250 mm in British Columbia, averaging 142 mm, the same as the present study predicts for the entire country. In spite of

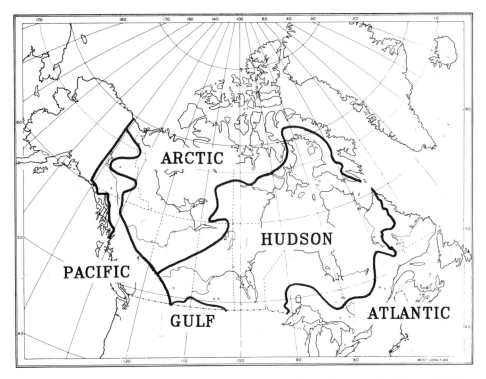

FIG. 11. Canada's major drainage regions. (Adapted from Greene and Godwin 1978)

increasing precipitation, its concentration during the winter, coupled with rising evapotranspiration, indicated higher probability of drought for the southeastern Canadian Prairies. This was an extension of a large area of increasing drought over most of the United States. The EPA model also indicated increases in the length of the growing season in southern Canada, varying from 45 d on the east coast to 25 d in the interior and 65 d on the west coast.

The above scenarios indicate that major changes are possible in the water balance of many regions of Canada. A substantial increase in winter precipitation, coupled with

TABLE 2. Present and predicted water balances of the major Canadian drainage basins using the UKMO GCM greenhouse warming climate scenario for 2085. Note that the data apply to only the Canadian portions of international basins.

Region	Area $\times 10^4$ (km^2)	Present			Predicted			Change		
		Prec.	Evap.	Runoff	Prec.	Evap.	Runoff	Prec.	Evap.	Runoff
		(mm • yr^{-1})			(mm • yr^{-1})			(%)		
Gulf	3	458	431	26	563	477	87	23	11	235
Pacific	94	1089	313	775	1342	344	998	23	10	29
Arctic	331	334	165	169	514	184	330	54	12	95
Atlantic	151	1022	399	623	1131	448	683	11	12	10
Hudson	356	559	272	287	652	308	344	17	13	20
Canada	935	607	259	348	749	291	459	23	12	32

possible reductions during the summer, will result in greater seasonal fluctuations in water levels in rivers and lakes. Increases in snowpack amounts, and earlier and more rapid springmelt, will have considerable effects on the rivers that drain the Rockies, for example. Changes in vegetation resulting from climate change will affect basin hydrological characteristics. We must not forget that if greenhouse warming does occur, it will also affect countries other than Canada, and effects elsewhere will have repercussions in Canada. For example, the expectation that the central United States would become considerably more arid would be bound to put pressure on Canada to export surplus water to the needy areas (see chapter by Bocking, this volume). While forecasts such as the above are not sufficiently accurate to be of much direct use to the water resource planner (Lettenmaier and Burges 1978), there are many useful measures that could be implemented now to help mitigate some of the undesirable effects of these changes in the future.

Canadian Water Quality

Any changes in the components of the water balance are bound to have repercussions on water quality, although they will likely be overshadowed by other concurrent changes due to population shifts and industrial expansion, for example. Some general inferences can be made, however, from the climate-change scenario described above: heavier winter precipitation and spring runoff will lead to greater sediment loads in rivers and streams, reduced summer precipitation will result in lower water quality because of reduced dilution effects, and generally higher temperatures will result in biological changes which will likely affect water quality.

Canadian Water Demand

Even in the absence of a change in climate, the demand for water will undoubtedly undergo considerable change during the next century (see chapter by Tate, this volume). This will be due not only to population growth and higher standard of living, but also to new patterns of use such as hydraulic mining and slurry pipelines. Many of the features of the changing climate may serve to exacerbate the problems created by increased demand due to other factors.

Pollution control measures are likely to require even more water for dilution purposes in future years, and this will aggravate even further the problem of low summer water levels mentioned above. On top of this, there will likely be increased demand for irrigation as farmers strive to increase production in spite of summer dryness.

The major influence on Canadian water demand during the twenty-first century may well be from outside our national borders. Many models predict reductions in precipitation and increases in evaporation for large areas of the United States. More plentiful supplies of water in Canada would undoubtedly lead to increased political pressure for large-scale intercountry water transfer, the northward shift of many agricultural activities, and perhaps even the migration of farm populations.

Impacts on Ecosystems and Society

Mankind has become aware of the inconstancy of climate through historical study, at first through oral and written records extending back less than a few hundred years, but more recently through geological evidence reaching millions of years into the past. The influence of variations in weather and climate on human affairs has been considerable, although it has not always been recognized by historians. While the impacts of severe storms and prolonged droughts have not escaped attention, more subtle changes have often been overlooked. Neumann (1976, 1978, and 1979) examined the role of weather in such historical events as the French Revolution and the Mongol invasions of Japan and noted that the exceptionally cold winter of 1657-58, during the "Little Ice Age," permitted the Swedish army to cross over frozen seas and defeat the Danes. Parry (1978) looked at the effects of changing climates over the past thousand years on: glacier advance and retreat, changes in natural vegetation, agricultural production, dates of wine harvests, and the establishment and abandonment of human settlements in Europe and North America. He also warned that "coincidence of climatic change and economic change [does not indicate] a causal connection" and advised the adoption of "a deductive approach, one which makes retrodictions about the past and tests them against historical actuality" (Parry 1981).

A careful study of past climatic variations and their societal impacts can be of great benefit in predicting impacts of future climatic change. However, if the proposed scenario for greenhouse warming is correct, the next century will see greater changes in climate than any experienced since the end of the last ice age, 12 000 yr ago. The impacts could far exceed any that have been experienced since the dawn of recorded history. The rapid changes in climates will tend to displace natural ecosystems. As Robinson (1981) has suggested, steppes may be transformed into forests or deserts, forests to deserts or bogs, and coniferous forests to deciduous forests. Mountain glaciers may expand or disappear, streams turn into rivers, and some rivers dry up. At the same time, other areas may be relatively unaffected. The direct effects on agricultural production and water supplies could well have severe social, economic, and political repercussions at the national and international levels. Rather than being "insulated" from climate by technology, agricultural systems in many instances may be just as vulnerable to climatic change as natural ecosystems (Barker et al. 1981). Human reactions to the uncertainty of changing climate will vary considerably. "Some will see great opportunity, some will see the need for careful planning, some will see impending disaster" (Ausubel 1983). The impacts will be greatest in areas that are climatically sensitive, such as those where temperature or moisture conditions are already marginal for plant or animal life. However, they may be felt first in those areas where resources are stressed the most. In Canada, the most climatically sensitive areas are the Arctic region and much of the Prairies, but impacts on these areas may not be noticed as soon as those in more heavily populated areas, for example. The likely impacts of our greenhouse warming scenario on Canadian ecosystems and society will be discussed in the following section.

What Can We Expect?

Our hypothetical climate change scenario for 2085 predicts a general temperature increase of 2–3°C nationally in summer and up to 6°C in central Canada in winter. Summer precipitation is expected to increase mainly in the Arctic and in southeastern Canada, while most of the remainder of the country will have no change or a reduction. Winter precipitation is expected to increase dramatically in the west, decreasing to no change on the east coast.

This will likely result in greater snowpack buildup in mountainous terrain and in the Arctic, particularly in the western section. However, higher temperatures will mean that more of the precipitation will likely fall as rain and that thaws will be more frequent and spring disappearance earlier. Thus, river flows on the Prairies, for example, will peak earlier and will likely be much reduced during the summer months. The central part of the country will have more snow and rain during the winter, but winter thaws, with snowmelt, will be more frequent. Summers will tend to be much drier, with a higher frequency of droughts over all of the country except the Arctic and the St. Lawrence–Atlantic region.

This scenario may be compared with estimates of Canadian climate during the Hypsithermal Period of 5000–8000 yr ago (Fig. 1), when global temperatures were about 2°C higher than at present. At that time the Arctic was more humid than now, and the boreal forest extended northward to the Arctic islands (McKay 1978). The Prairies were more arid, and the arid zone stretched as far eastward as the Great Lakes and over much of the United States (Butzer 1980).

Bernard (1980) expressed the view that historical climate analogues may provide better forecasts of the geographical and seasonal distributions of greenhouse warming than computer models. He suggested that the drought of the Thirties in the Great Plains of North America may provide some indication of what lies ahead for the region during the next century. Mean temperatures were as much as 1°C higher than at present, due mainly to warmer summers. Precipitation, on the other hand, was more variable, with summer amounts lower and winter amounts higher, than at present.

Summer aridity in central Canada would likely be accentuated due to the combination of higher temperatures (and hence higher evaporation rates), reduced summer precipitation, and longer growing season. However, this would be partially offset because of the increased water-use efficiency resulting from higher CO_2 concentrations (Rosenberg 1981). While natural vegetation would be exposed to the dry summer conditions for a longer period, agricultural crops could take advantage of the higher temperatures through earlier-maturing varieties, thus reducing water consumption.

The direct effect of increasing atmospheric CO_2 concentration on plant growth has been studied in the laboratory (Kimball 1983) and in the field (Rogers et al. 1983), and the subject has even spawned a Symposium with a published volume (Lemon 1983). Indications are that substantial gains in growth rate and yield, as well as water-use efficiency, are to be expected for many native and cultivated species of plants, especially those that employ the C_3 photosynthetic pathway (Ausubel 1983). This latter includes most temperate-zone species. Kimball's results for a wide range of crops showed an average yield increase of 33% in response to CO_2 doubling. Individual species estimates varied from 25% for barley and potatoes to 32% for wheat and 54% for legumes. It is interesting to note that the greatest percentage increases for wheat were found under conditions of water stress. Few CO_2-enrichment studies have spanned an entire growing season or longer, and there have been suggestions that plant response to CO_2 enrichment may be transient phenomenon. Jones et al. (1984) and Hanelka et al. (1984) applied $2\times$ and $4\times$ CO_2 enrichment to soybeans during the entire growing season and found up to 50% increases in yield and water-use efficiency for CO_2 doubling and 70% increase in yield for CO_2 quadrupling. Perhaps the most exciting evidence on the subject that has come to light so far relates to tree-ring studies of subalpine conifers growing in the western United States (LaMarche et al. 1984). They examined tree rings from a number of sites for the period from 1800 to the present and found a trend of increasing ring width index that could not be explained solely on the basis of climate changes. The hypothesis that this is due to CO_2 fertilization remains to be verified, but if it is true, then society has already been reaping benefits from CO_2 enrichment, in terms of increased plant growth.

The northward extension of agricultural regions because of greenhouse warming would be of little benefit across much of Canada, because of the lack of suitable soils (Stewart 1984). Only a few areas, such as the clay belt of northern Ontario and Quebec, and river valleys in the Yukon and Northwest Territories, would be able to take advantage of the higher summertime temperatures and longer frost-free season. The increase in winter precipitation would add to soil erosion and drainage problems, thus providing a challenge for agronomists to develop more effective methods of storing and conserving moisture for use during the arid summers. Another likely disadvantage of higher temperatures would be an increase in agricultural pests and diseases. The milder winters and longer growing seasons would increase the number of potential species and their reproduction rates, creating the need for the development of new control measures.

Prairie Agriculture

An increase in Prairie mean annual temperature of 3°C (Fig. 7b) would be equivalent to a southward shift of about 500 km, in terms of present climate. The length of the frost-free season could be expected to increase about 20–30 d and the time for a wheat crop to reach maturity (assuming a constant growing degree-day requirement) to decrease about 30% (Maybank 1985; Bootsma et al. 1984). Since these changes involve only temperature, they may ignore important effects of moisture, day-length, and soil properties. Also ignored are changes in the frequency of occurrence of severe weather, which may have much more limiting effects on agriculture than changes in mean values (Wigley 1985).

The expected increase in summer aridity would make the soil more susceptible to wind erosion (Wheaton 1984), which is already a major problem in the Prairies (Coote 1984). For plants the dryness would be at least partially offset by the improved water-use efficiency due to the higher atmospheric CO_2 concentration and the possibility of reducing seasonal crop water requirements through the use of more rapidly maturing varieties. This latter effect might reduce growing-season water use by as much as 10% in spite of higher evaporation rates (Terjung et al. 1984).

The Canadian Prairies are one example of a climatically marginal agricultural area (Treidl 1978). A comparison of grain-yield variability with weather in several countries (Sakamoto et al. 1980) showed that Canada had the greatest variability, as well as the highest probability of occurrence of consecutive years with poor yields. The possibility of crop losses from drought, frost, or hail is ever present. Since the choice of crops, varietal breeding, and agronomic practices have all been adapted to the existing conditions (including climate) in each agricultural area, it is reasonable to expect that any changes in these conditions will affect yields (Bolin 1980).In most cases, the immediate effects would be negative, although often countermeasures could be introduced to derive benefit from the changes. For example, an increase in precipitation might result in the loss of arable land due to flooding. If the problem were dealt with by appropriate water management techniques, the end result could well be an increase in yield (see chapter by deJong and Kachanoski, this volume).

A search of the literature has revealed relatively few quantitative studies of the impacts of greenhouse warming relevant to agricultural production on the Prairies. A comprehensive study by Bootsma et al. (1984) will be discussed, along with some "cooling" scenarios for Canada and "warming" scenarios for Alberta and United States, and an attempt will be made to interpret them in relation to the anticipated climatic effects of greenhouse warming on the Canadian Prairies.

Williams (1976) used a statistical model to assess the effects of temperature decreases on Prairie grain production. He assumed that the precipitation changes would alter yields, and that the temperature changes would alter the cropped area, through changes in the length of the growing season. For a 3°C decrease in temperature, he predicted that the barley-growing area of the Prairies would shrink by 30%. Ravenholt (1976) and Williams and Oakes (1978) found similar decreases in the wheat-growing area corresponding to a 1°C temperature decrease. For a reduction in precipitation of 30%, Williams found decreases of 11% in barley yield and 16% in wheat yield. An increase of 10% in precipitation had a negligible effect on barley yield and increased wheat yield 3%.

More detailed results were derived from another empirical study of wheat yields in several U.S. states (Thompson 1975). Those for North Dakota, adjoining the Canadian Prairies, show a positive response to precipitation increase, but a negative response to increasing temperature. A 2°C increase in temperature depressed yields 9% if the precipitation remained constant, or 2% if the precipitation increased 20%.

Bootsma et al. (1984) investigated the impacts of temperature changes ranging from −3 to +3°C, and precipitation changes up to ±40% of present values, on dry matter yields of a wide range of crops grown across Canada. They concluded that a rise in temperature without a change in precipitation would reduce crop yields, except in a few northern areas. A 20% increase in precipitation would increase yields between 0.5 and 0.8 t • ha^{-1}. A combination of higher temperatures and increased precipitation would result in increased yield in all areas, but especially in those with more arid climates.

Williams (1985) estimated the effect of CO_2 doubling on Alberta agricultural production as a whole, assuming no change in precipitation, would be a 16% increase. If the direct effect of CO_2 enrichment were included, he estimated the increase would rise to 54%.

Interpreting these results for our present scenario leads one to speculate that there would be a slight positive effect of the expected increase in annual precipitation, but very little change in yield due to the temperature rise and higher risk of summer drought. The main effects would likely be due to expansion of the area under cultivation (made possible by the milder temperature), and to CO_2 enrichment. These changes would provide challenges to agricultural scientists to maximize the benefits, principally in the areas of (1) improving the characteristics of many northern soils to make them more suitable for crop growth, (2) managing the increased winter precipitation, expected in many areas, so that loss of agricultural land due to excess moisture is minimized, (3) developing conservation practices to use excess winter moisture to compensate for the more frequent summer shortages, and (4) selecting and breeding crop varieties that are more suitable for the changed climate, and enriched CO_2, conditions.

The challenge to the Canadian Prairies would be especially important in view of the expected decreases in agricultural productivity due to increasing aridity in areas such as the United States and the southern USSR (although the northern USSR would also gain suitable land).

Great Lakes Watershed

The Great Lakes watershed has an area of about three-quarters of a million square kilometres and a population of over 35 million people, of which about 41% of the surface area, and 20% of the population, are in Canada.

The climate change scenario for CO_2 doubling, presented in this chapter, indicates an increase of about 100 mm $\cdot$ yr^{-1} in precipitation over the watershed, occurring mainly during the summer season. This would tend to spread out the present early spring runoff peak, making the streamflow more even throughout the year. Concurrently, higher temperatures would produce an increase in evapotranspiration, amounting to about 50 mm $\cdot$ yr^{-1}. The balance, 50 mm $\cdot$ yr^{-1}, would result in an increase in watershed runoff of about 38 km^3 $\cdot$ yr^{-1}, or 1200 m^3 $\cdot$ s^{-1}, i.e. 8% greater than the present runoff.

The present population and the high level of industrial activity are already placing heavy demands on the water resources of the region. During the next century, an additional 1000 m^2 $\cdot$ s^{-1} is likely to be required to satisfy increasing local needs, quite apart from the possibility of exportation of water to the U.S. High Plains region to ameliorate their increasing aridity.

A recent study by Cohen (1986) examined in detail the likely effects of CO_2 doubling on the water resources of the Great Lakes region. He considered that the changing climate would have direct effects on water flows and lake levels, water withdrawals, consumption, and external demand for water, electricity, oil, and gas demands, crop type and yield, recreation, storm damage, and road maintenance, and forestry and wildlife, not to mention a host of indirect effects involving all facets of society and the economy.

He suggested that alterations in the balance of inflows and outflows would first of all affect lake levels. Such fluctuations in the past have had significant effects on hydroelectric power generation and lake shipping, as well as causing considerable shoreline erosion.

Cohen's study considered several CO_2 warming scenarios, of which the results varied from a 28% decrease in basin water supply to a 7% increase, the latter being similar to the present study. Even the most optimistic of these would barely keep pace with the projected demand increases, and leave no surplus for export. The more pessimistic projections would result in large losses in hydroelectric power production,

166

which would be offset by reduced winter demand for heating, but augmented by increased summer demand for air conditioning.

In his concluding remarks, Cohen identified areas requiring further research. These included (1) information on weather variables other than temperature and precipitation, such as windspeed, humidity, cloudiness, etc., (2) information on aspects of the changes other than mean values, i.e. lengths of dry and wet spells and frequencies of occurrence of extreme events, (3) consideration of the effects of elevated CO_2 on plants and the resulting effects on the water balance, and (4) studies of the linkages of water supply with irrigation and energy demand.

A Warmer, Moister Arctic

During the Hypsithermal, the Canadian Arctic climate was still dominated by a residual ice sheet over northern Quebec and Labrador. There is evidence that winter temperatures in the region were 3-4°C higher than today, with those in the western Arctic perhaps as much as 11°C higher (Hengeveld 1984). Summer temperatures were 1-3°C higher generally, and 5° in the west. The climate was considerably moister than it is at present, and the tree line was about 250 km farther north.

A Hypsithermal analogue is perhaps not a bad estimate of the modelled conditions predicted for CO_2 doubling, although Fig. 7b indicates a little less warming than mentioned above. The warming scenario would probably result in more open water and higher water temperatures, providing more moisture to the overlying air, leading to the large increase in precipitation simulated in Fig. 8b.

The higher temperatures, together with the insulating effect of the increased winter precipitation (still mainly snow), would lead to the gradual reduction in the thickness of the permafrost layer. This would eventually result in increasing land instability, the development of thermokarst topography, and frequent mudslides (see chapter by Van Everdigen, this volume). At the same time, the higher snowfall could result in glacier advance and an increase in iceberg production.

Inland lakes and rivers would have a shorter season of ice cover by perhaps 2-3 wk (Hengeveld 1984), while a considerable retreat of the Arctic Ocean ice pack would be expected. Glacial ice might well increase in some regions due to the higher winter snowfall, and together with the longer summers might lead to increased iceberg production. Any sea-level change, resulting from thermal expansion of the oceans and possible glacier melt, is likely to be negligible in as short a time period as 100 yr.

The many changes in climate and physical environment, described above, would have profound effects on the vegetation, animals, and humans resident in the Arctic. The increases in temperature, CO_2 concentration, and precipitation will all aid agricultural development, at least in those areas with suitable soils. There will undoubtedly be changes in fish populations although these are very difficult to predict. Accessibility to fishing grounds will improve as ice cover diminishes.

Some of the effects of greenhouse warming on human life in the Canadian Arctic will likely include (1) a reduction in the hazards of living due to amelioration of the harsh environment, (2) reduced heating costs, (3) less suitable conditions for winter road use, offset somewhat by easier marine access, (4) increased damage to buildings and roads due to decaying permafrost, (5) greater agricultural potential, leading to lower food costs, and (6) substantial increases in forest productivity, but greater risk of forest fires because of summer dryness.

What Should We Do?

According to Fraser (1983), the "stakes are much too high for [us to play] wait and see" in the "game" of climate change. While some parts of the world climate puzzle are known with certainty (e.g. the present increase in atmospheric CO_2 concentration), others are as yet very much a guess (e.g. amounts and types of future energy sources). It may be decades before all the pieces fit together and we have incontrovertible proof

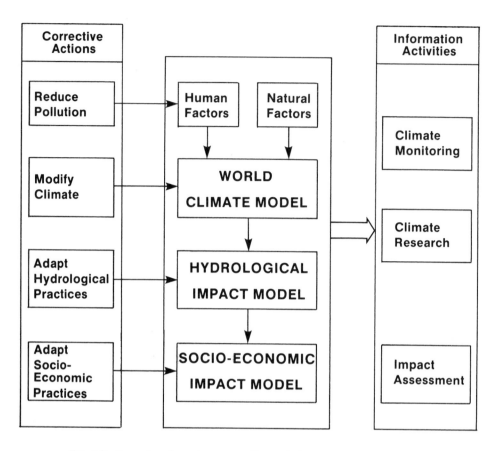

FIG. 12. Procedure for minimizing effects of climate change on society.

of greenhouse warming, and by that time it may be too late to deal with the changes without disastrous results.

One possible plan of action is outlined in Fig. 12. It involves the following: (1) continually working to improve forecasts of the nature, timing, and magnitude of climate change and its likely impacts on hydrological resources and society; this will require monitoring, research, and impact assessment activities (Hecht and Imbrie 1979; Kates and Hare 1980; Ausubel 1983); (2) devising ways to eliminate or reduce the potentially negative impacts, and to gain benefit from the positive impacts (Kellogg and Schneider 1974; Slater and Levin 1981); and (3) striving to prepare society for the change by developing and implementing an appropriate plan of action; if this were done gradually, it might avoid an abrupt and potentially catastrophic transition.

In 1975, the U.S. National Academy of Sciences published a report entitled "Understanding Climatic Change — A Program for Action" (Anonymous 1975) which recommended establishing programs for climatic monitoring, modelling, research, and palaeoclimatic study. All would be directed toward increasing our understanding of climatic variations, with the goal of improving our ability to predict them.

Information Acitivities

Through adequate programs of monitoring and research, we can hope to improve estimates of whether climate will change, by how much, and when the change will likely take place. This knowledge can then form the basis for the development of methodology for reducing negative impacts, taking advantage of positive impacts, and minimizing the disruption to society during the transition period.

In response to this challenge, the Canadian Climate Program was established during the late 1970s, having as its aims: to provide climatological information required to assist in meeting Canada's social, economic, and environmental objectives, to provide integration, stimulation, and an accountability for Canadian climatological activities, and to work along with, and in support of, the major goals of the World Climate Program and Canada's interests in related international programs (Anonymous 1983b).

As part of its mandate, the Canadian Climate Program carried out surveys and held a number of seminars and workshops across the country on a variety of climate-related topics. More information regarding the Canadian Climate Program, especially in relation to water resources, may be found in Anonymous (1982) and Anonymous (1983b).

Climate monitoring

Monitoring has two main aspects. The first involves regular climatological stations measuring standard meteorological parameters such as temperature, precipitation, wind, and pressure (Hogg 1982). These measurements must be continued, particularly at selected "baseline" stations that are situated away from local sources of heat and atmospheric pollution. Efforts must continue to be directed toward increasing the sparse oceanic, southern hemispheric, and upper air meteorological sampling networks and toward developing methods to produce better hemispheric temperature averages. The second aspect of monitoring involves atmospheric pollutants known, or suspected, to be agents of climatic change. These include not only CO_2 but carbon monoxide, methane, and the nitrogen and sulphur oxides among others. The need for this type of monitoring was first recognized internationally at the 1972 Stockholm Conference on the Human Environment. This led to the creation of the Global Environment Monitoring System which coordinates a number of monitoring programs around the world (Wallen 1980). The World Meteorological Organization's Background Air Pollution Monitoring Network had 109 stations operating in 41 countries in 1980 and saw the need to expand these to 200 stations in 70 countries. In Canada, CO_2 is now monitored on a routine basis at Sable Island, off the Nova Scotia coast, and at Alert, in the eastern High Arctic.

Such stations provide invaluable data for the development and verification of climate change models and, coupled with better estimates of CO_2 source and sink strengths and locations, will lead to better estimates of future atmospheric CO_2 concentration (Smagorinsky 1983). Kellogg (1983) has provided a good summary of the measurements needed for early identification of climatic change, together with the purpose, status, and priority of each. A key element that has received inadequate attention up to now is the solar radiation flux at the top of the atmosphere. This is the main energy source for the global climate system, and surprisingly few accurate data are available describing its variations.

Climate research and impact assessment

"We lack an entirely satisfactory theory of climatic trends and change. Without such a theory, prediction of future climate and hence availability of water supplies is impossible. Geophysicists, glaciologists, oceanographers, meteorologists, hydrologists, climatologists and specialists from many other fields are seeking to understand the earth's past changes so that they can better comprehend the present and the immediate future. A better understanding of climate could have inestimable benefits for the whole world" (Anonymous 1977).

The above statement clearly spells out the view of the World Meteorological Organization that the challenge of climatic change can only be met through coordinated research involving a large number of disciplines. A major thrust of this research will undoubtedly continue to be modelling of the carbon cycle and the atmospheric and oceanic components of the climate system (Hare 1981). Although at present the majority of this work is being carried out in the United States, the United Kingdom, and the USSR, recent improvements in Canadian climate computation facilities and budgeting bode well for increased activity in this country. Substantial work on the role of the oceans

169

in the CO_2 balance is being carried out at Canada's federal oceanographic institutes, and there is also highly significant work being done by many other federal, provincial, and university research groups. What is needed most, perhaps, is a focussing of effort, which is now being provided by the Atmospheric Environment Service's Canadian Climate Program.

Corrective Actions

While it is true that variation is a natural attribute of the earth's climate system, the changes expected during the next century, because of greenhouse warming, will far outweigh any known historical changes. While there is still a possibility that the climate system is more stable than we think, and will be able to absorb the greenhouse gases without major change (Hare 1981), mankind has too much at stake to sit back passively and wait. There are a number of things we can do to reduce the perturbation (CO_2 increase) itself, or reduce or ameliorate the impacts of the changes it is expected to cause, and we will consider a few of these now.

Reduce pollution

Goudriaan and Ketner (1984) used a carbon cycle model to address various management possibilities. They found that fixing the rate of deforestation, land reclamation, and rangeland burning at the 1980 level would reduce the projected atmospheric CO_2 concentration in 2030 by 12 vpm. Also, a fairly large reforestation program ($10 \text{ Mha} \cdot \text{yr}^{-1}$ in 1980, increasing annually by 1.5% would result in a 10 vpm reduction by 2030. In their words, "These simulation exercises demonstrate that the effect of environmental protection measures on the atmospheric CO_2 concentration is minor in comparison to the effect of changing energy consumption."

A report issued by the U.S. Environmental Protection Agency (Seidel and Keyes 1983) concluded that only a worlwide ban on coal combustion would have a substantial effect on the next century's greenhouse warming, reducing it by about 30%. They estimated that even a 300% tax on fossil fuels would only delay the inevitable a few years. Another report, by the U.S. National Research Council (Nierenberg 1983), was much less pessimistic, seeing fuel taxes as a potentially effective way of reducing emissions. The latter report did not think that the use of such strong measures was even warranted yet in view of the many uncertainties in the CO_2 issue. An analysis of the two reports by Hileman (1984) tended to favour the latter report, suggesting that "EPA's conclusions follow almost inevitably from its assumptions." Whichever scenario one prefers, it is safe to suggest that colossal pressure, whether environmental, economic, or political, would be required to reduce CO_2 emissions sufficiently to eliminate the threat of greenhouse warming.

Modify climate

There have been many schemes proposed over the years to modify climate to make it more suitable for human activities. These activities would be classed as "deliberate" climate modification mechanisms (if they worked!), in contrast with the inadvertent mechanisms discussed earlier in this chapter. In some cases the only difference between the two classes is the question of intent!

Kellogg and Schneider (1974) reviewed some of the proposed climate-change schemes that are applicable on a global or hemispheric scale. These included (1) installing pumps in the Bering Strait to pump Arctic water into the Pacific, thereby drawing Atlantic water into the Arctic Ocean and melting its ice pack (Borisov 1973), (2) achieving the same result by detonating nuclear devices in the Arctic Ocean, or by covering it with soot to absorb more radiation, (3) injecting dust or aerosol into the stratosphere to reduce surface temperatures, (4) diverting ocean currents, such as the Gulf Stream, and (5) creating artificial lakes by flooding low-lying land.

Some of the above have the potential to counteract part of the predicted changes in climate due to greenhouse warming, and the question will certainly arise some day, perhaps sooner rather than later, whether they should be used. It will be a difficult, if not impossible, question to answer, since not only are we unable to predict the outcome of such an action, but there almost certainly will be a conflict of interest regarding the desirability of the potential outcome. For example, greenhouse warming will benefit some countries to the detriment of others. Any operation to nullify the effects of the warming would almost certainly be opposed by the potential beneficiaries. Thus, any climate change operation that could have effects in more than one country would pose the question of how the decision would be made to carry it out, and how reparations would be paid to any party injured by the operation. With this view in mind, it will be essential to construct international mechanisms to guarantee that all new knowledge of the global climate system will be used only for constructive purposes.

Adapt hydrology

Since water resources are used for a multitude of purposes, including human consumption, industry, recreation, irrigation, and many energy processes, implications of any change in climate that affects the water cycle are obvious.

As better climate forecasts become available, hydrologists will have to use them with their basin models (Klemes 1982) to predict future water flows. This will provide some advance warning of potential trouble areas, with respect to water shortages and surpluses, and may lead to reallocation in areas of possible conflict.

The present greenhouse warming scenario predicts increases in annual precipitation for almost all parts of Canada (Fig. 8b). However, the uneven seasonal distribution is expected to result in a surplus in many areas during the winter and shortages during the summer. This will create a greater potential for flooding and land slides, particularly in northern British Columbia and the Yukon. Much of the central part of the country is expected to have warmer and drier summers, which is likely to increase the demand for recreational water use. This will have to be met by expansion of storage capabilities to preserve some of the surplus winter precipitation, so it can be used during the drier summers for recreation, irrigation, and power generation. Other appropriate actions might include tapping of additional aquifers, and interbasin transfers, although there will undoubtedly be some opposition to each of these from groups with conflicting interests (see chapter by Rosenberg et al, this volume). Major capital expenditures may be required for hydrologic structures designed to cope with the new climatic regime.

A Climate Change Seminar held in Regina in 1981 as part of the Canadian Climate Program received suggestions from provincial ministers from across the country regarding action priorities to meet the challenge of climatic change. Those most commonly heard included (Anonymous 1983b), increase the number of observing stations, and the relevance of the data collected, modify the design of hydrological works to make them more "climate proof", make better use of existing hydrological data, and develop analyses relevant to the climatic-change scenario, and keep the public and decision makers informed regarding climate-change scenario updates.

Adapt socio-economic systems

The final, and perhaps most difficult, link in the climate-impact chain (Fig. 12) is that involving society itself. It is probable that modern society is less able to adapt to a changing climate than any earlier societies because of its complexity and highly structured nature. On the other hand, it has more powerful tools than ever at its disposal and a greater capability to decide what is best to do than ever before. Of prime importance will be the continual updating of the climate-change forecast and its channelling to the public, politicians, and decision makers. This will engender an awareness which hopefully will ensure that adequate consideration is taken of climatic change in decisions regarding future structures, programs, and policies.

Although societies have had much experience in responding to short-term environmental crises, such as those caused by hurricanes, forest fires, volcanoes, and droughts, the changes associated with greenhouse warming "will be slow, pervasive environmental shifts. They will be imperceptible to most people from year to year because of the annual range of climatic variation. It would be prudent to begin thinking now about what the changes might be and how humankind might best avoid or ameliorate the unfavourable effects and gain the most benefit from the favourable ones." (Revelle 1982).

Conclusion

The climate of our planet has undergone many changes since its birth over a billion years ago. The appearance of life introduced a system of feedbacks which presumably has had a stabilizing influence. Even so, natural variations continue to occur and, as yet, are not fully understood. The rapid industrialization of the past century has released vast quantities of CO_2 and other pollutants to the atmosphere. This has raised the possibility of a major change in global climate which could have profound effects on the biosphere and man. Without adequate foreknowledge of these changes, the outcome could be disastrous. However, if society is made aware of them, and plans for them, it may be possible to meet the challenge and turn them to our advantage.

Only the passage of time will provide the definitive answer. In the meantime, we will have to continue picking and probing, refining our predictions with the addition of each hard-won piece of knowledge. That way we will be able to gain sufficient lead-time to permit us to develop strategies that will avoid the negative aspects of the changes, and capitalize on the positive.

References

AGEE, E. M. 1980. Present climatic cooling and a proposed causative mechanism. Bull. Am. Meteorol. Soc. 61: 1356-1367.

ANONYMOUS. 1972. World water balance. FAO-IASH-WMO, Proc. Reading Symposium, Gentbrugge, Paris, Geneva.
    1975. Understanding climatic change — A program for action. U.S. Committee for the Global Atmosphheric Research Program, National Academy of Sciences, Washington, D.C. 239 p.
    1977. Weather and water. World Meteorol. Publ. No. 463: 24 p.
    1982. Proceedings of the CCP Water Resources-Climate Workshop. Atmospheric Environment Service, Downsview, Ont. Workshop held in Edmonton, Feb. 28-29, 1980. 212 p.
    1983a. Understanding CO_2 and climate. Annual Report, Atmospheric Environment Service, Downsview, Ont., June. 13 p.
    1983b. Canada water year book 1981-1982, Water and the economy. Environment Canada, Ottawa, Ont. 105 p.

AUSUBEL, J. H. 1983. Can we assess the impacts of climatic changes? Clim. Change 5: 7-14.

BARKER, R., E. C. GABLER, AND D. WINKELMAN. 1981. Long-term consequences of technological change on crop yield stability; the case for cereal grain. *In* A. Valdea [ed.] Food security for developing countries. Westview Press, Boulder, CO.

BERNARD, H. W. 1980. The greenhouse effect. Ballinger Publ. Co., Cambridge, MA. 189 p.

BOLIN, B. 1980. Climatic changes and their effects on the biosphere. World Meteorol. Org. Publ. No. 542: 49 p.

BOLIN, B., E. T. DEGENS, S. KEMPE, AND P. KETNER [ED.]. 1979. The global carbon cycle. SCOPE 13, John Wiley & Sons, Chichester. 491 p.

BOOTSMA, A., W. J. BLACKBURN, R. B. STEWART, R. W. MUMA, AND J. DUMANSKI. 1984. Possible effects of climate change on estimated crop yields in Canada. Agriculture Canada, Research Branch, Tech. Bull. 1984-9E.

BORISOV, P. 1973. Can man change the climate? Progress Publishers, Moscow. 175 p.

BRAZEL, A. J., AND S. B. IDSO. 1979. Thermal effects of dust on climate. Ann. Assoc. Am. Geogr. 69: 432-437.

BROECKER, W. S., T. TAKAHASHI, H. J. SIMPSON, AND T. H. PENG. 1979. Fate of fossil fuel carbon dioxide and the global carbon budget. Science (Wash., DC) 206: 409-418.

BRUCE, J., AND H. HENGEVELD. 1985. Our changing northern climate. Geos 14: 1-6.

BRYAN, K., F. G. KOMRO, S. MANABE, AND M. J. SPELMAN. 1982. Transient climate response to increasing atmospheric carbon dioxide. Science (Wash., DC) 215: 56-58.

BRYSON, R. A., AND B. M. GOODMAN. 1980. Volcanic activity and climatic changes. Science (Wash., DC) 207: 1041-1044.

BURROUGHS, W. 1981. Climate and the earth's albedo. New Sci. 91: 144-146.

BUTZER, K. W. 1980. Adaptation to global environmental change. Prof. Geogr. 32: 269-278.

CALLENDAR, G. S. 1958. On the amount of carbon dioxide in the atmosphere. Tellus 10: 243-248.

CHARNEY, J. G., P. H. STONE, AND W. J. QUIRK. 1975. Drought in the Sahara: a biogeophysical feedback mechanism. Science (Wash., DC) 187: 434-435.

CHARTON, F. L. 1973. An additional comment on the La Porte precipitation anomaly. Bull. Am. Meteorol. Soc. 54: 26-28.

CLARK, D. L. 1983. Frozen ocean with the key to world climate. Geogr. Mag. Lond. 55: 352-356.

COHEN, S. J. 1986. Impacts of CO_2-induced climatic change on water resources in the Great Lakes. Clim. Change (In press)

COOTE, D. R. 1984. The extent of soil erosion in western Canada. Soil Erosion and Land Degradation, Proc. Second Annual Western Provincial Conference, Rationalization of Water and Soil Research and Management, Nov. 29 to Dec. 1, 1983. Saskatchewan Institute of Pedology, University of Saskatchewan, Saskatoon, Sask.

CUMBERLAND, J. H., J. R. HIBBS, AND I. HOCH [ED.]. 1982. The economics of managing chlorofluorocarbons: stratospheric ozone and climatic issues. Resources for the Future, Baltimore, MD.

DEGENS, E. T., H. K. WONG, AND S. KEMPE. 1981. Factors controlling global climate of the past and the future, Chap. 1, p. 3-24. In G. E. Likens [ed.] SCOPE 17. John Wiley & Sons, Chichester. 430 p.

DEN HARTOG, G., AND H. L. FERGUSON. 1978. Water balance — Derived precipitation and evapotranspiration, Plate 25, In Hydrological atlas of Canada. Fisheries and Environment Canada, Supply and Services, Ottawa, Ont.

ELSAESSER, H. W. 1984. The climatic effect of CO_2: a different view. Atmos. Environ. 18: 431-434.

ENGELMANN, R. J. AND G. A. SEHMEL. 1976. Atmosphere-surface exchange of particulate and gaseous pollutants (1974). Energy Research and Development Administration, NTIS, Springfield, VA. 988 p.

EPSTEIN, E. S. 1982. Detecting climate change. J. Appl. Meteorol. 21: 1172-1182.

ETKINS, R., AND E. S. EPSTEIN. 1982. The rise of global mean sea level as an indication of climatic change. Science (Wash., DC) 215: 287-289.

FERGUSON, H. L., A. D. J. O'NEILL, AND H. F. CORK. 1970. Mean evaporation over Canada. Water Resour. Res. 6: 1618-1633.

FLÖHN, H. 1982. Oceanic upwelling as a key for abrupt climatic change. J. Meteor. Soc. Jpn. 60: 268-273.

FOLLAND, C. K., D. E. PARKER, AND F. E. KATES. 1984. Worldwide marine temperature fluctuations, 1856-1981. Nature (Lond.) 310: 670-673.

FOWLER, W. B., AND J. D. HELVEY. 1974. Effect of large-scale irrigation on climate in the Columbia Basin. Science (Wash., DC) 184: 121-127.

FRASER, H. M. 1983. Making the most of climate change. Presented at Futurescan 83, Saskatoon, 9 June. 16 p. typescript.

GATES, W. L., AND K. H. COOK. 1980. Preliminary analysis of experiments on the climatic effects of increased CO_2 with the OSU atmospheric general circulation model. Rep. No. 14, Climate Research Institute, Oregon State University, Corvallis, OR. 33 p.

GILLILAND, R. L. 1982. Solar, volcanic, and CO_2 forcing of recent climatic changes. Clim. Change 4: 111-131.

GILLILAND, R. L., AND S. H. SCHNEIDER. 1984. Volcanic, CO_2 and solar forcing of northern and southern hemisphere surface air temperatures. Nature (Lond.) 310: 38-41.

GLANTZ, M. H., AND J. H. AUSUBEL. 1984. The Ogallala aquifer and carbon dioxide: comparison and convergence. Environ. Conserv. 11: 123-131.

GOUDRIAAN, J., AND P. KETNER. 1984. A simulation study for the global carbon cycle, including man's impact on the bioshpere. Clim. Change 6: 167-192.

GREENE, G. A. D., AND R. B. GODWIN. 1978. Annual large river flow, Plate 22, In Hydrological atlas fo Canada. Fisheries and Environment Canada, Supply and Services, Ottawa, Ont.

GRIBBIN, J. 1973. Planetary alignments, solar activity and climatic change. Nature (Lond.) 246: 453-454.

[ED.]. 1978. Climatic change. Cambridge University Press, Cambridge.

———. 1979. What's wrong with the weather? — The climatic threat of the 21st century. Charles Scribner's Sons, New York, NY. 172 p.

GRIFFITHS, R. W., AND P. F. LINDEN. 1985. Intermittent baroclinic instability and fluctuations in geophysical circulations. Nature (Lond.) 316: 801-803.

GROVEMAN, B. S., AND H. E. LANDSBERG. 1979. Simulated northern hemisphere temperature departures, 1579-1880. Geophys. Res. Lett. 6: 767-769.

HÄFELE, W. 1980. A global and long-range picture of energy developments. Science (Wash., DC) 209: 174-182.

HAMEED, S., R. D. CESS, AND J. S. HOGAN. 1980. Response of the global climate to changes in atmospheric chemical composition due to fossil fuel burning. J. Geophys. Res. 85: 7537-7545.

HANELKA, U. D., V. A. WITTENBACH, AND M. G. BOYLE. 1984. CO_2-enrichment effects on wheat yield and physiology. Crop Sci. 24: 1163-1166.

HANSEN, J. E. 1985. Global sea level trends. Nature (Lond.) 313: 349-350.

HANSEN, J. E., W-C. WANG, AND A. A. LACIS. 1978. Mount Agung eruption provides test of a global climatic perturbation. Science (Wash., DC) 199: 1065-1067.

HANSEN, J., G. RUSSELL, D. RIND, P. STONE, A. LACIS, S. LEBEDEFF, R. RUEDY, AND L. TRAVIS. 1983. Efficient three-dimensional global models for climate studies: models 1 and 2. Mon. Weather Rev. 111: 609-662.

HARE, F. K. 1981. Climate's impact on food supplies: can it be identified?, p. 9-21. In L. E. Slater and S. K. Levin [ed.] Climate's impact on food supplies. Westview Press, Boulder, CO.

HARE, F. K., AND J. E. HAY. 1974. The climate of Canada and Alaska, Chap. 2. In R. A. Bryson and F. K. Hare [ed.] Climates of North America. Elsevier, Amsterdam.

HARRISS, R. C., E. GORHAM, D. I. SEBACHER, K. B. BARTLETT, AND P. A. FLEBBE. 1985. Methane flux from northern peatlands. Nature (Lond.) 315: 652-654.

HAYS, J. D., J. IMBRIE, AND N. J. SHACKLETON. 1976. Variations in the earth's orbit: pacemaker of the ice ages. Science (Wash., DC) 1984: 1121-1132.

HECHT, A., AND J. IMBRIE. 1979. Toward a comprehensive theory of climatic change. Quat. Res. 12: 2-5.

HENDERSON-SELLERS, A., AND V. GORNITZ. 1984. Possible climatic impacts of land cover transformations, with particular emphasis on tropical deforestation. Clim. Change 6: 231-257.

HENGEVELD, H. 1984. About climate change and variability. Chinook 6: 52-53.

HILEMAN, B. 1984. Recent reports on the greenhouse effect. Environ. Sci. Technol. 18: 45A-46A.

HOGG, W. D. 1982. Climate monitoring, networks and specialized analyses to support the needs of the natural resource sector, p. 77-91. In Proceedings of the CCP Water Resources — Climate Workshop, Feb. 28-29, 1980, Edmonton. Atmospheric Environment Service, Downsview, Ont.

HUMPHREYS, W. J. 1920. Physics of the air. The Franklin Institute, Washington, DC.

HUNT, B. G. 1981. An examination of some feedback mechanisms in the CO_2 climate problem. Tellus 33: 78-88.

IDSO, S. B. 1980. The climatological significance of a doubling of earth's atmospheric carbon dioxide concentration. Science (Wash., DC) 207: 1462-1463.

———. 1983. Do increases in atmospheric CO_2 have a cooling effect on surface air temperature? Clim. Bull. 17: 22-26.

IDSO, S. B., AND A. J. BRAZEL. 1984. Rising atmospheric carbon dioxide concentrations may increase streamflow. Nature (Lond.) 312: 51-53.

JONES, P. D., T. M. L. WIGLEY, AND P. M. KELLY. 1982. Variations in surface air temperatures, Part 1 — Northern Hemisphere, 1881-1980. Mon. Weather Rev. 110: 59-70.

JONES, P., L. H. ALLEN, J. W. JONES, K. R. BOOTE, AND W. J. CAMPBELL, 1984. Soybean canopy growth, photosynthesis, and transpiration responses to whole-season carbon dioxide enrichment. Agron. J. 76: 633-637.

JUDAY, G. 1982. Climatic trends in the interior of Alaska: moving toward a high CO_2 world? Agroborealis, Jan. p. 10-15.

KANDEL, R. S. 1981. Surface temperature sensitivity to increased atmospheric CO_2. Nature (Lond.) 293: 634-636.

KATES, R. W., AND F. K. HARE. 1980. Improving the science of impact study. Int. Council of Scientific Unions, S.C.O.P.E., Paris. 19 p.

174

KEELING, C. D., R. B. BACASTOW, A. E. BAINBRIDGE, C. A. EKDAHL, P. R. GUENTHER, AND L. S. WATERMAN. 1976. Atmospheric carbon dioxide variations at Mauna Loa Observatory, Hawaii. Tellus 28: 538-551.

KEELING, C. D., R. B. BACASTOW, AND T. P. WHORF, 1982. Measurements of concentration of carbon dioxide at Mauna Loa Observatory, Hawaii, p. 377-385. In W. C. Clark [ed.] Carbon dioxide review: 1982. Oxford University Press, News York, NY.

KELLOGG, W. W. 1977. Effects of human activities on global climate — Part I. World Meteorol. Org. Bull. 26: 229-240.

1978. Effects of human activities on global climate — Part II. World Meteorol. Org. Bull. 27: 3-10.

1980. Modelling future climate: it is time to plan for a climate change and a warmer earth. Ambio 9: 216-221.

1983. Identification of the climate change induced by increasing carbon dioxide and other trace gases in the atmosphere. World Meteorol. Org. Bull. 32: 23-32.

KELLOGG, W. W., AND S. H. SCHNEIDER. 1974. Climate stabilization: for better or for worse? Science (Wash., DC) 186: 1163-1172.

KELLY, P. M., P. D. JONES, C. B. SEAR, B. S. G. CHERRY, AND R. K. TAVAKOL. 1982. Variations in surface air temperature, Part 2 — Arctic regions 1881-1980. Mon. Weather Rev. 110: 71-83.

KELLY, P. M., AND C. B. SEAR. 1984. Climatic impact of explosive volcanic eruptions. Nature (Lond.) 311: 740-743.

KERR, R. A. 1977. Carbon dioxide and climate: carbon budget still unbalanced. Science (Wash., DC) 197: 1352-1353.

KILLOUGH, C. G., AND W. R. EMANUEL. 1981. A comparison of several models of carbon turnover in the ocean with respect to their distributions of transit time and age, and responses to atmosphere CO_2 and ^{14}C. Tellus 33: 274-290.

KIMBALL, B. A. 1983. Carbon dioxide and agricultural yield: an assemblage and analysis of 430 prior observations. Agron. J. 75: 779-788.

KLEMES, V. 1982. Aspirations and realities, p. 24-41. In Proceedings of the CCP Water Resources — Climate Workshop, Feb. 28-29, 1980, Edmonton. Atmospheric Environment Service, Downsview, Ont.

KOENIG, L. R. 1974. A numerical experiment on the effects of regional atmospheric pollution on global climate. Report R-1429-ARPA, The Rand Corp., Santa Monica, CA. 77 p.

KONDRATYEV, K. YA. 1980. The greenhouse effect of minor constitutents in the atmosphere. Weather 35: 252-256.

KUKLA, G. J. 1975. Missing link between Milankovitch and climate. Nature (Lond.) 253: 600-603.

KUKLA, G., AND J. GAVIN. 1981. Summer ice and carbon dioxide. Science (Wash., DC) 214: 497-503.

LaMARCHE, V. C., H. C. FRITTS, AND M. R. ROSE. 1984. Increasing atmospheric carbon dioxide: tree ring evidence for growth enhancement in natural vegetation. Science (Wash., DC) 225: 1019-1021.

LANDSBERG, H. E., AND W. C. JACOBS. 1951. Applied climatology, p. 976-992. In Compendium of meteorology. American Meteorological Society, Boston, MA.

LANDSBERG, H. E., B. S. GROVEMAN, AND I. M. HAKKARINEN. 1978. A simple method for approximating the annual temperature of the Northern Hemisphere. Geophys. Res. Lett. 5: 505-506.

LARSEN, L. W., AND E. L. PECK. 1974. Accuracy of precipitation measurements for hydrologic modelling. Water Resour. Res. 10: 857-863.

LEMON, E. R. [ED.]. 1983. CO_2 and plants: the response of plants to rising levels of atmospheric carbon dioxide. Westview Press, 5500 Central Avenue, Boulder, CO 80301.

LETTENMAIER, D. P., AND S. J. BURGES. 1978. Climate change: detection and its impact on hydrologic design. Water Resour. Res. 14: 679-687.

LIST, R. J. 1966. Smithsonian meteorological tables. Smithsonian Institution, Washington, DC. 527 p.

LORENZ, E. N. 1976. Nondeterministic theories of climatic change. Quat. Res. 6: 495-506.

LOVELOCK, J. E. 1972. Gaia as seen through the atmosphere. Atmos. Environ. 6: 579-580.

1975. Natural halocarbons in the air and in the sea. Nature (Lond.) 256: 193-194.

LUTGENS, F. K., AND E. J. TARBUCK. 1979. The atmosphere: an introduction to meteorology. Prentice-Hall Inc., Englewood Cliffs, NJ.

MADDEN, R. A., AND V. RAMANATHAN. 1980. Detecting climate change due to increasing carbon dioxide. Science (Wash., DC) 209: 763-768.

MAKI, L. R., AND K. J. WILLOUGHBY. 1978. Bacteria as biogenic sources of freezing nuclei. J. Appl. Meteorol. 17: 1049-1053.

MANABE, S. 1983. Carbon dioxide and climatic change. Adv. Geophys. 25: 39-82.

MANABE, S., R. T. WETHERALD, AND R. J. STOUFFER. 1981. Summer dryness due to an increase of atmospheric CO_2 concentration. Clim. Change 3: 374-386.

MARLAND, G., AND R. M. ROTTY. 1980. Emissions from future fossils-fuel systems: impacts on atmospheric CO_2. Abstract of paper presented to American Chemical Society 182 (Aug.): 27.

MASON, B. J. 1976. Towards the understanding and prediction of climatic variations. Q. J. R. Meteorol. Soc. 102: 473-498.

MAYBANK, J. 1985. Climate change is food for you. Chinook 7: 20-22.

MCKAY, G. A. 1978. Climatic change and the Canadian economy, p. 151-169. In K. D. Hage and E. R. Reinelt [ed.] Essays in meteorology and climatology in honour of Richmond W. Longley. Univ. Alta. Stud. Geogr. Monogr. 3: 427 p.

MEEHL, G. A. 1984. Modeling the earth's climate. Clim. Change 6: 259-286.

MILANKOVITCH, M. 1920. Theorie mathematique des phenomenes thermiques produits par la radiation solaire. Université de Paris, Paris, France. 339 p.

MITCHELL, J. F. B. 1983. The seasonal response of a global circulation model to changes in CO_2 and sea temperatures. Q. J. R. Meteorol. Soc. 109: 113-152.

MITCHELL, J. M. 1976. An overview of climatic variability and its causal mechanisms. Quat. Res. 6: 481-493.

NAVARRA, J. G. 1979. Atmosphere, weather and climate: an introduction to meteorology. W. B. Saunders, Philadelphia, PA.

NEFTEL, A., E. MOOR, H. OESCHER, AND B. STAUFFER. 1985. Evidence from polar ice cores for the increase in atmospheric CO_2 in the past two centuries. Nature (Lond.) 315: 45-47.

NEUMANN, J. 1976. Great historical events that were significantly affected by the weather: 1. The Mongol invasions of Japan. Bull. Am. Meteorol. Soc. 56: 1167-1171.

1978. Great historical events that were significantly affected by the weather: 2. The year leading to the revolution of 1789 in France. Bull. Am. Meteorol. Soc. 58: 163-168.

1979. Great historical events that were significantly affected by the weather: 3. The cold winter of 1657-58, The Swedish Army crosses Denmark's frozen sea areas. Bull. Am. Meteorol. Soc. 59: 1432-1437.

NEWELL, R. E., AND T. G. DOPPLICK. 1979. Questions concerning the possible influence of anthropogenic CO_2 on atmospheric temperature. J. Appl. Meteorol. 18: 822-825.

NIERENBERG, W. A. [ED.]. 1983. Changing climate. NRC Carbon Dioxide Assessment Committee, National Academy Press, Washington, DC. 496 p.

OKE, T. R, AND G. B. MAXWELL. 1975. Urban heat island dynamics in Montreal and Vancouver. Atmos. Environ. 9: 191-200.

PARRY, M. L. 1978. Climatic change, agriculture and settlement. Dawson & Sons, Kent, England. 214 p.

1981. Evaluating the impact of climatic change, p. 3-16. In C. D. Smith and M. Parry [ed.] Consequences of climatic change. Department of Geography, University of Nottingham, England.

PETERSON, J. T., W. D. KOMYR, T. B. HARRIS, AND L. S. WATERMAN. 1982. Atmospheric carbon dioxide measurements at Barrow, Alaska, 1973-1979. Tellus 34: 166-175.

PLASS, G. N. 1956. Effect of carbon dioxide variations on climate. Am. J. Phys. 24: 376-387.

PREINING, O. 1983. The 1880 measurement of the average atmospheric CO_2 concentration by Ernst Lecher. Atmos. Environ. 17: 1595.

RAPER, S. C. B., T. M. L. WIGLEY, P. R. MAYES, P. D. JONES AND M. J. SALINGER. 1984. Variations in surface air temperatures, Part 3 — The Antarctic 1957-1982. Mon. Weather Rev. 112: 1341-1353.

RAPP, R. R., AND M. WARSHAW. 1974. Some predicted climatic effects of a simulated Sahara lake. Report R-1415-ARPA, The Rand Corp., Santa Monica, CA. 31 p.

RAVENHOLT, A. 1976. Who will grow grain for a world food bank? Field Staff Reports, American Universities Field Staff, Vol. 4, No. 1, Hanover, NH. 11 p.

REVELLE, R. 1982. Carbon dioxide and world climate. Sci. Am. 247: 35-43.

REVELLE, R. AND H. SUESS. 1957. Carbon dioxide exchange between atmosphere and ocean, and the question of an increase of atmospheric CO_2 during the past decades. Tellus 9:

18-27.

RIND, D., AND S. LEBEDEFF. 1984. Potential climatic impacts of increasing atmospheric CO_2 with emphasis on water availability and hydrology in the United States. Strategy Studies Staff, Office of Policy Analysis, Environmental Protection Agency, Washington, DC 96 p.

ROBINSON, J. 1981. Ecological implications of increased atmospheric CO_2. Fog's Edge Research. P.O. Box 330, Inverness, CA.

ROBOCK, A. 1978. Internally and externally caused climate change. J. Atmos. Sci. 35: 1111-1122.

———— 1982. The Russian surface temperature data set. J. Appl. Meteorol. 21: 1781-1785.

RODDA, J. C. 1971. The precipitation measurement paradox — the instrument measurement problem. Report 16, World Meteorological Organization, No. 316, Geneva.

ROGERS, H. H., J. F. THOMAS, AND G. E. BINGHAM. 1983. Response of agronomic and forest species to elevated atmospheric carbon dioxide. Science (Wash., DC) 223: 428-429.

ROSENBERG, N. J. 1981. The increasing CO_2 concentration in the atmosphere and its implication on agricultural productivity. I. Effects on photosynthesis, transpiration, and water-use efficiency. Clim. Change 3: 265-279.

ROTTY, R. M. 1980. Past and future emissions of CO_2. Experientia 36: 781-787.

SADLER, J. C., C. S. RAMAGE, AND A. M. HORI. 1982. Carbon dioxide variability and atmospheric circulation. J. Appl. Meteorol. 21: 793-805.

SAKAMOTO, C., S. LEDUC, N. STROMMER, AND L. STEYAERT. 1980. Climate and global grain yield variability. Clim. Change 2: 349-361.

SCHIFF, H. I. 1981. Review of the carbon dioxide greenhouse problem. Planet. Space Sci. 29: 935-950.

SCHLESINGER, M. E. 1983. A review of climate model simulations of CO_2-induced climatic change. Climate Research Institute Rep. 41, Oregon State University, Corvallis, OR. 135 p.

———— 1984. A review of climate model simulations of CO_2-induced climatic change. Adv. Geophys. 26: 141-235.

SCHNEIDER, S. H. 1984. Deforestation and climatic modification — An editorial. Clim. Change 6: 227-229.

SCHNEIDER, S. H., AND C. MASS. 1975. Volcanic dust, sunspots, and temperature trends. Science (Wash., DC) 190: 741-746.

SCHNEIDER, S. H., AND S. L. THOMPSON. 1979. Ice ages and orbital variations: some simple theory and modeling. Quat. Res. 12: 188-203.

SEIDEL, S., AND D. KEYES. 1983. Can we delay a greenhouse warming? U.S. Environmental Protection Agency, U.S. Government printing Office, Washington, DC.

SHUKLA, J., AND Y. MINTZ. 1982. Influence of land surface evapotranspiration on the earth's climate. Science (Wash., DC) 215: 1498-1501.

SIONIT, N., B. R. STRAIN, AND H. HELLMERS. 1981. Effects of different concentrations of atmospheric CO_2 on growth and yield components of wheat. J. Agric. Sci. Camb. 79: 335-339.

SLATER, L. E., AND S. K. LEVIN [ED.]. 1981. Climate's impact on food supplies: strategies and technologies for climate-defensive food production. Westview Press, Boulder, CO.

SMAGORINSKY, J. [ED.]. 1982. Carbon dioxide and climate — a second assessment. Climatic Review Panel, National Academy Press, Washington, DC. 72 p.

———— 1983. Climatic changes due to CO_2. Ambio 12: 83-85.

STEUDLER, P. A., AND B. J. PETERSON. 1984. Contribution of gaseous sulphur from salt marshes to the global sulphur cycle. Nature (Lond.) 311: 455-457.

STEWART, D. W. 1984. Atmospheric carbon dioxide and Canadian agriculture. Clim. Bull. 18: 3-14.

STEWART, R. W. 1978. The oceans, the climate and people (with a view from Mars). Atmosphere-Ocean 16: 367-376.

STOKER, H. S., AND S. L. SEAGER. 1976. Environmental chemistry: air and water pollution. 2nd ed. Scott. Foresman & Co., Glenview, IL. 231 p. + index.

STOMMEL, H., AND E. STOMMEL. 1979. The year without a summer. Sci. Am. 240(6): 176-186.

STOTHERS, R. B. 1984. The great Tambora eruption in 1815 and its aftermath. Science (Wash., DC) 224: 1191-1198.

STUIVER, M. 1978. Atmospheric carbon dioxide and carbon reservoir changes. Science (Wash., DC) 199: 253-258.

TERJUNG, W. H., H-Y. JI, J. T. HAYES, P. A. O'ROURKE, AND P. E. TODHUNTER. 1984. Actual and potential yield for rainfed and irrigated wheat in China. Agric. For. Meteorol. 31: 1-23.

THOMAS, M. K. 1975. Recent climatic fluctuations in Canada. A.E.S., Environment Canada, Cat. No. EN57-7/28. 92 p.

THOMPSON, L. M. 1975. Weather variability, climatic change, and grain production. Science (Wash., DC) 188: 535-541.

TREIDL, R. A. 1978. Climatic variability and wheat growing in the Prairies, p. 347-365. In K. D. Hage and E. R. Reinelt [ed.] Essays in meteorology and climatology in honour of Richmond W. Longley. Univ. Alta. Stud. Geogr. Monogr. 3: 427 p.

TYNDALL, J. 1981. Remarks on radiation and absorption. Philos. Mag. Ser. 4, 22: 377-378.

VOWINCKEL, E., AND S. ORVIG. 1982. On the evaluaiton of the influence of large-scale dam projects on weather and climate, p. 95-99. In Proc. CCP Water Res. — Clim. Wksp., Feb. 28-29, Edmonton. Atmospheric Environment Service, Downsview, Ont. 212 p.

WALLEN, C-C. 1980. Monitoring potential agents of climatic change. Ambio 9: 222-228.

WALSH, J. E. 1983. The role of sea ice in climatic variability: theories and evidence. Atmosphere-Ocean 21: 229-242.

WANG, W. C., J. L. YOUNG, A. A. LACIS, T. MO, AND J. E. HANSEN. 1976. Greenhouse effects due to man-made perturbations of trace gases. Science (Wash., DC) 194: 685-690.

WASHINGTON, W. M. AND G. A. MEEHL. 1983. General circulation model experiments on the climatic effects due to a doubling and quadrupling of carbon dioxide concentration. J. Geophys. Res. 88: 6600-6610.

WEERTMAN, J. 1976. Milankovitch solar radiation variations and ice age sheet sizes. Nature (Lond.) 261: 17-20.

WHEATON. E. 1984. Climatic change impacts on wind erosion in Saskatchewan, Canada. Tech. Rep. No. 153, Publ. E-906-16-B-84. Saskatchewan Research Council, Saskatoon, Sask., June. 27 p.

WIGLEY, T. M. L. 1985. Impact of extreme events. Nature (Lond.) 316: 106-107.

WIGLEY, T. M. L., AND P. D. JONES. 1981. Detecting CO_2-induced climatic change. Nature (Lond.) 292: 205-208.

    1985. Influences of precipitation changes and direct CO_2 effects on streamflow. Nature (Lond.) 314: 149-152.

WIGLEY, T. M. L., P. D. JONES, AND P. M. KELLY. 1980. Scenario for a warm, high CO_2 world. Nature (Lond.) 283: 17-21.

WILLIAMS, G. D. V. 1976. An assessment of the impact of some hypothetical climatic changes on cereal production in western Canada. Proc. Sterling Forest Conf., December 2-5, 1974, Sterling Forest Conference Center, Sterling Forest, New York.

    1985. Estimated bioresource sensitivity to climatic change in Alberta, Canada. Clim. Change 7: 55-69.

WILLIAMS, G. D. V., AND W. T. OAKES. 1978. Climatic resources for maturing barley and wheat in Canada, p. 367-385. In K. D. Hage and E. R. Reinelt [ed.] Essays on meteorology and climatology in honour of Richmond W. Longley. Univ. Alta. Stud. Geogr. Monogr. 3.

WILSON, A. T. 1964. Origin of ice ages: an ice shelf theory for Pleistocene glaciation. Nature (Lond.) 201: 147-149.

WOLLIN, G., G. J. KUKLA, D. B. ERICSON, W. B. F. RYAN, AND J. WOLLIN. 1973. Magnetic intensity and climatic changes 1925-1970. Nature (Lond.) 242: 34-37.

WONG, C. S. 1978. Atmospheric input of carbon dioxide from burning wood. Science (Wash., DC) 200: 197-200.

WOODWELL, G. M., AND E. V. PECAN [ED.]. 1973. Carbon and the biosphere. U.S. Atomic Energy Commission Report CONF-720510, NTIS, Springfield, VA. 392 p.

WOODWELL, G. M., R. H. WHITTAKER, W. A. REINERS, G. E. LIKENS, C. C. DELWICHE, AND D. B. BOTKIN. 1978. The biota and the world carbon budget. Science (Wash., DC) 199: 141-146.

CHAPTER 7

The Importance of Forests in the Hydrological Regime

E.D. Hetherington

Pacific Forestry Centre, Canadian Forestry Service, Victoria, B.C. V8Z 1M5

Introduction

Canada is a nation of vast forests as well as extensive water resources. These forests, which occupy almost half of the total land area, are unique associations of trees and soils that have developed together through geologic time. They exist where precipitation is sufficient for the development of trees and, during most years, is surplus to their growth needs. The stems, branches, and foliage of the trees within a forest make up a three-dimensional matrix that interacts with rain, snow, sunlight, and wind. Furthermore, the layer of soil occupied by the tree roots is a porous matrix that absorbs rain and snowmelt to delay and temporarily store the water en route to a visible stream channel. Roots also extract water from this layer, reducing the amount available for streamflow. Through this intimate association, water movement and storage are affected by forests (Fig. 1) and by changes in the forest cover.

The importance and influence of forests and forest management operations in relation to water in Canada are the subjects of this chapter. The focus is on knowledge gained from Canadian research and experience, whereas previous reviews (Freedman 1982; Jeffrey 1970; Moore 1983; Plamondon 1981) relied heavily on literature from other countries, particularly the United States.

Forest Hydrology in Canada

Canadian activity in forest hydrology[1] and watershed management[2] is of fairly recent origin. Active forest hydrology research began in the late 1950's in the Wilson Creek watershed in Manitoba and in the early 1960's in southern Alberta (Jeffrey 1967). The late W. W. Jeffrey was instrumental in initiating the Alberta watershed research program and promoting the expansion of forest hydrology activity in Canada during the 1960's. By the early 1970's, several experimental watershed studies had developed in other provinces, and additional research has since been carried out in a number of other watersheds (Fig. 2). These studies have been concentrated in southern Canada where commercial forests and man's use of water coincide. Forest hydrology programs are also now offered at four universities in Canada: British Columbia (Vancouver), Alberta (Edmonton), Laval (Quebec), and New Brunswick (Fredericton). With this increase in information, awareness, and trained personnel, water-related concerns and knowledge have been increasingly incorporated into forest management regulations and practices in most provinces. At present, watershed management in Canada consists mainly of putting constraints on forestry operations to maintain the water resource in its existing state. The one exception is in Alberta, where purposeful manipulation of forests with the objective of enhancing the water resource has been done experimentally (Golding 1981).

---

[1] Forest hydrology is the science of water-related factors influenced by the forest and forms the technical basis for watershed management.

[2] Watershed management denotes operational management of forest land for water-related purposes.

Popular Beliefs

Although there is a growing body of scientific information on forest-water relationships in Canada and elsewhere, several incorrect popular beliefs about these relationships persist. Many believe that forests "conserve water and provide maximum runoff" and "regulate streamflow by controlling spring snowmelt and sustaining summer flows." Others believe that forests "protect against floods" and that "the forest floor acts as a sponge that holds back water and reduces peak flows." Forests are thus commonly viewed as having positive effects on streamflow. In contrast, popular beliefs concerning the effects of forest harvesting on water are usually expressed in negative terms such as "logging dries up streams," "logging causes flooding," "clearcutting produces greater snowmelt runoff," and "logging silts up streams." Forest hydrology research has shown that most of these beliefs are either overly simplistic or false.

Issues and Concerns

Concerns about forest-water interactions arise from direct observation or perceptions of what constitutes desirable or adverse situations. However, conflicting mandates resulting from the separation of management responsibilities for these two resources are an

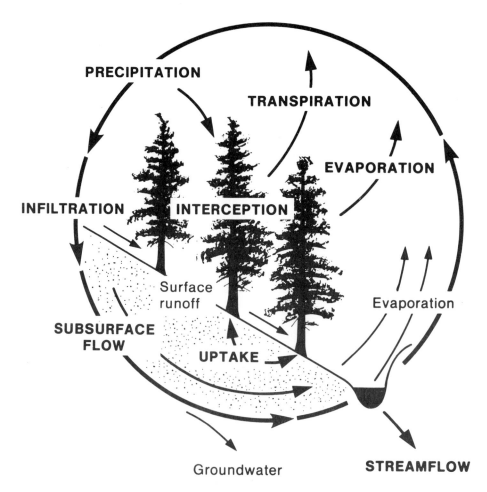

FIG. 1. Diagram illustrating the relative importance of components of the hydrologic cycle in forested areas.

180

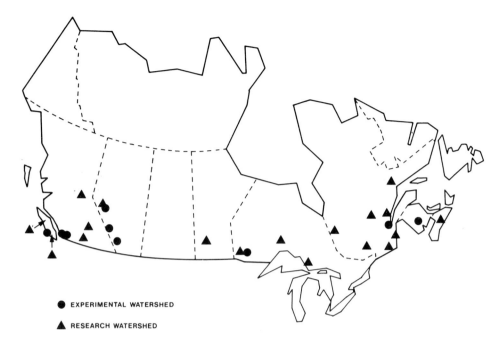

● EXPERIMENTAL WATERSHED

▲ RESEARCH WATERSHED

FIG. 2. Distribution of forest hydrology study watersheds in Canada. Experimental watershed studies involve the calibration of study streams against control streams prior to treatments such as logging. Research watershed studies are those that lack pretreatment calibrations. The research watersheds indicated on the map represent the principal locations of watershed-scale forest hydrology studies. There are also many individual study sites or watersheds where research related to forest hydrology has been conducted that are not shown.

underlying source of many problems. Forests and water in Canada are administered separately by both provincial and federal governments. Furthermore, forests are mostlly a provincial responsibility, while fish and their habitat in many streams and rivers, particularly those with anadromous fish such as salmon, fall under federal jurisdiction.

Most of the contentious issues stem from concerns about potential effects of forestry operations on the water resource. The degree of concern varies regionally and is greatest in provinces where forestry is important economically, and where people live close to current forest management activity. The one concern prevalent in all provinces involves the impacts of logging on fish and their habitat, which includes water quality and the physical attributes of the stream channel. However, fisheries impacts are a major issue only in British Columbia and the Atlantic provinces. An issue of growing importance in many areas of Canada is public concern over possible contamination of water and fish by herbicides used for forestry. Similar concerns over the impacts on water and fish of forest pesticide spraying to control spruce budworm outbreaks in eastern Canada have a lower profile than they once had.

Concerns over the effects of forest harvesting on water supplies are centered mainly on water used for domestic needs and irrigation in British Columbia and on managing forests to enhance water supplies in Alberta (Swanson 1978). Logging as an alleged cause of local flooding has occasionally been an issue in British Columbia, Alberta, and some of the Atlantic provinces. Forest management around waterways subject to recreational use is of some concern in parts of Ontario and Quebec. In addition to these identifiable regional issues, numerous local forestry-water problems and concerns exist in most parts of southern Canada, as illustrated by an earlier review of problems in British Columbia (Jeffrey 1968b).

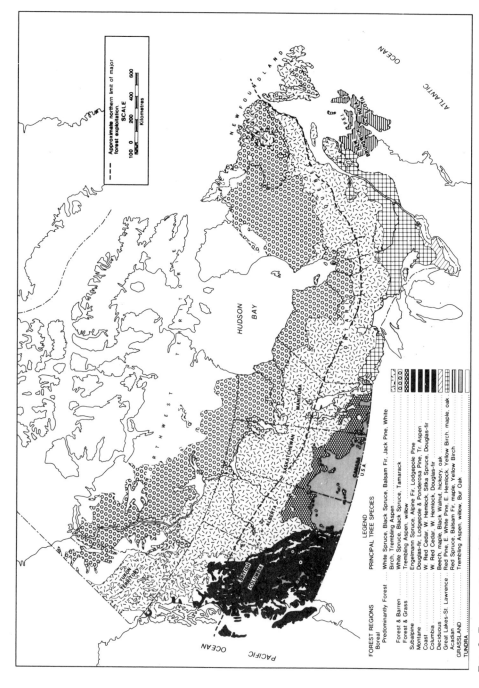

FIG. 3. Forest regions of Canada. (After Rowe (1972) with modifications by Weetman (1983). Reproduced by permission of the Minister of Supply and Services Canada.)

Regional Considerations

Canada is a country with great diversity of forests, soils, terrain, and climate (see chapter by A. H. Laycock in this volume). Although broad regional categorization of forests is possible (Fig. 3), forests within these categories are continuously changing due to natural causes and human exploitation. From 1977 to 1981, the area of forest cover on productive forest lands in Canada was reduced by 2.2 million ha annually, while satisfactory regeneration of new trees took place on 1.8 million ha (Honer and Bickerstaff 1985). The main causes of depletion of the forest cover were fire (43%), insects and disease (23%), and harvesting (34%). These changes affect the structure and species composition of Canada's forests and have resulted in a net increase in the area of land lacking an adequate forest cover. Fire, which is common in most of Canada except for the wetter parts of coastal British Columbia, not only kills trees but sometimes consumes the organic material of the forest floor. The most destructive forest pests, the spruce budworm in eastern Canada and the mountain pine beetle in British Columbia and southern Alberta, have killed trees over extensive areas (Kondo and Taylor 1984). These changes in the pattern of forest cover have marked effects on the water cycle (Swanson 1982).

Forestry operations are concentrated in the commercially productive forest lands across southern Canada (Fig. 3), and encompass a number of practices. The principal harvesting technique is clearcutting with log removal (yarding) by ground skidding (Weetman 1983). Cable yarding systems are used in the steeper terrain in British Columbia (Toews and Brownlee 1981). Conventionally, a watershed is cut in patches over a number of years, with varying limits on the size of clearcut patch and rate of harvest. In some areas, the rate of logging is accelerated to salvage trees damaged by insect infestations (Kondo and Taylor 1984). In most provinces, special attention is given to forest management along stream and lake margins to protect these water bodies from direct disturbance or debris inputs during logging, and maintain desirable fish habitat and water quality. This may involve leaving protective buffer strips of forest vegetation along water edges (Gouvernement du Québec 1977); Toews and Brownlee 1981) or special strategies for removing streamside trees (Moore 1983; Toews and Brownlee 1981).

Silvicultural and forest protection procedures employed to varying degrees across the country include scarification to expose mineral soil to provide for regeneration of the new forest, prescribed burning, mainly to reduce fire hazard and prepare sites for tree planting, herbicide application to control brush and permit conifers to become well established, insecticide spraying to control insect damage, aerial fertilizer application to stimulate better tree growth, and thinning and tree planting to promote growth and establishment of better forests. The impacts of forestry operations on the water resource depend on the extent and severity of change in the forest cover, exposure and compaction of mineral soil, and input of materials to water systems, plus the rate of recovery of soil and vegetation after disturbance.

Most of Canada has been glaciated, which has resulted in extensive wetlands in the boreal forest region and shallow soils, often less than 1 m deep, in most areas presently occupied by forests (Dickison et al. 1981; Hetherington 1976, 1982; Jablonski 1980; Nicolson et al. 1982). In this respect, Canada differs from many areas in the United States where much of the existing body of forest hydrology research has originated.

Influence of Forests and Forestry Operations on Microclimate

The microclimate of a forest stand is the local climate in, and immediately above, or adjacent to the forest. It is determined by the ways in which the forest interacts with energy from the sun, wind, and precipitation. Forest influences on the processes of energy and precipitation distribution affect rates of transpiration, evaporation and soil freezing, and patterns of snow accumulation and melt. Changes in the forest cover, whether due to human or natural causes, also modify these basic microclimatic processes.

Radiation and Energy Balances

The sun is the primary source of energy for microclimatic processes. Solar energy (shortwave radiation) received by the forest canopy is partly absorbed and partly reflected, while some passes directly to the ground. The forest also receives energy from the atmosphere (longwave radiation and sensible heat transfer) and condensation of moisture (latent heat), and it emits energy (longwave radiation) back to the atmosphere. Much of the net energy absorbed by the forest accounts for its microclimate and the remainder is used for tree growth, resulting in an overall energy balance.

The partitioning of incoming solar radiation is dependent on the density and type of forest canopy. The amount of sunlight reaching the forest floor decreases from over 80% for a 10% canopy cover to less than 12% for complete closure for conifers (Jeffrey 1968a). In summer, the amount of solar radiation reflected by a mature coniferous forest varies from about 7% for balsam fir (McCaughey 1981) to 13% for less dense black spruce stands (Haag and Bliss 1974). One mostly deciduous forest in eastern Canada was found to reflect at least 16% (O'Kane 1983).

In forests with closed canopies, most of the energy absorbed is retained within the canopy. The net radiant energy absorbed by the ground beneath the canopy was found to be only 5% of that at the top of the canopy in one study (Spittlehouse and Black 1981). The absorbed energy is used primarily for warming the surroundings (including the melting of snow) and for the processes of transpiration and evaporation of water (evapotranspiration). When the moisture supply is adequate, more energy goes into evapotranspiration. Under drier conditions, more energy is used to warm the environment. The amount of energy stored in the vegetation and soil and used for photosynthesis usually does not exceed 5-10% of the net radiant energy on a daily basis (McCaughey 1985).

Forest harvesting and silvicultural practices alter the radiant energy balances. Clearing the forest substantially increases solar radiation reaching the ground. Clearcut and nonforested tundra surfaces reflect 2-10% more solar radiation than adjacent forests (McCaughey 1981; Petzold and Rencz 1975), although the exposure of mineral soil by logging may initially decrease the amount reflected (O'Kane 1983). With these changes, the net radiant energy absorbed by the ground in open clearcuts (or tundra areas) is considerably greater than that beneath the forest canopy. However, the net energy absorbed by the canopy still remains about 10-20% greater than that absorbed in adjacent open clearcut (McCaughey 1981) or tundra areas (Rouse 1984). In winter, snow increases the reflection of solar radiation much more from open areas than from forest canopies. For example, the reflectivity of a northern black spruce forest under snow conditions was 32% compared with 78% from a snow-covered tundra surface (Rouse 1984). The use of absorbed radiation in open areas differs from that in forests, with less energy tending to go into evapotranspiration and more into warming the environment in the open.

Evapotranspiration

Evapotranspiration includes all evaporated or transpired water. Most of the net absorbed energy used for evapotranspiration in forests goes into transpiration of soil water and evaporation from wet vegetation; evaporation from the ground and water surfaces accounts for the remainder (Fig. 1). Transpiration rates are determined not only by radiant energy but also by the temperature and humidity of the air, by soil moisture, and by physiological controls imposed by the trees themselves. The highest reported estimates of daily transpiration by individual trees in Canada are 16 L for lodge-

Forests and water are intimately associated (R. Barry, Faculté de Foresterie et de Géodésie, Université Laval, Québec, Qué.)

pole pine (1544 stems/ha) in Alberta (Swanson 1975) and 24 L for young Douglas-fir (840 stems/ha) in British Columbia (Black et al. 1980). However, maximum single-tree transpiration volumes for Douglas-fir can exceed 100 L/d in stands of lower tree density (Black et al. 1980). The maximum transpiration rates from forest stands, expressed as depth of water over the area occupied by the stand, reach 0.3-0.5 mm/h and 3-5 mm/d for both conifers and hardwoods (McCaughey 1985; McNaughton and Black 1973). These rates usually occur on sunny days.

Total evapotranspiration from forests can be even higher when the vegetation is wet. Evapotranspiration rates from wet canopies of up to 0.8-1.0 mm/h and over 5 mm/d have been reported (McCaughey 1978; Singh and Szeicz 1979). The evaporation of precipitation intercepted by tree surfaces reduces the rate of transpiration, thereby reducing the loss of soil moisture by transpiration during wet conditions (Singh and Szeicz 1979; Spittlehouse and Black 1981). The proportion of time that canopies are wet can thus strongly influence total annual losses of water by evapotranspiration. The high wet-canopy evaporation rates reported for research plots, however, are caused by the evaporative power of drier air from nearby areas moving through the canopy in addition to net radiant energy. Whether or not wind-related evaporation forces are effective in increasing total evaporation over extensive areas of wet forest remains an unresolved question (Morton 1985). In returning water to the atmosphere, the processes of evapotranspiration affect streamflow by reducing the volume of water available for runoff and by increasing the ability of the soil to temporarily store and delay runoff.

Evapotranspiration rates vary according to vegetation type. Daily summer evapotranspiration of trees was found to be greater than that of other vegetation in three Canadian studies: 10-32% higher than grass and soil in Alberta (Cohen 1977), 6-58% higher than tundra in the north (Rouse 1976), and 36% higher than grasses and shrubs in Saskatchewan (Meyboom 1967). In contrast, under relatively dry conditions, transpiration by ground vegetation (salal) exceeded that of Douglas-fir in coastal British Columbia (Black et al. 1980). Furthermore, evidence from Oregon indicates that vigorous young deciduous vegetation uses more water than mature conifers (Harr 1983). On a seasonal or annual basis, however, total evapotranspiration from coniferous forests will tend to be higher than that from other vegetation in most situations, although there are suggestions that this might not always be the case (Morton 1984; Plamondon 1981). The reasons for greater evapotranspiration from forests include the more extensive root systems of trees, longer periods of transpiration, and greater evaporation of intercepted precipitation (Calder 1982).

Evapotranspiration during the winter is small in most parts of Canada because of the cold temperatures. Annual evaporation of snowfall in forested environments could be less than 3% (Jeffrey 1968a). One exception occurs in southern Alberta, where evaporation of snow during chinook winds can account for up to 20% of annual snowfall at lower elevations (Golding 1982).

Forest harvesting reduces evapotranspiration losses by eliminating transpiration and evaporation from the elevated canopy. In Ontario, for example, maximum hourly rates of evapotranspiration were 20% greater from the forest than from an adjacent clearcut (McCaughey 1985). One result of this is wetter soils in the harvested areas than in the forest (Kachanoski and de Jong 1982). This increase in soil water content in logged areas makes more water available for streamflow by reducing the ability of the soil to store rain and snowmelt.

Soil Frost

Soil freezing within forests is impeded by the insulating qualities of the forest canopy and organic forest floor. Even in the cold winters of central Canada, soil freezing within forests can be minor (Price and FitzGibbon 1982; Sahi and Courtin 1983). Snow is also a good insulator and can be more effective in reducing the degree of soil freezing than the forest. In Quebec, for example, soil frost was less severe in the open under a deep snowpack than in the forest where the snow was not as deep (Plamondon and Grandtner 1975). Soil frost has also been observed in the forested mountains of western Canada, although it frequently disappears before spring snowmelt due to thawing from below (Harlan 1969). Forest harvesting may either increase or decrease the severity and frequency of soil freezing, depending on whether snowpack depths are lower or higher after removal of the canopy (see section on snow accumulation and melt).

In the permafrost zone (see chapter by R. O. van Everdingen in this volume), insulation provided by the forest vegetation and organic forest floor reduces the depth of summer thaw in comparison with open areas. In some locations, such as the forested lowlands of northern Quebec, snow trapped by the forests may locally prevent permafrost formation (Granberg 1973). In general, however, the forest is more efficient in preventing thawing than preventing freezing. Forest removal can cause an expansion of permafrost formation to previously unfrozen ground and an increase in the depth of summer thaw in existing permafrost.

The effect of soil frost on runoff depends on whether the soil was saturated or not at the time of freezing. Saturated soils will freeze solid and cause surface runoff if the frost persists into the snowmelt period, whereas freezing of unsaturated soils may have little effect on infiltration or runoff (Price and FitzGibbon 1982).

Wind

Wind plays an important role in the processes of evapotranspiration, snow distribution, and snowmelt. Because of their great height and surface roughness, forests induce turbulence and reduce wind speed both over the canopy and at forest/clearing edges.

The reduced roughness of low ground vegetation or isolated trees in large open areas affect wind to a much lesser extent than fully forested areas. For example, wind speeds 10 m over a forest canopy are about 50% slower than 10 m over relatively smooth open ground (Silversides 1978). Beneath the forest canopy, wind speeds may be reduced to just a few percent of those in the open (Beaudry 1984) or above the canopy (Martin 1971). The denser the forest, the lower will be the relative wind speed (Szeicz et al. 1979).

Forest harvesting changes wind patterns and the effects of wind on the environment. After clearcutting, for example, wind speeds will increase at ground level (Meeres 1977), although the change may be small in openings less than six tree heights across (Swanson 1980). Once exposed to higher wind speeds, tress along clearcut edges are more susceptible to being blown down. This possibility is an important concern in the design of protective leave strips of trees along streams, lakes, and unstable areas.

Precipitation

The forest's direct influence on water begins with the onset of precipitation (Fig. 1). Through interception, forest canopies reduce the amount of rain and snow reaching the ground and alter its distribution. Forests further alter snow accumulation and melt patterns by sheltering snowpacks from sun and wind. These initial forest-precipitation interactions help determine the quantity and timing of inputs to runoff processes.

Interception

The capture of rain by forest canopies is controlled by the type and density of trees and by rainfall intensity. On a seasonal basis, conifer stands tend to intercept more rain than hardwoods, with mixed forests falling in between (Table 1). The interception values in Table 1 represent water lost to the atmosphere by evaporation. For individual storms, light showers might be almost totally intercepted, while interception might account for as little as 5% of heavy winter rainfall (McMinn 1960). Rain interception is determined partly by the storage capacity of the canopy for rainwater and partly by continual evaporation of intercepted water during rainfall. Total storm interception, for example, has been found to exceed 10 mm for a stand with a canopy storage capacity of 2.4 mm (Singh and Szeicz 1979). Interception losses are partly offset, however, by a concurrent reduction in transpiration when the foliage is wet, as already noted. Interception of fog or cloud droplets and consequent fog drip, as observed in coastal Oregon (Harr 1983), may be locally important in immediate coastal areas.

TABLE 1.　Summer rainfall interception as a percentage of total precipitation.

Forest type	Height (m)	Age (yr)	Density (no./ha)	Interception (%)	Reference
Conifers, B.C.	70	236	267	57	McMinn 1960
Conifers, B.C.	28	248	603	30	McMinn 1960
Balsam fir, Que.	15	50	4800	39	Plamondon et al. 1984b
Balsam fir, Que.		20		25	Frechette 1969
Red spruce, N.B.		46	4841	14	Mahendrappa and Kingston 1982
Hardwoods, Ont.	22	135	673	9	Foster and Nicolson 1986
Hardwoods, Que.				21	Frechette 1969
Birch, N.B.		61	4303	20	Mahendrappa and Kingston 1982
Aspen, N.B.		44	5649	8	Mahendrappa and Kingston 1982
Mixed wetland, Alta.	27	200	192	37	Rothwell 1982
Mixed, Que.		20		23	Frechette 1969

The amount of snow trapped by forest canopies, expressed in terms of its melted equivalent as water, will depend on the canopy density and whether the snow is dry or wet. Mature west coast forests can temporarily retain at least 16 mm water equivalent of wet snow (Beaudry 1984), although average values are likely much lower for the drier snow and smaller trees encountered in most other forest regions. Most of the intercepted snow is removed from forest canopies either by wind action, if the snow is relatively dry, or by melt processes when temperatures are mild. As already noted, the loss of snow by evaporation is small.

Clearcutting eliminates the elevated intercepting surfaces of the forest canopy. This permits rain or snow that would have been previously intercepted to reach the ground where it is less exposed to evaporative forces. Consequently, losses due to evaporation are reduced and more water made available for runoff processes. Even partial reductions of canopy or foliage density, such as that produced by spruce budworm defoliation of eastern conifers (Plamondon et al. 1984b) or forest fires, can reduce interception losses.

Snow Accumulation and Melt

The trees of a forest change the accumulation and distribution of snow on the ground by intercepting snow and altering wind speed and turbulence. The effect will vary depending on the characteristics of the forest canopy and the nature of the snow. Both depth and water equivalent of the snowpack are affected, although the water equivalent is more important in terms of runoff. Snow accumulation on the ground in forests usually decreases as the density of the canopy increases (Daugharty 1984). In areas with dry snow, however, more snow may accumulate in the forest than in the open due to trees trapping snow blown from adjacent open terrain (Payette et al. 1973). Leafless hardwood stands act as porous traps and accumulate more snow than coniferous forests (Daugharty 1984; Frechette 1968).

Removal of the forest cover, whether by harvesting or other causes, modifies snow accumulation patterns through elimination of interception losses and changes in redistribution processes. Small openings in the forest almost always accumulate more snow than the adjacent forest (Fig. 4). Maximum snowpack water equivalents in small clearings are usually less than 40% greater than in the forest (Golding 1982), although increases in clearings can be several hundred percent for shallow snowpacks (Jeffrey 1968a). In large open areas, late winter snowpacks are often smaller than within forests due to wind erosion of dry snow (Payette et al. 1973) or greater overwinter ablation by melting (Daugharty and Dickison 1982). In regions where milder winter temperatures prevail, snowpacks can also be greater in large openings if melt or evaporation from the canopy results in smaller snow accumulations in the forest (Beaudry 1984).

The major factors causing snow to melt are solar radiation, sensible heat and longwave radiation from the atmosphere or vegetation, and heat derived from the condensation of moisture. Forests moderate snowmelt rates by sheltering snowpacks from the direct effects of sunlight and heat transported by the wind. Consequently, melt rates of the snowpack increase with increasing exposure to sun and wind. For example, snowpacks under leafless hardwoods melt faster than those in coniferous stands (Daugharty and Dickison 1982), but snow can melt even faster while retained in the canopy (Beaudry 1984). In a lodgepole pine stand in Alberta, melting was actually slowest in very small openings about one tree height across, whereas melt rates in the forest were about the same as those in openings three tree heights in diameter (Golding 1981). Snowmelt rates were also found to be slower in small clearings in an aspen forest than beneath the canopy (Swanson and Stevenson 1971). Snow melts most rapidly, however, in large open areas (Plamondon et al. 1984a; Stanton 1966).

Removal of the forest cover by harvesting or fire usually accelerates snowmelt and runoff, and can advance the date at which the snowpack disappears. Snow in forest openings commonly disappears sooner than in the forest despite greater accumulations (Daugharty and Dickison 1982; Stanton 1966). Snowpacks in the forest have

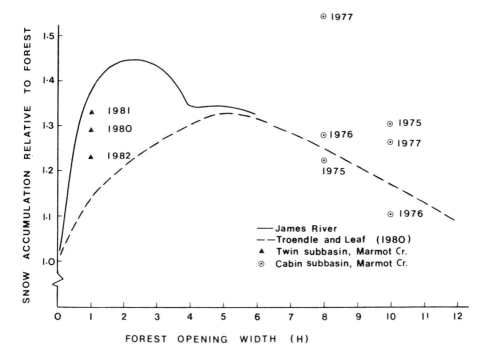

FIG. 4. Maximum snow accumulation (water equivalent) in forest openings at three locations in Alberta relative to accumulation in the uncut forest (Golding 1982). The broken line is based on data for the dry snow conditions of the subalpine zone in Colorado.

persisted for as much as 1–2 wk longer in Alberta (Stanton 1966) and 2–4 wk longer on well-shaded north aspects in Quebec (Plamondon et al. 1984a). On the other hand, because of factors such as snowmelt in the canopy or chinook wind-related processes, snow has been found to disappear sooner in lower elevation forests than in the open in coastal British Columbia (Beaudry 1984) and in Alberta (Jeffrey 1968a).

Influence of Forests and Forestry Operations on Runoff

Runoff is the product of several processes in which forest–water interactions play a key role. Following initial effects on microclimate and precipitation, forests next affect runoff by influencing the movement and storage of water on and within the soil mantle. Eventually, runoff waters move into groundwater aquifers or appear as streamflow in surface channels. Streamflow integrates the various effects of forests on runoff and also reflects changes in runoff processes caused by forestry operations.

Runoff Processes

On reaching the ground, most rain or snowmelt water will either move into and through the soil or over its surface. The processes of water movement and storage are thus conditioned by the characteristics of the soil mantle. Forest soils readily absorb water (Plamondon et al. 1972; Price and Hendrie 1983; Singh 1983), being highly permeable because of accumulations of organic matter, live tree roots, and decayed root channels and other voids (Chamberlin 1972; Foster and Nicolson 1984). As a result, surface runoff (also called overland flow by some authors) outside of stream channels is rare in most forested areas. The exceptions occur where infiltration is locally prevented because soils are saturated or frozen. Water moves more slowly through the soil than it does on the surface.

Forest soils are composed of a surface layer of organic material (called the forest floor) overlying a mantle of mineral soil, both of which can store water. The water storage capacity of a soil is the amount of space in the total soil volume not occupied by solids. Soil water storage opportunity is defined as the portion of this space not already filled with water and is created or increased by transpiration and vertical drainage. The water storage capacity of the upper 50 cm of mineral soil, which contains most of the tree roots, is typically about 250 mm. Water storage capacities of the forest floor are much lower, ranging from 13 to 23 mm (floor depth of 5-10 cm) in conifer stands in Alberta (Golding and Stanton 1972) and New Brunswick (Mahendrappa 1982), 12 to 38 mm (floor depth of 5-6 cm) in hardwood stands in New Brunswick (Mahendrappa 1982), and up to 81 mm (floor depth of 17 cm) in coastal British Columbia (Plamondon et al. 1972). Since soils under forest cover are seldom totally dry, the water storage opportunity in forest soils is usually considerably less than the maximum values for forest floors (Plamondon et al. 1972) and the mineral soil (Giles et al. 1985). The thin, very well-drained mineral soils in many forest regions in Canada further restrict soil water storage opportunity. Where the water table is close to the surface, transpiring forests also directly increase groundwater storage opportunity by lowering water levels (Rothwell 1982).

In most of the forested lands in southern Canada, peak streamflows from rain or snowmelt are generated either by subsurface flow into expanding surface channel systems (Cheng et al. 1975b) or surface runoff from zones of saturated soil near stream channels (Price and Hendrie 1983). These two processes are known as the "variable source area" and "partial area" concepts, respectively. Surface runoff on frozen ground during snowmelt is the dominant process in the permafrost zone (Price et al. 1978).

Forestry operations that remove the forest cover or disturb the soil also change runoff processes. Soil water storage opportunity is reduced by the harvesting of trees because of reduced transpiration losses of soil moisture. This results in wetter soils in cleared areas (Kachanoski and de Jong 1982), higher water tables in areas of shallow groundwater (Hetherington 1982), and increased zones of saturated soil near stream channels (Swanson and Hillman 1977). Fire may consume forest floor material, which reduces soil water storage capacity and exposes mineral soil to erosive forces (Smith and Wass 1982).

The pathways of water movement are also altered by soil disturbance. The compacted surfaces of haul roads and skid trails intercept seepage water and precipitation, creating surface runoff and speeding up water delivery to stream channels (Hetherington 1982). Yarding and scarification modify surface soil structures and block water entry to more open pathways such as decayed root channels. These changes can reduce infiltration and subsurface flow rates, cause local increases in groundwater tables during rain storms (Hetherington 1982), and could result in surface runoff. In the mountains of western Canada and particularly in coastal British Columbia, some of the steeper slopes are inherently unstable and landslides are a natural occurrence. However, the incidence of landslides can be increased by harvesting and road construction in steep terrain (Wilford and Schwab 1982). The landslide scars often become surface runoff channels, while landslide debris alters patterns of streamflow within channels.

Water Yield

Water yield is the total runoff from a drainage area through surface channels and by groundwater flow. However, it is usually taken to be the total measured streamflow over a given time period. This represents the amount of precipitation not lost to evapotranspiration or deep groundwater flow or retained in storage within the watershed. On an annual basis, water yields vary directly in relation to precipitation, since evapotranspiration is less variable than precipitation.

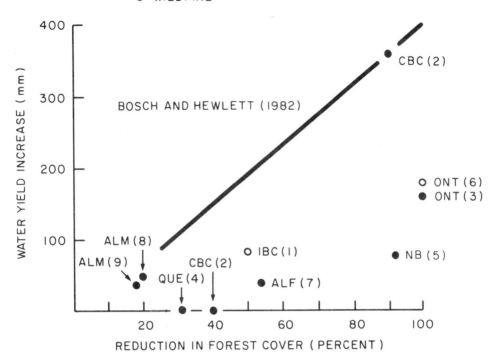

FIG. 5. Maximum annual or seasonal water yield increases within the first 4 yr following harvesting or wildfire in Canadian watersheds in comparison with worldwide averages. Values for 1, 3, and 7 are based on streamflow data for the ice-free season from spring to late fall. The regression line from Bosch and Hewlett (1982) is based on data from experimental watersheds in several countries. 1, J. D. Cheng, Ministry of Forests, Kamloops, B.C., pers. comm., the percent forest cover reduction is revised from Cheng 1980; 2, Hetherington 1982; 3, Nicolson et al. 1982; 4, Plamondon and Ouellet 1980; 5, R. B. B. Dickison and D. A. Daugharty, University of New Brunswick, Fredericton, N.B., pers. comm., a revision of data given by Dickison and Daugharty 1983; 6, Schindler et al. 1980; 7, Swanson and Hillman 1977; 8, R. H. Swanson, Canadian Forestry Service, Edmonton, Alta, pers. comm.; 9, Swanson and Golding 1982. ALF = Alberta foothills; ALM = Alberta Rocky Mountains; CBC = coastal British Columbia; IBC = interior British Columbia; NB = New Brunswick; ONT = Ontario; QUE = Quebec.

Forests reduce water yields because they are consumers of water. For the same total precipitation, a given area covered by trees will generally yield less annual runoff than an area covered with other types of vegetation, although there might be exceptions (Morton 1984). The lower yields from forests are due to greater evapotranspiration losses. However, where forest stands conserve water by trapping snow blown from open areas, forests can yield more water than adjacent treeless terrain. This occurs with drier snow in central and northern Canada.

Research studies in small (less than 25 km²) forested watersheds in Canada have shown that reductions in forest cover by harvesting or fire usually result in higher annual water yields (Fig. 5). The increases in runoff are generally in proportion to the amount of forest cover removed, and they vary with soil depth and precipitation. These changes are mainly attributable to reductions in evapotranspiration losses from the deforested areas. Except for coastal British Columbia, observed changes in water yield are lower on average than those found elsewhere (Fig. 5). This might be partly due to the limited soil water storage capacities and rapid drainage characteristics of the thin soils found in many forested areas of Canada. The magnitude of changes in water yield from insect-damaged forests in Canada could also be limited because of reductions in evapotrans-piration prior to deforestation. In areas where forests trap wind-blown snow, removal of forest stands will eliminate this trapping effect and may cause reductions in total runoff from the area affected. The rate of recovery of water yields is proportional to the rate of recovery of vegetation. With regrowth of forest vegetation and no further disturbance, water yields from deforested areas can return to predisturbance levels in only a few years on the humid west coast and in eastern Canada (Schindler et al. 1980), but recovery may take 30 yr or longer in central Canada (Swanson and Hillman 1977).

The processes that have caused observed changes in water yields in small water-sheds are likely to produce similar changes in large drainages. There is evidence from large watersheds (over 700 km^2) in the United States, for example, of increases in water yield following forest harvesting (Berndt and Swank 1970) and insect damage (Bethlahmy 1974) and reductions in water yield after reforestation (Schneider and Ayer 1961). The magnitude of the change will depend on the proportion of a watershed that is deforested or reforested.

Low Flows

Forest influences on summer low flows parallel those on water yields. Forest evapo-transpiration in summer can exceed precipitation inputs, thereby reducing the amount of soil water and groundwater available for streamflow. Daily flows in small streams can fluctuate in direct response to daytime water use by streamside vegetation, as observed in Washington State (Helvey 1972).

Removal of the forest cover will decrease evapotranspiration losses. Most research studies indicate that some of the extra water is translated into higher summer stream-flows. For example, flows in August and September increased 10–36% after a wildfire in interior British Columbia (Cheng 1980) and 133–318% after harvesting in New Brunswick (Dickison et al. 1981) and Ontario (Nicolson et al. 1982). Minimum daily flow increased by 78% after clearcutting in coastal British Columbia (Hetherington 1982), and clearcut plot runoff doubled in Saskatchewan (Kachanoski and de Jong 1982). However, no change was detected after partial clearcutting in Quebec (Plamondon and Ouellet 1980). Even small increases in low flows can be beneficial in terms of improved fish habitat and domestic water supply. The expected durations of changes in low flows are similar to those for water yields.

Under some circumstances, low flows might be reduced in small streams following logging or fire. After initial increases, flows in west coast or eastern Canadian streams could eventually be diminished below predisturbance levels by the vigorous transpira-tion of new streamside deciduous vegetation (Harr 1983). Reduced fog interception could locally reduce low flows in immediate coastal areas, as observed in Oregon (Harr 1983). Even if low flows are not reduced, buildup of gravel in stream beds, such as might occur after logging, could result in flow being entirely subsurface. This change in the channel would impair its use by fish. In general, however, the evidence indicates that low flows in most of Canada are more likely to increase rather than decrease after removal of the forest cover.

Peak Flows

Peak flows are maximum stream discharges generated primarily by short-duration, localized rain showers, more extensive long-duration rain storms, spring snowmelt, and rain-on-snow events. Floods represent extreme occurrences of peak flows produced by any of these sources of runoff. The effects of forests and forestry operations on peak flows can vary considerably, depending on the source of runoff.

TABLE 2. Changes in peak flows during rain storms following logging in Canadian watersheds (nd = not detected).

Location	Change in peak flow (%)	Comments	Reference
Carnation Creek, B.C.	nd	40% clearcut	Hetherington 1982
H Creek tributary	20	90% clearcut	Hetherington 1982
Near Haney, B.C.	−22	Soil 50% disturbed	Cheng et al. 1975a
Near Hinton, Alta.	nd–230	Avg. clearcut age 10 yr	Swanson and Hillman 1977
Experimental Lakes, Ont.	Increased	After 4 yr	Nicolson et al. 1982
Nashwaak Basin, N.B.	71	1979	Dickison and Daugharty 1983
Nashwaak Basin, N.B.	38	1980	Dickison and Daugharty 1983
Nashwaak Basin, N.B.	26	1981	Dickison and Daugharty 1983

Stormflow from Rainfall

Stormflow, the rapid runoff from storm rainfall, is defined by both the peak flow and the volume of runoff. In undisturbed forested lands, stormflow is mainly affected by available soil water storage opportunity and, to a lesser extent, by canopy interception of rainfall. The effect is greatest when soils are deep and/or saturated. In general, the influence of this storage is greatest during light showers, decreases as the duration and intensity of rainfall increases, and is least during major rain storms in late fall or winter. In coastal British Columbia, for example, the proportion of rainfall appearing as stormflow has ranged from as low as 3% for small summer storms to over 90% during major winter rains (Cheng et al. 1977; Hetherington 1982).

Peak flows during rain storms increased after forest harvesting in several research watersheds in Canada (Table 2). These results apply to storms of a magnitude that is likely to occur more often than once every 5–10 yr. Stormflow volumes also increased in the Alberta and New Brunswick watersheds. The stormflow changes in Alberta and eastern Canada occurred during summer storms and were caused mainly by reductions in soil water storage opportunity after logging. The production of surface runoff by roads caused the increase in coastal British Columbia and was a contributing factor in Alberta's increases. In the west coast Carnation Creek watershed, harvesting after road construction increased peak flows from one or two early fall storms, but had little detectable impact on most runoff events in this humid environment. Peak flows may thus be increased by water reaching the stream channel faster or simply by greater volumes of runoff. The decreased peak flows in the other west coast watershed were attributed to changes in subsurface flow rates caused by extensive soil disturbance and to flow delays by logging debris in the channel.

These results indicate that stormflow runoff peaks can potentially increase after harvesting for smaller storms in small watersheds in most forest regions in southern Canada. As the severity and duration of storm rainfall increases, the proportional increase in peak flow will diminish. Changes in stormflow caused by roads are likely to be long-lasting. Those changes associated with forest cover removal will diminish as vegetation regrows and take about the same length of time as water yields to return to prelogging levels.

| a. Low flow | b. Peak flow |

Forests influence both summer low flows and rainstorm peak flows (E. D. Hetherington, Canadian Forestry Service, Victoria, B.C.).

Snowmelt Runoff

Forests have a significant effect on snowmelt runoff through their influence on the processes of snow accumulation and melt, frost formation, and runoff generation. Peak flows from snowmelt are particularly affected by the rate and timing of melting and by soil water storage opportunity. By increasing storage opportunity through transpiration and slowing melt rates by sheltering snowpacks from the effects of sun and wind, forests help regulate the generation of runoff from snowmelt.

Forested areas composed of a mosaic of openings and tree cover, however, can generate different streamflow patterns than those arising from a complete forest cover by synchronizing or desynchronizing snowmelt (Federer et al. 1972). Peak flows are highest when heavy snowmelt runoff from all parts of a watershed is synchronized and arrives downstream at the same time. Because melt rates are higher in openings than beneath the forest canopy, the acceleration of spring snowmelt from openings may desynchronize runoff to the extent that downstream peak flows are smaller than would

TABLE 3. Changes in spring snowmelt runoff following forest harvesting and fire in Canadian watersheds (nd = no change detected).

	Change in runoff				
Location	Peak flow (%)	Volume (%)	Time of peak (d)	Treatment	Reference
Palmer Creek, B.C.	16	53[a]	− 13	50% burn	Cheng 1980
Marmot Creek, Alta. (Cabin Creek trib.)	nd	24[b]	− 14	20% patchcut	Golding 1980
Streeter basin, Alta.	78	41[a]	− 17[c]	Patchcut	Golding 1981
Near Hinton, Alta.	57[d]	59	nd	35–84% patchcut	Swanson and Hillman 1977
Experimental Lakes, Ont.		75[a]		Clearcut	Nicolson et al. 1982
Riv. des Eaux-Volées, Que.	nd	nd	nd	31% patchcut	Plamondon and Ouellet 1980
Nashwaak basin, N.B.	Lower	Lower	− 21 to − 4	92% clearcut	Dickison and Daugharty 1982

[a]Runoff for April and May.
[b]Runoff for May.
[c]Occurrence of 50% of total spring snowmelt runoff.
[d]Estimated from composite hydrograph.

occur from a completely forested watershed. Accelerated melt runoff from openings at high elevations could also augment the normally earlier melt at lower elevations to increase downstream peak flows. The location of openings in forested watersheds is thus critical in determining their influence on snowmelt streamflow.

Because of the major differences in snowmelt rates between forested and open areas and the possibilities for synchronizing or desynchronizing snowmelt runoff, the effects of forest cover removal on spring snowmelt runoff are variable (Table 3). Following harvesting and, in one case, fire, spring snowmelt peak flows increased, showed no change, and decreased. In the clearcut New Brunswick watershed, early-spring runoff was increased, but the reduction in late-spring runoff was of greater magnitude (Dickison and Daugharty 1982). Spring streamflow volumes were also higher in several streams, a probable result of reduced soil water storage opportunity in cleared areas. In the New Brunswick stream, the decreased spring runoff volume resulted from greater ablation of the snowpack during the winter in the open clearcut than in the forest. In four of the studies referred to, peak runoff occurred earlier due to faster melt in open areas. However, the dispersed clearcut pattern on relatively level terrain in the Hinton, Alberta, area resulted in little difference in the timing of peak flows (Fig. 6). In summary, complete clearcutting of a watershed will usually increase peak flows from snowmelt, except in areas such as New Brunswick where winter ablation reduces the snowpack prior to spring melt. To evaluate the probable effect on snowmelt peak flows of partially harvesting a watershed, both the distribution of openings and the proportion of the drainage area to be cleared must be taken into consideration. With no further disturbance of the forest cover, changes in snowmelt runoff may persist for 30 yr or longer in central Canada (Swanson and Hillman 1977) and possibly in eastern Canada, but recovery could occur sooner on the west coast.

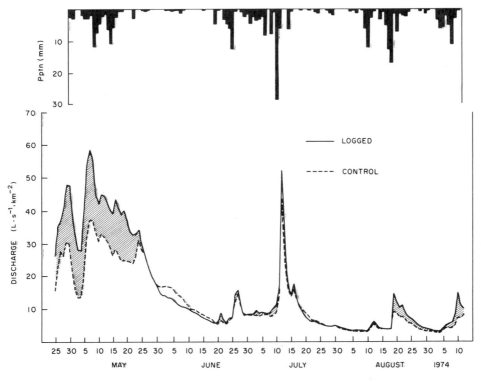

FIG. 6. Composite hydrograph for 1974 for logged and control catchments in the Hinton, Alberta, study area (Swanson and Hillman 1977). Note the increase in spring snowmelt runoff and the increases in peak flows from summer rains in the logged watershed.

Rain-on-Snow Runoff

Rain-on-snow runoff produces some of the largest peak flows in winter and spring on both the east and west coasts. However, the influence of forests on rain-on-snow runoff received little attention in Canada until recent expressions of concern over the effects of forest harvesting on peak flows in British Columbia (Toews and Wilford 1978). The presence or absence of snow in the canopy in addition to snow on the ground during rainfall can markedly affect the relative rates of runoff from forested versus open areas. With no snow in the forest canopy, rain-on-snow runoff rates and volumes were found to be higher in a clearcut than in the adjacent forest in coastal British Columbia (Beaudry 1984). When snow was present in the canopy during rainfall, peak runoff rates from the forest equalled those in the open. This result suggests that snow in the canopy melted faster than the snowpack in the open, and that temporary storage of rain and meltwater in the open snowpack exceeded that in the forest. The relative contribution of snowmelt to peak flows will diminish as the amount of storm rainfall increases. In the absence of snow in the canopy, the effects of harvesting on rain-on-snow peak flows should be similar to those for snowmelt without rain, including the duration of changes.

Floods

Floods are peak flows that rise above natural stream and river banks and flow onto adjacent areas. Damage caused by floods results from soil and debris carried by the floodwaters as well as by the water itself. In general, forests act to minimize flooding and flood damage by intercepting precipitation, increasing soil water storage opportunity through transpiration, moderating snowmelt rates, minimizing overland flow, erosion, and the effects of frost, retaining soil on steep slopes, and maintaining stream channel capacity for carrying peak flows. Forests may mitigate flood runoff from high-intensity summer showers (Anderson et al. 1976), but their influence on flood flow decreases as the duration and magnitude of rain storms increase. Snowmelt floods may also be moderated by the desynchronizing effects of accelerated melt from openings in the forest. Forest influences on extreme runoff caused by exceptional meteorological conditions, however, will be negligible (Teller 1968). In short, forests do not prevent floods, but they probably provide the best conditions for minimizing flood runoff and damage.

The effects of forest harvesting on floods have seldom been observed in experimental watershed studies and therefore must be inferred. Forestry operations that increase surface runoff through soil disturbance or more severe soil freezing can result in increased flood flows and erosion from all types of runoff. In small watersheds, flood peaks can be increased and flood damage aggravated by road-diverted runoff and localized transport of soil and debris into stream channels from landslides caused by logging operations. Direct inputs of logging debris can also create or destabilize debris jams in streams, increasing the risk of damaging surges of water and debris that can occur when the jams break up during flood flows. However, extensive and severe soil disturbance and reductions in infiltration and subsequent erosion, such as caused by cultivation or overgrazing, are not generally a problem in the forested areas of Canada.

Forest removal, apart from soil disturbance effects, might increase flood flows from high-intensity summer rain showers in smaller watersheds by reducing soil water storage opportunity. However, harvesting will have little effect on flooding caused by major long-duration rainstorms. Except for extreme events, harvesting could increase floods involving snowmelt runoff by accelerating snowmelt from clearcut watersheds or by synchronizing the melting of snow in different parts of partially cut watersheds. In most larger watersheds, however, changes in forest cover are unlikely to affect downstream flooding significantly due to natural diversity of terrain and cover conditions (Hewlett 1982). An important Canadian exception is that of basins with northward flowing rivers. Accelerated spring snowmelt following forest harvesting in the more southerly portions of such watersheds could aggravate downstream flooding if breakup has not occurred in downstream reaches (Swanson 1972).

Popular Beliefs Reviewed

The view of forests as having positive effects on water is generally valid for healthy young or mature forests complete with undisturbed forest floor and soil. However, there are some constraints. Forests do help "regulate streamflow" by moderating spring snowmelt, but their efficacy in minimizing snowmelt peak flows can be enhanced by appropriately spaced openings in the forest cover. Forests probably provide the best cover for minimizing flood flows and damage, but they are not as effective in "protecting against floods," in the sense of preventing them as this statement implies. The role of the forest floor as a "sponge that reduces peak flows" will be minor in most areas because of limited water storage opportunity in the forest floor at the time of major runoff. With the exception of situations where trees trap wind-blown snow, forests do not "provide maximum runoff" nor are they the best cover for "sustaining summer flows." Forests are consumers rather than "conservers" of water.

The negative views of logging impacts on streamflow are mostly not substantiated. Rather than "drying up streams," harvesting usually increases summer flows and total annual runoff. The effect of earlier disappearance of snow from cleared areas on summer flows is offset by higher soil moisture reserves resulting from reduced evapotranspiration losses. Furthermore, logging, in general, does not "cause flooding." Most floods are caused by exceptional meteorological conditions. Logging may aggravate flood flows and flood damage in small watersheds through accelerated additions of soil and debris to stream channels. Harvesting might increase flood flows because of "greater snowmelt runoff from clearcuts," but flood peaks could also be reduced if this process leads to desynchronization of snowmelt runoff. In summary, forest harvesting does not cause floods, and its effects on downstream flooding are usually minor and most likely to be noticeable only in small watersheds.

Influence of Forests and Forestry Operations on Water Quality

A forest influences water quality by first modifying the chemistry of precipitation as it passes through forest vegetation and then as it moves through the soil. Trees bordering streams further affect water quality by providing shade, which moderates stream temperatures and rates of in-stream biological activity, and by furnishing inputs of leaves and other organic material. Streams draining from naturally forested lands generally contain few sediments or harmful chemicals. In some areas, however, there may be naturally high levels of deleterious substances such as mercury, or biological contamination by wildlife. Mean concentrations of most dissolved nutrients in forest streams are usually low (Feller and Kimmins 1984; Hetherington 1976; Krause 1983; Nicolson 1975; Singh and Kalra 1977). The high quality of forest stream water results from protection against erosion by the forest floor and high soil infiltration rates, soil stability provided by tree roots, the ability of forest soils to retain nutrients, and the modest organic decomposition and mineralization rates beneath forest canopies. Tree roots and logs from fallen trees also help maintain good stream water quality and desirable aquatic habitat by stabilizing stream banks and maintaining a network of pools and riffles in the channel (Toews and Moore 1982).

Forestry operations can modify the quality of stream water through direct inputs of sediment, debris, fertilizers, or pesticides, changing the rate of leaching of nutrients and other dissolved substances from the soil, and removal of streamside vegetation which allows warming of the water and subsequent biological and chemical changes (Freedman 1982; Krause 1982). Changes in water quality are most noticeable in small streams.

Conventional clearcut harvesting in patches increases water yields and modifies snowmelt runoff (R. H. Swanson, Canadian Forestry Service, Edmonton, Alta.).

Forests as Filters of Airborne Contaminants

Forest canopies are effective in collecting airborne contaminants in precipitation and in dry fallout from the atmosphere by filtering particles or moisture from air moving through the canopy and by adsorption of gases. The contaminants of current concern in North America are primarily those causing acid precipitation and acidification of streams and lakes. These include sulphur dioxide, sulphate, oxides of nitrogen, especially nitrate, hydrogen ions, and ammonium (Bangay and Riordan 1983). Other contaminants such as heavy metals, trace elements, and organic micropollutants have received considerably less attention. Some elements, particularly nitrogen, are adsorbed by the foliage. The chemistry of precipitation is altered in passing through forest canopies by washing of deposited substances such as sulphate and leaching basic cations, primarily calcium, magnesium, and potassium, from the foliage (Morrison 1984). The resulting acidity of precipitation reaching the ground is lower in hardwoods than in the open (Foster 1984). Some conifers also lower the acidity of precipitation but to a lesser extent than most hardwoods (Mahendrappa 1983), while rainwater acidity remains unchanged or may be increased by other conifers (Foster 1984). The effect will vary with the acidity of the precipitation.

Forests soils are very important in determining the degree of protection from airborne contaminants provided water bodies by forests. Shallow, acidic soils with low basic cation content do little to modify the acidity of precipitation, while deeper soils rich in cations are highly effective in buffering or neutralizing water acidity (Foster 1984; Nicolson 1984). Soils neutralize acidic waters by exchanging basic cations for acidic cations. The result is increased leaching of calcium, magnesium, potassium, and sodium and enrichment of stream waters with these ions (Nicolson 1984). Sulphate plays an important role in cation leaching from soils low in iron and aluminum or high in organic matter, whereas soils high in iron and aluminum retain sulphate (Bangay and Riordan 1983; Foster and Nicolson 1984). As soils become more acidic, there is an increasing tendency for leaching of aluminum and manganese to occur (Bangay and Riordan 1983; Foster 1984). Nitrogen is retained by some forest soils as it is by forest canopies (Nicolson 1986).

Sediment

Increased sediment input to streams is probably the biggest change in water quality associated with forestry operations. The three main sources of sediment in streams are surface erosion, mass soil failures (landslides), and stream bank erosion. Mineral soil is exposed to accelerated erosion through removal of the forest floor during road construction, log skidding, prescribed burning, and scarification. The major sources of sediment in most regions, however, are roads and skid trails because of the associated increase in surface runoff that causes erosion. Maximum sediment loads usually occur during construction of roads, particularly in wet weather, and reported measurements have ranged from about 200 mg/L in main streams to over 8000 mg/L in small tributaries (Krause 1982; Ottens and Rudd 1977; Rothwell 1977). Even after construction, roads continue to be sources of sediment, and concentrations from 30 to over 500 mg/L have been reported (Rothwell 1983; Slaney et al. 1977). Sediment amounts derived from soil disturbed by other forestry operations are generally lower than those from roads. More frequent landslides following logging are a major periodic source of sediment in the mountains of British Columbia. In-stream sediment production can also increase as a result of bank erosion caused by additions of logging debris and altered debris structures in the channel (Toews and Moore 1982), or the greater erosive power of increased peak flows resulting from logging.

Sediment levels can be reduced or controlled by such management practices as the use of buffer strips of vegetation along the margins of water bodies. These keep logging equipment away from streams and lakes and trap soil eroded from roads or other exposed soil, except where tributary streams or roads cross the strips (Plamondon 1982; Plamondon et al. 1976). Erosion control rehabilitation measures, such as revegetation of forest roadsides, also help speed up the reduction of sediment inputs to streams (Carr and Ballard 1980). Sediment concentrations in stream waters normally decline with time and can return to predisturbance levels within a year (Hetherington 1976). Where roads are the primary sediment source, however, sediment inputs could persist almost indefinitely. Even small increases in sediment loads may have important cumulative impacts on aquatic habitat, such as the covering or plugging of spawning and rearing gravel beds, which are harmful to the early life stages of fish (Jablonski 1980; Sabean 1978). Sediment trapped in gravel beds may also persist for considerable lengths of time.

Relatively high, short-term inputs of sediment to forest streams are possible anywhere in Canada. The conclusions of Krause (1982), however, regarding the overall potentials for stream sedimentation in Canada seem reasonable: high in the (coastal) mountains of British Columbia and the Alberta foothills, intermediate in the Rocky Mountain region, comparatively low in the Canadian Appalachians, and very low within the Canadian Shield, except for areas with fine-textured soils. During most logging operations some soil is disturbed and some increase in stream sediment levels is usually unavoidable. The belief that "logging silts up streams" therefore bears some truth. Requirements for minimizing deterioration of stream water quality or aquatic habitat by sediment include sound planning, sufficient resource data, careful and appropriate logging operations, suitable attention to road maintenance, and application of rehabilitation measures where necessary. In reality, these objectives are not always met, although improved management practices have reduced the frequency and extent of the damaging stream sedimentation of the past.

Water Chemistry

The chemistry of stream water is altered following forest harvesting and, to a greater degree, after prescribed burning or wildfire. The uptake of nutrients in vegetation is substantially reduced, decomposition of the forest floor is accelerated, and more water passes through the soil to leach these extra elements into streams. Both increased sediment and decreased leaf inputs can also affect water chemistry, while additions of organic debris can modify water quality and decrease dissolved oxygen levels as the debris decays. Changes in concentrations and total amounts of the following dissolved nutrients and chemical parameters have been measured in forest streams after harvesting or burning: calcium, magnesium, sodium, potassium, nitrogen, phosphorus, sulphate, chlorine, bicarbonate, organic substances, dissolved oxygen, colour, and pH (Feller and Kimmins 1984; Hetherington 1976; Krause 1983; Nicolson 1975; Plamondon et al. 1982; Schindler et al. 1980; Scrivener 1982; Singh and Kalra 1977). The responses have included increases, decreases, and no change in the various parameters. The changes in water chemistry and the effects of these changes, however, are usually minor. Maximum values seldom exceed drinking water standards and, then, only briefly. Decreases in dissolved oxygen could adversely affect fish. On the other hand, observed increases in dissolved nutrients will benefit nutrient-poor streams, like those in coastal British Columbia, by enhancing their biological productivity. In general, these changes in water chemistry are short-lived. Predisturbance conditions usually return within about 3-5 yr, although recorded changes have lasted as little as 1 yr in northern Ontario (Nicolson 1975) and as long as 9 yr in coastal British Columbia (Feller and Kimmins 1984).

Krause (1982) has drawn the following conclusions regarding dominant changes in water chemistry to be expected after forest harvesting in Canada: (1) increases in concentrations of calcium, magnesium, and bicarbonate in areas with conifers and soils with low acidity in British Columbia and the foothills of Alberta; (2) increases in concentrations of calcium, magnesium, sodium, potassium, and dissolved organic substances and a lowering of pH in areas with conifers and acid soils, particularly across the Canadian Shield; and (3) significant increases in nitrate-nitrogen in areas with shade-tolerant hardwoods in eastern Canada. There can, however, be considerable variability in water chemistry responses to watershed disturbance within any given region. In one west coast watershed, for example, concentrations of some nutrients that had shown initial increases or no change actually fell below preharvesting and burning levels between 2 and 8 yr later (Feller and Kimmins 1984).

Water Temperature

Once the shade of streamside trees is removed by harvesting or fire, streams are more exposed to the warming effects of the sun. In Canada, maximum summer water temperatures in small streams have increased after harvesting by up to 15° C, although most increases have been less than 10° C (Feller 1981; Holtby and Newcombe 1982; Nicolson 1975; Plamondon et al. 1982; Sabean 1977; Toews and Brownlee 1981). Shallow, slow-moving streams are more vulnerable to temperature increases than larger, deeper streams with higher flows. During winter, exposed streams could experience increased freezing and ice formation in areas where they do not normally freeze over. On the other hand, increases in winter stream water temperatures have been measured after harvesting in watersheds on the south coast of British Columbia (Feller 1981; Holtby and Newcombe 1982). Changes in water temperatures resulting from stream exposure pose little problem for most users of the water except fish. Temperature increases could be harmful to fish if streams are overheated in hotter interior regions of Canada (Toews and Brownlee 1981), but can be beneficial for fish through increased biological productivity in cool, nutrient-poor streams like those along the west coast (Holtby and Newcombe 1982). Shade given by streamside buffer strips is effective in minimizing changes in water temperatures (Plamondon et al. 1976). Furthermore, water that has been warmed in the open may be cooled upon reentering the shelter of the forest,

Buffer strips protect streams during logging and help maintain good water quality and a natural environment for fish and human enjoyment. (Pacific Forestry Centre, Victoria, B.C.).

although the reduction in temperature appears to result more from additions of cooler groundwater than from the effects of shade (Sabean 1977).

Recovery of predisturbance stream temperatures will depend on the length of time it takes for streamside vegetation to grow high enough to shade the stream. In one west coast watershed, summer water temperatures in one stream returned to prelogging levels 6-7 yr after clearcutting, but showed little decline in another stream 7 yr after both clearcutting and slashburning (Feller 1981). Winter temperature changes in the same streams lasted only 1-2 yr.

Fertilizers

The principal fertilizers used in forestry are the nitrogen fertilizers urea and ammonium nitrate. Streams can be temporarily enriched with nitrogen by direct inputs of fertilizer during aerial application or by entry of fertilizer or its breakdown products of ammonium or nitrate through surface runoff or subsurface flow (Table 4). For the most part, observed concentrations of fertilizer nitrogen have remained below acceptable limits set for public water supplies (Hetherington 1985), except for brief peaks of ammonium and, in the Nashwaak basin, of nitrate (University of New Brunswick 1976). Excess nitrogen from streams draining fertilized forest lands has the potential to unduly enrich downstream lakes, but any effects of excess nitrogen can be offset by limitations in other nutrients such as phosphorus (Perrin et al. 1984). On the other hand, nitrogen increases can enhance the biological productivity of otherwise nutrient-poor streams such as those on the west coast. By avoiding direct inputs of fertilizer to open water during application, the amounts of fertilizer nitrogen entering streams and lakes can be considerably reduced (Perrin et al. 1984).

TABLE 4. Maximum measured nitrogen concentrations and additional nitrogen in streams following forest fertilization in Canada.

Location	Type	Fertilizer Rate[a] (kg N/ha)	Maximum concentration (mg N/L) Urea	NH$_4$-N[b]	NO$_3$-N[c]	Total N[d] (%)	Reference
Coastal B.C.							
Lens Creek	Urea	224	14.0	1.90	9.30	14.5	Hetherington 1985
Mohun Lake							
Tributary	Urea	200	57.6	4.78	0.79	5.2	Perrin et al. 1984
Tributary[e]	Urea	200	0.66	0.47	0.19	2.1	Perrin et al. 1984
Montmorency, Que.	Urea	150	15.0	3.5	1.3	1.3	Gonzalez and Plamondon 1977
Nashwaak basin, N.B.	AN[f]	110		5.5	11.5	22	Univ. of N.B. 1976

[a]Rate of fertilizer application.

[b]Ammonium.

[c]Nitrate.

[d]Percentage of total nitrogen applied on the watershed that was lost in streamflow.

[e]Nonfertilized strip of forest was left along streams.

[f]Ammonium nitrate.

The duration of changes in stream water nitrogen following fertilization is variable. Urea is normally present in stream water for only a few days because it usually is rapidly converted into ammonium and nitrate. In one west coast stream, however, urea was detected up to 20 wk after application due primarily to cold temperatures and sub-surface water flow through macrochannels in the soil (Perrin et al. 1984). Maximum ammonium concentrations are also usually observed at the time of fertilization, with lesser peaks occurring during rainstorms in the following few weeks. Ammonium increases lasted up to 6 wk in the two eastern streams and 12-20 wk on the west coast. Nitrate concentrations peaked during the first few rains in all streams listed in Table 4 and remained above background levels for up to 1 yr after fertilizer application in Lens Creek and the Nashwaak basin, but only for a few weeks in the other streams.

Pesticides

Forest pesticides are used primarily for suppression of unwanted deciduous vegetation (herbicides) to permit successful establishment of young conifers and for control of insects (insecticides) to protect existing forests. Pesticides are mostly applied by spraying from aircraft or from the ground, but some herbicides are injected directly into trees. The use of such chemicals often raises concerns over the risk of contaminating stream and potable waters. The most likely ways for pesticides to enter streams are by direct overspray and aerial drift during application. Secondary pathways or sources include entry by overland flow, subsurface leaching, leaf-fall, or accidental spills. Each chemical pesticide is unique in its properties, formulation, rate of breakdown, behaviour, and response in the environment. The other materials used in pesticide formulations may be much more harmful than the active ingredient itself. Any hazard associated with chemicals also results from both the amount (toxicity) and length of exposure to the chemical. Thus, while some generalities regarding pesticide impacts on the environment are possible, pesticides must be evaluated individually to obtain an accurate understanding of their effects and avoid erroneous conclusions. Because of environmental concerns about pesticides, strict federal and provincial regulations govern their registration, use, and application, including such measures as requiring unsprayed buffer zones around lakes, streams, and potable water sources.

Herbicides

The chemicals most commonly used now for forest weed control in Canada are phenoxy herbicides (e.g. 2,4-D) and glyphosate. These herbicides are applied infrequently on any given forest site, on average less than once every 40 yr. They may be used more often to clear roadside vegetation. Most herbicides decompose rapidly and do not accumulate in the environment (Toews and Brownlee 1981). Many are also strongly adsorbed on forest soils and are not readily leached.

Field evidence to date indicates that forestry uses of currently registered herbicides have not resulted in adverse effects on stream water or fish. Peak herbicide concentrations found in stream waters following operational applications have been low and transitory in nature (Wilson et al. 1983). Herbicides diminish to negligible levels within hours and disappear within days in stream water, although they may last from several days to several months on land before totally disappearing, depending on environmental conditions and chemical structure (Freedman 1982; Newton et al. 1984). Sublethal effects on fish such as impaired health, which might be caused by chronic exposure to low herbicide doses, have note been found. If streamside vegetation is killed by herbicides, aquatic habitat could be altered through changes in stream temperatures, reduction in terrestrial food supplies, and bank destabilization. However, precautions normally taken during spray applications minimize the risk of negative impacts of herbicides on aquatic resources (Freedman 1982).

Insecticides

Considerable research on the environmental effects of insecticides has been conducted in eastern Canada since the start of the spruce budworm spray program in the 1950's (Kingsbury 1984; Prebble 1975). The greatest use of insecticides has been in eastern Canada where fenitrothion, aminocarb, and the biological agent *Bacillus thuringiensis* var. *kurstaki (Btk)* are the principal insecticides employed for control of spruce budworm. Repeat application of these insecticides on any given area is much more likely than it is for herbicides. A few other chemicals are used to a lesser extent in British Columbia. Ongoing studies accompanying insecticide use have resulted in modifications to spray programs to minimize adverse effects. For example, both DDT, which was harmful to fish, and phosphamidon, which was harmful to birds, were eliminated from use during the 1960's and 1970's, respectively (Kingsbury 1984). Insecticides are normally distributed at very low concentrations (i.e. a few grams per hectare).

Field studies have indicated that operational spraying of fenitrothion and aminocarb has resulted in little effect on streams or fish. The greatest hazard is likely to come from mishandling of insecticides and their containers rather than from spraying. Concentrations of fenitrothion and aminocarb measured in streams after operational spraying have been substantially lower than levels that are toxic to fish (Morin et al. 1986). Both insecticides disappear rapidly from stream waters and usually disappear totally from the environment within a few days to a few weeks, depending on environmental conditions (Sundaram et al. 1984; Wilson et al. 1983). They also do not accumulate biologically. Where repeated insecticide applications take place, there is a risk of a gradual buildup of insecticides in the environment if the chemicals persist longer than the frequency of applications. However, no such long-term environmental accumulations have yet been documented for the insecticides in current use (Sundaram et al. 1984; Varty 1980).

Btk is a bacterial pathogen specific to caterpillars (Lepidoptera larvae), whose use in forest pest control has expanded rapidly in the 1980's. *Btk* is also applied in small amounts, and has been found to be of very low toxicity to fish and aquatic insects (Eidt 1985). It has lasted up to 100 d in water and several months in soil (Wilson et al. 1983), but is biologically inactive unless ingested by caterpillars. No adverse effects of *Btk* on aquatic life have been documented in the field.

Environmental Regulation of Forestry Operations

Forest management operations in Canada are governed by a variety of environmental regulations, many of which exist because of concern for water and fisheries resources. The primary objective of these regulations is to safeguard other resource values from adverse impacts resulting from forestry operations. The federal government, for example, is actively involved in enforcing sections of the *Federal Fisheries Act* aimed at protecting aquatic habitat and fish from possible damage resulting from such activities as forestry operations. All provinces have resource or environmental legislation which, while not necessarily addressed specifically at forestry, can be applied to forestry issues. The several laws and acts relating to stream protection in British Columbia are good examples (Dorcey et al. 1980; Toews and Brownlee 1981). At the operational level, regulations take the form of technical guidelines which specify the basic principles and procedures to be applied, such as those in Quebec (Gouvernement du Québec 1977). Resource agencies in most provinces have developed such guidelines which are implemented through interagency referral systems and incorporated into forestry use permits. In some areas, however, limitations in the capability of resource agencies to handle the large volume of forestry activity often restrict the effectiveness of these procedures. Although such practical limitations in applying some environmental regulations seem likely to persist, environmental concerns are being recognized and the level of dealing with them has improved across Canada in recent years.

Future Trends in Forest Hydrology in Canada

The next decade for forest hydrology in Canada is likely to be one of both consolidation and shift in emphasis. Forest hydrology research appears to be in a transition phase. At least two of the major experimental watershed projects are drawing to a close. The current level of research, including number of studies and researchers, will probably remain about the same, but with increasing emphasis on application of knowledge to operational watershed management (Swanson 1982). For the most part, activity will continue to focus on watershed protection. Fisheries–forestry interactions will continue to be a major focal point, primarily in relation to water quality and aquatic habitat. The demand for forest management techniques to enhance water supplies will increase because of increased need for supplies of high-quality water, increased awareness by water managers that forest and water management can be complementary, and more aggressive application of forest hydrology knowledge by hydrologists and watershed foresters.

The changing stress on forest lands will increase pressures on water resources and help focus forest hydrology attention and research. As forest harvesting continues to expand into increasingly more difficult terrain, related hydrologic problems will become more acute and the need for hydrologic input to management decisions ever more apparent. Logging in the steep, unstable mountainous terrain of British Columbia is one example. An upsurge in the management of wetlands for forestry, currently underway across central and eastern Canada, is another example. Drainage of these wetlands to improve the land for growing and planting trees significantly modifies their hydrologic regimes. Although wetlands are being studied (see chapter by T. H. Whillans in this volume), little is known about these changes and hydrologists will be increasingly called upon to investigate wetland hydrology.

A developing forestry initiative in Canada is renewal and rehabilitation of inadequately reforested land and more intensive management on highly productive forest lands. At the present time, about 10% of productive forest land in Canada is not satisfactorily reforested, and this amount has been increasing annually (Honer and Bickerstaff 1985). A renewal program has many implications for forest hydrology. Lands with minimal forest cover prolong hydrologic changes caused by harvesting or fire. Hydrologists will have an opportunity to apply their knowledge to help improve regeneration and growth of new forests (Swanson 1982) and recovery of hydrologic regimes. The concern over water quality will certainly increase, and with it, the need for hydrological input to help minimize undesirable impacts of forestry operations. Furthermore, more intensive forest management will mean more extensive road networks to maintain and a greater potential for stream sedimentation. The need to suppress or control brush in order to establish new coniferous forests on inadequately forested lands will be attended by a greater use of herbicides. Insecticides will continue to be used to ensure the survival of existing timber and of managed forest plantations. Greater use will be made of forest fertilizers to enhance tree growth in our more intensively managed forests. Fire suppression activity is also likely to increase to protect these expanded investments in Canada's forests. Each of these activities will provide an opportunity for forest hydrologists to help enhance compatible management and use of Canada's forest and water resources.

Conclusion

Our knowledge and understanding of forest hydrology and its application in Canada have advanced considerably since W. W. Jeffrey brought the subject to national attention in the 1960's. The information now available permits us to see more clearly the importance of forests in the hydrological regime. Forests and forestry operations have both positive and negative effects on stream water as perceived in relation to the fisheries resource and to human use and convenience. One of the chief values of forestry, which is often forgotten, is its role in maintaining and reestablishing healthy forests with all the positive benefits that this implies. For example, forest roads provide access for control of fire and insect damage, silvicultural improvement, recreation, and fish enhancement programs. Healthy forests with a mosaic of cover types enhance the value, use, and enjoyment of our water resources. The changing nature of our forested landscape will mean a need for continued and even expanded study and evaluation of forest-water relationships and a greater need to incorporate this knowledge into the management of both forests and water in Canada.

Acknowledgements

I gratefully acknowledge permission granted by my employer the Canadian Forestry Service to work on this chapter. The more detailed manuscript on which this chapter is based is planned for publication by the Canadian Forestry Service. I am indebted to R. B. B. Dickison, D. L. Golding, and R. H. Swanson for their reviews and to my many colleagues for their helpful comments.

References

ANDERSON, H. W., M. D. HOOVER, AND K. G. REINHART. 1976. Forests and water: effects of forest management on floods, sedimentation and water supply. U.S. For. Serv. Gen. Tech. Rep. PSW-18: 115 p.

BANGAY, G. E., AND C. RIORDAN [ED.]. 1983. United States - Canada memorandum of intent on transboundary air pollution. Impact Assessment Work Group I Final Report. Environment Canada, Ottawa, Ont. 626 p.

BEAUDRY, P. G. 1984. Effects of forest harvesting on snowmelt during rainfall in coastal British Columbia. M.F. thesis, Faculty of Forestry, University of British Columbia, Vancouver, B.C. 185 p.

BERNDT, H. W., AND G. W. SWANK. 1970. Forest land use and streamflow in central Oregon. U.S. For. Serv. Res. Pap. PNW-93: 15 p.

BETHLAHMY, N. 1974. More streamflow after a bark beetle epidemic. J. Hydrol. 23: 185-189.

BLACK, T. .A., C. S. TAN, AND J. U. NNYAMAH. 1980. Transpiration rate of Douglas fir trees in thinned and unthinned stands. Can. J. Soil Sci. 60: 625-631.

BOSCH, J. M., AND J.D. HEWLETT. 1982. A review of catchment experiments to determine the effect of vegetation changes on water yield and evapotranspiration. J. Hydrol. 55: 3-23.

CALDER, I. R. 1982. Forest evaporation, p. 173-193. In Proc. Can. Hydrol. Symp. '82. National Research Council of Canada, Ottawa, Ont.

CARR, W. W., AND T. M. BALLARD. 1980. Hydroseeding forest roadsides in British Columbia for erosion control. J. Soil Water Conserv. 35: 33-35.

CHAMBERLIN, T. W. 1972. Interflow in the mountainous forest soils of coastal British Columbia, p. 121-127. In H. O. Slaymaker and H. J. McPherson [ed.] Mountain geomorphology, B. C. Geographical Series No. 14, Tantalus Research Ltd., Vancouver, B.C.

CHENG, J. D. 1980. Hydrologic effects of a severe forest fire, p. 240-251. In Proc. Symp. Watershed Management, Boise, Idaho. American Society of Civil Engineers, New York, NY.

CHENG, J. D., T. A. BLACK, J. DE VRIES, R. P. WILLINGTON, AND B. C. GOODELL. 1975a. The evaluation of initial changes in peak streamflow following logging of a watershed on the west coast of Canada, p. 475-486. In Proc. Tokyo Symp. IASH Publ. No. 117.

CHENG, J. D., T. A. BLACK, AND R. P. WILLINGTON. 1975b. The generation of stormflow from small forested watersheds in the Coast mountains of southwestern British Columbia, p. 542-551. In Proc. Can. Hydrol. Symp. — 77. National Research Council of Canada, Ottawa, Ont.

　　　 1977. The stormflow characteristics of a small, steep and forested watershed in the Coast mountains of southwestern British Columbia, p. 300-310. In Proc. Can. Hydrol. Symp. — 77. National Research Council of Canada, Ottawa, Ont.

COHEN, S. J. 1977. Land use changes on micro-climate in central Alberta: an energy balance study. M.Sc. thesis, Department of Geography, University of Alberta, Edmonton, Alta. 263 p.

DAUGHARTY, D. A. 1984. The effects of topography and forest cover on snow depths and water equivalents in the Nashwaak experimental watersheds, New Brunswick. M.Sc.F. thesis, Faculty of Forestry, University of New Brunswick, Fredericton, N.B. 256 p.

DAUGHARTY, D. A. AND R. B. B. DICKISON. 1982. Snow cover distribution in forested and deforested landscapes, p. 10-19. In Proc. 50th Western Snow Conf., Reno, NV.

DICKISON, R. B. B., AND D. A. DAUGHARTY. 1982. The effects on snowmelt runoff of the removal of forest cover, p. 131-150. In Proc. IHP Fourth Northern Research Basin Symp. Workshop, Ullensvang, Norway. Norw. Natl. Comm. Hydrol. Rep. No. 12, Oslo, Norway.

_____ 1983. Climate and water balance, p. 8-14. In G. R. Powell [ed.] Nashwaak Experimental Watershed Project Annu. Rep. 1981-1982. Department of Natural Resources, Forestry Branch, Fredericton, N.B.

DICKISON, R. B. B., D. A. DAUGHARTY, AND D. K. RANDALL. 1981. Some preliminary results of the hydrologic effects of clearcutting a small watershed in central New Brunswick, p. 59-75. In Proc. 5th Can. Hydrotech. Conf., Fredericton, N.B. Canadian Society of Civil Engineers.

DORCEY, H. J., M. W. McPHEE, AND S. SYDNEYSMITH. 1980. Salmon protection and the B.C. coastal forest industry: environmental regulation as a bargaining process. Westwater Res. Centre, University of British Columbia, Vancouver, B.C. 378 p.

EIDT, D. C. 1985. Toxicity of Bacillus thuringiensis var. kurstaki to aquatic insects. Can. Entomol. 117: 829-837.

FEDERER, C. A., R. S. PIERCE, AND J. W. HORNBECK. 1972. Snow management seems unlikely in the northeast, p. 212-219. In Proc. Symp. Watersheds in Transition. American Water Resources Association, Urbana, IL.

FELLER, M. C. 1981. Effects of clearcutting and slashburning on stream temperature in southwestern British Columbia. Water Resour. Bull. 17: 863-867.

FELLER, M. C., AND J. P. KIMMINS. 1984. Effects of clearcutting and slash burning on streamwater chemistry and watershed nutrient budgets in southwestern British Columbia. Water Resour. Res. 20: 29-40.

FOSTER, N. W. 1984. Neutralization of acid precipitation within a deciduous forest, p. 143-148. In Proc. TAPPI Res. and Dev. Conf., Appleton, WI.

FOSTER, N.W., AND J. A. NICOLSON. 1984. Acid precipitation and water quality within a tolerant hardwood stand and soil, p. 337-342. In Speeches and Papers, Forest Resources Management, the Influence of Policy and Law. Int. For. Cong., Quebec, P.Q.

_____ 1986. Ion transfer through a tolerant hardwood canopy, Turkey Lakes watershed, Ontario. In Proc. Acid Rain and Forest Resources Conf., 1983. Quebec, P.Q. (In press).

FRECHETTE, J.-G. 1968. Accumulation de la neige sous divers types de couverts forestiers. Cah. Géogr. Qué. 12: 141-144.

_____ 1969. Interception de la pluie par une sapinière laurentienne. Nat. Can. 96: 523-529

FREEDMAN, B. 1982. An overview of the environmental impacts of forestry, with particular reference to the Atlantic provinces. Institute for Resources and Environmental Studies, Dalhousie University, Halifax, N.S. 219 p.

GILES, D. G., T. A. BLACK, AND D. L. SPITTLEHOUSE. 1985. Determination of growing season soil water deficits on a forested slope using water balance analysis. Can. J. For. Res. 15: 107-114.

GOLDING, D. L. 1980. Calibration methods for detecting changes in streamflow quantity and regime, p. 3-7. In Proc. Helsinki Symp. IAHS Publ. No. 130.

_____ 1981. Hydrologic relationships in interior Canadian watersheds, p. 107-115. In D. M. Baumgartner [ed.] Proc. Symp. Interior West Watershed Management. Coop. Ext., Washington State University, Pullman, WA.

_____ 1982. Snow accumulation patterns in openings and adjacent forest, p. 91-112. In Proc. Can. Hydrol. Symp. '82. National Research Council of Canada, Ottawa, Ont.

GOLDING, D. L., AND C. R. STANTON. 1972. Water storage in the forest floor of subalpine forests of Alberta. Can. J. For. Res. 2: 1-6.

GONZALEZ, A., AND A. P. PLAMONDON. 1977. Urea fertilization of natural forest: effects on water quality. For. Ecol. Manage. 1: 213-221.

GOUVERNEMENT DU QUÉBEC. 1977. Guide d'aménagement du milieu forestier. Ministère des Terres et Forêts, Québec, Qué. 158 p.

GRANBERG, H. B. 1973. Indirect mapping of the snowcover for permafrost prediction at Schefferville, Quebec, p. 113-120. In PERMAFROST: The North American Contribution to the Second International Conference, National Academy of Sciences, Washington, DC.

HAAG, R. W., AND L. C. BLISS. 1974. Functional effects of vegetation on the radiant energy budget of boreal forest. Can. Geotech. J. 11: 374-379.

HARLAN, R. L. 1969. Soil-water freezing, snow accumulation and ablation in Marmot Creek experimental watershed, Alberta, Canada, p. 29-33. In Proc. 37th Western Snow Conf., Salt Lake City, UT.

HARR, R. D. 1983. Potential for augmenting water yield through forest practices in western Washington and western Oregon Water Resour. Bull. 19: 383-393.

HELVEY, J. D. 1972. First-year effects of wildfire on water yield from mountain watersheds in north-central Washington, p. 308-312. In Proc. Symp. Watersheds in Transition. American Water Resources Association, Urbana, IL.

HETHERINGTON, E. D. 1976. Dennis Creek, a look at water quality following logging in the Okanagan Basin. Can. For. Serv. Inf. Rep. No. BC-X-147: 33 p.

    1982. Effects of forest harvesting on the hydrologic regime of Carnation Creek experimental watershed: a preliminary assessment, p. 247-267. In Proc. Can. Hydrol. Symp. '82. National Research Council of Canada, Ottawa, Ont.

    1985. Streamflow nitrogen loss following forest fertilization in a southern Vancouver Island watershed. Can. J. For. Res. 15: 34-41.

HEWLETT, J. D. 1982. Forests and floods in the light of recent investigation, p. 543-559. In Proc. Can. Hydrol. Symp. '82. National Research Council of Canada, Ottawa, Ont.

HOLTBY, B., AND C. P. NEWCOMBE. 1982. A preliminary analysis of logging-related temperature changes in Carnation Creek, British Columbia, p. 81-99. In G. F. Hartman [ed.] Proc. Carnation Creek Workshop, a 10 year review. Pacific Biological Station, Nanaimo, B.C.

HONER, T. G., AND A. BICKERSTAFF. 1985. Canada's forest area and wood volume balance 1977-1981: an appraisal of change under present levels of management. Can. For. Serv. Inf. Rep. BC-X-272: 84 p.

JABLONSKI, P. D. 1980. Pretreatment water quality of the TriCreeks experimental watershed. Alta. Energy Nat. Resour., For. Land Use Branch, Watershed Manuscr. Rep. No. 7: 67 p.

JEFFREY, W. W. 1967. Forest hydrology research in Canada, p. 21-30. In W. E. Sopper and H. W. Lull [ed.] Proc. Int. Symp. For. Hydrol. Pergamon Press, Oxford.

    1968a. Snow hydrology in the forest environment, p. 1-19. In Snow Hydrology, Proc. Workshop Seminar, Fredericton. Can. Natl. Comm. Int. Hydrol. Decade, Ottawa, Ont.

    1968b. Watershed management problems in British Columbia: a first appraisal. Water Resour. Bull. 4: 58-70.

    1970. Hydrology of land use. In D. M. Gray [ed.] Handbook on the principles of hydrology. Sect. XIII. 57 p. Can. Natl. Comm. Int. Hydrol. Decade, National Research Council of Canada, Ottawa, Ont.

KACHANOSKI, R. G., AND E. DE JONG. 1982. Comparison of the soil water cycle in clear-cut and forested sites. J. Environ. Qual. 11: 545-549.

KINGSBURY, P. D. 1984. Environmental impact assessment of insecticides used in Canadian forests, p. 365-376. In W. Y. Garner and J. Harvey, Jr. [ed.] Chemical and biological controls in forestry. American Chemical Society, Washingon, DC.

KONDO, E. S., AND R. G. TAYLOR. 1984. Forest insect and disease conditions in Canada 1983. Environment Canada, Canadian Forestry Service, Ottawa, Ont. 73 p.

KRAUSE, H. H. 1982. Effect of forest management practices on water quality — a review of Canadian studies, p. 15-29. In Proc. Can. Hydrol. Symp. '82. National Research Council of Canada, Ottawa, Ont.

    1983. Effect of clearcutting on stream-water quality and nutrient balance, p. 15-17. In G. R. Powell [ed.] Nashwaak Experimental Watershed Project Annu. Rep. 1981-1982. New Brunswick Department of Natural Resources, Forestry Branch, Fredericton, N.B.

MAHENDRAPPA, M, K. 1982. Effects of forest cover type and organic horizons on potential water yield, p. 215-224. In Proc. Can. Hydrol. Symp. '82. National Research Council of Canada, Ottawa, Ont.

    1983. Chemical characteristics of precipitation and hydrogen input in throughfall and stemflow under some eastern Canadian forest stands. Can. J. For. Res. 13: 948-955.

MAHENDRAPPA, M. K., AND D. G. O. KINGSTON. 1982. Prediction of throughfall quantities under different forest stands. Can. J. For. Res. 12: 474-481.

MARTIN, H. C. 1971. Average winds above and within a forest. J. Appl. Meteorol. 10: 1132-1137.

McCAUGHEY, J. H. 1978. Energy balance and evapotranspiration estimates for a mature coniferous forest. Can. J. For. Res. 8: 456-462.

    1981. Impact of clearcutting of coniferous forest on the surface radiation balance.

J. Appl. Ecol. 18: 815-826.

    1985. Energy balance storage terms in a mature mixed forest at Petawawa, Ontario — a case study. Boundary-Layer Meteorol. 31: 89-101.

McMINN, R. G. 1960. Water relations and forest distribution in the Douglas-fir region on Vancouver Island. Can. Dep. Agric. For. Biol. Div. Publ. 1091, Ottawa, Ont.

McNAUGHTON, K. G., AND T. A. BLACK. 1973. A study of evapotranspiration from a Douglas fir forest using the energy balance approach. Water Resour. Res. 9: 1579-1590.

MEERES, L. S. 1977. A preliminary study of changes in wind structures produced by a commercial-appearing cut in a mature forest, p. 220-236. In R. H. Swanson and P. A. Logan [ed.] Proc. Symp. Alberta Watershed Research Program. Can. For. Serv. Inf. Rep. NOR-X-176.

MEYBOOM, P. 1967. Groundwater studies in the Assiniboine River drainage basin Part II: Hydrogeologic characteristics of phreatophytic vegetation in south central Saskatchewan. Geol. Surv. Can. Bull. 139: 64 p.

MOORE, G. C. 1983. A report on the feasibility of harvesting timber within buffer zones: a literature review. New Brunswick Department of Natural Resources, Fredericton, N.B. 138 p.

MORIN, R., G. GABOURY, AND G. MAMARBACHI. 1986. Fenitrothion and aminocarb residues in water and balsam fir foliage following spruce budworm spraying programs in Quebec, 1979 to 1982. Bull. Environ. Contam. Toxicol. 36: 622-628.

MORRISON, I. K. 1984. Acid rain, a review of literature on acid deposition effects in forest ecosystems. Commonw. For. Bur. For. Abstr. 45: 483-506.

MORTON, F. I. 1984. What are the limits on forest evaporation? J. Hydrol. 74: 373-398.

    1985. What are the limits on forest evaporation — reply. J. Hydrol. 82: 184-192.

NEWTON, M., K. M. HOWARD, B. R. KELPSAS, R. DANHAUS, C. M. LOTTMAN, AND S. DUBELMAN. 1984. Fate of glyphosate in an Oregon forest ecosystem. J. Agric. Food Chem. 32: 1144-1151.

NICOLSON, J. A. 1975. Water quality and clearcutting in a boreal forest ecosystem, p. 733-738. In Proc. Can. Hydrol. Symp. — 75. National Research Council of Canada, Ottawa, Ont.

    1984. Ion concentrations in precipitation and streamwater in an Algoma maple-birch forest, p. 123-135. In Proc. Can. Hydrol. Symp. No. 15, 1984. National Research Council of Canada, Ottawa, Ont.

    1986. Ion movement in terrestrial basins in the Turkey Lakes forest watershed. In Proc. Acid Rain and Forest Resources Conf., 1983, Quebec, P.Q. (In press).

NICOLSON, J. A., N. W. FOSTER, AND I. K. MORRISON. 1982. Forest harvesting effects on water quality and nutrient status in the boreal forest, p. 71-89. In Proc. Can. Hydrol. Symp. '82. National Research Council of Canada, Ottawa, Ont.

O'KANE, T. 1983. The effect of clearcutting on the forest albedo..B.Sc.F. thesis, University of New Brunswick, Frederiction, N.B. 40. p.

OTTENS, J., AND J. RUDD. 1977. Environmental protection costs in logging road design and construction to prevent increased sedimentation in the Carnation Creek watershed. Can. For. Serv. Inf. Rep. BC-X-155: 28 p.

PAYETTE, S., L. FILION, AND J. OUZILLEAU. 1973. Relations neige-végétation dans la toundra forestière du Nouveau-Québec, Baie d'Hudson. Nat. Can. 100: 493-508.

PERRIN, C. J., K. S. SHORTREED, AND J. G. STOCKNER. 1984. An integration of forest and lake fertilization: transport and transformations of fertilizer elements. Can. J. Fish. Aquat. Sci. 41: 253-262.

PETZOLD, D. E., AND A. N. RENCZ. 1975. The albedo of selected subarctic surfaces. Arct. Alp. Res. 7: 393-398.

PLAMONDON, A. P. 1981. Écoulement et modification du couvert forestier. Nat. Can. 108: 289-298.

    1982. Augmentation de la concentration des sédiments en suspension suite à l'exploitation forestière et durée de l'effet. Can. J. For. Res. 12: 883-892.

PLAMONDON, A. P., T. A. BLACK, AND B. C. GOODELL. 1972. The role of hydrologic properties of the forest floor in watershed hydrology, p. 341-348. In Proc. Symp. Watersheds in Transition. American Water Resources Association, Urbana, IL.

PLAMONDON, A. P., A. GONZALEZ, AND Y. THOMASSIN. 1982. Effects of logging on water quality: comparison between two Quebec sites, p. 49-70. In Proc. Can. Hydrol. Symp. '82. National Research Council of Canada, Ottawa, Ont.

PLAMONDON, A. P., AND M. M. GRANDTNER. 1975. Microclimat estival d'une sapinière à Hylocomium de la forêt Montmorency. Nat. Can. 102: 73-87.

PLAMONDON, A. P., R. LEPROHON, AND A. GONZALEZ. 1976. Exploitation forestière et protection de quelques cours d'eau de la Côte Nord. Les Cahiers de Centreau. Vol. 1, No. 6. Université Laval, Québec, Qué. 43 p.

PLAMONDON, A. P., AND D. C. OUELLET. 1980. Partial clearcutting and streamflow regime of ruisseau Eaux-Volées experimental basin, p. 129-136. In Proc. Helsinki Symp. IAHS Publ. No. 130.

PLAMONDON, A. P., M. PREVOST, AND R. C. NAUD. 1984a. Accumulation et fonte de la neige en milieu boisé et déboisé. Géogr. Phys. 38: 27-35.

    1984b. Interception de la pluie dans la sapinière à bouleau blanc, forêt Montmorency, Can. J. For. Res. 14: 722-730.

PREBBLE, M. L. [ED.] 1975. Aerial control of forest insects in Canada. Environment Canada, Ottawa, Ont. 330 p.

PRICE, A. G., AND L. K. HENDRIE. 1983. Water motion in a deciduous forest during snowmelt. J. Hydrol. 64: 339-356.

PRICE, A. G., L. K. HENDRIE, AND T. DUNNE. 1978. Controls on the production of snowmelt runoff, p. 257-268. In S. C. Colbeck and M. Ray [ed.] Proc. Modeling of Snow Cover Runoff. U.S. Army Cold Reg. Res. Eng. Lab., Hanover, NH.

PRICE, J. S., AND J. E. FITZGIBBON. 1982. Winter hydrology of a forested drainage basin, p. 347-360. In Proc. Can. Hydrol.Symp. '82. National Research Council of Canada, Ottawa, Ont.

ROTHWELL, R. 1977. Suspended sediment and soil disturbance in a small mountain watershed after road construction and logging, p. 285-300. In R. H. Swanson and P. A. Logan [ed.] Proc. Symp. Alberta Watershed Research Program. Can. For. Serv. Inf. Rep. NOR-X-176.

    1982. Water balance of a boreal forest wetland site, p. 313-332. In Proc. Can. Hydrol. Symp. '82. National Research Council of Canada, Ottawa, Ont.

    1983. Erosion and sediment control at road-stream crossings. For. Chron. 59: 62-66.

ROUSE, W. R. 1976. Microclimatic changes accompanying burning in subarctic lichen woodland. Arct. Alp. Res. 8: 357-376.

    1984. Microclimate at arctic treeline. 1. Radiation balance of tundra and forest. Water Resour. Res. 20: 57-66.

ROWE, J. S. 1972. Forest regions of Canada. Can. For. Serv. Publ. No. 1300. Ottawa, Ont. 172 p.

SABEAN, B. 1977. The effects of shade removal on stream temperature in Nova Scotia. CAT/77/135/150. Nova Scotia Department of Lands and Forests, Halifax, N.S. 31 p.

    1978. Sediment: a preliminary investigation of the problem in Nova Scotia streams. Unpubl. Rep. Nova Scotia Department of Lands and Forests, Halifax, N.S. 12 p.

SAHI, S. V., AND G. M. COURTIN. 1983. Anthropogenic reduction in overstory tree canopy as a critical factor in cryopedogenesis, p. 177-179. In Proc. 40th Eastern Snow Conf., Toronto, Ont.

SCHINDLER, D. W., R. W. NEWBURY, K. G. BEATY, J. PROKOPOWICH, T. RUSZCZYNSKI, AND J. A. DALTON. 1980. Effects of a windstorm and forest fire on chemical losses from forested watersheds and on the quality of receiving streams. Can. J. Fish. Aquat. Sci. 37: 328-334.

SCHNEIDER, J., AND G. R. AYER. 1961. Effect of reforestation on streamflow in central New York. U.S. Geol. Surv. Water-Supply Pap. 1602: 61 p.

SCRIVENER, J. C. 1982. Logging impacts on the concentration patterns of dissolved ions in Carnation Creek, British Columbia, P. 64-80. In G. F. Hartman [ed.] Proc. Carnation Creek Workshop, a 10 year review. Pacific Biological Station, Nanaimo, B.C.

SILVERSIDES, R. H. 1978. Forest and airport wind speeds. Atmos.-Ocean 16: 293-299.

SINGH, B., AND G. SZEICZ. 1979. The effect of intercepted rainfall on the water balance of a hardwood forest. Water Resour. Res. 15: 131-138.

SINGH, T. 1983. A proposed method for preliminary assessment of erosion hazards in west-central Alberta. Can. For Serv. Inf. Rep. NOR-X-251: 17 p.

SINGH, T. AND Y. KALRA. 1977. Impact of pulpwood clearcutting on stream water quality in west central Alberta, p. 272-284. In R. H. Swanson and P. A. Logan [ed.] Proc. Symp. Alberta Watershed Research Program. Can. For. Serv. Inf. Rep. NOR-X-176.

SLANEY, P. A., T. G. HALSEY, AND A. F. TAUTZ. 1977. Effects of forest harvesting practices on spawning habitat of stream salmonids in the Centennial Creek watershed, British Columbia. B.C. Fish Wildl. Branch, Fish. Manuscr. Rep. No. 73. Vancouver, B.C. 45 p.

SMITH, R. B., AND E. F. WASS. 1982. Changes in ground-surface characteristics and vegetative cover associated with logging and prescribed broadcast burning, p. 100-108. In

G. F. Hartman [ed.] Proc. Carnation Creek Workshop, a 10 year review. Pacific Biological Station, Nanaimo, B.C.

SPITTLEHOUSE, D. L. AND T. A. BLACK. 1981. A growing season water balance model applied to two Douglas fir stands. Water Resour. Res. 17: 1651-1656.

STANTON, C. R. 1966. Preliminary investigation of snow accumulation and melting in forested and cut-over areas of the Crowsnest forest, p. 7-12. In Proc. 34th Western Snow Conf., Seattle, WA.

SUNDARAM, K. M. S., P. D. KINGSBURY, AND S. B. HOLMES. 1984. Fate of chemical insecticides in aquatic environments: forest spraying in Canada, p. 253-276. In W. Y. Garner and J. Harvey, Jr. [ed.] Chemical and biological controls in forestry. American Chemical Society, Washington, DC.

SWANSON, R. H. 1972. Forest hydrology in Canada — more water probably not wanted, p. 4301-4304. In Proc. 7th World For. Cong., Vol. III, Buenos Aires, Argentina.

1975. Water use by mature lodgepole pine, p. 264-277. In D. M. Baumgartner [ed.] Proc. Symp. Management of Lodgepole Pine Ecosystems, Vol. 1. Coop. Ext. Serv., Washington State University, Pullman, WA.

1978. Increasing water supply through watershed management. Can. Water Resour. J. 3: 85-93.

1980. Surface wind structure in forest clearings during a chinook, p. 26-30. In Proc. 48th Western Snow Conf., Laramie, WY.

1982. Problems and opportunities in Canadian forest hydrology, p. 1-13. In Proc. Can. Hydrol. Symp. '82. National Research Council of Canada, Ottawa, Ont.

SWANSON, R. H., AND D. L. GOLDING. 1982. Snowpack management on Marmot watershed to increase late season streamflow, p. 215-218. In Proc. 50th Western Snow Conf., Reno, NV.

SWANSON, R. H., AND G. R. HILLMAN. 1977. Predicted increased water yield after clearcutting verified in west-central Alberta. Can. For. Serv. Inf. Rep. NOR-X-198: 40 p.

SWANSON, R. H., AND D. R. STEVENSON. 1971. Managing snow accumulation and melt under leafless aspen to enhance watershed value, p. 63-69. In Proc. 39th Western Snow Conf., Billings, MT.

SZEICZ, G., D. E. PETZOLD, AND R. G. WILSON. 1979. Wind in the subarctic forest. J. Appl. Meteorol. 18: 1268-1274.

TELLER, H. L. 1968. Impact of forest land use on floods. Unasylva 22: 18-20.

TOEWS, D. A. A., AND M. J. BROWNLEE. 1981. A handbook for fish habitat protection on forest lands in British Columbia. Department of Fisheries and Oceans, Vancouver, B.C. 172 p.

TOEWS, D. A. A., AND M. K. MOORE. 1982. The effects of three streamside logging treatments on organic debris and channel morphology of Carnation Creek, p. 129-152. In G. F. Hartman [ed.] Proc. Carnation Creek Workshop, a 10 year review. Pacific Biological Station, Nanaimo, B.C.

TOEWS, D. A. A., AND D. WILFORD. 1978. Watershed management considerations for operational planning on T.F.L. #39 (Blk 6A), Graham Island. Fish. Mar. Serv. Manuscr. Rep. No. 1473. Vancouver, B.C. 32 p.

UNIVERSITY OF NEW BRUNSWICK. 1976. Nashwaak Experimental Watershed Project. Annu. Rep. 1975-1976, p. 17-19. Faculty of Forestry, Fredericton, N.B.

VARTY, I. W. 1980. Summary overview of environmental surveillance in 1978 and 1979, p. 1-4. In I. W. Varty [ed.] Environmental surveillance in New Brunswick 1978-1979: effects of spray operations for forest protection against spruce budworm. Faculty of Forestry, University of New Brunswick, Fredericton, N.B.

WEETMAN, G. F. 1983. Forestry practices and stress on Canadian land, p. 259-301. In W. Simpson-Lewis, R. McKechnie, and V. Neimanis [ed.] Stress on land in Canada. Environment Canada, Lands Directorate, Ottawa, Ont.

WILFORD, D. J., AND J. W. SCHWAB. 1982. Soil mass movements in the Rennell Sound area, Queen Charlotte Islands, British Columbia, p. 521-541. In Proc. Can. Hydrol. Symp. '82. National Research Council of Canada, Ottawa, Ont.

WILSON, R. C. H., W. R. ERNST, G. JULIEN, AND S. E. HALL. 1983. Brief to the Nova Scotia Royal Commission of Forestry. Vol. II, Sect. 3. Environment Canada, Environmental Protection Service, Halifax, N.S. 148 p.

CHAPTER 8

The Role of Grasslands in Hydrology

E. de Jong

Department of Soil Science, University of Saskatchewan, Saskatoon, Sask. S7N 0W0

and R. G. Kachanoski

Department of Land Resource Science, University of Guelph, Guelph, Ont. N1G 2W1

Introduction

The natural vegetation of the earth can be divided into grassland, forest, desert shrub, and tundra. Grasslands are believed to have occupied one fifth of the total land surface (Barnes 1948). Grasslands lie between forests and desert shrubs in terms of water supply and are divided into short grass or steppe, tall grass or prairie, and savanna. Short grass and savanna represent the bulk of the grasslands. Savannas are tropical grasslands whose water supply is limited by seasonal rainfall and high evaporation. Short grass is found in semiarid regions of the temperate zones where rainfall is very variable and insufficient to regularly wet the soil down to the ground water table. Tall grass occupies the subhumid area bordering the short grasslands. Near the transition zones between desert shrub, grassland, and trees a number of factors cause one biome to be favored over another. For example, the invasion of aspen (*Populus tremuloides*) into the prairie grassland during the last century is probably largely due to the absence of prairie fires and, to a lesser degree, overgrazing by buffalo (Bird 1961).

At the beginning of this century, most of the vast grassland area in Manitoba, Saskatchewan, and Alberta was uncultivated. Now, about 70 % of the area has been broken. This chapter gives a brief overview of the physical environment of the region, its agricultural development, and the current state of the water supply. The fluxes of the hydrological cycle and the effect of agricultural settlement on their quantity and quality are discussed. Finally, the demands for water and methods to augment the limited supply are reviewed.

Canada's Grassland

Environment, Agricultural Development

Canada's natural grasslands, which are part of a large North American area of short and tall grass prairie, are found mainly in the Prairie provinces (Fig. 1). The natural vegetation consists largely of mixed prairie (*Stipa-Bouteloua*) and fescue prairie (*Festuca scabrella*) invaded by aspen with only a relatively small area of true or tall grass prairie (*Stipa-Sporobulus*). The total area of grassland and interspersed Solonetzic soils in Canada is estimated at 541×10^3 km^2 (Agriculture Canada 1977), and represents approximately 6% of Canada's land surface.

Annual precipitation on the prairies ranges from 30 to over 50 cm and largely falls in early spring and summer (Fig. 2). In the mixed prairie region, precipitation is generally too low to wet the soil deeply (Barnes 1948). Going from the mixed to the tall grass region, potential evaporation decreases, and amount and dependability of precipitation increase. Growing season rain is usually substantially lower than potential evapotrans-

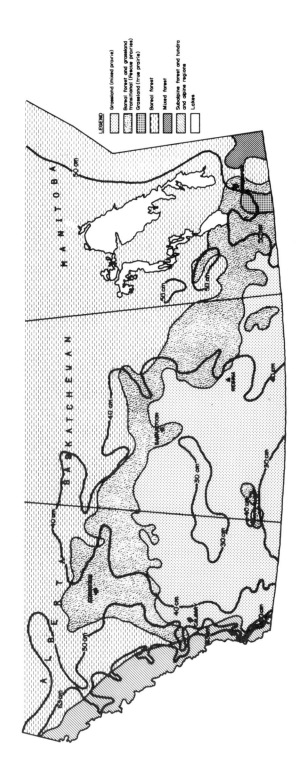

FIG. 1. Annual precipitation (Laycock 1965) and vegetation regions (Agriculture Canada 1977) of western Canada.

214

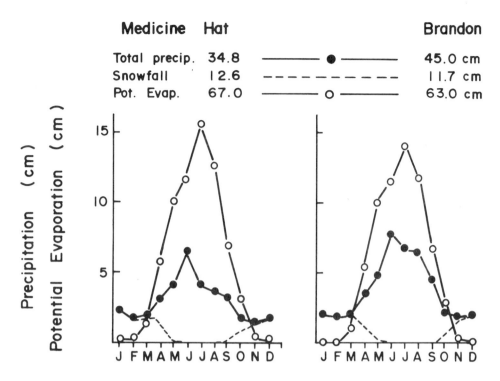

FIG. 2. Monthly precipitation (Environment Canada 1984) and potential evaporation (Coligado et al. 1968) for stations typical for the mixed (Medicine Hat) and tall grass (Brandon) area.

piration (Fig. 2), and water deficits severely limit plant growth. Water surpluses occur mainly in the spring as the snow metls and, to a lesser degree, in the autumn.

Most of Canada's prairie region lies in the Saskatchewan–Nelson basin and slopes from southwest to northeast (Fig. 3). A small area in southern Alberta and Saskatchewan drains southward to the Missouri system. The grassland area comprises most or all of the North Saskatchewan, Red Deer, Bow, Oldman, South Saskatchewan, Qu'Appelle, Souris, and Assiniboine drainage basins. Only small portions of the Red River and Lake Winnipeg basins are part of the grassland area. The gentle relief of the plains is interrupted by escarpments, hilly areas, and deeply incised glacial metlwater channels, some of which are still used by rivers. Between the major river valleys, surficial drainage is often poorly developed and numerous small depressions intermittently hold water. J. R. Millar (pers. comm.) counted up to 26 sloughs or potholes/km² in glacial till deposits and up to 14 sloughs/km² in lacustrine deposits. The sloughs were slightly larger in lacustrine than glacial till areas, but most were smaller than 0.04 km².

Potholes, or sloughs, are a focal point of the prairie hydrological cycle. They can be classified as recharge sloughs (water moves from the surface to the ground water), discharge sloughs (water moves from groundwater to the surface), and transitional sloughs (when water at different times of the year moves in different directions) (Lissey 1968). Sloughs retain runoff, stabilize streamflow, and are major breeding grounds for waterfowl (Simpson-Lewis et al. 1979; chapter by Whillans, this volume). Ludden et al. (1983) estimated that in an area where they occupied about 15% of the land surface, potholes could store on average 8.5 cm of runoff from the surrounding area. Evaporation and drainage often exceed the amount of water available, and the sloughs dry out during the summer. Many sloughs have been drained to increase the area of agricultural land and to facilitate cultivation.

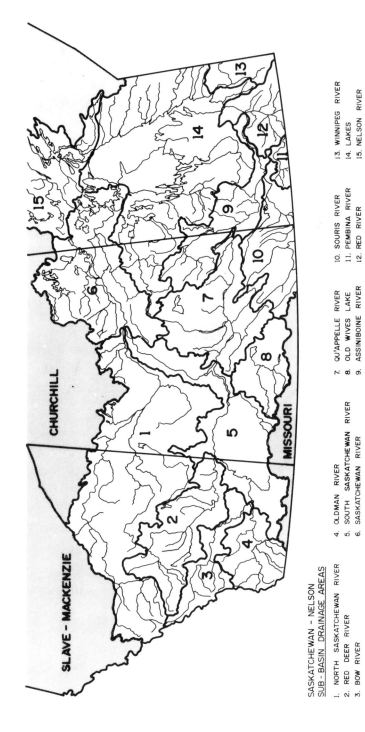

FIG. 3. Major drainage basins in the grassland area.

SASKATCHEWAN - NELSON
SUB - BASIN DRAINAGE AREAS

1. NORTH SASKATCHEWAN RIVER
2. RED DEER RIVER
3. BOW RIVER

4. OLDMAN RIVER
5. SOUTH SASKATCHEWAN RIVER
6. SASKATCHEWAN RIVER

7. QU'APPELLE RIVER
8. OLD WIVES LAKE
9. ASSINIBOINE RIVER

10. SOURIS RIVER
11. PEMBINA RIVER
12. RED RIVER

13. WINNIPEG RIVER
14. LAKES
15. NELSON RIVER

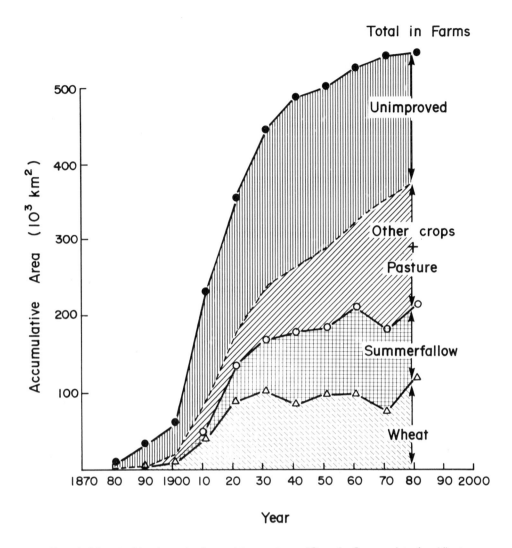

FIG. 4. Historical land use in the prairie provinces (Canada Census data for Alberta, Saskatchewan, and Manitoba).

Large areas of natural grassland have been broken and turned into cropland since the turn of this century (Fig. 4). The dependance on spring-seeded cereals, wheat in particular, and summerfallow is a major feature of the cultivated land use. Cultivation has not only decreased the total area of native grassland, but it has also utilized the better land. Most of the remaining natural grassland is found in Canada Land Inventory Classes (CLI) 5 and 6 (Horner et al. 1980) because of limitations due to climate (mainly too arid), soil (stony, sandy, salt-affected), or landscape (too rolling for cultivation). The CLI classification, which covers a larger area than the Census data in Fig. 4, indicates 316×10^3 km^2 of class 5 and 6 land. About half of this potential grazing land lies in the northern forest area, outside the boundaries of present farms.

Water Supply and Quality

The grassland area in the Prairie provinces is about 500×10^3 km^2. A crude water balance of the region can be calculated from Laycock's (1965) precipitation and evapotranspiration maps and river flow data. The mean annual input from precipitation is about 200 km³ (40 cm) of which 185 km³ (37 cm), or over 90%, is lost by evapotranspiration (see also chapter by Laycock, this volume). Streams rising in the Rocky Mountains feed about 15 km³ of water per year into the area at Edmonton, Red Deer, Bassano Dam, and Lethbridge (Environment Canada 1983), and about 19 km³ of water annually leaves the grassland area via the Saskatchewan River at Nipawin/Tobin Lake and the Assiniboine at Headingly. The difference represents about 1 cm/yr, and extrapolation to the whole grassland region indicates a net annual removal of 5 km³ of water by rivers.

Most of the river flow on the prairies originates in the Rocky Mountains and Foothills, with the grassland area contributing relatively little. In 1948-49, a dry year, the Plains region contributed about 10% of the flow of the Saskatchewan River (Laycock 1965). A major portion of the runoff never reaches the main rivers. Closed drainage basins, such as the Old Wives Lake basin (Fig. 3), occupy approximately 80 × 10³ km² (Lane and Sykes 1982), or about 15% of the total grassland area. The effective drainage area, the part of a drainage basin that might be expected to contribute to the main stream during a flood with a return period of 2 yr (Mowchenko 1976), averages about 50% of the gross drainage area.

Mountain rivers vary less in flow rate than those which originate on the prairies. For example, in very dry years the flow of the Oldman and Red Deer rivers is about 40% of the mean annual flow, whereas in the Souris and Qu'Appelle rivers it is about 3% of the mean annual flow (Lane and Sykes 1982). The salt content of the mountain rivers increases slightly as the smaller prairie rivers join them, but they show no serious quality deficiencies. There is no evidence that the water quality of the major rivers draining the grasslands has deteriorated (Fisheries and Environment Canada 1977).

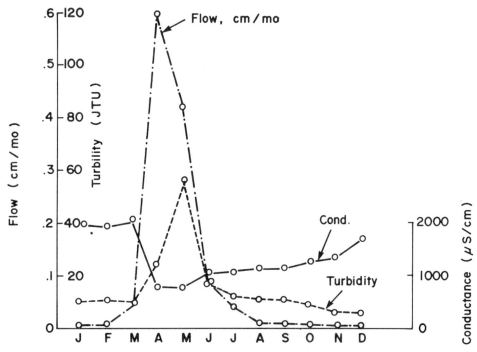

FIG. 5. Monthly variation in flow rate (over the gross drainage area) and water quality (specific conductance and turbidity in Jackson turbidity units) of the Souris River at Glen Ewan (data for 1974-82 from Environment Canada reports).

The rivers originating on the prairies show wide seasonal variations in quality (Fig. 5). Salt concentrations are lowest during the spring metl period and highest in winter. Prairie rivers tend to be high in suspended sediment (Stichling 1973), mainly derived from bank erosion (Hudson 1983). Turbidity is highest in spring and early summer and lags the occurrence of maximum spring flows (see also Chacko et al. 1981). This suggests that the greatest water erosion is associated with early spring rains, and not with snowmelt runoff (Burwell et al. 1975).

Most of the runoff on the prairies is stored in lakes and potholes. Major man-made reservoirs have a surface area of about 1000 km^2 (PPWB 1982), and small dams and dugouts probably create at least another 100 km^2 of storage space. Permanent lakes in the grassland region are usually saline, with freshwater lakes confined to uplands, drainageways, or the northern parkland (Hammer 1984). The size of the saline lakes varies, but the largest (Big Quill) has an area over 300 km^2 (Hammer and Haynes 1978).

Hammer and colleagues (Hammer 1978, 1984; Hammer and Haynes 1978) have summarized much of the existing knowledge on saline lakes in the grassland area. Many of the lakes are more saline than the oceans. Their salinity is highest during late fall or in winter under ice, and has tended to increase with time. Few data are available on the salt content of sloughs, but in the North Battleford area, specific conductances ranged from 200 to 20 000 μS/cm in the summer of 1984 (H. P. W. Rostad, pers. comm.).

Groundwater is an integral part of the hydrological cycle. The location of major surficial and bedrock acquifers has been summarized by Lane and Sykes (1982). Sustained annual yield of potable groundwater on the prairies has been estimated (PPWB 1982) at 8.3 km^3. Much of this water is in high yield aquifers in surficial deposits. The deeper, more mineralized groundwater represents a more reliable supply than the shallower groundwater (MacLeod 1977).

Effect of Cultivation on the Hydrological Cycle

This section emphasizes the effect of cultivation on the quantity and quality of the "fluxes" in the hydrological cycle. The information is largely based on measurements made on small research basins and plots in the grassland area of Canada and the United States.

Precipitation

Precipitation is highest in early summer and lowest during the winter months (Fig. 2). Most of the summer precipitation comes as showers. Local, high-intensity storms can cause high runoff from small areas, but their effect on streamflow rapidly diminishes downstream (Striffler 1969). These infrequent storms are the main cause of water erosion during the growing season (Toogood 1963).

On average, snow represents about 30% of the annual precipitation. Snow accumulates over the winter, and snow meltwater is a major contributor to the flow of rivers (Fig. 5) and recharge of soil water. At the time of melt, snow is rarely uniformly distributed over the landscape. Snow accumulates wherever surface obstructions encourage deposition. In the Bad Lake basin in Saskatchewan (Gray et al. 1979), grassland retained more snow than fallow fields but less than stubble. On native range, snow retention decreased from 100 to 20% as grazing intensity increased from zero to heavy (Wight et al. 1975). Snow water is lost by evaporation from the pack and from windblown snow particles; the latter depends largely on the distance travelled by the snow (Tabler 1975). Snow water losses can be especially high in southern Alberta and southwestern Saskatchewan where Chinook winds occur (Louie 1977; Steppuhn 1981).

The rate of snowmelt is one of the main factors determining whether the meltwater will infiltrate the soil or run off. Snowmelt is a complex process (Male and Gray 1981) that is largely controlled by the weather, but the nature of the vegetation also plays a role. Erickson et al. (1978) found that snowmelt rates on stubble and pasture fields were similar and slower than on fallow fields. Spring melt runoff peaks on small grass, stubble, and fallow basins near Swift Current, Saskatchewan, occurred within a few day of each other (W. Nicholaichuk, pers. comm.).

On the prairies, the natural precipitation should be slightly basic because of dust containing basic cations (Hammer 1980). Sulfur and nitrogen oxide emissions from the combustion of fossil fuels and the smelting of sulfide ores lower the pH of the precipitation. Such emissions are relatively low in western Canada. The volume-weighted pH of the precipitation on the prairies (Verry 1983) ranges from 5.3 in southern Manitoba to 6.5 for most of Saskatchewan. The pH of the snow is about one unit lower than that of the rain (Hammer 1980). In the grassland area, no acidification of the lakes or sloughs is anticipated due to their high buffering capacity (Hammer 1980). The same is also true for the small prairie streams, and to a lesser degree, the major rivers.

When potash mining started in Saskatchewan, concern was raised about possible effects of dust fallout on crops and soils, as studies showed detectable increases in soil chloride levels up to 6 km away from the refineries (Rennie 1983). Recent levels of dustfall within 2.5 km of the mines are about equivalent to the amount of potash removed by crops. It is not expected that such levels of dustfall would seriously affect water quality over the expected life span of most mines.

Interception

Part of the precipitation falling on a watershed is intercepted by, and held on, living vegetation and litter and is returned to the atmosphere by evaporation. Interception depends on the nature of the vegetation and precipitation. Interception losses by native grasses in southwestern Saskatchewan ranged from 21 to 32% of the growing season precipitation and were less on grazed than ungrazed prairie (Coutourier and Ripley 1973). In studies in the United States (Striffler 1969), grasses and natural mulches intercepted from 5 to 14% and from 3 to 8% of the annual precipitation, respectively. Water intercepted by a growing vegetation temporarily suppresses transpiration, while water intercepted by mulches represents a direct loss for the vegetation.

Cereals probably have similar interception characteristics as grasses, but cultivation reduces the buildup of crop residues. Mulches build up under chemical fallowing and zero-tillage, which may increase interception on crop land. Mulches also reduce evaporation and the net result might be a wetter topsoil.

Infiltration

Rain or snowmelt water that reaches the soil surface does not all infiltrate into the soil; some may accumulate as ponded water and some may be lost as runoff. The infiltration rate depends on several soil properties and is indirectly affected by vegetation and residues. Infiltration rates are generally much higher in unfrozen than frozen soils.

Several studies (see Branson et al. 1981) have shown that the infiltration rate of short and tallgrass range depends largely on the amount of vegetal cover, which protects the soil surface from rainfall impact and retards runoff. Grazing reduces vegetal cover, and as grazing intensity increases, infiltration decreases and runoff increases (Gifford and Hawkins 1978). The negative impact of grazing on infiltration could be due partly to soil compaction, particularly in areas of heavy traffic. Recovery from overgrazing varies with soil type and may take 10 yr or more. Overgrazing is not a recent problem; it also occurred before agricultural settlement (Bird 1961).

Prescribed fire is used to reduce the buildup of combustible residues, to control woody species and disease, and to improve rangeland condition. Excessive burning is, however, detrimental (Reeves 1977). Fire removes the mulch that protects the soil surface and is expected to decrease the infiltration rate; the effect may be compunded by condensation on the soil surface of hydrophobic substances released by the fire. Data on the impact of fire on the hydrological cycle are scarce (Branson et al. 1981).

Cultivation of grassland soils generally reduces their ability to take in water (Skidmore et al. 1975) due to an increase in bulk density, a decrease in aggregate size, and a decrease in biopores and continuous channels from the soil surface to the subsoil. The decrease in aggregate size can be retarded by reducing the amount of fallow in the rotation (Emmond 1971), or even partly reversed by returning to perennial grass. Although zero-tillage tends to increase the bulk density of the soil, it usually, though not always (Lindstrom and Onstad 1984), increases infiltration rates, since natural soil channels are left intact (Blevins et al. 1983).

The infiltration characteristics of frozen soils depend largely on the number and size of the ice-free pores, and the soil temperature (see chapter by van Everdingen, this volume). Post and Dreibelbis (1942) and Haupt (1967) found the lowest infiltration on burned, bare, or sparsely vegetated sites, while more rapid infiltration was found under denser cover. Gray et al. (1984) suggested three categories for snowmelt infiltration in seasonally frozen prairie soils: (1) restricted: infiltration negligible due to ice lenses at or near the surface (caused by rain or snowmelt just before freeze-up or melt/freeze events prior to the final melt); (2) limited: infiltration governed by the snow water equivalent and the water (ice) content of the 0- to 30-cm soil layer at time of melt (applies to most prairie soil); (3) unlimited: most or all of the metl infiltrate as the soil contains a high percentage of air-filled macropores (examples are dry, cracked, clays and dry, coarse sands).

Runoff

When the rate of rainfall or snowmelt exceeds the infiltration rate of the soil, water builds up on the soil surface. Overland flow and runoff start when the accumulated surface water begins to move downslope.

During the growing season, runoff losses are larger for cultivated than (native) grassland fields (e.g. Dragoun 1969). On the Canadian prairies, growing season runoff is small: for example, in Alberta, runoff losses from virgin sod, wheat, and fallow were less than 0.2, 0.8, and 1% of the 1950 growing season rainfall, respectively (Toogood 1963). Runoff losses from grassland increased with increased grazing intensity and percentage of bare soil (Branson et al. 1981).

Runoff losses from snowmelt, as a percentage of initial snow water equivalent, are higher from fallow than from stubble fields, as the wetter fallow fields have a lower infiltration capacity (Male and Gray 1981). Since fallow fields retain less snow than stubble or pasture, they may have the lowest snowmelt runoff. For example, over a 3-yr period, average spring runoff losses were 3.1, 3.2, and 0.7 cm for established grass, stubble, and fallow, respectively, from adjacent fields in southwestern Saskatchewan (W. Nicholaichuk, pers. comm.). Increased surface roughness (and possibly increased fall soil water evaporation) in fall-plowed fields may reduce snowmelt runoff compared with grass (Ayers 1964). On the Canadian prairies, autumn cultivation rarely increases overwinter soil water recharge (de Jong and Steppuhn 1983).

Overwinter soil water recharge depends on many factors: the amount of snow retained and the fraction of the meltwater that infiltrates, late fall and early spring rains, movement of water in the soil, and evaporation and transpiration losses (Gray et al. 1984). The overwinter gains shown in Table 1 reflect probable differences in snowcatch (buckbrush (*Symphoricarpos occidentalis*) versus native grass versus burned native grass) and/or frozen soil infiltration characteristics caused by differences in fall soil water content (stubble

TABLE 1. Overwinter soil water recharge on Sceptre heavy clay in southwestern Saskatchewan (unpublished data).

Cover	Overwinter recharge (cm)[a]	Period
Native grass	6.2 ± 2.4 (3)	1969-72
Burned native grass	2.9 (1)	1969-70
Buckbrush	10.4 ± 2.6 (2)	1969-71
Stubble	5.6 ± 7.1 (5)	1981-84
Fallow	−0.3 ± 4.0 (3)	1981-84

[a]Mean ± standard deviation (no. of winters in parentheses).

versus fallow). In Akron, Colorado, soil water recharge between October and April on winter wheat stubble was about 2 cm higher than on native grassland, possibly because the grass used some water (Greb 1980).

In the drier part of the Canadian prairies, over 80% of the runoff is derived from snowmelt (Card 1979; Nicholaichuk 1967; Hall and Langham 1970). High spring runoff is associated with above-average fall and winter precipitation and rapid melting of the snowpack. Median annual runoff over the effective drainage area ranges from less than 1 cm in southern Alberta and Saskatchewan to 5 cm and higher near the fringe of the grassland area (Fig. 6).

Few data are available on the effect of the increase in cultivated area on the runoff of prairie watersheds. Streamflow recording started as cultivation began, but early records are often incomplete and of doubtful quality. Durrant and Blackwell (1959) found that mean annual floods in the Assiniboine basin were generally lower for the 1911-56 period than for the 1941-58 period. To examine whether this might be due to the result of the increase in cultivated acreage, we compared flow records (Environment Canada 1983) for four streams draining areas with little cultivated land (Manyberries Creek, Lodge Creek, Horse Creek, and the Rolling River) with flow recrods for five streams draining areas where land use changed considerably (the Battle River at Ponoka, Moose Jaw Creek, Pembina River, Brokenhead River, and the Little Saskatchewan River). The location of the gauging stations is shown in Fig. 6. Streamflow was expressed in centimetres over the gross drainage area and divided into annual flow and flow after June 1. Streamflow after June 1 has a significant subsurface water component (e.g. see increases in salt content in Fig. 5) and is only partially caused by runoff.

The three streams in the largely uncultivated, drier part of the prairies (Manyberries Creek, Lodge Creek, and Horse Creek) show no systematic change in the long-term annual and seasonal discharge (Table 2; Fig. 7). Annual flows based on the effective drainage area are similar for all three streams and in about 10% of the years amount to 5 cm or more. Streamflow after June 1 represented about 10% of the annual flow. Card (1979) noted high summer flow in undisturbed basins in Alberta when June and July rainfall exceeded 20 cm, but did not distinguish between runoff and subsurface flow. The flow regime of these three streams appeared unchanged over the last 70 yr, indicating no changes in precipitation patterns. The apparent increase in summer and fall flow of the Rolling River (Table 2) is likely an artefact of the limited data. Spring flow accounted for 50-60% of the mean annual flow of the Rolling River.

In many areas, farmers have implemented surface drainage to reduce spring flooding of farmland. Such systems, often consisting of open ditches and drains, could have affected the streamflow patterns of the five cultivated basins. Records on the extent of surface drainage in these basins were not available; however, a recent review (Rubec and Rump 1984) indicates considerable conversion of wetlands to agricultural use in the grassland region.

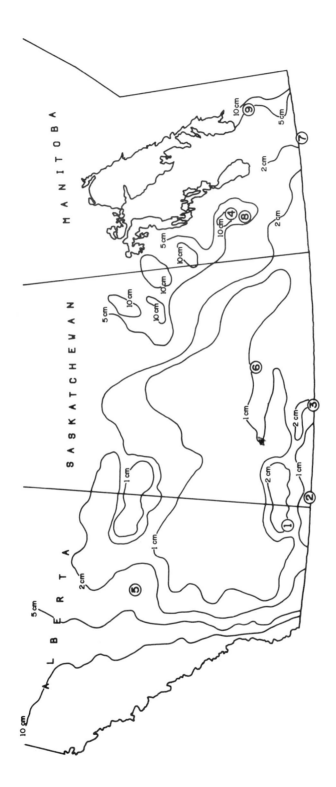

FIG. 6. Median annual runoff (over the effective drainage area) on the Canadian prairies (Mowchenko 1976) and location of the gauging stations in Table 2.

223

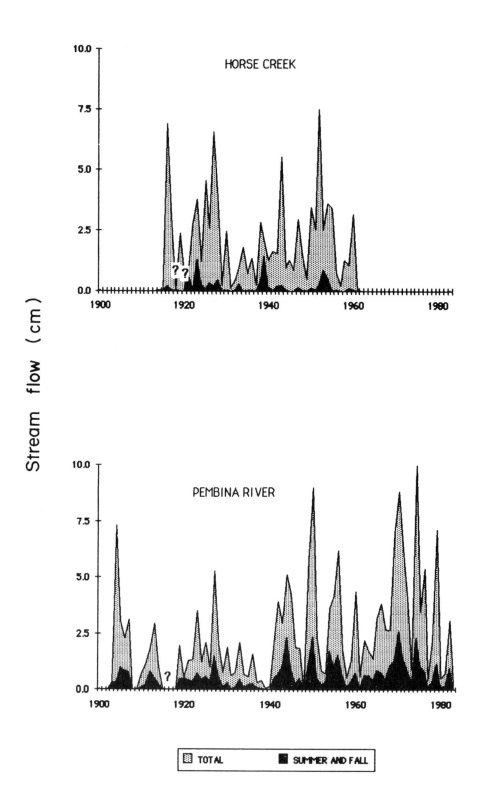

FIG. 7. Annual and summer plus fall flows of Horse Creek (watershed largely uncultivated) and the Pembina River (up to 75% of the watershed in field crops and fallow).

TABLE 2. Flow regime of some prairie rivers at different times.

Stream[a]	Drainage area (km²)[b]	Period	Flow (cm)[c] Annual	Flow (cm)[c] June–Dec.
		Largely uncultivated		
Manyberries Creek (1)[a]	339 (100)	1912–30	2.5 ± 1.8 (16)	0.4 ± 0.9 (19)
		1957–82	2.2 ± 1.6 (26)	0.2 ± 0.3 (26)
Lodge Creek (2)	1970 (89)	1910–31	1.9 ± 1.4 (19)	0.2 ± 0.3 (22)
		1932–51	1.1 ± 0.7 (20)	0.1 ± 0.2 (20)
Horse Creek (3)	196 (100)	1914–37	2.2 ± 1.9 (20)	0.2 ± 0.3 (24)
		1938–61	2.2 ± 1.9 (24)	0.2 ± 0.3 (24)
Rolling River (4)	759 (82)	1915–22	5.4 ± 1.8 (7)	2.2 ± 1.2 (8)
		1972–82	6.8 ± 3.4 (11)	3.3 ± 2.6 (11)
		Substantial increases in cultivated acreage		
Upper Battle River (5)	1840 (90)	1913–29	7.34 ± 4.4 (12)	3.1 ± 2.9 (17)
		1972–82	3.8 ± 2.9 (11)	1.7 ± 1.6 (11)
Moose Jaw Creek (6)	5240 (54)	1911–40	0.7 ± 0.7 (23)	0.1 ± 0.2 (28)
		1954–82	1.4 ± 1.6 (29)	0.2 ± 0.3 (29)
Pembina (7)	8470 (64)	1903–22	1.8 ± 1.7 (12)	0.7 ± 0.8 (14)
		1923–42	1.2 ± 1.0 (20)	0.4 ± 0.4 (20)
		1943–62	2.3 ± 1.7 (20)	0.9 ± 0.7 (20)
		1963–82	2.9 ± 2.4 (20)	0.9 ± 0.7 (20)
Little Saskatchewan River (8)	2620 (58)	1914–26	4.7 ± 2.8 (12)	2.3 ± 1.5 (13)
		1972–82	4.7 ± 2.9 (11)	2.0 ± 1.9 (11)
Brokenhead (9)	1538 (100)	1912–22	6.7 ± 2.8 (9)	4.1 ± 2.4 (11)
	1582 (100)	1957–82	8.2 ± 5.8 (23)	3.5 ± 3.4 (26)

[a]Location on Fig. 6 given in parentheses.
[b]Gross (effective as % of gross) (PFRA 1985).
[c]Based on gross drainage area, mean ± standard deviation (no. of years in parentheses).

Moose Jaw Creek receives about 35 cm of precipitation annually and has the greatest growing season water deficit of the five cultivated basins. The area of cultivated land has increased substantially since 1910. The annual flow appeared to have increased significantly, largely because of increased spring flow (Table 2). Summer and fall flow accounts for slightly less than 15% of the annual flow, which agrees with data for the nearby Davin watershed (Nicholaichuk 1967). Surface drainage has increased the effective drainage area by about 15% since 1950 (Thies 1985), which could be partly responsible for the increase in spring flows. Banga (1978) concluded that increasing the effective drainage area by 10% would increase the 1 in 2-yr flood by 13%, but would have less effect on rare floods at Moose Jaw.

Settlement in the Pembina basin started in the 1870's, and by 1900, most of the area had been homesteaded. The area has an annual precipitation of about 50 cm. Changes in the flow regime (Fig. 7; Table 2) are probably largely caused by increases in the area of field crops and fallow and surface drainage. Flow has varied considerably and was lowest in the dry 1930's. Spring flows amounted to two-thirds of the annual flow. Flows were higher after 1940, which could be the result of improved surface drainage.

The Little Saskatchewan River drains a slightly more humid area than the Pembina River, and summer and fall flow accounted for 40–50% of the annual flow (similar to the nearby Rolling River). The Little Saskatchewan River showed no significant changes in long-term annual streamflow with time (Table 2), despite increases in the area of field crops. A substantial portion of the basin is located in Riding Mountain Park, which could account for the absence of noticeable changes in the flow regime.

Flows of the upper Battle River appeared to have decreased (Table 2). Stolte and Herrington (1984) examined the flow records in detail and concluded that, except for spring flows, monthly flows were reduced significantly. They observed the same for the adjacent Blindman River and concluded that snow accumulation and snowmelt processes had not changed with time. The decreased flow from June to April was attributed to increased soil water use by fertilized crops (see Evapotranspiration). Improved surface drainage would also tend to decrease summer and autumn flows and increase spring flows. Stolte and Herrington (1984) found no changes in flow for the lower Battle basin, which is drier than the upper basin.

The Brokenhead River drains a mixture of grassland, forest, and organic soils. Increases in its flow regime (Table 2) were consistent with increased surface drainage.

The transformation from grass to cultivated land affects the quality of the runoff. Runoff and soluble nutrient losses from native prairie and cultivated fields occur mainly when the snowpack melts (Alberts et al. 1978; Burwell et al. 1975; Timmons and Holt 1977). Concentrations of soluble nutrients in runoff from prairie and cultivated plots are similar (White and Williamson 1973), but soil losses are greater on cultivated plots (Toogood 1963; Burwell et al. 1975). On grassland, nitrogen is mainly lost in soluble form whereas on cultivated fields, most of the nitrogen and phosphorus are lost as sediment. In the United States grassland area, most sediment loss from cultivated fields occurs after seeding (Burwell et al. 1975; Alberts et al. 1978). Figure 5 suggests that on the Canadian prairies, significant sediment losses occur during and shortly after the snowmelt.

Under normal management, nutrient losses from cultivated fields are low. For example, Burwell et al. (1975) found that nitrogen losses in runoff from cultivated watersheds were less than the amount of nitrogen in the precipitation, and in other experiments (Alberts and Spomer 1985; Alberts et al. 1978; Nicholaichuk and Read 1978), nutrient losses represented less than 5% of the applied fertilizer. Nutrient losses are low, since considerable deposition of sediment occurs in small depressions, as is illustrated in Fig. 8 for a small watershed near Saskatoon (Martz and de Jong 1985). Although nutrient losses in runoff are insignificant for agriculture, the runoff often exceeds water quality criteria.

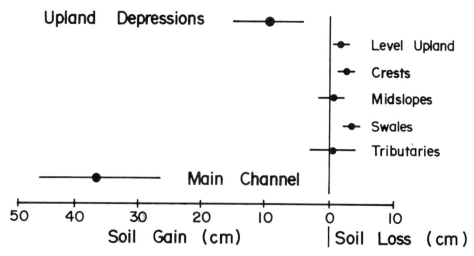

FIG. 8. Mean soil losses and gains from 1960 to 1982 for different landforms (mean and 95% confidence interval) (Martz and de Jong 1985).

Surface waters on the prairies have always been nutrient rich by virtue of the fertile land surrounding them. Erosion has probably slightly increased nutrient additions to surface water, although White and Williamson (1973) suggested that nutrient losses from cultivated land could be similar to those from periodically burned prairie. The latter was not uncommon before agricultural settlement (Bird 1961). Where livestock is concentrated, be it in feedlots or near the favorite watering places of buffalo, nutrients will likely enter surface water. Domestic and industrial wastes have also affected the surface water quality. For example, in 1970, 70% of the nitrogen and phosphorus in the Qu'Appelle River system came from the cities of Regina and Moose Jaw (Anonymous 1972), and wastes from a potash plant have practically eliminated the biota of Patience Lake near Saskatoon (Hammer 1984).

In summary, cultivation has no doubt increased runoff from individual fields, but the effect on streamflow appears to be small. Mean annual river flows have increased slightly in some areas, possibly, due to improved surface drainage and/or redistribution of snow due to summerfallowing. In most areas, streamflows from June to January have not changed significantly, and in a more humid area, a significant reduction was evident. The impact of the increased erosion on river water quality has been limited, as most of the sediment is trapped in the fields.

Evapotranspiration

Evapotranspiration includes water lost to the atmosphere by evaporation and by transpiration from living plants. Of all the fluxes in the hydrological cycle, transpiration has probably been changed most by the breaking of the native grassland. Annual potential evapotranspiration of forages exceeds that of spring cereals, although peak rates are similar (Pohjakas et al. 1967; Sonmor 1963). The practice of summerfallowing further reduces transpiration and increases evaporation. For example, on a heavy clay soil in southwestern Saskatchewan, evapotranspiration averaged 29.4 cm/yr on native grassland and 29.3 cm/yr for wheat that was grown on fields that had been summer-fallowed (de Jong and MacDonald 1975). The native grassland used about 90% of the annual precipitation and the wheat used about 45% of the precipitation over the 2-yr crop-fallow cycle. In long-term trials in the same area, evaporanspiration of wheat on fallow and stubble accounted for 37 and 63% of the precipitation, respectively (Cameron 1978). In Manitoba, under more humid conditions, water use was similar for spring wheat grown in 2-yr or longer rotations (Ferguson 1963; Michalyna and Hedlin 1961) and about 20% lower than water use of hay.

During the actual summerfallow year, fields conserve only a small portion of the precipitation (Table 3). More than 85% of the precipitation is lost by evaporation, runoff, and drainage. During the summer, runoff losses are small, while evaporation losses are large, as the light summer rains do not penetrate deeply in the soil (Cameron 1978). Drainage (see Groundwater Recharge and Discharge) increases as the climate becomes more humid. Most of the winter precipitation on fallow fields is lost as blowing snow or snowmelt runoff, while abundant spring rains can cause drainage losses. Despite its inefficiency in conserving precipitation, summerfallowing is used extensively. The average soil water gain of 4-5 cm in the drier areas increases wheat yields by about 500 kg/ha and stabilizes the production level. Summerfallowing also controls weeds and builds up nutrients in the soil.

Earlier studies (see de Jong and Steppuhn 1983) showed little difference in storage of plant available water between different fallow systems as long as weed growth was prevented. Recent studies in Alberta (Lindwall and Anderson 1981) indicate 0.5-1.0 cm more storage when some tillage operations are replaced by herbicides. The method of fallowing could alter the nature of the water losses. The better trash cover associated with chemical fallow may reduce evaporation and runoff during the summer, while drainage and spring runoff could increase.

TABLE 3. Gains in available soil water to 120 cm depth (ΔS) and water losses (cm) by evaporation, deep drainage, and runoff during the fallow period.

Measurement period	Location							
	SW Sask.[a]		S Alta.[b]		NE Sask.[c]		S Man.[d]	
	ΔS	Loss	ΔS	Loss	ΔS	Loss	ΔS	Loss
Harvest–fall	2.1	3.4					5.3	7.9
Fall–spring	3.5	7.5	7.9	10.1	2.2	12.0	4.1	13.2
Spring–fall	2.6	17.4	2.0	18.3	1.2	27.8	0.3	37.3
Fall–seeding	1.9	9.2	3.4	13.6	1.2	11.6	−1.0	18.3

[a]Staple and Lehane (1952).
[b]Lindwall and Anderson (1981).
[c]Bowren (1977).
[d]Siemens (1977).

Fertilization usually increases yields, as it increases efficiency of water use and, sometimes, extraction of soil water by the crop. Fertilization of native grass increased soil water extraction from greater depth; the effect persisted for 1–2 yr (Black 1968; de Jong and MacDonald 1975; Power 1980). Fertilization of cereals in the drier areas of the prairies had no effect on soil water extraction on light- and medium-textured soils (de Jong and Rennie 1969; Warder et al. 1963, but increased soil water extraction from heavy clays by about 2 cm (Warder et al. 1963). In wetter areas, fertilization increased soil water extraction on medium- and heavy-textured soils (Hoyt and Rice 1977), but not on light soils where all available water was extracted regardless of fertilization (Siemens 1977). In the drier areas, fertilization of continuously cropped and grassland fields only increases soil water extraction for a few years, as the amount of soil water that can be extracted on such fields becomes limited by the annual recharge. In or near the forest area, the total root zone is frequently recharged over the winter (Hoyt and Rice 1977), and fertilization increases evapotranspiration consistently on soils of moderate to high water-holding capacity.

The effect of grazing or fire on evapotranspiration is confounded by their effect on infiltration and the amount of snow that is trapped. Differences due to grazing seem to be least on light-textured soils (Branson et al. 1981). In one study, fire reduced overwinter soil water recharge and decreased evapotranspiration during the next growing season (de Jong and MacDonald 1975).

At least one third of the irrigated acreage in Alberta and Saskatchewan is devoted to growing hay, forage, or pasture (PPWB 1982). With adequate fertilization and irrigation to supply a total of 60–65 cm of water (Sonmore 1963; Pohjakas et al. 1967), yields are several times higher than on dryland. Although not commercially practiced, irrigation increases yield of native grassland, especially when combined with nitrogen fertilization (de Jong and MacDonald 1975; Smika et al. 1965). Under irrigation the possibilities of water loss by drainage and reduced absorption of snowmelt water are increased (Hobbs and Krogman 1971).

Groundwater Recharge and Discharge

Some of the precipitation that soaks into the ground will percolate through the root zone to the groundwater. Little is known about the rates of groundwater recharge or discharge (movement from the groundwater to the land surface). Laycock (1965) calculated water surpluses ranging from 2.5 cm or less in southern Alberta and Saskatchewan to 10 cm on the Canadian Shield and in the Foothills. Water surpluses occur during spring melt or during wet fall and spring periods and generate runoff and groundwater recharge. Direct runoff to streams is important in the effective area of a drainage basin; in the remaining area the water surplus should result in groundwater recharge and evapotranspiration of wetlands. Groundwater recharge utlimately causes base flow of rivers, but base flow is not an accurate indication of groundwater recharge (Meyboom 1967) because of the consumptive use by phraetophytes in the discharge areas. Rey (1983) estimated that over the settled area of Saskatchewan the recharge rate of shallow and medium-depth groundwater aquifers varied from 0.4 to 1.2 cm/yr.

In the drier areas, deep percolation is associated with periods of above-average precipitation. At Mandan, North Dakota, the top 180 cm of the soil under native grass occasionally wetted to field capacity when precipitation was above average for 2 yr or more (Power 1970); under annual crops the soil was wetted to 180 cm depth more frequently than under grass. Similarly it has been estimated that in southern Alberta, properly managed alfalfa would prevent groundwater recharge in 7 out of 10 yr whereas annual wheat would only do so in 1 out of 2 yr (Krogman 1973). Leaching of nitrate from wheat-fallow and continuous wheat rotations in southern Saskatchewan confirms that groundwater recharge occurs in cultivated fields in years of above-average precipitation or heavy spring rains (Campbell et al. 1983). Miller et al. (1981) estimated that under native vegetation, 1-4% of the annual precipitation percolated below the root zone versus 7-15% under crop/fallow systems.

The studies described above did not consider groundwater recharge through sloughs. In North Dakota, Shjeflo (1968) estimated an average annual seepage loss of 18 cm from potholes. This seepage represented an average annual groundwater recharge of 1.8 cm, since the potholes occupied about 10% of the drainage basin. Some of the seepage losses will be recovered as evapotranspiration by vegetation around and in the slough (Meyboom 1966). For example, Laksham (1971) estimated that in the summer of 1970 the weekly water loss of three sloughs in the Saskatoon area consisted of 1.5 cm of evaporation, 1.1 cm of seepage, and 2.2 cm of transpiration by vegetation around the sloughs. The sloughs held water for only part of the growing season.

Table 4 summarizes estimates of groundwater recharge and discharge in the prairie region. Most of the "measured" data are based on limited measurements of the hydraulic gradient and permeability at a few sites in large basins (Table 4, part I), Often, these "measured" data have been extended using models. The average recharge is low, with high recharge occurring in small parts of the basins such as sloughs or areas of coarse materials with a high water table (e.g. Freeze and Banner 1970; Rehm et al. 1982). The study by Rehm et al. (1982) illustrates the sensitivity of the predicted recharge rates to the estimated permeability: depending on the assumed permeability, the calculated recharge on the fine textured soils varied from 0.2 to 9 cm/yr.

Estimates of growing season drainage from computer models (Table 4, part II) are fairly accurate when the model runs are started each spring with measured soil water contents. These calculations indicate little or no drainage when a crop is actively growing. In one case (de Jong and Zentner 1985), an upward flux into the root zone of wheat grown on fallowed fields was predicted: unfortunately, the model was not run for the fallow years, when drainage is known to have occurred (Campbell et al. 1983). Drainage predictions by models run over a series of years without periodic calibration (Table 4, part II) must be treated with caution. The model predictions in Table 4, part II, do not include groundwater recharge through depressions. The higher predicted groundwater recharge under crops than under permanent grassland is confirmed by observations on the salt distribution in the soil. From the latter, Ferguson and Bateridge (1982) concluded that on cultivated sites, groundwater recharge in the fallow years was 2-3 cm/yr higher than on native grassland. Groundwater discharge areas were characterized by salt accumulation.

The major concerns with accelerated groundwater recharge under cultivated fields are nitrate contamination of the groundwater and increased salinity in discharge areas.

TABLE 4. Estimates of groundwater recharge $(+)$ or discharge $(-)$ by various authors.

Location	Flux (cm/yr)	Comments and source
I. Estimates for basins or part thereof		
Prairie provinces	1.7	Sustainable groundwater yield (PPWB 1982)
Saskatchewan	0.4–1.2	Rey 1983
Allan Hills	1.1	Measured (Meyboom 1966)
Arm River	0.3	Measured + model (Meyboom 1967)
Good Spirit Lake	0.2	Measured + model (Shiau 1978)
Sand and gravel plains, shallow water table	0.3–2.5	Measured (Freeze and Banner 1970)
Till plain, water table at 150 cm	<0.3	Measured (Freeze and Banner 1970)
Gravel plain, water table at 330 cm or deeper	0	Measured (Freeze and Banner 1970)
Alberta		
Dryland	<1	Measured + model (Chan and Hendry 1985)
Irrigation	0.6–5	Measured + model (Hendry 1981)
Irrigation	4–10	Measured + model (Hendry et al. 1982)
North Dakota Upland Basin	1–12	Measured + model (Rehm et al. 1982)
Sloughs	60	Measured + model (Rehm et al. 1982)
Exposed sand	13–71	Measured + model (Rehm et al. 1982)
Covered sand	6	Measured + model (Rehm et al. 1982)
Fine textured	0.2–9	Measured + model (Rehm et al. 1982)
Recharge areas for saline seeps		
North Dakota	1.4	Measured (Doering and Sandoval 1976)
Montana	4–4.5	Measured (Miller et al. 1981)
Discharge areas of saline seeps		
North Dakota	−1.2	Measured (Doering 1982)
Montana	−5.8	Measured (Miller et al. 1981)
Saskatchewan	−0.6 – −6	Measured (Henry et al. 1985)

Leaching of nitrate below the root zone occurs occasionally on cultivated fields (Campbell et al. 1983), but not on grassland (Power 1970; Sommerfeldt and Smith 1973). Losses of fertilizer nitrogen by leaching are affected by the rate of application and the amount of precipitation. In Iowa and Missouri, subsurface discharge accounted for 62–88% of the annual streamflow and 84–95% of the annual soluble nitrogen discharge from cultivated watersheds (Burwell et al. 1976; Alberts and Spomer 1985). Base flows on the Canadian prairies (Table 2) are generally less than 40% of the annual flow, and the balance between nutrient losses by runoff and leaching will favor the former.

Groundwater flow systems have been described as local, intermediate, and regional (Freeze and Cherry 1979). Regional systems develop in areas with negligible local relief and are characterized by slow flow rates, slow response to climatic changes, and high

TABLE 4. *(Concluded)*

Location	Flux (cm/yr)	Comments and source
II. Growing season estimates from models		
(a) Known starting soil moisture levels		
Southern Alberta		
Fallow	2.2	VSMB[a] (van Schaik et al. 1976)
Crop/fallow rotation	1.0	VSMB (van Schaik et al. 1976)
Continuous wheat	0.5	VSMB (van Schaik et al. 1976)
Grass	0	VSMB (van Schaik et al. 1976)
Southwest Saskatchewan		
Grass	0	VSMB (de Jong and MacDonald 1975)
Grass	0.06	de Jong and Hayhoe 1984
Wheat on fallow	− 1.2	de Jong and Zentner 1985
(b) No known starting soil moisture levels		
Assiniboine basin, crop/fallow rotation		
Southwest Saskatchewan	0.2	VSMB, 1944–60 (Meyboom 1967)
Southern Manitoba	3.8	VSMB, 1944–60 (Meyboom 1967)

[a]Versatile soil moisture budget.

mineralization of the water (MacLeod 1977). Local systems dominate in areas where the local relief is pronounced, the recharge occurs at a topographic high, and the discharge occurs in an adjacent low. These systems are shallow and respond quickly to climatic conditions. In glacial till, the salts in the water are mainly calcium and magnesium sulfate. In an intermediate groundwater system, recharge and discharge areas are separated by one or more topographic irregularities, each with its own shallow, local groundwater system. On the prairies, variations in stratigraphy lead to discharge–recharge relationships that are difficult to anticipate (Freeze and Whitherspoon 1967).

In the drier prairie region, permanent lakes are discharge areas as indicated by their salt content, while sloughs can be discharge and/or recharge areas (Meyboom 1966). Temporary sloughs are quite common. In the spring when filled with snowmelt water, flow is away from the slough. In the summer when the slough is dry, shallow groundwater flows to the slough, supporting a lush vegetation of willows and grasses. In the fall and winter the regional flow pattern dominates. Much of the spring melt groundwater recharge is depression focussed. Parts of the landscape that do not retain much snow may rarely or never contribute directly to groundwater recharge. Few estimates of groundwater discharge rate have been made (Table 4) despite its importance in soil salinization. Seepage rates of about 6 cm/yr into a saline area were observed in Montana (Miller et al. 1981) and 1.2 cm/yr in North Dakota (Doering 1982). These two examples refer to saline seeps which involve local groundwater systems. In Saskatchewan the upward flow from regional and intermediate aquifers below saline areas was estimated at 0.6–6 cm/yr (Henry et al. 1985). The salts in the discharging water cause the loss of arable land and the deterioration of surface and shallow groundwater resources (Miller et al. 1981).

The mechanism causing soil salinity must be understood before solutions can be proposed. In the case of saline seeps, excessive summerfallowing is often the cause of increased groundwater recharge. The most successful control practices are to grow deep-rooted perennial crops, drain upland recharge sloughs, and use less summerfallow (Miller et al. 1981). Special management practices, e.g. snow trapping (see Augmentation of the Water Supply), may be necessary to make continuous cropping feasible. Interception and removal of the incoming water by drainage systems is sometimes possible (Doering and Sandoval 1976). If the salinity problem is caused by intermediate or regional groundwater systems, reclamation may be possible through increased infiltration on the saline area. Studies in North Dakota suggest that summerfallowing, preventing runoff, and leaving a mulch to reduce evaporation can reduce salinity (Sandoval and Benz 1973; Sandoval et al. 1961). In some cases it may be possible to use water from the source aquifer for irrigation (Henry et al. 1985).

Water Demand and Possible Augmentation of the Supply

Demand for Water

Water demand studies for the Saskatchewan-Nelson basin (PPWB 1982) were used to estimate current water use on the prairies (Table 5). Water use is divided into gross, or total, use and net, or consumptive, use. Power generation accounts for nearly 90% of the gross water use, while agriculture is the largest net user of surface water (see also chapter by Tate, this volume). Table 5 does not include evapotranspiration from dryland crops and natural vegetation, which was estimated at 185 km^3/yr (see Water Supply and Quality).

TABLE 5.   Water use on the prairies in 1978.[a]

Category	Use (km^3) Gross	Net	Increase, 1951-78 (%/yr)	Comments
Municipal	0.50	0.10	7	87% surface water
Small communities	0.04	0.02	5	15% surface water
Industrial in 1976[a]	0.25	0.19	4	80% surface water
Agriculture				
Domestic	0.02	0.02	−2	Mainly groundwater
Livestock	0.15	0.15	2	Mainly groundwater
Irrigation				
District	1.41	0.95	27	Mainly surface water
Private	0.32	0.32	10	Mainly surface water
Agriculture total	1.90	1.44	11 (5)	11% increase in net use, 5% increase in gross use
Evaporation				
Large reservoirs	0.50	0.50		Approximately 25% from irrigation reservoirs
Small reservoirs	0.10	0.10		Agricultural use
Power	23.05	0.03	13	Net evaporation loss

[a]Estimated from PPWB (1982) report by considering only the grassland basins.

Irrigation is by far the greatest net user of surface water. The average net annual water use by irrigation is probably about 0.6 km^3/yr higher than shown in Table 5, as 1978 was a wet year; in 1976 the net water use of the irrigation districts was 1.58 km^3. Net water use by irrigation appears to include some recharge to groundwater. For example, it is unlikely that the net water use of private irrigation projects equals the gross use. Also, the net average water use in the districts was 68, 58, and 47 cm in 1961, 1971, and 1978, respectively (PPWB 1982), which exceeds the potential evapotranspiration of many irrigated crops (Pohjakas et al. 1967; Sonmor 1963) if one takes into account growing season rainfall and available soil water at seeding.

Most of the rivers on the prairies are used to some degree for municipal and industrial use, irrigation, and hydropower. The growing season water deficit on most of the prairies ranges from 20 to 40 cm, and Oosterveld and Nicholaichuk (1983) estimated that productivity per unit area can be increased up to four times by eliminating the water deficit. Currently, about 550 km^2 of land, about 1.5% of the area of improved farm land, is irrigated. Further extension of the irrigated area is constrained by water supply. In the Red Deer, Oldman, and Bow basins, water for irrigation is already limited during periods of low flow, and the same will be true in the foreseeable future for most of the other basins (Lane and Sykes 1982). Major demand conflicts occur when an abnormally dry growing season coincides with low river flows.

Augmentation of the Water Supply

Most of the user groups identified in Table 5 can improve their water supply through appropriate water management. Agriculture can have the biggest impact, but other users can also make important contributions.

Snow Management

Overwinter soil water recharge is extremely important for the water supply of the next crop. Table 3 summarized estimates of overwinter recharge for level stubble and fallow fields. From these and other data reported by de Jong and Steppuhn (1983), it is apparent that overwinter recharge is least efficient in the wetter regions. Table 6 shows some recent estimates of overwinter recharge on rolling, cultivated land in the grassland region. Tables 1, 3, and 6 show considerable differences in overwinter recharge between different slope positions and between fields with different amounts of vegetation cover. Redistribution of the overwinter precipitation in the landscape as blowing snow and/or snowmelt runoff is a major factor causing these differences in overwinter recharge. Snow management aims to minimize this redistribution and to capture blowing snow in preferred locations.

Currently, 5-10 cm of the overwinter precipitation is lost by sublimation and runoff or is blown into undesirable locations (de Jong and Cameron 1980). Sublimation/

TABLE 6. Average overwinter recharge (1982-83, 1983-84) under different vegetative cover (Kachanoski et al. 1985).

	No. of fields	Slope position (cm)		
		Upper	Middle	Lower
Stubble + snowtrap strips	26	3.5	4.3	6.7
Standing stubble	45	3.5	3.9	5.6
Cultivated stubble	17	2.8	3.1	5.2
Fallow	14	1.1	0.9	1.7

FIG. 9. Snow conserved by trap strips in a cultivated field.

evaporation losses could be reduced considerably if snow blowing was prevented. Leaving the stubble undisturbed or leaving trap strips (Fig.9) increased spring soil water supplies by 1–2 cm compared with cultivated stubble (Table 6) and wheat yields by about 10% (200 kg/ha) under good management (Kachanoski et al. 1985). If these gains could be achieved on all land cropped to cereals, snow management would yield 2–4 km^3 water/yr, which is equivalent to one to two times the amount of water now used by irrigation. Snow management increases the possibility of extended rotations, which would reduce the water losses that now occur during the summerfallow year (Table 3). Simple snow management procedures, such as leaving the stubble standing or stubble strips, are usually the most economic (Nicholaichuk 1980). On grassland, strips of unpalatable grasses might create the same effect (Wight et al. 1975).

Based on long-term climate records, the opportunity for snow management is better in the wetter than the drier areas of the prairies (de Jong et al. 1987). Farmers in the more humid areas are sometimes reluctant to adopt snow management because it conflicts with their preference for fall tillage. Alternative ways of utilizing snow should be sought in these areas, as growing season water deficits are common. Concentrating the snow and snow meltwater in large sloughs for use as supplementary irrigation, and mixing with groundwater of poorer quality, might be possible.

Snow management is of possible interest to increase the water supplies of small communities. The village of Nokomis, Saskatchewan, is experimenting with snowtrapping over a surficial aquifer to increase recharge.

The major remaining problem in snow management is to increase the infiltration of metlwater. Models predict that a large proportion of the extra snow held on stubble fields could be lost as runoff or drainage (de Jong et al. 1987). Recent studies on the effect of fall cultivation and subsoiling (Gray et al. 1984; Nicholaichuk 1984) are inconclusive, but earlier studies (de Jong and Steppuhn 1983) showed no consistent benefits.

Land Surface Shaping

In the 1930's, terracing increased wheat yields at Swift Current, Saskatchewan, by about 220 kg/ha (50%) on stubble and 360 kg/ha (26%) on fallow fields (Barnes

1938). The increased yields were probably largely the result of decreased losses of snow or snowmelt runoff. Numerous studies have been conducted in the United States on the effect of land surface modification on the soil water regime and yield of grassland (Branson et al. 1981; Wight 1976). Most studies pertain to conditions where runoff is mainly generated by summer rainstorms, whereas in the Northern Great Plains most of the runoff is from snowmelt. The data generally show increased overwinter soil water recharge after surface modification (Neff 1980; Haas and Willis 1971), but the effect is less on the Northern than the Central Great Plains (Wight et al. 1975).

The main purpose of slough consolidation and wetland drainage is to facilitate field operations and to increase the area of cultivated land. About 12×10^3 km^2 of prairie wetlands has been drained (Simpsons-Lewis et al. 1979), and in some areas more than half the wetlands has been lost (Rubec and Rump 1984). This reduction in water retention capacity must affect the quantity and quality of streamflows, but few data on this are available (see Runoff). The loss of wetland habitat and the possible reduction in groundwater recharge are additional factors that should be considered when drainage is contemplated. Consolidation of several small potholes into a larger one would create fewer but more permanent water bodies and would eliminate some of the adverse environmental impacts of drainage.

Vegetation Control

In areas of low precipitation, controlling shrubs and trees seems to have little effect on the hydrological regime of grasslands; in more humid areas, brush removal may increase water yield (Branson et al. 1981). Johnston (1970) found that water use increased in the order bare soil < grass < aspen. Aspen started to use water before grasses and had the lowest soil water contents in fall. Under buckbrush, fall soil water was about 2 cm lower than under native grassland (de Jong and MacDonald 1975), but increased overwinter recharge (Table 1) more than compensated for this and water use of buckbrush was about 4 cm higher than for native grassland.

The net effect of controlling woody vegetation or phraetophytes around sloughs is not clear. Eradication of such species may reduce evapotranspiration, but would also reduce the amount of snow trapped. Chemical control of woody species in grasslands could affect water quality. Properly applied, the concentration of the herbicides in runoff and drainage water should be low, and the low runoff and drainage during the growing season would further minimize adverse water quality effects on the Canadian prairies.

Most cereals on the Canadian prairies are spring seeded. Fall-seeded cereals probably use water more efficiently than spring-seeded varieties, as they take up water during autumn and early spring that might otherwise be lost by evaporation and drainage. Winter wheat is best suited for seeding in standing stubble and could reduce the dependence on the water-inefficient summerfallow system.

Increasing Efficiency of Water Use

High water use efficiency in dryland and irrigated agriculture depends on adequate soil fertility and weed and pest control. In irrigation, sufficient water should be applied to meet the needs of the crop and to provide enough drainage to prevent salt buildup. Drainage needs are lower for sprinkler systems where application rates are better controlled than for gravity systems. In the Vauxhall District of the Bow River project, average on-farm losses by runoff and drainage amounted to about 17 cm (Rapp et al. 1969), which is far in excess of any possible drainage requirements. Adaptation of improved irrigation methods would increase irrigation water use efficiency.

Currently, about 500 ha of land is irrigated with sewage effluent of small communities (PPWB 1982). Irrigation is an effective way of utilizing the nutrients in the effluent: however, the high salt content of most effluents imposes fairly high leaching requirements. On the prairies, effluent irrigation is limited by the need to store the effluent during the fall and winter periods.

Demand for irrigation, municipal, or other water, and the efficiency with which it is used, depends on its price. For example, per capita municipal water use is 10–30% higher in Alberta than in Saskatchewan and Manitoba (PPWB 1982). The difference is no doubt related to the pricing policy.

Others

Evaporation from reservoirs is equivalent to one third of the net offstream water use by agriculture (Table 5). Evaporation suppressants (Harbeck 1969) have shown savings of up to 20% on large reservoirs. Transpiration suppressants reduce plant water losses but often also reduce yields and are expensive (Branson et al. 1981). Agronomic practices such as varying seeding rates and row spacings may be more effective ways of stretching the supply of water available to the crop (de Jong and Cameron 1980).

Freeze-purification of brackish groundwater can increase the water supply of small communities (Husband and Spyker 1978). Geller (1962) has discussed the applicability of the technique for farm use, but no field tests have been performed in North America.

Summary

Before agricultural settlement, Canada's prairie grassland occupied an area of about 500×10^3 km^3. About two thirds of this land is now under crops. The annual precipitation varies from 30 to 50 cm and about one third falls as snow. Precipitation is considerably lower than potential evapotranspiration during the growing season. Water surpluses occur during the spring melt period and, to a lesser extent, in the fall and create streamflows equivalent to 1 cm/yr over the whole grassland area. Over 90% of the precipitation received in the grassland area is lost as evapotranspiration.

The change in land use has significantly altered the balance between evaporation and transpiration and has increased groundwater recharge. The increased groundwater recharge is blamed for an increase in soil salinity. If the current land use continues, soil degradation through salinization and erosion will increase. Excessive summer-fallowing, which leads to inefficient use of water and exposure of the soil to wind and water, is the major culprit. Cropping decisions based on soil water supplies at seeding, rather than on a fixed rotation, would be a first step towards more efficient utilization of the available water. Simple snow management procedures result in small gains in overwinter soil water storage which in some cases may be sufficient to enable continuous cropping.

Agricultural production in the grassland area could be increased severalfold with irrigation. Several users compete for the limited amount of surface water, and in the southern prairies, lack of water, and possibly economics, will limit further expansion of the irrigated area. Interbasin transfers could make more water available to the drier areas but would be very costly. Because of the limited surface water supplies, all users must improve their efficiency. Underground water supplies are vast, but not necessarily of good quality, and their use should be explored.

References

AGRICULTURE CANADA. 1977. Soils of Canada. Agriculture Canada, Ottawa, Ont.

ALBERTS, E. E., G. E. SHUMAN, AND R. E. BURWELL. 1978. Seasonal runoff losses of nitrogen and phosphorus from Missouri Valley Loess watersheds, J. Environ. Qual. 7: 203–208.

ALBERTS, E. E., AND R. G. SPOMER. 1985. Dissolved nitrogen and phosphorus in runoff from watersheds in conservation and conventional tillage. J. Soil Water Conserv. 40: 153–157.

ANONYMOUS. 1972. Report of the Qu'Appelle Basin Study Board. Queen's Printer, Regina, Sask.

AYERS, H. D. 1964. Effects of agricultural land management on winter runoff in the Guelph, Ontario region p. 167-182. *In* Hydrol. Symp. No. 4. Research Watersheds. National Research Council of Canada, Ottawa, Ont.

BANGA, A. B. 1978. Flood peak potential at Moose Jaw. Rep. No. HYD-5-47. Hydrology Branch, Saskatchewan Department of Environment, Regina, Sask.

BARNES, C. P. 1948. Environment of natural grassland, p. 45-49. *In* A. Stefferud [ed.] Grass, yearbook of agriculture. U.S. Department of Agriculture, Washington, DC.

BARNES, S. 1938. Soil moisture and crop production under dryland conditions in Western Canada. Publ. 595. Agriculture Canada, Ottawa, Ont.

BIRD, R. D. 1961. Ecology of the aspen parkland of Western Canada in relation to land use. Publ. 1066. Agriculture Canada, Ottawa, Ont.

BLACK, A. L. 1968. Nitrogen and phosphorus fertilization for production of crested wheat grass and native grass in northeastern Montana. Agron. J. 60: 213-216.

BLEVINS, R. L., M. S. SMITH, G. W. THOMAS, AND W. W. FRYE. 1983. Influence of conservation tillage on soil properties, J. Soil Water Conserv. 38: 301-305.

BOWREN, K. E. 1977. The effect of reducing tillage on summerfallow when weeds are controlled in the Black soils zone in Saskatchewan, p. 5-11. *In* Proc. 1977 Soil Fertility and Crops Workshop, Saskatoon, Sask. Univ. Sask. Ext. Div. Publ. No. 328.

BRANSON F. A., G. F. GIFFORD, K. G. RENARD, AND R. F. HADLEY. 1981. Rangeland hydrology. Range Sci. Ser. No. 1, Soc. Range Manage., Denver. Kendall/Hunt Publ. Co., Dubuque, IA.

BURWELL, R. E., G. E. SCHUMAN, K. E. SAXTON, AND H. G. HEINEMANN. 1976. Nitrogen in subsurface discharge from agricultural watersheds. J. Environ. Qual. 5: 325-329.

BURWELL, R. E., D. T. TIMMONS, AND R. F. HOLT. 1975. Nutrient transport in surface runoff as influenced by soil cover and seasonal periods. Soil Sci. Soc. Am. Proc. 39: 523-528.

CAMERON, D. R. 1978. Soil water and salt movement in the dryland farming area of southwest Saskatchewan, p. 22-56. *In* Dryland wheat production in southwestern Saskatchewan. Agriculture Canada, Research Station, Swift Current, Sask.

CAMPBELL, C. A., D. W. L. READ, V. O. BIEDERBECK, AND G. E. WINKLEMAN. 1983. The first 12 years of a longterm crop rotation study in southwestern Saskatchewan-nitrate-N distribution in soil and N uptake by the plant. Can J. Soil Sci. 63: 563-578.

CARD, J. R. 1979. Synthesis of streamflow in a prairie environment, p. 11-21. *In* The hydrology of areas of low precipitation. Proc. Canberra Symp. IAHS-AISH Publ. No. 128.

CHACKO, V. T., B. C. S. CHU, AND W. D. GUMMER. 1981. Seasonal variation of nitrogen and phosphorus species in the Red River. Inland Waters Directorate. Sci. Ser. No. 128, Environment Canada, Western and Northern Region, Regina, Sask.

CHAN, G. W., AND M. J. HENDRY. 1985. Impact of irrigation expansion on salinity adjacent to the Monarch branch canal. Phase II. Draft Rep., Irrig. Cons. Div., Alberta Agriculture, Lethbridge, Alta.

COLIGADO, M. C., W. BAIER, AND W. K. SLY. 1968. Risk analyses of weekly climatic data for agricultural and irrigation planning. Tech. Bull. 34 (Brandon) and 51 (Medicine Hat). Agrometeorology Section, Plant Research Institute, Agriculture Canada, Ottawa, Ont.

COUTOURIER, D. E., AND E. A. RIPLEY. 1973. Rainfall interception in mixed grass prairie. Can. J. Plant Sci. 53: 659-663.

DE JONG, E., AND D. R. CAMERON. 1980. Efficiency of water use by agriculture for dryland crop production. Prairie Production Symp., Saskatoon. Canadian Wheat Board, Winnipeg, Man.

DE JONG, E., R. G. KACHANOSKI, AND B. A. RAPP. 1987. Possibilities for snow management in Saskatchewan. *In* Symp. Snow Management for Agriculture, Swift Current, Sask. (In press)

DE JONG, E., AND K. B. MACDONALD. 1975. The soil moisture regime under native grassland. Geoderma 14: 207-221.

DE JONG, E., AND D. A. RENNIE. 1969. Effect of soil profile type and fertilizer on moisture use by wheat grown on fallow or stubble land. Can. J. Soil Sci. 49: 189-197.

DE JONG, E., AND H. STEPPUHN. 1983. Water conservation: Canadian Prairies, p. 89-104. *In* H.E. Dregne and W.O. Willis [ed.] Dryland agriculture. Agronomy 23. American Society of Agronomy, Madison, WI.

DE JONG, R., AND H. N. HAYHOE. 1984. Diffusion-based soil water simulation for native grassland. Agric. Water Manage. 9: 47-60.

DE JONG, R., AND R. P. ZENTNER. 1985. Assessment of the SPAW model for semi-arid growing conditions with minimal local calibration. Agric. Water Manage. 10: 31-46.

DOERING, E. J. 1982. Water and salt movement in prairie soils, p. 55-78. In Proc. 1st Annu. Western Prov. Conf. Rationalization of Water and Soil Res. and Manage., Soil Salinity. Alberta Agriculture, Lethbridge, Alta.

DOERING, E. J., AND F. M. SANDOVAL. 1976. Hydrology of saline seeps in the Northern Great Plains. Trans. ASAE 19: 856-861, 865.

DRAGOUN, F. J. 1969. Effects of cultivation and grass on surface runoff. Water Resour. Res. 5: 1078-1083.

DURRANT, E. F., AND S. R. BLACKWELL. 1959. The magnitude and frequency of floods on the Canadian Prairies, p. 101-156. In Proc. Hydrol. Symp. No. 1. Spillway Design Floods. National Research Council of Canada, Ottawa, Ont.

EMMOND, C. S. 1971. Effect of rotations, tillage treatments and fertilizers on the aggregation of a clay soil. Can. J. Soil Sci. 51: 235-241.

ENVIRONMENT CANADA. 1983. Historical streamflow summary: Alberta, Saskatchewan, Manitoba. Inland Waters Directorate, Environment Canada, Ottawa, Ont.

        1984. Principal station data. PSD/DSP-61 (Medicine Hat), and PSD/DSP-55 (Brandon). Atmospheric Environment Service, Environment Canada, Downsview, Ont.

ERICKSON, D. E. L., W. LIN, AND H. STEPPUHN. 1978. Indices for prairie runoff from snowmelt. Paper presented 7th Symp., Applied Prairie Hydrology. Water Studies Institute, Saskatoon, Sask.

FERGUSON, H., AND T. BATERIDGE. 1982. Salt status of glacial till soils of North-Central Montana as affected by the crop-fallow system of dryland farming. Soil Sci. Soc. Am. J. 46: 807-810.

FERGUSON, W. S. 1963. Effect of the intensity of cropping on the efficiency of water use. Can. J. Soil Sci. 43: 156-165.

FISHERIES AND ENVIRONMENT CANADA. 1977. Surface water quality in Canada — An overview. Inland Waters Directorate, Environment Canada, Ottawa, Ont.

FREEZE, R. A., AND F. BANNER. 1970. The mechanism of natural groundwater recharge and discharge. 2. Laboratory column experiments and field measurements. Water Resour. Res. 6: 138-155.

FREEZE, R. A., AND J. A. CHERRY. 1979. Groundwater. Prentice-Hall Inc., Englewood Cliffs, NJ. 604 p.

FREEZE, R. A., AND P. A. WITHERSPOON. 1967. Theoretical analysis of regional groundwater flow. 2. Effect of water-table configuration and subsurface permeability variation. Water Resour. Res. 3: 623-634.

GELLER, S. Y. 1962. Desalting of water by natural freezing for farm use. Izv. Akad. Nauk., SSR Ser. Geogr. 5: 71-77.

GIFFORD, G. F., AND R. H. HAWKINS. 1978. Hydrologic impact of grazing on infiltration; a critical review. Water Resour. Res. 14: 305-313.

GRAY, D. M., D. I. NORUM, AND R. J. GRANGER. 1984. The prairie soil moisture regime: fall to seeding, p. 159-207. In G. Hass [ed.] The optimum tillage challenge. Div. Ext. and Comm. Relations, University of Saskatchewan, Saskatoon, Sask.

GRAY, D. M., H.W. STEPPUHN, AND F. L. ABBEY. 1979. Estimating the areal snow water equivalent in the Prairie environment, p. 302-332. In Cold climate hydrology. Proc. Canadian Hydrology Symp, Vancouver. National Research Council of Canada, Ottawa, Ont.

GREB, B. W. 1980. Snowfall and its potential management in the semiarid Central Great Plains. Sci. Educ. Admin. ARM-W-18. U.S. Department of Agriculture.

HAAS, H. J., AND W. O. WILLIS. 1971. Water storage and alfalfa production on level benches in the Northern Plains. J. Soil Water Conserv. 26: 151-154.

HALL, P. L., AND E. J. LANGHAM. 1970. Drainage basin study: Progress Report No. 7 and program review. Publ. No. E70-4, Saskatchewan Research Council, Saskatoon, Sask.

HAMMER, U. T. 1978. The saline lakes of Saskatchewan. III. Chemical characterization. Int. Rev. Gesamten Hydrobiol. 63: 311-335.

        1980. Acid rain: the potential for Saskatchewan. Environmental Advisory Council, Province of Saskatchewan, Regina Sask.

        1984. The saline lakes of Canada, p. 521-540. In F. B. Taub [ed.] Lakes and reservoirs. Elsevier Science Publishers, Amsterdam.

HAMMER, U. T., AND R. C. HAYNES. 1978. The saline lakes of Saskatchewan. II. Locale, hydrography and other physical aspects. Int. Rev. Gesamten Hydrobiol. 63: 179-203.

HARBECK, G. E. 1969. Some recent research in evaporation suppression, p. 201-213. In A. H. Laycock [ed.] Water Balance in north America. Proc. Series No. 7, American Water Resources Association, Urbana, IL.

238

HAUPT, H. F. 1967. Infiltration, overland flow, and soil movement on frozen and snow-covered plots. Water Resour. Res. 3: 145-161.

HENDRY, M. J. 1981. Surface return flows under irrigated lands in the Bow River Irrigation District, Vol. 1. Irrig. Div., Alberta Agriculture.

HENDRY, M. J., D. GLOVER-PIKE, AND R. G. BURNETT. 1982. Surface return flows under irrigated lands north of Taber, Alberta, Vol. 1. Irrig. Div., Alberta Agriculture.

HENRY, J. L., P.R. BULLOCK, T. J. HOGG, AND L. D. LUBA. 1985. Groundwater discharge from glacial and bedrock aquifers as a soil salinization factor in Saskatchewan. Can. J. Soil Sci. 65: 749-768.

HOBBS, E. H., AND K. K. KROGMAN. 1971. Overwinter precipitation storage in irrigated and non-irrigated Chin loam soil. Can. J. Soil Sci. 51: 13-18.

HORNER, W. H., J. A. BROWN, J. C. GILSON, G. LEE, F. V. MACHARDY, H. D. MCRORIE, AND C. L. SIBBALD. 1980. Western Canadian agriculture to 1990. Canada West Foundation, Calgary, Alta.

HOYT, P. G., AND W. A. RICE. 1977. Effects of high rates of chemical fertilizer and barnyard manure on yield and moisture use of six successive barley crops grown on three Gray Luvisolic soils. Can. J. Soil Sci. 57: 425-435.

HUDSON, H. R. 1983. Erosional processes and sediment yield of a typical eastern slopes drainage basin in southwest Alberta, p. 81-96. In Soil erosion and land degradation, Proc. 2nd Annu. Western Prov. Conf. on Rationalization of Soil Res. Manage., Sask. Institute of Pedology, Saskatoon, Sask.

HUSBAND, W. H. W., AND J. W. SPYKER. 1978. Using the prairie climate to produce desalinated water for small communities by spray freezing, p. 89-95. In J. Whiting[ed.] Proc. 7th Symp., Applied Prairie Hydrology, Water Studies Institute, Saskatoon, Sask.

JOHNSTON, R. S. 1970. Evapotranspiration from bare, herbaceous, and aspen plots: a check on a former study. Water Resour. Res. 6: 324-327.

KACHANOSKI, R. G., E. DE JONG, AND D. A. RENNIE. 1985. The effect of fall stubble management on over-winter recharge and grain yield. p. 254-259. In Proc. Soils and Crops Workshop, Div. Ext. and Comm. Relations, University of Saskatoon, Sask.

KROGMAN, K. K. 1973. Cropping to control groundwater, p. 61-66. In Proc. Alta. Dryland Salinity Workshop, Plant Industry Division, Alberta Agriculture, Lethbridge, Alta.

LAKSHMAN, G. 1971. The water balance of shallow permanent of intermittent natural reservoirs as it affects ecnomy and wildlife. Prog. Rep. no. 6, Saskatchewan Research Council, Saskatoon, Sask. 80 p.

LANE, R. K., AND G. N. SYKES. 1982. Nature's lifelines: Prairie and northern waters. Canada West Foundation, Calgary, Alta. 467 p.

LAYCOCK, A. H. 1965. Geographic aspects of water and climate, p. 1-20. In Water and climate, a symposium. Water Studies Institute, Saskatoon, Sask.

LINDSTROM, M. J., AND C. A. ONSTAD. 1984. Influence of tillage systems on soil physical parameters and infiltration after planting. J. Soil Water Conserv. 39: 149-152.

LINDWALL, C. W., AND D. T. ANDERSON. 1981. Agronomic evaluation of minimum tillage systems for summerfallow in southern Alberta. Can. J. Plant Sci. 61: 247-253.

LISSEY, A. 1968. Surficial mapping of groundwater flow systems, with application to the Oak River Basin of Manitoba. Ph.D. thesis, University of Saskatchewan, Saskatoon, Sask. 141 p.

LOUIE, P. Y. T. 1977. Potential evaporative loss from snow in south-western Alberta. Proc. Can. Hydrol. Symp. 77: 25-32. National Research Council of Canada, Ottawa, Ont.

LUDDEN, A. P., D. L. FRINK, AND D. H. JOHNSON. 1983. Water storage capacity of natural wetland depressions in the Devil's Lake Basin of North Dakota. J. Soil Water Conserv. 38: 45-48.

MACLEOD, D. L. 1977. Drought in a prairie environment. Proc. Can. Hydrol. Symp. 77: 335-340. National Research Council of Canada, Ottawa, Ont.

MALE, D. H., AND D. M. GRAY. 1981. Snowcover ablation and runoff, p. 360-436. In D. M. Gray and D. H. Male [ed.] Handbook of snow: principles, processes, management and use. Pergamon Press Inc., Elmsford, NY.

MARTZ, L. W., AND E. DE JONG. 1985. The relationship between land surface morphology and soil erosion and deposition in a small Saskatchewan basin. Can. Soc. Civ. Eng. Annu. Conf., Saskatoon, Sask.

MEYBOOM, P. 1966. Unsteady groundwater flow near a willow ring in hummocky terrain. J. Hydrol. 4: 38-62.

1967. Estimates of groundwater recharge in the prairies, p. 128-153. In C. E. Dolman [ed.] Water resources of Canada, R. Soc. Can., Inland Waters Branch, Reprint Ser. No. 4, Department of Energy, Mines and Resources, Ottawa, Ont.

MICHALYNA, W., AND R. A. HEDLIN. 1961. A study of moisture storage and nitrate accumulation in soil as related to wheat yields on four cropping sequences. Can. J. Soil Sci. 41: 5-15.

MILLER, M. R., P. I. BROWN, I. J. DONOVAN, R. N. BERGATINO, J. L. SONDEREGGER, AND F. A. SCHMIDT. 1981. Saline seep development and control in the North American Great Plains: hydrogeologic aspects. Agric. Water Manage, 4: 115-141.

MOWCHENKO, M. 1976. Median annual unit runoff for the prairie provinces, p. 137-142. In J. Whiting [ed.] Proc. 7th Symp., Applied Prairie Hydrology. Water Studies Institute, Saskatoon, Sask.

NEFF, E. L. 1980. Snowtrapping by contour furrows in southeastern Montana. J. Range Manage. 33: 221-223.

NICHOLAICHUK, W. 1967. Comparative watershed studies in southern Saskatchewan. Trans. ASAE 10: 502-504.

    1980. Snow management to provide additional water for agriculture. Prairie Production Symp., Saskatoon. Canadian Wheat Board, Winnipeg, Man.

    1984. Tillage and snow management practices for the conservation of moisture, p. 75-90. In G. E. Hass [ed.] The optimum tillage challenge. Ext. Div. and Comm. Relations, University of Saskatchewan, Saskatoon, Sask.

NICHOLAICHUK, W., AND D. W. L. READ. 1978. Nutrient runoff from fertilized and unfertilized fields in western Canada. J. Environ. Qual. 7: 542-544.

OOSTERVELD, M., AND W. NICHOLAICHUK. 1983. Water requirements, availability and development restraint for increased crop productivity in Canada. Can. J. Plant Sci. 63: 33-44.

PFRA. 1985. Addendum No. 1 to Hydrology Rep. No. 104: The determination of gross and effective drainage areas for the Prairie provinces. Prairie Farm Rehabilitation Administration, Hydrology Division, Regina, Sask.

POHJAKAS, K., D. W. L. READ,, AND H. C. KORVEN. 1967. Consumptive use of water by crops at Swift Current. Can. J. Soil Sci. 47: 131-138.

POST, F. A., AND F. R. DREIBELBIS. 1942. Some influences of frost penetration and microclimate on the water relationships of woodland, pasture and cultivated soils. Soil Sci. Soc. Am. Proc. 7: 95-104.

POWER, J. F. 1970. Leaching of nitrate-nitrogen under dryland agriculture in the northern Great Plains, p. 111-122. In Relationship of agriculture to soil and water pollution. Cornell University Press, Ithaca, NY.

    1980. Response of semiarid grassland sites to nitrogen fertilization. I. Plant growth and water use. Soil Sci. Soc. Am. J. 44: 545-550.

PPWB. 1982. Water demand study: Historical and current water uses in the Saskatchewan-Nelson basin. Prairie Provinces Water Board, Regina, Sask. 147 p.

RAPP, E., J. C. VAN SCHAIK, AND N. N. KHAVAL. 1969. A hydrologic budget for a southern Alberta irrigation district. Can. Agric. Eng. 11: 54-57.

REEVES, H. C. 1977. Use of prescribed fire in land management. J. Soil Water Conserv. 32: 102-104.

REHM, B. W., S. R. MORAN, AND G. H. GROENEWOLD. 1982. Natural groundwater recharge in an upland area of central North Dakota, U.S.A. J. Hydrol. 59: 293-314.

RENNIE, D. A. 1983. Fate and impact of potash dust on the soil environment in Saskatchewan, p. 841-846. In R. M. McKercher [ed.] Potash '83, Intern. Potassium Technology Conf., Saskatoon, Pergamon Press, Toronto, Ont.

REY, T. W. 1983. Concerns associated with increased utilization of groundwater. In Proc. Annu. Meet. Can. Water Resour. Assoc., Applied Hydrology, Saskatoon, Sask.

RUBEC, C. D. A., AND P. C. RUMP. 1984. Prime wetland use monitoring activities and needs in southern Canada, p. 405-429. In Proc. 3rd Annu. Western Prov. Conf. on Rationalization of Water and Soil Research and Management, Agricultural Drainage, Winnipeg, Man.

SANDOVAL, F. M., AND L. C. BENZ. 1973. Soil salinity reduced by summerfallow and crop residues. Soil Sci. 116: 100-105.

SANDOVAL, F. M., C. W. CARLSON, R. H. MICKLESON, AND L. C. BENZ. 1961. Effects of run-off prevention and leaching on a saline soil. Can. J. Soil Sci. 41: 207-217.

SHIAU, S-Y. 1978. Watershed modelling in a subhumid prairie parkland basin — Good Spirit Lake Basin, Saskatchewan, p. 121-129. In J. Whiting [ed.] Proc. 7th Symp., Applied Prairie Hydrology, Water Studies Institute, Saskatoon, Sask.

SHJEFLO, J. M. 1968. Evapotranspiration and the water budget of prairie potholes in North Dakota. Geol. Surv. Prof. Pap. 585-B.

SIEMENS, L. B. [Chairman] 1977. Soils, p. 17-60. In Principles and practices of commercial farming. University of Manitoba, Winnipeg, Man.

SIMPSON-LEWIS, W., J. E. MOORE, N. J. POCOCK, M. C. TAYLOR, AND H. SWAN. 1979. Canada's special resources lands: a national perspective of selected land uses. Land Use Directorate, Environment Canada, Ottawa, Ont.

SKIDMORE, E. L., W. A. CARSTENSON, AND E. E. BANBURY. 1975. Soil changes resulting from cropping. Soil Sci. Soc. Am. Proc. 39: 964-967.

SMIKA, D. E., H. J. HAAS, AND J. F. POWER. 1965. Effects of moisture and nitrogen fertilizer on growth and water use by native grass. Agron. J. 57: 483-486.

SOMMERFELDT, T.G., AND A. D. SMITH. 1973. Movement of nitrate nitrogen in some grassland soils of southern Alberta. J. Environ. qual. 2: 112-115.

SONMOR, L. G. 1963. Seasonal consumptive use of water by crops in southern Alberta and its relationships to evaporation. Can. J. Soil Sci. 43: 287-297.

STAPLE, W. J., AND J. J. LEHANE. 1952. The conservation of soil moisture in southern Saskatchewan. Sci. Agric. 32: 36-47.

STEPPUHN, H. 1981. Snow and agriculture, p. 60-125. In D. M. Gray and D. H. Male [ed.] Handbook of snow: principles, processes, management and use. Pergamon Press Inc., Elmsford, NY.

STICHLING, W. 1973. Sediment loads in Canadian rivers, p. 38-73. In Proc. Hydrol. Symp. No. 9, Fluvial Processes and Sedimentation. National Research Council of Canada, Ottawa, Ont.

STOLTE, W. J. AND R. HERRINGTON. 1984. Changes in the hydrologic regime of the Battle River Basin, Alberta, Canada, J. Hydrol. 71: 285-301.

STRIFFLER, W. D. 1969. The grassland hydrologic cycle, p. 101-116. In R. L. Dix and R. G. Beidleman [ed.] The grassland ecosystem: a preliminary synthesis. Range Sci. Dept. Sci. Ser. No. 2. Colorado State University, Fort Collins, CO.

TABLER, R. A. 1975. Estimating the transport and evaporation of blowing snow, p. 85-104. In Symp. Snow Management on Great Plains, Great Plains Agriculture Council, Lincoln, NE.

THIES, D. 1985. Drainage project investigations and constructed works. Paper Moose Jaw River Watershed Manage., Saskatchewan Water Corp., Regina, Sask.

TIMMONS, D. R., AND R. F. HOLT. 1977. Nutrient losses in surface runoff from a native prairie. J. Environ. Qual. 6: 369-373.

TOOGOOD, J. A. 1963. Water erosion in Alberta. J. Soil Water Conserv. 18: 238-240.

VAN SCHAIK, J. C., D. S. CHANASYK, AND E. H. HOBBS. 1976. An assessment of the capability of a computer program to estimate fall soil moisture and deep percolation. Can. J. Soil Sci. 56: 357-362.

VERRY, E. S. 1983. Precipitation chemistry at the Marcell experimental forest in north-central Minnesota. Water Resour. Res. 19: 454-462.

WARDER, F. G. , J. J. LEHANE, W. C. HINMAN, AND W. J. STAPLE, 1963. The effect of fertilizer on growth, nutrient uptake and moisture use of wheat on two soils in southwestern Saskatchewan. Can J. Soil Sci. 43: 107-116.

WHITE, E. M., AND E. J. WILLIAMSON. 1973. Plant nutrient concentrations in runoff from fertilized cultivated erosion plots and prairie in eastern South Dakota. J. Environ. Qual. 2: 453-455.

WIGHT, J. R. 1976. Land surface modifications and their effects on range and forest watersheds, p. 165-174. In H. F. Heady et al. [ed.] Watershed management on range and forest lands. Utah Water Res. Lab, Utah State University Logan, UT.

WIGHT, J. R., E. L. NEFF, AND F. H. SIDDOWAY. 1975. Snow management on eastern Montana rangelands, p. 138-143. In Symp. Snow Management on the Great Plains, Great Plains Agriculture Council, Lincoln, NE.

CHAPTER 9

The Importance of Permafrost in the Hydrological Regime

Robert O. van Everdingen

Arctic Institute of North America, The University of Calgary, Calgary, Alta. T2N 1N4

Introduction

Permafrost

Permafrost is defined as ground (soil or rock) with temperatures remaining continuously below 0°C for 2 or more yr. Permafrost occurs in areas with mean annual air temperatures below 0°C and is generally overlain by a layer of ground of variable thickness, subject to seasonal freezing and thawing, known as the active layer (Fig. 1).

This chapter explores the influence of permafrost on the various terrestrial components of the hydrological regime. It includes a discussion of the effects of natural and man-induced changes in the extent of permafrost, and some speculation regarding the potential hydrological consequences of widespread changes in the extent of permafrost that might result from changes in climate (see also chapter by Ripley, this volume).

Distribution of Permafrost

Permafrost underlies about half the land area of Canada (Fig. 2) and also parts of the shallow offshore in the Canadian Arctic. It is discontinuous in the southern portion of the permafrost region and may occur as thin, isolated patches near the southern limit of this zone of discontinuous permafrost. As a consequence of decreasing mean annual air temperatures, permafrost gradually becomes thicker and more continuous at higher latitudes, but nonpermafrost "windows" can be found even in the zone of continuous permafrost.

The local distribution of permafrost is controlled by a number of factors, other than the climate, which influence the exchange of energy (heat) between the air and the ground. Relief and slope aspect affect the amount of incoming solar radiation; winter snow cover will provide insulation, but it can also reflect much of the solar radiation; vegetation cover can provide significant insulation, especially in dry peatlands (Brown 1970, p. 21). Bodies of water have a moderating effect on ground temperature variations because of their heat content and because of the latent heat of fusion of the water. Locally, moving groundwater can be a significant source of convective heat.

In the discontinuous zone, permafrost is most likely to be present under north-facing slopes with a thin winter snow cover. Permafrost is usually thinner and may even be absent under south-facing slopes with abundant vegetation and thick winter snow cover and beneath deep lakes and large rivers.

Hydraulic Properties of Frozen Ground

It is often assumed that frozen ground contains water only in the form of ice, that frozen ground will not allow the movement of water (i.e. that it is impermeable), and that permafrost is perennially frozen and, thus, impermeable. These assumptions are not always valid.

Frozen ground often contains a significant proportion of unfrozen water, even when the temperature is well below the ice-nucleation temperature of the ground (Fig. 3). This results from progressive lowering of the freezing point by the gradually increasing dissolved-solids content of the pore water which, in turn, is caused by the exclusion

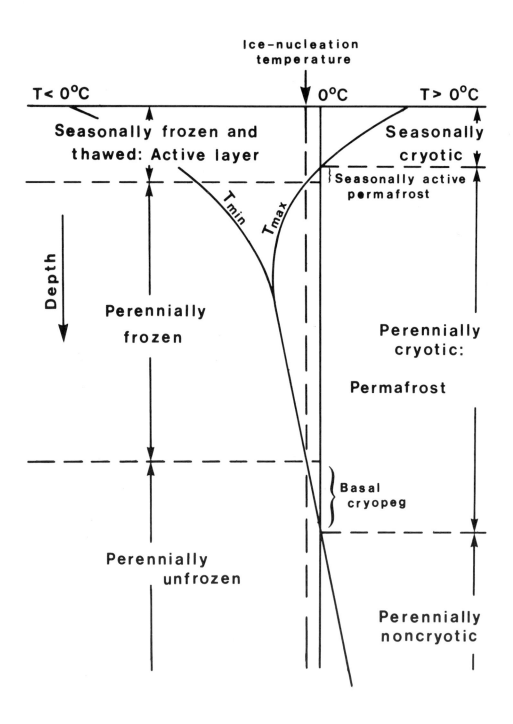

FIG. 1. Schematic representation of the variation of temperature with depth in a permafrost area, showing terms used to indicate the seasonal or perennial temperature relative to 0°C and terms indicating seasonal or perennial presence (or absence) of ice (modified after van Everdingen 1985, fig. 2).

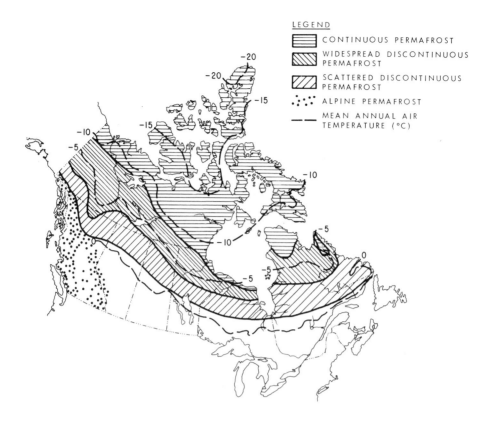

FIG. 2. Distribution of permafrost in Canada (from Johnston 1981, fig. 2.1).

of dissolved solids from the ice phase during freezing. In fine-grained sediments (with a high silt or clay content), some of the water adsorbed as thin films on the surfaces of the mineral particles can remain unfrozen even at temperatures below $-50°C$. The unfrozen water content of fine-grained frozen sediments is directly related to the temperature (Fig. 4), and it is nearly independent of the total water content (unfrozen water plus ice), except under very dry conditions.

Frozen ground is usually not completely impermeable. It is true, of course, that the formation of ice in pore spaces and fractures, and in the form of ice lenses, will significantly reduce the hydraulic conductivity of the ground. Freezing of clean, coarse-grained sediments and fractured bedrock may reduce their hydraulic conductivities to a small fraction of the original (unfrozen) values. Freezing of fine-grained sediment with a high clay content, on the other hand, may result in only a minor reduction in the hydraulic conductivity, especially if the temperature falls only slightly below $0°C$. The development of cracks, caused either by drying or by thermal contraction during periods of extremely cold weather, can greatly increase the bulk hydraulic conductivity of frozen ground.

Permafrost is not necessarily frozen at all. The temperature at which ice can start forming in the ground (the ice-nucleation temperature) may be depressed several degrees below $0°C$ (and below the actual temperature of the ground) by, for example, a high concentration of dissolved salts. Permafrost will affect the hydrological regime only if it is perennially frozen and if its hydraulic conductivity is much lower than it would be under unfrozen conditions. In that case the presence of permafrost will have perennial effects on the hydrological regime similar to the intermittent effects of seasonal freezing of the ground in areas without permafrost.

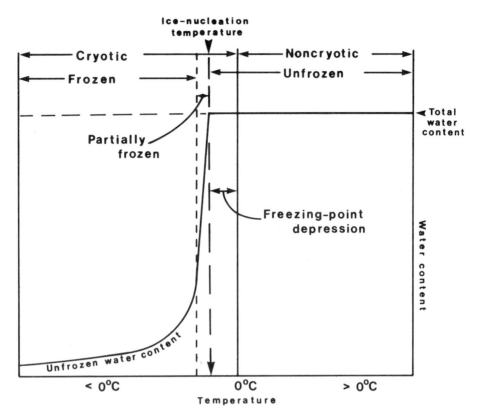

FIG. 3. Hypothetical graph of unfrozen water content vs. ground temperature, showing the relationship between terms indicating temperature relative to 0°C and those indicating phase conditions (modified after van Everdingen 1976, fig. 1).

Climate

The extreme climate of the permafrost region in Canada has several direct effects on the hydrological regime that may be just as significant as any of the effects related to the presence of permafrost.

Mean annual precipitation is relatively low over much of the region. It ranges from 500 to 600 mm • yr^{-1} locally in the zone of discontinuous permafrost to less than 100 mm • yr^{-1} over parts of the Arctic Islands. The low precipitation limits the amount of water available for either surface runoff or groundwater recharge.

Mean annual air temperatures as low as -25°C also impose severe limitations on the hydrological regime in the permafrost region. Air temperatures below 0°C may keep rivers, lakes, and the active layer frozen for up to 10 mo • yr^{-1}. Therefore, surface runoff and infiltration of snowmelt or rain to recharge groundwater aquifers may be possible for only 2 mo each year.

As indicated earlier, freezing affects different geological materials in different ways, leading to a great diversity of soil/water/ice conditions that are all covered by the terms permafrost and seasonal frost. The northern hydrologist is therefore generally interested not just in the distribution of permafrost, but in the complex distribution of both seasonal and perennial frozen conditions and their effects on the hydraulic conductivity of the ground.

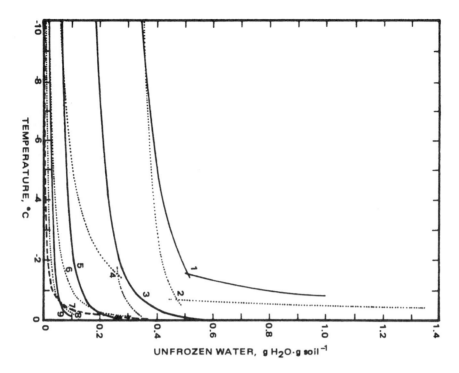

FIG. 4. Unfrozen water content vs. temperature for several soils and soil constituents (modified after Anderson and Morgenstern 1973, fig. 6 and 7). 1, Umiat bentonite; 2, Wyoming bentonite; 3, Hawaiian clay; 4, Kaolinite; 5, Suffield clay; 6, Dow Field silty clay; 7, Basalt; 8, E. Lebanon gravel < 100 μm; 9, Fairbanks silt.

Influence of Permafrost on Hydrological Processes

Availability of Data

Although the permafrost region includes approximately half the land area of Canada, it supports only a small fraction of the population. Compared with the rest of the country, economic development in the region is very limited and localized, and the need for hydrological data has therefore also been limited. During the past few years, efforts have been made to expand the northern networks for such observations as precipitation and surface runoff in advance of proposed or anticipated engineering developments such as mines, highways, and pipelines, which could affect, and could in turn be affected by, local hydrological processes. The number of weather stations north of 60°N increased from 84 in 1978 to 99 in 1983, while the number of prmanent stream-gauging stations increased from 159 to 204. These numbers are still small, however, when compared with those for the provinces, and the stations are distributed unevenly across the region.

A variety of research projects have been undertaken during the last 15 yr in connection with such planned and proposed developments as the Mackenzie and Dempster highways; the Mackenzie Valley, Alaska Highway, and Arctic Islands pipelines; and the Churchill/Nelson and James Bay hydropower developments. Many of those studies have not only lead to improved understanding of hydrological processes, but have also made valuable contributions to the available data base.

247

Precipitation

The quantity of precipitation falling on northern drainage basins is commonly underestimated for two reasons (see also chapter by Laycock, this volume). First, most precipitation gauges suffer from some degree of "undercatch" with respect to snowfall, especially under windy conditions. This shortcoming is particularly serious in the permafrost region, where snowfall generally constitutes between 40 and 80% of the total annual precipitation. Elimination of this shortcoming requires further improvements in gauge design. Second, weather stations in mountainous areas are almost without exception located in valleys, whereas precipitation is usually more frequent, heavier, and of longer duration at higher elevations. Significant improvements in this respect would require installation and operation of an increasing number of automatic weather stations at higher elevations. Such stations would have to incorporate special design features to enable them to operate reliably under extreme climatic conditions (low temperatures, high wind velocities, blowing snow, etc.).

An alternative to the direct measurement of snowfall is the determination of the depth and water content of snow cover on the ground in so-called snow courses. Again, most snow courses in mountainous areas are located in valleys, and the data they provide may therefore not be representative of the snowfall or snow cover on the surrounding area. In the Yukon, for example, the highest snow course lies at an elevation of 1465 m, along a tributary of the Duke River, where elevations exceed 2500 m within a radius of 5 km from the course.

Recent developments in the use of airborne radar and gamma-radiation surveys suggest that these may provide improved methods for the determination of the water content of snow cover over larger areas, at least in nonmountainous northern regions.

Evaporation

Available data (Hydrological Atlas of Canada 1978) indicate that direct evaporation can account for 25 to more than 50% of the total precipitation over a large portion of the permafrost region. In the High Arctic, total annual evaporation may account for as much as 80% of the total annual precipitation (Ohmura 1982). During the summer, the percentages and actual amounts are highest in areas of low relief where numerous lakes and ponds exist as a result of poor drainage due to the presence of perennially frozen ground.

Transpiration losses from vegetation tend to decrease with increasing latitude because of the decreasing density of the vegetation cover and the increasing proportion of mosses and lichens, which do not transpire. Little quantitative information is available on these losses. Loss of water from snow cover through sublimation appears to be small, but it may represent a significant fraction of the total snowfall in some areas.

Surface Runoff

Permanent stream-gauging stations in the north have traditionally been located on the larger rivers because those carry the bulk of the surface runoff and have the greatest potential for large-scale flooding. However, information on the frequency, magnitude, timing, and duration of extreme runoff events is not only needed for large rivers, but also for smaller streams with drainage areas of 500 km^2 or less, primarily for the proper design of culverts, bridges, streambank protection, and buried pipeline crossings. A recently added requirement is that of limiting the duration of extreme velocities in culverts on small streams through which fish must migrate.

Where streamflow data are lacking, regional flood-frequency analysis may provide relatively reliable estimates. Use of this method requires the transfer of information from large, gauged drainage basins to small, ungauged ones. This will often lead to underestimates of peak flows because the ratio between peak discharge rate and drainage area increases rapidly with decreasing basin size, particularly in mountainous areas.

Temporary gauging stations are now often established on smaller streams that will be affected by, for instance, highway construction. They are usually operated for periods of less than 5 yr, and only during the ice-free season. They are increasingly being replaced by permanent gauging stations.

Groundwater

Before 1970, the only information available on the occurrence of groundwater in the permafrost region in Canada consisted of observations of a few springs and data on a small number of widely scattered and commonly shallow wells (Beschel 1963; Brandon 1965). The Hydrological Atlas of Canada (1978) indicated the existence of only one groundwater observation well in the permafrost region, located in northern Alberta.

Recent studies related to proposed highway and pipeline projects (Hughes et al. 1973; Michel 1977; Rutter et al. 1973; van Everdingen 1974, 1975, 1981, 1982a; 1982b; van Everdingen et al. 1979) and an increasing number of water wells (Intera Environmental Consultants Ltd. 1975; Trimble et al. 1983) have greatly expanded the available data base. Additional information has also become available from mining operations that require mine-dewatering and/or surface disposal of mine and mill tailings.

Surface Runoff in Permafrost Areas

Snowmelt versus Rainfall Runoff

The chief contributions to surface runoff come from snowmelt and rainfall, primarily by direct surface flow, and also by "interflow" in the active layer once thawing has started. At higher latitudes an increasing percentage of the annual precipitation falls as snow, and the proportion of surface runoff contributed by snowmelt increases. In the High Arctic, snowmelt provides almost all of the annual runoff. The infrequent rainfall in this region is usually insufficient to generate any runoff, except immediately after the snowmelt period when the ground may still be saturated with water (Woo 1983).

In early spring, when the ground is still frozen, all of the snowmelt may become runoff. If the frozen ground contains little ice, and especially if it is fractured, a portion of the meltwater may also infiltrate and replenish soil moisture or recharge groundwater (Kane and Stein 1983). In many instances, however, ice lenses and layers forming at the base of the snowpack impede infiltration (Woo et al. 1983). Toward the end of the snowmelt period, when most of the ground is exposed, meltwater from the snow remaining in drifts and sheltered locations can increasingly infiltrate into the thawing active layer. The prediction of snowmelt-runoff patterns is extremely difficult in northern basins due to the great variability of snow cover and ground conditions.

In lower latitudes in the permafrost region, rainfall may generate more than half the annual surface runoff. Runoff-to-rainfall ratios for individual rainstorms generally show similar magnitudes and degree of variation as outside the permafrost region (see chapters by Hetherington and de Jong and Kachanoski, this volume). Nevertheless, the highest runoff yields have been observed in basins underlain by permafrost (Newbury 1974). The lower precipitation on more northern basins appears to be offset by lower evaporation losses during the shorter open-water season and by higher runoff percentages related to the presence of perennially frozen ground (Newbury et al. 1979).

Peak Flows

Annual peak flows in northern rivers may be caused by snowmelt, by rainfall, or by a combination of the two. In the southern part of the permafrost region and in the western mountain areas the annual peak flows are commonly caused by rainfall. Farther north, peak flows may be generated by rainfall one year and by snowmelt the next, depending on the amount of snow on the ground at the end of the winter and on the speed of melting. In the Mackenzie Delta region and on the Arctic Islands, annual peak flows are almost without exception generated by snowmelt. In the perma-frost region, as elsewhere, smaller basins tend to produce higher peak flows (up to $1.5 \ m^3 \cdot s^{-1} \cdot km^{-2}$) from rainfall, with relatively short response times between the start of precipitation and the occurrence of the associated peak runoff. In areas of low Arctic tundra, the water-retaining capacity of thick sod mats may slow the response, lowering and broadening the flood peak.

Water levels during peak-flow events can be dramatically increased by downstream ice jams, and also by icings that may cover river channels, and in some cases parts of the flood plain, with more than 2 m of ice (Fig. 5).

Baseflow

The baseflow (or low flow during the winter) in northern rivers that drain basins without significant lake storage decreases with increasing latitude and increasing extent of per-mafrost. Minimum mean monthly discharge rates range from about 0.002 to about $0.005 \ m^3 \cdot s^{-1} \cdot km^{-2}$ in the zone of discontinuous permafrost and approach zero in the zone of continuous permafrost; values exceeding $0.005 \ m^3 \cdot s^{-1} \cdot km^{-2}$ usually reflect discharge from large lakes (Williams and van Everdingen 1973).

The decrease in baseflow rates with increasing latitude is caused by the gradual decrease in the rates of groundwater discharge. However, a zero value for the baseflow at a particular gauging station does not always mean that there is no groundwater dis-charge in the basin. Examples of this are provided by Babbage River and Firth River in the Yukon, both of which have no streamflow at their respective gauging stations for 5.5–6 mo each year. During the winter, spring-fed discharge rates of 1.4, 0.9, and $0.5 \ m^3 \cdot s^{-1}$ have been measured in the upper Babbage River, in the upper Firth River, and in Joe Creek, a tributary of Firth River, respectively (van Everdingen 1974).

Baseflow rates are of interest because they indicate the magnitude of groundwater flow in drainage basins where no subsurface information is available from wells. More-

FIG. 5. Icing on the flood plain of Crow River, Yukon, showing layering of ice and development of a thermo-erosional niche (27 June 1973).

over, the open-water areas maintained in northern rivers during the winter by ground-water discharge play an important role in northern ecology because they provide over-wintering and/or spawning areas for a number of fish species (Bryan 1973).

Icings

During the winter, much of the groundwater discharged into northern rivers is stored as icings only a short distance (0.5-1 km) from the point of discharge. Some major exceptions exist, for instance in the South Fishing Branch in the basin of the Porcupine River, where perennial spring discharge of about 11 m^3 • s^{-1} maintains more than 30 km of open water during most winters (van Everdingen 1974).

Icings form not only in the permafrost region, but also in many streams farther south, especially in the mountains. The formation and subsequent melting of icings cause seasonal redistribution of surface runoff. For instance, the icings developed over a period of about 6 mo in the valleys of upper Babbage River, upper Firth River, and Joe Creek take about 3 mo to melt, increasing runoff in these streams by 1.0-2.8 m^3 • s^{-1} during the late spring and summer. In smaller streams in the zone of discontinuous permafrost, runoff from melting icings may be up to three times as high as the direct contribution from groundwater discharge.

Icings can also modify the dissolved-solids content of the streamwater because most of the dissolved salts in the discharging groundwater are precipitated during freezing. If those salts consist of poorly soluble calcium carbonate and calcium sulfate, only a fraction will redissolve in the meltwater from the icing. The remainder stays behind in the icing area untill it is washed away in suspension during the next period of high water (van Everdingen 1974).

The presence of icings usually reduces the carrying capacity of affected stream channels. This promotes lateral erosion during snowmelt runoff and may lead to the formation of a widened flood plain with braided stream channels.

Lakes

Numerous small lakes and ponds, as well as extensive bogs and marshes, are found in many areas in the permafrost region because the slow thawing of the active layer and the presence of perennially frozen ground tend to maintain a high water table, particularly in areas of low relief.

Lakes in the Mackenzie Delta area, initially developed on the outer portions of the growing delta, are maintained by low levees. Many delta lakes remain connected to active channels; drainage directions, into or out of such lakes, vary with the water levels in the delta channels. The low levees separating other lakes from the active channels are occasionally breeched during extreme high water. The presence of the lakes has a retarding influence on the development of permafrost in the underlying delta sediments (Mackay 1963).

In karst areas with sinkholes and underground drainage channels developed by dissolution of soluble rock strata (limestone, gypsum, or rock salt), temporary lakes and ponds form every year in some of the sinkhole depressions (Fig. 6). These lakes usually fill up with water within a few days during the snowmelt period, when the rate of input of surface runoff exceeds the capacity of the underground drainage system. Complete draining of the lakes may take until late September, especially if heavy rainfall occurs during the summer (van Everdingen 1981; Brook 1983).

The zonation of vegetation around the edges of larger, seasonally flooded karst depressions (Fig. 6) reflects, on the one hand, the relative duration of flooding (zones of sedges and grasses) and, on the other hand, the gradually decreasing thickness of the unfrozen layer (zones of willows and white spruce). This vegetation sequence forms a strong contrast with the black spruce and lichen forest, and the black spruce and sedge-tussock forest on the surrounding permafrost.

FIG. 6. Large sinkhole depression in the area of collapse karst west of Smith Arm, Great Bear Lake, N.W.T. (a) Drained depression, showing bedrock exposed at individual sinkholes (10 September 1975); (b) flooded depression (7 August 1976). Note contrast between vegetation zones along the edges of the depression and vegetation on the surrounding permafrost.

Groundwater Flow in Permafrost Areas

The occurrence of groundwater and the direction and rates of groundwater flow are in general controlled by the same physical parameters in areas with permafrost as in nonpermafrost areas (see chapter by Laycock, this volume). The seasonal and areal distribution of precipitation determines the amount of water available for groundwater recharge. Topography (or relief) provides the potential for groundwater movement under the influence of gravity. The distribution of rock types and structures in the subsurface determines the distribution of hydraulic conductivity and storage capacity, as well as locally preferred flow directions. In addition, groundwater regimes in the permafrost region are directly affected by decreased hydraulic conductivities wherever the ground is perennially frozen.

Recharge

The recharge of groundwater aquifers will be least affected by the presence of permafrost in areas where bedrock exposures present relatively large conduits for the inflow of water. Such conduits may consist of interconnected fracture and joint systems in sedimentary or volcanic rocks or they may be solution or collapse sinkholes formed as a result of underground dissolution of limestone, dolomite, gypsum, or salt beds. In areas covered with unconsolidated sediments (gravel, sand, silt, and clay), the influence of permafrost on groundwater recharge will vary widely with the grain size distribution and the unfrozen water content of the sediments. Unfrozen zones (taliks) that exist below many streams and lakes may penetrate the permafrost completely and present opportunities for the recharge of groundwater aquifers at greater depth.

The concentration of recharge in smaller areas and fewer points, as a result of the presence of permafrost, can lead to high local recharge rates. For instance, inflow into individual sinkholes in permafrost west of Great Bear Lake was found to exceed $1.0 \ \text{m}^3 \cdot \text{s}^{-1}$ during the snowmelt period (van Everdingen 1981).

Lateral Movement

The distribution of groundwater flow, from its point of recharge to the point where it is discharged, is strongly influenced by the presence of perennially frozen ground. In general, three types of aquifers, with different flow regimes, can be distinguished on the basis of their position relative to the permafrost (Fig. 7).

Suprapermafrost aquifers are found in the active layer and in closed taliks below many rivers and lakes. Significant groundwater flow can take place in the active layer only when it is thawed, whereas the flow in closed river taliks is often year-round. Unfrozen zones found along the base of some large alluvial fans (lateral taliks), and unfrozen zones with mineralized water in the lower portion of the permafrost (basal cryopegs), form intrapermafrost aquifers for the lateral movement of groundwater. Open taliks below large rivers and lakes, and the taliks associated with thermal and saline springs, form intrapermafrost conduits that carry recharge to, and discharge from, subpermafrost aquifers existing in the unfrozen ground below the permafrost.

Discharge

The concentration of discharge by the presence of perennially frozen ground is responsible for the high rates of discharge from individual springs that were mentioned earlier.

The discharge from subpermafrost aquifers is generally perennial. Discharge from lateral intrapermafrost taliks may stop before the end of winter, if decreasing flow rates from limited reservoirs allow the outlets to freeze. Seeps and springs discharging suprapermafrost water from the active layer generally do so only during the frost-free season, drying up shortly after the onset of winter.

The occurrence of groundwater discharge can be deduced from several natural phenomena other than springs and seeps. North of the tree line, well-developed willows and poplars are found almost exclusively in areas of groundwater discharge (Fig. 8). Winter baseflow in rivers, the presence of open-water areas in ice-covered rivers, and the occurrence of icings commonly indicate that a significant quantity of groundwater is being discharged upstream. Although such discharge phenomena can be used to obtain quantitative information on the amount of groundwater flow in various river basins, few attempts of this kind have been made in Canada to date.

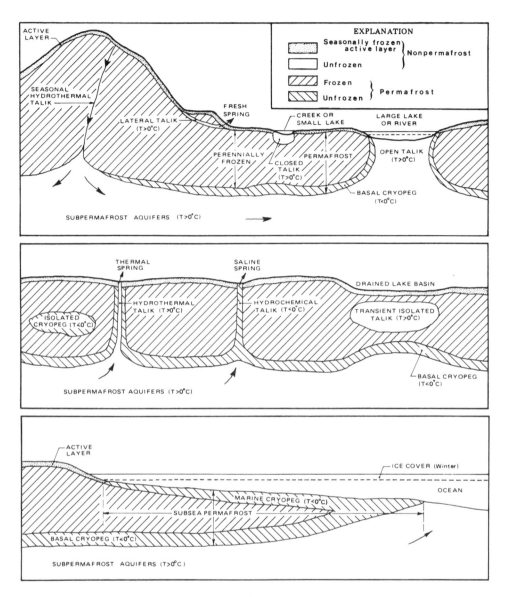

FIG. 7. Aquifers in permafrost areas. Suprapermafrost aquifers are found in the active layer and in closed river and lake taliks. Intrapermafrost aquifers occur in open, lateral, and transient isolated taliks and in isolated, basal, and marine cryopegs.

Flow Rates

The discharge from springs and seeps can provide reliable information on groundwater flow. The flow regime of a spring, and the temperature and chemical composition of its water, may give an indication of the source of the water (Michel 1977; van Everdingen et al. 1979). Perennial springs usually indicate discharge from subpermafrost aquifers or, less commonly, from intrapermafrost aquifers. All springs with discharge rates of more than 5 L • s^{-1}, or with water temperatures above 5°C, or with dissolved-solids concentrations of more than 1 g • L^{-1} likely derive their water from subpermafrost aquifers, via open taliks. Of course, springs with lower discharge rates or producing cooler or fresher water may also be discharging from subpermafrost

FIG. 8. Open water maintained by groundwater discharge in Babbage River, Yukon, showing contrast in vegetation between the discharge area (willows and poplars) and the surrounding tundra (photo by R. A. Mutch).

aquifers. Seasonal springs and seeps commonly discharge water from suprapermafrost aquifers.

A first approximation of the rate of groundwater discharge in a river basin can usually be derived from the baseflow rates in streams in the basin. In some cases it may be possible to derive such information from the seasonal variations in dissolved-solids concentration in the river water.

Icing volumes can often be used to evaluate the groundwater resources of an area because many significant icings represent groundwater discharge. In the permafrost region, the source of the groundwater has a significant bearing on the duration of icing growth. Icings fed by discharge of suprapermafrost water generally stop growing long before the end of the winter, whereas icings fed by discharge from intra- or subpermafrost aquifers will commonly continue to grow until mean daily air temperatures rise above 0°C in spring.

Icing volumes can be measured in the field at the end of the winter. Dividing the measured volume by the duration of the growth period will give the average groundwater discharge rate for the growth period. In the summer, after melting of the icings, it may still be possible to determine their approximate areal extent from a careful survey of the distribution of mineral precipitates left behind by the icings. These and other measurements of the areal extent of individual icings, for example on air photos or satellite imagery (van Everdingen 1975), can be converted into approximate icing volumes using empirical formulas developed for this purpose (Sokolov 1973).

Chemical Composition of Natural Waters

Effects of Low Temperatures

The presence of permafrost has probably only minor direct effects on the chemical composition of either surface water or groundwater. Under the lower temperatures prevailing in the permafrost region, the solubilities of the carbonate minerals in limestones and dolomites are somewhat increased because of the slightly increased solubility of

CO_2 gas in colder water. The solubility of gypsum (or anhydrite) also increases slightly, but the solubilities of most other minerals generally decrease with decreasing temperature.

The dissolution of minerals and any chemical reactions occurring in a groundwater, or between the groundwater and the aquifer materials, proceed more slowly at lower temperatures. The higher viscosity of water at lower temperatures, however, reduces the velocities of groundwater movement, increasing the subsurface residence time of the water and, therefore, the time available for dissolution and chemical reactions.

Groundwater Chemistry

The chemical composition of suprapermafrost groundwater reflects the influence of rainfall and snowmelt and a relatively short underground residence time. Dissolved-solids contents are commonly low, and calcium, magnesium, and bicarbonate are the main constituents. The chemical composition and dissolved-solids concentration may be quite different wherever the suprapermafrost water is affected by discharge from intra- or subpermafrost aquifers. In areas with extensive muskeg (peatland), suprapermafrost water is often high in organic (humic acid) content.

Intrapermafrost water in lateral taliks and in open taliks with downward flow usually resembles local suprapermafrost water, with slightly increased dissolved-solids concentrations. Intrapermafrost water in open taliks with upward flow, and in basal cryopegs, on the other hand, is often identical to local subpermafrost water in both composition and concentration.

The composition of subpermafrost water depends strongly on the underground residence time of the water and on the mineralogical composition of the rocks with which the water has been in contact. Subpermafrost waters can range from freshwater of the calcium/magnesium bicarbonate type, through sulfurous, brackish, and saline waters, to sodium chloride or calcium/sodium chloride brines.

Low dissolved-solids contents in subpermafrost water usually indicate proximity of a recharge area, or the existence of a relatively fast flow system in carbonate karst or in fractured bedrock. The concentrations of some ions, particularly bicarbonate and sulfate, may be unusually low where thawing occurs at the base of degrading permafrost. Low solution rates due to low temperature, and low solubility of calcite and dolomite in the absence of free CO_2, tend to limit redissolution of these minerals that were precipitated during the initial formation of the permafrost.

The chemical composition of spring waters commonly is a reliable indicator of the chemistry of intra- and subpermafrost groundwater, especially during the winter when dilution by water from the active layer is absent.

Surface-Water Chemistry

The chemical composition of surface waters in the permafrost region is affected primarily by the slightly increased solubilities of the carbonates and by somewhat decreased dissolution rates. In areas with extensive muskeg, surface waters are often high in humic acid, as indicated by their distinctive brown color.

The occurrence of groundwater discharge into northern rivers is reflected in the chemical composition of their waters (Brandon 1965; Hitchon et al. 1969; Reeder et al. 1972; Schreier 1979) and also in their isotopic composition (Hitchon and Krouse 1972). The chemical effects, which have been studied in more detail than the isotopic effects, show up in several ways. First, the chemical composition of the surface water may vary along the length of a stream because of local inflows of mineralized groundwater. This effect is illustrated in Fig. 9 for Vermilion Creek, a small tributary of the Mackenzie River, 38 km southeast of Norman Wells, N.W.T. Second, the chemical composition of river water at a particular point will vary seasonally because of changing contributions by snowmelt or rainfall, by discharge from the active layer, and by discharge from intra- and subpermafrost aquifers. This effect is illustrated in Fig. 10 for a station

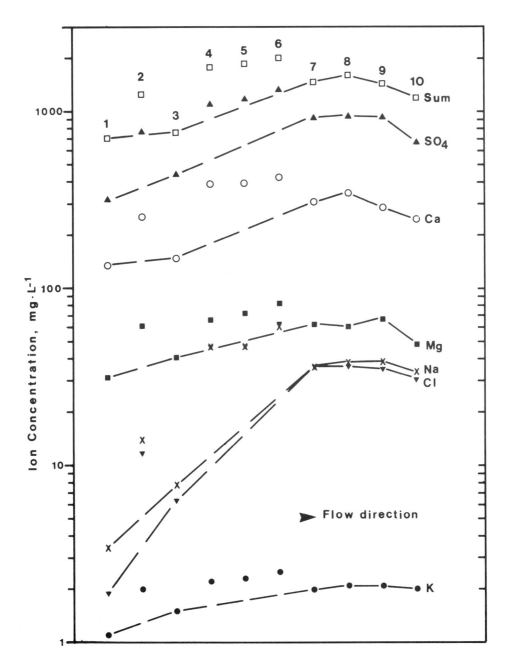

FIG. 9. Chemical composition of spring waters (2,4, 5,6) and variation of creek water chemistry along the lower 11 km of Vermilion Creek, N.W.T. (after van Everdingen 1974). No horizontal scale is implied.

on Ogilvie River, Yukon, below the confluence with Engineer Creek. Finally, as pointed out earlier, the storage of groundwater discharge during the winter, in the form of icings, tends to have a diluting effect on associated river water during melting of the icings.

Surface waters are also affected by natural acidic fallout in a few places in the Northwest Territories. Two different processes responsible for the generation of such

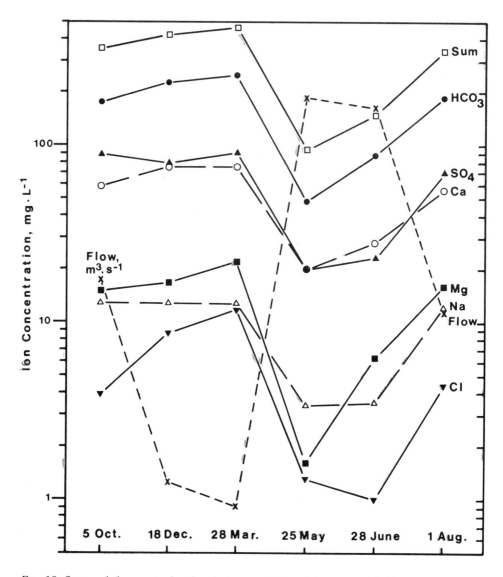

FIG. 10. Seasonal changes in the chemical composition of the water of Ogilvie River, Yukon, below Engineer Creek (data from Schreier 1979).

fallout have been identified: spontaneous combustion of pyrite and atmospheric oxidation of hydrogen sulfide.

Spontaneous combustion of pyrite occurs in several exposures of pyrite-bearing, organic mudstones along a 30-km stretch of sea cliffs in the Smoking Hills area on the western shore of Franklin Bay. Sulfur dioxide gas produced by the combustion process is converted into airborne sulfuric acid. As a result of the deposition of the sulfuric acid, 27 small ponds on the tundra within 1 km downwind from the burning cliffs show low pH values between 1.8 and 4.5, sulfate concentrations up to 17 000 mg • L^{-1}, and elevated concentrations of aluminium, iron, manganese, nickel, and cadmium (Havas and Hutchison 1983).

The Smoking Hills area thus provides a natural site for the study of the effects of acidic fallout on vegetation, soils, surface water, and aquatic life in the continuous permafrost zone. The presence of perennially frozen ground near the surface tends to restrict the infiltration of the fallout, and neutralization would have to take place in the relatively thin active layer than may have only a limited capacity for the neutralization of acidic water.

Acidic fallout on a much smaller scale has been studied at several spring locations in the Mackenzie River Valley, where escaping hydrogen sulfide gas is converted into airborne sulfuric acid. At these sites, much of the acid is neutralized upon contact with exposed carbonate rocks and carbonate-rich soil (van Everdingen et al. 1982).

Water Supply in Permafrost Areas

Usage of Surface Water and Groundwater

In the permafrost region, as elsewhere, water supplies are obtained from both surface waters and groundwater. Data on the approximate quantities taken from these two sources are available for the Yukon and the Northwest Territories (Table 1), but not separately for the areas underlain by permafrost in the provinces. For the purpose of this discussion it is assumed that the information on water use in the territories is representative of water use patterns in the permafrost region as a whole.

More than 63% of the population of the Yukon relies on groundwater for part or all of its water supply during part or all of the year. However, only about 22% of the total water used is obtained from groundwater sources because large quantities of surface water are used by the placer mining industry in the Yukon. In the Northwest Territories, on the other hand, less than 1% of the population relies on groundwater, but more than 48% of the total "water used" comes from groundwater sources. This reflects the extraction of large quantities of groundwater during mine dewatering by the mining industry in the Northwest Territories.

In general, the data given in Table 1 for the two sources of municipal and rural supplies reflect the greater availability of groundwater in the mountainous areas of the Yukon and the western portion of the District of Mackenzie, as compared with the areas of lower relief in the District of Keewatin, and areas with very thick permafrost in the Arctic Islands.

Limitations of Surface-Water Supplies

The reliability of surface-water sources is often limited by seasonal freezing. Many smaller streams and lakes freeze all the way to the bottom, and these can therefore provide seasonal supplies only. In larger rivers, where baseflow is sufficient to provide an adequate water supply, the quality of the water may vary widely with the seasons (Fig. 10), and the suspended-sediment content of the river water can become very high during spring breakup and peak-flow periods.

With increasing latitude, even larger rivers and lakes will freeze to the bottom, and surface-water supplies become increasingly unreliable. A case in point is presented by the 21 radar stations of the Distant Early Warning Line in the Yukon and Northwest Territories, all of which rely on surfce water. Some of the stations that use dependable lake sources have to haul their water over distances of several kilometres. Others have only a single source of supply that is limited by deep freezing during extremely cold winters. All would benefit from having a groundwater supply available as an alternative.

Limitations of Groundwater Supplies

Suprapermafrost aquifers in taliks below large rivers and deep lakes can provide reliable year-round supplies of good-quality water. Aquifers in the active layer will freeze at some time during the winter, and these can, therefore, only be used for part of the year. Intrapermafrost aquifers do not freeze, but their supply may still be exhausted

TABLE 1. Water use in Canada north of 60°N.[a]

	Yukon	Northwest Territories	North ot 60°N total
Surface water users[b]			
Municipal	7 538	37 451	44 989
Rural	958	8 008	8 966
Total	8 496	45 459	53 955
Groundwater users[c]			
Municipal	13 700	202	13 902
Rural	957	80	1 037
Total (and % of total population)	14 657 (63.3)	282 (0.6)	14 939 (21.7)
Total users	23 153	45 741	68 894
Surface water use ($1000 \text{ m}^3 \cdot \text{yr}^{-1}$)			
Municipal	1 922	3 520	5 442
Rural	67	560	627
Manufacturing	4	875	879
Mining	10 374	4 864	15 238
Total	12 367	9 819	22 186
Groundwater use ($1000 \text{ m}^3 \cdot \text{yr}^{-1}$)			
Municipal	3 494	19	3 513
Rural	67	6	73
Manufacturing	0	0	0
Mining	0	9 168	9 168
Total (and % of total water used)	3 561 (22.3)	9 193 (48.4)	12 754 (36.5)
Total water used ($1000 \text{ m}^3 \cdot \text{yr}^{-1}$)	15 928	19 012	34 940

[a]Data from P. J. Hess. (1984. Groundwater use in Canada. Internal report prepared for NHRI and WPMB, Inland Waters Directorate, Ottawa, Ont.).
[b]Using surface water exclusively.
[c]Relying on groundwater for part or all of their needs during part or all of the year.

before the end of the winter. Moreover, the dissolved-solids content in the water may be unacceptably high in open taliks that carry discharge from subpermafrost aquifers. It is also generally more difficult to find adequate groundwater supplies at higher latitudes because of the gradual decrease in the size of easily accessible unfrozen zones.

A special type of intrapermafrost aquifer, which has not yet been explored for water supply in northern Canada, is found in the isolated transient taliks that develop from originally closed taliks below recently drained lakes and abandoned river channels when permafrost starts penetrating from the surface down. Such aquifers will gradually decrease in size with time, and they may become artesian (Fig. 11). The dissolved-solids content of the remaining water in the shrinking talik will gradually increase, and the chemical composition of the water may change as a result of the precipitation of less-soluble salts. Initial above-zero temperatures in the remaining unfrozen zone will fall to slightly below zero before eventual freezing occurs.

Subpermafrost aquifers offer, in principle, the most reliable sources for groundwater supply but the mineralization of their water can be expected to be high, particularly in the low-relief portions of the Interior Plains in the zone of continuous permafrost.

FIG. 11. Artesian discharge from the transient isolated talik below a pingo on the Tuktoyaktuk Peninsula, N.W.T. (from Mackay 1978, fig. 7).

In areas of higher relief, with somewhat higher precipitation and shorter, faster flow systems, high-quality water can be found in subpermafrost bedrock aquifers, as shown by the artesian water supply for the settlement of Old Crow, Yukon, completed in 1982 (Trimble et al. 1983). In the zone of discontinuous permafrost, where the permafrost may be very thin, subpermafrost aquifers in unconsolidated deposits can also provide high-quality water supplies.

Karst development has created subpermafrost aquifers in areas with soluble bedrock in the British and Barn Mountains in the northern Yukon (Williams and van Everdingen 1973). In some parts of the Franklin Mountains and Colville Hills, N.W.T., where dissolution of gypsum and salt has occurred at depth, the chemical composition of groundwater discharge does not always reflect the sulfurous or saline nature of the karst. This suggests that some of the groundwater flow takes place through zones of collapsed strata above the soluble beds.

Lack of recharge likely prevents development of extensive groundwater flow systems in the mountains in the High Arctic. Some of the groundwater discharged from younger sedimentary deposits on the Artic Islands may be derived from the compaction of fine-grained sediments. Saline springs on Axel Heiberg Island (Beschel 1963) indicate that dissolution of salt beds is also taking place. In all these cases, the dissolved-solids concentrations are too high to allow use of the water for domestic supplies.

An additional factor to be considered during the exploration for groundwater supplies in the permafrost region is the contamination potential of various aquifer types because some aquifers can become contaminated almost as easily as rivers and lakes. In general terms, active-layer aquifers are unattractive because of their near-surface position, and karst aquifers are liable to contamination because of the lack of filtering during recharge and because the relatively high flow velocities in solution channels within the karst will cause rapid spreading of contamination in the subsurface.

The water in intra- and subpermafrost aquifers may be under sufficient pressure to produce artesian flow. If encountered during drilling, such flow can lead to a blowout (uncontrolled discharge) that may be difficult to bring under control because thawing of the surrounding frozen ground will accelerate cratering around the drill hole. In early winter, the water may also escape by channeling through the unfrozen portion of the active layer. It may then emerge onto the surface and form an icing, or cause the formation of frostmounds, at some distance from the drill site (Linell 1973).

Heat transmitted through the casing of a producing well can also cause thawing of frozen ground around the well bore. If the well penetrates ice-rich sediments, thawing may lead to instability of the well which can cause buckling and failure of the well casing and uncontrolled discharge outside the casing. Thawing around producing wells can generally be prevented by proper insulation of the casing. Insulation should also be installed in all wells in which heat cables are used to prevent freezing during interruptions of pumping or flow.

Thermal Springs and Geothermal Energy

More than 40 thermal springs with water temperatures ranging from 5 to 64°C have so far been found in the mountainous areas of the Yukon and the District of Mackenzie (Crandall and Sadlier-Brown 1976). Those with temperatures above 20°C are listed in Table 2. Several of these springs discharge enough warm or hot water to support recreational developments and to satisfy local space-heating requirements, but only one site, Takhini Hot Springs near Whitehorse, Yukon, has been developed for these purposes. The hot-spring water there is used in a swimming pool and to heat the buildings on the site. Most of the other hot-spring sites are too remote and inaccessible to attract such developments.

Low-grade geothermal energy is being used in Whitehorse and Mayo in the Yukon in the form of warm groundwater that is added to municipal surface-water supplies during the winter. This reduces the amount of fossil fuel that would otherwise have to be used for heating to prevent freezing of the water distribution systems.

Dependence of Fish Populations on Groundwater Discharge

During the early 1970's, fisheries studies in the northern Yukon and along the Mackenzie River and its tributaries established that several fish species depend on spawning grounds and overwintering areas that are kept from freezing by perennial discharge of groundwater. In the northern Yukon, Arctic char (*Salvelinus alpinus*) were found to use spawning areas immediately downstream from spring areas in Firth River, Joe Creek, Fish Hole Creek, and Fish Creek. Chinook salmon (*Oncorhynchus tshawytscha*) were found to spawn only in the open-water areas in the Miner River and chum (*O. keta*) and coho salmon (*O. kisutch*) only in an area just downstream from the large springs on the South Fishing Branch of the Porcupine River. Such areas where rates of groundwater discharge are high enough to maintain open-water reaches for spawning and overwintering grounds are much rarer in the low-relief portions of the Northwest Territories and in the High Arctic due to the progressive restriction of groundwater flow by the increasing extent of perennially frozen ground.

Because of the selective utilization of open-water areas by various species, engineering developments near any of these areas in northern rivers could have disastrous effects on some fish populations. Temporary effects may result from temporarily increased concentrations of suspended sediment, for instance during construction of a buried pipeline crossing. More serious, long-term effects can be expected to result if rates or locations of groundwater discharge are changed.

	Temperature (°C)	Discharge (L•s^{-1})	TDS (mg•L^{-1})	References[b]
British Columbia				
Liard Hot Spring 1	21.0	30	1100	1,2
Liard Hot Spring 2	54.0	40	670	1,2
Portage Brulé	48.0	5	1180	1,2
Northwest Territories				
Clausen Creek	36.7	30	5070	1
Wild Mint	29.0	30	473	2
Rabbitkettle Hot Springs	21.0	7	730	2
Hole-in-the-Wall	47.0	30	170	2
East Cantung	29.0	<1	234	2
West Cantung	41.0	<1	195	2
North Cantung	32.0	<1	137	2
Nahanni River North	58.0	40	260	2
Grizzly Bear	44.0	20	550	2
Broken Skull	45.0	35	945	2
Nahanni Headwater	64.0	50	254	2
Roche-qui-trempe-à-l'eau	31.3	15	12860	1,3
South Redstone Hot Springs	54.0	120	580	2,3
Ekwi River	46.0	30	16090	2
Deca East	22.0	10	2460	2
Yukon				
Larsen South	43.0	20	391	2
Larsen North	53.0	40	388	2
Pool Creek	54.5	3	454	2
Takhini Hot Spring	47.0	7	2625	1,2
McArthur Hot Springs	54.5	30	185	1,2

[a]For each site, maximum measured temperature, estimated combined discharge, and maximum dissolved-solids content are given.

[b]References: 1, Brandon (1965); 2, Crandall and Sadlier-Brown (1976); 3, Michel (1977).

Hydrological Effects of Disturbances of Permafrost Terrain

Any disturbance that affects the heat balance in permafrost terrain will result in either degradation or aggradation of permafrost. These in turn can give rise to detrimental ground settlement or frost heave, respectively. Either of these processes may cause local changes in surface and subsurface drainage patterns, which will further affect the heat balance.

Natural Disturbances

Natural processes that affect the heat balance at the surface of the ground include forest fires, erosion along riverbanks, lake shores, and coastlines, and draining of ponds and lakes.

Forest Fires

Destruction of the insulating vegetation cover by fire will allow deeper thawing during the summer, increasing the thickness of the active layer. In fine-grained sediments on slopes, such an increase in the depth of thaw may release enough water to cause slope failures, usually in the form of long, narrow mud flows (Fig. 12a). Incidentally, such active-layer failures can also be initiated by extremely heavy rains. In flat terrain where ice-rich sediment or layers of massive ice are affected by increased thaw depth, the

FIG. 12. Effects of increased thaw depth after forest fires. (a) Active-layer failure in the Grandview Hills, N.W.T. (Hughes et al. 1973, fig. 5); (b) thermokarst lake in the Takhini River valley, 38 km west of Whitehorse, Yukon (Klassen 1979, fig. 6).

ground will settle, usually accompanied by ponding of water in the resulting depressions. Such water-filled depressions, called thermokarst lakes (Fig. 12b), gradually expand by continued thawing of ice in the backshore materials. The expansion will stop when the ice or ice-rich sediment becomes insulated again by slumping of overburden and/or vegetation cover, or when all the ice has melted.

Active-layer failures will usually have negligible effects on the hydrology of the affected area. The development of thermokarst lakes, on the other hand, can significantly affect local hydrology because the lakes will provide increased storage for surface runoff.

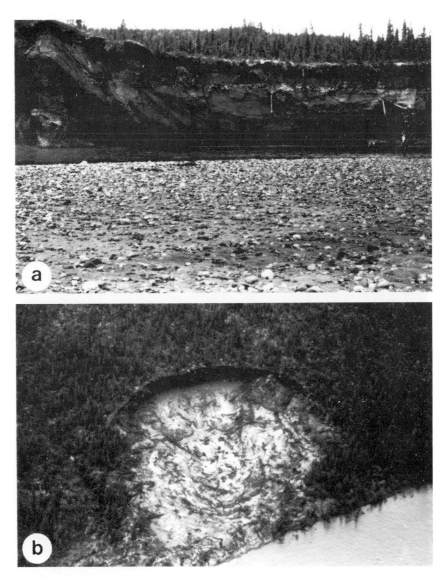

FIG. 13. Thermal erosion along rivers. (a) Thermo-erosional niche and bank collapse along Ram River, Mackenzie Mountains, N.W.T.; (b) thaw slump along South Nahanni River, Nahanni National Park, N.W.T.

Erosion and Sedimentation

Where ice-rich sediments are exposed by erosion of riverbanks, melting of the ice can significantly increase the rate of erosion. Development of a thermo-erosional niche by heat supplied by the river water will lead to eventual collapse of the overlying part of the bank (Fig. 13a). The relatively large quantities of silt- and clay-sized material released by this process can cause a large increase in the concentration of suspended sediment in affected streams. The process also encourages relatively rapid migration of river meanders.

Wherever significant thicknesses of ice-rich material or massive ice are exposed, rapid thawing may produce gradually expanding retrogressive thaw slumps (Fig. 13b) that

FIG. 14. Thaw slumping along the Yukon coast (photos by D. A. van Everdingen). (a) Massive ice exposed in the headwall of a slump at Kay Point; (b) ice wedges exposed along the top of the cliffs at King Point (mud flow in foreground).

will contribute further to the suspended sediment in rivers and lakes. Very large retrogressive thaw slumps have developed along portions of the Arctic coast, where wave erosion has exposed beds of massive ice (Fig. 14a) or numerous ice wedges (Fig. 14b). In these cases, slumping of the original cover of overburden and vegetation does not produce even a partial insulating cover on the thaw faces because the latter are too steep and slippery. Continued development of these very large thaw slumps may locally alter surface-drainage patterns near the coast.

Deposition of sediments on alluvial fans, inside river meanders, in cut-off meanders, and on river deltas will cause a gradual rise of the permafrost table into the newly deposited sediments. This expansion of permafrost will likely have some effect on the movement of groundwater, but not on surface runoff.

Lake Draining

Natural draining of a lake will occur when erosion creates an outlet with a lower elevation than the lake bottom. When this happens, exposure of the lakebed will allow development of permafrost in the unfrozen sediments of the lake talik. In an originally open talik that carries discharge of groundwater, the permafrost aggradation may stop the discharge. In an originally closed talik, the water remaining in the gradually decreasing unfrozen zone will come under increasing pressure, which may lead to the development of springs or to the growth of a pingo. (Fig. 15).

Man-Made Disturbances

Man-made disturbances of permafrost terrain are related to such activities as petroleum exploration, transportation, mining, and power generation, etc. Many of these disturbances lead to degradation of permafrost and to local changes in drainage patterns. For the following discussion, examples are grouped under three headings: limited-area disturbances, linear disturbances, and surface-water impoundments.

Limited-Area Disturbances

Local disturbances of permafrost terrain caused by drilling sites and buildings generally have negligible effects on hydrological processes. Improper disposal of drilling fluids, however, may cause serious permafrost degradation and erosion and subsequent contamination of surface waters.

Airstrips and runways on permafrost often have noticeable effects on surface drainage because most of the precipitation falling on their surface will drain towards the surrounding ground relatively rapidly. Unless proper drainage control is provided, erosion will occur along the edges of the runway, followed by degradation of the permafrost, which will further affect drainage patterns.

FIG. 15. Pingo in the valley of the Blackstone River upstream of Chapman Lake, Yukon.

Clearing and cultivation of areas with ice-rich permafrost or with polygonal networks of ice wedges will cause degradation of the permafrost and development of thermokarst topography. Melting of ice wedges will leave a topography of so-called cemetery mounds, with an intervening network of trenches. Such areas commonly have to be abandoned because they can no longer be worked by farm machinery. Usually only minor channeling of surface runoff occurs in the trenches, but they may store all of the local snowmelt.

Mining will in general have similar effects on hydrological processes in permafrost areas as in areas without permafrost. Both open-pit and underground mines commonly require some degree of dewatering to allow mining operations to proceed safely. The resulting lowering of water levels in the surrounding area will cause some decrease in the rates of natural groundwater discharge, and it may thus affect surface runoff to some extent. The associated changes in groundwater flow patterns will affect the heat balance in the subsurface, and this may locally cause either degradation or aggradation of permafrost, with the usual consequences.

Placer mining always has an effect on surface water, but the effect can be particularly severe in areas with ice-rich permafrost or massive ice in the overburden. In some instances, very large quantities of surface water are used to thaw and remove ice-rich overburden (Fig. 16), and it is often difficult to keep the resulting organic-rich slurry away from surface-water courses.

Linear Disturbances

Seismic lines, logging roads, and winter roads of various types all require clearing by removal of trees and shrubs and, locally, ground cover and soil. Such clearing and the compaction or destruction of the insulating vegetation mat by vehicle traffic usually lead to an increase in thaw depth. Where excess ice is present, the increased thaw depth will cause differential settlement of the ground and ponding of water on the right-of-way. Eventually, these features may develop into a drainage channel along the right-of-way, with associated erosion problems (Fig. 17a).

FIG. 16. Hydraulic stripping of ice-rich "muck" at a placer mine on Hunker Creek near Dawson, Yukon.

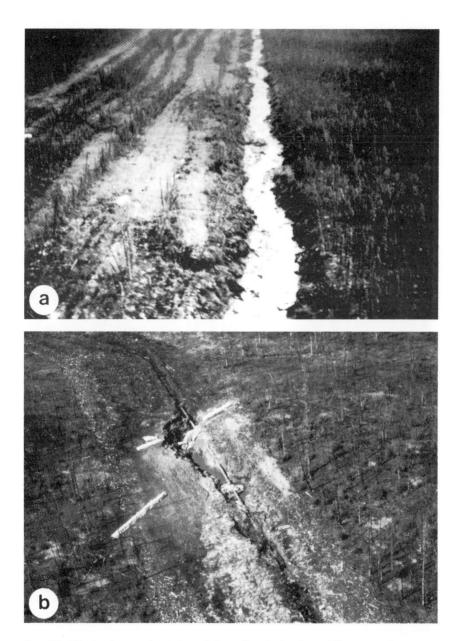

FIG. 17. Effects of permafrost degradation after destruction of the vegetation cover. (a) Stream channel developed along a seismic line southwest of Norman Wells, N.W.T. (from Rutter et al. 1973); (b) Norman Wells oil pipeline exposed by erosion near Bear Rock, N.W.T., as a result of channeling of surface drainage (photo by D.G. Harry).

All-weather highways and railroads in the permafrost region are now commonly constructed on a berm or embankment that should be sufficiently high to prevent thawing of the underlying permafrost. Cross-drainage by culverts is provided wherever there is any evidence of surface flow because obstruction of surface-water drainage by the embankment would result in erosion along the highway ditches or ponding of water, which could lead to permafrost degradation along the base of the embankment.

269

In places where a highway embankment is higher than required for the protection of permafrost, the permafrost table will gradually rise, possibly even into the embankment fill. Such excess embankment heights are often used at the crossings of small streams, where a rise of the permafrost table will restrict the movement of groundwater that is usually associated with small streams. The restriction will force the groundwater to discharge somewhere upstream from the highway crossing. In winter, the discharge will form icings that may completely plug the culverts, leading to flooding during the snowmelt period. Frostmounds with a core of massive ice may also form (van Everdingen 1982a). Such man-induced frostmounds and icings will commonly have detrimental secondary effects on vegetation in the affected areas.

Highway cuts that expose ice-rich sediments will experience thermal erosion and thaw slumping until a new insulating cover is established over the ice-rich material, either through slumping of the overlying vegetation cover or by artificial means. Where such cuts are made along the approaches to stream crossings, transport of the slumped material by surface drainage can contribute significantly to the suspended-sediment load in those streams during the thaw season.

Pipeline construction in the permafrost region can be expected to have a number of hydrological effects that are similar to those resulting from the clearing of seismic lines. Increased thaw depths may cause differential settlement and ponding, or erosion by water running along the right-of-way (Fig. 17b). In addition, groundwater flow may be concentrated in the backfill of the pipeline trench, possibly causing erosion, if the backfill remains unfrozen; the movement of groundwater across the trench may be blocked if the backfill freezes. Where a pipeline is installed in or on a berm, obstruction of cross drainage may also result in erosion.

Two further effects are related to pipeline operation (van Everdingen 1979). Where a hot oil pipeline, such as the trans-Alaska pipeline, is buried in ice-rich permafrost without being adequately insulated, thawing of the permafrost will lead to the earlier mentioned consequences, which may cause failure of the pipeline. To avoid such thaw problems, the pipeline can be elevated above the ground, as was done for large portions of the trans-Alaska pipeline. Vertical supports were provided with convective heat-exchange elements to make sure that the permafrost remains frozen. However, in some places where the pipes that serve as vertical supports extend through the base of the permafrost, they act as artesian wells (Fig. 18).

Buried gas pipelines can be cooled by refrigeration of the gas to keep the permafrost frozen, but significant frost heave may result from freezing of the trench backfill and the surrounding ground where such a line crosses areas of unfrozen ground. The induced freezing and related frost heaving may cause obstruction of groundwater movement and surface drainage, respectively.

Most of the undesirable hydrological consequences from limited-area and linear disturbances of permafrost terrain can be minimized by the use of special engineering designs, construction methods, and operational practices. In the case of the remaining group of disturbances, however, little can be done to minimize the consequential damage once the decision to proceed has been taken.

Surface-Water Impoundments

The impoundment of surface waters, for the creation of reservoirs for water supply, for hydroelectric power generation, or for the diversion of surface runoff from one drainage basin to another, is dealt with in detail by Newbury and Rosenberg elsewhere in this volume. Several hydropower and multipurpose impoundments, however, exist in the permafrost region in Canada, and their interaction with the permafrost terrain will be examined briefly here.

Where a surface-water reservoir is created, or the level of an existing lake raised, the mean annual ground-surface temperature of the submerged area will rise. If ice-rich permafrost is present, the increase in temperature may cause melting of the excess

FIG. 18. Vertical supports of the trans-Alaska oil pipeline discharging subpermafrost water. Note that water levels inside the pipes are indicated by condensation on the outside.

ice, leading to settling of the bottom of the reservoir and local increases in depth. Long-term shoreline erosion will occur where ice-rich, fine-grained sediments exist between the minimum and maximum water levels along the new shoreline. This phenomenon has been studied in detail at Southern Indian Lake in Manitoba, the level of which was raised by 3 m in 1976 to achieve a 75% diversion from the Churchill River drainage basin to the Nelson River drainage basin (Newbury et al. 1978; Newbury and McCullough 1984).

Rapid shoreline erosion, with rates of retreat as high as $12 \text{ m} \cdot \text{yr}^{-1}$, accompanied the rise in water level where flooding extended over permafrost-affected glacial, pro-glacial, and glaciolacustrine sediments. The rate of erosion appeared to be controlled by the rate of thaw of the frozen materials. The process caused greatly increased concentrations of suspended sediment in the lake water and some shallowing in nearshore areas. Whereas the preimpoundment annual sediment contribution from lakeshore erosion was small, the annual sediment input from eroding shorelines following impoundment exceeded 4×10^6 t. Five years after the start of impoundment, stabilization had occurred only where bedrock underlying the backshore became exposed at the water edge. Shoreline erosion is expected to continue for several decades everywhere else.

Although most of the preimpoundment predictions regarding the effects of impoundment on the physical environment were qualitatively correct, some unpredicted changes also occurred. The latter included a decrease in water temperatures and increases in mercury concentrations in fish (Hecky et al. 1984).

Potential Hydrological Consequences of the Effects of Climatic Changes on Permafrost

The general hydrological consequences of long-term changes in climate are discussed in detail by Ripley elsewhere in this volume. In the permafrost region, however, and particularly in areas with ice-rich sediments, a climatic change can be expected to have some consequences that would not occur elsewhere.

Several model studies have recently been undertaken to attempt prediction of the effect of climatic change on ground temperatures for selected sites in the permafrost region (Osterkamp 1983; Smith and Riseborough 1983; Goodwin et al. 1984). For the following discussion it is assumed that a change in climate will cause similar changes in mean annual ground-surface temperatures.

Consequences of Widespread Degradation of Permafrost

A general increase in mean annual ground-surface temperatures would cause a gradual decrease in the thickness and extent of permafrost, some northward displacement of the southern boundaries of the discontinuous and continuous permafrost zones and both intensification and wider occurrence of the degradation phenomena that were mentioned earlier: active-layer failures, thaw slumps, thermokarst development, and accelerated coastal, lakeshore, and riverbank erosion. All these processes would contribute to increased concentrations of suspended sediment in surface waters, which would have adverse effects on fish populations, and possibly also on hydroelectric power generation, through increased wear of turbines and reduction of reservoir capacity. In addition, thermal erosion of ice-rich highway cuts could be reactivated, pipeline trenches in ice-rich ground could become unstable, settlement and lateral cracking of highway embankments could be accelerated, and previously stable building foundations could become unstable. These effects would be most severe in areas underlain by extensive ice-rich permafrost.

Recharge and discharge of groundwater would likely become more active, occurring in more places. Icing activity could therefore increase, which would likely lead to additional icing problems along northern highways. Serious flooding problems could result at stream crossings with culverts, if increased snow cover causes higher runoff rates during the snowmelt period. Many culverts could turn out to be inadequate if increased total precipitation were to lead to increased peak-flow rates.

Changes in drainage patterns, and particularly changes in the timing of freeze-up and breakup, could also interfere with migration and spawning habits of various fish species.

Consequences of Widespread Aggradation of Permafrost

A general lowering of mean annual ground-surface temperatures would cause a gradual increase in the extent and thickness of permafrost, some southward displacement of the southern boundaries of the discontinuous and continuous permafrost zones, and some decrease in the intensity of current phenomena of permafrost degradation.

The hydrological effects of widespread aggradation of permafrost would likely consist of a decrease in the total quantity of groundwater recharge and the further concentration of discharge in fewer and smaller discharge areas. This would in turn reduce baseflow rates in many streams, possibly eliminating some of the spawning and overwintering areas currently used by various fish species. Reduced thickness and slower thawing of the active layer would somewhat increase runoff rates during the snowmelt period, but this effect could be offset by a decrease in the thickness of snow cover.

Conclusions

Although much has been learned, a variety of studies are still required to improve our understanding of the interactions between permafrost and various hydrological processes, and also to provide the basic data required for the proper management and protection of the water resources of the permafrost region. A number of subjects for further studies are listed below without any implied order of priority.

The effects of natural, long-term changes in thermal regime on the distribution of permafrost, on surface runoff, and on the groundwater regime should be monitored at carefully selected sites, and models for the prediction of such effects should be further refined.

Studies should also be directed towards quantitative prediction and measurement of the hydrological effects of man-made changes in the thermal regime and of the thermal effects of man-made changes in the hydrological regime (such as river diversions, mine dewatering, and draining or infilling of muskeg areas). On a local scale, the intentional modification of groundwater and soil-moisture movement should be investigated to determine its potential usefulness in minimizing the severity of icing problems and frost-heave damage.

Regional studies of the chemistry of both surface waters and groundwaters, and of the seasonally varying influence of groundwater chemistry on the quality of surface waters, should provide estimates of the groundwater contribution to surface runoff in the permafrost region. Surveys of icing locations and dimensions, using airphotos, satellite imagery, and field measurements, could provide similar information, as well as estimates of the contributions to summer streamflow from melting of the icings. The effects of icings on the incidence of extreme water levels and peak-flow rates, and on lateral erosion and streambed scour, should also be investigated.

The various methods available for estimating flood frequency and magnitude in small, ungauged drainage basins should be further refined to allow the selection of optimum sizes for culverts and other drainage structures. This would also require further development of the ability to predict the potential occurrence of ice jams and icings and the magniture of their effects on stream hydraulics.

The economic potential of groundwater as a source of low-grade thermal energy in the permafrost region should be assessed.

Several of the above studies would benefit from further development of surface geophysical methods for the delineation of the frozen ground that forms seasonal and perennial low-permeability boundaries for groundwater flow and further development of airborne methods for the determination of the water content of snow cover. Some of those studies will also require the development of low-cost instrumentation for the measurement, or long-term automatic recording, of total and unfrozen water contents in soil and of changes in phase (water or ice), water level or pressure, and water chemistry in situations where the sensing elements will be subjected to seasonal freezing.

The study of permafrost hydrology, and of the potential interactions between permafrost, engineering developments, and hydrological processes, would also benefit from the establishment of a central clearing house for the exchange, compilation, and reporting of all types of northern hydrological data. Such a clearing house would have to be provided with relevant data by municipal, territorial, provincial, and federal agencies, by environmental and engineering consultants, and by various industries.

Acknowledgements

The following people have provided illustrative material for this chapter: D. G. Harry, O. L. Hughes, and R. W. Klassen, Geological Survey of Canada; G. H. Johnston, National Research Council of Canada; J. R. Mackay, University of British Columbia; R. A. Mutch, University of Calgary; N. W. Rutter, University of Alberta; and D. A. van Everdingen, Memorial University. Prints from color photography were prepared by the Photography Laboratory of the Institute of Sedimentary and Petroleum Geology. All these contributions are gratefully acknowledged. Careful reviews and helpful comments were received from O. J. Ferrians, Jr., U.S. Geological Survey; R. Janowicz, Water Resources Division, DIAND; G. K. McCullough, Freshwater Institute; J. B. Wilson, Environmental Protection Service; and M. K. Woo, McMaster University.

References

ANDERSON, D. M., AND N. R. MORGENSTERN. 1973. Physics, chemistry, and mechanics of frozen ground: a review, p. 257-288. In Permafrost, North American Contribution, Second International Conference. National Academy of Sciences, Washington, DC.

BESCHEL, R. E. 1963. Sulphur springs at Gypsum Hill. Preliminary Report, 1961-1962, p. 183-187. Axel Heiberg Island Research Reports, McGill University, Montreal, Que.

BRANDON, L. V. 1965. Groundwater hydrology and water supply in the District of Mackenzie, Yukon Territory, and adjoining parts of British Columbia. Geol. Surv. Can. Pap. 64-39: 102 p.

BROOK, G. A. 1983. Hydrology of the Nahanni, a highly karsted carbonate terrain with discontinuous permafrost, p. 86-90. In Proceedings, Permafrost, Fourth International Conference, Vol. 1. National Academy Press, Washington, DC.

BROWN, R. J. E. 1970. Permafrost in Canada — Its influence on northern development. University of Toronto Press, Downsview, Ont. 234 p.

BRYAN, J. E. 1973. The influence of pipeline development on freshwater fishery resources of northern Yukon Territory. Environmental-Social Program Northern Pipelines, Rep. No. 73-6: 63 p.

CRANDALL, J. T., AND T. L. SADLIER-BROWN. 1976. Data on geothermal areas, cordilleran Yukon, Northwest Territories and adjacent British Columbia. Report prepared for Department of Energy, Mines and Resources, Contract No. ISQ5-0136.

GOODWIN, C. W., J. BROWN, AND S. I. OUTCALT. 1984. Potential responses of permafrost to climatic warming, p. 92-105. In J. H. McBeath [ed.] The potential effects of carbon dioxide-induced climatic changes in Alaska. University of Alaska, Fairbanks, Misc. Publ. 83-1.

HAVAS, M., AND F. C. HUTCHISON. 1983. The Smoking Hills: natural acidification of an aquatic ecosystem. Nature (Lond.) 301(5895): 23-27.

HECKY, R. E., R. W. NEWBURY, R. A. BODALY, K. PATALAS, AND D. M. ROSENBERG. 1984. Environmental impact prediction and assessment: the Southern Indian Lake experience. Can. J. Fish. Aquat. Sci. 41: 720-732.

HITCHON, B., AND H. R. KROUSE. 1972. Hydrogeochemistry of the surface waters of the Mackenzie River drainage basin, Canada — III. Stable isotopes of oxygen, carbon and sulphur. Geochim. Cosmochim. Acta 36: 1337-1357.

HITCHON, B., A. A. LEVINSON, AND S. W. REEDER. 1969. Regional variations in river water composition resulting from halite solution, Mackenzie River drainage basin, Canada. Water Resour. Res. 5(6): 1395-1403.

HUGHES, O. L., J. J. VEILLETTE, J. PILON, P.T. HANLEY, AND R. O. VAN EVERDINGEN. 1973. Terrain evaluation with respect to pipeline construction, Mackenzie Transportation Corridor, Central Part. Environmental-Social Program Northern Pipelines, Rep. No. 73-37: 74 p.

HYDROLOGICAL ATLAS OF CANADA. 1978. Department of Supply and Services, Ottawa, Ont.

INTERA ENVIRONMENTAL CONSULTANTS LTD. 1975. Groundwater management study of the Yukon Territory. Report prepared for Department of Indian and Northern Affairs, Ottawa, Ont. 122 p.

JOHNSTON, G. H. [ED.] 1981. Permafrost: engineering design and construction. John Wiley & Sons, Canada Ltd., Toronto, Ont. 540 p.

KANE, D. L., AND J. STEIN. 1983. Field evidence of groundwater recharge in interior Alaska, p. 572-577. In Proceedings, Permafrost, Fourth International Conference, Vol. 1. National Academy Press, Washington, DC.

KLASSEN, R. W. 1979. Thermokarst terrain near Whitehorse, Yukon Territory. Geol. Surv. Can. Curr. Res. Pap. 79-1A: 385-388.

LINELL, K. 1973. Risk of uncontrolled flow from wells through permafrost, p. 462-468. In Permafrost, North American Contribution, Second International Conference. National Academy of Sciences, Washington, DC.

MACKAY, J. R. 1963. The Mackenzie delta area, N.W.T., Canada. Dep. Mines Tech. Surv. Geogr. Branch Mem. No. 8: 202 p.

1978. Sub-pingo water lenses, Tuktoyaktuk Peninsula, Northwest Territories. Can. J. Earth Sci. 15: 1219-1227.

MICHEL, F. A. 1977. Hydrogeologic studies of springs in the central Mackenzie Valley, Northwest Territories, Canada. M.Sc. thesis, University of Waterloo, Waterloo, Ont. 185 p.

NEWBURY, R. W. 1974. River hydrology in permafrost areas, p. 31-37. In Permafrost Hydrology, Proceedings of Workshop Seminar. Canadian National Committee, International Hydrological Decade, Environment Canada, Ottawa, Ont.

NEWBURY, R. W., K. G. BEATY, J. A. DALTON, AND G. K. McCULLOUGH. 1979. A preliminary comparison of runoff relationships and water budgets in three experimental lake basins in the continental, sub-arctic and arctic climatic regions of the Precambrian Shield, p. 516-535. In Proceedings Canadian Hydrology Symposium '79 — Cold Climate Hydrology. National Research Council of Canada, Ottawa, Ont.

NEWBURY, R. W., K. G. BEATY, AND G. K. McCULLOUGH. 1978. Initial shoreline erosion in a permafrost affected reservoir, p. 834-839. In Proceedings, Third International Conference on Permafrost, Vol. 1. National Research Council of Canada, Ottawa, Ont.

NEWBURY, R. W., AND G. K. McCULLOUGH. 1984. Shoreline erosion and restabilization in the Southern Indian Lake reservoir. Can. J. Fish. Aquat. Sci. 41: 558-566.

OHMURA, A. 1982. Evaporation from the surface of the Arctic tundra on Axel Heiberg Island. Water Resour. Res. 18(2): 291-300.

OSTERKAMP, T. E. 1983. Potential impact of a warmer climate on permafrost in Alaska, p. 106-113. In J. H. McBeath [ed.] The potential effects of carbon dioxide-induced climatic changes in Alaska, University of Alaska, Fairbanks, Misc. Publ. 83-1.

REEDER, S. W., B. HITCHON, AND A. A. LEVINSON. 1972. Hydrogeochemistry of the surface waters of the Mackenzie River drainage basin, Canada — I. Factors controlling inorganic composition. Geochim. Cosmochim. Acta 36: 825-865.

RUTTER, N. W., A. N. BOYDELL, K. W. SAVIGNY AND R. O. VAN EVERDINGEN. 1973. Terrain evaluation with respect to pipeline construction, Mackenzie Transportation Corridor, Southern Part. Environmental-Social Program Northern Pipelines, Rep. No. 73-36: 135 p.

SCHREIER, H. 1979. Winter water quality sampling and its use in determining hydrological conditions in the Ogilvie River system in the Yukon Territory, p. 559-569. In Proceedings, Canadian Hydrology Symposium '79 — Cold Climate Hydrology. National Research Council of Canada, Ottawa, Ont.

SMITH, M. W., AND D. W. RISEBOROUGH. 1983. Permafrost sensitivity to climatic change, p. 1178-1183. In Proceedings, Permafrost, Fourth International Conference, Vol. 1. National Academy Press, Washington, DC.

SOKOLOV, B. L. 1973. Regime of naleds, p. 99-108. In Permafrost, U.S.S.R. Contribution, Second International Conference. National Academy of Sciences, Washington, DC.

TRIMBLE, J. R., J. M. GRAINGER, AND P. K. GLEN. 1983. A sub-permafrost groundwater supply for the community of Old Crow, Yukon. Proceedings, Cold Regions Environmental Engineering Conference, University of Alaska, Fairbanks, May 1983.

VAN EVERDINGEN, R. O. 1974. Groundwater in permafrost regions of Canada, p. 83-93. In Permafrost Hydrology, Proceedings of Workshop Seminar. Canadian National Committee, International Hydrological Decade, Environment Canada, Ottawa, Ont.

    1975. Use of ERTS-1 imagery for monitoring of icings, N. Yukon and N.E. Alaska, p. 75-87. Summaries of progress and short research reports. Hydrology Research Division, Inland Waters Directorate, Environment Canada, Rep. Ser. No. 42.

    1976. Geocryological terminology. Can. J. Earth Sci. 13: 862-867.

    1979. Potential interactions between pipelines and terrain in a northern environment. National Hydrology Research Institute, Environment Canada, NHRI Pap. No. 8: 7 p.

    1981. Morphology, hydrology and hydrochemistry of karst in permafrost terrain near Great Bear Lake, Northwest Territories. National Hydrology Research Institute, Environment Canada, NHRI Pap. No. 11: 53 p.

    1982a. Management of groundwater discharge for the solution of icing problems in the Yukon, p. 212-226. In The Roger J. E. Brown Memorial Volume. Proceedings of the Fourth Canadian Permafrost Conference. National Research Council of Canada, Ottawa, Ont.

    1982b. Frost blisters of the Bear Rock spring area near Fort Norman, N.W.T. Arctic 35(2): 243-265.

    1985. Unfrozen permafrost and other taliks. Workshop on Permafrost Geophysics. U.S. Army Cold Regions Research and Engineering Laboratory, Spec. Rep. 85-5: 101-105.

VAN EVERDINGEN, R. O., F. A. MICHEL, H. R. KROUSE, AND P. FRITZ. 1979. Hydrochemical and isotope analysis of groundwater flow systems, Franklin Mountains, District of Mackenzie, N.W.T., Canada. National Hydrogeological Conference, Canadian National Chapter, International Association of Hydrogeologists, Edmonton, Alta. p. 166-178.

VAN EVERDINGEN, R. O., M. A. SHAKUR, AND H. R. KROUSE. 1982. Isotope geochemistry of dissolved, precipitated, airborne and fallout sulfur species associated with springs near Paige Mountain, Norman Range, N.W.T. Can. J. Earth Sci. 19: 1395-1407.

WILLIAMS, J. R., AND R. O. VAN EVERDINGEN. 1973. Groundwater investigations in permafrost regions of North America: a review, p. 436-446. In Permafrost, North American Contribution, Second International Conference. National Academy of Sciences, Washington, DC.

WOO, M. K. 1983. Hydrology of a drainage basin in the Canadian High Arctic. Ann. Assoc. Am. Geogr. 73(4): 577-596.

WOO, M. K., P. MARSH, AND P. STEER. 1983. Basin water balance in a continuous permafrost environment, p. 1407-1411. In Proceedings, Permafrost, Fourth International Conference, Vol. 1. National Academy Press, Washington, DC.

CHAPTER 10

Urbanization and the Hydrological Regime

F. B. MacKenzie

AJAMAC Management, 6408-128th Street, Edmonton, Alta. T6H 3X5

Introduction

Water in the urban environment has been referred to as a resource out of place because of the impacts of urbanization on the hydrological cycle (McPherson 1969). In fact, of all land use activities, urbanization probably has the greatest impact per unit area on the hydrological regime. These impacts are well documented and include modifications to the local climate, increased flooding potential, reduced water quality, and increased erosion and sedimentation.

In the initial stages of urban development much of the natural vegetation is removed and soil surfaces are exposed. The interception of precipitation by foliage and the evapotranspiration from plants are thus drastically reduced. As well, the bare soil becomes highly susceptible to erosion. Streets and parking lots are then paved and closely spaced buildings are constructed reducing the surface permeability which allows for little moisture to infiltrate into the soil. This may result in a lowering of groundwater levels but baseflow in urban streams is often increased through industrial discharges and water distribution system leakage.

As a result of these land surface changes, additional runoff is created in the urban environment as more of the precipitation is dispensed through surface runoff. Furthermore, the time of concentration of water in the basin is reduced because water runs off faster from streets and roof tops than from natural vegetated surfaces and because the number and total length of drainage channels is increased through the installation of ditches, culverts, and complex underground sewer networks. Sewer conduits are also much more hydraulically efficient than natural stream channels. The general result is to greatly accelerate surface runoff.

The urban hydrological cycle is further complicated by the high population densities found in cities. The air pollution that develops over cities has notable climatic effects in that the amount of solar radiation reaching the surface is reduced, mean air temperatures are generally increased, and total precipitation may be increased (Landsberg 1981). In addition, substantial water withdrawals are required to satisfy the large domestic, commercial, and industrial demands for potable water in cities. In some regions, the demand for potable water actually exceeds the sustainable withdrawals from the immediate urban vicinity and this has necessitated the search farther afield for adequate supplies. The city of Regina, Saskatchewan, is a good example of this, as provisions are being made to pipe water from Lake Diefenbaker on the South Saskatchewan River to the city (a distance of approximately 200 km) because local supplies are no longer adequate.

The problem of supplying water to urban areas is exacerbated by the reduction in quality the water undergoes through its use within the city. The quality of the water returned to the basin after use is often significantly diminished, even after treatment. The disposal of wastewater is a major concern for most municipalities. Finally, stormwater runoff, which is normally routed directly to receiving water bodies without treatment, is usually highly polluted because of the great quantities of pollutants it collects in washing off the streets, parking lots, and sidewalks. How to deal with poor stormwater quality is another pressing issue in municipal water pollution abatement programs.

The hydrological changes associated with urbanization are numerous and varied. Most of these impacts are amply discussed in the literature. In this chapter, the objective is to review and elaborate upon the most significant problems of the urban hydrological cycle, to outline current trends, and to speculate on future prospects. Canadian examples are highlighted wherever possible. For specific information the reader is referred to the literature, of which many examples are listed in the bibliography.

Erosion and Sedimentation

The natural pattern of erosion and deposition in a river basin is disrupted to the greatest extent during the initial construction stage in urbanization when soil surfaces are exposed to the environment and, following development, when the local stream channels have to adjust to the larger, more frequent peak flows that accompany urban development. The consequences of accelerated erosion and deposition during urbanization can be serious, since sediments may clog channels, fill reservoirs, increase flooding, promote the growth of algae, reduce the recreational potential of water bodies, and increase the costs of water treatment.

Wolman (1967) postulated that erosion and sedimentation in urban stream channels follows an evolutionary path involving three basic stages. Stage one is prior to development during which equilibrium conditions prevail. Only small changes in inflow of water and sediment occur from year-to-year and channel form and gradient may remain relatively stable. Land use in the drainage basin may be primarily agricultural or may be dominated by natural grasslands and forests. The sediment yields may range from 17.5 to 2800 kg $\cdot$ ha^{-1} $\cdot$ yr^{-1}. Stage two is the period of construction during which bare land is exposed to erosion and the sediment yields may become as great as 490 000 kg $\cdot$ ha^{-1} $\cdot$ yr^{-1} (Wolman 1967). The final stage is the mature urban environment during which the landscape is dominated by impermeable surfaces and sediment yields may fall to approximately 14.3 kg $\cdot$ ha^{-1} $\cdot$ yr^{-1}. These sediment yields are based on Wolman's observations in the Middle Atlantic region of the United States.

Similar variations in sediment yield during urban development have been noted by others. Graf (1975) reported that sediment yields in the Denver, Colorado, area are thirty times greater during suburban development than before construction. Vice et al. (1969) claimed that the sediment yield per acre for an average storm event in construction areas in Virginia is about 10 times greater than for agricultural areas, 200 times greater than for grassland, and 2000 times greater than for forested lands. Wolman (1975) suggested that construction within a basin can increase sediment yield by five to several hundred times although sediment traps and seasonal construction can reduce sediment loss.

In addition to the noted increases in sediment yield during construction, several authors have documented a tendency toward channel enlargement after urbanization is completed. Hammer (1972) found that channel enlargement associated with urbanization in Pennsylvania varies with the extent of impervious land uses and the length of time that impervious development has been in existence. He reported an average channel enlargement of about 2.5 times after urbanization. Hollis and Luckett (1976) claimed that complete urbanization of a catchment in England in which 50% of the basin is paved would result in a channel enlargement of about 4.9 times.

Added to the above problems is the reduction in stream channel aesthetics that often accompanies urbanization. Many channels are either straightened or paved to accommodate peak flows, and low-order streams may be completely filled in. Leopold (1968) also complained that urban stream channels tend to collect the "artifacts of civilization" such as beer cans, old tires, oil drums, and other rubbish. Recently, however, more effort has been expended in maintaining urban streams and ravines for recreational purposes. In Toronto and Edmonton, for example, most development along local streams is limited to recreational use. Parks, bicycle paths, and trails for hiking, cross-country skiing, and nature appreciation dominate the landscape in the ravines and along the stream channels in these two cities.

Precipitation

That metropolitan regions influence their climate has long been recognized. There is an abundance of literature on the subject which has become a very distinct branch of climatological research. Several bibliographies, each citing several hundred references, have been published (e.g. Chandler 1970; Changnon et al. 1977; Oke 1974, 1979). In addition, there is at least one textbook available that deals with urban climatology (Landsberg 1981).

The influence of urban areas on climate has both direct and indirect impacts on the hydrological cycle. Anomalies in the precipitation regime are a direct impact, while the interception of solar radiation, higher ambient air temperatures resulting in the so-called "urban heat island," and altered atmospheric conditions such as local wind systems indirectly affect the hydrological cycle. In this section, only the anomalies dealing with precipitation will be discussed.

There is strong evidence that precipitation is enhanced by urban areas. It is also apparent that not only is total precipitation increased but the frequency of thunderstorms and hailstorms is enhanced as well. Studies in many of the major United States cities, and elsewhere, have shown rainfall to be 5-25% greater than surrounding rural areas, while the frequency of thunderstorms can be up to 40% greater and hailstorms over 400% more frequent (Changnon 1968,1969, 1973, 1976; Landsberg 1956, 1962, 1981). These effects, however, appear to be most notable in the downwind regions of the metropolitan areas (e.g. Changnon 1968, 1976).

It is believed that three factors contribute to the modification and enhancement of rainfall in metropolitan regions. First, the urban heat island, although more dominant in winter, may lead to vertical air movement and subsequent convective precipitation. Any additional rainfall as a direct result of this is likely to occur in the downwind areas away from the city. Second, the aerodynamic roughness of urban development may impede the progress of weather systems (Landsberg 1981). Cyclonic activity thus delayed in passage over a city may produce more precipitation than normally might occur but only very tall buildings are likely to create this effect. Finally, the abundance of aerosols in the urban atmosphere act as condensation nuclei for raindrop formation and may speed up the development of rain producing clouds. This would also likely have a greater impact on downwind regions.

No firm evidence exists with respect to any possible augmentation of snowfall by cities (Oke 1978) but the increased air temperatures in cities may be important in reducing the length of permanent snowcover. For example, Hage (1975) reported that during the primary snowmelt period in Edmonton it is not uncommon to have temperatures above freezing at night in the city and below freezing in the countryside, thus enhancing the snowmelt rate within the city. Snow clearing in cities probably also has the effect of decreasing albedos compared with rural environs, thus allowing more heat to be available for melting snow.

Changes in Runoff Characteristics

Time Distribution and Shape of Urban Storm Hydrographs

The impact of urbanization on the time distribution and shape of urban storm hydrographs is well documented (e.g. Anderson 1963; Bras and Perkins 1975; Gregory 1974; Hollis 1974, 1975; Leopold 1968, 1973). The reduction in runoff response time, or lag time, is particularly significant. Lag time may be defined as the elapsed time between the peak rainfall intensity and the associated runoff peak flow. Lag times are decreased because of the higher flow velocities water attains in running off streets and roof tops as opposed to natural vegetated surfaces but also because the installation of sewer networks greatly increases the drainage density of a basin. Drainage density is the sum of the length of all drainage channels divided by the basin area. For instance, Roberts (1972) reported that the drainage density of a basin in suburban Bloomington, Indiana, was increased by a factor of 2.4 times with sewering. Graf (1977) claimed that an increase of 1.5 times in the basin drainage density occurred with the development of South Branch, Iowa.

Another factor in reducing basin lag time is that sewers are more hydraulically efficient than natural stream channels. The smoother surfaces of these artificial channels offer less resistance to flow than the rougher, natural streambeds.

The changes in lag time caused by urbanization are striking. Anderson (1963) reported that lag time can be decreased by 85% for a basin with complete sewering. Martens (1968) claimed a reduction in lag time of about 75% for basins that are almost completely developed for urban use. Gregory (1974) reported a decrease of 49% in the lag time for a basin in which 12.2% of the land is affected by urban development. In one other study, Hollis (1974) described how the urbanization of 80% of a catchment in England has caused a 56% reduction in lag time.

The effect of urbanization on storm runoff peak flows was also reported by numerous authors (e.g. Anderson 1970; Carter and Thomas 1968; Gregory 1974; Hollis 1974, 1975; James 1965; Leopold 1968; Stall et al. 1970; Taylor 1977). In most of these studies, emphasis is placed upon the changes in peak flow as related to the percentage of a basin covered with impervious surfaces and the storm recurrence interval. For example, the impact of urbanization on the mean annual flood peak is frequently cited. The mean annual flood is simply the average of the annual peak flows and is found to have a recurrence interval of about 2 yr (Gregory and Walling 1973). It has been demonstrated that the peak flow of the mean annual flood after urbanization can be eight times higher than the peak flow prior to urbanization (Anderson 1970; Carter and Thomas 1968; Leopold 1968; Stall et al. 1970). The exact increase depends upon the level of imperviousness and the extent of sewering.

It was also reported in the above studies that the impact of urbanization on peak flows and total runoff is greatest for the frequent storm events while the infrequent, large-magnitude events are not appreciably affected. The reason for this is that basins under rural land use conditions behave in a similar fashion to highly impervious basins during the low frequency events. The high rainfall intensities normally associated with infrequent storm events tend to exceed the soil infiltration capacities in an undeveloped basin resulting in significant surface runoff much the same as in a highly impervious urbanized basin. For example, both James (1965) and Anderson (1970) claimed that the effect of urbanization upon the 100-yr flood peak flow is insignificant, while others reported increases in peak flow of about two times for the 100-yr event (Carter and Thomas 1968; Hollis 1975; Reimer and Franzini 1971; Stall et al. 1970). Storm events with recurrence intervals greater than 100 yr are probably not materially affected by urban development.

This direct relationship between the impact of urbanization on peak flows and the frequency of storm events has serious implications because it is the very frequent events that are affected the most. The area affected by frequent flooding is increased and storm sewers have to be designed to accommodate volumes greater than peak flows that would occur from undeveloped basins. This is one reason that municipalities can only afford to install sewers capable of accommodating flows from the 5-yr, and in some cases the 10-yr, flood. Sewers capable of handling the flow volumes from larger events are not practical from both a physical and economic perspective.

Changes in Groundwater Yield

Intuitively, the yield from groundwater should be diminished with urbanization because the reduced infiltration rates that occur in urban areas should result in less groundwater recharge. This has been the case in many areas. James (1965) reported that stream baseflow in an urbanized basin in California is only 0.7 times the rural value and attributed this to the lower urban infiltration rates. Franke (1968) reported an average drop of 2.1 m in the water table in Nassau Country, Long Island, New York because of sewering. This has produced a 20% decline in the groundwater yield for that region. In contrast, there are studies which indicate that increases in baseflow can occur with urban development (Miller 1966; Sawyer 1963). These examples of conflicting reports led the American Society of Civil Engineers Task Committee on the Effect of Urbanization (1975) to conclude that urban development can either decrease or increase the low flow of streams.

This dichotomy is the result of several factors. On the one hand, groundwater recharge from natural sources is probably reduced because of the lower infiltration rates. The exception may be snow dumps which can act as local recharge sites. In addition, sewers with loose fitting joints, especially in areas of high water tables, can act as weeping tile in draining areas and lower the water table. On the other hand, there is significant leakage from urban water distribution systems and this leakage probably contributes to groundwater recharge. Indeed, it seems that municipal water distribution systems have a propensity for leakages. Howe (1971) reported that for a sample of 91 cities in the United States during the period 1964-69, the mean loss of treated water production was 12%. Tate and Lacelle (1978) reported a mean loss of about 19% of production for municipal water treatment systems across Canada. System losses are so common that a municipal engineering rule of thumb states that systems that lose less than 15% of total production are in good shape (Howe 1971). Some of the above losses constitute unreported usage, but still a great deal of water is lost through leakage and this can contribute to groundwater recharge. Another source of recharge may be lawn overwatering. Most of the surplus water above plant needs may percolate to the water table.

The impact of urbanization on groundwater supplies is obviously very complex and no one definitive statement can be applied for all regions or individual localities. The impact has to be evaluated on an individual basis.

Annual Yield Modifications

There are a few studies that document increases in total annual water yield with urban development. Using the Stanford Watershed Model to simulate flows with various levels of urbanization in the Morrison Creek watershed in California, James (1965) found that over a 10-yr period the annual runoff with complete urbanization was 2.29 times its rural value. Harris and Rantz (1964) employed a double-mass analysis to determine the effect of urban development on the streamflow regimen of Permanente Creek, California. They found that the ratio of stream outflow to rainfall inflow increased by 44% from 1945 to 1958. During that same period, the amount of impermeable surfaces in the basin grew from 4 to 19%. Waananen (1969) reported that, on the basis of mean annual flows, the average annual yield from urban areas in Connecticut was twice that from undeveloped areas. Hartmann (1972) employed the Thronthwaite climatic water budget analysis to determine the effect of urbanization on annual yields in the Chester Creek watershed in southeastern Pennsylvania. Using climatic data for the period 1936-68, Hartmann determined that the annual yield had increased by about 17%. This corresponded to an increase from 4.9 to 12.1% in the impervious cover with no noticeable change in the annual rainfall totals. Muller (1967, 1969) also employed the Thornthwaite procedures in two separate studies to determine the water balance patterns of two river basins in New Jersey. In the earlier study, Muller reported a 48% increase in the South Raritan River annual flow after 50% of the basin had been covered by impervious surfaces. In the later study, he found that the average annual runoff from subdivisions in Middlesex County increased about 29% after 25% of the area was paved. This additional surplus is sufficient to meet the domestic water requirements of about 36 000 people according to Muller, notwithstanding quality limitations.

Two other studies dealing with urban-induced yields are more specific in terms of cause and effect. Hollis (1977) claimed that the total annual yield from the Canon's Brook catchment in Harlow, England, increased by 75% from 1950 to 1968 with 16.7% of the basin having been paved during that period. Hollis (1977) also reported that the increase in total annual flow was closely related to the percentage of the basin paved but also to the amount of rainfall in the preceding summer. The proportional increase in annual yield was greatest for those years when rainfall was reduced in the preceding summer. In other words, there is an inverse relationship between rainfall and the effect of urbanization on annual water yield.

Laycock and MacKenzie (1984) also found that greater proportional yields are produced by urban areas in semi-arid and subhumid regions than by urban areas with more humid climates. In their study, water balances were computed for eight cities in Canada that represent different climatic regions. They found that Ottawa and Vancouver, two cities with humid climates, experience higher annual yields by about 19 and 13% respectively, over their environs. In contrast, the proportional increase in surplus is very striking in the drier regions. Saskatoon and Medicine Hat, for example, have urban-induced annual yields that are over five times and almost 10 times greater, respectively, than would occur under rural land use conditions.

In summary, there is not a strict causal relationship between the increase in annual yield and the percentage of impervious surfaces in urban areas. Although the percentage imperviousness is no doubt an important variable, the annual rainfall regime is also very important. In humid regions, where soils remain close to saturation and annual surpluses are high, the proportional increase in yield caused by urbanization is relatively small. In drier climates, where much of the annual precipitation normally infiltrates into the soil, the effects of urbanization on annual yields are much greater.

Water Quality

There are three major areas of concern when dealing with the water quality changes induced by urbanization: (i) wastewater disposal, (ii) stormwater runoff, and (iii) combined sewer overflows. Wastewater disposal refers to all domestic and industrial sewage produced in municipalities. In Canada, most cities provide some level of treatment for their sewage but there are some notable exceptions such as St. John's, Halifax, and Quebec City. In most of the newer areas of cities, stormwater is collected in separate sewers and disposed of in local water bodies. It is now recognized that stormwater runoff represents a major source of water pollution in most municipalities. In the older portions of most cities, stormwater and domestic sewage are collected in the same sewer system and directed toward a treatment facility. During rainfall and snowmelt periods, however, the flows may exceed the capacity of the treatment facility and some sewage may be discharged to a local water body directly without any treatment.

Wastewater Disposal

Sanitary sewage disposal is a major concern for most municipalities because of the large volumes of water involved and the need for costly treatment to ensure that the effluent does not exceed the assimilative capacity of the receiving stream. The magnitude of the problem can be appreciated when it is realized that wastewater flows in municipalities represent a very high proportion of the domestic water withdrawals.

Baumann and Dworkin (1978) reported that the water collected as sewage usually amounts to nearly 65% of the volume of water supply. Tate and Lacelle (1978) claimed that 60% of Canadian municipal water supplies is returned as wastewater. In Canada in 1976 this represented approximately 7360 million litres per day (MLD) of wastewater directed to treatment plants (Tate and Lacelle, 1978).

The costs of treating wastewater are very high and depend upon the level of treatment provided, the ease of sludge disposal, and the volume of sewage treated. There are significant economies of scale in treating wastewater. For example, Bauman and Dworkin (1978) claimed that the cost per 1000 U.S. gallon of sewage receiving secondary treatment is almost twice as high for a plant treating 1 million U.S. gallons per day as for one treating 10 million U.S. gallons per day, and almost three times greater than in a 100 million U.S. gallon per day plant. Furthermore, sewage treatment is energy intensive and there are large energy input increases as treatment levels increase.

In Canada, wastewater treatment is a municipal responsibility controlled by provincial regulations and guidelines that vary from province to province. In 1984, about 85% of the urban population in Canada was served by sewers but only about 57% was served by some form of sewage treatment (Environment Canada 1982). The breakdown by province is provided in Table 1. In Quebec, a very small percentage of the urban population was served by sewage treatment but a large assistance program is underway to upgrade municipal sewage treatment facilities in that province (Inquiry on Federal Water Policy 1985b).

The level of wastewater treatment provided in the major Canadian cities in 1978 is listed in Table 2. The disparate level of treatment provided in these cities is revealing. Cities that dispose of their wastewater into the ocean such as St. John's, Charlottetown, Halifax-Dartmouth, Vancouver, and Victoria provide little or no treatment. In Quebec, Quebec City discharges raw sewage directly to the St. Lawrence River, and in Montreal only a very small proportion of the wastewater is treated. Most of that city's sewage is dumped untreated into the St. Lawrence. The much stronger environmental controls exercised in Ontario are evident in the high level of wastewater treatment provided by that province's major cities. The remaining cities across the country provide secondary treatment and most have some form of sludge processing.

TABLE 1. Population served by wastewater systems in Canada. Source: Environment Canada (1982, data updated to 1985).

Province or territory	Total population surveyed	Percentage served by sewers	Percentage served by sewage treatment
Newfoundland	497 018	60.0	12.8
Prince Edward Island	57 587	100.0	94.4
Nova Scotia	536 604	83.3	21.4
New Brunswick	409 900	91.8	60.7
Quebec	6 685 434	81.9	6.2
Ontario	7 641 607	86.1	83.5
Manitoba	839 158	94.9	94.7
Saskatchewan	611 072	99.0	99.0
Alberta	1 852 714	99.3	99.3
British Columbia	2 175 754	77.5	77.5
Yukon Territory	21 888	88.0	86.2
Northwest Territories	43 953	91.6	52.5

TABLE 2. Municipal wastewater treatment in the major Canadian cities, 1978. Source: Environment Canada (1978, table 18, p. 87).

City	Percentage of population served by sewers	Percentage served by sewage treatment	Disinfection	Primary	Secondary	Tertiary	Sludge processing
St. John's	92.0	0					
Charlottetown	83.3	83.3	×	×			
Halifax–Dartmouth	97.4	0					
Saint John	90.9	28.4	×	×	×		×
Fredericton	94.8	94.8	×	×	×		×
Quebec City	100	0					
Montreal (metro)	98.1	5.3	×	×	×		×
Ottawa–Carleton	n/a	96.7	×	×	×	×	×
Toronto (metro)	99.6	99.6	×	×	×	×	×
Hamilton	93.5	93.5	×	×	×	×	×
London	98.3	98.3	×	×	×	×	×
Windsor	79.0	79.0	×	×	×	×	×
Winnipeg	100	100		×	×		×
Saskatoon	100	100	×	×	×		×
Regina	100	100	×	×	×	×	×
Calgary	100	100		×	×		×
Edmonton	79.2	84.2		×	×		×
Vancouver (metro)	99.0	90.1	×	×			
Victoria	92.8	81.9	×				

Stormwater Runoff

Pollution problems associated with stormwater runoff are particularly worrisome. It is a nonpoint source of pollution and, therefore, difficult to characterize let alone control. Furthermore, it is now evident that stormwater runoff is a very significant source of stream pollution and contains many diverse pollutants. The sources of these pollutants include airborne materials (e.g. dust and lead) that have settled to the surface or been washed out of the atmosphere by precipitation, rubber particles from automobile tires, asbestos fibres from brake linings, animal and bird droppings, lawn fertilizers, leaves and lawn clippings, salt, and general litter.

Several Canadian studies (e.g. Luckmann 1979; Mills 1977; James F. MacLaren Ltd. 1980a, 1980b; Marsalek 1979; Sullivan et al. 1978; Swain 1985) reported finding high levels of total suspended solids, dissolved solids, inorganic and organic carbon, phosphorous, phenols, hydrocarbons, coliform bacteria, and heavy metals in urban runoff. A profusion of studies from the United States and elsewhere reported similar findings. A summary of the composition of pollutants in urban runoff is presented in Table 3.

A problem common to most Canadian municipalities is the impact on local streams of salt from road deicing. For example, James F. MacLaren Ltd. (1980a) reported chloride concentrations as high as 3200 mg/L in runoff from local streets in North York, Toronto, following salt applications. However, according to Scott (1979), these very high salt concentrations do return to so-called "normal" levels shortly after runoff events.

In addition to the concentration of pollutants in stormwater runoff, the time distribution of the quality parameters is very important. Because of the accumulation of materials

TABLE 3. Characteristics of urban stormwater. Source: Roesner (1982, table 2, p. 163).

Characteristic	Range of values
Biochemical oxygen demand (mg/L)	1-700
Chemical oxygen demand (mg/L)	5-3100
Total suspended solids (mg/L)	2-11 300
Total solids (mg/L)	450-14 600
Volatile total solids (mg/L)	12-1600
Settleable solids (mg/L)	0.5-5400
Organic nitrogen (mg/L)	0.1-16
Soluble phosphate (mg/L)	0.1-10
Total phosphate (mg/L)	0.1-125
Chlorides (mg/L)	2-25 000[a]
Oils (mg/L)	0-110
Phenols (mg/L)	0-0.2
Lead (mg/L)	0-1.9
Total coliforms (no./100 mL)	$200-146 \times 10^6$
Fecal coliforms (no./100 mL)	$55-112 \times 10^6$
Fecal streptococci (no./100 mL)	$200-1.2 \times 10^6$

[a]With highway deicing.

on the streets and in the sewers the concentration of pollutants peaks in the stormwater runoff very early in a storm and normally precedes the discharge peak. This is often referred to as the "first flush effect" (Lazaro 1979) and can have serious consequences because it produces an initial slug of highly polluted water. After the first flush, the concentration of pollutants tends to decline to a constant level. Frequent street sweeping and sewer flushing can reduce the impact of the first flush.

Stormwater pollution is such a serious problem that a parallel is often drawn between the quality of urban runoff and secondary treated wastewater effluent (Table 4). The summary data in Table 4 are an indication that the BOD of stormwater is very high relative to secondary treated effluent, and the suspended solids in urban runoff can exceed those found in secondary effluent. The quantity of nitrogen and phosphorous in stormwater is also very high.

The recognition that uncontrolled urban storm runoff is highly polluted has led many people, especially in the United States, to question the justification for massive expenditures of public funds for tertiary treatment of municipal wastewater when the highly polluted stormwater is allowed to pass untreated to the same receiving water body as

TABLE 4. Comparison of annual pollution loadings from stormwater and secondary waste treatment plant effluents. Source: Environment Canada (1978, table 16, p. 91).

Population density (no./ha)	Source	Annual loadings (kg/ha)			
		BOD	Suspended solids	Nitrogen	Phosphorous
60	Stormwater	34	392	9	1.6
60	Secondary effluent	164	246	39	27
150	Stormwater	90	560	11	3.4
150	Secondary effluent	410	615	87	68

the treated effluent. The improvement is water quality that can be anticipated in the receiving stream with tertiary treatment of wastewater, opponents of such treatment argue, would be inconsequential and very expensive because stormwater runoff is as highly polluted as secondary treated effluent. Now that most municipalities have secondary treatment facilities, treatment of stormwater appears to be more desirable than further treatment of wastewater. Unfortunately, there are some major obstacles, mostly economic, in attempting to treat urban runoff. It represents a very significant volume of water that occurs on an irregular basis. Pisano (1975) claimed that the 1975 costs of controlling stormwater runoff in the United States are $235 billion. The annual control costs for storm areas in 56 urbanized areas in Ontario have been calculated to be approximately $168/ha for 50% control and $445/ha for 75% control (Sullivan et al. 1978). No estimates were completed for 100% control but the magnitude of the costs involved is apparent.

Measures to control stormwater pollution are similar to treating wastewater, that is, storage, treatment, disinfection usually with chlorine, and then discharge. In many new suburban developments, urban runoff pollution is being partially controlled through the use of stormwater impoundments. These artificial lakes are used primarily to reduce the stormwater runoff peak flows by storing the water and releasing it slowly at a later time, but they also act as settling basins and some improvement in water quality has been noted (Michaels et al. 1985). These impoundments have other advantages in addition to flood control and improved water quality. The lakes are often popular recreational areas, their presence can enhance local real estate values, and they may serve as a habitat for waterfowl and wildlife. Unfortunately, stormwater lakes are not a practical alternative for the older portions of most cities. In those areas, control of stormwater would probably require some form of subsurface storage which could be very costly.

There is little evidence that urban stormwater pollution is a major concern in Canada except perhaps to municipal officials. There is no mention of this problem in either the Participation Paper or the Public Hearings Synthesis of the Inquiry on Federal Water Policy (1984, 1985a) and only brief mention of it in the Final Report (Inquiry on Federal Water Policy 1985b). Nor did Foster and Sewell (1981) discuss urban runoff in their book on emerging water problems in Canada. The attitude appears to be that this is a local pollution problem that should be dealt with by the municipalities. On the contrary, it is a widespread and pervasive pollution problem that is difficult and costly to control.

Combined Sewer Overflows

The older portions of most Canadian cities are serviced by combined sanitary/stormwater sewer systems. Wastewater and stormwater, collected in the same sewer network, are directed for treatment to wastewater plants. If, however, the flows exceed the capacity of the treatment plant, the excess is routed directly to the receiving water body without any treatment. These discharges of raw sewage combined with stormwater runoff are, needless to say, a significant source of pollution. LaGrega and Keenan (1975), for example, claimed that the organic loadings of combined sewer overflows average about one half the concentration of untreated sanitary sewage. More importantly, these overflows, like stormwater runoff, produce a shock loading on the receiving water body because they generally occur only during rainfall and snowmelt events. Depending upon the climatic region, these overflows may exceed 100 occurrences per year. A characterization of the pollutant concentrations in combined sewer overflows is presented in Table 5.

The resolution to the problem of combined sewer overflows has traditionally been in-system storage or separation of the sanitary and stormwater systems with the stormwater sewers discharging directly to local water bodies. By this means, all the raw sewage

TABLE 5. Characteristics of combined sewer overflows. Source: Roesner (1982, table 1, p. 163).

Characteristic	Range of values
Biochemical oxygen demand (mg/L)	30-600
Total suspended solids (mg/L)	20-1700
Total solids (mg/L)	150-2300
Volatile total solids (mg/L)	15-820
Settleable solids (ml/L)	2-1550
Organic nitrogen (mg/L)	1.5-33.1
Soluble phosphate (mg/L)	0.1-6.2
Total coliforms (no./100 mL)	20 000-90 × 10^6
Fecal coliforms (no./100 mL)	20 000-17 × 10^6
Fecal Streptococci (no./100 mL)	20 000-2 × 10^6

receives treatment. The actual drawbacks of the combined system may be less than assumed, however, because of three factors. First, stormwater is itself highly polluted. Second, stormwater runoff receives treatment with combined sewers during rainfall and snowmelt events that produce runoff in quantities that are below the sewer and treatment plant capacities. Third, separate sewer systems are more costly than combined systems, especially in the older sections of the cities where retrofitting is required. Wiggers and Bakker (1978) reported that the annual average pollution load with the separate system in the Netherlands is roughly 10 times greater than with the combined system. Furthermore, the separate system in that country is approximately 50% more expensive to install than the combined system.

Water Supply

Municipal withdrawals are a very significant water use. Of all the major water withdrawals in Canada, municipal water use represents the third largest user (behind thermoelectric and manufacturing) and the third largest consumptive use (behind agriculture and mining) (Environment Canada 1980). With urbanization and rising standards of living, the per capita demand for potable water has grown steadily over the past several decades in North America. In the United States the daily municipal per capita usage was 340 L in 1900, 530 L in 1940, 590 L in 1965, and 715 L in 1970 (Baumann and Dworkin 1978). There are no comparable figures available for Canada but it is safe to assume that the pattern has been the same in this country. There is evidence, however, that this upward trend in per capita use has slowed in the past decade as new sources of supply have become increasingly expensive to develop because of scarcity and/or poor quality of sustainable withdrawals.

Current Trends

Total municipal and rural residential water withdrawals grew in Canada from 9956.5 MLD in 1974 to approximately 12 410 MLD in 1980, an increase of about 24.6% (Table 6). This corresponds fairly closely to the growth in population over that period. The per capita demand in both 1974 and 1980 was about 520 L/d. The rate of consumptive use for the individual provinces is not available for 1974 but the average value for the country was estimated to be approximately 20%, representing about 1990 MLD (Environment Canada 1975). The rates of consumptive use in 1980 ranged from 10.7% for the Atlantic region to 33.3% for the Yukon and Northwest Territories. The value for the north is believed to be so high because of the need to allow running water during the winter months to prevent pipes from freezing. The average rate of consumptive use for the country in 1980 was 16.7%, or approximately 2075 MLD

287

TABLE 6. Canadian municipal and rural domestic water use (in millions of litres per day, MLD). Sources: Environment Canada (1975, 1976, 1983).

Withdrawals	1974		1980	
Atlantic	772.8		1 443	
Quebec	3240.4		4 510	
Ontario	3554.0		3 615	
Prairie Provinces	1226.0		1 739	
British Columbia	1136.9		1 076	
Territories	26.4		27	
Total	9956.5		12 410	

Consumption		% of withdrawals		%
Atlantic	n/a		155	10.7
Quebec	n/a		637	14.1
Ontario	n/a		522	14.4
Prairie Provinces	n/a		470	27.0
British Columbia	n/a		282	26.2
Territories	n/a		9	33.3
Total	1991.3 (estimated)	20.0	2 075	16.7

(Environment Canada 1983; see also chapter by Tate, this volume). For comparative purposes, the rate of consumptive use for municipal and rural domestic use in the United States was about 23.5% in 1978 (U.S. Water Resources Commission 1978).

Future Trends

Projecting the future requirements for municipal water use is a perilous task. In the past it was sufficient to calculate a value for per capita demand and multiply it by projected population values. An allowance was often incorporated to account for an increase in per capita demand. Currently, however, there are indications that the growth in per-capita demand for municipal water has peaked and is levelling off. For example, the U.S. Water Resources Commission (1968), in its First National Assessment, projected that by the year 2000 approximately 192 007 MLD would be required to satisfy municipal demands for water in the United States (up 136% from 1965), with the rate of consumptive use rising from 22% in 1965 to 37.5% by the year 2000. No apparent consideration was given to the fact that the United States does not have sufficient freshwater supplies to meet this projected demand. In contrast, in the Second National Assessment, the U.S. Water Resources Commission (1978) drastically reduced the projected long-term growth rate in municipal water demand. They now project a demand for municipal water by the year 2000 of about 105 679 MLD, or 45% less than the previous projection. Furthermore, the anticipated rate of consumptive use for the United States in the year 2000 is 23.8%, only a marginal increase from current levels. The projected demand for water by municipalities in the United States is currently expected to increase 32% by the year 2000.

Unfortunately, a regular, comprehensive assessment of current and future water demands for municipal use, like the National Assessments of the U.S. Water Resources Commission, is lacking in Canada. Nevertheless, the Inland Waters Directorate of Environment Canada (1980) predicted that the Canadian demand for municipal and rural domestic use in the year 2000 will be approximately 13 865 MLD, an increase of about

11.7% from 1980. The rate of consumption use is expected to remain at about 18% for this same period.

Regardless of the exact increase in demand, there remains the problem of satisfying these future water requirements. Traditionally, increased water demand was met by increasing supply but, as previously mentioned, the increasing scarcity of useable supplies plus the escalating costs of new developments, treatment, and distribution will limit our economic capacity to continue expansion according to past trends. Even now there are areas in Canada where current demands for water exceed the reliable minimum local supplies. Six drainage basins were identified by Environment Canada (1983) where current and future demands either exceed, or are near, the reliable minimum monthly flow. They are the Okanagan, Milk, North and South Saskatchewan, Red-Assiniboine, and southern Ontario drainage basins. With the exception of southern Ontario, these basins are located in the drier areas of the contry. Whereas southern Ontario has the advantage of proximity to the Great Lakes, considerable capital will be required to construct water distribution networks as demands exceed available local supplies. Poor water quality is also a major concern in southern Ontario.

A practical alternative to increasing water supplies is controlling demand (see chapter by Tate, this volume) Demand control, which is gaining in popularity throughout North America, usually takes the form of metering and pricing policies. Traditional thought has been that the demand for municipal water is inelastic, that is, not responsive to price changes. It was assumed that because water is a unique commodity and a necessity for modern life it must be provided almost without regard for cost. Water demands, consequently, were viewed as immutable requirements which increase with economic and population growth and, therefore, must be satisfied by constructing larger supply facilities (Zamora et al. 1981). Furthermore, because water demand was considered to be price inelastic, pricing policies were geared toward recovering the costs of providing water services to different customer services. No value was placed on the resource. Rate structures usually consisted of a fixed service charge per billing period regardless of total individual consumption or in some cases a declining block rate charge in which lower rates were charged as total individual consumption increased.

Numerous studies have demonstrated that the demand for municipal water is actually price elastic. Grima (1979) stated "the literature on metering and pricing is extensive and the results of research are clear: metering and pricing achieve substantial reductions in municipal water use in general and residential water use in particular." The best Canadian example of this is probably the contrasting water demand patterns in Calgary and Edmonton. Although both cities have about the same population, water demand in Calgary historically has been 30-40% greater than in Edmonton (Gysi 1981a). The discrepancy is believed to be largely the result of an almost completely metered population in Edmonton, while in Calgary very few of the residential water users are metered. Further, the nonmetered users in Calgary are charged a fixed monthly rate, regardless of consumption. The unmetered residential customers in Calgary have a daily per capita demand that is approximately 300 L greater that the residential demand in Edmonton, but the per capita demand for the metered residential customers in Calgary is about the same as in Edmonton (Gysi and Lamb 1977). There is some debate as to whether the major difference between the two cities is due to higher leakage in Calgary's distribution system but there are data which show that the average per capita consumption dropped by 30-50% in several Canadian cities after complete metering programs were carried out (Gysi 1981b).

There are other pricing policies that have been proposed, and implemented in some cases, to control municipal water demand. These include incremental block pricing, summer peak use rates, and differential rates that are based on geographic location within the municipality. Incremental block pricing refers to water rates that increase as usage increases. Peak use rates may be applied because peak demands for water, which usually occur during the summer lawn watering period, can result in system pressure loss or reduced reserves. Consequently, higher rates can be imposed during this period in order to lessen peak demands. Differential rates based on geographic location within the municipality are designed to offset the higher costs of delivering water to areas that are isolated from the water treatment plant or to areas that may be significantly higher in elevation than the treatment plant.

One other form of demand control is restricting water usage. These restrictions may be either voluntary or imposed. An advantage to the use of restrictions is their flexibility. They may be based on hours of use or types of activity or they may be confined to peak demand periods. Voluntary restrictions, however, will only be accepted by the public if people are convinced that a crisis situation exists (Baumann and Dworkin 1978).

Conclusion

The future trends with respect to the impacts of urbanization on the hydrological cycle are difficult to predict. The major problem areas are likely to be the ongoing pollution of water bodies by wastewater and stormwater disposal and the continuing need to supply large per capita demands for high quality water. There are impending water shortage problems in the Prairies because of the lack of supplies and in southern Ontario because of the contamination of useable supplies. Solutions to these problems will no doubt follow past trends and involve attempts at augmenting and improving supplies, but management of the demand will have to be a component of any strategy.

A partial solution to the problems of urban runoff pollution and the scarcity of useable water supplies may be alternative stormwater management techniques. The stormwater runoff produced in municipalities represents a large unappropriated source of water. There are undeniably quality and temporal constraints associated with this water but the quantity available should be put to the best practical use. Stormwater has been viewed as a liability in most municipalities. Traditional thought in stormwater management has been to collect, route, and dispose of stormwater runoff as quickly as possible without regard to any potential uses of this water. The current trend in stormwater management, however, is towards the development of surface detention basins or stormwater lakes which do have many positive benefits. However, with the apparent increasing acceptance of these lakes as an effective stormwater management tool, other tools and uses may be overlooked.

A radical change is required in our approach to municipal water management. Rather than classifying water supply, drainage, and wastewater treatment as separate entities, an approach that views the entire spectrum of municipal water management activities as a complex system should be employed. This "total management concept" would be one that presents all water-related functions of a municipality in such a way that not only is each individual component identified, but also the very real and significant interactions between components are realized. This holistic philosophy is based on two simple fundamentals: first, all water, regardless of the quantity and quality, must be viewed as an asset, and second, the conservation of water is an integral part of each component of the total system. Stormwater is a component of the system and should be managed in this context. In other words, the capture and reuse of stormwater runoff should be a strong objective in municipal water management. Probably the greatest potential for stormwater use is lawn watering. This water use accounts for a significant portion of the water consumed in domestic use and is a primary factor in the peak summer demands for water. The direction of runoff from rooftops onto lawns or into onsite storage facilities for later use could be employed. Homeowners could be provided the incentives of a lower water rate or lower taxes because of the decrease in runoff from individual lots and the reduced demand for potable water. Many other uses of this water are available and should be considered.

References

ANDERSON, D. G. 1963. Effect of urbanization of floods in northern Virginia. U.S.G.S. Prof. Pap. 475-A: A69.

1970. Effects of urban development on floods in northern Virginia. U.S.G.S. Water Suppl. Pap. 2001-C.

BAUMANN, D., AND D. DWORKIN. 1978. Water resources for our cities. Resour. Pap. College Geogr. No. 78-2. Association of American Geographers.

BRAS, R. L., AND F. E. PERKINS. 1975. Effects of urbanization on catchment response. J. Hydraul. Div. Proc. ASCE 101(HY3): 451-466.

CARTER, R. W., AND D. M. THOMAS. 1968. Flood frequency in metropolitan areas, p. 56-67. In P. Cohen and M. N. Francisco [ed.] Proceedings of the fourth American Water Resources Conference. AWRA Proc. No. 6.

CHANDLER, T. J. 1970. Selected bibliography on urban climate. WMO Publ. No. 276.

CHANGNON, S. A. 1968. The LaPorte weather anomaly — fact or fiction? Bull. Am. Meteorol. Soc. 49(1): 4-11.

1969. Recent studies of urban effects on precipitation in the United States. Bull. Am. Meteorol. Soc. 50(6): 411-421.

1973. Inadvertent weather and precipitation modification by urbanization. J. Irrig. Drain. Div., Proc. ASCE 99(IR1): 27-41.

1976. Inadvertent weather modification. Water Resour. Bull. 12: 695-718.

CHANGNON, S. A., F.A. HUFF, P. T. SCHICKEDANZ, AND J. L. VOGEL. 1977. Summary of METROMEX. Vol. 1. Weather anomalies and impacts. Ill. State Water Surv. Bull. 62.

ENVIRONMENT CANADA. 1975. Canada Water Year Book 1975.

1976. Canada Water Year Book 1976.

1978. Canada Water Year Book 1977-1978.

1980. Water and the Canadian economy. Inland Waters Directorate.

1981. Canada Water Year Book 1979-1980.

1982. National inventory of municipal waterworks and wastewater systems in Canada 1981. Data updated to January 1985.

1983. Canada Water Year Book 1981-1982.

FOSTER, H. D., AND W. R. D. SEWELL. 1981. Water: the emerging crisis in Canada. James Lorimer & Co., Toronto, Ont.

FRANKE, O. L. 1968. Double-mass-curve analysis of the effects of sewering on ground water levels on Long Island, New York. U.S.G.S. Prof. Pap. 600-B: B205-B209.

GRAF, W. L. 1975. The impact of suburbanization on fluvial geomorphology. Water Resour. Res. 11(5): 690-692.

1977. Network characteristics in suburbanizing steams. Water Resour. Res. 13(2): 459-463.

GREGORY, K. J. 1974. Streamflow and building activity, p. 107-122. In K. J. Gregory and D. E. Walling [ed.] Fluvial processes in instrumented watersheds. Spec. Publ. No. 6, Institute of British Geographers, London.

GREGORY, K. J., AND D. E. WALLING. 1973. Drainage basin form and process: a geomorphological approach. Edward Arnold Ltd., London.

GRIMA, A. P. 1979. Municipal water demand management. Water Spectrum 11(3): 27-35.

GYSI, M. 1981a. The cost of Peak Capacity Water. Water Resour. Bull. 17(6): 956-961.

1981b. Demand management of water — constraints for consumption. Paper presented to the CWRA annual conference, Banff, Alta.

GYSI, M., AND G. LAMB. 1977. An example of excess urban water consumption. Can. J. Civ. Eng. 4(1): 66-71.

HAGE, K. D. 1975. Urban-rural humidity differences. J. Appl. Meteorol. 14(7): 1277-1283.

HAMMER, T. R. 1972. Stream channel enlargement due to urbanization. Water Resour. Res. 8(6): 1530-1540.

HARRIS, E. E., AND S.E. RANTZ. 1964. Effect of urban growth on streamflow regimen of Permanente Creek, Santa Clara County, California. U.S.G.S. Water Suppl. Pap. 1591-B.

HARTMANN, B. J. 1972. The effect of urbanization on annual water yield. Publ. Climatol. Lab. Climatol. Centerton, NJ. 25(1): 42-73.

HOLLIS, G. E. 1974. The effect of urbanization on floods in the Canon's Brook, Harlow, Essex, p. 123-139. In K. J. Gregory and D. E. Walling [ed.] Fluvial processes in instrumented watersheds. Spec. Publ. No. 6, Institute of British Geographers, London.

1975. The effect of urbanization on floods of different recurrence interval. Water Resour. Res. 11(3): 431-435.

1977. Water yield changes after urbanization of the Canon's Brook Catchment, Harlow, England. Hydrol. Sci. Bull. 22(1): 61-75.

HOLLIS, G. E., AND J. K. LUCKETT. 1976. The response of natural channels to urbanization: two case studies from southeast England. J. Hydrol. 30(4): 351-363.

HOWE, C. W. 1971. Savings recommendations with regard to water system losses. J. Am. Water Works Assoc. 63(5): 284-286.

INQUIRY ON FEDERAL WATER POLICY. 1984. Water is a mainstream issue. Participation paper, Inquiry on Federal Water Policy, Ottawa, Ont.

1985a. Hearing about water. A synthesis of public hearings of the Inquiry on Federal Water Policy, Ottawa, Ont.

1985b. Currents of change. Final report, Inquiry on Federal Water Policy, Ottawa, Ont.

JAMES, L. D. 1965. Using digital computers to estimate the effects of urban development on flood peaks. Water Resour. Res. 1: 223-234.

JAMES F. MACLAREN LTD. 1980a. Brucewood urban test catchment. Research program for the abatement of municipal pollution within the provisions of the Canada-Ontario Agreement on Great Lakes water quality. Res. Rep. No. 100, Environment Canada, Ottawa, Ont.

1980b. Stormwater management technology systems demonstration in the city of St. Thomas. Canada Housing and Mortgage Corp., Ottawa, Ont.

LAGREGA, M. D., AND J. D. KEENAN. 1975. Characterization of water quality from combined sewer discharges, p. 110-119. In W. Whipple Jr. [ed.] Urbanization and water quality control. AWRA Proc. No. 20, Minneapolic, MN.

LANDSBERG, H. E. 1956. The climate of Towns, p. 584-606. In W.L. Thomas [ed.]. Man's role in changing the face of the earth. Vol. 2. The University of Chicago Press, Chicago, IL.

1962. City air — better or worse? Air over cities. SEC Tech. Rep. A62, Public Health Service, Cincinnati, OH. p. 1-22.

1981. The urban climate. Academic Press Inc., New York, NY.

LAYCOCK, A. H., AND F.B. MACKENZIE. 1984. The increase in water yield with urbanization in the Canadian Prairies. Can. Water Resour. J. 9(1): 83-90.

LAZARO, T. R. 1979. Urban hydrology. A multidisciplinary perspective. Ann Arbor Science Publishers Inc., Ann Arbor, MI.

LEOPOLD, L. B. 1968. Hydrology for urban land planning — a guidebook on the hydrologic effects of urban land use. U.S.G.S. Circ. 554.

1973. River channel change with time: an example. Geol. Soc. Am. Bull. 84(6): 1845-1860.

LUCKMANN, B. H. [ed.]. 1979. Urban runoff and water quality monitoring in the Carling Street catchment, London, Ontario. The University of Western Ontario Geographical Paper No. 39, London, Ont.

MARSALEK, J. 1979. Malvern urban test catchment. Vol. 2. Research program for the abatement of municipal pollution within the provisions of the Canada-Ontario Agreement on Great Lakes water quality. Res. Rep. No. 95, Environment Canada, Ottawa, Ont.

MARTENS, L. A. 1968. Flood inundation and effects of urbanization in metropolitan Charlotte, North Carolina. U.S.G.S. Water Suppl. Pap. 1591-C.

MCPHERSON, M. B. 1969. The nature of changes in urban watersheds and their importance in the decades ahead, p. 157-164. in W. L. Moore and C. W. Morgan [ed.] Effects of watershed changes on stramflow. The University of Texas Centre for Research in Water Resources, Water Resources Symposium No. 2, Austin, TX.

MICHAELS, E. A., E. A. MCBEAN, AND G. MULAMOOTTIL. 1985. Canadian stormwater impoundment experience. Can. Water Resour. J. 10(2): 46-55.

MILLER, E. G. 1966. Effect of urbanization on low flow. U.S.G.S. Prof. Pap. 550-A: A166.

MILLS, W. G. 1977. Water quality of urban storm water runoff in the Borough of East York. Research program for the abatement of municipal pollution within the provisions of the Canada-Ontario Agreement on Great Lakes Water Quality. Res. Rep. No. 66, Environment Canada, Ottawa, Ont.

MULLER, R. A. 1967. Some effects of urbanization on runoff as evaluated by Thornthwaite water balance models, p. 127-136. In M. N. Francisco [ed.] Proceedings of the Third Annual American Water Resources Conference, Ser. No. 3. AWRA, Urbana, IL.

1969. Water balance evaluation of effects of subdivisions on water yield in Middlesex County, New Jersey. Proc. Am. Assoc. Geogr. 1: 121-126.

OKE, T. R. 1974. Review of urban climatology, 1968-1973. WMO Publ. Tech. Note 134.

——— 1978. Boundary Layer climates. Methuen, London.

——— 1979. Review of urban climatology. WMO Publ. Tech. Note 169.

PISANO, W. C. 1975. Cost-effective approach for combined and storm sewer cleanup, p. 219-230. In W. Whipple Jr. [ed.] Urbanization and water quality control. AWRA Proc. No. 20, Minneapolis, MN.

REIMER, P. O., AND J.B. FRANZINI. 1971. Urbanization's drainage consequence. J. Urban Plann. Dev. Div. Proc. ASCE 97(UP2): 217-237.

ROBERTS, M. C. 1972. Watersheds in the rural-urban fringe, p. 388-393. In S. C. Csallany, T. G. McLaughlin, and W. D. Striffler [ed.] Watersheds in transition. AWRA Proc. Ser. No. 14, Minneapolis, MN.

ROESNER, L. A. 1982. Quality of urban runoff, p. 161-187. In D. F. Kibler [ed.] Urban stormwater hydrology. Am. Geophys. Union Water Resour. Monogr. No. 7.

SAWYER, R. M. 1963. Effect of urbanization on storm drainage and ground water recharge in Nassau County, New York. U.S.G.S. Prof. Pap. 475-C: C185-C187.

SCOTT, W. S. 1979. Road de-icing salts in an urban stream and flood control reservoir. Water Resour. Bull. 15(6): 1733-1742.

STALL, J. B., M. L. TERSTRIEP, AND F. L. HUFF. 1970. Some effects of urbanization on floods. Meeting Reprint No. 1130, ASCE National Water Resources Engineering Meeting, Memphis, TN. ASCE, New York, NY.

SULLIVAN, R. H., W. D. HURST, T. M. KIPP, J. P. HEANEY, W. C. HUBER, AND S. NIX. 1978. Evaluation of the magnitude and significance of pollution loadings from urban storm water runoff in Ontario. Research program for the abatement of municipal pollution under provisions of the Canada-Ontario Agreement on Great Lakes water quality. Res. Rep. No. 81, Environment Canada, Ottawa, Ont.

SWAIN, L. 1985. Stormwater management — the next step. Can. Water Resour. J. 10(1): 47-68.

TATE, D., AND D. LACELLE. 1978. Municipal water use in Canada. Can. Water Resour. J. 3(2): 61-78.

TAYLOR, C. H. 1977. Seasonal variation in the impact of suburban development on runoff response time: Peterborough, Ontario. Water Resour. Res. 13(2): 464-468.

U.S. WATER RESOURCES COMMISSION. 1968. The nation's water resources. The first national assessment. U.S. Water Resources Commission, Washington, DC.

——— 1978. The nation's water resources 1975-2000. Second national assessment. Vol. 1, summary report. U.S. Water Resources Commission, Washington, DC.

VICE, R. B., H. P. GUY, AND G. E. FERGUSON. 1969. Sediment movement in an area of suburban highway construction, Scott Run Basin, Fairfax County, Virginia, 1961-64. U.S.G.S. Water Suppl. Pap. 1591-E.

WAANANEN, A. O. 1969. Urban effects on water yield, p. 169-182. In W. L. Moore and C. W. Morgan [ed.] Effects of watershed changes on streamflow. The University of Texas Center for Research in Water Resources, Water Resources Symposium No. 2, Austin, TX.

WIGGERS, J. B. M., AND K. BAKKER. 1978. In what way is water pollution influenced by sewerage systems? Hydrol. Sci. Bull. 23(2): 257-266.

WOLMAN, M. G. 1967. A cycle of sedimentation and erosion in urban river channels. Geogr. Ann. 49A(2-4): 385-395.

——— 1975. Erosion in the urban environment. Hydrol. Sci. Bull. 20(1): 117-125.

ZAMORA, J., A. V. KNEESE, AND E. ERICKSON. 1981. Pricing urban water: theory and practice in three southwestern cities. Southwest. Rev. 1(1): 89-113.

CHAPTER 11

Freshwater Fish and Fisheries of Canada

Henry A. Regier

Department of Zoology, University of Toronto, Toronto, Ont. M5S 1A1

Introduction

The variety of human interests served by fish can be classified under three main headings: *material well-being* as when fish are caught to be traded commercially by participants in the fishing industry and used as food for humans, domestic animals, and pets; *cultural opportunity* as when catching and perhaps eating the fish is part of a recreational or cultural experience as by anglers; and *environmental harmony* as when preservationists participate in non-consumptive contemplation of fish, or when sensitive fish species are used and protected as indicators of ecosystem health (see Loftus et al. 1978; Regier 1981, Livingston 1981; Evernden 1985). Somewhat different scientific and scholarly concepts and methods are used to serve these different interests (see Regier 1981). The relevant considerations simply cannot be captured by a single ecological measure (such as biomass produced annually) or economic measure (such as dollars in the context of gross national product). Such oversimplification would obscure rather than enlighten. Different kinds of qualities and quantities are associated with the three categories of interests.

For the three categories, appropriate interdisciplinary information services are now best developed for *material well-being*, less well developed for *cultural opportunity*, and even less developed for *environmental harmony*. This bias is now less pronounced than it was a decade ago because of growing recognition that all three interests are legitimate and important. Also, the relative number of Canadians with personal involvement with these interests has been shifting with a relative decrease in the first and a relative increase in the second and third. Development of appropriate information services for the latter two has lagged somewhat behind social and political changes.

As with almost any simple classification, the one sketched above should be viewed as a first approximation. An individual human (such as the late R. Haig-Brown 1982) may in fact recognize the legitimacy of all three interests, but then generally exhibit a bias to a particular one. Not infrequently, an individual may be strongly committed to a particular interest and be hostile to the others (for example Livingston 1981). The perspective of the present chapter is that all three interests, with their associated values, are legitimate if practiced responsibly and that priority rankings in a political context vary between different locales of Canada.

For the sake of brevity the three main types of interests will be characterized by a particular activity as follows: material well-being by commercial food fisheries; cultural opportunity by angling; and environmental harmony by preservation. These activities are intended to be illustrative of the more complete concepts implied by each interest and should be so interpreted.

Valuation of Freshwater Fish

Different values predominate within different interest groups and each kind of value has associated with it particular concepts and techniques related to "valuation." Valuation has been developed furthest with respect to commercial food fisheries (i.e. related to material well-being) in the context of conventional fisheries biology and fisheries economics. This kind of valuation will be reviewed first after which the other two kinds will be contrasted with the first.

Common names are used for fish species throughout this chapter. Scientific names correpsonding to these common names and much additional information on these species may be found in Scott and Crossman (1973).

The state of fisheries in Canada a quarter of a century ago was described by Larkin (1961) and Ricker (1961). Table 1 shows 1980 landings of freshwater fish by commercial food fisheries across Canada. More detailed summaries are provided for Ontario (Table 2) as an example; for similar data concerning the Prairie Provinces, see Ayles (1983) and Atton (1984).

Many species are harvested from Canadian freshwaters, each with its own set of commercial roles. For example, Lake Superior lake herring may be dressed and sold fresh, frozen or smoked; their roe may be processed and sold as caviar; the "waste" may be processed to make food for pets and other animals. Freshwater commercial fisheries are usually small enterprises based on specialized knowledge and skills concerning a variety of species, locales, techniques, and products. Altogether these fisheries, the related commercial services, and the government administrative activities comprise a very complicated system. For example, high-quality yellow perch or lake trout from the Great Lakes may be netted and placed on ice (but not frozen), transported quickly to an international airport, shipped by air freight to Europe, and appear in a restaurant meal within 2 or 3 d of harvest.

Estimates of the number of fish caught and kept by anglers in 1980 are shown in Table 3. Comparable data for Quebec were not available but it is known that Quebec's totals are much less than Ontario's totals. Some general patterns are apparent. Salmonines (chars, trouts, salmon, grayling) are particularly important in a broad margin around much of the country; west, north, east, and also in the Great Lakes. Percids (yellow perch and walleye) and esocids (especially northern pike) dominate catches in the middle of the country and in shallow parts of the Great Lakes. Centrarcids (black bases and sunfishes) are commonly caught in southern Ontario, including the edges of the lower Great Lakes. Across the country, anglers take relatively few of the abundant coregonines (lake whitefish, lake herring, etc.), cyprinids (carp, creek chub), catostomids (suckers), gadids (ling), or ictalurids (bullheads). Europeans who have angled for some of these species while in Europe generally do not show interest in them in Canada. Recent immigrants from Southeast Asia have taken up fishing for some of these.

Based on questionnaire data from the 1980 Survey of Sportfishing in Canada, Tuomi (1985) inferred that expenditure and investment of $1.7 billion, in 1980 dollars, was directly attributable to sport fishing in Canada. These expenditures were distributed by province as shown on the bottom line of Table 3.

Anglers also use small fish, mostly cyprinids and catostomids, as bait to help catch larger fish. In Ontario in 1980, some 1732 bait fish harvest licences were issued, and the bait fishermen harvested altogether some 180 million small fish. Total revenue from sales was $4.3 million (Ridgely 1981b).

The data given above indicate something about the qualitites and quantities of various kinds of Canadian freshwater food and angling fisheries. It is difficult to get comprehensive and reliable numerical data with respect to quantities of fish caught and much more difficult to determine what economic value(s) are attribuable to these catches. If accurate data existed, of the type appropriate for the conventional national accounts (e.g. as in gross national product), the total for freshwater fisheries, including recreational, commercial food, domestic bait fisheries, would not exceed 1% of the GNP. But for small communities, the welfare of which is a national concern, the importance of small commercial food or commercial angling fisheries is not meaningfully reflected in the statement that the relative contribution of freshwater fisheries to the conventional accounts is less than 1% (Regier 1976).

TABLE 1. Landed catches and landed values of freshwater commercial fisheries in Canada in 1980, by species or species groups. Statistics from the Department of Fisheries and Oceans, courtesy of V. Lasker. Catch data are in metric tonnes and value data in thousands of dollars.

Species	New Brunswick	Quebec	Ontario	Manitoba	Saskatchewan	Alberta	Northwest Territories	Total catches in Canada	Total value of catches
Sturgeon	22	70	22	—	—	—	—	130	394
Alewife	522	—	—	—	—	—	—	522	80
Shad	9	6	2	—	—	—	—	17	13
Salmon[a]	0	—	16	—	—	—	—	16	34
Arctic char	—	—	—	—	—	—	81	81	365
Lake trout	—	—	280	60	740	—	134	1 214	1 140
Whitefish	—	7	1 771	4 527	1913	837	1208	10 263	9 533
Tullibee[b]	—	—	1 926	306	50	43	—	2 325	2 240
Smelt	—	0	11 426	—	—	—	—	11 426	2 499
Northern pike	—	3	269	2 562	1035	248	184	4 301	2 490
Sucker (mullet)	—	—	572	4 721	231	—	—	5 543	941
Carp	—	36	122	778	586	—	—	1 522	331
Catfish	—	137	328	—	—	—	—	476	402
Eel	16	176	169	—	—	—	—	361	1 063
Burbot	—	—	109	—	—	—	—	109	21
Tomcod	—	80	—	—	—	—	—	80	50
White bass	—	—	911	—	—	—	—	911	938
Rock bass	—	0	49	—	—	—	—	49	94
Sunfish	—	7	119	—	—	—	—	126	101
Yellow perch	—	113	6 344	65	—	4	—	6 526	9 009
Sauger	—	—	17	1 708	—	—	—	1 725	2 759
Walleye	0	4	1 353	3 316	756	60	44	5 533	13 340
Other fish	7	29	896	43	25	21	66	1 041	515
Total landed catch	576	668	26 701	18 086	5336	1213	1717	54 297	
Total landed value	177	1339	23 644	16 591	3794	1014	1793		48 352

[a]Pacific salmon species introduced into the Great Lakes and taken incidentally in commercial gear.
[b]Lake herring, chub, and cisco, i.e. small coregonines.

TABLE 2. Commercial food fishery landings in Ontario in 1980 (from Ridgely 1981a). Data are in kilograms and dollars; average prices are used to calculate total values.

Species	Lake Ontario	Lake Erie	Lake St. Clair	Lake Huron	Georgian Bay	North Channel	Lake Superior	Northern inland waters	Southern inland waters	Total landings	Total value
Sturgeon	366	279	—	1 669	644	4 249	—	12 100	2 227	21 534	104 015
Bowfin	54	6 918	2 124	—	—	—	—	—	35	9 131	1 334
Shad	189	360	—	1 436	—	—	—	—	—	1 985	556
Pacific salmon	—	9 703	—	6 408	26	52	—	—	1 073	17 262	31 746
Lake trout	5	359	—	27 458	385	5 681	212 089	32 069	—	278 046	499 155
Splake	—	—	—	32	36 630	2 280	—	—	—	38 942	85 034
Lake whitefish	4 100	851	—	523 958	88 215	73 467	225 474	765 369	—	1 681 434	2 819 422
Lake herring (cisco)	5 269	—	—	2 563	19 661	1 231	1 281 615	153 874	—	1 464 213	1 383 228
Chub (deepwater cisco)	—	—	—	101 279	77 064	645	239 878	27 430	—	446 296	684 092
Round whitefish	26	—	—	5 607	14 922	1 274	53 528	—	—	75 357	62 435
Smelt	22 224	11 296 440	—	305	—	—	16 632	—	171	11 335 772	2 498 886
Northern pike	19 857	9 581	1 062	116	2 929	5 808	1 188	225 971	153	266 665	311 640
Sucker (mullet)	4 880	30 623	4 498	18 352	40 784	34 461	77 312	336 996	15 375	563 281	151 368
Carp	—	10 162	14 103	15 060	7 095	3 217	—	—	71 042	120 679	62 487
Bullhead	165 244	16 821	996	140	626	435	—	3 315	53 193	240 570	197 658
Catfish	10 731	39 361	6 570	24 456	—	—	—	—	3 685	84 833	95 251
Eel	164 031	57	—	—	—	7	—	—	3 548	167 643	454 645
Burbot (ling)	—	—	—	—	9 837	1 651	4 898	91 028	268	107 682	21 304
White perch	54 956	—	—	—	—	—	—	—	2 151	57 107	36 374
White bass	3 293	887 667	1 726	4 482	2 940	2 720	—	631	437	903 896	937 989
Rock bass	6 141	9 191	810	90	1 234	439	—	30 008	—	47 913	94 273
Sunfish	72 170	17 052	—	—	—	—	—	—	28 769	117 991	93 942
Crappie	8 205	14 315	—	—	—	—	—	—	4 473	26 993	31 761
Yellow perch	265 463	5 674 037	172	201 139	36 402	7 340	37 942	2 936	10 742	6 236 173	8 709 114
Sauger	1	113	—	—	—	—	—	16 752	—	16 866	37 240
Walleye	56 933	800 152	—	123 854	11 262	13 133	348	336 937	—	1 342 619	4 192 795
Freshwater drum	25	151 308	4 721	21 671	—	—	—	187	824	178 736	29 524
Animal food	27 050	514 151	—	3 308	10 409	—	—	44 536	35 958	635 412	16 994
Total landings	891 213	19 489 501	36 782	1 083 383	361 095	158 090	2 150 904	2 080 139	234 124	26 485 231	
Total landed value	1 147 054	14 024 375	23 123	2 368 984	505 434	279 477	2 605 932	2 532 623	157 260		23 644 262

TABLE 3. Number of freshwater fish caught and retained in 1980 by anglers by province and territory and funds spent and invested during 1980; data are in thousands. Totals may vary due to rounding. Comparable data from Quebec were not available due to differing survey coverage and methods used. A dash implies that the relevant datum is insignificantly small. Source: Nationally Coordinated Survey of Sportfishing in Canada (see Tuomi 1985).

Species	B.C.	Alta.	Yukon	N.W.T.	Sask.	Man.	Ont.	N.B.	N.S.	PEI	Nfld.	Canada[a]
Pacific salmon	1935	—	—	—	—	—	—	—	—	—	—	1 935
Rainbow trout	4177	1423	7	—	54	66	1 064	—	245	25	379	7 440
Atlantic salmon	—	—	—	—	—	—	—	46	19	—	124	189
Arctic char	2	—	1	27	—	—	—	—	—	—	10	41
Brook trout	432	118	—	—	27	48	2 418	3388	2009	540	6032	15 012
Lake trout	380	197	38	55	145	129	1 318	—	—	—	—	2 262
Other trout	1067	507	1	5	26	4	834	217	63	111	689	3 524
Freshwater salmon	1523	—	11	—	—	—	313	51	—	—	82	1 983
Lake whitefish	288	1265	4	6	25	57	514	—	—	—	—	2 159
Arctic grayling	51	158	83	35	9	3	1	—	—	—	—	340
Rainbow smelt	88	—	—	—	—	9	24 415	1504	2542	82	14	28 654
Northern pike	2	2808	—	58	1922	1734	5 640	113	34	—	1	12 165
Black basses	3	—	—	—	—	136	10 060	113	—	—	—	10 386
Yellow perch	13	2995	—	—	1635	475	22 442	169	1	—	—	27 730
Walleye	1	891	—	34	1602	2510	14 959	61	1	—	—	20 059
Other	8252	227	2	38	57	410	9 586	336	511	154	226	19 630
Angler number												
Resident	489	252	6	6	134	133	1 737	98	67	15	98	3035
Nonresident	95	11	8	8	47	47	635	8	2	2	4	867
Dollars spent and invested in 1980[b]	408	155	6	11	70	84	709	33	20	3	28	1 527

[a]Not including data from Quebec.
[b]Of the British Columbia total, $209 million was devoted to tidal fisheries; the estimate for Quebec is $236 million.

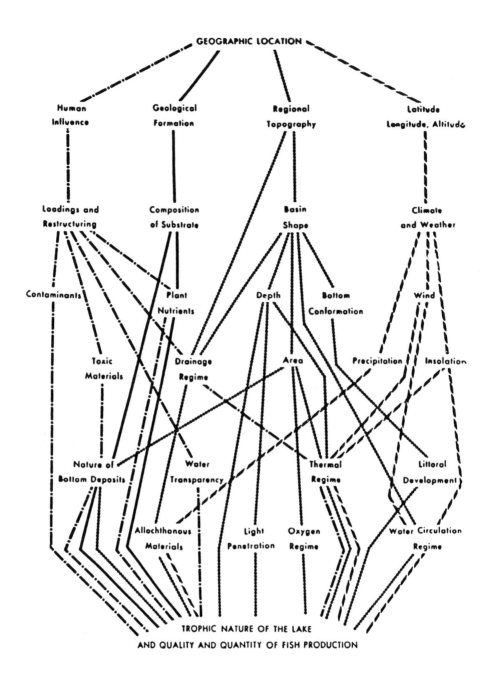

FIG. 1. Various human and abiotic factors that influence the biotic characteristics of lakes, especially with respect to the ecological productivity of fish of various species. This is a modified version of D. S. Rawson's diagram (see Larkin 1974a, p. 51); the modifications involve primarily a further elaboration of the human master factor, which was sketched in a very simple form by Rawson.

Conventionally, the political importance of freshwater food fisheries has often been related to the special economic benefits of foreign trade. Much of the market is foreign, mostly in the United States but now also in Europe and Japan. Such importance has also been argued with respect to commercial angling services that cater to foreign tourists, who are almost exclusively American.

Indication of the importance that Canadians ascribe to sport fishing may be seen in data on the number of Canadians who participate annually in the sport and the money they expend for that purpose (Table 3; see also Tuomi 1982). Their political interests are focussed through the Canadian Wildlife Federation as well as in angling organizations not affiliated with it; these organizations have large and active memberships.

Data similar to the above (landings, earnings, expenditures, participants) are not available for the third interest group, those related to environmental harmony or "preservation." Many people of this group would hold that economic and numerical information would not be relevant and would be contradictory to the values that motivate them (Evernden 1985). Nevertheless, some indication of the social importance of this interest group may be reflected in the breadth and intensity of public interest and involvement in the work of such groups as the Canadian Nature Federation and its affiliated organizations, the Sierra Club and Greenpeace.

The Ecology of Fish Production

As with values in the preceding section, the ecology of fish will be viewed here from three perspectives: food fishing, angling, and "preservation."

Figure 1 is based on a diagram developed by D. S. Rawson, whose approach might be termed "classical comparative fisheries limnology" (Rawson 1939, 1958). His schema related in the first instance to the general nature of a lake and to the quality and quantity of fish production from commercial food fisheries. P. A. Larkin, a student of Rawson, sketched his mentor's approach as involving four groups of factors that are instrumental in determining the levels of productivity of lakes and streams: *edaphic*, soil or chemical fertility factors; *morphometric*, physical dimensions of the aquatic environment; *climatic*, such as length of growing season; and *biotic*, which relate to species present in waters that may have "a deficient complement of freshwater organisms... because of geologically recent glaciation" (Larkin 1961). In another version of Rawson's perspective (see Fig. 1, which is based on a diagram published by Larkin 1974a) the *biotic* factors are deleted and a *human cultural* group is included instead. Of course, humans may be seen as a special subset of the biota from a comprehensive ecosystem perspective.

Figure 2 illustrates a relationship between some of the factors shown in Fig. 1, as discovered by R. A. Ryder who based his approach on that of Rawson (see Ryder et al. 1974). Put somewhat simplistically, the morphoedaphic index combines the two variables *plant nutrients* and *depth*, shown near the centre of Fig. 1. The second is a morphometric factor and the first an edaphic factor (see above); thus, Ryder called the new combined factor a morphoedaphic index (MEI). As used by Ryder (1965), *morphoedaphic* refers explicitly to only two of the four or five groups of factors that Rawson considered important determinants of the productivity of freshwaters. But the distinction between the two sets of data in Fig. 2, each fitted by a different line, includes some consideration of human cultural phenomena, a third generic factor. Although it is an obvious abbreviation, *morphoedaphic* seems preferable to *morphoedaphoclimatobiotocultural*, except perhaps to a German.

Rawson's shema refers to *ecological productivity* of fish in a lake in the first instance, but he then used a measure based on *economic landings* by commercial food fisheries as related to productivity. Rawson interpreted such landings data in the light of his direct expertise with those fisheries; a somewhat more formal version of this method is still used as shown in Fig. 2 and 3 (see Matuszek 1978 for a discussion of this). More formal analytical methods have become available since Rawson started his work (Ricker 1978);

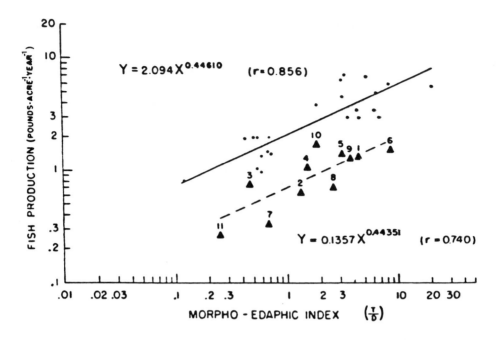

FIG. 2. Logarithmic plots of fish production (1 lb. • acre^{-1} = 1.12 kg • ha^{-1}) and the morphoedaphic index (MEI) for 34 lakes of North America. Here, MEI is the ratio of the concentration of total dissolved solids and mean depth of the lake. The continuous regression line relates to 23 moderately to intensively fished lakes. The broken regression line is for the 11 lakes with restricted fisheries or incomplete records (numbered triangles). (From Ryder 1965)

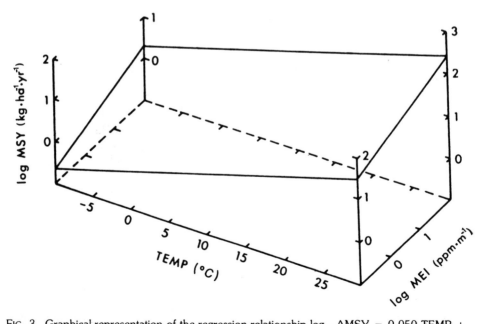

FIG. 3. Graphical representation of the regression relationship $\log_{10}$ AMSY = 0.050 TEMP + 0.280 $\log_{10}$ MEI + 0.236. Here, AMSY is aggregated maximum sustainable yield, TEMP is mean annual air temperature, and MEI is Ryder's morphoedaphic index. (From Schlesinger and Regier 1982)

302

and the estimates in Fig. 4 were obtained by more recent methods. The data in Fig. 4 confirm with respect to production (and presumable productivity) what has long been speculated from observations of temperatures, of preferred habitats, and from laboratory studies of growth as a function of temperature: walleye thrive in warmer temperatures than do northern pike, which in turn thrive in warmer water than do lake whitefish.

Relationships have been discovered between "fisheries yield" and measures of various subsets of the factors shown in Rawson's schema for sets of comparable lakes, reservoirs, etc. (see Jenkins 1968; Ryder et al. 1974; Ryder 1982; Oglesby 1982). Where possible, estimates of maximum sustainable yield (MSY) are derived separately for the most preferred species in a lake. These estimates may then be combined (not necessarily as a simple sum) to obtain an aggregated maximum sustainable yield (AMSY). An estimate of the latter quantity may alternatively be obtained more directly from a study of the data on annual aggregated catches.

Over the past two decades an analytical trophodynamic approach has been under development in the hope of providing better estimates of ecological productivity and then of maximum sustainable yields under different kinds of fish harvest regimes. The study of trophodynamics starts with a description and quantification of trophic or feeding interactions, extending from the chemicals in the water, through plankton, eventually to fish, and perhaps to their nonaquatic predators such as fish-eating birds and humans. Trophodynamic ecologists have encountered severe difficulties in explicating fish production — analytically and deterministically, within particular ecosystems — say, in terms of plankton production as a driving variable. The higher the behavioural capabilities of the organisms within a trophic level, the more difficult it is to characterize them realistically within trophodynamic models. Fisheries ecologists are now discussing the relative importance of "bottom-up" versus "top-down" controls. Bottom-up control would presume that the productive capability of the phytoplankton was a master variable driving zooplankton and fish production, whereas top-down control would presume that the nature of the association of the fish and other large aquatic animals and their external predators largely determines fish production, through such avenues as the effects of the larger fish species on the smaller fish and of the smaller fish on the zooplankton (Kerr 1974; Kerr and Ryder 1977). Trophodynamic ecologists are increasingly studying the inferences and generalizations of the comparative fisheries limnologist and are finding useful hypotheses to test within the borders of their own speciality.

Figure 5 depicts the "realizable or potential productivity" of natural waters of North America in which neither the fish resources nor the aquatic habitats have been abused by humans and in which none of the natural "Rawson factors" are seriously suboptimal. The realizable productivity is in the sense of AMSY of the set of species most preferred by commercial fisheries. The isolines are based on regression relationships for three master variables in the context of Rawson's schema (Fig. 1): the morphometric, the edaphic, and the climatic. Effects of the fourth master variable, the cultural, are avoided in the sense that no serious human effects (positive or negative) are assumed to have occured in the lakes for which these estimates apply. The numbers on the isolines relate to AMSY (kilograms per hectare per year) *under optimal combinations of the first two variables*, i.e. when the MEI of Ryder (1982) is about 40 units (see Schlesinger and Regier 1982). The regression relationship on which Fig. 5 is based is graphed in Fig. 3. It should be noted that yield (AMSY) and the morphoedaphic variables are transformed to logarithms, while the climatic temperature variable is on an arithmetic scale. This implies that AMSY is likely to drop rapidly with respect to the MEI's deviation from the optimal, *ceteris paribus*. The form of relationship shown in Fig. 3 was selected (or "discovered") through stepwise multiple regression procedures and not as a result of informed judgement based on a sufficient understanding of the actual interactions of factors as they contribute to "yield."

The scatter of points around each regression line (Fig. 2-4) is quite extensive, plus or minus a factor of 2 or even 3. Thus, any estimate of the AMSY productivity using

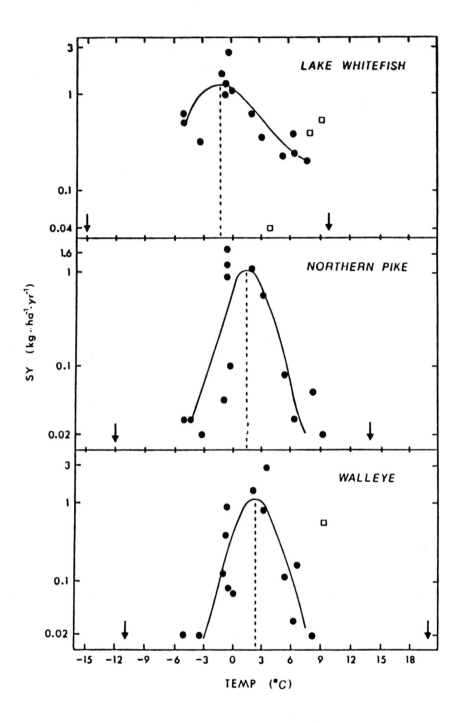

FIG. 4. Relationship between maximum sustainable yield and mean annual air temperature for three species based on a preliminary study (Schlesinger and Regier 1983). Squares represent points that were not included in the final curve-fitting. Arrows at the abscissae indicate the approximate TEMP range for native populations of each species.

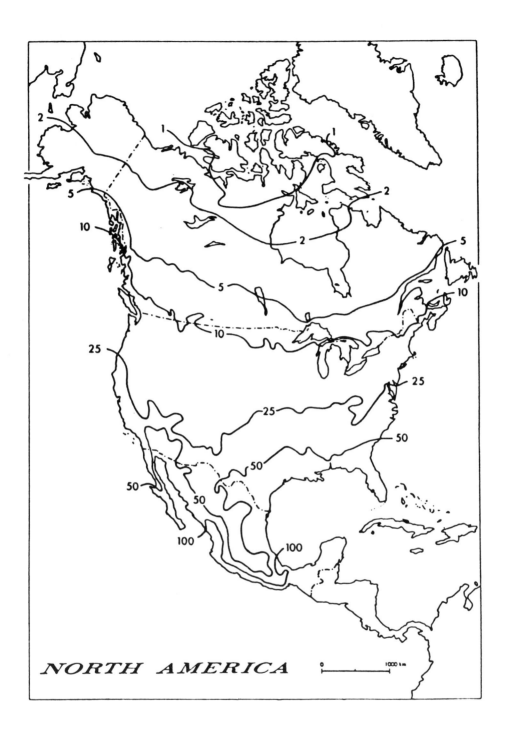

FIG. 5. Isolines showing theoretical upper limits to maximum sustainable fish yield (expressed as kg • ha^{-1} • yr^{-1}) in North America, based on the relationship graphed in Fig. 3 and an inference that a morphoedaphic index value of 40 approximates optimal conditions for fish production. Localized altitudinal differences in mountainous areas, which are not depicted on the map, also would affect climatic patterns and potential fish yields. (From Schlesinger and Regier 1982)

such a relationship will be only a "ball-park figure," though much more precise than if based only on an arithmetic mean and not on any king of "Rawson regressions." Again these regression estimates are likely to be useful for regional planning with sets of lakes rather than for detailed management of individual lakes. Additional and other information will be necessary for the latter purpose.

The information shown in Fig. 5 may be related to the north-south zonation based on a "nordicity index" by Hamelin (1972) (Fig. 6). In 1966 about 83% of Canada's population lived in the Main Ecumene of the Near North, 16% in the rest of the Near North, 1% in the Middle North, 0.1% in the Far North, and 0.001% in the Extreme North. Regier (1976) and Christie (1978) have related fisheries issues to Hamelin's zones.

The "Rawson productivity" of Canadian waters decreases as temperature decreases northward. The edaphic factor has low values on the Canadian Shield that takes up so much of Canada's geography. The morphometric factor, as characterized by the inverse of the mean depth of the water body, is suboptimal with respect to AMSY for many mountain lakes and most of our large lakes, including the Great Lakes that we share with the United States. There are climatic complications with those shallow waters which otherwise should be quite productive. Shallow waters of the south may become anoxic under ice cover and shallow waters in the north may be frozen to the bottom; such complications are not reflected in the regression relationship depicted in Fig. 3

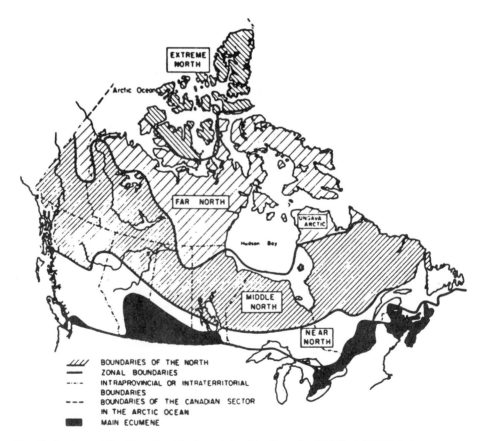

FIG. 6. Regions of the Canadian North as defined by Hamelin (1972). A VAPO index measures "nordicity" (see text), and numerical values of the index were used by Hamelin to locate the VAPO isolines shown here.

because data on very shallow waters were excluded because of unsolved problems in quantifying winter kill.

The fourth master variable of Rawson's schema (Fig. 1) is human cultural influences. With respect to abusive influences the overall intensity decreases from south to north; in other words the degree of deviation from optimality is the opposite direction to that of the climatic variable. More about this follows in the next section.

A fifth master variable (Larkin 1961) concerns the fish (or biotic) association; Larkin surmised at the time that some waters may be relatively depauperate due to the lingering effects of recent glaciation. Within the Main Ecumene, many fish species have been transplanted or introduced (Table 4) so that it is now questionable whether those waters are still depauperate. (Entry of exotics into a fish community may well shift it from a "harmonic" to an "astatic" state; see Kerr and Ryder 1977.) Farther north there may be cases of overly simple fish associations, but little is now understood about such matters. The number of fish species is strongly correlated with latitude so that any judgement of depauperate status must be made with respect to relevant ecological baseline considerations.

Of course, the actual species of fish available to food fishermen is about as important to them as is some measure of AMSY. Figure 7 depicts the extreme geographic ranges of three percid species, and Fig. 8 describes the geographic distribution in Canada of a centrarchid in somewhat more detail. In Fig. 7, the large horseshoe-shaped region that encircles the percid domain to the north and west within Canada is dominated by the salmonids: salmonines plus coregonines. Along the coasts the rivers are used for spawning by sea-run salmon and char species. The salmonids also occur (or once occurred), although usually to a less dominant extent, in thousands of lakes within what is shown as the range of the percids in Fig. 7.

The main fisheries, of all types, are off our sea coasts, in the St. Lawrence Basin (the lakes, River, and Gulf), and in the larger more southerly lakes elsewhere across Canada. These are generally the most productive waters (Fig. 5) and are nearest the people (Fig. 6), thus limiting transportation costs.

How does one describe and measure productivity with respect to fish in the context of recreational potential and *cultural opportunity*, or the productivity of aquatic ecosystems in supporting the continued existence or preservation of endangered species, with respect to *environmental harmony*?

TABLE 4. Dates of initial appearance or introduction of "successful" populations of non-native species in Canadian waters of the Great Lakes where "success" in indicated by some subsequent natural reproduction by the population.[a]

Species	Lake Ontario	Lake Erie	Lake Huron	Lake Superior
Sea lamprey	ca. 1890	1921	1937	1946
Alewife	1873	1931	1933	1953
Pink salmon	1979	1979	1971	1956
Coho salmon	1968	1967	1967	1966
Kokanee salmon	—	—	1964	—
Chinook salmon	1970	—	1969	1967
Rainbow trout	1922	1921	1890	1883
Brown trout	1929	1913	1929	1930
Rainbow smelt	1931	1935	1925	1930
Goldfish	ca. 1880	ca. 1880	ca. 1880	ca. 1880
Carp	1896	1891	1900	1915
White perch	1952	ca. 1970	ca. 1980	—

[a]For references, see Berst and Spangler (1973), Christie (1973), Crossman (1984), Emery (1981), and Hartman (1973).

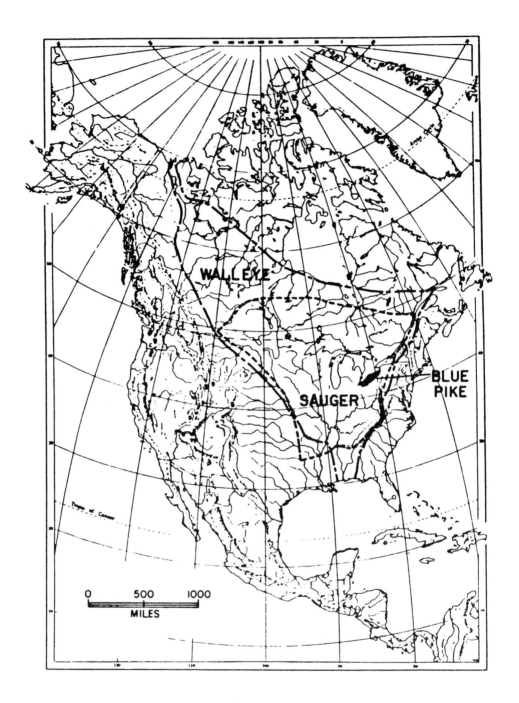

FIG. 7. Distribution of walleye, blue pike, and saugers (modified from Trautman 1957). The eastern limit of the walleye does not include rivers flowing into the Atlantic in which the species may have been introduced. Blue pike in Lake Ontario may have been migrants from Lake Erie; in any case, this taxon is now extinct (McAllister et al. 1985)

There have always been anglers whose joys were directly proportional to the total number or aggregated weight of preferred species in their creels, tubs, or gunnysacks. For these the concepts and methods sketched with respect to commercial food fisheries are adequate. But for many anglers the recreational experience is related to a variety of other considerations, especially the *qualities* of the environment, the fish available, their gear, the company of others. Killing of harvesting of fish may be limited to the needs of a shore lunch. Commercial wilderness angling services are finding that prospects of long-term success may depend on "low-kill fishing" so that large fish of the most preferred species may be caught, and released relatively unharmed, to suppress the populations of low-valued competitive species and then be available to be caught again.

What is the analogue of AMSY with respect to low-kill trophy fisheries? Let us call that analogue maximum sustained recreational benefits (MSRB). On which natural factors does it depend? Suppose it is the same set of master factors as with Rawson's approach to AMSY. The coefficients and perhaps the mathematical form of the regression relationships will presumably be quite different when MSRB is the dependent variable rather than AMSY.

In the absence of commercial and/or domestic food and/or angler "meat" fisheries, and of environmental abuses from many causes, the MSRB productivity of Canadian waters was high. Prior to intensive settlement, all waters of Canada, except some saline or shallow lakes in the prairies, provided exceptionally fine fishing for anglers, from all accounts. Currently in waters that are not abused environmentally, if high-kill fisheries of all types could be focussed very selectively on fish species that compete with the trophy-sized fish of preferred species, then both kinds of interests would reap benefits. Insufficient attention has yet been directed to this option in Canada; it is time that fisheries managers take it more seriously, initially on an experimental basis. Multiple-use productivity can be increased considerably in many of the more intensively used waters but only with a reduction of environmental abuses and with development of new "symbiotic" multiple harvest strategies that require that each use be quite selective. Various strategies and techniques already in use could be organized to interrelate properly in order to achieve this purpose.

Finally, consider productivity as it relates to *environmental harmony*, seen as a good in itself, sometimes to the exclusion of killing any fish of a species when its survival is endangered (McAllister et al. 1985). If "productivity" were to be assigned some meaning in this context, then it would presumably relate to the factors that are particularly influential to the survival of the endangered species.

Uses and Abuses of Fish and Fish Habitat

In Canada the issue of who may use the components and processes of aquatic ecosystems in particular ways has always been contentious. The scale and intensity of the consequent allocation struggles have increased progressively so that now the problem is pervasive across much of the country and extends throughout our "exclusive economic zone" in the oceans. The productive potential of fish can be allocated to high-kill fishermen to harvest the resource, or to low-kill fishermen to play the fish (though the fish may not fancy this kind of play), or to the interests of environmental harmony and "preservation" as with some endangered species or populations. The productive potential may also be tacitly allocated (through preemption) to other uses of the aquatic habitat such as the dilution and inactivation of harmful pollutants, harbour and marina construction, removal of much of the water for irrigation, etc. The other uses generally act to reduce the ecosystem's productivity at least with respect to the preferred species (Regier and Grima 1985).

Historically, different uses of the aquatic ecosystem first became intense at the sites of the larger human settlements and then expanded up or down the tributaries or rivers, along shores, and toward the centres of lakes (Regier and Loftus 1972). Progressively more degrading uses — characteristic of conventional industrialization, commercialization, and urbanization — cascaded over each other in these locales and expanded hundreds of kilometres downstream and into the lakes and estuaries. A kind of domino effect occured in which the more degrading abuses displaced the less degrading which then expanded outward to displace some even less degrading use, etc. (Loftus et al. 1982). By the mid-1970's the aquatic ecosystems of almost the entire area shown as the Main Ecumene in Fig. 6 was in a state of severe degradation (e.g. see Johnson 1978 for Saskatchewan information). The most intense sites of degradation in the Great Lakes came to be called "areas of concern" (IJC/WQB 1985).

Under intense abuse of many kinds acting separately or jointly, the overall effect is to engender a kind of general distress syndrome in the biotic parts of the ecosystem (Paloheimo and Regier 1982; Rapport et al. 1985) (Table 5). One feature of this syndrome is that large, nearshore, bottom-oriented fish species are suppressed to the advantage of small, offshore, midwater species.

Much degradation of freshwater ecosystems occured in spite of the *Canada Fisheries Act*, which has long forbidden uses of the aquatic habitat that adversely affect fish. Part of the problem is related to the constitutional division of power which accorded the federal government responsibility for fish and the provincial governments responsibility for the water, except where the water crosses interjurisdictional boundaries in which case the responsibilities are shared (see also chapter by Thompson, this volume). Provincial water laws (Beerling 1984) generally do not rate fish production as a high priority; hence, provincial political support for enforcement of the Act has generally not been strong. Parts of the federal responsibility for administering fisheries have been delegated to some of the provinces, especially to the Prairies Provinces and Ontario, and this has also weakened the force of the Act. But this may be changing gradually.

FIG. 8. Distribution of smallmouth bass in Canada (Shuter et al. 1980). The hatched areas indicate regions where the species is widespread. The stippled areas indicate regions where the species is restricted to isolated bodies of water. Also included on the map are the 16.6 and 18.3°C mean July air temperature isotherms. From a simulation model using a variety of data, mean winter survival was "predicted" to be essentially zero in lakes typical of areas where the mean July air temperature is 16.6°C or less. As the mean July air temperature drops below 18.3°C, there is a rapid increase in the frequency at which very low values for winter survival occur in typical lakes. There is also a steady increase in the relative number of lakes where average winter survival is too low to permit the existence of self-reproducing populations.

In 1984 the then Federal Minister of Fisheries, John Fraser, reasserted the importance of the Act and restated a policy of habitat protection and habitat rehabilitation (DFO 1984).

Strictly within Canada the issue of what may rightfully be done in and to aquatic ecosystems is a contentious issue for which no definitive, consistent statement of practical principles has yet emerged. This is also the case between national jurisdictions, although some progress in enunciation of principles has been registered in a variety of international negotiations including the 1909 Boundary Waters Treaty between the United States and Canada, the 1972 Stockholm Conference of the Human Environment, and the Law of the Sea negotiations, the latter not yet ratified by some countries (see Caldwell 1984). With respect to rightful use of shared freshwater resources in the Great Lakes area, there is a growing sense in Canada that the Americans have progressively come to enjoy a greater share of the combined benefits of these ecosystems than can be justified on the basis of simple interjurisdictional equity. With respect specifically to Great Lakes fishery resources the issue of what would be rightful sharing is very complex, inasmuch as interjurisdictional differences in the intensity of the abuse of the shared habitat as well as the intensity of the use of the shared fish resources must be included in the considerations.

The widespread deposition of acidic and toxic substances from the atmosphere is leading to the acidification and contamination of many waters to the detriment of their fish and fisheries. These substances are loaded deliberately into the atmosphere through venting of waste gases with their associated particles through chimneys, exhaust pipes, etc. Some substances such as PCB's enter the atmosphere by volatilization from dump sites, sewage treatment plants, etc., and then may enter the water by exchange from the atmosphere.

Much research is underway in Canada on the effects of acid rain on fish (e.g. see Harvey et al. 1981; Harvey and Lee 1982; Harvey 1982; Beggs et al. 1985; Minns and Kelso 1985). Figure 9 (from McAllister et al. 1985) illustrates the approximate sensitivity of four fish species to the acidity of water, with lake trout more sensitive than yellow perch. The sensitivity of other species has been documented by Harvey (1982): pumpkinseed sunfish are about as tolerant as yellow perch and the common shiner somewhat more sensitive than lake trout.

Beggs et al. (1985) have inferred that the level of acidity with a pH of 5.2 is critical for lake trout. Minns and Kelso (1985) judged that populations of lake trout, and of some other salmonines, east of the Manitoba–Ontario border and south of 55°N tend to occur in waters that are naturally acid sensitive in that their alkalinities are so low as to have only moderate to low buffering capacity. The interacting set of factors that determines ecosystemic sensitivity to acid loading is quite complex, but perhaps not as complex as Fig. 1 (see also chapter by Hamilton et al., this volume).

Beggs et al. (1985) judged that some 10 000 North American lakes until recently contained natural populations of lake trout. This includes some 2200 lakes in Ontario, or about 1% of Ontario's 225 000 total. Lake trout populations have been rendered extinct in some 3% of Ontario's lake trout lakes, and about 16% of lake trout lakes are currently acidified or are extremely sensitive to acidic precipitation. Strong circumstantial evidence indicates that most of the extinctions that have so far occurred were due primarily to acid pollution from the smoke stacks at Sudbury.

Of course, populations of fish species other than lake trout have been extinguished in many lakes (see references cited above). But extinctions of lake trout and other salmonids (i.e. salmonines plus coregonines) are of particular concern (Maitland et al. 1981; Ryder and Edwards 1985; McAllister et al. 1985). These taxa have produced quite diverse populations that appear to be particularly closely adapted to their native lakes. Once extinguished, recovery of the native salmonids with comparable local adaptions may be slow and erratic, as experience in the Great Lakes demonstrates (Fetterolf 1985).

ACIDITY
PH

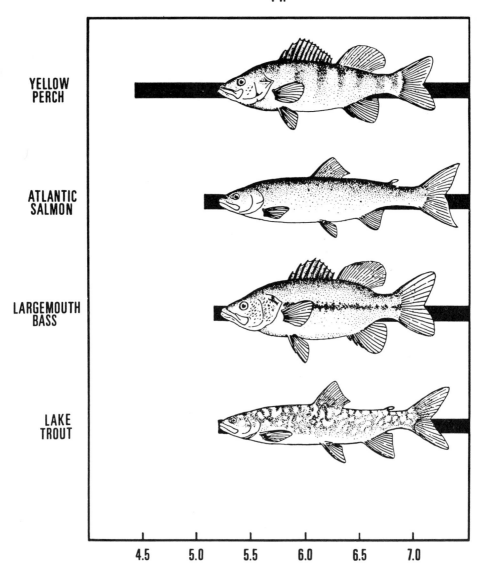

FIG. 9. pH scale from 0 to 14 measures acidity, with 0 being highly acidic, 7 indicating neutrality, and 14 caustic. Each unit indicates a 10-fold increase. The end of each bar indicates where adverse effects appear. (From McAllister et al. 1985)

It should be emphasized that the nature of the ecosystemic response to acidification from atmospheric sources does not resemble closely the responses to other abuses as sketched in Table 5. Why it sould be anomalous is not clear.

We turn now to consideration of allocation between fishermen on extant fish resources and set aside the issue of environmental abuses. The process according to which particular fish species and populations are assigned and apportioned to different groups of fishermen, or even to different individual fishermen, has become very complex (e.g. see Pearse 1982; Kirby 1982; Regier and Grima 1985). The political process of

TABLE 5. A general syndrome of ecological degradation of Canadian waters as caused by conventional uses and abuses by humans bent on economic development in the usual manner (based on Paloheimo and Regier 1982 and Francis et al. 1985).

The major ecological stresses associated with human uses as conventionally practiced often act synergistically so as to exacerbate each other's adverse effects and seldom act antagonistically so as to cancel out adverse effects

The stresses separately and jointly act to alter the fish association from one that is dominated by large fish usually associated with the bottom and edge of the water body to one characterized by small short-lived midwater species. A similar change happens with respect to vegetation; firm-rooted aquatic plants near shore that were there orginally give way to dense suspensions of open-water plankton algae. Further, the association of relatively large benthic invertebrates directly on bottom (such as mussels and crayfish) is supplanted by small burrowing insects and worms (such as midge larvae and sludge worms). Broadly similar changes occur in the flora and fauna of the wetlands and nearshore areas bordering these waters

With the above changes, year to year variability in abundance of particular species increases, especially of landings of different fish species by anglers and commercial fishermen. Fluctuations are also more pronounced in the species associations of wetland, benthic, and pelagic areas

The shift from large organisms associated with the edges and the bottom to small organisms in the bottom mud and in midwater is not accompanied by a great increase in the total biomass of living material, certainly not of the most preferred species

Market and sport value per unit biomass is generally much lower with small midwater fish species than with large bottom species, and processing costs are higher. Similarly, the aesthetic value to recreationists of the rooted plants nearshore is higher than the value of a pea-soup-like mixture of suspended algae and pollutants

The overall effect on fisheries is that nearshore labour-intensive specialized fisheries (sport and commercial) tend to disappear though highly mechanized, capital-intensive offshore enterprises may persist, if the combined stresses do not become excessive and if the fish are not so contaminated as to become a health threat for those who would eat them. Boaters may quickly sail from polluted marinas through the foul nearshore water to less polluted offshore waters. Beaches are posted as hazardous to health

The combined effect is one of debasement and destabilization of these ecosystems with respect to the features of greatest value to humans

identifying certain species or populations as sport fish to be reserved for anglers began over a century ago. Thus, the *Fisheries Act (of Canada)* of 1885 stated in Section 9.2 that "No one shall at any time, fish for, catch, or kill [speckled] trout by other means than angling by hand with hook and line, in any inland lake, river or stream, except in tidal waters." But a subsequent section (9.4) permits use of small trout for baiting traps and codfish hooks and permits incidental catch of trout when fishing fo lake herring and whitefish.

The black basses, smallmouth and largemouth bass, were reserved for anglers long ago, e.g. in southern Ontario starting in 1900 or 1901. Prior to that time these species were actively fished commercially. Table 6 shows the commercial catches of black bass from some Canadian waters of the Great Lakes for the period for which relatively raliable commercial catch statistics are available for this genus, i.e. between 1873 and 1902. Since 1902 these basses have thrived in some areas where they were protected from commercial fishing, such as Long Point Bay of Lake Erie (Whillans 1985), but not in others such as the Toronto area where almost all black bass habitat was destroyed or degraded (Whillans 1979).

TABLE 6. Commercial landings of the black basses (*Micropterus* spp.) in Canadian waters of the Great Lakes (from Ford 1943). There were some inconsistencies in the manner of collecting data, as described in the publication cited. Data are in metric tonnes or 10^3 kg.

Year	Lake Ontario	Lake Erie	Lake St. Claire	Lake Huron	Georgian Bay	North Channel
1875	29.9	19.8	3.5	2.1	0	0
1880	26.1	32.2	15.2	3.4	0	0
1885	112.7	50.1	4.7	4.3	0	0
1890	6.4	61.1	8.0	4.2	51.7	0
1895	22.6	20.6	16.6	15.1	2.0	0.5
1900	2.1	17.1	1.8	0	0	<0.1
1901	2.5	4.6	0.9	0	<0.1	6.8
1902	0	0	0	0	0	0

Various commercial services for anglers are now proliferating, and the political power of the entrepreneurs and their clients is growing. Anglers are often hostile toward food fishermen, for a mixed set of reasons, only some of which are rationally defensible. In freshwaters, as in some coastal marine waters, food fishermen are gradually being squeezed out by angling interests, often under leadership from the more commercialized parts of the angling interest group. Partial compensation for loss of livelihood is now commonly offered by governments to food fishermen that in such situations are prevented from fishing. That anglers and food fishermen might be able to fish to mutual *advantage*, given an appropriate allocative and management regime, has not yet been addressed seriously in Canada. When anglers remove the large sport fish, less desirable species often become more abundant and may compete with recruitment of sport fish. Commercial fishermen might be subsidized to harvest those competitor species, if a market could be developed for them.

Correction of Abuses and Enhancement of Uses

It came to be widely recognized in the 1970's that increased catches of preferred species, some of which had waned markedly, would only follow correction of excessive fishery exploitation and of fish habitat degradation (e.g. see Johnson 1978 for the view of Saskatchewan experts). At least for the short run, some enhancement of fishery catches could be achieved even in relatively abused ecosystems with "enhancements" by stocking the young of preferred species that could not reproduce in these waters as successfully as was desired. For the longer term, rehabilitation of habitats and stocks was also necessary (Larkin 1974b; Loftus 1976; Loftus et al. 1978).

The Salmon Enhancement Program (SEP) was undertaken in British Columbia with an enthusiasm due in part to Larkin (1974b). SEP was formally begun in 1977 and Phase I then continued until 1983, when a Transition Phase was initiated which was to run until 1986 (see SEP 1983). Numerous facilities were constructed, numerous community development projects were organized, and many "small projects" were undertaken. As they were operating in 1983, several of the largest facilities were each expected to contribute about a quarter million mature salmon to the fisheries, while some of the smallest of the small projects might each produce only one mature fish.

By 1983 some $150 million had been devoted to SEP and about 85% of the original target had been achieved, according to the judgement of SEP officials. Some components of the SEP program are generally considered to be fully successful, while both the original inclusion and performance of others, especially of the larger facilities, have been criticized.

314

The complicated developments with respect to aquaculture cannot be considered in this paper: see Larkin (1982) and Pearse (1982) for some recent information.

In Ontario, the Strategic Plan for Ontario Fisheries (Loftus et al. 1978; also Loftus 1976) called for a primary emphsis on rehabilitation in southern Ontario, where many aquatic ecosystems had become degraded. Numerous small rehabilitation projects have been undertaken in recent years (Biette 1983) in the context of the Community Fisheries Involvement Program which was designed on the basis of experience in British Columbia with SEP.

A much more comprehensive program of ecosystem "restoration" has been under-way in the Great Lakes under the 1972 and 1978 Great Lakes Water Quality Agreements overseen by the International Joint Commission (IJC 1984; IJC/WQB 1985; IJC/SAB 1985; NRC/RSC 1985). The eutrophication process due to phosphate enrichment has been reversed to some extent and the levels of toxic substances in the biota have been reduced, but not yet to levels that are acceptably low. Because several major sources of toxic contaminants have not yet been effectively stopped, a risk of increasing contamination exists, though prospects are good that they will come under better control within the next 5–10 yr. Meanwhile, binational efforts to rehabilitate the fisheries are progressing under the general auspices of the Great Lakes Fishery Com-mission (Francis et al. 1979; Loftus 1984; Fetterolf 1985). The sea lamprey, as an exotic invader, is now under partial control to the advantage of lake trout, lake whitefish, and other species. Walleye populations in various bays and shallow parts of the lakes had dwindled under a number of abuses, but are now recovering. Originally introduced on a put-grow-and-take basis to prey upon and reduce small pelagic exotic species of fish which had become pests in the lakes and to enhance the fishery resource for anglers, Pacific salmon are doing well in this chain of lakes with some natural reproduc-tion in evidence. Some size ranges of some species of fish in some locales are con-taminated to levels that make them legally unfit for sale; health advisories are updated and reissued annually to warn anglers specifically where the dangers lie.

Current politics being what they are, one cannot now be confident that various ecosystem abuses will not recur, perhaps in some modified form, so as to reverse the trend in the Great Lakes back in the direction of degradation rather than recovery (NRC/RSC 1985). Some parts of the Great Lakes came through the abuses of recent decades in reasonably healthy state. Surprisingly this also occurred in parts of Lake Erie, which was thought to be "dying" 25 yr ago. The Long Point ecosystems has sur-vived and efforts are now underway to ensure its future preservation in a healthy state (Francis et al. 1985).

Husbandry of Aquatic Ecosystems

As fisheries and environmental researchers and managers were learning how to cor-rect improprer fisheries practices and environmental abuses of local to regional scales during the 1960's and 1970's, environmental insults of a regional to hemispheric scale were mounting. These included airborne pesticides, acids, and other contaminants, the latter two due to the primitive technology still associated with the combustion of fossil fuels. No comprehensive assessment of the impacts of these vast processes of degradation is available.

Meanwhile, consensus is growing that the global climate is beginning to change due to increased content of carbon dioxide and methane at high levels of our atmosphere (see chapter by Ripley, this volume). It seems likely that a warming is underway that may become empiricaly detectable within two or three decades. A very preliminary assessment of impacts of possible climate warming on Great Lakes fish and fisheries (Meisner et al. 1987) focussed on the likely increase in productivity (Fig. 5) and a change in the geographic distribution of fish species (Fig. 6) due to a temperature increase.

Until recently, economic development reigned supreme and fish and fisheries were progressively sacrificed to uncaring industrialization, commercialization, and urbanization and a rather transitory form of wealth and superficial well-being. Fortunately our culture also contains traditions of sensitivity and husbandry towards other species of our man-nature ecosystems. Those traditions have gradually mobilized during the past two decades to play a larger role in our postindustrial culture. Consistent with this, some researchers and managers are clarifying and implementing the idea that thriving highly valued fish species should serve as both the objective of rehabilitation and enhancement of the fish as resources and the measure of proper husbandry of the aquatic ecosystem with respect to a wide range of human uses and interests in these environments. In Canada, the salmonines are particularly appropriate integrative indicators of ecosystem quality for the more oligotrophic waters (Maitland et al. 1981; Ryder and Edwards 1985). The walleye may be most appropriate for mesotrophic waters and the black basses for the inshore areas of eutrophic and mesotrophic waters in warmer parts of the country.

Fisheries professionals now know that they must concern themselves not only with the management of direct use as in harvesting by various kinds of fishermen, but also with indirect abuse that destroys the habitat and thus undercuts the productivity of the resource. Canada's Minister of Fisheries has reemphasized this responsibility (DFO 1984). In trying to implement this broadened realization, fisheries workers are coming to appreciate anew the comparative fisheries limnolgy of D.S. Rawson and of his student colleagues of 50 yr ago, F. E. J. Fry, R. B. Miller, and W. E. Ricker (Fry and Legendre 1963).

Acknowledgements

My thanks to the following for help with this chapter: F. M. Atton, G. B. Ayles, J. L. Goodier, H. H. Harvey, M. C. Healey, S. R. Kerr, G. D. Koshinsky, J. G. I. Lark, V. Lasker, C. K. Minns, J. Tilt, A. L. W. Tuomi, F. Ward, and T. H. Whillans. J. Retel and G. Rania helped with the manuscript. Some of the contents are based on discussions at the Rawson Workshop convened in Niagara-on-the-Lake, Ontario, in October 1983: my thanks to those who participated in that Workshop. Financial support for this work came from the Max Bell Foundation.

References

ATTON, F. M. 1984. Fish resources and the fisheries industry of the Canadian Plains. Prairie Forum 9(2): 315-325.

AYLES, G. B. 1983. Fisheries of the Canadian Prairies. Department of Fisheries and Oceans, Winnipeg, Man. 20 p. (Reproduced).

BEERLING, C. W. 1984. Water allocation law: potential impacts in economic development. Can. Water Resour. J. 9(1): 46-58.

BEGGS, G. L., J. M. GUNN, AND C. H. OLIVER. 1985. The sensitivity of Ontario lake trout (*Salvelinus namaycush*) and lake trout lakes to acidification. Ont. Minist. Nat. Resour. Fish. Tech. Rep. Ser. No. 17: iii + 24 p.

BERST, A. H., AND G. R. SPANGLER. 1973. Lake Huron, the ecology of the fish community and man's effects on it. Great Lakes Fish. Comm. Tech. Rep. 21: v + 41 p.

BIETTE, R. M. 1983. Fisheries habitat rehabilitation in Ontario. N. Am. Lake Manage. Soc. Lake Line 3(2): 6-7.

CALDWELL, L. K. 1984. International environmental policy: emergence and dimensions. Duke University Press, Durham, NC. xv + 367 p.

CHRISTIE, W. J. 1973. A review of the changes in the fish species composition of Lake Ontario. Great Lakes Fish. Comm. Tech. Rep. 21: v + 65 p.

1978. A study of freshwater fishery regulation based on North American experience. Rome, Food and Agricultural Organization of the United Nations. FAO Fish. Tech. Pap. 180: 46 p.

CROSSMAN E. J. 1984. The introduction of exotic fishes into Canada, p. 78-101. In W. R. Courtenay, Jr. and J. R. Stauffer, Jr. [ed.]. Distribution, biology and management of exotic fishes. The Johns Hopkins University Press, Baltimore, MD. xv + 430 p.

DFO. 1984 Submission to the Inquiry on Federal Water Policy. Department of Fisheries and Oceans, Ottawa, Ont. iii + 27 p. + 5 appendices. (Reproduced)

EMERY, L. 1981. Range exetension of pink salmon (Oncorhynchus gorbuscha) into the lower Great Lakes. Fisheries 6(2): 7-10.

EVERNDEN, N. 1985. The natural alien: humankind and environment. University of Toronto Press, Downsview, Ont. x + 160 p.

FETTEROLF, C. M. JR. 1985. Lake trout futures in the Great Lakes, p. 163-171. Proceedings of Wild Trout Three Symposium, Yellowstone National Park, Wyoming, Sept. 24-25, 1984. Federation of Fly Fishers and Trout Unlimited, Washington, DC. 192 p.

FORD, M. A. 1943. Annual landings of fish on the Canadian side of the Great Lakes from 1867 to 1939 as officially recorded. Board of Inquiry for the Great Lakes Fisheries. Ottawa, King's Printer. 91 pp.

FRANCIS, G. R., A. P. GRIMA, H. A. REGIER, AND T. H. WHILLANS. 1985. A prospectus for the management of the Long Point ecosystem. Great Lakes Fish. Comm. Tech. Rep. 43: 108 p.

FRANCIS, G. R., J. J. MAGNUSON, H. A. REGIER, AND D. R. TALHELM. 1979. Rehabilitating Great Lakes ecosystems. Great Lakes Fish. Comm. Tech. Rep. 37: ii + 99 p.

FRY, F. E. J., AND V. LEGENDRE. 1963. Ontario and Quebec, p. 487-519. In D. G. Frey [ed] Limnology in North America. University of Wisconsin Press, Madison, WI.

HAIG-BROWN, R. 1982. Writings and reflections. McClelland and Stewart, Toronto, Ont. 220 p.

HAMELIN, L. E. 1972. L'ecoumene du Nord Canadien, p. 25-40. In W. C. Wonders [ed] The North, Studies in Canadian geography. University of Toronto Press, Downsview, Ont. 151 p.

HARTMAN, W. L. 1973. Effects of exploitation, environmental changes and new species on the fish habitats and resources of Lake Erie. Great Lakes Fish. Comm. Tech. Rep. 22: v + 43 p.

HARVEY, H. H. 1982. Population responses of fishes in acidified waters, p. 227-242. In R. E. Johnson [ed] Acid Rain/Fisheries, Proceedings of an International Symposium on Acidic Precipitation and Fishery Impacts in Northeastern North America, Ithaca, NY. August 2-5, 1981. American Fisheries Society, Bethesda, MD. viii + 357 p.

HARVEY, H. H., AND C. LEE. 1982. Historical fisheries changes related to surface water pH changes in Canada, p. 45-55. In R. E. Johnson [ed] Acid Rain/Fisheries, Proceedings of an International Symposium on Acidic Precipitation and Fishery Impacts in Northeastern North America, Ithaca, NY, August 2-5, 1981. American Fisheries Society, Bethasda, MD. viii + 357 p.

HARVEY, H. H., R. C. PIERCE, P. J. DILLON, J. R. KRAMER AND D. M. WHELPDALE. 1981. Acidification in the Canadian aquatic environment. Publ. NRCC No. 18475. Environmental Secretariat, National Research Council of Canada, Ottawa, Ont. 369 p.

IJC. 1984. Second Biennial Report under the Great Lakes Water Quality Agreement of 1978. Ottawa, Ontario, and Washington, DC. 12 p.

IJC/SAB. 1985. 1985 Annual Report. Science Advisory Board, International Joint Commission, Windsor, Ont. xiii + 40 p.

IJC/WQB. 1985. Report of the Great Lakes Water Quality Board for 1984. Great Lakes Regional Office, Windsor, Ont. xii + 212 p.

JENKINS, R. M. 1968. The influences of some environmental factors on standing crop and harvest of fishes in U.S. reservoirs, p. 298-321. In Reservoir Fishery Resources Symposium, Athens, GA, April 1967. Southern Division, American Fisheries Society, Bethesda, MD.

JOHNSON, R. P. 1978. Land use and fish. Government of Saskatchewan, Saskatoon, Sask. Water Studies Int. Rep. No. 13: 92 p. (Reproduced)

KERR, S. R. 1974. Theory of size distribution in ecological communities. J. Fish Res. Board Can. 31: 1859-1862.

KERR, S. R., AND R. A. RYDER. 1977. Niche theory and species community structure. J. Fish. Res. Board Can. 34: 1952-1958.

KIRBY, M. J. L. 1982. Navigating troubled waters: a new policy for the Atlantic fisheries. Report on the Task Force on Atlantic Fisheries. Ministry of Supply and Services Canada, Ottawa, Ont. Cat No. CP32-43/1983-1E: x + 152 p.

LARKIN, P. A. 1961. The effect of man-made changes on the environment of fisheries. Proceedings of the Conference on Resources for Tomorrow, Oct. 23-28, 1961, held in Montreal, Vol. 2: 785-795. Queen's Printer, Ottawa, Ont. Cat No. R29-5261/2.

    1974a. Freshwater pollution Canadian style. McGill-Queen's University Press. 132 p.

    1974b. Play it again Sam — a perspective on salmon enhancement in Canada. J. Fish. Res. Board Can. 31: 1433-1456.

    1982. Aquaculture in North America: an assessment of future prospects. Can. J. Fish. Aquat. Sci. 39: 155-156.

LIVINGSTONE, J. 1981. The fallacy of wildlife conservation. McClelland and Stewart, Toronto, Ont.

LOFTUS, K. H. 1976. Science for Canada's fisheries rehabilitation needs. J. Fish. Res. Board Can. 33: 1822-1857.

    1984. Fisheries: past, present, and future in the Great Lakes. J. Great Lakes Res. 10: 164-167.

LOFTUS, K. H., A. S. HOLDER, AND H. A. REGIER. 1982. A necessary new strategy for allocating Ontario's fishery resources, p. 255-264. In J. H. Grover [ed.] Allocation of fishery resources. Proc. FAO Tech. Consult., Vichy France, 20-23 April 1980. Food and Agricultural Organization, Rome. 623 p.

LOFTUS, K. H., M. G. JOHNSON, AND H. A. REGIER. 1978. Federal-provincial strategic planning for Ontario fisheries: management strategy for the 1980's. J. Fish. Res. Board Can. 35: 916-927.

MAITLAND, P. S., H. A. REGIER, G. POWER, AND N. A. NILSSON. 1981. A wild salmon, trout, char watch: an international strategy for salmonid conservation. Can. J. Fish. Aquat. Sci. 38: 1882-1888.

MATUSZEK, J. E. 1978. Empirical predictions of fish yields of large North American lakes. Trans. Am. Fish. Soc. 107: 385-394.

MCALLISTER, D. E., B. J. PARKER, AND P. M. MCKEE. 1985. Rare, endangered and extinct fishes in Canada. Nat. Mus. Can. Syllog. No. 54: 192 p.

MEISNER, J. D., J. L. GOODIER, H. A. REGIER, B. J. SHUTER, AND W. J. CHRISTIE. 1987. An assessment of the effects of climate warming on Great Lakes Basin fishes. J. Great Lakes Res. 13. (In press)

MINNS, C. K., AND J. R. M. KELSO. 1985. Estimates of existing and potential impact of acidification of the freshwater fishery resources and their use in eastern Canada. Department of Fisheries and Oceans, Burlington, Ont. (Paper presented to Conference at Muskoka, Ont., in September 1985)

NRC/RSC. 1985. The Great Lakes Water Quality Agreement: an evolving instrument for ecosystem management. National Research Council of the U.S. and the Royal Society of Canada. National Academy Press, Washington, DC. xix + 224 p.

OGLESBY R. [ed.]. 1982. The morphoedaphic index — concepts and practices. Trans. Am. Fish. Soc. 111: 133-175.

PALOHEIMO, J. E., AND H. A. REGIER. 1982. Ecological approach to stressed multispecies fisheries resources. In M. C. Mercer [ed.] Multispecies approaches to fisheries management advice. Can. Spec. Publ. Fish. Aquat. Sci. 59: 127-132.

PEARSE, P. H. 1982. Turning the tide, a new policy for Canada's Pacific fisheries. Final Report of the Commission on Pacific Fishery Policy. Ministry of Supply and Services Canada, Ottawa, Ont. Cat. No. Fs 23-18/1982: xii + 292 p.

RAPPORT, D. J., H. A. REGIER, AND T. C. HUTCHINSON. 1985. Ecosystem behavior under stress. Am. Nat. 125: 617-640.

RAWSON, D. S. 1939. Some physical and chemical factors in the metabolism of lakes. In Proceedings of the Symposium on Problems of Lake Biology. Am. Assoc. Adv. Sci. Publ. 10: 9-26.

    1958. Indices to lake productivity and their significance in predicting conditions in reservoirs and lakes with disturbed water levels. H. R. MacMillan Lectures in Fisheries, University of British Columbia, Vancouver, B.C. p. 27-42.

REGIER, H. A. 1976. Science for the scattered fisheries of the Canadian interior. J. Fish. Res. Board Can. 33: 1213-1232.

    1981. Optimum use, fisheries management. Fisheries 6(5): 4-7.

REGIER, H. A., AND A. P. GRIMA. 1985. On the allocation of fish to fishermen. Can. J. Fish. Aquat. Sci. 42: 845-859.

REGIER, H. A., AND K. H. LOFTUS. 1972. Effects of fisheries exploitation on salmonid communities in oligotrophic lakes. J. Fish. Res. Board Can. 29: 959-968.

RICKER, W. E. 1961. Productive capacity of Canadian fisheries — an outline. Resources for Tomorrow Conf. Background Papers, Vol. 2: 775-784. Queen's Printer, Ottawa, Ont.

    1978. Computation and interpretation of biological statistics of fish populations. Fish. Mar. Serv. Bull. 191: xviii + 382 p.

RIDGELY, J. 1981a. Bait fish summary by region for 1980. Ontario Ministry of Natural Resources, Toronto, Ont. (Unpublished)

    1981b. Commercial fish industry, commercial fish harvests for 1980. Ontario Ministry of Natural Resources, Toronto, Ont. (Unpublished)

RYDER, R. A. 1965. A method for estimating the potential fish production of north temperate lakes. Trans. Am. Fish. Soc. 94: 214-218.

    1982. The morphoedaphic index — use, abuse and fundamental concepts. Trans. Am. Fish. Soc. 111: 154-164.

RYDER, R. A. AND C. J. EDWARDS. 1985. A conceptual approach for the application of biological indicators of ecosystem quality in the Great Lakes Basin. Report to the Great Lakes Science Advisory Board, International Joint Commission, Windsor, Ont. xxix + 169 p.

RYDER, R. A., S. R. KERR, K. H. LOFTUS, AND H. A. REGIER. 1974. The morphoedaphic index, a fish yield estimator — review and evaluation. J. Fish. Res. Board Can. 31: 663-688.

SCHLESINGER, D. A., AND H. A. REGIER. 1982. Climatic and morphoedaphic indices of fish yields from natural lakes. Trans. Am. Fish. Soc. 111: 141-150.

    1983. The relationship between environmental temperature and yields of subarctic and temperate zone fish species. Can. J. Fish. Aquat. Sci. 40: 1829-1837.

SCOTT, W. B., AND E. J. CROSSMAN. 1973. Freshwater fishes of Canada. Bull. Fish. Res. Board Can. 184: 966 p.

SEP. 1983. 1983 Annual report summary of the Salmonid Enhancement Program. Department of Fisheries and Oceans, Vancouver, B.C. 16 p.

SHUTER, B. J., J. A. MACLEAN, H. A. REGIER, AND F. E. J. FRY. 1980. Stochastic simulation of temperature effects on first year survival of smallmouth bass. Trans. Am. Fish. Soc. 109: 1-34.

TRAUTMAN, M. B. 1957. The fishes of Ohio. University of Ohio Press, Columbus, OH. xviii + 683 p.

TUOMI, A. L. W. 1982. The role and place of sportfishing in water-based recreation. Can. Water Resour. J. 7(3): 53-68.

    1985. Angling in Canada. In Proc. of the First World Angling Conf., Cap d'Agde, France, Sept. 12-18, 1984.

WHILLANS, T. H. 1979. Historic transformations of fish communities in three Great Lakes bays. J. Great Lakes Res. 5: 195-215.

    1985. Related long-term trends in fish and vegetation ecology of Long Point Bay and marshes, Lake Erie. Ph.D. thesis, Department of Zoology, University of Toronto, Toronto, Ont. viii + 331 p.

CHAPTER 12

Wetlands and Aquatic Resources

Thomas H. Whillans

Environmental and Resource Studies, Trent University, Peterborough, Ont. K9J 7B8

Introduction

Wetlands, of all aquatic environments, are the least readily recognized as aquatic. Yet they can rank amongst the most productive, diverse, and interesting. Perhaps better known popularly as marshes, swamps, bogs, and sloughs, wetlands are enigmas in the spectrum of Canadian aquatic resources. Judging from the impoverished literature on wetlands in journals of aquatic science, their actual nature must be relatively obscure to mainstream aquatic scientists. Perhaps the position of many wetlands, at the margins of waterbodies, has led to ambiguous interpretations of their relationships to these waterbodies. Certainly their natural incompatibility with and sensitivity to most intensive human uses and their convertibility to terrestrial-like conditions have caused some ambivalence about their natural value. This situation has often resulted in wetlands protection and conservation receiving equivocal support from wetland owners and governments. Recently, however, Canadian wetlands have begun attracting more attention from aquatic scientists, resource stewards, and the public.

The intent of this chapter is to expose current scientific understanding of Canadian freshwater wetlands. It emphasizes the relevance to "traditional" aquatic science. Terrestrially oriented issues, brackish estuaries, and maritime salt marshes will not be addressed in detail. Wetland functions, the status of wetlands, and wetlands management in Canada will also constitute major foci of the chapter. The chapter is not comprehensive in detail. It cannot be, as will be plain in the section on information needs. It is relatively comprehensive in scope in that all major aspects of wetland science are addressed.

The title of this chapter implies that the two are somehow linked. In fact the title contains a redundancy. Wetlands are aquatic resources. It illustrates the common perception that wetlands are environments in which fish, wildlife, desirable plants, peat, and other valuable resources may be obtained — like marketplaces for resources somewhat removed from their hinterland. A more realistic view of wetlands is that they are hydrophylic systems of which each of these so called resources is an integral part. In general terms a wetland is a system in which standing water or saturated soil is a dominant feature at least seasonally, hydrophytes (water-adapted plants) are the predominant rooted macrophytes, soils are wet intrazonal (explainable by global geoclimatic conditions) or peaty azonal (explainable by local conditions), and secondary and tertiary production depend heavily on and affect any or all of the above.

Typology and Distribution of Wetlands

There are a number of types of wetland. An understanding of these types and their distribution is necessary before addressing wetland functions and status.

Types of Wetlands Development

Classification has progressed in a steadier, more systematic manner than perhaps any branch of wetland science in Canada. Early efforts of Tarnocai (1970), Jeglum et al. (1974), Zoltai et al. (1973), Millar (1976), and others paved the way for the reasonably comprehensive scheme of the National Wetland Working Group (WWG) of the Canada Committee on Ecological Land Classification (National Wetland Working Group 1981).

The WWG identified five basic types of wetland: bog, fen, marsh, swamp, and shallow water. Bogs and fens are wetlands in which peat has accumulated to considerable depth. Hydrologic flow is often internal and retarded in comparison with other wetland types. Bog waters are nutrient poor and very acidic whereas fens are supplied with some nutrients from nearby mineral soils. Fens are circumneutral; if acidic, they are at least an order of magnitude less so. Bogs are dominated by mosses, especially *Sphagnum*, and shrubs of the family Ericaceae; fens, on the other hand, are dominated by sedges, *Carex* in particular, and mosses, but they can also be well endowed with wildflowers of the orchid (Orchidaceae), buckbean (Menyanthaceae), and other families. Black spruce (*Picae mariana*) is the common tree species in bogs, and black spruce and tamarack (*Larix laricina*) are found in fens, although the latter may also support cedars (*Thuja*), alders (*Alnus*), and other deciduous trees. The high acidity of bogs is not conducive to much zoological activity. Fens, however, may be inhabited by fish, herptiles (reptiles and amphibians), birds, mammals, and a variety of invertebrates.

Marshes, swamps, and shallow-water wetlands are nutrient-rich, low acidic systems containing standing water seasonally to a depth of about 2 m maximum. Swamps are the least tolerant of long periods (multiyear) of standing water; the water table is normally at or slightly below the soil surface throughout much of the year. Peat may accumulate, often initially remnant from a rich fen and later from the abundant trees that characterize swamps. Swamp trees include black ash (*Fraxinus nigra*), white cedar (*Thuja occidentalis*), and red maple (*Acer rubrum*). Bird and mammal species such as prothonotary warbler (*Prothonotari citrea*), great crested flycatcher (*Myiarchus crinitus*), and beaver (*Castor canadensis*) abound, at least locally, and depend heavily on this habitat. Fish diversity is not usually high although species such as central mudminnow (*Umbra limi*) and northern redbelly dace (*Chrosomos eos*) do well.

Marshes and shallow-water wetlands are characterized by dense submerged, floating-leaf, and/or emergent vegetation. Emergents are prominent in marshes. Nutrient-rich mineral water flows freely among stands of vegetation, usually but not always preventing accumulation of much peat. Marshes contain vegetation such as cattail (*Typha*), bulrushes (*Scirpus*), reeds (*Phragmites*), coontail (*Ceratophyllum*), and white water lily (*Nymphaea odorata*). Both marshes and shallow-water wetlands are inhabited by plants of the milfoil (*Myriophyllum*) and (*Potamogeton*) genera. Often, shrubs such as willows (*Salix*) and sweetgale (*Myrica*) predominate along the edges of marshes and shallow-water wetlands or on alluvial deposits where seasonal currents prevent establishment of other plants. Marshes and shallow-water wetlands support rich assortments of fish and waterfowl. Marshes may additionally contain dense populations of small mammals and herptiles.

Wetlands develop initially because either surface drainage is poor or surface water is in nearly constant supply. The key causal factors are thus climate and physiography. Once wetlands are established, other factors begin to influence their development. Stability of environmental conditions and successional tendencies, for example, become important.

Relatively few studies have focussed on the dynamics of wetland systems — how and why they develop and the rate at which they develop. Such process-oriented studies would reveal much about differences and commonalities among wetland systems, the remarkable productivity of some wetland systems, and the potential for recovery after human disturbance. Much of the understanding of wetland development derives from studies outside of Canada (Mitsch et al. 1982). The models may apply to Canadian wetlands, but they require quantification and calibration.

A typical model might portray successional change in a marsh-swamp-shallow-water wetland continuum in response to fluctuating water levels. This might be depicted as a simple sequence of changes in the abundance of dominant plant species. By focussing on form, this type of model tends to hide the complex processes which are necessary to explain wetland development. Consider, for example, two situations. In the first, periodic water level changes influence wetland development by stimulating increased biotic diversity and productivity; hence, the wetland system is recognized as a pulse stable system (Odum 1971). In the second situation, water levels are relatively constant, yet wetland development occurs endogenously as peat accumulation, decomposition, and acidity influence each other. Tarnocai (1978), for example, observed that water level is not the only factor which can influence peat deposition. Understanding of relatively complex processes is necessary to explain the differences between these two situations. Although quantified predictive models may be achievable, they are rare.

Existing models of wetlands tend to be applied to vegetation, climate, and soil. The zoological component is conspicuously underrepresented. The model presented in Fig. 1 was developed for bog- and fen-dominated ecosystems in northern Ontario (Jeglum and Cowell 1982). It is an attempt to conceptualize and begin to quantify relationships among wetland types. It captures the sense that wetlands are dynamic systems and the potential of modeling as an explanatory tool. The five wetland types are in fact representative "snapshots" in time or space. They can only truly be understood in terms of each other, history, and potential.

Extent of Canadian Wetlands

The surface area of Canada, excluding marine waters, is approximately 9.98×10^6 km^2 (Simpson-Lewis et al. 1979). The surface area of lakes is about 7.56×10^5 km^2 (Simpson-Lewis et al. 1979). Various authors have estimated total wetland area in Canada and some of its more renowned wetland regions (Table 1). A detailed province-by-province summary is presented in National Wetland Working Group (1985); this will likely serve as a benchmark for most future discussions of wetland area. It is doubtful that any of these estimates include all deepwater marshes, historic wetland losses, and aerially indistinguishable wetlands. The National Wetland Working Group (1985) figure represents about one seventh of the area of Canada. Estimates of wetland area in Canada range from 1.05×10^6 to 2.30×10^6 km^2.

Based on the relative abundance of the five types of wetland, variability in Canadian climate and physiography, and density of wetlands, the National Wetland Working Group (1981) has delineated wetland regions of Canada (Fig. 2). Six major types of region are represented. Each of these may be further subdivided into as many as 11 subordinate regions.

Regional traits range from permafrost-dominated fen systems typical of the Arctic Wetland Region to the marsh, swamp, bog, fen, and shallow-water mosaic of the Temperate Wetland Region. Characteristics of Canada's wetland regions are summarized in Table 2.

Wetlands are unevenly distributed within wetland regions and across Canada. The National Wetland Working Group (1981) has summarized relative area covered by wetlands (Fig. 3). The highest percent cover of wetlands in Canada occurs in the Hudson Bay lowland. This results from low relief, moist climate, and relatively recent successional response to postglacial isostatic rebound. Low relative area of wetland reflects climatic and physiographic extremes, especially cold temperatures, and rough terrain. In areas such as southern Ontario, low percent cover likely also reflects conversion of wetlands to other human uses.

Special Functions of Wetlands

Reviews of wetland functions and values have been undertaken recently by Greeson et al. (1978), Herdendorf and Hartley (1980), Richardson (1981), Patterson (1984),

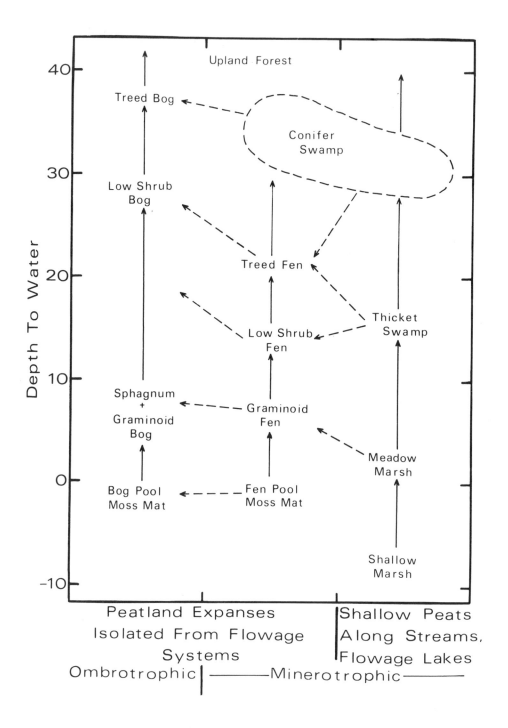

FIG. 1. Model of relationships among trophic status, depth to water (below soil surface), and wetland types. Redrawn from Jeglum and Cowell (1982).

TABLE 1. Selected estimates of wetland area in Canada.

Geographic extent of estimate	Estimated area (km^2)	Reference
Canada	1.27×10^6	National Wetland Working Group 1985
	1.30×10^6	Rennie 1977
	1.70×10^6	National Wetland Working Group 1981
	1.50×10^6 to 1.70×10^6	Zoltai 1979
	1.05×10^6 to 2.30×10^6	D. Marchand pers. comm.
Pacific region		
B.C. interior	9.32×10^3	British Columbia Ministry of Agriculture and Food 1981
S.W. Fraser R. lowland	2.35×10^2	Pilon and Kerr 1984
Prairie region		
Battle River Basin	3.03×10^2	Rittner 1980
Alberta parkland	2.58×10	Schick 1972
South Saskatchewan River Basin	4.71×10	Schmitt 1980
Minnedosa District Newdale Basin	4.48	Adams and Gentile 1978
Black soil zone	1.85×10	Goodman and Pryor 1972
Ontario		
S. Ontario	1.13×10^4 to 1.29×10^4	Cox 1972
	6.07×10^3	Laidlaw 1978
Hudson Bay lowland	2.10×10^3	Riley 1981
Northern clay belt	2.60×10^4	Riley 1981
Ontario	2.58×10^5	Monenco Ontario Ltd. 1981
Quebec		
St. Lawrence River	5.54×10^2	Canadian Wildlife Service 1978
Atlantic region		
Nova Scotia salt marshes	1.50×10^2	Hatcher and Patriquin 1981
New Brunswick	8.87×10^2	Airphoto Analysis Associates Consultants Limited-Toronto 1975

Sather and Smith (1984), and Office of Technology Assessment (1984). This review focusses on some major functions, emphasizing Canadian literature.

Primary Productivity

Some types of wetland contain plant communities that are among the most productive in the world. For example, reed marshes are the most productive major natural communities of vegetation, having a 5% efficiency of conversion of photosynthetically available radiation to energy in the form of dry matter. In comparison, agricultural grasses, considered to be highly productive crops, have a 2% conversion efficiency (Etherington 1983). Such efficiencies are difficult to estimate; hence, comparative data are rare.

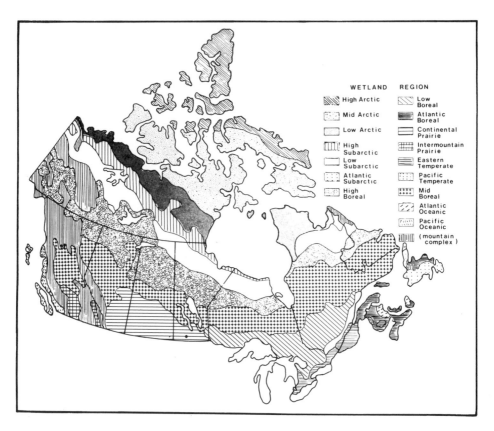

FIG. 2. Wetland regions of Canada. Based on a map by the National Wetland Working Group (1981) and reproduced in Tarnocai (1984).

A more widely employed measure of productivity is change in biomass per unit area. Even so, although attempts at standardizing measurement of productivity have been made (Milner and Elfyn-Hughes 1968), methods are not standard. Moreover, estimates of aerial productivity are more numerous than below ground productivity. All methods are tedious; consequently, understanding of variation in primary productivity across Canada is limited. Examples of aerial primary productivity suggest the range of values that might be expected. Reader (1978) estimated productivity of leatherleaf (*Chamaedaphne calyculata*) in a subarctic bog-marsh complex to be 227 g $\cdot$ m^{-2} $\cdot$ yr^{-1} (dry wt.); the same species in a boreal bog produced 370 and in a boreal fen 1020 g $\cdot$ m^{-2} $\cdot$ yr^{-1} (dry wt.) (Richardson 1978). The sedge *Carex rostrata* produced 116 g $\cdot$ m^{-2} $\cdot$ yr^{-1} (dry wt.) in a subarctic bog-marsh complex (Reader 1978) and 515 in a mountain marsh-shallow water-sedge meadow complex (Bernard and Gorham 1978). These trends are consistent with the expectation of increases in productivity in more southern wetland regions and mineral-trophic wetland types. In fact, average primary productivity for all species in one subarctic bog-marsh complex was estimated by Reader (1978) to be 169 g $\cdot$ m^{-2} $\cdot$ yr^{-1} (dry wt.); the value for a single species, the yellow water lily *Nuphar polysepala*, in a mountain marsh-shallow water-sedge meadow complex was 1121 (Davies 1970). Interpretation of figures such as these is further clouded by community relationships. Gosselink and Turner (1978), for example, reported that cattail (*Typha*) productivity ranged in one study from over 1000 g $\cdot$ m^{-2} $\cdot$ yr^{-1} (dry wt.) when the species was alone to less than 50 when four other species were present.

TABLE 2. Wetland regions of Canada (National Wetland Working Group 1985).

Wetland region		Key word description
1st order	2nd order	
Arctic	High Mid Low	Cold climate; arid, continuous permafrost; common wetlands: fens, lowland polygonal bogs, peat mound bogs, marshes, and shallow waters
Subarctic	High Low Atlantic	Cool summers, cold winters; low to moderate precipitation; discontinuous permafrost; common wetlands: fens, bogs, marshes, and shallow waters
Boreal	High Mid Low Atlantic	Mild to cold winters, warm to moderately cool summers; moderate to high precipitation; discontinuous permafrost; common wetlands: fens and bogs
Prairie	Continental Intermountain	Cold winters, hot summers; semiarid; common wetlands: marshes and shallow waters
Temperate	Eastern Pacific	Mild winters, warm summers; moderate to high precipitation; common wetlands: swamps and marshes in east, bogs and fens in west
Oceanic	Atlantic Pacific	Cool summers, cold (east) or mild (west) winters; high precipitation; common wetlands: bogs, fens, and marshes
Mountain	Coastal Interior Rocky Eastern	Cool to cold climate; precipitation moderate with many exceptions; discontinuous permafrost; common wetlands: bogs, fens, and some marshes

Production below ground as a proportion of aerial production ranges approximately from 0.13 to 4.0 (bog), 0.30 to 0.50 (fen), 0.10 to 0.20 (*Carex* marsh), and 0.66 to 0.30 (cattail marsh) (Richardson 1978), although most of these figures are not for Canadian wetlands. Measurement of below-ground primary production has rarely been published for Canadian wetlands.

In general, marshes represent the most productive type of wetland system. Highest production has been recorded for cattail-dominated marshes. Shoot development in late winter and early spring (Bayly and MacLatchy 1982), high leaf area (Richardson 1978), prolific vegetative reproduction, minimal water stress, and minimal nutrient limitation contribute to this productivity (Wetzel 1983).

Among the primary producers which attract most attention are those which are either harvestable or pests. Two of these are especially noteworthy: wild rice (*Zizania*) and eurasian watermilfoil (*Myriophyllum spicatum*).

Wild rice is a reed marsh emergent which is harvested for human consumption. It is gathered from wild stands or cultivated. In Canada the yield has been substantially from wild stands, although many small-scale attempts at cultivation have been made (Plansearch 1980). Saskatchewan's natural rice was of inadequate quantity for commercial harvest, but introductions in the 1930s and subsequent unassisted spread of the rice have resulted in commercially harvestable rice.

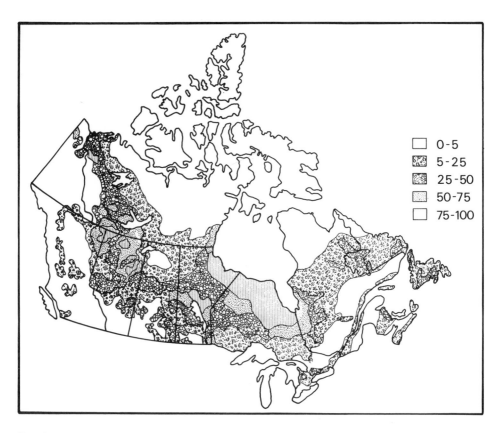

FIG. 3. Percentage cover of wetlands in Canada. The key contains categories of the percentage of the Canadian land surface covered by wetland. Based on Zoltai and Pollett's (1983) summary of National Wetland Working Group (1981).

There are now eight large wild rice processing plants in Canada, most in Manitoba (Winchell and Dahl 1984). Other smaller ones exist. Production of wild rice in the main producing provinces from 1980 to 1982 was (× 1000 kg) (Winchell and Dahl 1984).

	Saskatchewan	Manitoba	Ontario	Total
1980	58	254	194	506
1981	93	82	137	312
1982	95	75	34	204

Productivity ranges from about 630 to 1450 $g \cdot m^{-2} \cdot yr^{-1}$ (Wetzel 1983). Natural factors such as climate, water level in particular, cause considerable variation in rice production. Also a factor in Canada is conflict between potential producers about who should have rights to the resource (Hartt 1978).

In contrast with wild rice, eurasian watermilfoil is a nuisance. Introduced to the Chesapeake Bay area in about 1902 (Wile and Hitchin 1977), the species has slowly invaded lakes throughout North America. At some point during its invasion of a water body the eurasian milfoil characteristically develops a dense population, outcompeting other plant species and creating a problem for water-based recreation and transport.

In Canada, eurasian watermilfoil has notably been a problem in the Kawartha Lakes (Ontario) and Lake Okanagan (British Columbia). Standing biomass in the Kawarthas in the mid-1970s exceeded 200 g • m^{-2}, more than half of the total macrophyte biomass (Wile and Hitchin 1977). Mechanical harvesters and herbicides have been used to control the milfoil. For example, in the Kawarthas, mechanical harvesting between 1973 and 1976 yielded over 10 000 metric tons (t) of vegetation from about 1170 ha (Wile and Hitchin 1977). In the Kawarthas, however, as elsewhere, the eurasian water-milfoil population collapsed after a period of dominance (Wile et al. 1979) although nuisance patches remain.

Natural wetlands are used for agriculture. Drainage enables intensive cultivation of wetland areas; however, once drained, "wetland" is a misnomer. Natural wetland productivity, mainly grasses, is utilized on the prairies for grazing livestock (Rakowski and Jurick 1984). In some areas the fringe of grasses surrounding potholes and other small wetlands is a prime source of forage. The water retained by such wetlands can be invaluable for livestock and crops in arid areas such as Pallister's Triangle (Prairie Association of Water Management 1984).

Forestry practices have been expanding into wetland areas in recent years. Sheltered, fertile slope fens are most suitable for afforestation although generally they are small. Larger domed (raised) bogs are also possible sites (Wells and Pollett 1983). Experimental partial drainage of suitable sites for afforestation is currently underway in northern Ontario and Newfoundland.

Harvest of black spruce in northern wetlands has been undertaken for some time. Black spruce forest stands, for example, cover about 10% of the area of Ontario (Rennie 1977). There is, however, some doubt about whether these sites can regenerate naturally. Site damage during harvesting and successional alteration of seedling habitat may require site preparation similar to that for afforestation (Rubec and Lynch-Stewart 1985).

Conventional forest harvesting has been undertaken in hardwood and cedar swamps. In southern Quebec and southern Ontario especially, selective harvest of species such as silver maple (*Acer saccharinum*) and white cedar has proved profitable.

Animal Productivity and Diversity

Wetland Birds

A generally conspicuous function of wetlands, because of its recreational significance, is the production of fish and wildlife. This is rather difficult to define precisely. Relatively few of the well known game and nongame species are permanent residents in wetlands. Some are only present for a few critical days or weeks of a year and the activities of many in wetlands are incompletely understood. Both permanent and transient components of wetlands merit attention.

Waterfowl are perhaps the animals most closely identified with wetlands. They are generally divided into those which breed in wetlands and those which stage or moult there. Dabbling ducks (Anatinae), diving ducks (Aythyinae), ruddy and masked ducks (Oxyurinae), geese (Anserinae), and swans (Cygninae) are the prominent wetland waterfowl.

The most important wetlands for waterfowl breeding in North America are located in the prairie and parkland biomes of the prairie provinces (Table 3). Some 5-6 million pond wetlands (potholes, sloughs, etc.) provide breeding habitat for as many as 50% of North American ducks (Neave 1983). In Canada alone this represents about 1.8 × 10^7 birds (Hochbaum 1983). Averaging about 0.4 ha in area (Jahn 1969) the pond wetlands occur at spring densities of 10-15 • km^{-2} in the parkland and 5-10 • km^{-2} in the prairie (Hochbaum 1983). Although about eight species of dabbling duck and eight species of diving duck breed in the region (Wishart et al. 1983), mallards are the most important, representing about 4.3 × 10^6 birds and breeding densities ranging from 2 to 50 pairs • km^{-2} (Hochbaum 1983). Pond and bird densities are subject to high variation because of periodic drying (Jahn 1969).

TABLE 3. Duck breeding and broods in Ducks Unlimited projects in various Canadian biomes. Data were obtained from Wishart et al (1983). Based on various surveys between 1978 and 1981. The numbers of breeding species for the deciduous Ontario and Great Lakes - St. Lawrence biomes were pooled.

Number of breeding species				Breeding pair densities per hectare of wetland			Number of broods per hectare of wetland	
Dabblers	Divers	Biome	N	X̄	(± SE)		X̄	(± SE)
8	8	Parkland B.C.	18	2.4	(0.3)		0.8	(0.1)
7	8	Forested Alberta	30	1.2	(0.2)		0.6	(0.1)
8	8	Parkland Alberta	10	3.5	(0.6)		1.6	(0.4)
8	6	Prairie Alberta	20	4.0	(0.5)		0.9	(0.2)
7	7	Parkland Sask.-Man.	37	1.7	(0.2)		1.0	(0.2)
7	5	Prairie Sask.-Man.	34	4.2	(0.4)		3.8	(0.6)
6	2	Deciduous Ontario	3	0.8	(0.2)		0.5	(0.1)
6	2	Gt. Lakes–St. Lawrence	8	2.3	(0.4)		3.3	(1.2)
6	1	Annapolis Valley	13	0.4	(0.1)		0.6	(0.1)
7	3	Coastal Maritimes	18	1.3	(0.2)		0.8	(0.1)
6	2	Forested Maritimes	6	2.5	(0.8)		2.9	(1.2)

In Ontario, Quebec, and the Atlantic provinces the density of breeding pairs and species diversity are much reduced (Wishart et al. 1983). Bluewing teal (*Anas discors*) in Ontario and black ducks (*Anas rubripes*) there and eastward assume greater importance, although evidence is accumulating that mallards are succeeding the other species where habitat is unstable (Collins 1974; Dennis 1974a; Laperle 1974). Northward in boreal forests of eastern and western Canada, black ducks, Canada geese (*Branta canadensis*), and diving ducks are important, with densities of breeding pairs ranging from 0 to 6 • km^{-2} (Dennis 1974b; Gillespie and Wetmore 1974; Hochbaum 1983). Canada geese, lesser snow geese (*Anser c. caerulescens*), and whistling swans (*Olor columbianus*) are important in the Arctic (Gillespie and Wetmore 1974; Hochbaum 1983).

Larger wetlands are particularly important for staging ducks, geese, and swans during spring and autumn migration. Four major migratory flyways exist in North America (Jahn and Kabat 1984). Waterfowl rest, feed, and moult in and adjacent to major wetlands, marshes especially, along the flyways. Of particular importance are the marshes in large lakes along the edge of the Canadian Shield (Bailey 1983), in prairie river deltas (Canadian Wildlife Service 1979), along the St. Lawrence River (Sarrazin et al. 1983), and in coastal estuaries (Hunter and Jones 1982; McAloney 1981).

Autumn migrations generally involve greater use of wetlands by waterfowl because the urgency of southward movement is less pronounced than immediately before breeding. Also, food is more readily available and moulting requires some time. Along the Great Lakes, for example, the number of waterfowl using staging areas during autumn exceeds that in spring by about a factor of 4 (Dennis and Chandler 1974).

Wetlands also produce birds other than waterfowl. Shorebirds both breed and stage in wetlands. Some, such as the semipalmated sandpiper (*Calidris pusilla*), are widespread; others, such as the hudsonian godwit (*Limosa haemastica*), are concentrated in one or a few areas (Morrison 1983). Particularly important for breeding is the transition (ecotone) between boreal forest and tundra systems (Howe 1983). Understanding of nongame birds is limited in comparison with that for game birds. Isolated studies reveal the potential value of wetlands. McCracken et al. (1981), for

TABLE 4. Numbers of waterbirds harvested ($\times 1000$) in 1982 in Canada (data obtained from Metras 1984).

Taxon[a]	North	Pacific	Prairie	Ontario	Quebec	Atlantic	Total
American widgeon	0	24	38	10	6	5	83
Black duck	0	0	1	87	118	132	338
Blue-winged teal	0	tr	26	32	15	6	79
Gadwall	0	6	74	11	7	0	98
Green-winged teal	0	15	43	37	66	41	202
Mallard	3	91	686	334	95	4	1213
Pintail	0	14	56	10	20	4	104
Shoveler	0	3	30	1	3	0	37
Wood duck	0	1	tr	91	21	3	116
Ruddy duck	0	0	1	2	0	0	3
Bufflehead	1	2	5	24	4	2	38
Common merganser	0	tr	1	4	16	7	28
Hooded merganser	0	tr	17	10	tr	0	27
Red-breasted merganser	0	1	1	7	10	0	19
Canvasback	0	12	6	1	0	0	19
Redhead	0	tr	13	12	2	0	27
Ring-necked duck	0	2	14	75	21	12	124
Lesser scaup	0	2	48	40	20	1	111
Greater scaup	0	tr	6	17	16	2	41
Goldeneye	0	1	5	41	21	23	91
Oldsquaw	0	0	0	4	10	4	18
Common scoter	0	0	0	1	6	0	7
Surf scoter	0	tr	tr	1	15	12	28
White-winged scoter	0	1	1	2	8	7	19
Harlequin	0	0	0	tr	0	9	9
Common eider	0	1	0	3	18	0	22
King eider	0	0	0	0	tr	0	0
Geese	0	18	428	72	74	33	625
Total	4	194	1500	929	592	307	3526

[a]American widgeon (*Anas americana*), black duck (*A. rubripes*), blue-winged teal (*A. discors*), gadwall (*A. strepera*), green-winged teal (*A. crecca*), mallard (*A. platyrhynchos*), pintail (*A. acuta*), shoveler (*A. clypeata*), wood duck (*Aix sponsa*), ruddy duck (*Oxyura jamaicensis*), bufflehead (*Bucephala albeola*), common merganser (*Mergus merganser*), hooded merganser (*Lophodytes cucullatus*), red-breasted merganser (*Mergus serrator*), canvasback (*Aythya valisineria*), redhead (*A. americana*), ring-necked duck (*A. collaris*), lesser scaup (*A. affinis*), greater scaup (*A. marila*), goldeneye (*Bucephala*), oldsquay (*Clangula hyemalis*), common scoter (*Melanitta nigra*), surf scoter (*M. perspicillata*), white-winged scoter (*M. deglandi*), harlequin (*Histrionicus histrionicus*), common eider (*Somateria mollissima*), king eider (*S. spectabilis*), geese (*Anserinae*).

example, identified 102 bird species other than waterfowl breeding in the wetlands and immediate vicinity at Long Point, Lake Erie. At the same location about 300 species of migratory bird have been identified (McKeating 1983).

The value of wetlands for bird production is difficult to measure in human terms. Between 1976 and 1981 the number of duck and goose hunters in Canada averaged about 390 000 roughly 90% of whom were residents. Approximately 3.5×10^6 ducks and 6.0×10^5 geese were harvested annually on average (Cooch 1984). The importance of mallards and black ducks is evident in Table 4. Most of the harvest of waterfowl bred in Canada takes place in the United States.

TABLE 5. Sport fishing catch of wetland-associated freshwater fish in 1980 in Canada (× 1000 fish) (data from K. W. Brickley, Department of Fisheries and Oceans, pers. comm.).

Taxon[a]	Yukon	N.W.T.	B.C.	Alta.[b]	Sask.	Man.	Ont.	N.B.	N.S.	P.E.I.	Nfld.	Total
Goldeye	0	0	0	188	59	0	0	0	0	0	0	247
Northern pike	23	230	0	2808	3298	4 321	13 308	0	0	0	0	23 988
Bullhead	0	0	0	0	0	0	1 356	0	0	0	0	1 356
Catfish	0	0	0	0	0	0	2 216	0	0	0	0	2 216
White perch	0	0	0	0	0	0	0	102	0	11	0	113
White bass	0	0	0	0	0	0	669	0	0	0	0	669
Black bass	0	0	21	0	0	248	10 832	172	0	0	0	11 273
Pumpkinseed	0	0	0	0	0	0	4 139	0	0	0	0	4 139
Rock bass	0	0	0	0	0	0	6 864	0	0	0	0	6 864
Crappies	0	0	0	0	0	0	1 101	0	0	0	0	1 101
Sheepshead	0	0	0	0	0	0	558	0	0	0	0	558
Yellow perch	0	0	0	2995	2060	0	33 575	295	0	0	0	38 925
Walleye	0	57	0	891	2158	3 756	22 246	126	0	0	0	29 234
Other	1	24	835	25	27	1 855	12 693	0	0	0	0	15 435
Total	24	311	856	6907	7602	10 180	109 557	695	0	11	0	136 143

[a]Goldeye (Hiodon tergisus), northern pike (Esox lucius), bullhead (Ictalurus other than I. punctatus), catfish (I. punctatus), white perch (Morone americana), white bass (M. chrysops), black bass (Micropterus), crappies (Pomoxis), pumpkinseed (Lepomis gibbosus), rock bass (Ambloplites rupestris), sheepshead (Aplodinotus grunniens), yellow perch (Perca flavescens), walleye (Stizostedion), "other" are mostly littoral fish.
[b]Harvest.

Wetland Fish

The use of wetlands by fish is far less well understood than waterfowl use. For various reasons, pelagic, profundal, and stream salmonidae — salmon, trout, char, and whitefish in particular — have received more scientific attention in Canada than coolwater and warmwater littoral fish. The regular occurrence of some species such as northern pike (*Esox lucius*), largemouth bass (*Micropterus dolomieui*), and bowfin (*Amia calva*) is well documented (Scott and Crossman 1973). Many, in fact most other Canadian fish species, make at least short seasonal or life stage specific appearances in wetlands. It is generally accepted that families such as the pike (Esocidae) and sunfish (Centrarchidae) use wetlands for many purposes: spawning, nursery, foraging, and cover. Other families, perch (Percidae) and catfish (Ictaluridae), for example, depend on wetlands for nursery and foraging. Yet others such as salmon, trout, and whitefish (Salmonidae) benefit more from wetland-dependent food organisms than wetland conditions directly.

These and other relationships between fish communities and wetlands are documented qualitatively (Liston et al. 1983; Whillans 1979, 1985). Quantitative relationships, however, between wetland condition and fish abundance or survival have not yet been determined.

The best Canada-wide reflection of the importance of wetlands to fisheries pertains to the sport fishery. This focusses more intensively on littoral fish than the commecial fishery, is spatially more diffuse, and is more thoroughly documented than the domestic fishery. It is clear from Table 5 that the major wetland-associated freshwater sport fishery is located in central Canada, Ontario in particular. Salmon, trout, char, and whitefish are not tabulated here. This does not mean that they do not benefit from freshwater wetlands, only that they are usually not considered part of wetlands. The value of wetland-associated freshwater sport fish is not easily estimated. Direct expenditure on all Canadian sport fisheries is about 1.1×10^9 annually (Department of Fisheries and Oceans 1984). This compares with a direct value of 2.5×10^7 for 13 888 t of wetland-associated comercial freshwater fish (Statistics Canada 1978).

The difference between actual and potential yields has long been a subject of interest to fishery scientists. Most of the large number of small ponds, potholes, sloughs, etc., in the prairie provinces — so valuable for duck breeding — are not utilized for fish production to their full potential. The Province of Saskatchewan in particular has been attempting to change this by culturing rainbow trout in underutilized waters on private land (Saskatchewan Tourism and Renewable Resources 1980).

Licences to undertake such activity numbered 1888 in 1982; all but 52 of these were for domestic (unmarketed, home-consumption) purposes (Sawchyn 1984). The harvest is largely uncensussed. The presence of a large commercial market (Saskatchewan Research Council 1984) has prompted development of more efficient culture techniques. Timing of culture (the full ice-free period), supplementary feeding, and the use of cages for 100% harvest suggest that an 18- to 25-fold weight gain could be obtained in one season. Ultimate trout densities of about 14–20 kg $\cdot$ m^{-3} could be achieved (Sawchyn 1984).

Wetland Mammals

Mammals also make use of wetlands. Although many species will venture into wetlands periodically for forage, cover, drink, or other purposes, only a few are truly wetland species. The four in Table 6, beaver, mink, muskrat, and otter, are heavily dependent on wetlands. Regionally they are most frequently captured in central Canada. Value of the three most valuable pelts amounts to about 4.5×10^7 for mink (including ranch mink), 3.24×10^5 for beaver, and 6.0×10^6 for muskrat (Statistics Canada 1985). Not reported here are domestic harvests. Some 12 other mammal species make substantial use of wetlands in Canada, including economically important ones such as moose (*Alces alces*), red fox (*Vulpes vulpes*), and white-tailed deer (*Odocoileus virginianus*) (Hummel 1981).

TABLE 6. Canadian harvests of mammals heavily dependent on wetlands ($\times 1000$ animals) (data from Statistics Canada 1985).

Species[a]	North	Pacific	Prairie	Ontario	Quebec	Atlantic	All
Beaver	0	13	131	118	48	11	325
Mink	2	4	21	16	8	12	63
Muskrat	160	25	462	448	222	92	1409
Otter	0	1	3	7	3	1	15

[a]Beaver (*Castor canadensis*), mink (*Mustela vision*), muskrat (*Ondatra zibethicus*), otter (*Lontra canadensis*).

Wetland Reptiles and Amphibians

Reptiles and amphibians are among the least thoroughly studied wildlife residents and transient users of wetlands. Some are integral components of wetlands during their whole life cycles; others are present only for critical purposes such as breeding. Herptiles utilizing wetlands in Canada are summarized in Table 7. As in the cases of fish and mammals, central Canada contains high diversity and presumably productivity of wetland-associated herptiles. About nine of these are currently of concern to the Amphibian and Reptile Subcommittee of the Committee on the Status of Endangered Wildlife in Canada (Cook 1985).

Reptiles and amphibians, unlike waterfowl, fish, and mammals, are not subjected to heavy consumptive exploitation. Like all conspicuous animals, however, they are subject to considerable nonconsumptive use (viewing, photography, etc.). Consumptive wildlife-related activities in general interested about 27% of Canadians in a 1981 survey; the figure was 83% for nonconsumptive use (Filion et al. 1983).

Modification of Surface and Groundwater Flow

It is the nature of wetlands to be tied to the hydrologic regime. Winter (1981) observed that the link may involve receiving groundwater discharge, conducting water, or recharging groundwater. The circumstances and degree of these conditions are subjects of some debate (Barnes 1984). It is generally accepted, though, that hydrology is a major determinant of wetland characteristics (Gosselink and Turner 1978). The influence of wetlands on hydrology is less clear.

The controversy centres around whether wetlands function as "sponges," conserving water in effluent streams until dry periods, or if they actually cause excessive water to be lost to the atmosphere through evapotranspiration (Barnes 1984). The "sponge" concept is based on the porous nature and hydraulic conductivity of peat. The mean volume of water at saturation in peat in Canada is reportedly in the order of 92% of the total volume of peat. Normally, 50-80% of the water is loosely contained in peat pores and available for release from new peat and 10-15% from well-decomposed peat (Walmsley 1977). Contrasting with this potential for temporary storage and slow release of water, some data reveal situations in which wetlands did little to alter the flow of water other than through friction on the plants (Taylor 1982; Woo and Valverde 1982).

It is likely that differences of opinion arose from the lack of a sufficient data set for comparative study. For example, it is known that hydraulic conductivity tends to decrease with increased depth (and bulk density) in a peat profile (Romanov 1961). The upper "hydrologically active layer" is that which potentially determines the fate of water as it passes through a wetland. The height of the water table determines the actual effect of the active layer. In regions with high precipitation, during spring melt,

TABLE 7. Herptiles utilizing wetlands in Canada (compiled from Cook 1984; Behler and King 1979).

Taxon[a]	Presence in region						Wetland types	Use[b]
	B.C.	Prairie	Ont.	Que.	Atl.	North		
Mudpuppy		×	×	×			Marshes, shallow water	R,F
Fowler's toad			×	×			Fen, marsh	R
Cricket frog			×				Shallow open water	R
Spring peeper		×	×	×	×		Swamps, marshes	R
Bullfrog			×	×	×		Shallow water, marshes	R,F
Green frog			×	×	×		Shallow open water	R
Pickerel frog			×	×	×		Shallow water	R
Leopard frog	×	×	×	×	×	×	Marshes, shallow water	R
Mink frog			×	×	×		Shallow water, bog	R
Wood frog	×	×	×	×	×	×	Swamps	R
Tailed frog	×						Swamp	R
Pacific tree frog	×						Marshes, shallow water	R
Eastern fox snake			×				Shallow water, marshes	R,F
Eastern ribbon snake			×				Shallow water, swamps, bogs	R,F
Northern water snake			×	×			Shallow water, marshes	R,F
Massasauga			×				Swamps, bogs	R,F
Midland painted turtle	×		×	×	×		Marshes, shallow water	R,F
Spotted turtle			×				Swamps, marshes, fens	R,F
Wood turtle			×	×	×		Swamps	R,F
Blanding's turtle			×	×			Shallow water, marshes	R,F
Musk turtle			×				Shallow water, marshes	R,F
Snapping turtle		×	×	×	×		Marshes, shallow water	R,F
Eastern newt			×	×	×		Shallow water, marshes	R
Roughskin newt	×						Shallow water, marshes	R

[a]Mudpuppy (*Necturus maculosus*), fowler's toad (*Bufo woodhousei fowleri*), cricket frog (*Acris crepitans*), spring peeper (*Hyla crucifer*), bullfrog (*Rana catesbeiana*), green frog (*R. clamitans*), pickerel frog (*R. palustris*), leopard frog (*R. pipiens*), mink frog (*R. septentrionalis*), wood frog (*R. sylvatica*), tailed frog (*Ascaphus truei*), Pacific tree frog (*Hyla regilla*), eastern fox snake (*Elaphe vulpina gloydi*), eastern ribbon snake (*Thamnophis sauritus*), northern water snake (*Nerodia sipedon*), massasauga (*Sistrurus catenatus*), midland painted turtle (*Chrysemys picta*), spotted turtle (*Clemmys guttata*), wood turtle (*C. insculpta*), blanding's turtle (*Emydoidea blandingi*), musk turtle (*Sternotherus odoratus*), snapping turtle (*Chelydra serpentina*), eastern newt (*Notophthalmus viridescens*), roughskin newt (*Taricha granulosa*).
[b]Compiled from Cook (1984) and Behler and King (1979).

or after excessive rainfall when the water table is above the active layer, a wetland is preempted from sponge-like behaviour (Barnes and Rogerson 1985; Taylor 1982; Verry and Boelter 1981). When the water table is below the active layer the depth of that layer and gradient in bulk density determine the potential response of a wetland to an influx of water (Barnes 1984; FitzGibbon 1982). Freezing effectively increases the height of the water table in organic soils (Price and FitzGibbon 1982).

It is thus apparent that wetlands can act as sponges during dry periods. Otherwise their hydrologic influence depends on inflow, evapotranspiration, and friction (Rothwell 1982; Woo and Valverde 1981). If vegetation is relatively thick they experience lower evapotranspiration than would exposed water (Bertulli 1981; Munro 1982).

Wetlands thus function on balance to moderate effluent stream flow, though sometimes to a very temporary degree (Larson 1981; Novitzki 1978; Woo and Valverde 1982). They thus contribute to flood control. In Canada it is estimated conservatively that wetlands are storage areas for some 1.5×10^3 km^3 of water (Barnes 1984).

Wave Attenuation

Shallow-water wetlands and marshes tend to develop in protected locations, so in this sense their potential value in protecting shorelines from waves is limited. Nevertheless, it is known, mainly from studies outside of Canada, that aquatic vegetation protects shorelines from erosion (Sather and Smith 1984).

According to Dean (1978) the protection results from four mechanisms: (a) binding of sediment by roots, (b) diffusion of erosion wave energy by plants, (c) sedimentation because of reduced wave velocity, and (d) establishment of protective sand dunes because of all of the above. In general, woody vegetation such as willow is hardiest. Its roots make it most useful in shoreline protection (Allen 1978). Cattail and other robust emergents can also be very important.

The economic value of wetland vegetation in wave attenuation has not been fully quantified (Jaworski et al. 1980). The practicality of protecting shorelines by enhancing riparian wetlands has, however, been widely demonstrated by the U.S. Army Corps of Engineers (Garbisch 1977).

In Canada, comparable work is rare. The recent situation in Rondeau Bay, Lake Erie, is an exception. In that case, denudation of much of a shallow-water wetland in the 1970s left an expanse of exposed sediment and a decimated largemouth bass (*Micropterus salmoides*) fishery. Subsequent wave action resuspended sediment and the turbidity prevented natural establishment of vegetation. Replanting costs for 1200 ha were projected at $395 000 (Hanna and Associates 1984).

A corollary of wave (or current) attenuation is the influence on water temperature. The reduction of circulation by vegetation tends to cause increased water temperatures in littoral waters (Dale and Gillespie 1977). In swamps, however, with dense tree crowns the shade serves to reduce water temperatures.

Uptake and Burial of Pollutants

High productivity and potential for assimilating nutrients are reasons for the contention that wetlands could function in the treatment of wastewater or the rehabilitation of lake and stream water quality. Natural and artificial wetlands have been proposed as useful in these respects. Possibilities have been explored in much more detail in the United States and Europe than in Canada (Sather and Smith 1984).

Kadlec and Kadlec (1978) reported that wetlands tend to alter the characteristics of any water that passes through them. The means and degree of alteration vary with wetland type. Richardson et al. (1978) compared nutrient assimilation in bogs, fens, swamps, and marshes. Concentrations of nutrients were generally lowest in bog plants. This reflects the low availability. In other types of wetland, soil constitutes a source. For example, in fens the concentration of N and P in stems and leaves increased throughout the growing season. It was noted that substantial translocation of nutrients from plants did not occur as a function of shedding of leaves and fruit. This helps to explain why Richardson et al (1978) found that in fens more than 97% of N, P, and Ca was tied up in peat (the remainder was in living organisms and water).

To estimate the potential of wetlands for trapping nutrients, it would be necessry to understand hydrology, concentration of constituents, mass flow, seasonality, and other factors. Generally, however, wetlands affect nutrients and other microcontaminants of inflowing water by (Kadlec 1981) (a) straining through litter, (b) contributing detritus to a finite suspended load, (c) sorption of dissolved material on sediment surface, (d) precipitation of dissolved material and settling of suspended material to the sediment surface, (e) bacterial denitrification on sediment surface, and (f) incorporation of material by organisms during periods of growth (and vectoring of material to sediment as excretia or dead tissue).

Richardson et al. (1978) observed that episodic nutrient export from wetlands can be high because of hydrology and other factors. This raises the question of whether natural wetlands are sinks or net exporters of nutrients. The answer is indefinite. Valiela and Teal (1978) have noted, for example, that riverine wetlands tend to be sinks for N, but not for P. Phosphorus is generally available in the sediments and is thus not limiting. There is no question, however, that at least seasonally, when peat accumulates, and in definable hydrologic conditions, wetlands do function as sinks for microcontaminants.

Natural wetlands are of limited value for dependable wastewater treatment because of the need to control hydrology and seasonal exports (van der Valk et al. 1978). In Canada, various wetlands have received wastewater. Recently, efforts have been made to identify situations in which natural wetlands can serve in sewage treatment. The Ontario Ministry of the Environment has since 1981 been testing the performance of a natural 1400-m^2 marsh near Bradford, Ontario (Black 1983). Although natural marshes have some utility, they may also experience undesirable changes in productivity and species composition when used (Benforado 1981; Guntenspergen and Stearns 1981). Hydraulic control and harvest of aerial material would minimize recycling of nutrients to effluent water (Sloey et al. 1978; Wile et al. 1981); however, further alteration of natural characteristics should be expected.

Most wetland wastewater treatment facilities are artificial marshes (Brennan and Garra 1981). The main problems in sewage treatment are P, N, biochemical oxygen demand (BOD), and suspended solids (SS). An artificial wetland at Listowel, Ontario, reduced these components by 60-90% (SS), 50-96% (BOD), 10-90% (total P), and 30-98% (total N) (Black 1983). All but the latter were generally better than the objectives for conventional secondary treatment of sewage. The Bradford natural wetland achieved results which were comparable with those for Listowel, with the exception that less P was removed (Black 1983). Wetland sewage treatment facilities have also been undertaken at Cobalt, Ontario (Black 1983), and Humboldt, Saskatchewan (Saskatchewan Research Council 1984), and are planned for other Canadian locations. The practice is not nearly as widespread as it is in the United States (Garbisch 1977).

The use of wetlands for treatment of toxic wastes and control of pathogenic organisms has been postulated, but seldom implemented (Black et al. 1981). At the Listowel and Cobalt marshes, bacterial levels were reduced substantially (total coliforms, fecal coliforms, *Escherichia coli*, fecal streptococci, and *Pseudomonas aeruginosa*). The Listowel facility was older and considerably more efficient (Desjardins and Seyfried 1984).

Treatment of potentially toxic waste using wetlands has received considerable attention recently. The natural immobilization of contaminants such as heavy metals and pesticides has been observed (Kadlec and Kadlec 1978). Organic soils are known to attenuate a variety of contaminants, including Mg, Fe, and Mn (Creasy et al. 1981) and pesticides (Horstman 1976). Correlations of contaminant concentration in aquatic plant tissue (Pb, Ni, Cu, and Cr) with that of Ca implies that calcite on the surface of vegetation plays a role (Mudrock 1981). Highest concentrations of metals in plant tissue have been found in the roots (Behan et al. 1979; Taylor and Crowder 1983). This and the tendency for submerged vegetation to register higher concentrations than emergent vegetation (Mudrock 1981) suggests an aquatic route of uptake, possibly related to sediment conditions.

Accumulation of contaminants in wetland vegetation is useful if this renders the contaminants less dangerous. Tests on duckweed in leacheate waters revealed bioconcentration of As and Cu to 100–2600 times the level in the leacheate (Hanna 1984). Pesticide levels are known to undergo bioaccumulation in wetlands (Horstman 1976). Giblin (1985), however, observed that the uptake of metals (and likely other non-nutrient contaminants) tends to be a significant portion of the available metals only when the amount of available metals is small. Giblin (1985) argued, however, that the small amount of metals uptaken by vegetation may still merit attention because of the potential for transfer into herbivorous animals.

Bioaccumulation has proven useful in another sense. Plants, mosses especially, in ombrotrophic bogs (bogs receiving nutrients solely from precipitation) have served as monitors of atmospheric contaminants. Monitoring networks have developed globally (Tyler 1977). Concerted Canadian interest has been more recent (Glooschenko and Capoblanco 1982).

Benforado (1981) has summarized the mechanisms by which wetlands remove contaminants in wastewater (Table 8). It is important to note that although vegetation per se plays a key role, the ultimate effect of wetlands is the result of their total biological–physical–chemical environment.

Peat Accumulation

When decomposition of organic matter is retarded by water as typically occurs in wetlands, then peat begins to accumulate. Both plant and animal matter are normally involved, though the latter tends to be a minor component (Moore and Bellamy 1974). Canada's peat reserves represent 9000–10 000 yr of large-scale accumulation in northern maritime areas that were cool and moist immediately after the melting of the Wisconsinian glaciers (MacPherson 1982; Tarnocai 1978; Zoltai and Pollett 1983). Local lacustrine deposits may have occurred previously (Heinselman 1970). In other areas, cool and moist periods initiated major periods of peat accumulation some 2000–3000 yr later (Heinselman 1970; MacPherson 1982; Terasmae 1977; Tarnocai 1978; Zoltai and Pollett 1983).

Peat development is relatively slow. In peatlands with rapid peat accumulation, rates of 0.2–1.0 cm $\bullet$ yr^{-1} have been measured (Moore and Bellamy 1974). Tarnocai's (1978) identification of time as the dominant factor in peat development is thus well taken. Other key determinants include climate, nutrients, physiography, vegetation, and biotic community (Jeglum and Cowell 1982; Tarnocai 1978).

Earlier studies of peat development referred to phases in development (Dansereau and Segadas-Vianna 1952). Contemporary reports distinguish two underlying processes of peat development (Glaser 1983; Tarnocai 1980). One is a succession from a predominance of shallow water to vegetation-covered water tracks. The other is the raising or buildup of peat on level or gently sloping terrain. The former tends to produce deeper peat (Tarnocai 1980), though not always (Tarnocai 1983). Wells and Pollett (1983) broke the second process into two: primary establishment of peat-producing plants on moist soils and lateral or vertical expansion of peat (paludification). Others refer to only one basic process of peat development (Heinselman 1963).

It is clear that peat forms distinct patterns in association with water availability, some of which represent the elevating influence of peat accumulation on the water table (Damman 1979; Glaser et al. 1981). In fact, changes in water supply cause rather rapid change in cover of wetland vegetation (Heinselman 1970; Jeglum 1975).

338

TABLE 8. Utility of wetlands for removal of contaminants from wastewater (taken from Benforado 1981).

Mechanism	Contaminant affected[a]								Description
	Settleable solids	Colloidal solids	BOD	Nitrogen	Phosphorus	Heavy metals	Refractory organics	Bacteria and virus	
Physical									
Sedimentation	P	S	I	I	I	I	I	I	Gravitational settling of solids (and constituent contaminants) in pond/marsh settings
Filtration	S	S							Particulates filtered mechanically as water passes through substrate, root masses, or fish
Adsorption		S							Interparticle attractive force (van der Waals force)
Chemical									
Precipitation					P	P			Formation of or coprecipitation with insoluble compounds
Adsorption					P	P	S		Adsorption on substrate and plant surfaces
Decomposition							P	P	Decomposition or alteration of less stable compounds by phenomena such as UV irradiation, oxidation, and reduction
Biological									
Bacterial metabolism[b]		P	P	P			P		Removal of colloidal solids and soluble organics by suspended, benthic, and plant-supported bacteria. Bacterial nitrification/denitrification
Plant metabolism[b]				S			S	S	Uptake and metabolism of organics by plants. Root excretions may be toxic to organisms of enteric origin
Plant absorption				S	S	S	S		Under proper conditions significant quantities of these contaminants will be taken up by plants
Natural die-off								P	Natural decay of organisms in an unfavourable environment

[a]P = primary effect, S = secondary effect, I = incidental effect (effect occurring incidental to removal of another contaminant).
[b]The term metabolism includes both biosynthesis and catabolic reactions.

339

Standing reserves of peat in Canada have been estimated at 3.0×10^8 m^3. The harvestable portion is approximately equivalent to an energy value ranging from 6.7×10^{21} to 5.7×10^{20} J (Tarnocai 1984; Zoltai and Pollett 1983). The peat reserves thus represent an energy source similar in size to Canada's coal reserves (Tarnocai 1984). Distribution of peat is concentrated in boreal and subarctic zones (Fig. 4). Boville et al. (1983) estimated that peat contains over half of the major nonliving organic carbon reservoir in Canada.

Peat is not harvested commercially for energy purposes in Canada (Zoltai and Pollett 1983), although experimental projects have been undertaken (Newfoundland Department of the Environment 1984). In fact, only about 1.3% of Canada's peatlands have been surveyed (Monenco Ontario Ltd. 1981). Peat is sold for horticultural purposes, about 4.88×10^5 t in Canada (Monenco Ontario Ltd. 1981).

Wetlands Status in Canada

Wetlands "status" here refers to two things. It has a comparative meaning; in this case the current condition of wetlands is compared with that historically. It also refers to the way in which wetlands are regarded in Canada — formally and from a utilitarian perspective. Status reports are meaningful only in context and generally have a limited period of relevance.

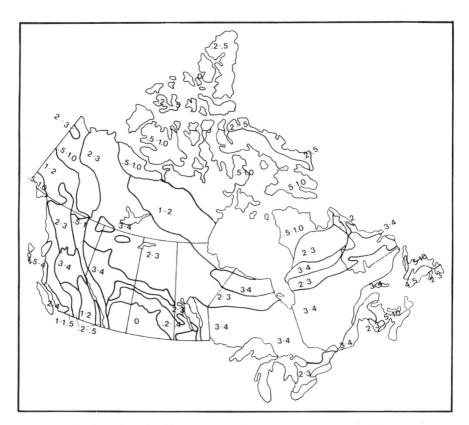

FIG. 4. Peat depth in Canada. The numbers refer to the average depth of peat in metres in wetlands within each region delineated. Adapted from Tarnocai (1984).

Wetland Abuses and Losses

Some of the ways in which wetlands are used in Canada have already been mentioned in the discussion of wetland functions. These and other uses exploit natural products and capabilities of wetlands. However, wetlands are also used in ways that exceed their natural limitations. Both the suitable and unsuitable types of use can in fact be abuses if they cause wetland conditions to change irreversibly or for long periods.

The most widely cited abuses of wetlands in Canada have been agricultural. Drainage of excess water or dyking to exclude water improves the capability of land for agriculture. These are irreversible practices because the organic soils deteriorate quickly when exposed to air and wetland biota are eliminated by agricultural activities.

Although increased productivity is the major reason for converting wetlands to agricultural use, simple convenience resulting from the elimination of obstacles is frequently cited (Desjardins et al. 1984). Losses of crops to wetland-associated animal pests and the nuisance created by recreational uses of wetlands are also given as justification (Jones 1981).

In British Columbia, both drain and dyke construction have increased by about 200% since the early 1970s (Bertrand 1985). Only a little more than 6.0×10^6 m of tile drains have been installed in Alberta (Tupper 1985). Manitoba has about 2.7×10^7 m of drains (Weber 1985). In Ontario, municipal drains alone affected some 2200 ha between 1970 and 1978 (Bardecki 1984). Dyking has enclosed about 30% of former wetland in the area of Lake St. Clair, Ontario, since 1965 (McCullough 1985).

Other common practices in preparation for agriculture in areas of thin peat include harvest of peat, burn-off of peat, and deep ploughing. These practices are prevalent on the prairies.

Agriculture also encroaches on wetlands through the clearing of wetland margins during dry years, in spite of probable low crop yields during wet years. On the prairies, for example, about 61% of wetland margins have been impacted in this manner (Brace and Pepper 1985).

Agriculture is generally practicable on organic soils which have been prepared by drainage, dyking, removal of peat, and encroachment. Yields tend to be high initially, after which fertilizer and organic inputs become economically limiting. A wide variety of crops have been grown in these areas of Canada (Stewart 1977). The less arable the initial conditions, the more irreversible the changes needed. Wetlands, once exploited in this manner, cannot recover their original characteristics. However, destruction of wetlands by agriculture is often incomplete.

It is perhaps unfair to brand agriculture as the major villain in wetland losses without qualification. Certainly, expectations of increased productivity and improvements in convenience and farming efficiency are major impetuses for conversion of wetlands to agricultural use. However, these must be weighed carefully against high levels of subsidization of drainage and taxes on natural wetlands which are "unimproved" for agriculture (Rakowski and Jurick 1984).

Regional distribution of agricultural and other abuses of wetlands in Canada are summarized in Table 9. The north has not suffered wetland loss. In all other regions agriculture is an extremely important cause of wetland loss.

Urbanization is the other major cause of wetland loss that is active in all regions except the north. The losses are much more restricted in area than those associated with agriculture. The widespread impacts of both urbanization and agriculture suggests a similar set of values for wetlands among both types of land user or land use regulator. In a number of urban areas there are indications that protection of wetlands has occurred in recent years, notably Vancouver, Calgary, Winnipeg, Hamilton, London, and Halifax (Rubec and Lynch-Stewart 1985). The other causes of wetland loss listed in Table 9 are more regional.

Information on losses of wetlands in Canada is accumulating from regional studies. Lynch-Stewart (1983) summarized many of these; Rubec and Lynch-Stewart (1985)

TABLE 9. Major causes of wetland degradation and loss in Canada by region (compiled from various sources).

Factor	North	B.C.	Prairie	Ont.	Que.	Atlantic
Urbanization						
Residential		×	×	×	×	×
Industrial		×		×	×	×
Agriculture						
Drain and dyke			×	×	×	×
Encroachment		×	×			
Peat elimination			×	×		
Grazing		×	×			
Peat harvest						
Horticultural		×		×		×
Energy						×[a]
Damming			×		×	
Recreational landfill				×		
Transportation					×	
Dredging		×				
Logging		×				

[a]Immediate potential.

updated the findings. Wetland losses are summarized in Fig. 5. Most of the losses are located in heavily settled or farmed areas. Where land-use activity has been most intense, losses of upward of 70% of the "original" wetland area have occurred. Intense but less conflicting land use has commonly resulted in areal losses of 40-60%. Even relatively rural locations have undergone losses of up to 30%. All regions except for the north have experienced losses.

A striking feature of the pattern of loss is its regional variation. No two regions have similar sets of causes of loss. This may reflect regional differences in predominant wetland types as much as it does regional differences in human activity.

Evaluating Losses

Trends in wetland loss have stimulated concern about the meaning of losses in terms of resources. This has been accompanied by growing managerial and public interest in wetlands. Consequently, resource managers across Canada have increasingly been faced with difficult decisions on the allocation of wetland resources to various uses. One response has been the development of evaluation methodologies.

A variety of types of wetland evaluation methodology have been applied, representing a variety of purposes (Table 10). The types are generally linked conceptually to each other, but tend to have different qualitites. Some of the methodologies cited in Table 10 resemble inventories more than evaluations. Nevertheless it is clear that evaluation represents a major thrust in wetland management now in Canada.

Few of the existing evaluation methodologies are being applied Canada-wide. Wetlands are provincial resources. This and the high time and cost involved in applying

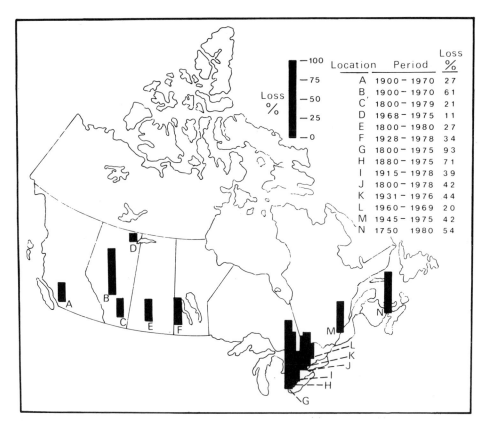

Location	Period	Loss %
A	1900 – 1970	27
B	1900 – 1970	61
C	1800 – 1979	21
D	1968 – 1975	11
E	1800 – 1980	27
F	1928 – 1978	34
G	1800 – 1975	93
H	1880 – 1975	71
I	1915 – 1978	39
J	1800 – 1978	42
K	1931 – 1976	44
L	1960 – 1969	20
M	1945 – 1975	42
N	1750 1980	54

FIG. 5. Example of wetland losses in Canada. Histograms depict the percent of wetland area lost in each region during a historic period. The exact percentages and time periods are summarized in the key. Compiled from various sources cited in Lynch-Stewart (1983) and Rubec and Lynch-Stewart (1985).

the methodologies may preclude Canada-wide evaluation in the future unless satellite imagery can be utilized.

Legislation and Government Policy

Widespread and dramatic losses of wetlands have occurred across Canada, in part because of fragmentation of responsibility. Wetlands tend to be addressed in legislation in terms of components such as fish, migratory birds, and fuel peat, but they are not addressed as systems. This is true of other resources; however, for wetlands, fragmentation is extreme annd responsibilities are especially unclear. For example, Rakowski and Jurick (1984) have identified for the prairies some 21 federal Acts, 12 Manitoba Acts, 17 Saskatchewan Acts, and 24 Alberta Acts that apply to wetlands. None refer directly to natural wetlands. Legislation is sufficient to enable many government departments to address wetlands; however, it limits drastically the extent to which wetlands protection may be pursued and does not facilitate resolution of conflicts between interests.

The United States has a *National Wetlands Act*. The most important pieces of legislation in Canada that pertain to wetlands are the *Canada Fisheries Act*, *Migratory Birds Convention Act*, *Canada Wildlife Act*, provincial planning Acts, provincial environmental assessment Acts, and provincial water management Acts. The *Canada Fisheries Act*

TABLE 10. Selected methodologies for evaluating Canadian Wetlands.

Region	Focus	Reference
B.C. interior	Landuse capability	Runka and Lewis 1981
B.C. interior	Wildlife capability	Moon and Selby 1983
B.C. coast	Estuary habitat	Hunter et al. 1983
Alberta	Agricultural drainage	Leskiw et al. 1984
Prairies	Waterfowl habitat	Gilmer et al. 1978
Manitoba	Peat quality	Mills 1983
Manitoba	Agricultural capability	Smith 1974
S. Ontario	Multipurpose planning	Patterson 1984
S. Ontario	Multipurpose planning	Ontario Ministry of Natural Resources and Environment Canada 1983
Quebec	Ecological classification	Environnement Illimité 1982
New Brunswick	Multipurpose resources	Monenco Ltd. 1983
New Brunswick	Multipurpose planning	Hudgins 1983
Nova Scotia	Peat utilization	Anderson and Broughm 1980
Newfoundland	Agricultural capability	Nowland 1980
Canada	Landuse capability	Canada Land Inventory 1970
North America	Duck productivity	Ducks Unlimited 1985

specifies the need to protect spawning sites and other habitat from contamination or disturbance. The strength of the *Migratory Birds Convention Act* is its international basis. It does not, however, have a strong focus on bird habitat and waterfowl managers prefer to use the *Canada Fisheries Act* when possible. The *Canada Wildlife Act* enables regulation of hunting and acquisition of wildlife areas. Provincial planning Acts enable the provinces or municipal designates to zone landuse and thus routinely protect valuable areas such as wetlands. Environmental assessment Acts allow governments to address large scale projects which would be difficult to assess under other Acts. Provincial water management Acts, such as created Ontario's Conservation Authorities, facilitate integrated wetlands management.

Potentially antagonistic legislation exists in the *Drainage Act* and provincial mining Acts. These encourage activities which could ultimately destroy a wetland. Policies such as that of the Wheat Board, which ties grain quotas to area of land "improved," and regional development grants for drainage are also antagonistic.

Perhaps the most encouraging development recently has been policy formulation at the level of departments. Environment Canada recently adopted a policy on land use (Government of Canada 1984). It provides a mandate to prevent damage to sensitive and key habitats. The Department of Fisheries and Oceans has proposed a policy on fish habitat (Department of Fisheries and Oceans 1985). Among other things, it expounds the concept of "no net loss" of habitat: useful if habitat destruction is inevitable; dangerous if interpreted to mean that natural wetland can be replaced. At the provincial level, Manitoba includes in its formal policy for crown land under agriculture a clause stating that no wetland may be drained or filled without approval (Manitoba Crown Land Classification Committee 1984). New Brunswick's recently developed Land Use Policy provides machanisms to resolve conflicts over ecological effects of land use (Macenko and Neimanis 1983). Ontario's Ministry of Natural Resources has policy in the form of guidelines leading toward effective provincial wetlands management. It specifies inventory and evaluation schemes, schedules, and money available to carry out intentions (Ontario Ministry of Natural Resources 1984). Provincial habitat acquisition programs have also proven useful for protecting key wetlands.

In short, what is missing in terms of wetland legislation is compensated for by less formal policy. Some private organizations have bolstered this by developing semi-formal working relationships with governments. Ducks Unlimited, for example, has signed major heritage marsh agreements with all provinces. These require governments to manage or allow Ducks Unlimited to manage certain key wetlands in a prescribed manner in return for guaranteed expenditures on those wetlands (about $5 million per agreement).

Wetlands management in Canada and the prevention of future losses do not at the moment appear to be as limited by inadequate legislation or a lack of models for informal institutional arrangements as they are by the failure to fully use existing legislation and arrangements. This does not mean that it could not be facilitated by changes such as explicit wetlands legislation. Governments and interest groups have, however, only recently begun considering or undertaking concerted management of wetlands. As the ongoing phase of wetlands evaluation produces results, it seems likely that facilitating mechanisms will be developed.

Nationally Significant Wetlands

Partly throught design and in part inadvertantly a national system of wetlands is developing. These are wetlands which have been identified as special areas and designated as national and provincial parks, Ramsar Convention wetlands of international importance, International Biological Program (IBP) sites, national wildlife areas, migratory bird sanctuaries, and Biosphere reserves or prospective ones (Fig. 6).

Parks and national wildlife areas are protected by legislation, the former in order to preserve nationally unique landscapes and the latter to protect significant wildlife

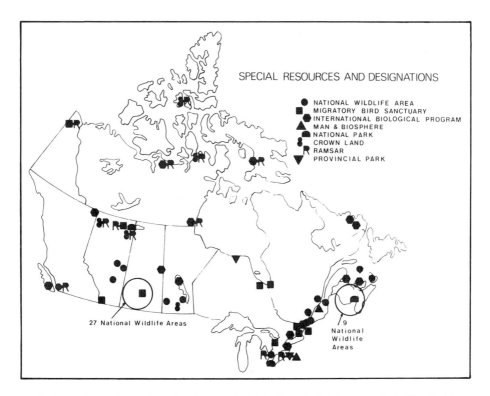

FIG. 6. Special area designations involving wetlands in Canada (major examples). Sites that have received more than one designation are identified by a horizontal line connecting the relevant symbols. Compiled from various sources.

habitat. Acquisition of areas such as Point Pelee National Park, Akimiski Island Migratory Bird Sanctuary, and Tintamarre National Wildlife Area assures that at least uses incompatible with wetland preservation will be excluded. The attraction of large numbers of sightseers to the areas requires that access be carefully controlled.

IBP sites were selected on the merit of scientifically interesting components. These components were identified for consideration in any decision that might affect the site. Many are wetland sites. A more precise designation of this type is that of wetlands of international importance. Eighteen wetlands representing 106 513 km^2 were so designated as of February 1985 (Canadian Wildlife Service 1985). The effectiveness of this designation remains to be judged.

Biosphere reserves are the most recent development in special area designation in Canada. These are areas that are representative of biogeographical provinces, that have undergone intensive management of human activity, and that are represented by a management committee. Two have existed for some time in Canada and contain minor amounts of wetland (Francis 1982). Two more were approved early in 1986 (G. Francis, University of Waterloo, pers. comm.). One of these, Long Point, Lake Erie, is a major wetland system (Francis et al. 1985).

These types of special area designation are complemented by provincial conservation areas, areas of natural and scientific interest, and other regional and local designations. They function not only as protective mechanisms and management foci, but also as ways of assembling comparative information on Canadian wetlands. To encourage further collection of such information, a Canadian Wetland Registry has been established (Tarnocai 1979). The National Wetland Working Group developed a method of collecting basic data on wetlands in a uniform manner for assembly in a central facility.

Wetlands Creation

Perhaps the most proactive means of demonstrating recognition of the value of wetlands would be to create or rehabilitate them. In fact, this is quite common. In Canada the intentional creation of wetlands is not as widespread as Garbisch (1977) reported for the United States. It is likely that interruption of hydrologic patterns by roads and other human-engineered structures has inadvertently created an undetermined but large number of small wetlands. The same types of activity have also potentially altered the conditions of existing wetlands. Jeglum (1975), for example, documented a shift in wetland type in response to road implacement.

Wetlands creation in Canada has more often taken the form of enhancement or rehabilitation. The Ontario Ministry of Natural Resources has recently been attempting to reestablish a littoral wetland in Lake Erie to stabilize sediments (Hanna and Associates 1984). The Ontario Ministry of the Environment transplanted wetland plants from one location to another to create a sewage treatment facility (Wile et al. 1979). Attempts to expand the area of wild rice production have been frequent (Plansearch 1980). Rehabilitation of degraded urban wetlands has also received some attention, for example in Second Marsh, Ontario (Cecile 1983), Rattray Marsh, Ontario (Hanna 1979), and the Fraser River Estuary (O'Riordan and Wiebe 1984).

The most widespread form of wetlands enhancement in Canada has involved impoundment. Predictable successional responses of wetland systems to water level regime have enabled marsh managers to maintain "optimal" conditions for desirable components such as ducks. Ducks Unlimited is generally acknowledged as most experienced in this field. The 48-yr-old organization has developed 746 218 ha of waterfowl habitat in over 2000 projects across Canada (Ducks Unlimited 1984). Most activity has taken place in the prairie provinces where dyking projects have ranged in size from less than 4 ha to over 200 000 ha. Although dyking is the most frequently used basic method of marsh enhancement, other techniques have been employed to reset succession and maintain a desirable balance between open water and plant species composition (Ball 1985). A problem with dyking is that it enhances duck productivity at the expense of other wetland quality. Fisheries, for example, can suffer dramatically (Snyder and Johnson 1984). The other techniques such as blasting, controlled fire, and harvesting can alleviate these problems.

Wetlands Future

A chapter such as this one which assembles previously diffuse information on wetlands and wetlands management in Canada could leave the impression that much is known and much has been achieved. It is clear that the scientific, managerial, and public focus across Canada on wetlands has intensified markedly over the past 15 yr. The vast majority of references cited in this chapter were published in that period. Nevertheless, in comparison with other major types of ecosystem, lakes and upland forests for example, wetlands are relatively poorly understood. Losses of wetland have been substantial in all regions of the country, except the north. These will doubtless continue for some time. Management is fragmented, with no immediate prospects for comprehensiveness, but with considerable potential for useful action on priority items. These can be categorized as information needs and management needs.

Information Needs

Information needs on Canadian wetlands are of two basic types: (a) theoretical information on wetland functions and characteristics and (b) empirical information on distribution and variation in such attributes.

Among the most important of the theoretical requirements are integrative models of wetland systems. Ultimately these would be predictive models of wetland development. Also required is understanding of responses of wetlands to nutrient loading and contaminants, especially as affected by temperature, snow cover, and other climatic variables. Long-term agricultural, aquacultural, and forest productivity on organic soils merits attention. Sustainability should be a focus.

Systemic empirical information is required on differences in primary, secondary, and tertiary productivity amongst wetland types and regions. Of all wetland types, shallow-water wetlands are least well served by comparative surveys. A rapid methodology for remote sensing is of prime importance in this regard. Although wetland inventories and estimates of losses of other wetland types are widely underway, what is missing is assessment of the meaning of wetland losses in terms of lost productivity, irreplaceable functions, system stability, and socioeconomic values. Finally, it is not now possible to compare the values of wetlands at a national scale because of the wide variation in evaluation methodologies. Even local transboundary comparisons are difficult. The basis for national evaluation is thus weak.

Management Needs

To overcome deficiencies in understanding and to implement programs to use the understanding, some adjustments in institutions and approaches for wetlands research and management are recommended. A prime requirement is the creation of financial incentives or at least removal of disincentives for private wetlands protection. Closely linked to this would be the development of public (private land owner) understanding of wetlands and involvement in wetlands conservation.

Clear, direct statements of institutional responsibilities are necessary. An explicit legal or semiformal basis for government involvement in wetlands management is required. This could take the form of regular preparation of statistical summaries of the condition of wetland resources.

A broader information base could result from the promotion of wetland registries or related means. Also useful in this regard would be an exchange of information between nongovernmental and governmental organizations. Some nongovernmental organizations maintain data banks of great public utility. Governments vary in their willingness to relinquish control of data. These data banks could be made accessible by developing agreements in which nongovernmental organizations would release information and provide expertise in exchange for funding, open access to government information, and effective input to wetlands management.

It should be recognized that interests in wetlands that are currently allied in their request for action on wetlands will be in conflict as wetland resources are developed (e.g. peat harvest and wildlife interests). Other interests already conflict. An approach should be developed for resolution of conflicts amongst renewable and nonrenewable uses of wetlands.

Finally, many Canadian wetlands have an international profile or potential for one. Attention should be directed at promotion of an interational wetlands network, including centres of expertise and preserves of key migratory habitat.

Acknowledgements

Preparation of this chapter was facilitated by funding for related research from the Environmental Conservation Service of Environment Canada. Research assistance was provided by P. Van Wyck and G. Balch. Discussions were conducted with scientists from many public and private organizations representing all regions of Canada, including members of the National Wetland Working Group. Opinions and data were freely offered. To all of these persons and organizations I extend my heartfelt appreciation.

References

ADAMS, G. D., AND G. C. GENTILE. 1978. Spatial changes in waterfowl habitat, 1964-1974, on two land types in the Manitoba Newdale Plain. Can. Wildl. Serv. Occas. Pap. No. 38: 29 p.

ALLEN H. H. 1978. Role of wetland plants in erosion control of riparian shorelines, 403-414. In P. E. Greeson, J. R. Clark, and J. E. Clark [ed.] Wetland functions and values: the state of our understanding. American Water Resources Association, Minneapolis, MN.

ANDERSON, A. R., AND W. BROUGHM. 1980. Preliminary assessment of the peatland resources of Southwest Nova Scotia. Nova Scotia Department of Mines. 50 p.

AIRPHOTO ANALYSIS ASSOCIATES CONSULTANTS LIMITED-TORONTO. 1975. Wetlands peatlands resources: New Brunswick. For New Brunswick Department of Natural Resources, Fredericton, N.B. 105 p. + maps.

BAILEY, R. O. 1983. Use of southern boreal lakes by moulting and staging diving ducks, p. 54-59. In H. Boyd [ed.] First Western Hemisphere Waterfowl and Waterbird Symposium, Canadian Wildlife Service, Ottawa, Ont.

BALL, J. P. 1985. Marsh management by water level manipulation or other natural techniques: a community approach. In H. H. Prince and F. M. D'Itri [ed.] Coastal wetlands. Lewis Publishers Inc., Chelsea, MI.

BARDECKI, M. J. 1984. Wetland conservation policies in southern Ontario: a Delphi approach. York Univ. Geogr. Monogr. No. 16: 275 p.

BARNES, J. L. 1984. The hydrology and water resources of peatlands in Canada and the role of the inquiry on federal water policy. Peat News 6(3): 1-18.

BARNES, J. L., AND R. J. ROGERSON. 1985. The hydroclimatological characteristics of a peatland-dominated catchment in Newfoundland. Presented at Canadian Waters, the State of the Resource Symposium, Toronto, May 26-29: 20 p. Available from J. L. Barnes, Newfoundland and Labrador Hydro, P.O. Box 9100, St. John's, Nfld.

BAYLY, I., AND I. MACLATCHY. 1982. Nutrient behavior and abscisic acid content during fall and winter in a Typha glauca community, p. 35-52. In M. J. Bardecki. Wetlands research in Ontario. Occasional Paper, Department of Applied Geography, Ryerson Polytechnical Institute, Toronto, Ont.

BEHAN, M. J., E. B. KINRAIDE, AND W. L. SELSER. 1979. Lead accumulation in aquatic plants from metalic sources including shot. J. Wildl. Manage. 43(1): 240-244.

BEHLER, J. L., AND F. W. KING. 1979. The Audubon Society field guide to North American reptiles and amphibians. Alfred A. Knopf, New York, NY. 744 p.

BENFORADO, J. 1981. Ecological considerations in wetland treatment of municipal wastewater, p. 307-324. In B. Richardson [ed.] Wetland values and management. Freshwater Society, Navarre, MN.

BERNARD, J. M., AND E. GORHAM. 1978. Life history aspects of primary production in sedge wetlands, p. 39-51. In R. E. Good, D. F. Whigham, and R. L. Simpson [ed.] Freshwater weltands: ecological processes and management potential. Academic Press, New York, NY.

BERTRAND, R. A. 1985. Overview of agricultural land drainage in B.C. In Provincial Proceedings of 3rd Western Conference on Rationalization of Water and Soil Research and Management, Manitoba Water Resources Branch, Winnipeg, Man.

BERTULLI, J. 1981. Influence of a forested wetland on a southern Ontario watershed, p. 33-47 In A. Champagne [ed.] Proceedings of the Ontario Wetlands Conference. Federation of Ontario Naturalists, Toronto, Ont.

BLACK, S. A. 1983. The use of marshlands in wastewater treatment — Ontario's research programs. MS Research Advisory Committee, Ontario Ministry of the Environment, 1 St. Clair Ave. W., Toronto, Ont.

BLACK S. A., I. WILE, AND G. MILLER. 1981. Sewage effluent treatment in an artificial marshland. Ontario Ministry of the Environment, 1 St. Clair Ave. W., Toronto, Ont.

BOVILLE, B. W., R. E. MUNN, AND F. K. HARE. 1983. The storage of non-living organic carbon in boreal and arctic zones — Canada. Institute for Environmental Studies, University of Toronto. Prepared for U.S. Department of Energy.

BRACE, R. K., AND G. W. PEPPER. 1985. Impact of agricultural drainage on wildlife, p. 451-482. In Proceedings of 3rd Western Provincial Conference on Rationalization of Water and Soil Research and Management, Manitoba Water Resources Branch, Winnipeg, Man.

BRENNAN, K. M., AND C. G. GARRA. 1981. Wastewater discharges to wetlands in six midwestern states, p. 285-294. In B. Richardson [ed.] Wetland values and management. Freshwater Society, Navarre, MN.

BRITISH COLUMBIA MINISTRY OF AGRICULTURE AND FOOD. 1981. Management and improvement of organic wetlands in the interior of B.C. Central Interior Forage Extension Committee. 12 p.

CANADA LAND INVENTORY. 1970. Land capability classification for wildlife. Report No. 7, Lands Directorate, Ottawa, Ont. 30 p.

CANADIAN WILDLIFE SERVICE. 1978. Riverine habitats on the St. Lawrence: thirty years of change (1950-1978). Canadian Wildlife Service, Quebec Region, Environment Canada. 56 p.

1979. Migratory bird habitat priorities — prairie provinces. Habitat Management Section, Western and Northern Region. 104 p.

1985. Canadian sites dedicated as wetlands of international importance. Dated May 24, 1982 with list of designated wetlands updated to February 1985, unpaginated. Canadian Wildlife Service, 1725 Woodward Dr., Ottawa, Ont.

CECILE, C. P. 1983. Oshawa Second Marsh baseline study. Environment Canada, Ottawa, Ont. 158 p.

COLLINS, J. M. 1974. The relative abundance of ducks breeding in southern Ontario in 1951 and 1971, p. 32-44. In H. Boyd [ed.] Waterfowl studies. Can. Wildl. Serv. Rep. Ser. No. 29.

COOCH, F. G. 1984. The kill of ducks and geese in Canada by non-resident hunters. Progress Notes No. 147, Canadian Wildlife Service, Environment Canada. 15 p.

COOK, F. R. 1984. Introduction to Canadian amphibians and reptiles. National Museums of Canada, Ottawa, Ont. 200 p.

1985. Amphibian and reptile subcommittee report to COSEWIC annual meeting, 2-3 April 1985. 5 p. (Available from National Museum of Natural Sciences, Ottawa, Ont.)

COX, E. T. 1972. Estimates of cleared wetlands in southern Ontario. MS, Ontario Department of Lands and Forests. 5 p. Available from Ontario Ministry of Natural Resources, Wildlife Branch, Whitney Block, Queen's Park, Toronto, Ont.

CREASY, D. E. J., R. J. PATTERSON, AND W. A. GORMAN. 1981. A hydrogeochemical study of contaminant attenuation and remobilization in the Big Swamp overburden near Picton, Ontario. Scientific Series No. 119. Inland Waters Directorate, Environment Canada. 53 p.

DALE, H. M., AND T. J. GILLESPIE. 1977. The influence of submerged aquatic plants on temperature grandients in shallow water bodies. Can. J. Bot. 55: 2216-2225.

DAMMAN, A. W. H. 1979. Geographic patterns in peatland development in eastern North America, p. 42-57. *In* Proceedinngs of the International Symposium on Classification of Peat and Peatlands. International Peat Society, Helsinki, Finland.

DANSEREAU, P. AND F. SEGADAS-VIANNA. 1952. Ecological study of the peat bogs of eastern North America — I. structure and evolution of vegetation. Can. J. Bot. 30: 490-520.

DAVIES, G. S. 1970. Productivity of macophytes in Marion Lake, British Columbia. J. Fish. Res. Board Can. 27: 71-81.

DEAN, R. G. 1978. Effects of vegetation on shoreline erosional processes, p. 415-426. *In* P. E. Greeson, J. R. Clark, and J. E. Clark [ed.] Wetland functions and values: the state of our understanding. American Water Resources Association, Minneapolis, MN.

DENNIS, D. G. 1974a. Breeding pair surveys of waterfowl in southern Ontario, p. 45-52. *In* H. Boyd [ed.] Waterfowl studies. Can. Wildl. Serv. Rep. Ser. No. 29.

1974b. Waterfowl observations during the nesting season in Precambrian and clay-belt areas of north-central Ontario, p. 53-57. *In* H. Boyd [ed.] Waterfowl studies. Can. Wildl. Serv. Rep. Ser. No. 29.

DENNIS, D. G., AND R. E. CHANDLER. 1974. Waterfowl use of the Ontario shorelines of the southern Great Lakes during migration, p. 58-65. *In* H. Boyd [ed.] Waterfowl studies Can. Wildl. Serv. Rep. Ser. No. 29.

DEPARTMENT OF FISHERIES AND OCEANS. 1984. Summary of data from national survey of sport fishing in Canada. MS, Department of Fisheries and Oceans, Ottawa, Ont.

1985. Proposed policy and procedures for fish habitat management. Communications Directorate, Department of Fisheries and Oceans, Ottawa, Ont.

DESJARDINS, R., K. MACDONALD, AND D. WUTZKE. 1984. The economics of drainage in Alberta, p. 261-326. *In* Proceedings of the 3rd Western Provincial Conference on Rationalization of Water and Soil Research and Management, Manitoba Water Resources Board, Winnipeg, Man.

DESJARDINS, R. M., AND P. L. SEYFRIED. 1984. Effects of metals from mine tailings on the microflora of a marsh treatment system, p. 449-483. *In* Ontario Ministry of the Environment Technology Transfer Seminar.

DUCKS UNLIMITED. 1984. Ducks Unlimited Canada 1984 annual report. Winnipeg, Man. 26 p.

1985. Wetland inventory and rating forms. Mimeographed. 9 p. Available from Ducks Unlimited, Winnipeg, Man.

ENVIRONNEMENT ILLIMITÉ. 1982. Vol. 1: Élaboration et validation des clés d'interprétation des cartes écologiques synthèse.

ETHERINGTON, J. R. 1983. Wetland ecology. The Institute of Biology's Studies in Biology No. 154. Edward Arnold, London, England, p. 67.

FILION, F. L. ET AL. 1983. The importance of wildlife to Canadians: highlights of the 1981 National Survey. Canadian Wildlife Service, Ottawa, Ont. 40 p.

FITZGIBBON, J. E. 1982. The hydrologic response of a bog-fen complex to rainfall, p. 333-346. *In* Proceedings of the Symposium on Hydrological Processes of Forested Areas, National Research Council, NRCC No. 20548.

FRANCIS, G. R. 1982. Biosphere reserves. Canadian Man and the Biosphere Committee, Ottawa, Ont. 10 p.

FRANCIS, G. R., A. P. GRIMA, H. A. REGIER, AND T. H. WHILLANS. 1985. A prospectus for the management of the Long Point ecosystem. Great Lakes Fish. Comm. Tech. Rep. No. 43: 109 p.

GARBISCH, E. W. 1977. Recent and planned marsh establishment work throughout the contiguous United States: a survey and basic guidelines. U.S. Army Engineers, Dredged Material Research Program. Contract Rep. D-77-3: 42 p.

GIBLIN, A. E. 1985. Comparisons of the processing of elements by ecosystems, II metals, p. 158-177. *In* P. J. Godfrey, E. R. Kaynor, S. Pelczarski, and J. Benforado [ed.] Ecological considerations in wetlands treatment of municipal wastewaters. Van Nostrand Reinhold Co., New York, NY.

GILLESPIE, D. I., AND S. P. WETMORE. 1974. Waterfowl surveys in Labrador-Ungava, 1970, 1971, 1972, p. 8-17. *In* H. Boyd [ed.] Waterfowl studies. Can. Wildl. Serv. Rep. Ser. No. 29.

GILMER, D. S., J. E. COLWELL, AND E. A. WORK. 1978. Use of landsat for evaluation of waterfowl habitat in the Prairie Pothole region. Pecora IV Symposium on application of remote sensing data to wildlife management, Sioux Falls, South Dakota.

GLASER, P. H. 1983. Vegetation patterns in the North Black River peatland, northern Minnesota. Can. J. Bot. 61: 2085-2104.

GLASER, P. H., G. H. WHEELER, E. GORHAM, AND H. E. WRIGHT. 1981. The patterned mires of the Red lake peatland, northern Minnesota: vegetation, water chemistry annd landforms. J. Ecol. 69: 575-599.

GLOOSCHENKO, W. A. AND G. J. A. CAPOBLANCO. 1982. Trace element content of northern Ontario peat. Environ. Sci. Technol. 16: 187.

GOODMAN, A. S.,, AND S. P. PRYOR. 1972. A preliminary study of the methods and rates of alteration of waterfowl habitat in the black soil zone of western Canada. MS, Canadian Wildlife Service, Edmonton, Alta. 55 p.

GOSSELINK, J. G., AND R. E. TURNER. 1978. The role of hydrology in freshwater wetland ecosystems, p. 63-78. In R. E. Good, D. F. Whigham, and R. L. Simpson [ed.] Freshwater wetlands: ecological processes and management potential. Academic Press, New York, NY.

GOVERNMENT OF CANADA. 1984. Federal policy on land use. Lands Directorate, Ottawa, Ont. 10 p.

GREESON, P. E., J. R. CLARK, AND J. E. CLARK. 1978. Wetland functions and values: the state of our understanding. American Water Resources Association, Minneapolis, MN. 675 p.

GUNTENSPERGEN, G., AND F. STEARNS. 1981. Ecological limitations on wetland use for wastewater treatment, p. 273-284. In B. Richardson [ed.] Wetland values and management. Freshwater Society, Navarre, MN.

HANNA, E. 1979. Battles for wetlands: Rattray Marsh a pyrric victory. Ont. Nat. 19(2): 19-23.
        1984. Heavy metal mobilization and biological uptake: cobalt mine tailings, p. 537-560. In Ontario Ministry of the Environment Technology Transfer Symposium.

HANNA AND ASSOCIATES. 1984. A management strategy for the restoration of aquatic vegetation in Rondeau Bay, Lake Erie. Prepared for Southwestern Region, Ontario Ministry of the Environment, London, Ont.

HARTT, E. P. 1978. Interim report of the Royal Commission on the Northern Environment, Toronto, Ont.

HATCHER, A., AND D. G. PATRIQUIN [ed.] 1981. Salt marshes in Nova Scotia: a status report of the salt marsh working group. Dalhousie University (N.S.), Institute for Resource and Environmental Studies and Department of Biology. 70 p.

HEINSELMAN, M. L. 1963. Forest sites, bog processes, and peatland types in the glacial Lake Agassiz region Minnesota. Ecol. Monogr. 33: 327-374.
        1970. Landscape evolution, peatland types, and the environment in the Lake Agassiz Peatlands Natural Area, Minnesota. Ecol. Monogr. 40(2): 235-261.

HERDENDORF, C. E., AND S. M. HARTLEY [ed.] 1980. A summary of the knowledge of the fish and wildlife resources of the coastal wetlands of the Great Lakes of the United States. Vol. 1: Overview. U.S. Fish and Wildlife Service, Twin Cities, MN. 468 p.

HOCHBAUM, G. S. 1983. Waterfowl of the prairie provinces and Northwest Territories: population, habitat and management, p. 34-37. In H. Boyd [ed.] First Western Hemisphere Waterfowl and Waterbird Symposium. Canadian Wildlife Service, Ottawa, Ont.

HORSTMAN, L. P. 1976. Current pesticide use in Alberta: fish and wildlife concerns and recommendations. Alberta Recreation, Parks and Wildlife, Fish and Wildlife Division, Wildlife Branch. 149 p.

HOWE, M. A. 1983. Breeding ecology of North American shorebirds: patterns and constraints, p. 95-100. In H. Boyd [ed.] First Western Hemisphere Waterfowl and Waterbird Symposium, Canadian Wildlife Service, Ottawa, Ont.

HUDGINS, E. J. 1983. Workbook for the initiation of the New Brunswick freshwater wetlands inventory. Canadian Wildlife Service, Sackville, N.B. 84 p.

HUMMEL, M. 1981. Wetland wildlife values, p. 27-32. In A. Champagne [ed.] Proceedings of the Ontario Wetlands Conference. Federation of Ontario Naturalists, Toronto, Ont.

HUNTER, R. A., AND L. E. JONES. 1982. Coastal waterfowl and habitat inventory program: summary report. B.C. Ministry of Environment and Ducks Unlimited (Canada). 28 p.

HUNTER, R. A., L. E. JONES, M. M. WAYNE, AND B. A. PENDERGAST. 1983. Estuarine habitat mapping and classification system manual. B.C. MOE Manual No. 3. British Columbia Ministry of Environment. 33 p.

JAHN, L. R. 1969. Summary [of the significance of small wetlands], p. 29-34. In Saskatoon Wetlands Seminar, Can. Wildl. Serv. Rep. Ser. No. 6.

JAHN, L. R., AND C. KABAT. 1984. Origin and role [of Flyway Councils], p. 374–385. *In* A. S. Hawkins et al [ed.] Flyways: pioneering waterfowl management in North America. U.S. Fish and Wildlife Service.

JAWORSKI, E., C. N. RAPHAEL, P. J. MANSFIELD, AND B. B. WILLIAMSON. 1980. Impact of Great Lakes water level fluctuations on coastal wetlands, p. 103–297. *In* C. E. Herdendorf and S. M. Hartley [ed.] A Summary of the knowledge of the fish and wildlife resources of the coastal wetlands of the Great Lakes. Vol. 1. U.S. Fish and Wildlife Services, Division of Ecological Services, Region 3, Twin Cities, MN.

JEGLUM, J. K. 1975. Vegetation–habitat changes caused by damming a peatland drainageway in northern Ontario. Can. Field-Nat. 89(4): 400–412.

JEGLUM, J. K., A. N. BOISSONNEAU, AND V. F. HAAVISTO. 1974. Toward a wetland classification for Ontario. Canadian Forestry Service, Sault Ste. Marie, Ont. Rep. 0-X-215: 54 p.

JEGLUM, J. K., AND D. W. COWELL. 1982. Wetland ecosystems near Kinoje Lakes, southern interior Hudson Bay lowland. Nat. Can. 109: 621–635.

JONES, R. 1981. The viewpoint from the farm, p. 152–153. *In* A. Champagne [ed.] Proceedings of the Ontario Wetlands Conference. Federation of Ontario Naturalists, Toronto, Ont.

KADLEC, R. H. 1981. How natural wetlands treat wastewater, p. 241–254. *In* B. Richardson [ed.]. Wetland values and management. Freshwater Society, Navarre, MN.

KADLEC, R. H., AND J. A. KADLEC. 1978. Wetlands and water quality, p. 436–456. *In* P. E. Greeson, J. R. Clark, and J. E. Clark [ed.] Wetland functions and values: the state of our understanding. American Water Resources Association, Minneapolis, MN.

LAIDLAW, S. A. 1978. A report concerned with the loss of wetlands in southern Ontario. MS, Environmental Protection Service, Environment Canada, Toronto, Ont. 118 p.

LAPERLE, M. 1974. Effects of water level fluctuation on duck breeding success, p. 18–30. *In* H. Boyd [ed.] Waterfowl studies. Can. Wildl. Serv. Rep. Ser. No. 29.

LARSON, C. L. 1981. Effects of wetland drainage on surface runoff, p. 177–120. *In* B. Richardson [ed.] Wetland values and management. Freshwater Society, Navarre, MN.

LESKIW, L. A., M. ANDERSON, AND A. R. V. RIBEIRO. 1984. Farmland drainage in central and northern Alberta. Farming for the Future, Research Project No. 82-0070: 151 p.

LISTON, C. R., ET AL. 1983. Environmental baseline studies of the St. Mary's River near Neebish Island, Michigan, prior to proposed extension of the navigation season. Mich. State Univ. Dep. Fish Wildl. Rep. U.S. Fish Wildl. Serv. FWS IOBS-80162.2: 316 p.

LYNCH-STEWART, P. 1983. Land use change on wetlands in southern Canada: review and bibliography. Canada Land Use Monitoring Program, Lands Directorate, Environment Canada. 115 p.

MACENKO, S. L., AND V. P. NEIMANIS. 1983. An overview of Crown lands management in Canada. Land Policy and Research Branch, Lands Directorate, Environment Canada. 78 p.

MACPHERSON, J. B. 1982. Postglacial vegetation history of the eastern avalon peninsula, Newfoundland, and holocene climatic change along the eastern Canadian seaboard. Geogr. Phys. Quater. 36(1-2): 175–195.

MANITOBA CROWN LAND CLASSIFICATION COMMITTEE. 1984. Policy and procedures manual. Manitoba Crown Land Classification Committee, Department of Agriculture, Natural Resources, and Municipal Affairs, Winnipeg, Man.

MCALONEY, K. 1981. Waterfowl use of Nova Scotian salt marshes, p. 60–66. *In* A. Hatcher and D. G. Patriquin [ed.] Salt Marshes in Nova Scotia. Institute for Resource and Environmental Studies, Dalhousie University, Halifax, N.S.

MCCRACKEN, J., M. S. W. BRADSTREET, AND G. L. HOLROYD. 1981. Breeding birds of Long Point, Lake Erie. Can. Wildl. Serv. Rep. Ser. No. 44: 74 p.

MCCULLOUGH, G. B. 1985. Wetland threats and losses in Lake St. Clair, p. 201–208. *In* H. H. Prince and F. M. D'Itri [ed.] Coastal wetlands. Lewis Publishers Inc., Chelsea, MI.

MCKEATING, G. 1983. Management plan: Long Point National Wildlife Area. Environment Canada, Canadian Wildlife Service, London. 72 p.

METRAS, L. 1984. Migratory birds killed in Canada during the 1982 season. Progress Notes No. 143, Canadian Wildlife Service, Environment Canada. 39 p.

MILLAR, J. B. 1976. Wetland classification in western Canada. Can. Wildl. Serv. Rep. Ser. 37: 38 p.

MILLS, G. F. 1983. Peatland inventories in Manitoba, p. 35–50. *In* S. M. Morgan and F. C. Pollett [ed.] Proceedings of a Peatland Inventory Methodology Workshop. Land Resource Research Institute, Ottawa, Ont.

MILNER, C., AND R. ELFYN-HUGHES. 1968. Methods for the measurement of the primary production of grasslands. I.B.P. Handbook No. 6, Blackwell Publications, Oxford.

MITSCH, W. J., J. W. DAY, J. R. TAYLOR AND C. MADDEN. 1982. Models of North American freshwater wetlands, p. 5-32. In Ecosystem dynamics in freshwater wetlands and shallow water bodies. Scientific Committee on Problems in the Environment, United Nations, Moscow.

MONENCO LTD. 1983. A wetland evaluation system for New Brunswick. Vol. 1 of 2, File ENB 7975-2. Prepared for Environment New Brunswick. 30 p.

MONENCO ONTARIO LTD. 1981. Evaluation of the potential of peat in Ontario. Ont. Minist. Nat. Resour. Miner. Resour. Branch Occas. Pap. No. 7: 193 p.

MOON, D. E., AND C. J. SELBY. 1983. Wetland systems of the Caribou-Chilcotin region of B.C., p. 54-74. In S. M. Morgan and F. C. Pollett [ed.] Proceedings of a Peatland Inventory Methodology Workshop. Land Resource Research Institute, Ottawa, Ont.

MOORE, P. D., AND D. J. BELLAMY. 1974. Peatlands. Springer-Verlag, New York, NY. 217 p.

MORRISON, R. I. G. 1983. A hemispheric perspective on the distribution and migration of some shorebirds in North and South America, p. 84-94. In H. Boyd [ed.] First Western Hemisphere Waterfowl and Waterbird Symposium. Canadian Wildlife Service, Ottawa, Ont.

MUDROCH, A. 1981. A study on selected Great Lakes coastal marshes. Scientific Series No. 122. Inland Waters Directorate, Environment Canada. 44 p.

MUNRO, D. S. 1982. Hydroclimatological investigations in the Beverly Swamp, Ontario, p. 87-104. In M. J. Bardecki [ed.] Wetlands Research in Ontario. Occasional Paper, Department of Applied Geography, Ryerson Polytechnic Institute, Toronto, Ont.

NATIONAL WETLAND WORKING GROUP. 1981. Wetlands of Canada. Map, Ecological Land Classification Series No. 14, 1:7,500,000. Environment Canada, Lands Directorate, Ottawa, Ont.

1985. Canada — distribution of wetlands. Map 1:7M, National Atlas of Canada, Draft.

NEAVE, D. J. 1983. Waterfowl and wetlands: problems and programs in Alberta, p. 8-12. In H. Boyd [ed.] First Western Hemisphere Waterfowl and Waterbird Symposium. Canadian Wildlife Service, Ottawa, Ont.

NEWFOUNDLAND DEPARTMENT OF THE ENVIRONMENT. 1984. A preliminary report on the hydrological investigations at the fuel peat demonstration project near Bishop's Falls. Water Resources Division, Energy, Mines and Resources Canada.

NOVITZKI, R. P. 1978. Hydrologic characteristics of Wisconsin's wetlands and their influence on floods, stream flow and sediment, p. 377-388. In P.E. Greeson, J. R. Clark, and J. E. Clark [ed.] Wetland functions and values: the state of our understanding. American Water Resources Association, Minneapolis, MN.

NOWLAND, J. L. 1980. Interpretation of organic soil properties for agriculture, p. 69-98. In C. Tarnocai [ed.] Proceedings of a Workshop on Organic Soil Mapping and Interpretation. Land Resource Research Institute, Ottawa, Ont.

ODUM, E. P. 1971. Fundamentals of ecology. W. B. Saunders, Philadelphia, PA.

OFFICE OF TECHNOLOGY ASSESSMENT. 1984. Wetlands: their use and regulation. U.S. Congress, Washington, D.C. OTA-0-206: p. 208.

ONTARIO MINISTRY OF NATURAL RESOURCES. 1984. Guidelines for wetlands management in Ontario. Whitney Black, Queen's Park, Toronto, Ont. 3 p.

ONTARIO MINISTRY OF NATURAL RESOURCES AND ENVIRONMENT CANADA. 1983. An evaluation system for wetlands of Ontario south of the Precambrian Shield. 50 p. + appendices.

O'RIORDAN, J., AND J. WIEBE. 1984. An implementation strategy for the Fraser River estuary management programme. Fraser River Estuary Management Prog. Review Comm., Environment Canada and B.C. Ministry of the Environment.

PATTERSON, N. J. 1984. An approach to wetland evaluation and assessing cultural stresses on freshwater wetlands in southern Ontario. M.Sc. thesis, Department of Geography and the Institute for Environmental Studies, University of Toronto, Toronto, Ont. 192 p.

PILON, P., AND M. A. KERR. 1984. Land use change on wetlands in the southwestern Fraser lowland British Columbia. Working Paper No. 34. Lands Directorate, Environment Canada, Vancouver, B.C. 31 p.

PLANSEARCH. 1980. Feasibility of wild rice production in Nova Scotia. Report to the Nova Scotia Department of Agriculture and Marketing. 96 p.

PRAIRIE ASSOCIATION OF WATER MANAGEMENT. 1984. Submission to the Inquiry on Federal Water Policy. Environment Canada, Ottawa, Ont. 7 p.

PRICE, J. S., AND J. E. FITZGIBBON. 1982. Winter hydrology of a forested drainage basin, p. 347–360. In Proceedings of the Symposium on Hydrological Processes of Forested Areas. National Research Council, NRCC No. 20548.

RAKOWSKI, P. W., AND D. R. JURICK. 1984. Alternatives for protecting habitat in the Minnedosa Pothole region of Manitoba. Canadian Wildlife Service, Winnipeg, Man. 55 p.

READER, R. J. 1978. Primary production in northern bog marshes, p. 53–62. In R. E. Good, D. F. Whigham, and R. L. Simpson [ed.] Freshwater wetlands: ecological processes and management potential. Academic Press, New York, NY.

RENNIE, P. J. 1977. Forests, muskeg and organic terrain in Canada, p. 167–207. In N. W. Radforth and C. P. Brawner [ed.] Muskeg and the northern environment in Canada. University of Toronto Press, Downsview, Ont.

RICHARDSON, C. J. 1978. Primary productivity values in freshwater wetlands, p. 131–145. In P. E. Greeson, J. R. Clark, and J. E. Clark [ed.] Wetland functions and values: the state of our understanding. American Water Resources Association, Minneapolis, MN.

RICHARDSON, B. [ed.] 1981. Selected proceedings of the Midwest Conference on Wetland Values and Management, St. Paul, MN.

RICHARDSON, C. J., D. L. TILTON, J. A. KADLEC, J. P. M. CHAMIE, AND W. A. WENTZ. 1978. Nutrient dynamics of northern wetland ecosystems, p. 217–241. In R. E. Good, D. F. Whigham, and R. L. Simpson [ed.] Freshwater wetlands: ecological processes and management potential. Academic Press, New York, NY.

RILEY, J. L. 1981. Evaluation of northern Ontario wetlands, p. 128–132. In A. Champagne [ed.] Proceedings of the Ontario Wetlands Conference. Federation of Ontario Naturalists, Toronto, Ont.

RITTNER, A. 1980. Wetland reconnaissance within Battle River basin: 1979 summary. MS, Ducks Ulimited Canada, Edmonton, Alta. 25 p.

ROMANOV, U. U. 1961. Hydrophysics of bog. Translated from Russian by N. Kaner, U.S. Department of Agriculture. 299 p.

ROTHWELL, R. L. 1982. Water balance of a boreal forest wetland site, p. 313–332. In Proceedings of the Symposium on Hydrological Processes of Forested Areas. National Research Council, NRCC No. 20548.

RUBEC, C. D. A., AND P. LYNCH-STEWART. 1985. Wetland utilisation in Canada: conservation and exploitation. Ms, Lands Directorate, Environment Canada, Hull, Que.

RUNKA, G. G., AND T. LEWIS. 1981. Preliminary wetland managers manual: Cariboo Resource Management Region. 1st ed. APD Tech. Pap. No. 5. Assessment and Planning Division, B.C. Ministry of the Environment. 113 p.

SARRAZIN, R., M. CANTIN, A. GAGNON, C. GAUTHIER, AND G. LEFEBVRE. 1983. La protection des habitats fauniques au Québec. Le Group de Travail pour Protection des Habitats, Ministère du Loisir, de la Chasse et de la Pêche du Québec. 256 p.

SASKATCHEWAN TOURISM AND RENEWABLE RESOURCES. 1980. Aquaculture in Saskatchewan. Fish Management Division, Fisheries and Wildlife Branch. 18 p.

SAKATCHEWAN RESEARCH COUNCIL. 1984. The 1983 Annual Report. Saskatchewan Research Council, Saskatoon, Sask. 55 p.

SATHER, J. H., AND R. D. SMITH. 1984. An overview of major wetland functions and values. U.S. Fish Wildl. Serv. FWS/OBS-84/18: 68 p.

SAWCHYN, W. W. 1984. Intensive cage culture of rainbow trout in farm dugouts. Sask. Res. Counc. Tech. Rep. No. 161.

SCHICK, C. D. 1972. A documentation and analysis of wetland drainage in the Alberta Parkland. MS, Western and Northern Region, Canadian Wildlife Service, Edmonton, Alta. 15 p.

SCHMITT, K. 1980. Wetland reconnaissance within the South Saskatchewan River basin. MS, Ducks Unlimited Canada, Winnipeg, Man. 29 p.

SCOTT, W. B., AND E. J. CROSSMAN. 1973. Freshwater fishes of Canada. Bull. Fish. Res. Board Can. 184: 966 p.

SIMPSON-LEWIS, W., J. E. MOORE, N. J. POCOCK, M. C. TAYLOR, AND H. SWAN. 1979. Canada's special resource lands: a national perspective of selected land uses. Map Folio No. 4, Lands Directorate, Ottawa, Ont. 232 p.

SLOEY, W. E., F. L. SPANGLER, AND C. W. FETTER. 1978. Management of freshwater wetlands for nutrient assimilation, p. 321–340. In R. E. Good, D. F. Whigham, and R. L. Simpson [ed.] Freshwater wetlands: ecological processes and management potential. Academic Press, New York, NY.

SMITH, R. E. 1974. Soil capability classification, p. 47-68. *In* J. H. Day [ed.] Proceedings of the Canada Soil Survey Committee Organic Soil Mapping Workshop. Canada-Manitoba Soil Survey.

SNYDER, W. S., AND D. L. JOHNSON. 1984. Fish community structure, movements, and reproduction in controlled and uncontrolled Lake Erie wetlands. Paper presented at the Great Lakes Coastal Wetlands Colloquium, East Lansing, November 1984.

STATISTICS CANADA. 1978. Fisheries Statistics of Canada, Canada summary 1978. Ottawa, Ont. 28 p.

_____ 1985. Fur production, season 1983-84. Catalogue 23-207.

STEWART, J. M. 1977. Canadian muskegs and their agricultural utilization, p. 208-220. *In* N. W. Radforth and C. O. Brawner [ed.] Muskeg and the northern environment in Canada. University of Toronto Press, Downsview, Ont.

TARNOCAI, C. 1970. Classification of peat landforms in Manitoba. Agriculture Canada, Research Station, Pedology Unit, Winnipeg, Man. 45 p.

_____ 1978. Genesis of organic soils in Manitoba and the Northwest Territories, p. 453-470. *In* Quaternary soils: Proceedings of the 3rd York Quaternary Symposium. Geo. Abstracts, Norwick.

_____ 1979. Canadian wetland registry, p. 9-38. *In* C. D. A. Rubec and F. C. Pollett, [ed.] Proceedings of a Workshop on Canadian Wetlands. Environment Canada, Lands Directorate. Ecological Land Classification Series No. 12.

_____ 1980. Development, age and classification of Canadian peatlands, p. 3-13. *In* C. Tarnocai [ed.] Proceedings of a Workshop on Organic Soil Mapping and Interpretation. Land Resource Research Institute, Ottawa, Ont.

_____ 1983. Peatland inventory methodology used in soil survey, p. 13-22. *In* S. M. Morgan and F. C. Pollett [ed.] Proceedings of a Peatland Inventory Methodology Workshop. Land Resource Research Institute and Newfoundland Forest Research Centre.

_____ 1984. Peat Resources of Canada. National Research Council of Canada, Peat Energy Program, NRCC No. 24140: 17 p.

TAYLOR, C. H. 1982. The hydrology of a small wetland catchment near Peterborough, Ontario, p. 105-130. *In* M. J. Bardecki [ed.] Wetlands research in Ontario. Occasional Paper, Department of Applied Geography, Ryerson Polytechnical Institute,, Toronto, Ont.

TAYLOR, G. J., AND A. A. CROWDER. 1983. Copper and nickel tolerance in *Typha latifolia* clones from contaminated and uncontaminated environments. Can. J. Bot. 62: 1304-1308.

TERASMAE, J. 1977. Postglacial history of Canadian muskeg, p. 9-30. *In* N. W. Radforth and C. P. Brawner [ed.] Muskeg and the northern environment in Canada. University of Toronto Press, Downsview, Ont.

TUPPER, D. A. 1985. Overview of Legislative and Programme Frameworks for Drainage in Alberta, p. 25-35. *In* Proceedings of 3rd Western Provincial Conference on Rationalization of Water and Soil Research and Management. Manitoba Water Resources Branch, Winnipeg, Man.

TYLER, G. 1977. Historical monitoring — ecological documentation of historical changes in the level of environmental pollutants, particularly heavy metals. Monitoring Assessment Research Center, Chelsea College, London, Internal Report 037: 14 p.

VALIELA, I., AND J. M. TEAL. Nutrient dynamics: summary and recommendations, p. 259-266. *In* R. E. Good, D. F. Whigham, and R. L. Simpson [ed.] Freshwater wetlands: ecological processes and management potential. Academic press, New York, NY.

VAN DER VALK, A. G. , C. B. DAVIS, J. L. BAKER, AND C. E. BEER. 1978. Natural fresh water wetlands as nitrogen and phosphorus traps for land runoff, p. 457-467. *In* P. E. Greeson, J. R. Clark, and J. E. Clark [ed.] Wetland functions and values: the state of our understanding. American Water Resources Association, Minneapolis, MN.

VERRY, E. S., AND D. H. BOELTER. 1981. Peatland hydrology, p. 121-122. *In* B. Richardson [ed.]. Wetland values and management. Freshwater Society, Navarre, MN.

WALMSLEY, M. E. 1977. Physical and chemical properties of peat, 82-129. *In* N. W. Radforth and C. O. Brawner [ed.] Muskeg and the northern environment in Canada. University of Toronto Press, Downsview, Ont.

WEBER, T. E. 1985. Agricultural land drainage in Manitoba, p. 38-45. *In* Proceedings of the 3rd Western Provincial Conference on Rationalization of Water and Soil Research and Management. Manitoba Water Resources Branch, Winnipeg, Man.

WELLS, E. D., AND F. C. POLLETT. 1983. Peatlands, p. 207–265. *In* G. R. Smith [ed.] Biogeography and ecology of the Island of Newfoundland. W. Junk Publishers, The Hague, The Netherlands.

WETZEL, R. G. 1983. Limnology. W. B. Saunders Co., Philadelphia, PA. 743 p.

WHILLANS, T. H. 1979. Historic transformations of fish communities in three Great Lakes bays. J. Great Lakes Res. 5: 195–215.

    1985. Related long-term trends in fish and vegetation ecology of Long Point Bay and marshes, Lake Erie. Ph.D. dissertation, Department of Zoology, University of Toronto, Toronto, Ont. 331 p.

WILE, I., AND G. HITCHIN. 1977. An assessment of the practical and environmental implications of mechanical harvesting of aquatic vegetation in southern Chemung Lake. Ontario Ministries of the Environment and Natural Resources. 180 p.

WILE, I., G. HITCHING, AND G. BEGGS. 1979. Impact of mechanical harvest on Chemung Lake, p. 145–159. *In* J. E. Breck et al. [ed.] Proceedings of Conference on: Aquatic Plants, Lake Management and Ecosystem Consequences of Lake Harvesting, Madison, WI.

WILE, I., G. PALMATEER, AND G. MILLER. 1981. Use of artificial wetlands for wastewater treatment, p. 255–272. *In* B. Richardson [ed.] Weltand values and management. Freshwater Society, Navarre, MN.

WINCHELL, E. H., AND R. P. DAHL. 1984. Wild rice production, prices and marketing. Agric. Exp. Stn. Univ. Minn. Misc. Publ. 29: 35 p.

WINTER, T. C. 1981. The hydrology of prairie lakes and wetlands, p. 113–116. *In* B. Richardson [ed.] Wetland values and management. Freshwater Society, Navarre, MN.

WISHART, R. A., P. W. HERZOG, P. J. CALDWELL, AND A. J. MACAULAY. 1983. Waterfowl use of Ducks Unlimited projects across Canada, p. 24–32. *In* H. Boyd [ed.] First Western Hemisphere Waterfowl and Waterbird Symposium. Canadian Wildlife Service, Ottawa, Ont.

WOO, M-K., AND J. VALVERDE. 1981. Summer streamflow and water level in a midlatitude forested swamp. For. Sci. 27(1): 177–189.

    1982. Ground and water temperatures of a forested mid-latitude swamp, p. 301–312. *In* Proceedings of the Canadian Hydrology Symposium, Fredericton. National Research Council of Canada.

ZOLTAI, S. C. 1979. An outline of the wetland regions of Canada, p. 1–18. *In* Proceedings of a Workshop on Canadian Wetlands. Ecological Land Classification Series No. 12. Environment Canada, Lands Directorate.

ZOLTAI, S. C., AND F. C. POLLETT. 1983. Wetlands in Canada: their classification, distribution and use, p. 245–268. *In* A. J. P. Gore [ed.] Mires: swamp, bog, fen and moor. Elsevier Scientific Publishing Co., Amsterdam, The Netherlands.

ZOLTAI, S. C., F. C. POLLETT, J. K. JEGLUM, AND G. D. ADAMS. 1973. Wetland classification, p. 497–511. *In* B. Bernier and C. H. Winget [ed.] Proceedings of the 4th North American Forest Soils Conference. Les Presses de L'Université Laval, Quebec.

CHAPTER 13

Major Aquatic Contaminants, Their Sources, Distribution, and Effects

R. D. Hamilton, J. F. Klaverkamp, W. L. Lockhart, and R. Wagemann

Department of Fisheries and Oceans, Freshwater Institute, 501 University Crescent, Winnipeg, Man. R3T 2N6

Introduction

The title of this chapter suggests a very broad definition of a very major issue and we cannot begin without defining what we can address and what we cannot. We have chosen to not address biological contaminants and will confine ourselves to chemical issues. The definition we have adopted for "toxic chemicals" is that used by the Treasury Board of Canada in a recent review of the problem:

> A toxic chemical is a substance which may, through inadvertent exposure, produce adverse, acute or chronic effects in humans, animals or other biota.

This, in its turn, is a very broad definition but is at least one that more than 20 federal departments could accept when it was proposed (Treasury Board 1983). Thus, the topic is tremendously complex and is becoming more Byzantine by the moment, for we humans have been creating these chemical problems since about 1900 at rates that have been increasing in a manner similar to compound interest funding. Actually, one can go further back in time and cite the following (Exodus 7, 17-18):

> The waters shall be turned to blood and the fish that is in the river shall die, and the river shall die and the Egyptians shall loath to drink the water.

In between that remote time and today, we note, with some sense of déjà vu, the comment of a fisheries inspector, in what was the District of "Manitoba and Northwest Territories" in 1885, in reporting to the then Minister of Marine and Fisheries: "Two instances of pollution of water, resulting in the destruction of fish have come under my observation; one being that of the Red River, through deleterious substances from the gas works and public sewers of the city of Winnipeg; the other, that of waters at the mouth of the Winnipeg River, near Fort Alexander, where saw-dust has proved destructive to fish. I have taken steps to abate the latter nuisance, but scarcely know how to overcome the former. A large number of dead fish have been noticed by settlers along the banks of the Red River, between Winnipeg and Selkirk, whose destruction I can attribute to no other cause than that just mentioned." (Annual Report 1886).

Thus, we cannot pretend to cover the whole of this ancient, pervasive problem. We therefore chose to select a few major issues. We attempt not to provide a comprehensive picture of even these few selected issues but to provide an overview and, through the use of selected references, to provide the interested reader with food for thought regarding some key issues.

In the far past (3000-4000 B.C.) until the industrial revolution of A.D. 1800-1900, society only produced chemicals; they were not managed in any way save in an economic sense. That is to say production, distribution, and use were based solely upon return on investment; safe disposal of the chemicals themselves or the chemical consequences of production and use were not a matter for consideration; nor for that matter were considerations of health, safety, and impact on the environment.

In the immediate past, management came to be thought of as a matter for gorvern-ment concern, and governments, while acknowledging new political realities, responded in a typically political manner; that is to say they responded to specific immediate issues and had little overt concern with or commitment to the resolution of the long-term issues. Moreover, their responses typically focussed on but a single stage in the use of a single chemical. Of course, responsibilities for various aspects became the sole problem of various agencies of government, or the same problem became the shared responsibility of several agencies.

One resultant of such activities has been the appearance of a wide variety of legislative instruments, and at almost every level of government. While this might be seen as positive action by some people, it cannot be denied that there has been created an unfortunately complex tangle of acts, regulations, and rules of conduct. While efforts are being made to streamline and coordinate these legal instruments, much work needs to be done, for many of such instruments have proven to be ineffective or counter-productive when placed before the courts.

In the past 15 yr, attempts have been made to rationalize both the legal and manage-ment atmosphere such as to obtain overall management which is truly more effective and which effects economies for the control agencies, users, and producers. However, constraints within governments, between governments, and between governments and users or producers remain very real. In fact, to deal with but one sector of manage-ment, it was recently noted "As far as the Federal Government as a whole is concerned however, no overall strategy or policy respecting toxic chemicals is in place…" (Auditor General 1983).

The existence of these management constraints and this lack of comprehensive policy and strategy must concern anyone who realizes that they impede our ability to avoid damage to our freshwaters and the biota they support. They constitute a very key issue indeed: the political or public understanding and acceptance of the problem. Simply stated, three major problem areas must be faced: (1) manufacture/extraction, use, disposal, and dispersion of specific chemicals, (2) industrial and domestic effluents which are, by nature, of mixed, variable, and changing composition, and (3) focussing political will so as to reach mutual agreement on the resolution of such issues (Hamilton 1976).

In the case of specific chemicals it is possible to employ fairly specific figures. Approxi-mately two million chemical compounds have been synthesized by man, at a present rate of approximately 25 000/yr. Ten thousand compounds or more are now in com-mercial production, about 1000 of which are prepared in 60 000 different formula-tions. Few of the compounds in commercial production have been tested for even such an important aspect as carcinogenicity, fewer still for their impact on the environment. The testing of compounds to date presents some curious anomalies: (1) why was nitrilotriacetate (a replacement for phosphate in detergents) tested so thoroughly when vinyl chloride (a precursor of certain plastics) was not, when the production of the latter is many times that of the former and (2) why was the lampricide TFM accepted for use in the Great Lakes system without any significant mammalian testing being done until many years later?

Even the quality of testing is in some doubt under certain circumstances. While the rate of synthesis is quite high, the rate at which new compounds come into the market is much lower. However, that rate suggests that decisions regarding the acceptability of new compounds would have to be made once every 10 d if governments wished to accept that responsibility. Clearly this is impossible and priorities will have to be set in a rational manner and testing imposed in only selected cases or only to a certain degree. It is not clear that we have comprehensive mechanism in place to accomplish these ends.

The problem of industrial and domestic effluents is harder to quantify but it is clear that despite our best efforts we have a serious problem. It can be shown that man has become a geological agent! That is to say our efforts now transport much more of certain chemicals to the world ocean and the atmosphere than do geological processes. It is clear that some marine areas are now little more than saline sewage treatment ponds (parts of the North Sea, the New York Bight, the Gulf of St. Lawrence, the Pacific off the California coast, etc.), while others are in danger of the same fate (the Baltic, the Mediterranean, the Gulf of Mexico, etc.). These factors are certainly influencing our fisheries, and similar discharges to the atmosphere are threatening our air quality. Moreover, we are now discharging materials to our surface and groundwaters that prejudice their future use as potable or even in some cases as industrial process waters. It is important to Canada that these resources remain useable or perhaps even marketable as some have suggested. While it is not acceptable to place undue constraints upon our industries or communities, some accommodation must be sought and it is not clear that a mechanism exists for attaining that goal or for enforcing it when and where required.

We must be forced by such considerations to address the basic question: what are we to do about our chemical dependency and our chemical by-products? How are we, as a community, to manage this problem? The problem is an old one and even then encompassed many aspects of society, for example food supply, drinking water sources, agricultural practise, urban sewage disposal, and the environment with its attendant aesthetic considerations.

It is hoped that the following consideration of four areas that we saw as major, yet general, problems with regard to our aquatic resources will serve as some food for thought. In this text we often range far beyond strictly aquatic issues. This is because the problems posed need a total systems response, and few decisions in these regards can be taken in isolation. Thus, lessons learned from one isolated circumstance or environment must be considered and the experience used in the context of many other situations.

Organic Chemicals

Specific Compounds of Concern

Organic compounds frequently dominate lists of substances raising environmental concern. For example the United States Environmental Protection Agency list of 129 "priority pollutants" included 31 "purgeable organics," 46 "base/neutral extractable organic compounds," 11 acid extractable organic compounds," and 26 "organic pesticides" (Keith and Telliard 1979). The agreement between Canada and the United States regarding water quality in the Great Lakes listed over 150 organic compounds regarded as either "hazardous polluting substances" or "potential hazardous polluting substances" (Nriagu and Simmons 1984).

Substances appear and disappear from lists such as these in accordance with our changing perceptions of hazard and the changing state of control actions. Generally, such lists include a number of chlorinated hydrocarbons, pesticides, and aromatic hydrocarbons and a smaller number of other compounds.

The properties of compounds on priority lists vary greatly but information used in deriving priority lists tends to include chemical stability leading to persistence in the environment, the quantities in use, the types of use, the tendency to accumulate in living organisms, and the biological effects associated with acute or chronic exposure.

Some materials that seem likely to be subjects of ongoing and future debate in Canada include compounds that appear as components of toxaphene and trace materials such as chlorinated dibenzodioxins and dibenzofurans. No doubt we shall continue to be perplexed by unanticipated consequences such as the appearance of chlordane in arctic fish and fish-consuming mammals (Norstrom et al. 1985). Older, proscribed materials such as DDT, PCBs, and Mirex will continue to generate concern as long as significant residues are reported. The fate and biological implications of organo-sulphur compounds (originating from coal- and petroleum-based process) in the environment are poorly understood. Polyaromatic hydrocarbons are known to be widely distributed and have been implicated in the increased incidence of tumors in certain fish populations (Black 1983). The use of methoxychlor to control blackfly larvae in western Canadian rivers remains controversial, as do aerial applications of synthetic pyrethroid insectides which are very toxic to fish, the growing use of herbicides in forest management, and the use of chlorophenols as wood preservatives. An emerging issue may be that we have few if any techniques for evaluating the impact of new control agents provided through "biotechnology."

Sources

Efforts to understand and to control toxic chemicals ultimately lead to a consideration of sources. Often the simplest cases are point sources where control technology can be applied to limit discharge of the material in question. Keith and Telliard (1979) provided a list of 21 types of industrial activity in the United States for which regulation of chemical discharges was most urgent. Many of these are equally relevant to Canada such as processing of wood products, petroleum refining, manufacturing of chemicals, rubber processing, paint and ink formulation, leather tanning and finishing, and coal mining. Given appropriate technology and incentive, control of point sources can reduce chemical emissions with relatively few corporate decisions.

Diffuse sources are much more difficult to control partly because the number of individual or corporate decisions and actions required is so much larger. The disposal practises of millions of individual farmers and consumers with traditions of independent decision making are virtually impossible to regulate; the most promising influence is through education.

With some chemicals the contribution by natural sources can be significant. For example, materials like polyaromatic hydrocarbons (PAHs) are formed during the incomplete combustion of carbon-based materials and so natural events such as forest or peat fires contribute to the environmental load of these compounds. These natural inputs contribute as much of certain PAH compounds as do human activities (Ontario Ministry of the Environment 1979).

Major Impacts

Generally the primary concern that is expressed is with the direct effects of chemicals on the health of people or on the economic or cultural well-being of people who depend upon the productivity of animals and plants. These include occasional mass mortalities of people (Bhopal), involving an exotic compound, and fishes (Placentia Bay), which involved the direct death of the animals or the accumulation of substances (such as mercury or radionuclides) in biota such that their use as human food must be prohibited or restricted, with a resulting impact on the grower, harvester, or processor. Many other examples can be cited such as the accumulation of stable contaminants like Mirex in fish so that tolerances set to protect human health are exceeded (St. Lawrence river eels) or the accumulation of compounds like petroleum oils which impart offensive tastes or odours so that fish products cannot be used (e.g. The Athabasca delta winter fishery following an under ice spill of refinery effluent).

Massive mortalities in fish and invertebrate populations due to chemical exposure have often been observed (Holden 1973; Kerswill 1967; Nimmo 1979). In some cases certain sublethal responses seen in the laboratory have been seen in the affected populations, especifically in those cases involving exposure to pesticides. For example, fish and many other organisms have a striking ability to metabolize foreign compounds using a group of enzymes called mixed function oxidases (MFO). This system is said to be inducible in that fish exposed to certain organic compounds (e.g. PCBs, petroleum) synthesize enhanced quantities of these enzymes in an apparent effort to metabolize the compounds to new, more easily excreted metabolites. Enhanced levels of such enzymes can be detected easily in confined and natural populations and have been interpreted as responses to chemical stress (e.g. Khan et al. 1979, Lech et al. 1982; Payne 1984; Stegeman et al. 1986). While elevated levels of such enzymes are presupposed to indicate chemical stress, it is not clear what the ecological cost is to the stressed animals. Presumably there would be some energy cost in synthesizing and maintaining an NADPH-dependent system. Consistent with an energy cost, Roesijadi and Anderson (1979) reported reduced growth in aquatic animals exposed chronically to sublethal quantities of petroleum oils. These enzymes are also implicated in the regulation of plasma levels of androgens, estrogens, and corticoids, since treatment to increase the enzymes results in a corresponding drop in plasma steroids (Sivarajah et al. 1978).

Distribution

Our knowledge of distribution relates directly to our increasing sophistication in detection. Current techniques allow us to detect chemicals that would not have been sensed even a few years ago. Our ability to predict that organic compounds will become concentrated in aquatic organisms has been improving recently, based upon concepts such as exchange equilibria (Hamelink et al. 1971) and bioenergetics (Norstrom et al. 1977). Models describing the fate of chemicals in different environmental compartments of ecosystems which are based on partition phenomena are now becoming commonplace (Kenaga and Goring 1980), and they are being improved continuously.

Improved analytical methods have allowed surveys of organic pollutants in fish, and other environmental samples have shown that very large scale movements of pollutants do take place. Such pollution cannot be considered as a site-specific issue any longer. It is now well recognized that the dispersal of many compounds can often be atmospheric and therefore much more widespread than recognized earlier (e.g. DDT in Antarctic snow; Peterle 1971). In the case of PCBs there is an elegant example of the pervasiveness of atmospheric transport in the study of Lake Superior sediments by Eisenreich et al. (1979) and especially from the studies of Isle Royale (isolated from human habitation since the 1940s) by Swain (1978) and Murphy (1984). Czuczwa et al. (1984) have reported dibenzo-p-furans from the same areas. PAHs have been reported from arctic air samples (Daisey et al. 1981) and arctic sediments (Stich and Dunn 1980). To stress the global extent of the atmospheric dispersal of toxic substances, we note that many inorganic compounds also are widely transported in this manner (Rahn and Lowenthal 1985.) Thus, many pollutants, having reached the waters, volatilize from the water to the air (Mackay and Wolkoff 1973; Harrison et al. 1975) where they are dispersed for deposition elsewhere.

An aspect not usually considered, but one that is of especial interest to Canada, is the role played by ice and snow. Pollutants do accumulate in such materials and are realeased as a "pulse" when the ice or snow melts, thus inflicting an insult to biota that is momentarily manyfold that of the so-called "annual deposition rate." Moreover, some materials such as volatile components of petroleum that are normally lost to the atmosphere within a short period of time will persist and move with water currents under ice cover, as was demonstrated in the Lodgepole gas well blowout (Crowther et al. 1984).

Specific Regional Impacts

Chlorinated dibenzodioxins and dibenzofurans have been recorded in the Great Lakes' water and sediments (Stalling et al. 1983) as have a wide variety of other organic pollutants, notably Toxaphene (Nriagu and Simmons 1984). Very remote areas of Canada are demonstrably contaminated, albeit at low levels (Wagemann and Muir 1984; Norstrom et al. 1985). The problem is both severe and widespread (Ng and Patterson

1981). It affects water and its associated aquatic life throughout Canada and is locally quite severe from time to time.

It must be recognized that such pollution is not simply a scientific curiosity, for it has very distinct practical consequences. For example, shipments of Canadian eels from the St. Lawrence River were recently rejected by European markets by reason of their Mirex levels. Court records reflect that release of "oil and grease" under the ice of the Athabasca River resulted in oil moving down the river to its delta where it caused the closure of the winter whitefish fishery in 1982 due to "tainting" problems. Other examples of impacts of chemicals on our fisheries have been reviewed by Gilbertson (1984).

Implications of Continued Loading

Continued exposure of animals to some chemicals can alter the animals. Fish from eastern Lake Erie and the upper Niagara river do have a very high incidence of tumours and these have also been induced in fish (bullheads) by exposing them to extracts of Niagara River sediments (Black 1983). Moreover, people fishing for certain fish in the Great Lakes have been formally advised not to eat them on a routine basis.

Animals consuming contaminated fish (and this includes man) have been shown to suffer a wide variety of consequences. Possibly the first example of this was the decline in the western grebe (a bird) population at Clear Lake (California) following the use of DDD for insect control (Hunt and Bischoff 1960). Another convincing example is that of the effect of feeding contaminated coho salmon from Lake Michigan to mink with the resulting mortality and reproductive failures (Aulerich et al. 1973), apparently as a result of PCB contamination of the fish. Other fish-eating mammals have shown abnormalities in reproduction which were thought to be related to contamination with organochlorine compounds (DeLong et al. 1973), perhaps a warning of events such as the low human birth weights associated with residence near a chemical dump site (Vianna and Polan 1984).

With regard to populations of aquatic animals and plants as measured by methods designed to estimate distribution, abundance, productivity, and species diversity, it is not clear what specific impacts can be associated directly with the input of detectable organic compounds. Signs of stress can be detected at the ecosystem level (Rappaport et al. 1985) but it has proven very difficult to link them specifically to inputs of particular chemicals. The best examples are related to stable organochlorine compounds in that accurate residue measurements can be used to compare natural with experimental populations. In the case of certain unstable compounds, especially those that produce identifiable metabolic products, this mode of approach shows some promise in defining the consequences of continued loading (Krahn et al. 1986). The studies of Murchelano and Wolke (1985) of carcinomas in winter flounders serve, however, to illustrate the continuing difficulty in relating exposure to effect. The current trend to overcome these problems seems to be to use large-scale experimental enclosures or "mesocosms" such as those described by Odum (1984), although he was by no means the first person to discuss such arrangements.

Metals

Specific Compounds of Concern

Interest in metals in aquatic ecosystem has arisen largely because many metals, especially those not essential for life, are toxic even at comparitively low levels, yet are ubiquitously present in aquatic systems (Luoma 1983).

Metals fall in to two general categories: those that are essential for the proper development and growth of organisms and those that are unessential. However, even the essential elements are essential only within fairly narrow concentration ranges and if present in concentrations exceeding these ranges are also toxic.

The ubiquity of metals in aquatic systems arises because the surface of the earth is mainly composed of minerals which are largely a mix of metalloids and especially a mix of metals combined with oxygen and sulphur. Thus, these minerals form the lining of water bodies and their catchment areas. A delicate balance has been established over time between these surfaces and the water they enclose producing a concentration range of metals in the water that is normally within the proper range for the maintenance of biota. It is when this balance is altered through man's intervention that the elevated levels of metals appear which we call "pollution."

Historically, the acceptable or maximum permissible limits of metals in water were established with regard only to man and little if any attention was paid to the health of aquatic biota. Such limits were first published in Canada the "Canadian Drinking Water Standards and Objectives, 1968" (Department of National Health and Welfare 1969). They were based on scientific information on long-term effects on human health of various toxic substances, the "Public Health Sercvice Drinking Water Standards" (U.S. Department of Health, Education and Welfare, Public Health Service 1962), and the "International Standards for Drinking Water (World Health Organization 1963). The Canadian standards were revised and republished in 1979 as the "Guidelines for Canadian Drinking Water Quality, 1978" (Federal-Provincial Working Group on Drinking Water of the Federal-Provincial Advisory Committee on Occupational Health 1979).

The concentrations permitted in livestock and wildlife watering sources are somewhat higher than the standards for human consumption. The limits for fish are less precisely defined in any set of water standards, but in Canada fish are also protected under the *Canada Fisheries Act*. This Act essentially states that any deleterious substance may not be deposited in waters frequented by fish or in areas where it is likely to enter such waters. The determination of what is "deleterious" and the translation of this into a limiting concentration in waters is difficult except in the most obvious case where death is the outcome of exceeding a limit. At the sublethal level, the determination of what is and what is not deleterious to biota has not been codified in law except in the context of a U.S.-Canada agreement (International Joint Commission 1977). It is only fairly recently that aquatic organisms other than fish have been studied with regard to the impact of elevated metal concentrations.

Sources

The elements of environmental concern are generally zinc, copper, selenium, mercury, cadmium, lead, arsenic, and possibly silver. This is due either to their inherent toxicity or because they are the most prevalent by-products of various industries that discharge to the environment. Common point sources of some of these metals are mining industries, various refining, smelting, and manufacturing industries, cement plants, and sewer systems serving domestic and industrial sources. Non-point sources include the burning of fossil fuels, incineration of municipal refuse, and metal-processing operations generally. The deliberate use of toxic metals in the environment (e.g. copper as an algicide and as a control agent for other aquatic "weeds") has been discussed widely in recent literature and therefore will not be addressed here, save for the peculiar problem of arsenic which is discussed later.

All of the above elements can be found in mine discharge waters but their concentrations will vary from mine to mine, depending upon the type of ore being extracted and the control measures that are in place (Clarke 1974). Moreover, there were 14 large smelters operating in Canada at last count, 6 of which were aluminium smelters. Stack emissions from smelters do measurably pollute aquatic environments in their vicinity (Franzin 1984; Freedman and Hutchinson 1980) and, depending upon the physical state of their discharges, can sometimes contribute to very long-range effects. Such long-range transport is particularly associated with the more volatile elements such as lead, mercury, and cadmium. Indeed, lead from such sources is now a global problem (Schaule and Patterson 1981) although it is particularly evident in the Northern Hemisphere (Rahn 1981).

Arsenic is found in discharges from gold and silver mines and in stack effluents from the smelting of most metallic ores, particularly copper, zinc, and lead. Also, arsenic has been used deliberately in the environment for such purposes as controlling aquatic weeds, as a wood preservative, and as general herbicides. While such uses are largely behind us, the legacy of prior use is still with us. The demand for arsenic is much less than its production as a by-product of the production of other metals, although demand for it as a wood preservative is increasing. Consequently, its storage (as arsenic trioxide) can create short and very long term local environmental problems.

Selenium, since it has a strong affinity for sulphur and since we are largely mining sulphide ores, is also largely produced as a by-product of the extraction of other metals and the production of sulphuric acid. However, this element also occurs naturally in high concentrations in certain soils. Again, man intervenes and introduces irrigation practices to previously undisturbed soils which inevitably result in drainage waters high in selenium that have been recorded as harming biota.

Lead oxide (Demayo et al. 1980) was the most important ingredient of paints prior to 1945, when it was largely replaced by titanium dioxide (which is now a suspected carcinogen). Consequently, old painted surfaces remain a major source of lead in certain areas but it would appear that leaded gasoline is now the major source. Inputs of lead to only two of the Great Lakes have been estimated to be 700-900 t/yr (Swanson 1976), a major part of this input being atmospheric in origin (Kemp and Thomas 1976). Lead isotope profiles within sediment cores from Lake Michigan have been correlated with the increased use of such fuels and with the combustion of coal (Edgington and Robbins 1978); however, it must be noted that lead concentrations in rainfall over the Great Lakes are extremely variable in nature (Environment Canada 1978). As one might expect, urban stormwater runoffs (washing the highways and their verges) have lead concentrations 20 times greater than rainwater (Wilber and Hunter 1977).

Cadmium is a by-product of the mining and refining of zinc, copper, and lead. In the latter part of this century the use of cadmium has increased by 300- to 500-fold over previous demand. Total cadmium emissions to the atmosphere are increasing at a rate of 4% annually. (Flinn and Reimers 1974) and it can be reasonably presumed that this increased loading is finding its way into our waters. Another major avenue for the widespread distribution of cadmium is the use of phosphate fertilizers which can contain tens of parts per million of this element. This loading is finding its way into agricultural products (Kjellstrom et al. 1975) and into man (Kjellstrom 1979).

Important sources of mercury pollution in our aquatic environments were chloralkali plants associated with pulp and paper operations (Armstrong 1979). While alternatives to mercury cells were known for some time, the plants in Canada chose to continue using the old process for reasons that appear even economically unsound. In any event we are still living with the effects of irresponsible corporate decisions. In fact it took federal intervention to change the picture through the imposition in 1972 of specific regulations under the *Canada Fisheries Act*, and the number of plants using such cells in Canada was reduced from 16 to a few (Sherbin 1979). A more diffuse source of mercury contamination of freshwater has been associated with the "acid rain" phenomenon and presumably reflects the simple fact that acid rain per se is not the entire problem; it is the "long-range tranpsort of atmospheric pollutants."

Major Impacts

Metals exert lethal effects on stages of fish development not normally tested (Giles and Klaverkamp 1982) and can exert subtle, yet major sublethal effects such as affecting migratory behaviour prior to spawning (Sprague et al. 1963). Thus, we see everything from outright kills due to mine effluents (Clarke 1974) through gradual population reductions over long periods of time (Somers and Harvey 1984). Food organisms such as zooplankton can also be affected (Marshall et al. 1981). While it is known that exposure to gradually increasing concentrations of metals increase the tolerance of fish to metals (Duncan and Klaverkamp 1983) and increases certain defense mechanisms such as enhanced metallothionein production (Klaverkamp et al. 1985), the energy cost to individual animals or whole populations is unknown.

The impact that any metal has on the biota is dependent upon its chemical species (Florence and Batley 1980). For example, it is now generally accepted that uncomplexed, or "free," metals are more toxic than those complexed with such natural substances as humic acids, fulvic acids, or other strong complexing agents (Andrew 1976; Magnuson et al. 1979). It would also appear that the more acidic the water, the more pronounced the toxicity of the metal; however, this is not a simple situation, for the hydrogen ion itself influences the toxicity of metals (Peterson et al. 1984). Increased acidification is known to increase the concentration of metals in the water (Schindler and Turner 1982; Forester 1980; Sloan and Schofield 1983), thus contributing to the complexity of the acid rain problem and to our attempts to understand metal toxicity issues.

It should be noted that while chemical analyses of water have a definite purpose to serve in developing our understanding of such issues and implementing control strategies, it is only in severe cases of pollution that persistently elevated levels of an element can alert us to a problem. What is more frequent are situations involving persistent low levels of contamination, involving a mixture of metals, that result in little overt concentration increases in the water but that produce marked results in the biota. Mercury is a classis case (Armstrong and Hamilton 1973; Rudd et al. 1983).

Thus, one must look beyond the undoubtedly complex question of water chemistry and elemental mass balances within water systems, even though these are important in their own right, in coming to grips with these problems. For example no one could drive through the Sudbury area during the last 40 yr and not understand that smelter operations were a problem. After all, when one goes from areas with heavy forest to a lunar landscape and back again wherein there is but one change — smelters — one cannot help coming to certain conclusions. Latterly we have learned that a less obvious resource (fisheries) was also gone from the area. Now we know that the effect is not truly confined to that area but extends for thousands of kilometres in varying degrees of severity. This single lesson suggests that we should look to indications of stress produced in the biota for answers both in the short and long term.

In Manitoba another smelter, at Flin Flon, has been a major source of heavy metal deposition for over 55 yr (Franzin 1984). Fish from lakes subject to such fallout are demonstrating higher tissue levels of certain metals than the same species of fish from lakes in the same area that are not so critically exposed (McFarlane and Franzin 1980). Moreover, such fallout does stress young fish and impair reproductive capacity in adults (Franzin and McFarlane 1980).

One unexpected impact due to metals occurs during the creation of new reservoirs. The problem relates mainly to markedly elevated mercury levels in fish and has been recorded throughout the world but is perhaps best documented in studies of the Southern Indian Lake situation in northern Manitoba (Bodaly et al. 1984).

Arsenic originating from mine effluents has largely eliminated certain taxa of invertebrates from affected lakes (Wagemann et al. 1978). Elevated concentrations of copper and zinc in rivers in the mining areas of New Brunswick have adversely affected migration and spawning of Atlantic salmon (Surface Water Quality in Canada 1977).

Invertebrates are important food organisms for fish, but in assessing the impact of toxicants upon such biota, one must usually look to population rather than individual effects. One exception to this are certain studies on a group of insects (Chironomidae) wherein specific morphological changes have been recorded from Lake Ontario (Warwick 1980a) and other lakes (Warwick 1980b).

Where pollution involves several metals, the effect seems to be a nonspecific decrease in species richness and population sizes. This has lead to the use of various "environmental quality indices" based upon the analysis of such parameters (e.g. Molitor and Vanhooren 1984). This appears to be a useful approach but a costly one, which may explain the dearth of such data in the literature.

Enhanced metal levels are now being found in aquatic mammals although the sources and effects are unknown (Wagemann and Muir 1984). Finally, of course, one must come to the ultimate consumer — man; noting that while total cadmium emissions are increasing at a significant rate annually (Flinn and Reimers 1974), the body burden of that element in man has increased by four to five times during the last century (Drasch 1983).

Distribution

The distribution of metals presents very curious and very complex problems due to the fact that these elements form a wide variety of compounds, complexes, and "species." The composition of any one of these and the relationship between them is governed by a number of factors; for example, complex formation depends upon the magnitude of "formation constants," the concentration of the metal, the concentration of the complexing agent, the concentration of other metals, the pH of the environment, the oxygen content of the environment, and other factors.

Each of the above-mentioned factors has an effect on the distribution of metals in the aquatic environment and its biota. For example the simple presence or absence of oxygen or the presence or absence of sulphide may either liberate metals to the biota or almost irreversibly bind them to the sediments. Also, intervention by man tends to cause metals to move through their intricate chemical webs in ways that are very hard to predict. However, some effects are painfully obvious such as our enhanced metal loading to closed aquatic systems (indeed some of our loadings to the world ocean are much greater than any geological process) or the enhanced productivity of such systems due to our input of sewage which results in oxygen depletion or hydrogen sulphide production. Another example is acidification which is known to promote mobilization of metals from inert or stable forms into the biota in amounts or at rates such as to cause serious damage (Hakanson 1980).

Thus, the distribution of metals presents a complex set of problems but there are natural controls which limit their movement into biota; there are also interventions by man that enhance that movement. There is some predictability about the control mechanisms and therefore some ability to predict the results of perturbation. The speciation of metals in aqueous systems has been modelled even in multicomponent systems with some very real success. Such models or predictions utilize extensive thermodynamic data bases and complex computer programs which consider hundreds of differing chemical species (Ball et al. 1979; Nordstrom et al. 1979; Wagemann et al., unpubl. data; McDuff and Morel 1973). Further modelling efforts attempt to address movements of metals between compartments in aquatic systems (Fontaine 1984; Luoma 1977; Tiwari and Hobbie 1976) but the reliability of these models is severely restricted due to the paucity of reliable rate constants for the transfer of metals between "compartments." A further impediment to understanding these processes is the paucity of information regarding the efficiency of biotic extraction of metals, for example by fish from their food (such as that provided by Pentreath 1976; Merlini et al. 1976) or directly from water (Willis and Sunda 1984; Klaverkamp et al. 1983).

To use but a single example of the complexity of problems involving a metal, one might use mercury: its uptake by fish is quite different depending upon whether the animal is presented with metallic mercury in solution/suspension or with methyl mercury. The element is introduced into waterbodies by man and by the environment. Native ore deposits, hot springs, volcanoes, and even the world ocean place mercury in the atmosphere from which it returns to the land in precipitation; add to this hospitals, small industries, and chloralkali plants. The common form of mercuric sulphide can be solubilized by bacteria (Fagerstrom and Jernelov 1971) and reduced by bacteria (Wood 1974). Elemental mercury is quite volatile even from relatively cold water bodies and can be reintroduced to the atmosphere. As a personal detoxification mechanism, many bacterial species are capable of converting mercury to methyl mercury which is readily accumulated by other biota and which is a potent neurotoxin for mammals. Still other microbes reduce such compounds to elemental mercury and methane. Thus, in trying to understand the distribution of metals, we are faced not only with chemical complexities but biological complexities as well.

Specific Regional Impacts

While few incidences of metal pollution come to the attention of the public or even the science community, the problems are pervasive in the extreme. Mercury is lost daily from health care facilities, titanium dioxide is used and disposed in large quantities as paints, and arsenic is stored at various gold extraction plants (e.g. Yellowknife) and will ultimately have to be disposed of safely. Heavy metals are introduced into our freshwater and marine environments, including the arctic, through mining and ore-concentrating operations. Extractions of asbestos has become recognized as a health problem in at least two provinces. High levels of radioactivity from radon gas have been measured in various parts of the country. The single largest source of environmental lead pollution is still our use of leaded gasoline. Thus, we have many scientific and political judgements to make in the near future.

Implications of Continued Loading

Demographic distributions of elevated concentrations of metals and the increased incidence of human diseases have been studied and correlations have been established. However, distinct cause-and-effect relationships have been established in only a few cases.

In the physical world things are more certain and it is now known that the natural distribution of lead has been disturbed on a worldwide scale. It is probable that the same statement can be made with regard to mercury and could be proposed with regard to cadmium and other metals.

In that it has been adequately demonstrated that such increased loadings to the aquatic environment usually manifest themselves in effects on man, two factors become obvious: the first being that we should drastically curtail our input of metals to the environment and the second being that if we do not adopt such a course of action we can expect to experience some very personal problems.

Fossil Fuel Consumption

Specific Compounds of Concern

Although there are major knowledge gaps regarding the environmental and human health effects associated with fossil fuel consumption, a wealth of information does exist on its impact on aquatic resources. Such information can be found in reports (National Research Council of Canada 1981; National Academy of Sciences 1981), conferences proceedings (Alberta Environment 1982; Toriba et al. 1980; Johnson 1982), review articles (Fromm 1980; Spry et al. 1981; Haines 1981; Dillon et al. 1983), a nine-volume series on acid precipitation (Teasley 1984), and numerous government reports and popular magazine articles. Moreover, a recent comprehensive document produced by the Standing Committee on Fisheries and Forestry (1981) describes major industrial sources, effects produced on terrestrial and aquatic systems and on human health, emission trends, legal contexts and implications, and the economic costs of acidification to Canada.

Thus, we will deal only with highlights of this very important problem and will focus particularly on the combustion of fossil fuels, placing lesser emphasis here upon issues relating to drilling, mining, refining, and transport of such energy sources. For a much more detailed review the reader is referred to Campbell (1981).

The major combustion products are the oxides of carbon (CO_2), nitrogen (NO_x), and sulphur (SO_2) and the associated emissions of metals and trace elements such as mercury, selenium, tin, chromium, nickel, lead, cobalt, copper, cadmium, zinc, vanadium, and iron (Nriagu 1979; Goldberg et al. 1981; Landheer et al. 1982).

Nitrogen oxides originate due to the oxidation of nitrogen in the air used in the combustion process, while sulphur oxides result from the oxidation of sulphur impurities in the fuel. Paradoxically, more efficient combustion processes produce greater amounts of NO_x. Approximately one half of the CO_2 produced is transferred to the ocean; the remainder goes to the atmosphere. Present atmospheric concentrations of CO_2 are about 350 ppm, while it is estimated that such concentrations prior to widespread use of fossil fuel were only about 220 ppm (see also chapter by Ripley, this volume; Gribben 1984a, 1984b).

The ease of metal volatilization during fossil fuel combustion is considered to be a good indicator of the distance over which such materials can be transported by atmospheric processes (Bertine and Goldberg 1971). Considering only the elemental state, the order of volatility is: $Hg > As > Cd > Zn > Sb > Mn > Ag > Sn > Cu$. For oxides, sulphates, carbonates, silicates, and phosphates the order is $As + Hg > Cd > Pb > Ag + Zn > Cu > Sn$, while for sulphides (the principal form found in fossil fuels) the order is $As + Hg > Sn + Cd > Sb + Pb > Cu + Fe + Co + Ni + Mn + Ag$.

The preferential long-range transport of mercury, cadmium, arsenic, lead, zinc, and copper suggested by these data is supported by a recent analysis of global metal transport (Lantzy and Mackenzie 1979). These investigators assessed the relative importance of atmospheric and stream processes in the transport of 20 metals. Two categories were found to exist: (1) those metals undergoing greater transport to oceans by streams than by the atmosphere and (2) those moving mainly via atmospheric transport. The first group included aluminum, mangnese, cobalt, chromium, vanadium, and nickel, while the second group included mercury, selenium, arsenic, lead, and tin.

Drilling, mining, refining, and transport can also produce environmental damage and have serious impacts upon aquatic resources. For example, the physical disturbance of lands associated with coal and tar sands mining can give rise to watercourse alterations, increased sediment loading, and increased leaching of chemicals from newly exposed terrestrial systems. Liquid, solid, and gaseous discharges such as mine waste waters and brines, washing and cooling waters, drilling fluids, hydrocarbon spills, elemental sulphur and hydrogen sulphide releases are also associated with such activities.

Sources

Base metal smelters are, by far, the largest source of sulphur dioxide emissions in Canada today. Summers and Whelpdale (1976) estimated that smelting produces some 63% of such emissions in Canada; the Ontario Ministry of the Environment (1980) report on acidic precipitation places the percentage at "over 40% of the Canadian total"; and the Memorandum of Intent (February 1981) on Transboundary Air Pollution between the United States and Canada states that "smelting accounts for 45% of the total sulphur emissions" in Canada. Their table 1 provides current estimates of emissions of sulphur and nitrogen oxides in the United States and Canada, while their table 2 provides projected estimates for SO_2 emissions in Canada. The third table identifies the major base metal smelters as of March 1976, and there is little reason to believe that the situation has changed. Environmental impacts of metals emitted from smelters are discussed by Nriagu (1984) and elsewhere in this chapter.

While Canada generates about 5.5 million tons of sulphur dioxide annually (24% from the Sudbury, Ontario area alone), the United States produces 29 million tons (Ontario Ministry of the Environment 1980). Power plants produce about 19.5 million tons in the United States whereas equivalent Canadian sources produce 0.8 million tons. It seems that most of the SO_2 produced in Canada and one quarter of that produced in the United States is deposited in Canada. Most of the Canadian emissions of SO_2 and NO_x originate east of the Manitoba–Saskatchewan border and are deposited in that area. In the case of NO_x emissions, approximately 22 million tons are produced in the United States versus 2 million tons in Canada. About one half of the Canadian emissions are due to automobile exhausts and this amount is projected to increase by at least 50% by the year 2000 (Ontario Ministry of the Environment 1980).

Major Impacts

Such discharges are certainly resulting in some of the most severe environmental changes ever produced by man.

These changes include the destruction of freshwater ecosystems through the deposition of sulphur and nitrogen oxides, the contamination of aquatic organisms with heavy metals, and perhaps profound climatic changes. In 1980 there were recorded some 180 lakes without fish in Ontario, mostly around Sudbury (Ontario Ministry of the Environment 1980), and nine rivers in Nova Scotia no longer supported salmon populations (Anonymous 1981). If current trends continue, it is projected that some 48 000 lakes in Ontario are threatened as are 11 more rivers in Nova Scotia. All other Canadian provinces as well as the Territories can expect similar problems (Anonymous 1981).

Atmospheric transport of heavy metals and other trace elements, their residence times, and their environmental dispersal are largely determined by their state (i.e. gaseous, fine particles, coarse particles, etc.). Estimates of the amounts of these substances emitted are variable, but more importantly, there is increasing evidence that they excessively contaminate resources used for human consumption before the resources themselves are actually destroyed.

Increases of atmospheric concentrations of CO_2 in excess of 25% above historic levels may have already occurred (Anonymous 1984). The rate of increase is increasing. It is suggested that increased CO_2 loadings will produce increases in climatic temperatures (the "greenhouse effect"). Impacts are projected to include flooding of coastal areas, redistribution of biota, both wild and domesticated, desiccation of shallow lakes in southern latitudes, and changes in permafrost boudaries (see chapter by Ripley, this volume).

Other major effects of fossil fuel combustion include human health problems, reduction of agricultural and forest productivity, and structural damage to buildings (Anonymous 1981; Environment Canada 1981; Royal Society of Canada 1984).

Distribution

Atmospheric processes such as wind speeds and directions, planetery boundary layers, solar radiation, and complex diurnal phenomena control the dispersal of pollutant gases (Whelpdale 1978; National Academy of Sciences 1981). Distance of transport seems largely dependent upon physical state: metals associated with heavy dusts (greater than 20 μm) or coarse particles (2-20 μm) have atmospheric residence times measured in hours or days at most and are transported from hundreds of metres to a few hundred kilometres from a point source (National Academy of Sciences 1981). Some oxides of nitrogen also appear to fall into this category because of the loading levels at which most are emitted. Fine particles (less than 2 μm) and gases have residence times measured in months or years and therefore are capable of continental or even global distribution.

Deposition of pollutants is dependent upon climatic conditions (Whelpdale 1978; National Academy of Sciences 1981). During dry periods, the deposition of various sized particles and the surface adsorption of gases is continual and extremely difficult to measure. During periods of precipitation, particles and gases are scavenged from the atmosphere and such events are relatively easy to assay. In spite of the time differential between processes, they are judged to be equivalent in effect on an annual basis.

Aquatic systems are important sinks for pollutants produced by fossil fuel combustion. None of the world's aquatic systems are free from such impacts and the problem can be seen as truly global in nature. It is a major instance of how the actions of one nation can affect the productivity of another, and little further research is required to demonstrate the obvious; what is required instead are greater efforts to stop such transboundary violations. As has been demonstrated in Europe, this will require political will and a change in ethics; neither is likely to occur until the producing nations realize that they are damaging their own territories, if the European example is to be believed.

Specific Regional Impacts

Acidification affects at least 5 million km^2 of North America (National Research Council of Canada 1981). Most of this area is located in the northeastern quadrant of the continent although major vulnerable areas now are recognized in the south and midwest. These latter areas are not associated with the geology of the Canadian Shield which was until recently seen as the only vulnerable area.

Specific regional impacts in Canada occur either in areas where there are major regional discharges or where atmospheric factors introduce such pollutants from sources to the south. It is not surprising that Ontario produces most of the emissions and suffers most of the damage. What is not readily apparent is that it receives a lot of material from the United States and exports much to Quebec and the Maritime provinces. Moreover, certain areas in the Northwest Territories are experiencing "acidic rain"; the source of this is problematical but it is most likely northern Europe.

An analysis of the amounts of sulphur and nitrogen oxides produced in Canada is available (Standing Committee on Fisheries and Forestry 1981). It is interesting to note that discharges from some relatively small operations become very significant when combined. For example, from their table 6 it can be seen that the SO$_2$ dishcarges from five natural gas processing plants in Alberta and British Columbia exceed a major point source in Flin Flon, Manitoba. In toto, these five closely related geographical sources become the fourth largest source in Canada (see their table 4).

Implications for Continued Loading

However much one would like to cite ecological issues, conservation of what may be only partially renewable resources such as fisheries and forestry, or even aesthetic issues, the real and very practical issue is economics (see, however, chapter by Vallentyne and Hamilton, this volume). In addition to human health (a datum never satisfactorily quantified by economists) there are four major economic areas affected by acidification), namely tourism, forestry, agriculture, and building materials. Tourism in northern Ontario is estimated to be worth 1 billion dollars annually. This is largely based upon the attraction of the fisheries which are threatened, and to this figure one could add at least another 50% for loss of commercial fisheries. In sum, the total loss *to Ontario alone can be estimated to be in excess of 1.5 billion dollars.* Certainly these figures can be argued; perhaps quoting them in this context will lead to reasoned argument and therefore reasoned estimates. In Nova Scotia the loss of nine salmon rivers is a genetic loss upon which we cannot place a price, but the economics of the situation suggest an annual revenue loss of $300 000. There simply are no clear estimates of the loss to the forestry sector in Canada but the United States loss is estimated to be in the "hundreds of millions of dollars." The real and potential loss to Canada could be as high or higher by reason of the distribution of the pollutants; the relative impact on the GNPs of the two countries are not comparable.

Economic losses to Canadian agriculture have not yet been defined adequately but one would naturally suspect very reduced productivity of the delicate crops typical of southern Ontario. It is widely suggested that the productivity of sugar maple groves can be directly or indirectly reduced.

Estimated health benefits through a 20% reduction of acid pollution range from 3 to 32 billion dollars annually in the United States (0.3–3.2 billion in Canada?). Certainly the Standing Committee on Fisheries and Forestry (1981) would agree with our estimate for Canada in that they suggested a saving of 0.5 billion dollars if acid rain is stopped.

The effects of acid precipitation on our buildings, monuments, and statues are causing serious loss. Several examples come to mind such as the Parthenon, Roman structures, the Notre Dame Cathedral in France, St. Peter's Basilica in Italy, the Washington Monument in the United States, and the Parliament Buildings in Canada. While there are many other specific estimates from Scandanavia, the cost of structural deterioration in the United States alone is estimated to be some 2 billion dollars annually.

Radionuclides

Specific Issues of Concern

This topic encompasses a variety of concerns ranging from scientific issues to those of economic import and in this case especially, political problems.

Canada has very substantial reserves of suitable ores. Exploitation of those reserves poses serious decisions in science, engineering, economics, and politics. What may very well be crucial is that said exploitation, especially the safety of exploitation, may now become dependent upon foreign expertise due to a variety of economic factors.

Also, Canada operates several reactors and these reactors produce waste products that must be safely stored for very long periods of time. This storage cannot be avoided. It is as much a fact of life as the fact that we use leaded gasoline with a certain penalty and pesticides with yet another penalty. The most promising radioactive waste storage areas seem to involve the extremely stable geological formation called the Precambrian Shield. Much of this formation is either in Canada or so near to it (in the United States) as to be of no practical difference to Canadians. Thus, we must face not only the question of storage of our own wastes but also issues with regard to the problems of our prolific neighbour to the south, a neighbour who seems to be planning to store very near to our border. Much basic engineering research into safe storage is underway but these by-products of an energy-producing technology that cannot be ignored are very exotic indeed. Knowledge of their behaviour, should they enter into Canadian aquatic systems, is deficient (National Research Council of Canada 1983). Few resources are available for the necessary studies and political will to provide such data is simply not extant. What is readily apparent is the NIMBY syndrome (Not In My Back Yard). Unfortunately, it must be in someone's backyard, for the future use of nuclear power is essential given that fossil fuels are a finite resource. Fortunately, there are scenarios that would appear to make such disposal practices both safe even attractive under certain circumstances.

Radioactive materials, if handled intelligently, are no more of a threat than some other materials such as the pesticides moving in transcontinental transport today. Like the pesticides, it is only when some authority, corporation, or individual decides to cut corners that irresponsibility emreges and a threat is manifest. For example, it is not generally recognized that many of our current activities release radioactive materials. An obvious one would be testing of nuclear bombs, but one has to wonder if the average person has been made aware that the use of certain fertilizers, the use of certain cements in their basements, the burning of coal, oil, and natural gas, and certain mining efforts all release radionuclides to the aquatic environment and to their own home. Moreover, effects such as eutrophication and "acid rain" have more impact on releases of radionuclides to the environment than did COSMOS 954, the Russian nuclear-powered satellite that was gratuitously gifted to us.

Reason seldom seems to enter the debate on the use of nuclear energy, either from the public, industry, or the government. Ignorance of issues and genuine concerns can be seen on each side of the problem. We suggest that some reasoned discussion occur, and occur quickly, for our present stockpiles of wastes and our present nuclear policies must be rationalized. We must ensure that this problem does not become another "Love Canal," and in this regard all parties must realize that there are three sides to every story: your side, my side, and the truth.

Sources

Mining of radioactive ores began in Canada in 1932; indeed we supplied the minerals for early experiments in the United States (Johnson 1975; RECS 1978). The initial site ceased production some 30 yr later but to date the tailings are still of concern. High-grade uranium ores exist in every province of Canada save Prince Edward Island. These ore bodies either have been exploited since the 1940s, are being exploited, or will be exploited. Canada possesses about 25% of known world reserves (Runnalls 1972). Environmental problems associated with production are still an issue (RECS 1978; Moffett and Tellier 1977, 1978; Ruggles and Rowley 1978) and the economics are still as problematical as are those of oil reserves.

Extraction of uranium ores and milling procedures are designed to deliberately isolate certain isotopes. Radioactive materials that are of little value are lost to wastes: liquid, solid, gases, and dust. Like base metal processing sites the sites of such processes have levels of contaminants that are well above "background" (Moffett and Tellier 1978; Rugges and Rowley 1978). Also like certain base metal operations these produce sulphidic waste products and the resultant sulphyric acid simply adds to the mobility of the mineral wastes. Thus, the environmental problems associated with extraction and processing are not very different from those associated with the extraction and processing are not very different from those associated with the extraction and processing of nonradioactive materials. In this case, however, the problem combines chemical toxicity with radiochemical toxicity and thus the threat to man and his environment begins to approximate what we have already done with our waste chemical dumps in North America.

A third source of radiochemical wastes are our reactors which have been in operation since 1946 (Merritt and Patrick 1960). The statistics regarding power production and the associated wastes are far more readily available than those for fossil fuel power production (e.g. Grisak and Jackson 1978). Given that no decision has been reached with regard to the reasoned disposal of associated wastes, highly radioactive materials are now being stored next to the reactors. Politically this seems feasible but practically it is not unlike storing the gasoline for the snowblower next to your barbeque; sure you are not going to make a mistake but if you do it is going to be spectacular!

Finally, a potential source of radioisotopes is the fuel reprocessing industry which is an almost inevitable outgrowth of the nuclear power industry. The specific environmental problems seem to relate largely to such "transuranic" isotopes such as plutonium and americium (Hohenemser et al. 1977).

Major Impacts

While there are distinct environmental issues, there can be no doubt whatsoever that *the* major concern of the public is human health.

These elements and their isotopes are both chemically and radiochemically toxic. They must be extracted, manipulated, used, and stored with extreme caution. The hazards to human health have been thoroughly documented (National Research Council of Canada 1983, the references therein). There is no point in repeating this or other such compilations here.

However, there are less direct issues, issues that involve the aquatic environment. Such matters have been explored in detail by Millard and Stannard (1977) and by the National Research Council of Canada (1983). Concerns range from the distribution of materials within food chains to the susceptibility of various biota to radiochemical threats. The problem is even more complex than that outlined for metals because of the nature of the materials and because of the dual threat; chemical and radiochemical toxicity.

Moreover, nuclear power production results in several associated environmental problems. Waste heat from reactors has been a problem from the beginning and its discharge to aquatic environments has caused both damage and benefit. The production of spent ores from mining and their long-term storage is still an issue, despite such recent examples as the spill in Saskatchewan in 1984. Uranium mining produces sulphidic by-products which are currently of no commercial value and are therefore perceived to be wastes.

Storage of the inevitable wastes is undoubtedly a major impact. Efforts are being made to establish safe methodology (Boulton 1980; Cherry and Gale 1979) but there is a very strong public reaction to even the most technically sound proposals.

In some contrast with the other toxic chemicals with which we deal, there is a strong sociological impact in this case. It is so strong as to lead to decisions which are less than rational but which are a fact of life. Any professional dealing with the question of using radionuclides, even for such beneficial purposes as health care, must be fully prepared to deal with such honest concerns.

Distribution

The problems surrounding our knowledge or lack of knowledge of the mechanisms of distribution of radionuclides within our own environment and the aquatic environment are essentially the same as those developed for metals. However, to those arguments posed earlier one might also call the reader's attention to the threats posed by both stable and radioactive elements to our marine and especially our estuarine systems (National Research Council of Canada 1983).

In the specific case of storage, most plans call for the use of geological structures which are deemed to be very stable. In the sense of long-lived radionuclides, stability is a relative term, as groundwater is thought to move within such structures albeit very slowly. Currently, public pressures are making it very difficult to even explore such hypotheses.

Specific Regional Impacts

Impacts from mining can be expected in northern Saskatchewan, the western part of the Northwest Territories, parts of Labrador, and, perhaps, British Columbia. Saskatchewan and Ontario have current operations and have already experienced problems. Unless there is a drastic reduction in the market for such ores, a change that might occur in the short term but cannot last in the long term, we can expect future development in these areas and future problems associated with such developments.

With regard to reactors, at the present time we face problems sited mainly in Ontario, Quebec, and the Maritimes. We may also be experiencing some impact from the many reactors located in the United States but close to the border. Current Canadian reactors do release low levels of radioactive materials to the atmosphere and the hydrosphere. What is particularly worrying is the fact that we have yet to establish suitable waste repositories and that wastes are currently being stored on-site in the interim.

The impact of deliberate disposal or storage seems to focus now on the Precambrian Shield because of its geological stability. Moreover, there are specific geological structures within the Shield that are especially favoured. It would appear, then, that current consideration is confined to northwestern Ontario, Manitoba, Quebec, and the Territories.

The question of fuel reprocessing is even more problematical than that of spent fuel storage. Specific regional impacts cannot be reasonably discussed at this time.

General Comments on the Problem

In certain instances it is easy to make value judgements that exposures to chemicals should be limited; these choices usually involve a clear threat to animals (obvious "kills") or more frequently a threat to humans (obvious "spills").

In the longer term the concern is for the integrity of organisms and ecosystems. It is in this area that value judgements regarding what is or is not acceptable become problematical, and these values vary greatly within and among societies (Goodman 1974). Many different value statements are often expressed but two major ones can be used to illustrate the conflict: firstly the need to apply chemicals (fertilizers, pesticides) to ecosystems in order to maximize production of crops for human use, and secondly the need to conserve and observe naturally diverse ecosystems as reservoirs of genetic diversity and as indicators of ecological stress. An important value statement has been made with regard to the preservation of certain natural stocks of plants in critical areas (Wilkes 1977), and similar reasoning surely applies to aquatic resources (see also chapter by Vallentyne and Hamilton, this volume).

The direct effects of various concentrations of chemicals in waters on fish have been investigated frequently in the laboratory, especially at concentrations producing lethal results. Such results are customarily reported as LC50 values or that concentration required to kill one half the test animals in a specified period of time; they are commonly refered to as "sudden death" assays or, more properly, as "acute lethal" assays.

Such data vary widely depending upon the substance being assayed, the fish being tested, the water being used for test purposes, and several other factors. However, they are commonly taken to be an indication of relative toxicity at least and have been used to establish guidelines for certain industrial effluents. Thus, to a limited extent they are useful, particularly in establishing the upper limits of acceptability with regard to undeniably toxic elements. Such data are available in various publications regarding water quality criteria. There is little doubt that these criteria are frequently exceeded in certain political and industrial practises.

Scientists tend to avoid such simplistic assays nowadays in preference to more subtle tests for the impact of specific chemicals on aquatic life using a wide variety of laboratory exposures. However, the linkages between the effects on individuals or small groups in the laboratory and those seen in populations and communities in the field are still unclear and deserving of further work. Within the laboratory, exposure conditions can be manipulated and known accurately, but even there surprises occur. For example, rainbow trout are almost always much more sensitive to pesticides than channel catfish. In the case of the fungicide Benomyl, however, the channel catfish are about five times more sensitive to the pesticide itself than rainbow trout and about 20 times more sensitive to its metabolites (Johnson and Finley 1980). Attempts to reproduce natural expo-

374

sure conditions have led to a number of techniques for acute and chronic test procedures in efforts to find concentrations that should be safe for aquatic life. Kenaga (1982) has tabulated the differences between tests of acute and chronic exposures, and most chemicals (90%) produce chronic effects within a concentration range down to 125 times lower than that producing acute effects. However, with some materials the difference was much greater; for example the herbicide "Propanil" was reported to cause chronic effects at concentrations 18 000 times lower than concentrations producing acute effects. In trying to generalize from the laboratory to natural settings where there is always much less known about exposure and where there is little or no opportunity to control other variables, the opportunities for surprise become manifest.

A promising way to begin linking laboratory and field studies is through the various signs of stress or overt pathology that can be related to exposure in laboratory experiments. Such signs can then be sought in natural populations known or thought to have been exposed to the chemicals in question. Application of some of these signs and symptoms has been reviewed by Pickering (1981) and Cairns et al. (1984). Key unanswered questions concern the results of exposure to combinations of chemicals.

Unfortunately, even when a sublethal response is directly related to a certain compound or class of compounds, the interpretation of the response is not always straightforward. For example, Malathion inhibits the function of cholinesterase enzymes in fish (Coppage et al. 1975), and in that these enzymes are critical to nerve function, such inhibition should be "bad" for the affected animals. However, Lockhart et al. (1985) demonstrated a loss of over 70% of such activity in fish exposed to a mosquito spraying program, yet the population response was only a brief cessation of growth. Thus, much more work needs to be done to link the biochemical, cellular, behavioural, and physiological responses observed in the laboratory to those seen, *or not seen*, in the field, much the same situation as human toxicology in its infancy.

Hypotheses linking one response to another are inherently difficult to prove or disprove rigorously. For example, it is reasonable to suggest that exposure to chemical stress might cause increased incidence of fish diseases (Sindermann 1978). A striking coincidence of chemical pollution (fuel oil) and fish diseases has been reported in a population of bluegills in Lake Winona, Minnesota (Fremling 1981). Similarly, comparing the incidence of diseases in fish from polluted versus nonpolluted areas might be expected to show the anticipated result, as indeed they do (e.g. Brown et al. 1977). Unfortunately, "cause and effect" is not strictly demonstrated by a survey approach, and few convincing experimental studies have appeared.

Specific Comments

With regard to the acid rain problem, sensitivity of aquatic and terrestrial ecosystems is now defined using relatively simple chemical parameters, notably alkalinity, bicarbonate, calcium, conductance, and pH. Of these the apparently simple parameters of alkalinity and pH are particularly subject to error in measurement and interpretation. Due to the wide variety of monitoring networks now in place in North America and Europe, it is essential to agree on standard methods and to mount interlaboratory comparisons (check sample programs). Galloway et al. (1979) and Herczog et al. (1985) discussed some of the issues.

Since the very early days of analytical chemistry, chemists have pleaded with their counterparts who sent them samples for analysis that the objectives of the laboratory and the agency requesting the analyses must be seen as interdependent. Even today this is seldom the case and one could be discussing an analytical laboratory dedicated to environmental science or to a hospital. This is a common complaint of the chemist and we can do it no more justice here save to voice it once again: to those of you who want good analyses done on your samples, please consult your chemists before, not after, your collection! These and additional points were discussed by Keith et al. (1983).

The statistical design of sample collection is commonly ignored. In this day and age, analyses cost any project much of its resources. It is therefore difficult to understand why this aspect is so frequently ignored by the "purchaser" of the analytical information. The "more is better" or "we're here, why not collect it" approach must be abandoned, for a single datum is seldom useful even if it is recovered from the North Pole at significant expense. What is essential is time series data, for this is so crucial to both the biological and the chemical analysts who might be assisting that collector.

Radionuclides pose a very specific problem in that we are dealing with very small amounts of materials, indeed, which could be dispersed in very large amounts of water. Indeed, even the collection of that water (groundwater) in proposed disposal areas is a major problem. At present the detection issue seems capable of solution using existing (but very expensive) technology. The collection of meaningful water samples remains problematical.

With specific regard to trace metals, one should note that we live in a metal-contaminated environment. Consequently, the chemist is faced with the task of assessing such elements within a laboratory that must be regarded as suspect on first principals. The first comparable example that comes to mind is an operating theater, but in this case the question of contamination is much more serious and very much more invidious than microbial occurrences. While the lack of state-of-the-art instrumentation is a major problem for most analysts in the world, perhaps the key requirements are the financial support required to produce ultraclean reagents and especially to maintain clean facilities. To the lay reader this might seem to be an issue easily resolved; one can only respond by saying that an operating room is a *contaminated area* compared with an analytical facility for the determination of trace metals in our current environment! For example, the accuracy of most data with regard to lead in the environment has been seriously questioned (Patterson 1974) and in many instances has been shown to be essentially of little utility (Patterson and Settle 1976; Settle and Patterson 1980). The problem in each case was contamination, and this "case history" has made it clear that the techniques are sound but that the provision of what are now known as "ultraclean" facilities is simply an essential requirement. Needless to say, such requirements extend to the collection, storage, and transportation of samples for analysis, but that must be the subject of another article.

One must also question the very reliability of reported chemical and biological measurements. Of particular concern is the fact that many of these results are reported in nonrefereed fora (e.g. consulting reports, unedited computer data banks, unrefereed "journal," etc.). Therefore, a lack of confidence in their reliability is almost implicit, unfortunately.

Future Research or Management Efforts

Activity in this area falls into two broad categories: (1) documentation of past problems and, (2) efforts to predict and prevent future problems.

Documentary types of research have proven to be formidably complex in that the linkage between the laboratory and the field, between the cause and the effect, between the specific chemical and the specific symptom, or more often between the group of chemicals and the general malaise have been difficult to prove to the people that matter — the body politic and their representatives, the politicians. Moreover, obtaining such proofs is expensive in time, monies, and skilled manpower. Even when all necessary resources are marshalled (usually around some current "scare" problem) one seldom finds the integrated and interdisciplinary effort that is required to make really new advances. Our addiction to fighting of forest fires leads neither to clear documentation of the phenomenon nor to clear preventative soluations for future problems.

Preventive or predictive efforts which include laboratory investigations on toxicological mechanisms and a variety of efforts devoted to modelling or forcasting the fate of toxicants in the environment are required. In the many chemical examples one can cite, there has seldom been the opportunity to validate the predictions with actual environmental studies. Such opportunities are appearing, however, and the results of verification in one environment or with one animal can be extrapolated to many species and many environments. One type of study without the other (laboratory versus field) might be politically and economically acceptable in the short term but can be potentially dangerous in the long term and may well be irrational under most circumstances.

There are very compelling reasons, economic, political, and scientific, to define better the kinds of observations needed to monitor problem issues in order to link the two approaches such as to better define the degree to which laboratory results are actually predictive of what happens in the environment. This key issue impacts on our management of all chemical related issues, be they mining developments, approvals to utilize new pesticides, or the storage of nuclear wastes. A more detailed discussion of such issues is to be found in a recent National Research Council of Canada (1986) document.

Within these various approaches to some sort of more improved management of toxic chemicals in the environment a number of suggestions have been made by a variety of agencies/authorities. For example only:

(1) With regard to use of fossil fuels it has been suggested (National Research Council of Canada 1981) that three critical questions need to be addressed: (a) What are the quantitative relationships between release of materials to the atmosphere and impact on the aquatic environment? (b) What is the rate of acidification of the aquatic environment? (c) What is the total effect of acidic deposition on aquatic ecosystem? The authors of this key article went on to make very specific suggestions as to how these questions might be addressed, if not answered.

(2) Similarly, in the case of radionuclides, another study by the National Research Council of Canada (1983) suggested a set of questions: (a) Current data on both anthropogenic and natural sources of radionuclides in Canadian environments are both limited and of limited value. Adequate data in these regards on the materials from all sources must be obtained (power plants, mining, fertilizers, etc.) (b) All laboratories within Canada should be involved in an intercomparison program such that their data interrelate both nationally and internationally. (c) Canada is neglecting the impact it may be having upon marine environments via air and the freshwater environments. In fact, it depends largely upon other countries for such information. This stance must be changed. (d) There is a set of problems with regard to the movement, partition, and concentration of radionuclides that is essentially similar to that associated with metals. In this case, however, even less is known, particularly of circumstances uniquely Canadian. (e) There is little research on the long-term effects of chronic, low-level radiation on aquatic organisms as opposed to high-level radiation. Methods for determining such somatic and genetic changes are needed.

(3) The questions surrounding use and distribution of metals exhibit common, fundamental problems. For example again: (a) Metal concentration data in water, air, and animal tissues are available but their accuracy is now suspect due to improvements in analytical and collection techniques. These data must be verified and reviewed. (b) Information on tissue damage and metabolic disruptions are presumed from field data but are still controversial in many cases. Moreover, there are massive gaps in our understanding of the combined effects of the metals and organics introduced into the environment through our activities. (c) Metal loadings are known to fluctuate seasonally but the corresponding fluctuations in concentrations in tissues are largely unknown as are the consequences of such fluctuating loadings. (d) Again, the lack of basic analytical facilities would be paramount; these include the resources to produce clean reagents

required for analyses, to provide truly clean facilities, and to provide similarly rigorous collection and storage of samples from even the most remote areas of this country. Few agencies besides government can provide the wherewithall.

(4) The control and management of organic chemicals has remained a problematical issue in the minds of decision-makers since the early 1960s in Canada. It is only recently though that they have become more than slightly aware that a problem exists which is national in scope and regional in its specifics. (a) This issue has been raised within the science community and by senior levels of government as cited earlier in this chapter. It is also an international problem of major scope and consequences. To date, one sees a fair amount of well-intentioned science, some well-intended management, and a lot of dithering at the policy level (Auditor General 1983). Surely it is safe to recommend once more that either some action be evidenced or that a drastic rethinking of our whole approach is required. (b) The clean-up of old dumping grounds is something that cannot be avoided. While this is generally true, it is specific to water issues, for we have examples of the consequences of the predeliction to bury mistakes that range from the arsenicals associated with mining of gold to the truly horrendous problems that our industries have created for us in the area of the Great Lakes. Provoking such corrections and preventing future problems will require a thorough and very public documentation of the problems. One could be totally pragmatic and note that should we chose to sell water, we cannot sell a soiled product. One could also consider the Canadian need to use the resource and suggest that such a soiled resource is equally unacceptable. (c) In our discussion above we have mentioned the link between laboratory and field biological observations as being seen as especially costly. Modern advances may alter that case but here there is quite real major cost and that is the one of modern instrumentation. In reading either popular or science literature it is common to note references to very small amounts of chemicals which have been linked to effects upon humans or their environment. Such instrumentation is not readily available and is becoming so costly that some integration of effort between users is as crucial as that between the users of the big machines associated with modern physics. This is not a new observation; it it is an old observation that has never been pursued in detail. Even in the early 1900s it was being cited as a matter-of-substance with regard to the training of students.

Conclusions

Effects upon biological components within ecosystems are very difficult to address both scientifically and economically. Much of the uncertainty expressed earlier in this article relates to inabilities to detect *early* adverse effects on various ecosystems and upon man. Society cannot afford to continue to wait until entire populations of animals are destroyed in order to decide that there is a problem. While progress is being made (Cairns et al. 1984; Sheehan et al. 1984; National Research Council of Canada 1986), methods are urgently required for detecting the early symptoms of chemically induced stress that have credence with both the scientific and, more especially, the political community.

One major problem that contributes to uncertainty is that both the biologists and the chemists are faced with problems of detection. For example, an analytical chemist, faced with analyzing a sample from a human cadaver, could easily conclude that said sample was largely composed of an oxide of hydrogen but certainly was contaminated with trace organic and inorganic compounds. It is only the exercise of skills that would determine the origin of the object, its function, and the manner in which it functions from the analysis of those "contaminants." People faced with environmental issues confront very similar problems, yet they must consider a variety of target species and whole target systems.

Thus, one must conclude that three major problems emerge: (1) the skill and training provided to those determining the situations (= support to our universities), (2) the level of support provided to universities and to fundamental research, including the provision of state-of-the-art equipment required to support a case in the literature or in court (support by governments and business to academic and government science), and (3) a basic political will to deal with such problems in the long term rather than only those of immediate political or economic importance. Simply stated, we need an investment plan that considers the long term and that is, in fact, executed.

References

ALBERTA ENVIRONMENT. 1982. Symposium/workshop Proceedings: Acid forming emissions in Alberta and their ecological effects. Alberta Department of the Environment, Canadian Petroleum Association, Oil Sands Environmental Study Group; March 9-12, Edmonton, Alta. 648 p.

ANDREW, R. W. 1976. Toxicity relationships to copper forms in natural waters, p. 127-145. In R. W. Andrew, P. V. Hodson, and O. E. Konasewich [ed.] Toxicity to biota of metal forms in natural water. International Joint Commission, Ottawa, Ont.

ANONYMOUS. 1981. How many more lakes have to die? Canada Today 12(2): 11.

1984. Trends in atmospheric composition. CO-2/climate report. Fall 1984. Climate Program Office, Atmospheric Environment Service, Environment Canada, Downsview, Ont.

ANNUAL REPORT. 1986. Second annual report, Department of Marine and Fisheries, Ottawa, Ont. p. 330-335.

ARMSTRONG, F. A. J. 1979. Mercury in the aquatic environment, p. 84-100. In Effects of mercury in the Canadian environment. NRCC No. 16739, National Research Council of Canada, Ottawa, Ont.

ARMSTRONG, F. A. J., AND A. L. HAMILTON. 1973. Pathways of mercury in a polluted northwestern Ontario lake, p. 131-156. In P. C. Singer [ed.] Trace metals and metal-organic interactions in natural waters. Ann Arbor. Science Publishers Inc., Ann Arbor, MI.

AUDITOR GENERAL. 1983. Report of the Auditor General of Canada to the House of Commons. Canadian Government Publishing Center, Supply and Services Canada, Ottawa, Ont. 65 p.

AULERICH, R. J., R. K. RINGER, AND S. IWAMOTO. 1973. Reproductive failure and mortality in mink fed on great lakes fish. J. Reprod. Fert. Suppl. 19: 365-376.

BALL, J. W., E. A. JENNE, AND D. K. NORDSTROM. 1979. WATEQ2-A computerized chemical model for trace and major element speciation and mineral equilibria of natural waters, p. 815-835. In E. A. Jenne [ed.] Chemical modelling in aqueous systems. American Chemical Society, Washington, DC.

BERTINE, K. K., AND E. D. GOLDBERG. 1971. Fossil fuel combustion and the major sedimentary cycle. Science (Wash., DC.) 173: 233-235.

BLACK, J. R. 1983. Field and laboratory studies of environmental carcinogensis in Niagara River fish. J. Great Lakes Res. 9: 326-334.

BODALY, R. A., R. E. HECKY, AND R. J. P. FUDGE. 1984. Increases in fish mercury levels in lakes flooded by the Churchill River diversion, northern Manitoba. Can. J. Fish. Aquat. Sci. 41: 682-691.

BOULTON, J. 1980. Second Annual Report of the Canadian Nuclear Fuel Waste Management Program. Atomic Energy of Canada, Ltd. Publ. No. AECL-6804.

BROWN, E. R., T. SINCLAIR, L. KEITH, P. BEAMER, J. J. HAZDRA, V. NAIR, AND O. CALLAGHAN. 1977. Chemical pollution in relation to diseases in fish. Ann. NY Acad. Sci. 298: 535-546.

CAIRNS, V. W., P. V. HODSON, AND J. O. NRIAGU [ED.] 1984. Contaminant effects on fisheries. Advances in environmental science and technology. Vol. 16. John Wiley & Sons, New York, NY.

CAMPBELL, P. 1981. Horizons of fisheries habitat research. Department of Fisheries and Oceans, Freshwaters Institute, Winnipeg, Man. 131 p.

CHERRY, J. A., AND J. E. GALE. 1979. The Canadian program for a high-level radioactive waste repository: a hydrological perspective. Geol. Surv. Can. Pap. 79-10: 35-44.

CLARKE, R. McV. 1974. The effects of effluents from metal mines on aquatic ecosystems in Canada. A literature review. Can. Fish. Mar. Serv. Res. Dev. Tech. Rep. No. 488: 150 p.

COPPAGE, D. L., E. MATTHEWS, G. H. COOK, AND J. KNIGHT. 1975. Brain acetylcholinesterase inhibition in fish as a diagnosis of environmental poisoning by malathion. Pestic. Biochem. Physiol. 5: 536-542.

CROWTHER, R. A., B. F. BIETZ, M. E. LUOMA, B. M. SHELAST, M. J. STAITE, AND J. L. DAVIS. 1984. Lodgepole gas well blowout: sources, fate and biological effects of petroleum hydrocarbons in the freshwater environment. International Conference on Oil and Freshwater. Edmonton, Alberta. (Abstracts: 26-27)

CZUCZWA, J. W., B. D. McVEETY, AND R. A. HITES. 1984. Polychlorinated dibenzo-p-dioxins and dibenzofurans in sediments from Siskiwit Lake, Isle Royale. Science (Wash., DC) 226: 568-569.

DAISEY, J. M., R. J. McCAFFREY, AND R. A. GALLAGHER. 1981. Polycyclic aromatic hydrocarbons and total extractable particulate material in the Arctic aerosol. Atmos. Environ. 15: 1353-1363.

DeLONG, R. L., W. G. GILMARTIN, AND J. G. SIMPSON. 1973. Premature births in California sea lions: association with high organochlorine pollutant residue levels. Science (Wash., DC) 181: 1168-1170.

DEPARTMENT OF NATIONAL HEALTH AND WELFARE. 1969. Canadian drinking water standards and objectives, 1968. Department of National Health and Welfare, Ottawa, Ont. 39 p.

DEMAYO, A., M. C. TAYLOR, AND S. W. REEDER. 1980. Guidelines for surface water quality. Vol. 1. Inorganic chemical substances — lead. Water Quality Branch, Inland Waters Directorate, Ottawa, Ont. 45 p.

DILLON, P. J., N. D. YAN, AND H. H. HARVEY. 1983. Acidic deposition: effects on aquatic ecosystems. CRC Crit. Rev. Environ. Control 13: 167-194.

DRASCH, G. A. 1983. An increase of cadmium body burden for this century — an investigation of human tissues. Sci. Total Environ. 26: 111-119.

DUNCAN, D. A., AND J. F. KLAVERKAMP. 1983. Tolerance and resistance to cadmium in white suckers (Catostomus commersoni) previously exposed to cadmiun, mercury, zinc, or selenium. Can. J. Fish. Aquat. Sci. 40: 128-138.

EDGINGTON, D. N., AND J. A. ROBBINS. 1978. Records of lead deposition in Lake Michigan sediments since 1800. Environ. Sci. Technol. 10: 266-274.

EISENREICH, S. J., G. H. HOLLOD, AND T. C. JOHNSON. 1979. Accumulation of polychlorinated biphenyls (PCBs) in surficial Lake Superior sediments. Atmospheric deposition. Environ. Sci. Technol. 13: 569-573.

ENVIRONMENT CANADA. 1978. National Water Quality Data Bank — NAQUADAT. Water Quality Branch, Inland Waters Directorate, Environment Canada, Ottawa, Ont.

1981. Downwind: the acid rain story. Information Directorate, Environment Canada, Ottawa, Ont. 19 p.

FAGERSTROM, T., AND A. JERNELOV. 1971. Formation of mercury from pure mercuric sulfide in aerobic organic sediments. Water Res. 5: 121-122.

FEDERAL-PROVINCIAL WORKING GROUP ON DRINKING WATER OF THE FEDERAL-PROVINCIAL ADVISORY COMMITTEE ON OCCUPATIONAL HEALTH. 1979. Guidelines for Canadian drinking water quality, 1978. Canadian Government Publishing Center, Hull, Que. 77 p.

FLINN, J. E., AND R. S. REIMERS. 1974. Development of predictions of future pollution problems. U.S. Environmental Protection Agency. EPA Soc. Econ. Ser. Rep. EPA-600/5-74-005: 219 p.

FLORENCE, T. M., AND G. E. BATLEY. 1980. Chemical speciation in natural waters. CRC Crit. Rev. Anal. Chem. 9: 219-296.

FONTAINE, T. D. III. 1984. A non-equilibrium approach to modeling metal speciation in acid, aqueous systems: theory and process equations. Ecol. Model. 21: 287-313.

FORESTER, A. J. 1980. Monitoring the bioavailability of toxic metals in acid stressed shields lakes used pelecypod molluscs (clams, mussels), p. 142-147. In D. D. Hemhill [ed.] Proceedings of the University of Missouri Ann. Conf. on Trace Substances. (Trace Substances and Environmental Health Vol. 14).

FRANZIN, W. G. 1984. Aquatic contamination in the vicinity of the base metal smelter at Flin Flon, Manitoba, Canada — a case history, p. 523-550. In J. O. Nriagy [ed.] Environmental impact of smelters. Wiley Interscience, Toronto, Ont. (Adv. Environ. Sci. Technol. Vol. 15)

FRANZIN, W. G., AND G. A. McFARLANE. 1980. Fallout, distribution and some effects of Zn, Cd, Cu, Pb, and As in aquatic ecosystems near a base metal smelter on Canada's PreCambrian Shield, p. 302-303. In D. Drablos and A. Tollen [ed.] Ecological Impact of

Acid Precipitation, Proceeds of an International Conference, Sandefjord, Norway, March 11-14, 1980. SNSF Project, Oslo, Norway.

FREEDMAN, B. AND T. C. HUTCHINSON. 1980. Smelter pollution near Sudbury, Ontario, Canada and effects on forest litter decomposition, p. 395-434. In T. C. Hutchinson and M. Havas [ed.] Effects of acid precipitation on terrestrial ecosystem. Plenum Press, New York, NY.

FREMLING, C. R. 1981. Impacts of a spill of No. 6 fuel oil on Lake Winona. Proc. Oil Spill Conference. American Petroleum Institute Publ. No. 4334: 419-421.

FROMM, P. O. 1980. A review of some physiological and toxicological responses of freshwater fish to acid stress. Environ. Biol. Fishes 5: 79-93.

GALLOWAY, J. N., B. J. CROSBY, JR., AND G. E. LIKENS. 1979. Acid precipitation: measurement of pH and acidity. Limnol. Oceanogr. 24: 1161-1165.

GILES, M. A., AND J. F. KLAVERKAMP. 1982. The acute toxicity of vanadium and copper to eyed eggs of rainbow trout (Salmo gairneri). Water Res. 16: 885-889.

GILBERTSON, M. 1984. Need for development of epidemiology for chemically induced fish diseases in Canada. Can. J. Fish. Aquat. Sci. 41: 1534-1540.

GOLDBERG, E. D., V. F. HODGE, J. J. GRIFFIN, M. KOIDE, AND D. N. EDGINGTON. 1981. Impact of fossil fuel combustion on the sediments of Lake Michigan. Environ. Sci. Technol. 15: 466-471.

GOODMAN, G. T. 1974. How do chemical substances affect the environment? Proc. R. Soc. Lond. Ser. B 185: 127-148.

GRIBBEN, J. 1984a. Another hot year in the greenhouse. New Scientist, Mar. 15, 1984: 5.
          1984b. Meteorology blows hot and cold. New Scientist, Sept. 6, 1984: 17-20.

GRISAK, G. E., AND R. E. JACKSON. 1978. An appraisal of the hydrogeological processes involved in shallow subsurface radioactive waste management in Canadian terrain. Environment Canada, Inland Waters Directorate Sci. Ser. No. 84: vii + 194 p.

HAINES, T. A. 1981. Acidic precipitation and its consequences for aquatic ecosystems: a review. Trans. Am. Fish. Soc. 110: 669-707.

HAKANSON, L. 1980. The quantitative impact of pH, bioproduction, and Hg-contamination on the Hg-content of fish (pike). Environ. Pollut. (B) 1: 285-304.

HAMELINK, J. L., R. L. WAYBRANT, AND R. C. BALL. 1971. A proposal: exchange equilibria control the degree chlorinated hydrocarbons are biologically magnified in lentic environments. Trans. Am. Fish. Soc. 100: 207-214.

HAMILTON, R. D. 1976. Aquatic environmental quality: toxicology. J. Fish Res. Board Can. 33: 2671-2688.

HARRISON, W., M. A. WINNIK, P. K. KWONG, AND D. MACKAY. 1975. Crude oil spills: disappearance of aromatic and aliphatic components from small sea-surface slicks. Environ. Sci. Technol. 9: 231-234.

HERCZOG, A. L., W. S. BROECKER, R. F. ANDERSON, S. L. SCHIFF, AND D. W. SCHINDLER. 1985. A new method for monitoring temporal trends in the acidity of fresh waters. Nature (Lond.) 315: 133-135.

HOHENEMSER, C. R. KASPERSON, AND R. KATES. 1977. The distrust of nuclear power. Science (Wash., DC) 196: 25-34.

HOLDEN, A. V. 1973. Effects of pesticides on fish, p. 213-253. In C. A. Edwards [ed.] Environmental pollution by pesticides. Plenum Press, New York, NY.

HUNT, E. G., AND A. I. BISCHOFF. 1960. Inimical effects on wildlife of periodic DDD applications to Clear Lake. Calif. Fish Game 46: 91-106.

INTERNATIONAL JOINT COMMISSION. 1977. New and revised Great Lakes water quality objectives, Vol. II: An IJC Report to the Governments of the United States and Canada, October 1977. 155 p.

JOHNSON, L. J. 1975. Great Bear Lake: its place in history. Arctic 28(4): 230-244.

JOHNSON, R. E. [ED.] 1982. Acid rain/fisheries. Proceedings of an International Symposium on Acidic Precipitation and Fishery Impacts in northeastern North America. Cornell University, Ithaca, NY, August 2-5. American Fisheries Society, Bethesda, MD. 357 p.

JOHNSON, W. W., AND M. T. FINLEY. 1980. Handbook of acute toxicity of chemicals to fish and aquatic invertebrates. U.S. Department of the Interior, Fish and Wildlife Service, Resource Publ. 137: 98 p.

KEITH, L. H., W. CRUMMETT, J. DEEGAN JR., R. A. LIBBY, J. K. TAYLOR, AND G. WENTLER. 1983. Principles of environmental analysis. Anal. Chem. 55: 2210-2218.

KEITH, L. H., AND W. A. TELLIARD. 1979. Priority pollutants I — a perspective view. Environ. Sci. Technol. 13: 416-423.

KEMP, A. L. M., AND R. L. THOMAS. 1976. Impact of man's activities on the chemical composition in the sediments of Lakes Ontario, Erie and Huron. Water Air Soil Pollut. 5: 469-490.

KENAGA, E. E. 1982. Predictability of chronic toxicity from acute toxicity of chemicals in fish and aquatic invertebrates. Environ. Toxicol. Chem. 1: 347-358.

KENAGA, E. E., AND C. A. I. GORING. 1980. Relationship between water solubility, soil sorption, octanol-water partioning and concentration of chemicals in biota, p. 78-115. In J. G. Eaton, P. R. Parrish, and A. C. Hendricks [ed.] Aquatic toxicology. American Society for Testing and Materials (ASTM) Publ. STP 707.

KERSWILL, C. J. 1967. Studies on the effects of forest sprayings with insecticides, 1952-63, on fish and aquatic invertebrates in New Brunswick streams: introduction and summary. J. Fish. Res. Board Can. 24: 701-708.

KHAN, M. A. Q., F. KORTE, AND J. F. PAYNE. 1979. Metabolism of pesticides by aquatic animals, p. 191-220. In M. A. Q. Khan [ed.] Pesticides in aquatic environments. Plenum Press, New York, NY.

KJELLSTROM, T. 1979. Exposure and accumulation of cadmium in people from Japan, U.S.A. and Sweden. Report of a three-year co-operative research project. Environ. Health Perspect. 28: 169-198.

KJELLSTROM, T., B. LIND, L. LINNMAN, AND C. G. ELINDER. 1975. Variation of cadmium concentration in wheat and barely in Sweden. Arch. Environ. Health 30: 321-328.

KLAVERKAMP, J. F., W. A. MACDONALD, L. J. WESSON, AND A. LUTZ. 1985. Metallothionein and resistance to cadmium toxicity in white suckers (*Catostomus commersoni*) impacted by atmospheric emissions from a base-metal smelter (abstract and resume), p. 163-164. In P. G. Wells and R. F. Addison [ed.] Proceedings of the Tenth Annual Aquatic Toxicity Workshop, November 7-10, 1983, Halifax, Nova Scotia. Can. Tech. Rep. Fish. Aquat. Sci. 1368: 475 p.

KLAVERKAMP, J. F., M. A. TURNER, S. E. HARRISON, AND R. H. HESSLEIN. 1983. Fates of metal radiotracers added to a whole lake: accumulation in slimy sculpin (*Cottus cocnatus*) and white sucker (*Catostomus commersoni*). Sci. Total Environ. 28: 119-128.

KRAHN, M. K., L. D. RHODES, M. S. MYERS, L. K. MOORE, W. D. MACLOED JR., AND D. C. MALINS. 1986. Associations between metabolites of aromatic compounds in bile and the occurence of hepatic lesions in english sole (*Parophrys vetulus*) from Puget Sound, Washington. Arch. Environ. Contam. Toxicol. 15: 61-67.

LANDHEER, F., H. DIBBS, AND J. LAUDA. 1982. Trace elements in Canadian coals. Can. Environ. Protect. Ser. Rep. EPS 3-AP-82-6: 41 p.

LANTZY, R. J., AND F. T. MACKENZIE. 1979. Atmospheric trace metals: global cycles and assessment of man's impact. Geochim. Cosmochim. Acta 43: 511-525.

LECH, J. J., M. J. VODICNIK, AND C. R. ELCOMBE. 1982. Induction of monooxygenase activity in fish, p. 107-148. In L. J. Weber [ed.] Aquatic toxicology. Raven Press, New York, NY.

LOCKHART, W. L., D. A. METNER, F. J. WARD, AND G. M. SWANXON. 1985. Population and cholinesterase responses in fish exposed to malathion sprays. Pestic. Biochem. Physiol. 24: 12-18.

LUOMA, S. N. 1977. The dynamics of biologically available mercury in a small estuary. Estuarine Coastal Mar. Sci. 5: 643-652.

1983. Bioavailability of trace metals to aquatic organisms — a review. Sci. Total Environ. 28: 1-22.

MACKAY, D., AND A. W. WOLKOFF. 1973. Rate of evaporation of low-solubility contaminants from water bodies to atmosphere. Environ. Sci. Technol. 7: 611-614.

MAGNUSON, V. R., D. K. HARRIS, M. S. SUN, D. K. TAYLOR, AND G. E. GLASS. 1979. Relationships of activities of metal-ligand species to aquatic toxicity, p. 635-657. In E. A. Jenne [ed.] Chemical modelling in aqueous systems. American Chemical Society, Washington, DC. (American Chemical Society Symposium Series, Vol. 43)

MARSHALL, J. S., D. L. MELLINGER, AND J. I. PARKER. 1981. Combined effects of cadmium and zinc on a Lake Michigan zooplankton community. J. Great Lakes Res. 7: 215-223.

McDUFF, R. E., AND F. M. MOREL. 1973. Description and use of the chemical equilibrium program REDEQL2. Keck Lab. Tech. Rep. EQ-73-02. California Institute of Technology, Pasadena, CA. 75 p.

MᴄFᴀʀʟᴀɴᴇ, G. A., ᴀɴᴅ W. G. Fʀᴀɴᴢɪɴ. 1980. An examination of Cd, Cu, and Hg concentrations in livers of northern pike, *Esox lucius*, and white sucker, *Catostomus commersoni*, from five lakes near a base metal smelter at Flin Flon, Manitoba. Can. J. Fish. Aquat. Sci. 37: 1573–1578.

Mᴇʀʟɪɴɪ, M., G. Pozzɪ, A. Bʀᴀᴢᴢᴇʟʟɪ, ᴀɴᴅ A. Bᴇʀɢ. 1976. The transfer of Zn-65 from natural and synthetic foods to a freshwater fish. *In* E. D. Cushing [ed.] Radioecology and energy resources. Ecological Society of America, Stroudsburg, PA. (Ecol. Soc. Am. Publ. 1)

Mᴇʀʀɪᴛᴛ, W. F., ᴀɴᴅ P. Pᴀᴛʀɪᴄᴋ. 1960. Radionuclides present in cooling water from the NRX reactor. Atomic Energy of Canada, Ltd. Publ. AECL-1177.

Mɪʟʟᴀʀᴅ, M. W., ᴀɴᴅ J. N. Sᴛᴀɴɴᴀʀᴅ. 1977. Environmental toxicity of aquatic radionuclides: models and mechanisms. Ann Arbour Science Publishers Inc., Ann Arbour, MI. 333 p.

Moffett, D., ᴀɴᴅ M. Tᴇʟʟɪᴇʀ. 1977. Uptakes of radioisotopes by vegetation growing on uranium mine tailings. Can. J. Soil Sci. 57: 417–424.

    1978. Radiological investigations of an abandoned uranium mine tailings area. J. Environ. Qual. 3: 310–314.

Mᴏʟɪᴛᴏʀ, A. M. M., ᴀɴᴅ G. Vᴀɴʜᴏᴏʀᴇɴ. 1984. L'analyse hydrobiologique cours d'eau au Benelux. Sci. Total Environ. 32: 167–181.

Mᴜʀᴄʜᴇʟᴀɴᴏ, R. A., ᴀɴᴅ R. E. Wᴏʟᴋᴇ. 1985. Epizootic carcinoma in the winter flounder, *Pseudopleuronectes americanus*. Science (Wash., DC) 228: 587–589.

Mᴜʀᴘʜʏ, T. J. 1984. Atmospheric inputs of chlorinated hydrocarbons to the Great Lakes. Adv. Environ. Sci. Technol. 14: 53–80.

Nᴀᴛɪᴏɴᴀʟ Aᴄᴀᴅᴇᴍʏ ᴏғ Sᴄɪᴇɴᴄᴇ. 1981. Atmosphere-biosphere interactions: toward a better understanding of the ecological consequences of fossil fuel combustion. National Academy Press, Washington, DC. 263 p.

Nᴀᴛɪᴏɴᴀʟ Rᴇsᴇᴀʀᴄʜ Cᴏᴜɴᴄɪʟ ᴏғ Cᴀɴᴀᴅᴀ. 1981. Acidification in the Canadian aquatic environment: scientific criteria for assessing the effects of acidic deposition on aquatic ecosystems. NRCC Publ. 18475, National Research Council of Canada, Ottawa, Ont. 369 p.

    1983. Radioactivity in the Canadian aquatic environment. NRCC Publ. 19250, National Research Council of Canada, Ottawa, Ont. 292 p.

    1986. The role of biochemical indicators in the assessment of ecosystem health: their development and validation. NRCC Publ. 24371, National Research Council of Canada, Ottawa, Ont. 119 p.

Nɢ, A., ᴀɴᴅ C. Pᴀᴛᴛᴇʀsᴏɴ. 1981. Natural concentrations of lead in ancient arctic and antarctic ice. Geochim. Casmochim. Acta 45: 2109–2121.

Nɪᴍᴍᴏ, D. R. 1979. Pesticides: their impact on the estuarine environment, p. 259–270. *In* W. B. Vernberg, F. P. Thurberg, A. Calabrese, and F. J. Vernberg [ed.] Marine pollution: functional responses. Academic Press, New York, NY.

Nᴏʀᴅsᴛʀᴏᴍ, D. K., ᴇᴛ ᴀʟ. 1979. Comparison of computerized chemical models for equilibrium calculations in aqueous systems, p. 857–892. *In* E. A. Jenn [ed.] Chemical modelling in auqeous systems. Am. Chem. Soc. Symp. Ser. 93. American Chemical Society, Washington, DC.

Nᴏʀsᴛʀᴏᴍ, R. J., A. E. Mᴄᴋɪɴɴᴏɴ, ᴀɴᴅ A. S. W. DᴇFʀᴇɪᴛᴀs. 1977. A bioenergetics based model for pollution accumulation by fish. J. Fish. Res. Board Can. 33: 248–267.

Nᴏʀsᴛʀᴏᴍ, R. J., D. C. G. Mᴜɪʀ, ᴀɴᴅ R. E. Sᴄʜᴡᴇɪɴsʙᴜʀɢ. 1985. Long-range transport and accumulation of organochlorine pollutants in arctic marine mammals. Presented at International Conference on Arctic Water Pollution Research: Applications of Science and Technology, Yellowknife, N.W.T.

Nʀɪᴀɢᴜ, J. O. 1979. Global inventory of natural and anthropogenic emissions of trace metals to the atmosphere. Nature (Lond.) 279: 409–411.

    [ED.] 1984. Environmental impact of smelters. Wiley Interscience, Toronto, Ont. 608 p. (Adv. Environ. Sci. Technol. 15)

Nʀɪᴀɢᴜ, J. O., ᴀɴᴅ M. S. Sɪᴍᴍᴏɴs. 1984. Toxic contaminants in the Great Lakes, p. 483–491. Appendix I. Wiley Interscience, New York, NY. (Adv. Environ. Sci. Technol. 14)

Oᴅᴜᴍ, E. P. 1984. The mesocosm. Bioscience 34: 558–562.

Oɴᴛᴀʀɪᴏ Mɪɴɪsᴛʀʏ ᴏғ ᴛʜᴇ Eɴᴠɪʀᴏɴᴍᴇɴᴛ. 1979. Polynuclear aromatic hydrocarbons — a background report including available Ontario data. ARB-TDA Rep. No. 58-79, Ontario Ministry of the Environment, Toronto, Ont. 193 p.

    1980. The case against acid rain. A report of acidic precipitation and Ontario programs for remedial action. Ontario Ministry of the Environment, Toronto, Ont. 24 p.

PATTERSON, C. 1974. Lead in seawater. Science (Wash., DC) 183: 553-554.

PATTERSON, C., AND D. SETTLE. 1976. The reduction of orders of magnitude errors in lead analyses of biological materials and natural waters by evaluating and controlling the extent and sources of industrial lead contamination introduced during sample collecting, handling and analysis, p. 321-351. In P. D. Lafleur [ed.] Accuracy in trace analysis sampling, sample handling and analysis. Proceedings of Seventh Materials Research Symposium, U.S. National Bureau of Standards, Gaithersburg, MD. U.S. Government Printing Office, Washington, DC. (NBS Spec. Publ. 422)

PAYNE, J. F. 1984. Mixed function oxygenase in biological monitoring programs: review of potential use in different phyla of aquatic animals, p. 625-655. In G. Personne, E. Jaspers, and C. Claus [ed.] Ecotoxicological testing for the marine environment. Vol. I. State University of Ghent and Institute of Marine Scientific Research, Bredene, Belgium.

PENTREATH, R. J. 1976. The accumulation of mercury from food by the plaice, Pleuronectes platessa. J. Exp. Mar. Biol. Ecol. 25: 51-61.

PETERLE, T. J. 1971. DDT in antarctic snow. Nature (Lond.) 224: 620.

PETERSON, H. G., F. P. HEALEY, AND R. WAGEMANN. 1984. Metal toxicity to algae: a highly pH dependent phenomenon. Can. J. Fish. Aquat. Sci. 41: 974-979.

PICKERING, A. D. 1981. Stress and fish. Academic Press, Toronto, Ont. 367 p.

RAHN, K. A. 1981. Atmospheric, riverine and oceanic sources of seven tarce constituents to the Arctic Ocean. Atmos. Environ. 15: 1507-1516.

RAHN, K. A., AND D. H. LOWENTHAL. 1985. Pollution aerosol in the northeast: northeastern-midwestern contributions. Science (Wash., DC) 228: 275-284.

RAPPAPORT, D. J., H. A. REGIER, AND T. C. HUTCHINSON. 1985. Ecosystem behaviour under stress. Am. Nat. 125: 617-640.

RECS. 1978. Monitoring program design recommendations for uranium mining localities. Environ. Can. Environ. Protect. Serv. Rep. EPS 5-NW-78-10.

ROESIJADI, G., AND J. W. ANDERSON. 1979. Condition index and free amino acid content of Macoma inquinata exposed to oil-contaminated marine sediments, p. 69-83. In W. B. Vernberg, F. P. Thurnberg, A. Calabrese, and F. J. Vernberg [ed.] Marine pollution; functional responses. Academic Press, New York, NY.

ROYAL SOCIETY OF CANADA. 1984. Long-range transport of airborne pollutants in North America. A peer review of Canadian Federal research. Atmospheric Environment Service, Environment Canada, Downsview, Ont. 115 p. + four appendices.

RUDD, J. W. M., M. A. TURNER, A. FURUTANI, A. L. SWICK, AND B. E. TOWNSEND. 1983. The English-Wabigoon River system: I. A synthesis of recent research with a view toward mercury amelioration. Can. J. Fish. Aquat. Sci. 40: 2206-2217.

RUGGLES, R. G., AND W. J. ROWLEY. 1978. A study of water pollution in the vicinity of the Eldorado Nuclear LTD Beaverlodge Operation 1976 and 1977. Environ. Can. Environ. Protect. Serv. Rep. EPS 5-NW-78-10.

RUNNALLS, O. J. C. 1972. Uranium...the future supply and demand. Geos. Fall 1972. Energy, Mines and Resources Canada. Ottawa, Ont.

SCHAULE, B. K., AND C. C. PATTERSON. 1981. Lead concentrations in the northeast Pacific: evidence for global anthropogenic perturbations. Earth Planet. Sci. Lett. 54: 97-116.

SCHINDLER, D. W., AND M. A. TURNER. 1982. Biological, chemical and physical responses of lake to experimental acidification. Water Air Soil Pollut. 18: 259-271.

SETTLE, D., AND C. C. PATTERSON. 1980. Lead in albacore: guide to lead pollution in Americans. Science (Wash., DC) 207: 1167-1176.

SHEEHAN, P. J., D. R. MILLER, G. C. BUTLER, AND P. BOURDEAU [ED.] 1984. Effects of pollutants at the ecosystem level. John Wiley & Sons, Toronto, Ont. 443 p. (SCOPE 22)

SHERBIN, I. G. 1979. Mercury in the Canadian environment. Environ. Can. Environ. Protect. Serv. Rep. EPS 3-EC-79-6.

SINDERMANN, C. J. 1978. Pollution-associated diseases and abnormalities of fish and shellfish: a review. Fish. Bull. 76: 717-749.

SIVARAJAH, K., C. S. FRANKLIN, AND W. P. WILLIAMS. 1978. The effects of polychlorinated biphenyls on plasma steroid levels and hepatic microsomal enzymes in fish. J. Fish Biol. 13: 401-409.

SLOAN, R., AND C. L. SCHOFIELD. 1983. Mercury levels in brook trout (Salvelinus fontinalis) from selected acid and limed Adirondack lakes. Northeast. Environ. Sci. 2: 165-170.

SOMERS, K. M., AND H. H. HARVEY. 1984. Alteration of fish communities in lakes stressed by acid near Wawa, Ontario. Can. J. Fish. Aquat. Sci. 41: 20-29.

SPRAGUE, J. B., P. F. ELSON, AND R. L. SANDERS. 1963. Sublethal copper and zinc pollution in a salmon river, a field and laboratory study. Adv. Water Pollut. Res. 1: 61-82.

SPRY, D. J., C. M. WOOD, AND P. V. HODSON. 1981. The effects of environmental acid of freshwater fish with particular reference to the softwater lakes in Ontario and in the modifying effects of heavy metals. A literature review. Can. Tech. Rep. Fish. Aquat. Sci. 999: 145 p.

STALLING, D. M., L. M. SMITH, J. D. PATTY, J. W. HOGAN, J. L. JOHNSON, C. RAPPE, AND H. R. BUSER. 1983. Residues of polychlorinated di-benzo-p-dioxins and dibenzofurans in Laurentian Great Lakes fish, p. 221-240. In R. E. Tucher, A. L. Young, and A. P. Gray [ed.] Human and environmental risks from chlorinated dioxins and related compounds. Plenum Press, New York, NY.

STANDING COMMITTEE ON FISHERIES AND FORESTRY. 1981. Still waters: the chilling reality of acid rain. A report by the Sub-committee on Acid Rain, House of Commons, Supply and Services Canada, Ottawa, Ont. 150 p.

STEGEMAN, J. J., P. J. KLOEPPER-SAMS, AND J. W. FARRINGTON. 1986. Monooxygenase induction and chlorobiphenyls in the deep sea fish *Coryphaenoides armatus*. Science (Wash., DC) 231: 1287-1289.

STICH, H. F., AND B. P. DUNN. 1980. The carcinogenic load of the environment: benzo(a)pyrene in sediments of arctic waters. Arctic 33: 807-814.

SUMMERS, P. W., AND D. M. WHELPDALE. 1976. Acid precipitation in Canada. Water Air Soil Pollut. 16: 447-455.

SWAIN, W. R. 1978. Chlorinated organic residues in fish, water and precipitation from the vicinity of Isle Royale, Lake Superior. J. Great Lakes Res. 4: 398-407.

SWANSON, D. L. 1976. Atmospheric input of pollutants to the upper Great Lakes, p. 197-206. In D. O. Hemphill [ed.] Proc. Univ. Missouri Annual Conf. on Trace Substances in the Environment. Environ. Health X.

TEASELY, J. I. [SER. ED.] 1984. Acid precipitation series, 9 volumes. Butterworth Publishers, Toronto, Ont.

TIWARI, J. L., AND J. E. HOBBIE. 1976. Random differential equations as models of ecosystems: Monte Carlo simulation approach. Math. Biosci. 28: 25-44.

TORIBA, T. Y., M. W. MILLER, AND P. E. MORROW [ED.] 1980. Polluted rain. Proceedings of the 12th Rochester International Conference on Environmental Toxicity, Rochester, NY, May 21-23, 1979. Plenum Press, New York, NY. 502 p.

U.S. DEPARTMENT OF HEALTH, EDUCATION AND WELFARE, PUBLIC HEALTH SERVICE. 1962. Public Health Service drinking water standards, rev. 1962. U.S. Public Health Service Publ. 956. Government Printing Office, Washington, DC. 61 p.

VIANNA, N. J., AND A. K. POLAN. 1984. Incidence of low birth weight among Love Canal residents. Science (Wash., DC) 226: 1217-1219.

WAGEMANN, R., AND D. C. G. MUIR. 1984. Concentrations of heavy metals and organochlorines in marine mammals or northern waters: overview and evaluation. Can. Tech. Rep. Fish. Aquat. Sci. 1279: v + 97 p.

WAGEMANN, R., N. B. SNOW, D. M. ROSENBERG, AND A. LUTZ. 1978. Arsenic in sediments, water and aquatic biota from lakes in the vicinity of Yellowknife, Northwest Territories, Canada. Arch. Environ. contam. Toxicol. 7: 169-191.

WARWICK, W. F. 1980a. Paleolimnology of the Bay of Quinte, Lake Ontario: 2800 years of cultural influence. Can. Bull. Fish. Aquat. Sci. 206: 117 p.

1980b. Pasqua Lake, Southeastern Saskatchewan: a preliminary assessment of trophic status and contamination based on the Chironomidae (Diptera), p. 255-267. In D. A. Murray [ed.]CHIRONOMIDAE, ecology, systematics, cytology and physiology. Pergamon Press, Oxford.

WHELPDALE, D. M. 1978. Atmospheric pathways of sulfur compounds. Chelsea College, University of London, England. 39 p. (MARC Rep. No. 7)

WILBER, W. G., AND J. V. HUNTER. 1977. Aquatic transport of heavy metals int he urban environment. Water Resour. Bull. 13: 721-734.

WILKES, G. 1977. The world's crop plant germplasm — an endangered resource. Bull. Atom. Sci. 33: 8-16.

WILLIS, J. N., AND W. G. SUNDA. 1984. Relative contributions of food and water in the accumulation of zinc by two species of marine fish. Mar. Biol. 80: 273-279.

WOOD, J. M. 1974 Biological cycles for toxic elements in the environment. Science (Wash. DC)
183: 1049-1052.
WORLD HEALTH ORGANIZATION. 1963. International standards for drinking water. 2nd ed. World
Health Organization, Geneva, Switzerland.

CHAPTER 14

Groundwater Occurrence and Contamination in Canada

John A. Cherry

Institute for Groundwater Research, University of Waterloo, Waterloo, Ont. N2L 3G1

Introduction

Groundwater exists everywhere beneath the land surface. In southern Canada, the top of the groundwater regime, referred to as the water table, generally lies within 20 m of the surface. In mountainous regions it can be much deeper. Above the water table the ground commonly contains water, but the soil pores or fissures are less than totally saturated. This shallow, partially saturated domain is the vadose zone. Nearly all of the water in the vadose zone and in the groundwater zone is continually in motion, travelling from one component of the hydrologic cycle to another.

Geologic formations that are permeable enough for water well yields to be useful are known as aquifers. Water-containing formations that are less permeable are known as aquitards (Freeze and Cherry 1979). Shallow aquifers, generally less than 50-100 m in depth, generally contain freshwater (< 1500 mg total dissolved solids/L). Deeper aquifers generally contain water with a higher content of natural dissolved constituents that makes them unfit for consumption by humans or livestock but sometimes suitable for industrial use. At depths greater than about 500 m, groundwater in Canada is generally as saline or more saline than seawater. The volume of freshwater in aquifers in Canada is unknown but it is certainly much larger than the total volume of water in the Great Lakes.

Although the volume of the fresh groundwater resource in Canada is huge, it is not distributed uniformly across the country. Most major cities are not located on or close to large aquifers, but many towns, villages, and farms are situated on aquifers that are large enough to supply all of their water needs. Few major cities in Canada use groundwater because of the abundance of surface water. Cities can afford to draw water a considerable distance by pipeline from rivers or lakes.

The use of groundwater in Canada is increasing. Surveys of groundwater use indicate that Canadian reliance on groundwater has grown from about 10% of the population in the 1960's to about 26% in 1981 when the most recent estimate was made (Hess 1986). Nearly 40% of the municipalities in Canada, which comprise about 2.2 million people, rely on groundwater for a significant portion of their water supply. In some regions more than half of the population uses groundwater in the home. About two thirds of Canada's rural population depends on groundwater because well supplies are generally more reliable and are generally less expensive than water obtained by pipeline from ponds or small streams. Also, aquifers are generally less contaminated than most streams or lakes. Groundwater provides nearly all of the water used for the production of livestock in Canada (Hess 1986). Usage of groundwater varies regionally. In Ontario and Quebec, municipal use is greatest whereas in the Prairies, agricultural use is largest. Industrial use is predominant in British Columbia. Groundwater is exceptionally important in Prince Edward Island where more than 99% of the population uses groundwater.

Groundwater contamination can render well water unsuitable for use. Furthermore, contaminants may migrate along subsurface paths, many of which lead to streams, marshes, or lakes. The extent of contamination of Canada's groundwater resource is unknown. Predictions of future contamination have not been made on national or even provincial scales. However, many causes of groundwater contamination are known from hundreds of site-specific investigations. Although our knowledge of the behavior of contaminants in the groundwater zone has some large gaps, this knowledge is nevertheless considerable.

Where groundwater is available from an aquifer, the cost of extracting the water is generally low. If the aquifer becomes contaminated, the cost of acquiring a new water supply is commonly large. First recognition of groundwater contamination normally occurs after water-well users have been exposed to a potential health risk. Once contamination is recognized, cleaning the aquifer sufficient for use to recommence is often not feasible. Little groundwater protection is achieved in Canada with existing government policies. In this chapter, groundwater contamination problems existing at present and expected to increase in the future are explored to form a basis for consideration of strategies for research, planning, and regulation that places emphasis on protection of the groundwater resource.

Groundwater Occurence

Nearly all groundwater used in Canada is extracted from wells in sand or gravel deposits or from wells in fissured bedrock. Several types of aquifers are illustrated in Fig. 1. Most sand and gravel deposits were laid down in rivers or lakes formed from melting Pleistocene glaciers. Unconfined aquifers do not have cover deposits of less permeable silty or clayey materials. The water table in an unconfined aquifer is the

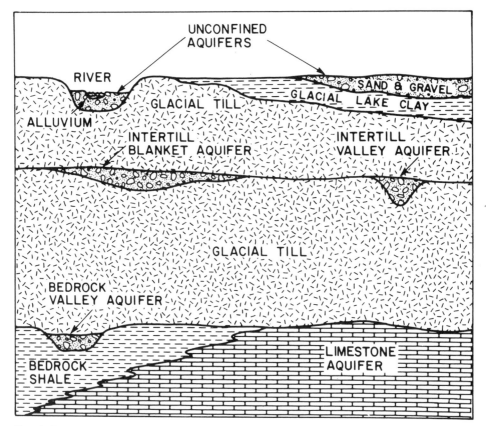

FIG. 1. Typical types of aquifers in Canada (adapted from Saskatchewan — Nelson Basin Board 1973).

top of the saturated zone. The water table declines when the aquifer is pumped. The aquifer is recharged primarily by water that infiltrates directly from the ground surface. Unconfined aquifers are therefore most prone to contamination, particularly where the water table is within 5 or 10 m of ground surface, which is common. Unconfined aquifers commonly exist in river valleys or as blankets of sand or gravel on flat or gently rolling plains.

Confined aquifers in glaciated terrain are formed of similar material but these materials are covered by silty or clayey deposits. These aquifers are saturated from top to bottom and the water table occurs in the overlying silty or clayey material. These aquifers are

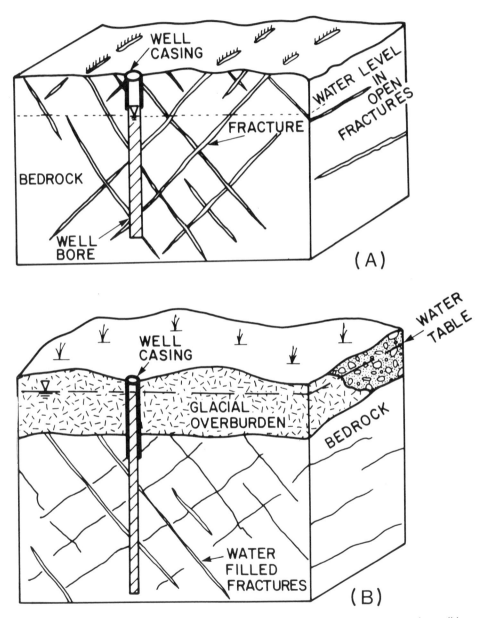

FIG. 2. Water wells in fractured bedrock showing (A) how fractures can intersect the well bore below the well casing and act as conduits to carry contaminants from the surface into the well and (B) how glacial deposits can provide a cover on the fissured bedrock.

less prone to contamination because the less permeable silty or clayey material provides protection against infiltration of overlying contaminated water. This protection may be short term, for only a few years, or it may last for centuries or longer, depending on the thickness and on the degree of weathering of the silty or clayey deposits.

Unconfined and confined aquifers are also present in bedrock areas, as illustrated in Fig. 2. Wells in bedrock derive water almost exclusively from fissures, except in some types of sandstone where the pore space between the sand particles as well as the fissures yield water. If the fissured bedrock is exposed at ground surface (Fig. 2A) or if it is convered only by thin permeable deposits, such as sand or gravel, easy entry of many types of contaminants may occur. Some bedrock aquifers are deeply buried beneath thick layers of glacial deposits (Fig. 1). The nature of aquifers and groundwater conditions across Canada are described in the volume edited by Brown (1967) and in a volume on the Hydrogeology of North America to be issued by the Geological Society of America in 1987. More detailed descriptions of groundwater conditions in many areas are provided in reports and maps published by provincial government agencies. The most comprehensive coverages are for the Prairie provinces, where systematic mapping of groundwater occurrence has been undertaken for more than two decades by the Saskatchewan and Alberta Research Councils and by the Manitoba Department of Natural Resources.

Sand and gravel aquifers vary greatly in size and shape. If located several metres or more below the water table, a sand or gravel deposit only a metre or two thick and only a few hectares or tens of hectares in area is commonly large enough to provide a sustainable yield of several litres per minute or more. This is an adequate supply for a household or for small livestock operations. The sand or gravel deposits act as a storage reservoir for water that is recharged by infiltration of rain and snowmelt. As water is pumped from the aquifer, it is normally replenished at a more rapid rate than under natural conditions because more of the surface water is able to recharge the aquifer. In addition to these small aquifers that are ubiquitous in many regions, there are also many thousands of larger sand or gravel aquifers in Canada that are individually capable of supplying water at rates in the order of tens to hundreds of litres per minute. The small aquifers and these intermediate-size aquifers commonly do not appear on maps of geologic deposits or of groundwater resources because their areal extent is rarely known.

At the other end of the scale there are large regional aquifers, the locations and sizes of which are generally known. In the southern populated or developed regions of Canada, nearly all aquifers in this category have already been located by the water-well industry or by groundwater exploration programs undertaken by provincial governments during the past 30 yr. Unlike petroleum exploration in Canada, the era of major groundwater exploration on a regional scale is over.

Regional aquifers are capable of yielding thousands or millions of litres per minute from wells distributed across them. Many of the largest of these aquifers in Canada are situated where the population density is low and where few water-consuming industries exist. This is particularly the case in the Prairie provinces where most of the large aquifers are located far from the few major cities in this region (Fig. 3). Winnipeg is an exception. One of the largest regional freshwater aquifers in Canada is the Paleozoic carbonate-rock aquifer in Manitoba. At its southern extremity, it provides most of the industrial water used in the city of Winnipeg (Render 1970). The aquifer extends northward over a large areas where the population is small and very little of the potential yield of the aquifer is used. A second major city in the Prairie region that is a large user of groundwater is Regina, which derives about one third of its municipal water supply from wells. The aquifer in Regina is not large enough for the city to avoid having to bring river and lake waters through a 80-km-long pipeline to the city. This surface water is contaminated, and before entering the municipal water distribution system, contaminants are removed in a large water treatment facility.

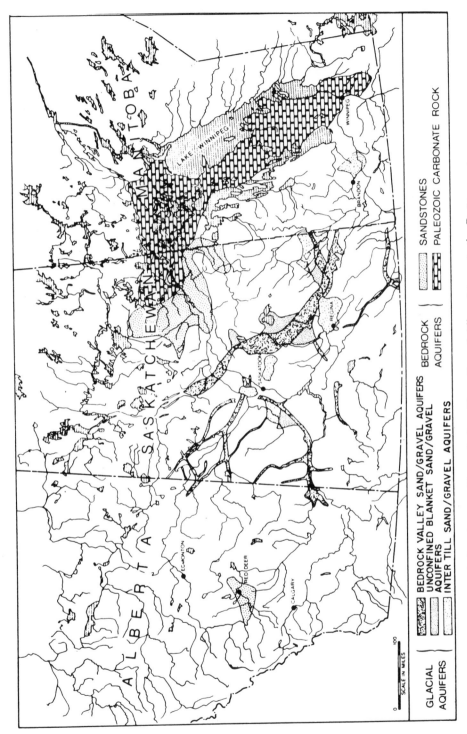

FIG. 3. Distribution of large, high-yield aquifers of different types in the Prairie provinces.

391

Natural Causes of Poor Quality Groundwater

The natural salinity of many aquifers in Canada is sufficiently high to limit the use of the groundwater. Salinity due to high sulfate concentrations causes many aquifers in the Prairie provinces to be unsuitable for household or agricultural uses. Much of this sulfate formed during oxidation of organic sulfur and mineral sulfur in shallow clay-rich glacial deposits. Much of this oxidation took place soon after the glaciers disappeared from the landscape about 10 000 yr ago (Hendry et al. 1986) when the climate was warmer and the water table was deeper. In some parts of the Prairies, sulfate salts from bedrock formations have been transported by groundwater flow into surficial aquifers. If it were not for the limitations caused by excessive sulfate, the potential for groundwater to support large-scale irrigation in the Prairie provinces would be much greater. The sulfate, which is very soluble, is gradually being flushed from the groundwater system but the flushing rate is too slow to result in significant amelioration of groundwater quality for present use.

In parts of southern Alberta, nitrate concentrations much above the limit specified for drinking water are common in shallow wells in sand seams within clay-rich glacial till. This nitrate is natural, having been derived by the oxidation of exchangeable ammonium on soil particles in the vadose zone of the till (Hendry et al. 1985). This oxidation occurred during the warm dry postglacial period, probably between 9000 and 3000 yr ago.

There are other geologically derived chemical constituents that cause the quality of groundwater to be poor in some regions. In a few areas in the Prairies, fluoride or iodine occurs naturally in groundwater at levels high enough to render the water unfit for human consumption. Limitations on groundwater use owing to naturally poor water quality are generally less widespread outside of the Prairie region, but nevertheless they are significant locally in some provinces. For example, in some counties of New Brunswick, the concentrations of natural arsenic in 10–20% of water wells commonly exceed the limits specified for drinking water. This arsenic is derived from local geologic sources. In York and Charlotte counties of New Brunswick, nearly half of the water wells in the Harvey Formation, which is composed of volcanic rocks, yield arsenic-rich water (Peters 1977). The Nova Scotia Department of Health (1983) has published a map of Nova Scotia showing locations where arsenic has been detected near or above the maximum acceptable concentration in drinking water. About one fifth of the total area of Nova Scotia has bedrock that locally can contribute excessive arsenic levels to groundwater.

Spotted across Canada are many small areas where concentrations of natural iron or manganese are high. These elements are derived from the minerals that comprise the particles or rock that form the matrix of aquifers. They are not detrimental to the health of well-water users but they render the water unsuitable for other household uses.

Along parts of the east coast, and beneath islands off the coast of British Columbia, saline water occurs in some aquifers. In some places this salinity is caused by present-day intrusions of seawater, and in others, such as in parts of the Gulf Islands of British Columbia, the salinity is due to relic salt leached slowly by groundwater circulation through bedrock shale and sandstone (Dakin et al. 1983). The intrusion of seawater is coastal aquifers is particularly important in the Maritime provinces because much of the population of these provinces lives within several kilometres of the sea.

Groundwater in the Precambrian Shield region of Canada is saline at depths generally greater than a few hundred metres. This deep saline groundwater in the Precambrian Shield was originally rain and snow that long ago infiltrated into the subsurface. The origin of the salt in this deep water is less clear. It may represent relic salt from seas that invaded the continent after the Precambrian Shield was formed or it may be salt that has been released over geologic time from some of the minerals that comprise the Precambrian rock. Nearly all groundwater at depths less than 100 m on the

Precambrian Shield is very fresh, but little of it is used for water supply because of low population densities in this region.

In some areas, groundwater is not suitable for household use because of a foul odor given off by the water. This odor is caused by natural hydrogen sulfide gas. Very small amounts of hydrogen sulfide can cause this odor. This problem is most common in petroleum-producing areas where shaley bedrock strata contribute the gas to the water.

Natural radioactive constituents are a problem in some regions primarily because of dissolved uranium or dissolved radon gas. According to the *Canadian Drinking Water Guidelines*, uranium should not exceed 0.02 mg/L (Health and Welface Canada 1978). This limit is based on the chemical toxicity of this element rather than effects related to radioactivity. Uranium in groundwater tends to be highest where the water has high carbonate alkalinity and dissolved oxygen. Parts of the southern interior of British Columbia and of Nova Scotia and New Brunswick are particularly prone to this type of natural contamination. The limit for uranium in drinking water indicated above was specified federally in 1978. The previous limit for uranium was 25 times higher. Prior to the 1978 guideline, it is likely that very few water-supply wells in Canada exceeded the guideline. This is probably not the case when the 1978 guideline is applied.

Because of its radioactivity, radon gas in groundwater poses a health hazard locally in many parts of Canada, but particularly in the Maritime provinces. This gas is acquired by groundwater as a result of the radioactive decay associated with uranium and thorium in bedrock or in rock particles in glacial deposits. The health hazard associated with radon does not result from drinking the water but rather from inhalation of air containing the radon. Radon from groundwater enters basements and contaminates the air in houses, or it may occur at excessive concentration in the air when showers are being taken, if the shower water comes directly from a well without prior aeration. Routine testing of well water for radon is uncommon in most regions of Canada, and therefore, well users generally are unaware of the radon levels in their water supply. Radon is formed as a result of the radioactive decay of radium-226. This radioactive constituent has an extremely low drinking water limit in both federal and provincial guidelines. Little is known regarding the occurrence of radium-226 in well water in Canada.

Susceptibility of Aquifers to Contamination

In unconfined sand or gavel aquifers in Canada the water table is generally between 2 and 10 m below ground surface. These aquifers are most easily contaminated because many types of contaminants travel quickly downward from surface or near-surface sources to the water table. After arrival at the water table, contaminants spread laterally and at the same time often migrate deeper in the aquifer. Lateral rates of groundwater flow in sand and gravel aquifers are generally in the range of 0.1-3 m/d (Mackay et al. 1985). Many contaminants travel as fast or nearly as fast as the groundwater, whereas others move at a slower rate because of chemical interactions with soil particles in the aquifer. Some contaminants that enter the groundwater zone are attenuated in this zone by microbiological processes.

In permeable fractured rock, groundwater flow can be much faster than the rates presented above for sand or gravel aquifers. The fractures may form relatively direct conduits for water and contaminants, particularly in limestone, dolomite, and some types of crystalline Precambrian rocks such as granite. The capacity for these rocks to retard contaminant movement or to attenuate contaminant concentrations is generally very low. In areas of Canada where the glacial deposits overlying the rock are thin or absent, wells in bedrock are generally very susceptible to contamination, particularly if the wells are not deep, as illustrated in Fig. 2A. These bedrock conditions are common in the Maritime provinces, parts of Quebec, northern Ontario, central Manitoba, and much of British Columbia.

Most of the silty or clayey layers that overlie confined aquifers in Canada were deposited during Pleistocene time by glaciers (glacial till) or as sediments on the bottom of glacial lakes (glaciolacustrine deposits) or on the bottom of inland intrusions of the ocean (glaciomarine sediments). Desaulniers et al. (1986) have shown that thick deposits of clay-rich glacial till overlying aquifers in parts of southwestern Ontario contain groundwater flowing at rates that are less than a few millimetres per year. Studies of the age of this water based on carbon-14, oxygen-18, and deuterium indicate that at depths below about 15 or 20 m, the groundwater is about 10 000 to 15 000 yr old. Much of this water originated during Pleistocene time when the last glaciers were advancing and then receding northward. Thus, in areas of southern Ontario where surficial glacial till or clayey glaciolacustrine deposits are greater than about 10–15 m thick, aquifers are isolated from the sources of municipal, industrial, and agricultural contaminants on top of the clay. Where clayey glaciomarine deposits are thick in parts of the St. Lawrence lowland in Quebec, there is a similar degree of isolation (Desaulniers 1986). Where the aquifers that occur beneath these thick clayey deposits are extensive, they receive much of their recharge from outlying areas where the clayey cover is thin. Contaminants may enter these aquifers in these outlying areas.

Long-term aquifer protection is not provided where clayey deposits are thin or where fissures go deep. In many areas of southern Ontario and Quebec the surficial clayey deposits on top of aquifers are less than 5 or 10 m thick. Weathering features such as fissures and root holes are common to depths of 4 m and in some areas to 7 m or even deeper. These features provide pathways for rapid downward migration of contaminants. In the Prairies, open fissures commonly extend to much greater depths than in Ontario and Quebec. Keller et al. (1986) described a site near Saskatoon where open fissures penetrate downward through a 20-m-thick clayey deposit to a large sand aquifer. In Winnipeg an extensive limestone and dolomite aquifer is covered by a 13- to 20-m-thick deposit of glaciolacustrine clay. The clay has vertical fractures that go deep, perhaps all the way to the bottom of the clay. These fractures provide much more rapid downward vertical flow of groundwater than would be the case if they were absent (Day 1978). The rate at which contaminants from surface sources move downward through these fractures is unknown.

In addition to the use of geologic and land-use information, tritium (^3H) in groundwater can be used as a means of identifying aquifers most susceptible to contamination. Tritium has a half life of radioactive decay of 12 yr. The concentrations of tritium in rain and snow rose markedly in 1953. The rise occurred as a result of large aboveground tests of nuclear bombs by the United States and the Soviet Union. These large tritium-producing tests began in 1952 and ended in 1967. In the past 20 yr, tritium levels in rain and snow have declined but they are still much above the pre-1953 level.

All potable groundwater in aquifers in Canada originated as rain and snowmelt. Rain and snowmelt that recharged the groundwater zone prior to 1953 is nearly devoid of tritium because the tritium levels in rain and snow prior to 1953 were low and because radioactive decay has caused nearly all of the tritium that was present in the water prior to 1953 to disappear due to radioactive decay. Groundwater that originated after 1953 can be identified by its considerable tritium content. Tritium is a good tracer because it does not adsorb on geologic materials, nor is it influenced by chemical or biochemical reactions. Tritium in unconfined sand and gravel aquifers is commonly found to depths between about 5 and 15 m and deeper where water-supply wells have caused hydraulic disturbance. Many of the industrial and agricultural chemicals that commonly cause groundwater contamination entered widespread use several years after the end of World War II. Thus, young groundwater identified by the presence of tritium is water that originated in this modern industrial period. Therefore, aquifer zones containing tritium are particularly prone to contamination.

Sources of Contamination

The causes of groundwater contamination in Canada are generally similar to those affecting groundwater quality in the United States and Europe although the relative importance of each cause may differ. Pye et al. (1983) provided an extensive overview of groundwater contamination in the United States. Pupp (1985) gave a brief description of the nature and causes of groundwater contamination in Canada. Beak Consultants Limited (1986) assembled an annotated compilation of cases of groundwater contamination in Canada obtained from the files of provincial agencies and augmented by cases contributed by several consulting firms.

The main sources of groundwater contaminants in Canada are listed in Table 1. The sources are in two categories: local and distributed. At local sources the contaminants are derived by leaching of wastes or from spills or leakages of industrial or agricultural chemicals in relatively small areas. For example, household septic systems for sewage disposal and leaky gasoline tanks are local sources. Industrial waste loggons and municipal landfills are also local sources even though they are much larger than a septic system or a gasoline tank. They are local in the sense that they are distinct sources that cover only a small percentage of the land area of most aquifers. In contrast, distributed sources are those that contribute contaminants over much larger areas. Pesticides applied to agricultural fields are an example of a distributed source.

TABLE 1. Sources of contaminants that can cause groundwater contamination.

Point sources

On-site septic systems
Leaky tanks or pipelines containing petroleum products
Leaks or spills of industrial chemicals at manufacturing facilities
Municipal landfills
Livestock wastes
Leaky sewer lines
Chemicals used at wood preservation facilities
Mill tailings in mining areas
Fly ash from coal-fired power plants
Landfarms and sludge disposal areas at petroleum refineries
Land spreading of sewage or sewage sludge
Graveyards
Road-salt storage areas
Wells for disposal of liquid wastes
Runoff of salt and other chemicals from roads and highways
Spills related to highway or railway accidents
Coal tar at old coal gasification sites
Asphalt production and equipment cleaning sites

Distributed sources

Fertilizers on agricultural land
Pesticides on agricultural land and forests
Contaminants in rain, snow, and dry atmospheric fallout

Septic Systems

The nature of some of the sources on the list needs no description. Others, such as septic systems and leaks of petroleum products or of dense industrial organic liquids, have features that need greater explanation. These contaminant sources have one aspect in common: they are small in size but are large in number.

The sewage from nearly 2.5 million rural households (Beak Consultants Limited 1986) in Canada and from hundreds of thousands of summer cottagers and campsites is disposed of in on-site septic tanks that feed drain tiles buried in blankets of sand or gravel. All provinces have regulations governing the construction of such septic systems. In a properly situated and designed septic system, the effluent from the tank feeds the tiles from which it drains freely through the sand or gravel blanket into the underlying geologic material. In most cases the effluent has nowhere to go but to the groundwater zone. This is the intent of a properly designed septic system in that direct discharge from the tile beds to streams, lakes, or ditches is avoided. Once installed, septic tanks and piping are rarely inspected for failure. They are not replaced unless problems become noticeable.

In the septic system some of the contaminant mass in the sewage is degraded in the tank, and some is degraded or absorbed in the sand or gravel surrounding the tile drains and in the subsoil beneath the drains. Even in the best of systems, however, a portion of the mass of some contaminants makes its way to the water table. The contaminants may be bacteria, viruses, or chemical compounds. In the United States, septic tanks and cesspools are the most frequently reported causes of groundwater contamination (Canter and Knox 1985). The manner in which household septic systems cause contamination of well waters is illustrated in Fig. 4.

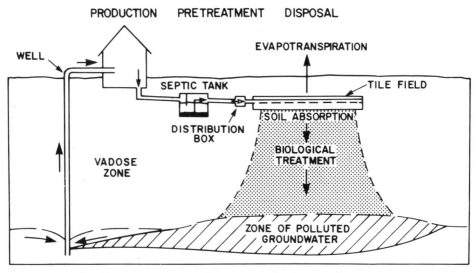

FIG. 4. Example of how contamination from a septic field can find its way into the domestic water well system.

Septic systems allow a variety of chemical contaminants to enter the groundwater zone, most notably nitrate (NO_3^-), ammonium (NH_4^+), some aromatic hydrocarbons, and some organic compounds derived from household or personal cleaning products and from septic-system cleanser. In the early 1960's, septic systems were recognized as a major cause of groundwater contamination in the United States. This recognition derived from the fact that in the late 1940's, manufacturers of detergents began to use alkyl benzene sulfonate (ABS) as a surfactant in the detergents. This surfactant proved to be mobile in groundwater and resistant to biodegradation in the septic system and in the groundwater zone. When groundwater contaminated with ABS was pumped from water-supply wells, the presence of ABS was commonly evident by the presence of foam. Thus, because of the foam, ABS served as a direct indicator to well users of septic-system contamination.

In the mid-1960's, the use of ABS in detergents was discontinued in favor of alternative compounds that are much more rapidly biodegraded in septic systems and in soil. The new compounds have caused little groundwater contamination. A lack of a distinct indicator of septic-system contamination easily recognized by foaming or taste, however, means that contamination in many wells near septic systems may go unrecognized.

There is a paucity of information on the behavior and effects on groudnwater of modern chemicals such as new detergent components and of petroleum-based or chlorinated organic cleansers. What little is known is cause for concern. Recently, a study of raw domestic sewage in the United States (U.S. Environmental Protection Agency 1985) showed the presence of many hazardous volatile organic pollutants. Further, there was little evidence of removal of these compounds during the typical 2-d retention in a septic tank. These pollutants showed higher levels on the weekend, probably reflecting household repair and cleaning activities including those using paint thinners and grease removers. When the volatile organic compounds are not degraded in the septic tank, they may enter the groundwater zone after passing through the drain tiles and drain bed. Of the various major causes of groundwater contamination, septic systems are the least monitored and they have been subjected to the least groundwater-related research. Perhaps this is because they have been such a common and seemingly innocuous household facility for so many decades. Unfortunately, it is commonly believed that septic systems constructed according to standards specified by government do not represent a hazard to groundwater quality. Therefore, vigilance that should be practiced by water-well users and government health and environmental agencies in septic tank areas is lacking. The analyses of water samples for detection of the most important types of organic contamination from septic systems are expensive and rarely done in routine assessments of well-water quality.

Petroleum Products

Another ubiquitous source of contamination to groundwater is underground storage tanks for petroleum products. Nearly all tanks installed prior to 1980 were constructed of steel. Steel tanks corrode. In the state of Maine it is estimated that 50% of the steel tanks now in the ground leak by the time they are 15 yr old (Maine Department of the Environment 1985). There are about 200 000 underground petroleum storage tanks in Canada, of which about 70 000 are located at retail gasoline outlets. The remaining 130 000 tanks in the ground are owned by transportation users, by industrial, commercial, and agricultural consumers of petroleum products and by heating oil consumers (Edgett and Coon 1986). It was only after the large number of steel tanks installed during the 1950's and 1960's exceeded their 10- to 20-yr service life that the vast scale of the petroleum leakage problem was recognized. In limited surveys, about 20-25% of storage tanks at petroleum retail outlets in Canada were found to be or suspected to be leaking (Pupp 1985). In estimates by the U.S. Environmental Protection Agency of tank leakage in the United States, 10% of the tanks were judged to be leaky.

Gasoline contains many organic compounds that are sufficiently soluble in water to pose a major threat to groundwater quality. Examples of the solubility of toxic chemicals in gasoline are as follows (milligrams per litre) (Brookman et al. 1984): benzene, 65; toluene, 34; o-xylene, 5.4; p-xylene, 13.8; 2-methylbutane, 3.7. The U.S. Environmental Protection Agency has classified both benzene and toluene as hazardous contaminants. The maximum permissible concentration for benzene in drinking water proposed by the U.S. Environmental Protection Agency is zero. Therefore, even small spills of gasoline have the potential to render large portions of the water in an aquifer unfit for drinking. For example, benzene constitutes about 1-2% of gasoline. Using 5 μg/L as the detection limit for benzene and therefore as a criterion for contamination, 1 L of gasoline has the potential to contaminate 4×10^6 L of groundwater.

The nature of gasoline contamination of groundwater changes from one decade to the next as the chemical nature of gasoline changes. For example, since 1980, methyl tertiary butyl ether (MTBE) has become a common component of gasoline in the United States. It is added as an octane booster and constitutes up to 10% of some premium gasolines. MBTE is exceptionally solluble in water and travels more rapidly in aquifers than other gasoline-derived contaminants. MBTE affects the behavior of other contaminants derived from the gasoline. Garrett et al. (1986), who have described several cases of MBTE contamination of groundwater in Maine, referred to this compound as a "pollution enchancer."

The mechanism of groundwater contamination caused by spills or leakages of petroleum products is shown in Fig. 5. The petroleum product floats on the water table and normally does not travel more than about 100 or 200 m from the spill. By contact with the floating mass, the groundwater that passes beneath it acquires dissolved constituents such as benzene (Ptacek et al. 1986). Although the main mass of the petroleum product tends to remain close to the spill or leakage site, many mobile dissolved contaminants such as benzene, which are influenced only weakly by adsorption of aquifer material, may be transported many kilometres from the site. The detailed behavior of some petroleum-derived contaminants in a shallow aquifer near Alliston, Ontario, is described by Patrick and Barker (1985).

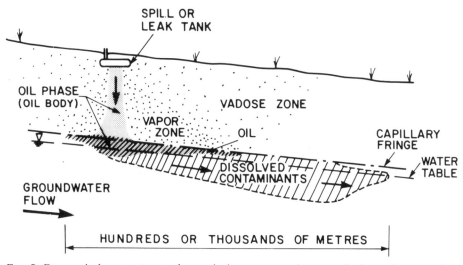

FIG. 5. Dispersal of contaminants, from a leaky storage tank or a spill of petroleum products, into the soil and groundwater.

Contamination of groundwater by gasoline and other, petroleum products is particularly severe in the Maritime provinces. Edgett and Coon (1986) presented a summary of statistics for confirmed incidences of this type of contamination in New Brunswick. In 1979, 19 spills or leaks caused contamination of 35 water-supply wells. The annual rate of spills and numbers of additional contaminated wells recognized have increased since 1979. In 1984, 90 more spills or leaks caused contamination of 100 additional wells. Many cases of contamination from petroleum products probably go unnoticed by users of well water because the taste and odor threshold concentrations for hazardous compounds such as benzene in water (Alexander et al. 1982) are much higher than the maximum permissible concentration specified for drinking water. Unless a chemical analysis is done, severe benzene contamination can therefore pass unnoticed by people drinking the water. This problem is also common to some of the other important contaminants derived from petroleum products.

Dense Organic Liquids

Since the 1940's, a group of halogenated organic chemicals that are liquids at ambient temperature have been produced in large volume by the chemical industry. These liquids, many of which are chlorinated solvents, are used for a variety of purposes such as dry cleaning, wood preservation, microelectronics manufacturing, machining, automotive production and repair, asphalt operations, aviation equipment maintenance, and munitions factories. The liquids are much like oil in appearance and are generally as insoluble as oil or gasoline in water. They are referred to as dense nonaqueous phase liquids (DNAPLs).

The U.S. Environmental Protection Agency has issued a list of chemicals that are hazardous to human health when consumed in drinking water and that are abundant or widespread in their industrial usage or as components of waste. Table 2 indicates the common heavy organic solvents included on this list. These solvents have been widely distributed for more than half a century, but only in the past few years have their exceptionally deleterious effects on groundwater become recognized. Feenstra (1982), Schwille (1984), and Villaume (1985) have provided descriptions of the movement of these dense liquids in groundwater and of the contamination that they cause.

DNAPLs are generally so much heavier than water that, when spilled on unconfined aquifers or on fractured or fissured aquitards, they sink, unaffected by groundwater flow, below the water table. Also, many of these liquids, primarily the chlorinated solvents, are less viscous than water. Because of their high density and low viscosity, these liquids sink quickly. The solubilities of nearly all of these compounds in water are in the range of about 100–8000 mg/L. However, for many of these compounds, safe concentrations in drinking water are very low, generally several orders of magnitude less than their solubilities. Drinking water limits or guidelines for these compounds have not yet been established in Canada. New federal guidelines that will include many of them will probably be issued this year or next. The maximum permissible concentration for total chlorinated solvents in drinking water specified by the European Economic Community is 0.025 mg/L. The maximum drinking water concentrations proposed by the U.S. Environmental Protection Agency for several common chlorinated solvents are "below detection," which in practice is a few micrograms per litre or less (Table 2). Maximum concentrations for these organic compounds are also included in the *Criteria for Potable Water* issued by the state of New York (Table 2).

TABLE 2. Common chlorinated solvents designated as priority pollutants by the U.S. Environmental Protection Agency and maximum permissible concentrations (MPCs) specified for drinking water by the State of New York and by the U.S. Environmental Protection Agency.* BD = below detection.

Solvent and MPC (μg/L)		MPC not available
Chloroform	100	Chlorobenzene
Carbon tetrachloride	5, BD*	1,2-Dichlorobenzene
1,1,1-Trichloroethane	10, 200*	1,3-Dichlorobenzene
Vinyl chloride	5, BD*	1,1-Dichloroethane
Trichloroethylene	10, BD*	Ethylene dichloride
1,2-Dichloroethane	BD*	Trans-1,2-dichloroethylene
		Methyl chloroform
		1,1,2-Trichloroethane
		1,1,2,2-Tetrachoroethane

The nature of groundwater contamination caused by dense solvents is illustrated in Fig. 6. In Fig. 6A, the volume of spilled liquid is small and all of it is held in the pore spaces in the vadose zone. None of the liquid penetrates to the water table. However, water from rain and snowmelt infiltrating through the liquid held in the vadose zone

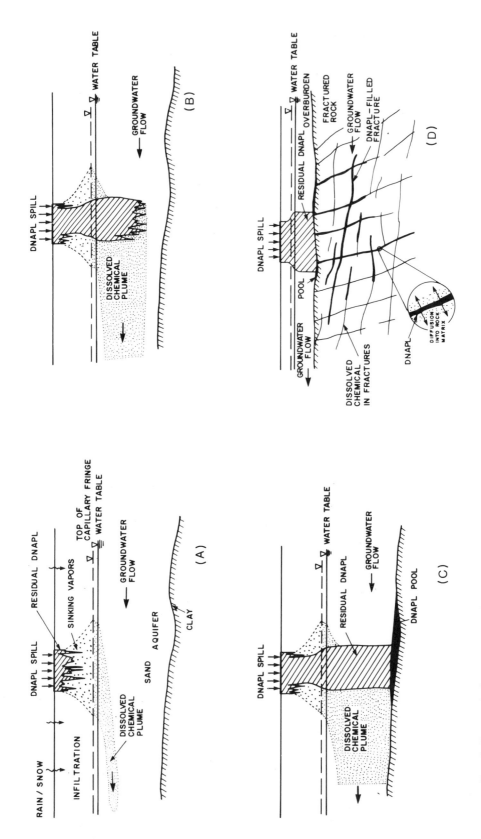

FIG. 6. Dispersal of contaminants from various sized spills of a dense nonaqueous phase liquid into a sand aquifer and into fissured bedrock.

dissolves some of it and carries these dissolved contaminants to the wate table. In Fig. 6B, the spill is larger and direct penetration of the DNAPL into the aquifer occurs. The DNAPL in the aquifer occupies only a small percentage of the multitude of open pore spaces amongst the mineral and rock particles that comprise the aquifer. Groundwater can flow freely through these open pore spaces. Dissolved contaminants acquired by this flow causes a plume of contamination to spread in the aquifer. In Fig. 6C, the spill is sufficiently large for free DNAPL to reach the bottom of the aquifer. In Fig. 6D, the DNAPL penetrates through the sand aquifer and then into the fissured rock below.

Contamination caused by these heavy organic liquids is generally much more difficult to cope with than that caused by leaks or spills of petroleum products. Even if the exact surface location of the spill or leak is known, the location of the main mass of the liquid in the aquifer is normally difficult to determine. The organic liquid may also move laterally large distances along the top of silty or clayey seams within the aquifer, or along the sloping bottoms of aquifers. The pools of organic liquid spread at a diminishing rate and become immobile in the aquifer after a few days or weeks. This immobilization occurs in much the same way as when a glass of water is spilled on a large floor. The spilled water spreads across the floor, leaving behind an immobile film or thin veneer of water on the floor. When the volume of water in this film equals the volume of the spill, water movement ceases. But, in an aquifer the immobile organic liquid on the bottom of the aquifer would normally be a spread-out pool rather than a water-like film. This immobile mass than slowly dissolves into the flowing groundwater over time periods of decades or centuries. In fissured bedrock, it may flow from one fissure to the next, and so on, as it moves farther from the spill site. In each fissure, it leaves behind a film of organic liquid on the surfaces of the fractures. The organic liquid as an immiscible phase is rarely found in water-supply wells, but the dissolved fraction derived from the organic liquid is found with increasing frequency in such supplies and is the cause of contamination.

Widespread and long-lasting contamination of groundwater can result from even very small spills of DNAPL. As an example, consider trichloroethylene (TCE), which is one of the most common industrial solvents. A single drum of this liquid (about 200 L or 45 gal) dissolved in water and diluted to a maximum concentration permissible in drinking water of 0.005 mg/L produces 60×10^9 L (16×10^9 gal) of contaminated water. This volume of water is equivalent to the total cumulative yield from one well pumping at 100 gal/min for 250 yr. One hundred gallons per minute is suffcient water to supply a small village.

For most of the common industrial solvents the maximum concentrations specified in guidelines or standards for drinking water are much less than the odor and taste thresholds for these compounds in water (Alexander et al. 1982). Except for the water supplies of large cities, drinking waters are rarely tested for these compounds. Until recently, spills or leakages of industrial solvents normally went unnoticed or otherwise unabated by industries and governments. Because only small volumes of these liquids can contaminate large volumes of groundwater, because migration in groundwater of the dissolved contaminants from the spill sites to wells or surface waters takes many years or decades and because the locations and magnitudes of most spills or leakages that took place prior to the 1970's or 1980's are unknown, this type of contamination is a problem of increasing magnitude. Cleanup of much of this contamination is not feasible with existing technology.

Case Studies of Groundwater Contamination

To illustrate the mechanisms causing groundwater contamination and the variety of problems and implications that derive from groundwater contamination, several case studies will now be described. Although information on hundreds of cases of groundwater contamination exists in the files of provincial government agencies, cases with detailed information are rare (Beak Consultants Limited 1986). The studies described here were selected primarily from the few Canadian cases for which intensive research-oriented investigations are described in the scientific or technical literature.

Municipal Landfills on Sand Aquifers

Contaminated sand aquifers at three municipal landfills in Ontario have been the focus of intensive research for many years. The largest of the three landfills, which occupies 32 ha, is situated on the outskirts of the city of North Bay. The other two, each covering about 4 ha, are located near Alliston and Kitchener-Waterloo. Cherry (1983) has described the geologic and hydrologic features of these three sites. At each of these sites hundreds of monitoring wells have been used to determine the extent and nature of the contamination. This discussion will focus on the North Bay landfill because it presents the most difficult problems pertaining to government policies and remedial action.

The North Bay landfill is situated on a shallow unconfined sand aquifer overlying Precambrian bedrock. The aquifer is 10 m thick near the landfill and between 10 and 20 m thick in most of the contaminated area away from the landfill. The landfill has received municipal refuse from North Bay since 1962. The refuse is deposited in layers and each layer is covered with a veneer of local sand.

Rain and snowmelt infiltrate through the sand and through the refuse. Infiltration through refuse and its cover layers occurs at most landfills in Canada. This infiltration produces leachate that seeps from the bottom of the landfill. At the North Bay site the leachate-contaminated groundwater migrates southwest, down the water table slope (Fig. 7A). This migration created a plume of contamination (Fig. 7B) extending from the landfill to an area of springs discharging to a small stream named Chippewa Creek about 700 m from the landfill. Investigations of the chemical nature of this plume are described by Buszka (1982) and Reinhard et al. (1984).

The front of the plume arrived at the spring long before studies of the plume began in 1979. The average rate of groundwater flow from the landfill to the springs is about 150-200 m/yr. It takes about 4 yr for the most mobile contaminants in the plume to travel the 700-m distance from the landfill to the spring (Moore 1986). Therefore, the springs have been discharging contaminated water to Chippewa Creek since about 1966. Chippewa Creek flows through the North Bay and then into Lake Nipissing.

The plume contains both inorganic and organic contaminants. The pattern of chloride contamination shown in Fig. 7B is generally similar to the patterns of sodium, calcium, magnesium, ferrous iron, and bicarbonate but the concentrations of these constituents are somewhat lower than chloride. The chloride pattern is also similar to that of dissolved organic carbon (DOC).

From maps and vertical profiles of concentrations in the plume, Moore (1986) obtained estimates of the total mass of chloride and DOC in the plume of 70 and 60 tonnes (t), respectively. The annual contributions of the landfill to the plume are about one quarter of these amounts. The plume is in a near-steady-state condition and therefore the annual mass inputs are approximately equal to those that discharge to the springs each year. The chloride and DOC contributed to Chippewa Creek from the landfill as well as contributions of these constituents from many other urban sources in North Bay are carried by the creek into Lake Nipissing.

To put the landfill-derived loading of chloride and DOC to surface water in perspective, the contributions of chloride from winter road salting and DOC from municipal sewage to Lake Nipissing are considered. North Bay applies an average of 3600 t of chloride salt to its roads each year. It is reasonable to expect that nearly all of this salt is transported in surface runoff and in the sewage system to Lake Nipissing. The load of chloride contributed by the landfill to Lake Nipissing via Chippewa Creek is therefore an insignificant fraction of this total contribution to the lake from the urban environment.

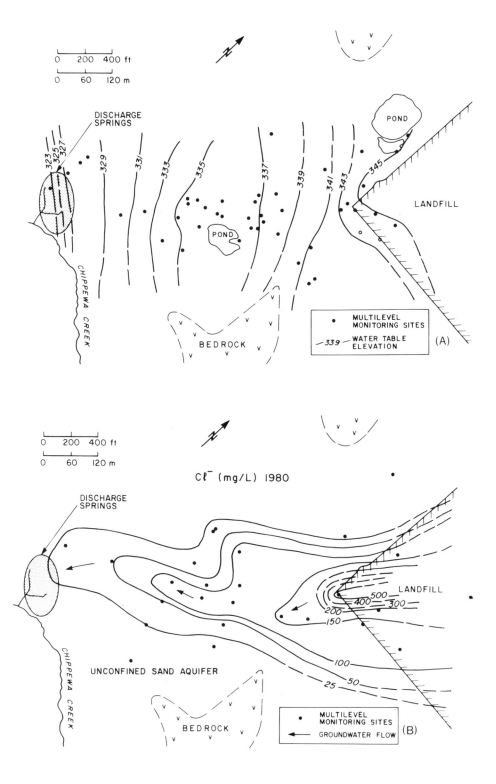

FIG. 7. (A) Water table slope and (B) plume of contamination originating from a landfill near North Bay, Ontario (adapted from Moore 1986).

An obvious contributor of DOC from the urban environment to Lake Nipissing is the North Bay sewage treatment facility. The annual load to Lake Nipissing from this source is about 117 t, which is about eight times larger than the contribution from the landfill. Lake Nipissing is a source of municipal water for North Bay and it is an important recreational resource. When judged on the basis of salt and total DOC, the landfill contributions are insignificant.

This case study illustrates several aspects of groundwater contamination at municipal landfills. The contamination that is most readily detected by visual inspection or by routine chemical analyses of well samples is relatively nonhazardous to humans and to the aquatic environment to which the plume discharges. Ferrous iron gives the water a murky brown color as it oxidizes to ferric iron on exposure to air. Chloride, total DOC, and other standard water quality indicators of gross contamination are not useful for basing judgement on the potential hazards posed to human health or to the ecological system.

TABLE 3. Concentration of chloride, total DOC, and selected trace organics in five monitoring wells at the North Bay landfill (from J. F. Barker, written communication).

	Outside of plume (background)		Within the contaminant plume		
Distance from landfill (m)	—	5	5	400	700
Depth (m)	5	6	15	13	5
Chloride (mg/L)	2.4	377	100	175	53
Total DOC (mg/L)	3.3	176	38	81	17
Volatile chlorinated organics (μg/L)					
1,1,1-Trichloroethane		0.01	0.03	0.0	0.0
Trichloroethylene		0.0	1.6	0.0	0.0
Aromatic hydrocarbons (μg/L)					
Benzene	0.3	29	108	71	3.9
Toluene	0.2	0.27	7.4	0.64	0.14
Ethylbenzene	0.1	5.4	6.8	14	0.03
m/p-Xylene	<0.1	12	22	12	0.16
o-Xylene	<0.05	4.3	18	2.3	0.12
1,2,4-Trimethylbenzene	<0.05	7.8	11	20	0.21
Naphthalene	<0.05	2.7	1.5	2.7	0.0
Chlorinated benzenes (μg/L)					
Chlorobenzene	0.3	4.3	0.5	11	2.1
1,2-Dichlorobenzene	0.2	0.18	0.26	1.0	0.61
1.4-Dichlorobenzene	<10.1	6.1	1.5	5.7	2.8

It is the many specific organic compounds in the plume (Table 3), such as chlorinated organic compounds and aromatic hydrocarbons, that present the greatest potential problems. The North Bay landfill receives small but significant quantities of industrial wastes that may impart many of these organic contaminants to the groundwater. These types of contaminants are found at most municipal landfills that have been investigated in Ontario. Prior to about a decade ago, municipal landfills were used for disposal of both normal municipal wastes and most types of hazardous industrial wastes. In recent years nearly all municipal landfills have been prohibited by provincial regulations from accepting most types of hazardous industrial wastes. Although restrictions have been placed on municipal landfills, alternative disposal sites for hazardous industrial wastes have not yet been established in any regions other than southwestern Ontario, where only one such disposal facility exists.

The North Bay case study illustrates some common features of groundwater contamination: the degradation of the aquifer is most severe near the source and the plume attains a maximum areal extent limited by location of surface waters into which the groundwater discharges. The North Bay landfill has rendered the stretch of aquifer between it and the springs unsuitable for potable use. To bring the water in this aquifer back to a drinkable quality would require application of major engineering measures to the landfill and to the aquifer. Tens of millions of dollars would probably be required. The aquifer is already severely contaminated relative to drinking water standards. The severity is unlikely to increase substantially even if the landfill is operated for many more years. Without engineering intervention, the aquifer will probably remain contaminated far into the next centrury. A choice exists between (a) discarding the aquifer permanently as a source of potable water supply or (b) allocating large amounts of tax revenue to clean the aquifer. The nature of aquifer contamination at many landfills and at numerous other contamination sources is such that these two options are generally the main ones available.

The aquifer at the North Bay landfill is small. The population that now lives near the aquifer is also small and draws its water from other sources. Future homeowners near the landfill could be prevented by municipal restrictions from installing water wells that may be affected by the plume. The plume does minimal harm to lake water. The dilemma in North Bay posed by these options has not yet been resolved, but there is little doubt that the noncleaning option will prevail. However, there are probably many waste disposal sites in Canada where, in the next few decades, expensive cleanup will be required.

Industrial Waste Disposal in a Gravel Pit

Across Canada many pits remaining from quarrying operations for sand or gravel have been used for waste disposal. Sand and gravel pits commonly occur on unconfined aquifers where the water table is close to the bottom of the pits. The potential for rapid contamination of groundwater is therefore great. To illustrate the nature of problems associated with the use of these pits for waste disposal, a case study of a site near Montreal is described. This description is based primarily on the work by Poulin (1977) and Poulin et al. (1983).

Between 1968 and 1972 a company in the business of transporting used oil disposed of about 40 000 m^3 of liquid wastes into lagoons in an old gravel pit area near the town of Mercier about 20 km south of Montreal (Fig. 8). The wastes originated mostly from chemical and petrochemical industries. When disposals ceased in the fall of 1972, the lagoons contained liquid equivalent to about half of the original volume. Most of the missing volume had seeped downward into the sand and gravel aquifer. In October 1971, nearby water-supply wells were found to be contaminated.

Shortly after this discovery, the local residents were supplied with aquaduct water drawn from a well located in a town 10 km away. The plume of contamination in the aquifer was subsequently discovered a much greater distance from the lagoons. The aquaduct was extended in 1976 and again in 1982 to supply water to the entire region between the town of Mercier and the town of Ste. Martine.

Investigations conducted between 1973 and 1983 have shown that contaminants from the lagoons caused a plume to develop in both the sand and gravel aquifer and in an aquifer of fissured dolomitic bedrock beneath the sand and gravel. In places the plume is at least 30 m thick. The average lateral velocity of groundwater flowing southwest from the lagoons was estimated to be about 100 m/yr in the sand and gravel and about 500 m/yr in the bedrock.

Zones of contamination within the plume and the areal extent of the plume which spanned a longitudinal distance of 7 km in 1981 are displayed in Fig. 8. The principle contaminants in each zones are listed in Table 4. Zone 1, in which more than 80 organic

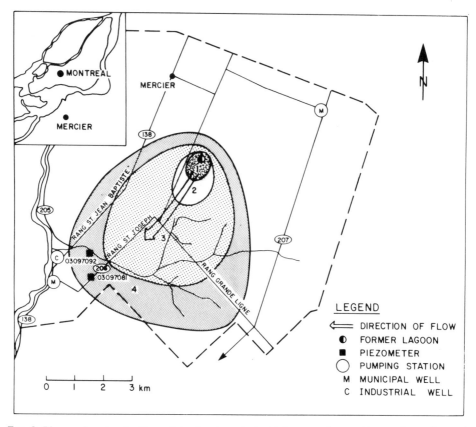

FIG. 8. Plume of contamination originating from industrial wastes dumped into a disused gravel pit near Mercier, Quebec.

compounds have been identified, is highly contaminated with toxic compounds of common industrial origin. Concentrations of individual contaminants vary within each zone because contaminants are transported by the groundwater at different rates due to different affinities for adsorption on soil particles in the aquifer and because some contaminants are diminished in concentration by degradation or transformation by natural bacteria in the aquifer. The processes of adsorption, degradation and transformation, however, have not been sufficiently strong to prevent hazardous contaminants

TABLE 4. Principal organic contaminants and concentration ranges in the various zones shown in Fig. 8 in the Ville Mercier plume (from Poulin et al. 1983).

	Contaminant range and zone		
	>20 μg/L	0.01–20 μg/L	Nondetectable
Chloroform	1, 2	3	4
Dichloroethylene	1, 2	3, 4	
Trichloroethylene	1	2	3, 4
Dichloroethane	1, 2	3, 4	
Trichloroethane	1	2	3, 4
Chlorobenzene	1	2, 3	4
Phenols	1, 2, 3	4	
PCBs	1, 2	1, 2	3, 4

from migrating far from the lagoons. Such extensive migration of industrial organic contaminants in aquifers has been observed at many other sites of industrial waste disposal in North America and Europe.

By 1975, part of the residual mass of wastes in the lagoons had been destroyed in an incinerator constructed at the site. By 1981, the remainder of the waste residual was removed from the lagoons, treated, and buried nearby in clay. In 1983, a program of aquifer restoration began, consisting of new wells for pumping contaminated water from the aquifer. This water is cleaned in a treatment plant and then discharged at the surface. The pumping was in progress at the time of writing of this chapter. When the restoration program began, it was thought that pumping for 5 yr would accomplish the cleanup of the aquifer; however, it is possible that near the lagoons, immiscible liquids (DNAPLs) exist in the sand and gravel. To clean this part of the aquifer, pumping for many decades or longer may be necessary. If all pumping ceases without having cleaned this zone, the plume will grow once again.

Several thousand residents of the Ville Mercier area whose water supply was changed from local wells to aqueduct water were subjected to inconvenience related to this change. Some of these residents were subjected to potential health risks due to consumption of contaminated water prior to recognition of the contamination after which time well usage ceased. By 1986, the direct cost of this groundwater-contamination case was about 10 million dollars, which includes 5 million for the change in water supply, 3 million for construction of the pumping and water treatment system, and about 0.5 million/yr for operation of this system (Poulin et al. 1983). Even the expenditure of many more millions of dollars may not ensure that this aquifer will once again yield potable water.

Wood Preservatives in Unconfined Aquifers

For more than a century in Canada, wood has been treated with hazardous chemicals to enhance its capability to resist rot and insects. Hundreds of these wood preserving facilities across Canada have at one time or another been in operation, although at present the number of active facilities is a small fraction of the total. Wood preserving generally consists of two steps. The first is a conditioning or moisture removal step. The second is the actual treatment of the wood with toxic chemicals serving as preservatives or insecticides. Wood preservation plants use one of three types of chemical treatments involving either creosote, chlorophenolic compounds, or metal salts.

The management of these chemicals at wood preservation plants has, until recently, included few precautions to prevent entry of spilled chemicals or contaminated waste fluids into the soil. By deduction from studies conducted at wood preservation sites in the United States, it is expected that at least some degree of groundwater contamination (but not necessarily aquifer contamination) occurs at most wood preservation sites in Canada.

Cresote gained widespread use for wood preservation in the first half of this century. Creosote is a black tar-like liquid that is heavier than water. It is a DNAPL and as a separate liquid phase in groundwater it flows in soil or fissured rock or clay in the same manner as the chlorinated solvents described above or the polychlorinated biphenyl (PCB) rich oils described below, but more slowly. Creosote in the subsurface generally does not travel far from locations where it is spilled or leaked but it can cause extensive degradation of groundwater quality because groundwater flowing past the creosote in aquifers will acquire organic contaminants by dissolution. The groundwater contaminants dissolved from creosote are a class of compounds known as polynuclear aromatic hydrocarbons (PAHs). Little information is available on the mobility of these compounds in the types of aquifers common in Canada.

Another type of chemical treatment for wood preservation makes use of chlorinated phenolic compounds. This process is used at many wood preservation plants in Canada, particularly in British Columbia, where air-dried lumber for export is routinely treated with chlorophenolic compounds to prevent staining due to fungal growth.

Patterson and Liebscher (1986) described a study at a saw mill site near Penticton, British Columbia, where groundwater contamination from chlorophenols was severe. In October 1978, 18 000 L of chlorophenol solution leaked from dip tanks at this site. This resulted in contamination of an unconfined sand aquifer used for domestic and industrial water supply. The spill occurred within 120 m of the Okanagan River. The aquifer connects to this river. Studies of the effects of the spill on groundwater began 1 mo after the spill took place. The solution was a high-pH water mixture of sodium tetrachlorophenate and sodium pentachlorophenate at a concentration of about 7500 mg total chlorophenate/L. When this solution entered the aquifer, mixing with groundwater caused a decline in pH which is believed to have caused conversion of some of the chlorophenate to chlorophenol. The water table at the spill site was 5-5.5 m below ground surface. The lateral groundwater velocity towards the Okanagan River was about 2-3 m/d. The chlorophenate/phenol contamination travelled only slightly slower than groundwater velocity.

In December 1978, production wells were installed at the front of the plume where it approached the river. With wells pumping at a combined capacity of 200 m^3/d (31 gal/min), the front of the plume was drawn to the wells, thereby preventing seepage of contaminated groundwater into the river. The contaminated water from the wells was conveyed to the Penticton sewage treatment plant, first by tanker truck and later by a pipeline. The water was passed through carbon filters to remove most of the contamination before discharge to the sewage treatment plant. The pumping system was operated for 89 d. About 40% of the spilled contaminants were recovered. At the end of the pumping, relatively low but environmentally significant concentrations remained in the pumped groundwater (1-2 mg/L). Thus, after 89 d, the recovery operation had become very inefficient. This low recovery rate was attributed to conversion within the aquifer of the chemical form of the contamination. It is believed that chemical precipitation converted dissolved contaminants in the groundwater to an immobilized solid form. Unfortunately, subsequent slow dissolution of this precipitate has continued to cause the groundwater flowing to the Okanagan River to contain dissolved contaminants. This situation was found to persist 6 yr after the spill when the last sampling was reported.

This case study demonstrates that once dissolved contaminants enter an aquifer, it may not be feasible to remove them quickly or completely. Further, chemical or biochemical processes may intervene in a way that greatly diminishes the rate of recovery. In this case study, chemical precipitation followed by slow dissolution caused the slow recovery rate. Other types of contaminants can be influenced by other chemical or physical processes to cause persistence of contamination even when aquifer-cleaning operations are applied.

A third type of wood preservation process relies on metal salts rather than organic compounds. Salts of chromium, arsenic, and copper are most common. Chromium has the greatest potential to cause extensive contamination of groundwater. Copper is generally immobile in soil. Arsenic may be mobile but generally much less so than chromium. In subsurface environments containing oxygen, which are common in shallow unconfined aquifers, chromium tends to travel nearly as fast as the groundwater. Where oxygen is lacking, chromium is generally insoluble and therefore immobile. Calder (1986) provided a review of the behavior of chromium in groundwater. A case study of a large plume of chromium caused by a small wood preservation facility on a sand aquifer in Michigan is described in French et al. (1985). No groundwater studies at wood preservative plants using metal salts in Canada have been reported in the literature.

A PCB Spill on Clayey Deposits

PCBs are hazardous industrial compounds that have an exceptional tendency to accumulate in the food chain. These compounds are frequently mentioned in the news media and the public is well aware that they are a threat to the environment and to human health. The industrial use of PCBs ceased in 1976 but the spread of these compounds in the environment continues.

PCBs were commonly used industrially in a liquid form similar to mineral oil in which the PCBs were generally mixed with other chlorinated organic compounds such as trichlorobenzene. These industrial liquids are much more dense than water and relatively insoluble. Thus, they are DNAPLs. Their solubility is low but it is sufficient for dissolved PCB concentrations in water to greatly exceed the extremely low levels permissible in drinking water. A major use of these PCB-rich industrial liquids was in manufacture of electrical transformers and capacitors. Spills of these liquids at manufacturing plants were common. The disposal of PCB-rich liquids at municipal landfills and even at Canada's only waste disposal site used exclusively for hazardous waste (near Sarnia) is prohibited and incineration is not yet permitted in Canada. Therefore, old transformers and capacitors containing PCBs are being stored temporarily at many sites across Canada, some of which are unsuitable for such storage. Leakage of these liquids from old transformers at temporary storage sites is now a cause of entry of PCBs to soil. The case study presented below illustrates some of the problems associated with spills of PCBs in their DNAPL form.

Between 7000 and 21 000 L of oil composed primarily of PCBs and chlorobenzenes was spilled when an underground pipe broke in 1976 at a transformer manufacturing plant in Regina, Saskatchewan. Subsurface conditions at the site were studied intensively in 1979 by Roberts et al. (1982) and Schwartz et al. (1982). Contrary to initial expectations, large quantities of PCBs were found to have migrated both vertically and horizontally at the site. PCBs moved downward into Regina clay through a 1-m-thick surficial layer of sandy fill material. The thin sandy layer was placed on top of the Regina clay to form a pad on which the building in which manufacturing takes place was constructed.

PCBs in the subsurface at the site occur in two liquid phases: a dissolved aqueous phase and a heavy oily liquid phase. Contaminant migration in shallow perched groundwater along the interface of the clay and the sandy pad probably accounts for the extensive near-surface lateral movement.

Six major geologic units were identified from detailed test drilling and sampling. From the surface downward these are the thin sandy pad, the Regina clay, the Condie silt, a glacial till, a deep silt, and the Regina aquifer system. This deep aquifer system, which is a large deposit of sand and gravel, contributes about one third of Regina's municipal water. The permanent water table is situated near the bottom of the Condie silt at a depth of about 15 m below ground surface. The top of the Regina aquifer is 40 m below ground surface.

When it became public knowledge in 1978 that large quantities of a PCB-rich liquid had leaked at the site, reporting in the news media caused fear in Regina of PCB contamination of the Regina aquifer. This lead to the appointment of a group of scientists (I was a member of this group) directed by the National Research Council (NRC) to study the movement of the PCBs at the site. This group concluded that the dense oily liquid had moved downward in many small vertical fractures and that some of the liquid had moved in these fractures all the way through the 9-m-thick Regina clay into the Condie silt beneath the clay. No evidence was found to suggest that PCBs had moved deeper than the Condie silt. This is also the finding of more recent studies of the site. The geological history of the Regina area and drilling records suggest that vertical fractures exist in all layers between the Regina clay and the aquifer. Whether or not any of these fractures are open much below the permanent water table is not known. It was recognized by the NRC investigators that there exists a theoretical possibility that some of the dense oily liquid may have travelled down fractures all the way to the aquifer, but the main conclusion was that this, very likely, has not taken place and that the aquifer is not and will not soon become contaminated. Probabilities of PCB travel to the aquifer based on any rigorous form of quantitative evaluation could not be specified. The NRC group also concluded that the most significant hazard represented by the spill is the potential for contamination of water-supply wells that could be placed on the site in future decades. Drilling in the future could provide contaminant pathways to the aquifer if boreholes or wells are not properly sealed (Schwartz et al. 1982).

A total of more than 1 million dollars was spent on the NRC study and on other studies of the site. The fact that there was significant uncertainty associated with the main conclusion was disappointing in light of public and governmental expectations. The study showed that determination of the maximum depth of penetration of DNAPLs in fractured clayey and silty deposits is an extremely difficult task. Short of removing part of the industrial plant, which would cause major economic loss, and of excavating major amounts of contaminated soil that cannot presently be destroyed or disposed of in Canada, the task probably cannot be accomplished with certainty. The value of the benefits that would accrue from more certain knowledge of PCB migration, when considered in relation to the large expenditures that would be necessary to acquire this knowledge, is very difficult to evaluate.

Deep penetration of open fractures in clayey deposits in the Prairie provinces is common (Grisak et al. 1976). The types of difficulties that lead to the uncertainties in the conclusions obtained in this case study will occur at many other sites in the Prairie region when downward migration of contaminants is investigated. The case study described above is the most detailed study ever of a point source of contamination conducted in the Regina area. Relative to the many other potential sources of ground-water contamination in this area, the PCB site may not be important. It was selected for an exceptionally intense investigation because, under scrutiny of the news media, it became a focus of public concern. The same can be said about some of the other sites in Canada that have been subjected to high-priority investigations.

Inactive Mine Tailings

A cornerstone of the Canadian economy is the mining industry, the principle products of which are base metals, uranium, and potash. This industry produces vast quantities of mill tailings. These tailings in addition to the large volumes of mine rock produced at open-pit mining operations are deposited on the landscape. This discussion focuses on mill wastes because of their considerable potential to cause adverse impact on ground-water and surface water.

Nearly all mill wastes are referred to as tailings. These are the materials that remain after the economic minerals are extracted from ores. Tailings are normally sent from the mill via pipeline to disposal areas. Most disposal areas are situated in valleys or other lowlands around which or across which dams or dikes are constructed to form enclosed basins (tailings impoundments). The waste entering these impoundments is normally a slurry of water and sediment.

At metal and uranium mines the sediment is almost entirely sand and silt-sized particles derived from the ore with small amounts of chemical additives from the mill. The process water contains rock-derived and mill-derived contaminants. The continuous feed of water from the mill and additional water from rain and snow result in a controlled discharge of tailings leachate from the impoundments to local streams or lakes. The quality of this effluent comes under the scrutiny of provincial or federal agencies and generally must meet surface water quality objectives or standards. At many mine sites this effluent passes through treatment plants for pH adjustment and removal of metals or radionuclides.

These activities occur when the mine and mill are operating. When the ore body is exhausted or alternatively when economic conditions cause termination of mining, the impoundments become a focus of postmining reclamation and maintenance to minimize long-term adverse impact on the aquatic environment. When mines closed prior to the 1970's, the impoundments were normally abandoned entirely or otherwise left with minimal maintenance. After the mine and milling operations cease, only rain and snow contribute water to the tailings impoundments. The bottoms of tailings impoundments are to some degree permeable. The continual additions of water to the top of the tailings cause leachate to leak downward into the underlying geologic domain. The leachate contains contaminants, the most important of which are metals at base metal and precious metal mines and radionuclides as well as metals at uranium mines. At potash mines the leachate is very saline water.

Very few tailings impoundments at metal or uranium mines in Canada are situated on aquifers currently used for water supply or even on aquifers likely to be used in the future. Mining in Canada rarely occurs in areas where there is a lack of abundant surface water for consumption. Entry of leachate into the groundwater zone is nevertheless cause for environmental concern at many old tailings sites because contaminants are transmitted along groundwater pathways into streams or lakes. The first arrival of these contaminants may not occur until long after the mine is abandoned. The time scale is such that if the company that derived profits from the mining is small, it will probably not be in existence when expenditures for water quality protection are necessary.

The long-term problem posed by old tailings at metal and uranium mines is compounded by the tendency for the water within many tailings to gradually evolve from an alkaline, low-metal content state to a very acidic, high-metal state. This normally occurs in tailings containing more than about 1% by weight of sulfide minerals and a much lesser amount of carbonate minerals. The sulfide minerals react with oxygen and water to produce sulfuric acid and other forms of acidity. The low pH causes release of various contaminants from the tailings solids to the leachate. Although the scenario outlined above occurs at hundreds of old tailings impoundments across Canada, little research has been directed at this problem. Few metal and uranium mines in Canada are situated close to urban centres or main highways. For nearly all Canadians, tailings are an unseen by-product of the industry.

Much of what is known about the details of old acid-producing tailings is drawn from studies in two sulfide mineral rich areas: an old tailings impoundment at a closed base metal mine in northeastern New Brunswick (Boorman and Watson 1976) and five old uranium tailings impoundments in the Elliot Lake district of northeastern Ontario (e.g. Morin et al. 1982; Feenstra et al. 1981; Dubrovsky et al. 1984, 1985). The base metal mining industry in northern New Brunswick is located within watersheds that include important spawning grounds for Atlantic salmon. The Elliot Lake area drains into the Serpent River flowing to Lake Huron.

The tailings studied by Boorman and Watson (1976) were deposited in 1957–58 and 1962–63, so there was an opportunity for geochemical change to have occurred by the time the field investigation began in 1974–75. These tailings have a maximum thickness of 10 m. Core samples were collected and trenches inspected. Boorman and Watson identified three chemical–physical environments in the tailings: an oxidation zone extending to a maximum of 0.5 m beneath surface, a hard pan up to 0.15 m thick beneath the oxidation zone, and a reduction zone in the underlying unaltered tailings. In the oxidation zone the pH is low and concentrations of dissolved copper, zinc, lead, and iron in the water are high. Although chemical reactions in the hard pan serve to decrease the concentrations of some of the metals, very deleterious levels remain throughout the underlying tailings and, therefore, also in the leachate emanating from the tailings. Although the permeability of these tailings is low, it is sufficient to allow much of the total annual precipitation, which is approximately 1 m/yr, to permeate into the tailings where it produces leachate. The tailings area was resampled in 1985 and 1986 by D. W. Blowes (written communication). The adverse geochemical conditions observed by Boorman and Watson in 1974-75 also existed in 1985-86. Without prevention of oxidation or infiltration in the tailings, these conditions are expected to persist far into the next century. The mining compoany that now owns these tailings is preventing impact to the aquatic environment from exceeding acceptable limits by operating an effluent treatment system primarily involving neutralization with lime. The annual cost of this treatment is large and is slowly increasing.

The next case study site, the Nordic Mine tailings, is situated in the uranium mining district of Elliot Lake. Investigations of the geochemistry of inactive pyrite-rich (FeS_2) uranium tailings in this mining district have focused on the Nordic tailings area, where two impoundments are located in natural topographic valleys (Dubrovsky et al. 1985). The tailings are 8-12 m thick and overlie a localized deposit of glaciofluvial sands (Fig. 9). Analyses of the solid, liquid, and gas phases in the vadose zone of the tailings show that gas-phase oxygen levels drop rapidly within 0.7-1.5 m of the surface. This indicates rapid oxygen consumption during pyrite oxidation. Oxidation during the past 15-20 yr has caused a marked depletion of pyrite near the surface of the tailings.

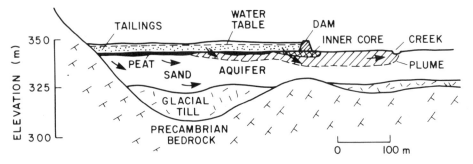

FIG. 9. Geological structure and contaminant plume in an aquifer impacted by tailings leachate near the Nordic uranium mine.

The oxidation of pyrite in the vadose zone imparts to infiltrating precipitation high concentrations of iron, sulfate (SO_4^{2-}), and various heavy metals and a pH generally between 1.5 and 4.0. The acidic water infiltrates downward towards the bottom of the tailings at a rate of 0.2-2.0 m/yr, displacing high-pH groundwater that originated as process water discharged from the mill. The acidic water now occupies the entire tailings thickness over a small area of the tailings.

At one location a well-defined plume of tailings-derived groundwater containing high ferrous iron concentrations has developed in the sand aquifer beneath and adjacent to the tailings. The plume has a core, where the pH is low and levels of various metals and radionuclides are high. Although the average velocity of groundwater flowing laterally beneath the dam is about 440 m/yr, reactions such as mineral dissolution, precipitation, and coprecipitaiton retard the migration of the front of the core of the plume, producing an observed frontal migration rate of this core of only 1 m/yr.

Groundwater from the outer zone of the plume flows laterally towards a small stream, where a portion of it is now discharging into the steambed. The discharge results in the precipitation of amorphous ferric hydroxide in the stream which imparts a brown color to the bed. Most of the acid produced by ferrous iron precipitation is buffered, and only a moderate decrease in stream pH is observed. The core of the plume of groundwater contamination will not reach the stream for a long time, but nevertheless the outer zone will probably cause a progressive decline in the quality of the stream water in the next few decades.

Although the rate of pyrite oxidation in the Nordic Mine tailings has been decreasing, there is sufficient pyrite in the tailings to generate groundwater high in ferrous iron for several decades or more. Calculated rates of downward migration in the tailings indicate that in the next few decades, acidic groundwater will occupy the entire tailings thickness over most of the tailings area. This will cause an increase in the total flux of contaminated groundwater into the underlying aquifer.

Based on their study of the tailings at the closed base metal mine in northeastern New Brunswick, Boorman and Watson (1976) stated that "early consideration must be given to developing suitable techniques" for minimizing the production of low-pH, metals-rich water from this type of tailings. The Elliot Lake tailings are also in this category. They concluded that appropriate reclamation procedures for these tailings dumps should be directed at limiting the entry of oxygen into the tailings. Ten years have passed and no affordable means available to the mining industry have been developed to adequately limit the entry of water or oxygen to tailings.

There are 10 potash mines in Saskatchewan. Most of these mines are situated on or near potable aquifers, but none are near heavily used aquifers. At each mine, tailings in the form of chloride salts exist in lagoons and as large solid mounds. The lagoons are part of the ongoing effluent management system. The mounds increase in size as mining proceeds and are the permanent waste deposits. It is not possible to prevent salt from the lagoons and mounds from entering the groundwater zone. Engineering controls can be used to minimize the spread of the salt where it is necessary, but these controls will have to be actively maintained for as long as it is desired to minimize the spreading. The salt mounds will continue to contribute salt to the groundwater zone for centuries or longer.

Distributed Sources of Groundwater Contamination

Probably one of the most common contaminants in shallow groundwater in Canada is nitrate, which has a maximum permissible limit of 10 mg/L (as N) in drinking water. It is toxic to infants but not to noninfants. Nitrate is a common contaminant because it is derived from many types of sources, such as fertilizer application, septic tanks, and livestock, and because it is very mobile in groundwater. It does not adsorb on aquifer materials and it is very soluble. The only natural mechanism that can remove nitrate from groundwater is transformation to nitrogen gas, which is a harmless substance in water. The biological transformation of nitrate to nitrogen gas is referred to as denitrification.

Gillham and Cherry (1978), Hendry et al. (1983), and Trudell et al. (1986) have studied nitrate contamination in unconfined sand aquifers in southern Ontario. Widespread nitrate at shallow depth in many of these aquifers is attributed to the use of fertilizers on agricultural land overlying these aquifers.

An example of the vertical distribution of nitrate in one of these aquifers is shown in Fig. 10. In this aquifer, nitrate only occurs in the shallow zone very near the water table. At depth, nitrate is absent as a result of denitrification, which occurs where there is sufficient organic carbon to act as a source of energy and cell carbon for natural bacteria in the aquifer. This process prevents nitrate from penetrating to the lower part of the aquifer. Farm wells draw water primarily from this lower zone. In unconfined sand aquifers and in fractured bedrock aquifers where denitrification in not an active process, nitrate commonly penetrates much deeper. It is not known whether the organic matter that enables denitrification to proceed is solid-phase organic matter incorporated into the sand when it was deposited 10 000 yr ago or whether particulate organic carbon has been recently supplied from the land surface as a result of continual disturbance by cultivation. Thus, it is unknown whether denitrification will be sustainable in the long term.

Until recently, it was generally assumed that pesticides are not an important cause of groundwater contamination. It was thought that pesticides are adequately degraded biologically in surficial soil to prevent pesticide residuals from penetrating to the groundwater zone. However, studies in the United States, Europe, and most recently in Canada

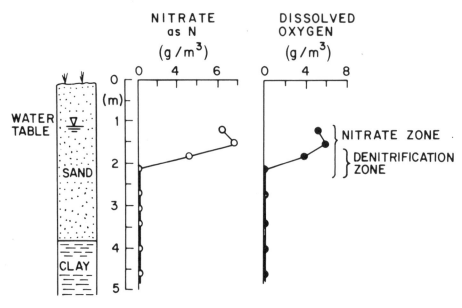

FIG. 10. Distribution of nitrate and dissolved oxygen in an unconfined sand aquifer near Rodney, Ontario (adapted from Trudell et al. 1986).

have now established that this is not the case in many areas. Aquifers particularly susceptible to contamination by pesticides are those in fractured rock where the overburden is thin. Organic pesticides have been found in groundwater supplies in agricultural areas of Prince Edward Island. Organic chemicals sprayed for vegetation control along transmission corridors for power lines have contaminated bedrock wells in Quebec. The search for pesticides in groundwater in Canada is still in its earliest stage.

It is well known that "acid rain" is causing widespread impacts on lakes in central and eastern regions of Canada. Little is known, however, regarding impacts of acid rain and other atmospheric contaminants on groundwater. Robertson and Cherry (1986a, 1986b) have investigated two shallow unconfined sand aquifers, one in an area about 90 km east of Winnipeg, Manitoba, and one near North Bay, Ontario. At each site the water table is within 2 m of ground surface. Atmospheric-derived sulfate attributed to combustion of fossil fuels is found in the aquifer to a depth of 10–15 m below the water table. Tritium dating of the groundwater and results of mathematical modelling of groundwater flow indicate that anthropogenic sulfate began penetrating the aquifer more than about 40 or 50 yr ago. At these sites the low pH of precipitation is not causing a significant decline in groundwater pH. The elevated sulfate levels are well below the limits recommended for drinking water. Whether or not atmospheric contaminants in addition to sulfate occur in groundwater at these two sites or in groundwater at any other sites in Canada remains to be determined.

Cole and Taylor (1986) studied 17 public water-supply wells in Pennsylvania and, for two of the supplies, found strong evidence that the water has become more acidic during the past 20 yr. They attributed this acidification to the effects of acidic pecipitation. They also cited studies in southern Sweden indicating progressive acidification of well water over a 30-yr period. In Canada, acidification of shallow well water is most likely to occur where the bedrock is not sedimentary in origin, such as in much of the Precambrian Shield region, where the acid-buffering capacity of the glacial overburden and bedrock is generally lowest.

The Current Situation and Prognosis

At nearly all of the sites in Canada where groundwater contamination has been discovered, the contamination was identified only after visual or odor impacts on the water become apparent or after recognition of the existence of nearby surface contaminant sources (Beak Consultants Limited 1986). These known sites of contamination probably comprise only a small fraction of the total number of sites where groundwater contamination now occurs. It is rare that aquifer monitoring has provided advance warning of impending contamination of water-supply wells because detailed monitoring of aquifers is minimal. Contamination of water-supply wells is usually unexpected. Once contamination of a well water is recognized, it is often impossible to deduce how long the well has been contaminated, what exposure people have sustained, or what health effects, if any, have been suffered.

Groundwater problems are increasing primarily because the number of organic compounds used in industry and agriculture is very large and because many of these compounds tend to be mobile and persistent in the groundwater zone. Of the many hazardous chemicals found in groundwater, few can be detected by taste or odor at the very low concentrations which may pose serious health problems over periods of long exposure. Many of these compounds are quite soluble relative to the low concentrations that are permissible in drinking water. Many laboratories for water quality testing have not yet acquired capabilities for analysis of the important organic contaminants at the very low but potentially hazardous concentration levels. To detect these compounds at these concentration levels, sophisticated sampling and sample processing techniques are required. Consequently, when a user of well water requests a laboratory to conduct a routine assessment of water quality, the chemical analyses that are conducted include few, if any, industrial organic compounds.

About 6.5 million Canadians obtain their household water supply from wells. About 4 million of these Canadians are rural residents using on-site household wells.

Wells that are most likely to be contaminated are those that have one or two of the following features: (a) shallow depth, generally less than about 20-30 m in unconfined sand or gravel aquifers; (b) fractured rock where there is little or no overburden on top of the rock; (c) improper well construction that allows surface water to drain down openings between the well casing and the borehole wall. Even very deep wells in fractured rock can be prone to contamination in areas where the overburden is thin because the well casing is normally very short. It is commonly sealed only a small distance into the rock (Fig. 2). Thus, although the depth of the hole is deep, the depth at which contaminants can first enter the pumped zone in the rock is shallow.

Quantitative information is lacking on the land use or contaminant sources near millions of household wells, but it is reasonable to expect that many of these wells are affected by contaminants from septic systems, buried fuel oil or gasoline tanks, disgarded motor oil, road salt, road water runoff, fertilizers, pesticides, livestock wastes, or other common contaminant sources. The health of a million or more Canadians dependent on well water may be affected due to consumption of contaminated well water. Based on studies in New Brunswick, Edgett and Coon (1986) stated that discovering that a well is polluted can be a devastating experience for the well users, an experience that has already marked the beginning of real hardship for a growing number of people in New Brunswick.

The perception of the nature of groundwater contamination outlined in this chapter provides the basis for the following prognosis for groundwater contamination in Canada in the next few decades.

(1) Aquifer contamination that already exists will, in many cases, gradually spread.

(2) Many water-supply wells that are not presently known to be contaminated will be identified as being contaminated.

(3) The number of contaminating compounds observed in wells will increase and new contaminants will be identified.

(4) Many aquifers that are not now contaminated will become contaminated.

(5) An increase in monitoring of wells using modern analytical methods to detect industrial organic contaminants will show that groundwater contamination is generally more widespread and deeper than previously thought.

(6) The discharge of contaminated groundwater into wetlands, streams, and lakes will increase.

(7) An increasing number of water supply wells in which contaminants are identified will be shut off and the former users of these wells will be supplied at much higher cost with water form other sources.

(8) There will be an increase in the number of sites where attempts will be made to remove contamination from aquifers but, for some time, successes will be few and costs will be large because appropriate technology has not yet been developed or tested.

(9) Public concern and fear with regard to the effects of waste disposal sites, pesticides, and industrial spills into groundwater will increase. This trend will be fueled by the seemingly unexpected occurrences of contamination and the inability of government and industry to predict trends or to solve the problem.

Monitoring

The prognosis just presented prompts one to ask whether there are ways to ameliorate the developing situation. Some steps have been taken in this direction. For example, monitoring wells are installed at nearly all new landfills in Canada to provide early warning of impending offsite contaminant migration in the groundwater zone. With an early indication of contaminant migration, remedial actions can be taken before major problems arise. This is, of course, an appropriate course of action, but current monitoring networks involve only a small number of sites. All sites for radioactive waste disposal are scrutinized intensely through groundwater monitoring, but the number of these sites, including uranium tailings impoudments, is less than 25. Monitoring is also undertaken on a regular basis at four sites in Canada where PCBs are known to have been released into the groundwater zone as a result of major spills or leakage from storage facilities. In a few local areas, rural water supply wells are being monitored to determine whether or not they are being affected by distributed sources of contamination, primarily agricultural chemicals.

The monitoring directed at new landfills, radioactive waste sites, and major spills of PCBs shows that provincial and federal governments recognize to some degree that groundwater contamination is an important issue. However, this "high-profile" monitoring pertains only to an insignificant fraction of the total number of potential sources of groundwater contamination and relates to sites that are generally not located on major aquifers or close to water-supply wells.

Groundwater monitoring involves two major steps: first, the selection of the number of wells and their locations and, second, the selection of parameters or constituents for analysis. The geology at many sites of actual or potential contamination is complex, which makes the design of networks of monitoring wells difficult and site specific. The probability of a particular monitoring-well network at a site adequately serving the intended purpose, such as early warning of contaminant migration, is generally not known in quantitative terms. The selection of parameters or constituents for analysis in well samples is a simple task at sites of PCB spills or radioactive waste disposal. This is not the case at many other contaminant sources. Traditional parameters for indentification of gross contamination such as chloride, total dissolved solids, total DOC nitrogen compounds, and coliform bacteria may suggest that the water is potable even though the water in fact contains harmfull industrial or agricultural chemicals. To analyze individually for the multitude of organic and inorganic contaminants that may occur in groundwater costs many hundreds or even thousands of dollars per sample. To select a small, less expensive subset of trace constituents as an indicator of significant contamination is feasible at some sites but at others the selection may not reveal the presence of some important contaminants. Proven strategies for cost-effective monitoring involving trace constituents are lacking.

Some Research Issues

Groundwater research in Canada is conducted primarily by several universities and by Atomic Energy of Canada Limited, Environment Canada, the Ontario Ministry of the Environment, Ontario Hydro, the research councils of Alberta and Saskatchewan, and the Alberta and federal Departments of Agriculture. A major groundwater research effort exists in only one topic area: subsurface storage and disposal of radioactive wastes. For several years, this research effort included, to a major degree, topics pertaining to uranium mill tailings, but now this is no longer the case. Th nuclear fuel cycle is primarily a federal regulatory issue and this research is a federal initiative. Most of the groundwater research pertaining to radioactive wastes focuses on very deep storage in Precambrian rock of spent nuclear fuel and other highly radioactive wastes. Not much of the knowledge that is gained from this research is directly applicable to many of the other types of actual or potential groundwater problems in Canada.

Research efforts, primarily in Ontario, are directed at the effects on groundwater of municipal landfills and of hazardous waste disposal facilities but these efforts are much less intensive than those directed at issues pertaining to radioactive wastes. In the Prairie provinces, substantial research is directed at the effects of groundwater on salinization of agricultural land and on wetlands and on the effects of coal mining and of tar-sands mining on groundwater. Small research programs in the Prairies and the Maritime provinces pertain to the effects of agriculture (nitrate, pesticides) on water quality in aquifers used for drinking water.

Techniques and strategies for groundwater monitoring have received scant attention, as is the case for problems pertaining to the rehabilitation of contaminated aquifers. Only a very small portion of groundwater research in Canada involves nonpriority research, including groundwater circulation, age and replenishment times, and natural geochemical and biochemical processes in the groundwater zone. Nonpriority research of this nature is needed to provide the basic framework of knowledge within which results of research specific to contaminant sources and effects are best utilized.

Considering that about one quarter of the population of Canada depends on water from aquifers, the total Canadian research effort directed at groundwater issues is extremely small. This is probably a reflection of the fact that each identified case of groundwater contamination potentially affects only a household or a few households. Rarely does a large urban population feel threatened, as was the case when the Regina PCB spill was investigated. At present, most rural Canadians that may be drinking contaminated groundwater are unaware of their situation and therefore public pressure for research pertaining to groundwater contamination is not coming from the part of the population that would likely benefit most quickly from the research.

Groundwater resources are owned by the provinces. Boundary issues, either interprovincial or international, rarely involve groundwater. For these reasons the federal government apparently has fostered few research initiatives directed at groundwater contamination beyond issues specific to the nuclear industry. This is unfortunate because groundwater contamination is a national problem, many aspects of which are beyond the provincial perspective. Results of research on topics such as the behavior of industrial and agricultural chemicals in groundwater, on the circulation and natural geochemistry and microbiology of groundwater, and on methods for cleaning con-

417

taminated parts of aquifers will assist in the management and protection of ground-water resources in all provinces. With the exception of Ontario, the provinces do not have underway any major research programs focused on groundwater contamination, either in their universities or in provincial agencies.

Planning and Regulation

In many circumstances, aquifers that lose their potability because of contamination cannot be returned to a potable state even after large expenditures over many years. Even the prevention of the spread of zones of contamination without ultimate cleaning of these zones can be difficult and costly. Most contaminated but unpumped aquifers eventually discharge contaminants in seepage to rivers, lakes, or wetlands. This may persist for decades or even centuries. A major goal of planning and regulation should therefore be prevention of aquifer contamination. Pursuit of this goal will provide long-term benefits. At many locations in Canada, shallow groundwater is already con-taminated or is becoming contaminated due to existing sources. Therefore, a second major goal of planning and regulation should be to prevent use of household or municipal wells that are already contaminated or that soon will become contaminated.

At present, most household consumers of groundwater draw their water from wells located on their property. The location and depth of household wells are selected by the property owner or the well driller. Guidelines in some provinces specify that household wells must be at least 15 or 30 m from septic systems. This distance is arbi-trary and does not depend on site-specific geologic or hydrologic factors. Drillers are mostly concerned with issues pertaining to well yield, the taste of the water, and esthetic water quality factors such as the iron content and hardness of the water. The problem is further complicated by the fact that early testing of the quality of the well water could give good results, with subsequent deterioration of the quality being caused by the drawing of contaminant plumes towards the well due to continual pumping. In some provinces, the driller is requested or advised to submit a water sample from each new household well to a provincial laboratory for assessment of a few water quality parameters. Following the initial assessment of the water quality of a new well, the onus is on the well owner to determine whether the well continues to show no indica-tion of contamination.

Planning and regulation by governments in Canada at present contribute little towards aquifer protection or prevention of use of contaminated well water. The existence of more than a million water-supply wells and of thousands of aquifers makes these goals difficult to achieve. However, there are many useful steps that could be taken imme-diately, and undoubtedly there are many others that would become evident after research and planning efforts are directed at these goals.

There is an urgent need to develop reliable national estimates of the numbers of households that are currently using contaminated groundwater and of the adverse health effects or risks to health caused by this contamination. To develop a statistical basis, provincial of federal agencies should periodically sample rural wells in representative areas to determine numbers and types of contaminated wells and the nature of the contamination. Areas in which natural constituents in groundwater pose a health hazard to well-water users should be identified in much more detail than at present. Of par-ticular importance are natural constituents such as uranium, radon, radium, lead, selenium, and arsenic. Nearly all municipal wells should be sampled periodically for analyses of various water quality indicators including many organic industrial compounds.

Whether or not it is the responsibility of government to monitor the quality of household wells on private property is debatable. Government should, however, make an effort to minimize use of contaminated groundwater by its citizens regardless of well ownership.

Wells that yield water containing tritium, which identifies groundwater of young age, are more likely to become contaminated than wells showing no tritium. Regulations could require a tritium analysis to be done on water from new wells shortly after they are put into production. Owners of tritium-producing wells would have particular reason for frequent monitoring of the quality of their well water. To achieve periodic testing of wells, provincial and municipal agencies should promote accress by well owners to diagnostic water quality analyses and tritium analyses at an affordable price. Reliable information on the effectiveness of household water purification systems for removal of contaminants should be readily available to users of well water. Possibilities for successfully replacing shallow wells prone to contamination with deeper wells should also be made known to well users. In some areas, deeper potable water is abundant and in other areas it is not. The possibilities are dependent on the local geological conditions.

For most parts of the habitated regions of Canada, maps of topography, soil conditions, and geology exists. Using these maps as well as borehole and water-well data on file with provincial agencies, terrain can be categorized with regard to its general susceptibility for aquifer contamination. In the United States a formal information and logic system for ranking areas for susceptibility to aquifer groundwater contamination has recently come into use by some state and local governments (Aller et al. 1985). An adaptation of this system may prove useful in Canada for identification of areas where priority efforts would best be directed for prevention of groundwater contamination or for assessment of the quality of water supplied by existing wells.

It is appropriate in areas susceptible to contamination for government regulations to require new wells to meet specific siting and design criteria. In these areas it is also appropriate for regulations governing septic systems, livestock operations, storage of gasoline and heating fuels, and proximity to roads and sewage lines to be particularly demanding.

In 1985, federal and provincial legislation requiring stringent precautions during the transportation of hazardous substances was enacted in Canada. The legislation is the result of a cooperative effort among federal, provincial, and territorial governments and industry. The goal of these regulations, which pertain to transportation on roadways, railways, and navigable waters, is to prevent public harm from explosions, gas releases, and liquid spills. These regulations will minimize possibilities for these occurrences to cause groundwater contamination along transportation routes.

There is also a need for stringent regulations pertaining to the storage and handling of industrial and agricultural chemicals and of petroleum products on the land where these substances are manufactured or used. These regulations should apply to the chemicals in their liquid product form and as liquid waste. In areas where there is significant susceptibility for aquifer contamination, these regulations should include a focus on preventation of spills or leakages into soil. Spills or leakages can be reduced to very rare occurrences if all storage tanks are double walled with associated leak controls and collectors and if all piping at manufacturing or retail facilities is placed in corridors, either above or below ground. These corridors should provide access for visual inspection and they should have leak detection and leak collection systems. At points where trucks or rail cars discharge or take on these substances, it should be made nearly impossible for spilled liquids to flow to exposed soil.

A few large corporations in Canada and in other countries, after recognizing the need for groundwater protection, recently constructed chemical storage and piping systems that include leak and spill prevention features of the type mentioned above. Thus, good examples of this approach to groundwater protection are now available.

Many industries in Canada producing hazardous liquid wastes that cannot be released legally to municipal sewers are in a difficult situation because possibilities for destruction or disposal of these wastes are poor. Only one integrated industrial waste processing and disposal facility exists in Canada. It is located near Sarnia, Ontario. By 1990, a second facility is to begin operation in Ontario. A facility is expected to be operating in Alberta in 1987. Other provinces ar making little or no progress (Ohlendorf 1986).

It is resonable to expect that this lack of facilities for hazardous waste disposal in all parts of Canada except the Sarnia area is a contributor to groundwater contamination. Long transportation distances to disposal facilities produce high disposal costs. Only a small percentage of the hazardous liquid industrial waste produced in Canada is disposed of at the Sarnia site. A minor amount is sent to disposal sites in the United States. The rest is (1) stored on-site where the wastes are produced, (2) disposed of in municipal landfills, or (3) dumped on the property of the waste generator, or along public roads or elsewhere. The situation cannot improve until facilities for disposals of these wastes exist in many parts of Canada and until the cost of disposal of hazardous liquid wastes is generally affordable by industry.

Deep pits excavated in thick unweathered clay-rich deposits offer good possibilities for safe burial of processed of solidified liquid industrial wastes of nearly all types. Deposits of this type occur in large areas in southern Ontario, southern Quebec, and the Prairie provinces. The disposal site near Sarnia is situated on a thick clay deposit. The site selected for the second disposal facility in Ontario is also located on thick clay. In a geologic perspective, Canada should be more able to develop well-distributed disposal facilities for disposal of hazardous wastes than nearly any other industrialized country. In addition to favorable geologic conditions, there are low population densities near most industrial centres in Canada. Except for British Columbia and the Maritime provinces, where rainfall is very high and geologic conditions are generally poor, the hydrogeologic impediments to the establishment of the necessary number and distribution of disposal facilities for hazardous wastes in Canada are minimal.

For several years now there has been much public awareness and concern regarding the contamination potential of landfills, both municipal and industrial. This general recognition that landfills are a potential cause of groundwater contamination has lead several provinces to implement searches for abandoned landfills. Provincial and municipal records pertaining to landfills abandoned prior to the 1970's are scanty or nonexistent. To assemble a comprehensive inventory of old landfill sites will be difficult, but it is a task that should be pursued.

In the Regional Municipality of Waterloo, Ontario, coal tar residuals from an old coal gasification plant were recently discovered during excavation at a new building site overlying a shallow confined aquifer. This potential cause of groundwater contamination, which has persisted out-of-sight beneath the urban landscape, was created about 85 yr ago. Nearly all of the water used in this municipality is well water. The municipality is now conducting a search of records for locations of other coal tar sites.

Attempts to locate old landfills and coal tar sites are desirable, but even more necessary are searches for all possible sources of potential groundwater contaminants in areas where such sources could cause contamination of aquifers containing potable groundwater or where groundwater contamination could eventually cause contaminant inputs to surface water. In relation to many other sources of groundwater contaminants, landfills and coal tar sites may not be particularly important.

When information on locations and nature of potential sources of groundwater contamination is combined with information on the hydrogeology and on well locations and depths, it will be possible to develop priorities for action. There are various possibilities for action such as intensive groundwater monitoring, removal of some of the contaminant sources, or localization of groundwater contamination by installation of impervious cutoff walls in the aquifer or operating of control wells. In some cases the nature of groundwater contamination and of the existing contaminant sources is such that removal of the sources and cleaning of the aquifer would be very expensive, on the order of many millions or tens of millions or more dollars. Where alternative sources of water supply can be developed, it may be appropriate to leave the contaminant sources in place and permanently write the aquifer off as a source of potable water.

420

In other cases, cleanup will be necessary to avoid growth of the contamination to an unacceptable degree. These will be difficult choices and for many years they will be hindered by the lack of an adequate research base.

Groundwater provides most of the drinking water of Europe: 65% in the Netherlands and Czechoslovakia, 71% in West Germany, East Germany, Switzerland, France, and Belgium, 93% in Italy, and nearly 100% in Austria and Denmark. Much of this groundwater usage is provided from large municipal wells or clusters of wells drawing from major aquifers. Thus, there has been much impetus for aquifer protection in European contries. The European approach involves the identification and delineation of major aquifers containing potable water and the establishment of protection zones on the terrain overlying these aquifers. This strategy pertains primarily to public wells, which are those wells operated by a city, municipality, or water company and which supply of a number of households. Milde et al. (1983) indicated that the protection zone concept implies numerous protection requirements. With decreasing distance from public water supply wells, land use regulations specify an increasing number of bans and restrictions. Inherent in this management approach is the expectation that contaminants travelling along a long or slow groundwater flow path to public water-supply wells represent less hazard to the well water than those travelling along shorter or faster paths. This approach is designed primarily to minimize occurrences of bacterial or viral contamination of wells. Table 5 indicates the industrial, municipal, and agricultural activities that are banned from the regulated zones in West Germany. Zone I is closest and Zone III is farthest from the well. The technical specifications for each zone vary from country to country, but not the overall framework of the management scheme.

TABLE 5. Land usages and facilities not acceptable in the three classification zones in the Federal Republic of Germany (from Milde et al. 1983). WES = water endangering substances.

Zone III	Zone II	Zone I
IIIA	Constructions, plants, workshops	Pedestrian traffic
		Vehicle traffic
Storage and commercial use of WES[2] and water endangering pesticides	Building sites and stocks	Pesticides
	Roads and railways	Manure
Mass livestocks	Transfer points, parking lots	
Hospitals, sanatoriums	Farms, stables, sheds	(Usually fence-protected property of water utility company)
Urbanizations	Intensive grassing	
New cemetaries	Fishponds	
Military/airport facilities	Sport/camping facilities	
Shunting stations	Allotments	
Waste sites	Cemetaries	
Sewage plants	Removal of surface layers	
Waste water treatment, sewage frams	Mining, blowups	
Essential removal of surface layers and drilling	Fuel storage	
Road construction with WES	Transport of WES	
	Sewage passage	
IIIB		
Oil refineries, chemical plants, smelting works		
Waste water injections		
Deposition/underground storage of WES		
Pipelines for WES		

For the protection of aquifers in Canada, this European approach in its present form is not generally suitable. Most of our major aquifers are confined. Public water-supply wells do not yet exist in many areas where major aquifers occur. There is a need for protection of unused aquifers for future use, and therefore, distance to wells cannot be a criterion in relatively unused aquifers. Many hazardous contaminants common in groundwater do not diminish appreciably in concentration even if they travel along groundwater flow paths for a long time and/or a long distance. The main lesson for Canada in the European effort is importance that these countries attach to land use regulation as a factor in groundwater protection.

Nearly 50% of the population of the United States uses groundwater in the household. New cases of well-water contamination are being reported at an alarming rate. Many new federal and state regulations pertaining to groundwater have been enacted in the past 7 yr. Much of this legislation focuses on municipal landfills, hazardous waste disposal sites, and idustrial waste storage facilities. The so-called Superfund Program is part of this legislative framework. The Superfund legislation pertains primarily to designated inactive waste disposal sites. It is retroactive in that it places financial responsibility for corrective action on the corporations that contributed the hazardous materials to the sites, regardless of whether or not these corporations were abiding by the regulations of the day or were using the generally accepted disposal practices of the time. In some cases, corporations are being held responsible for waste disposal or spills that occurred prior to the beginning of this century.

The Superfund Program now calls for corrective action at nearly 1000 major sites in the United States. It is expected that the number of designated sites will increase to about 5000 in the next few years. At many of these sites, aquifers are contaminated. The Superfund legislation calls for aquifer cleanup. The legislation has been in existence for more than 6 yr now and no contaminated aquifers have been returned to potability. At many Superfund sites the cost estimates for soil and aquifer cleanup are enormous. The program operates largely within an adversarial framework of litigation with very little awareness of the fact that it will not be feasible to return many of the aquifers at Superfund sites to a permanent potable condition. Much of the difficulty with aquifer cleanup at Superfund sites is caused by the DNAPLs that have penetrated deep into aquifers. Vast financial and technical resources are being squandered at many sites where the return of the aquifer to potability is not feasible, where the groundwater resource had very little value even before the contamination occurred, and where actual or future risks to the health of humans are very small. These resources are being consumed in a manner which will probably provide few long-term groundwater benefits, but which will detract from the application of major resources in the broader task of groundwater protection. It would be unwise for Canada to develop a program similar to the Superfund Program.

Some very recent federal legislation in the United States pertains specifically to aboveground and underground storage tanks and piping for hazardous industrial and agricultural chemicals and oil and gasoline. The tank design and monitoring aspects in these regulations are very stringent. In Florida, a new regulation requires that at least one groundwater-monitoring well be placed at every underground storage tank used for petroleum products or other hazardous liquids. This will result in the drilling of tens of thousands of new monitoring wells. These federal and state regulations pertaining to tanks and their associated piping will probably make a much greater contribution to groundwater protection than will the Superfund Program.

If governments in Canada should decide to pursue policies for groundwater protection, approaches specific to the Canadian conditions and needs rather than simply importation of American or European approaches will be necessary to get the most benefit from the effort.

Nearly all Canadian cities with a population greater than 100 000 derive their water supply from rivers or lakes. Much of our sewage, generally in a partially treated form and, in some cases, without treatement, is discharged to rivers and lakes. Many of our industries discharge waste effluents to rivers or lakes. Thus, except in norhtern Canada, our rivers and large lakes are contaminated with industrial and agricultural chemicals and with products of human wastes. Many but not all of these contaminants are removed in municipal water treatment plants. Other organic contaminants are created in the water as it passes through these treatment plants. It is generally acknowledged that it is not yet possible to make reliable assessments of the long-term chronic health effects of low doses of the chemicals commonly found in the treated drinking water obtained from many of our rivers and lakes. For some communities located near major aquifers, particularly confined aquifers, use of aquifer water that is many decades old and that does not contain industrial or agricultural chemicals is an alternative to the use of contaminated river or lake water. It may be that aquifer water in many parts of Canada has more value than previously thought.

Acknowledgements

I thank the following individuals for their very helpful comments on manuscript drafts: Robert Gillham, Michael Healey, Gordon Hodgson, Richard Jackson, Ronald Nicholson, Sherry Schiff, Peter Sly, Marcel Sylvestre, and Hennie Vedlhuizen. Assistance was also provided by Fred Baechlor and Bob Gunn who sent me information on groundwater contamination problems in the Maritimes. These acknowledgements do not imply that the individuals mentioned above agree with the opinions expressed in this chapter. I also thank Stephanie O'Hannesin who provided considerable editorial assistance, Nadia Bahar who drafted the diagrams, and Marilyn Bisgould who typed too many drafts of this manuscript.

References

ALEXANDER, H. C., W. M. McCARTY, E. A. BARTLETT, AND A. N. SYVERUD. 1982. Aqueous odor and taste threshold values of industrial chemicals. J. Am. Water Works Assoc. 595-599.

ALLER, L., T. BENNETT, J. H. LEHR, AND R. J. PETTY. 1985. DRASTIC: a standardized system for evaluating ground water pollution potential using hydrological settings. U.S. Environmental Protection Agency, Kerr Environmental Research Laboratory, EPA/600/2-85/018. 163 p.

BEAK CONSULTANTS LIMITED. 1986. Groundwater contamination in Canada: selected cases, potential sources and protection strategy. Final report, DDS File No. 5255.KE-145-5-0138. (Available from Environment Canada, Hull, P.Q.)

BOORMAN, R. S., AND D. M. WATSON. 1976. Chemical processes in abandoned sulphide tailings dumps and envrionmental implication for northeastern New Brunswick. Can. Inst. Mining Bull. 86-96.

BROOKMAN, G. T., M. FLANAGAN, AND J. O. KEBE. 1984. Solubilities of hydrocarbon compounds of gasoline. TRC Environmental Consultants Inc., East Hartford, CT. (Available from The American Petroleum Institute)

BROWN, I. C. [ED.]. 1967. Groundwater in Canada. Geol. Surv. Can. Econ. Geol. Rep. No. 24: 228 p.

BUSZKA, P. M. 1982. Hydrogeochemistry of contaminated groundwater in a sand aquifer at a landfill near North Bay, Ontario. M.Sc. project, University of Waterloo, Waterloo, Ont. 92 p. (Available from Department of Earth Sciences, University of Waterloo, Waterloo, Ont.

CALDER, L. M. 1986. Chromium in groundwater. In J. O. Nriagu [ed.] Chromium in the environment. John Wiley & Sons, New York, NY.

CANTER, L. W., AND R. C. KNOX. 1985. Septic tank system effects on ground water quality. Lew Publishers inc., Chelsea, MI. 336 p.

CHERRY. J. A. 1983. Occurrence and migration of contaminants in groundwater at municipal landfills on sand aquifers, p. 127-147. In C. W. Francis, S. I. Auerback, and V. A. Jacobs [ed.] Environment and solid wastes. Butterworths, Boston, MA.

COLE, C. A., AND F. B. TAYLOR. 1986. Possible acidification of some public groundwater supplies in Pennsylvania. Water Quality Bulletin, World Health Organization, Collaborating Centre On Surface And Groundwater Quality. Vol. II, p. 123-130, 171.

DAKIN, R. A., R. N. FARVOLDEN, J. A. CHERRY, AND P. FRITZ. 1983. Origin of dissolved solids in groundwaters of Mayne Island, British Columbia. J. Hydrol. 63: 122-270.

DAY, M. J. 1978. Movement and hydrochemistry of groundwater in fractured clayey deposits in the Winnipeg area. M.Sc. thesis, University of Waterloo, Waterloo, Ont.

DESAULNIERS, D. 1986. Ground water origin, geochemistry and solute transport in three major glacial clay plains of east-central North America. Ph.D. thesis in Earth Sciences, University of Waterloo, Waterloo, Ont. 455 p.

DESAULNIERS, D. E., R. S. KAUFMANN, J. A. CHERRY, AND H. W. BENTLEY. 1986. ^{37}Cl-^{35}Cl variations in a diffusion-controlled groundwater system. Geochim. Cosmochim. Acta 50: 1757-1764.

DUBROVSKY, N. M., J. A. CHERRY, E. J. REARDON, AND A. J. VIVYURKA. 1984. Geochemical evolution of inactive pyritic tailings in the Elliot Lake uranium district: 1. The groundwater zone. Can. Geotech. J. 22: 110-128.

DUBROVSKY, N. M., K. A. MORIN, J. A. CHERRY, AND D. J. SMYTH. Uranium tailings acidification and subsurface contaminant migration in a sand aquifer. Water Pollut. Res. J. Can. 19: 55-89.

EDGETT, J. AND D. COON. 1986. Petroleum on tap: the legacy of leaking underground storage tanks. Conservation Council of New Brunswick, Fredericton, N.B. 45 p.

FEENSTRA, S. 1982. Subsurface contamination from spills of dense non-aqueous phase liquid (DNAPL) chemicals. Proc. Second Annual Technical Seminar on Chemical Spills, sponsored by Environment Canada, Feb. 5-7 Montreal, P.Q.

FEENSTRA, S., R. D. BLAIR, J. A. CHERRY, J. C. CHAKRAVATTI, AND E. LaROCQUE. 1981. Hydrogeochemical investigations of two inactive tailings areas in the Elliot Lake uranium district, Ontario, Canada, P. 367-379. In Proceedings, Fourth Symposium On Uranium Mill Tailings Management, Fort Collins, CO.

FREEZE, R. A., AND J. A. CHERRY. 1979. Groundwater. Prentice-Hall Inc., Englewood Cliffs, NJ. 604 p.

FRENCH, W. B., M. GALLAGHER, R. N. PASSERO, AND W. T. STRAW. 1985. Hydrogeologic investigation for remedial action related to chromium-arsenic-copper discharge to soil and groundwater, p. 209-229. In Glysson, Swan, and Way [ed.] Innovations in water and wastewater fields. Ann Arbor Press, Ann Arbor, MI.

GARRETT, P., M. MOREAU, AND J. D. LAWRY. 1986. Methyl tertiary butyl ether as a groundwater contaminant. Proceedings of the Conference on Petroleum Hydrocarbons in Ground Water Conference. National Water Well Association, November 1986, Houston, TX.

GILLHAM, R. W., AND J. A. CHERRY. 1978. Field evidence of denitrification in shallow groundwater flow systems. Water Pollut. Res. Can. 13: 53-71.

GRISAK, G. E., J. A. CHERRY, J. A. VONHOF, AND J. P. BLUMELE. 1976. Hydrogeologic and hydrochemical properties of fractured till in the interior plains region, p. 304-335. In R. F. Leggett [ed.] Glacial till, Proceedings of a Symposium, Ottawa 1975. R. Soc. Can. Spec. Pub. No. 12.

HEALTH AND WELFARE CANADA. 1978. Guidelines for Canadian drinking water quality. Canadian Government Publishing Centre, Supply and Services Canada, Hull Que. 79 p.

HENDRY, M. J., J. A. CHERRY, AND E. I. WALLICK. 1986. Origin of groundwater sulfate in a fractured till in an area of southern Alberta. Can. Water Resour. Res. 22: 45-61.

HENDRY, M. J., R. W. GILHAM, AND J. A. CHERRY. 1983. An integrated approach to hydrogeological investigations: a case history. J. Hydrol. 63: 211-232.

HENDRY, M. J., R. G. L. McREADY, AND W. D. GOULD. 1985. Distribution, source and evolution of nitrate in a glacial till of southern Alberta, Canada. J. Hydrol. 70: 177-198.

HESS, P. J. 1986. Groundwater use in Canada, 1981. Inland Waters Directorate Bull. No. 140. Environment Canada, Ottawa, Ont. 43 p.

KELLER, C. K., G. VAN DER KAMP, AND J. A. CHERRY. 1986. Fracture permeability and groundwater flow in clayey till near Saskatoon, Saskatchewan. Can. Geotech. J. 23: 229-240.

MACKAY, D. M., P. V. ROBERT, AND J. A. CHERRY. 1985. Transport of organic contaminants in groundwater. Environ. Sci. Technol. 19: 384-392.

MAINE DEPARTMENT OF THE ENVIRONMENT. 1985. Survey of existing underground oil storage tanks: a report to the Main legislature, October 1985, Maine.

424

MILDE, G., K. MILDE, P. FRIESEL, AND M. KIPER. 1983. Basis and new development of the groundwater quality protection concepts in Central Europe, p. 287-296. *In* Proc. International Conference on Groundwater and Man, Sydney, Australia.

MOORE, M. B. 1986. An investigation of the permeability of a land aquifer to determine the contaminant mass flux of selected contaminants from a landfill plume. M.Sc. thesis, University of Waterloo, Waterloo, Ont. 305 p.

MORIN, K., J. A. CHERRY, T. P. LIM, AND A. VIVYURKA. 1982. Contaminant migration in a sand aquifer near an inactive uranium tailings impoundment, Elliot Lake, Ontario. Can. Geotech. J. 19: 49-62.

NOVA SCOTIA DEPARTMENT OF HEALTH. 1983. Naturally occurring arsenic contamination of water wells in Nova Scotia. (Map obtainable from the Nova Scotia Department of Health).

OHLENDORF, P. 1986. Waste not, want not. Report on Business Magazine, August, p. 42-49.

PATRICK, G. C., AND J. F. BARKER. 1985. A natural-gradient tracer study of dissolved benzene, toluene and xylenes in groundwater. 2nd Annual Canadian/American Conference on Hydrogeology: Hazardous Waste in Ground Water — a soluble dilemma, June 25-29, 1985, Banff, Alta. Sponsored by Alberta Research Council and NWWA.

PATTERSON, R. J., AND H. M. LIEBSCHER. 1986. Laboratory simulation of pentachlorophenol/phenate behaviour in an alluvial aquifer. Water Pollut. Res. J. Can. Special Issue on Groundwater Pollution and Remediation.

PETERS, L. P. 1977. Summary of the occurrence of arsenic and selected heavy metals in rural domestic water supplies in Fork and Charlotte counties of the Province of New Brunswick, Canada. Department of the Environment, Water Resources Branch, Fredericton, N.B. 27 p.

POULIN, M. 1977. Groundwater contamination near a liquid waste lagoon, Ville Mercier, Quebec. M.Sc. thesis, University of Waterloo, Waterloo, Ont. 158 p.

POULIN, M., G. SIMARD, AND M. SYLVESTRE. 1983. Pollution des eaux souterraines par les composes organiques a Mercier, Québec. Proceedings, Geological Association of Canada, Annual Meeting, 12 p.

PTACEK, C. J., J. A. CHERRY, AND R. W. GILLHAM. 1986. Mobility of dissolved petroleum-derived hydrocarbons in sand aquifers. *In* J. H. Vandermeulen and S. E. Hudrey [ed.] Oil and freshwater: chemistry, biology and technology. Pergamon Press Canada Ltd., Willowdale, Ont.

PUPP, C. 1985. An assessment of ground water contamination in Canada, Part 1. (Available from Environmental Interpretation Division, Environment Canada, Hull, P.Q.)

PYE, V. I., R. PATRICK, AND J. QUARLES. 1983. Groundwater contamination in the United States. University of Pennsylvania Press, Phyladelphia, PA. 315 p.

REINHARD, M., N. L. GOODMAN, AND J. F. BARKER. 1984. Occurrence and distribution of organic chemicals in two landfill leachate plumes. Environ. Sci. Technol. 18: 953-961.

RENDER, F. W. 1970. Geohydrology of the metropolitan Winnipeg area as related to groundwater supply and construction. Can. Geotech. J. 7: 243-274.

ROBERTS, J. R., J. A. CHERRY, AND F. W. SCHWARTZ. 1982. A case study of a chemical spill: polychlorinated biphenyls (PCBs). 1. History, distribution, and surface translocation. Water Resour. Res. 18: 525-534.

ROBERTSON, W. D., AND J. A. CHERRY. 1986a. Sulfate in shallow groundwater in the recharge area of a sand aquifer: relation to fossil fuel combustion. (Unpublished manuscript)
    1986b. Groundwater sulphate for identification of historical trends in precipitation pH; phase 2: Stratigraphy, groundwater, flow patterns, and preliminary hydrochemistry of the Sturgeon Falls field site. A progress report, submitted from the University of Waterloo. 45 p. (Available from Inland Waters Directorate, Environment Canada, Hull, P.Q.)

SASKATCHEWAN - NELSON BASIN BOARD. 1973. Other considerations — Groundwater. Tech. Rep. Append. 7.

SCHWARTZ, F. W., J. A. CHERRY, AND J. R. ROBERTS, 1982. A case study of a chemical spill: polychlorinated biphensyls (PCBs). 2. Hydrogeological conditions and contaminant migration. Water Resour. Res. 18: 535-545.

SCHWILLE, F. 1984. 3. Migration of organic fluids immiscible with water in the unsaturated zone, p. 28-48. *In* B. Yaron, G. Dagan, and J. Goldshmid [ed.] Pollutants in porous media. Spinger-Verlag, New York, NY.

TRUDELL, M. R., R. W. GILLHAM, AND J. A. CHERRY. 1986. An in-situ study of the occurrence and rate of denitrification in a shallow unconfined sand aquifer. J. Hydrol. 83: 251-268.

U.S. Environmental Protection Agency. 1985. Determination of toxic chemicals in effluent from household septic tanks. Rep. PB85-196798/AS,NTIS, 5285 Port Royal Rd., Springfield VA.

Villaume, J. F. 1985. Investigations of sites contaminated with dense, nonaqueous phase liquids (NAPLS). Ground Water Monit. Rev. (Spring) 60-74.

CHAPTER 15

Who Controls the Aquatic Resources?

Andrew R. Thompson

Professor of Law and Director, Westwater Research Centre,
The University of British Columbia, Vancouver, B.C. V6T 1W5

Some Fundamental Questions

There is a certain presumption in the topic — that mankind can control nature. Manipulate, maybe, but control, except in an ephemeral time scale, no! In the end, nature marches according to its own rules and rythms. The aquatic resource, by its floods and droughts, reminds us of vulnerability to nature's whims.

Yet we try to discipline the aquatic resources to serve human needs, and, within our ability to judge, we enjoy considerable success. We contain the floods, harness the water's energy, transport its life-giving nourishment to drought-stricken fields, and harvest its living resources. It is likely we will keep on trying to bend water to more and more beneficial uses according to our perceptions of needs. So the question "who controls aquatic resources?" is a central one, for its answer will determine whose perceptions and whose needs will govern our development plans.

There are those who resist this frankly anthropocentric view of nature. In the extreme they argue for a purely ecosystem-centred approach wherein nature is left undisturbed and mankind must limit expectations to needs that can be fulfilled without harm to nature. A more human-centred approach is one that accepts the inevitability of impacts on natural systems but calls for restraint, arguing that trees and animals should have rights as well as people — rights of preservation and of compensation in kind when damage occurs.

In the sense of fundamental human rights, such as the rights of persons entrenched in the Canadian Charter of Rights, trees and animals are outlaws. But that is not to say that mankind's governance systems ignore intrinsic values in nature altogether. The *National Parks Act* in Canada expressly requires protection of park values for future generations. The *Fisheries Act* imposes legal duties to protect fish from deleterious substances. Fish habitat is protected as well. A policy of the Department of Fisheries and Oceans says that where fish habitat is disturbed, compensation shall be provided in kind. "No net loss" is the goal, to be accomplished by requiring the restoration or replacement of habitat that is lost.

An example of an explicit regime for protecting water resources in their natural state is to be found in the *Northern Inland Waters Act*, the federal statute that governs water use in the Yukon and Northwest Territories. Under this statute a "conservation use" can be licensed to preserve water in its natural state. A Hunters and Trappers Association, or a fish camp operator, can apply to the Water Board for a licence that would identify a body of water and impose terms and conditions restricting its use so that its natural conditions of flow and quality are maintained. A similar provision is contained in the Alberta *Water Resources Act* but not in other provincial water legislation. Recently, applications have been made for conservation use licensing to protect a bog (Alberta) and a domestic water supply (Yukon), but these applications are without precedent and are presenting the licensing tribunals with difficult policy and legal questions related to the degree of protection and priority that should be given them.

Note that in all these examples, human beings must act to assert the protected status, and human values and needs are the measures of the protections given. People, then, are the fundamental subjects and objects of the governance system; it is people who are given rights to use natural resources, and it is people who are given duties to protect them. Consequently, answering the question "who controls aquatic resources?" must start with an examination of individual legal rights to use water. We must ask who owns the water resources, what rights does an owner have, and what constraints bind the owner?

An Illustration

Let me begin with an illustration of how complicated the answers can be, seen in the simple case of dumping rock into a river. I am going to describe the "twin-tracking" dispute in British Columbia and list the legal questions that arise in that situation. CN Rail has been given authority by the federal government to build a second track for its mainline through British Columbia. For about 700 km this mainline follows the valleys and canyons of the Thompson and Fraser rivers. Building a road-bed for the second track can be most efficiently accomplished in most locations by dumping rock into the river to "rip-rap" the banks. The Fraser and Thompson rivers are principle migration rivers for salmon and they provide spawning and rearing habitat for fish as well. Some 36 Indian bands representing three separate Indian nations have been allocated reserves under the *Indian Act* along the banks of the rivers. Many of these reserve allotments include express rights to fish in the rivers.

Seldom are there property rights in the Canadian legal system that are not complicated by the twists and turns of our history as a nation. The building of Canada's first transcontinental railway, the CPR, was a precondition of British Columbia joining the federation, and special land arrangements were necessary. Consequently, while British Columbia retained ownership and control of its natural resources, including water, under its Act of Union with Canada in 1871,[1] it was required to transfer the "Railway Belt"[2] — a 20-mile strip on each side of the CPR line through the province — to Canada. The Railway Belt provided the Dominion with lands with which it could honour the subsidy commitments made to the railway entrepreneurs. The remnants of this Railway Belt — those lands that had not been transferred to the CPR or to private persons by Canada — were returned to British Columbia when the federal government relinquished ownership and control of natural resources to the western provinces by the Natural Resource Transfer Agreements of 1930.[3] Consequently, two distinct periods of time must be considered in reviewing any land ownership question in this corridor of British Columbia, a federal ownership period prior to 1930 and a provincial period thereafter. The significance of this history for the twin-tracking dispute is that most of the Indian reserve rights and CN Rail rights of way in this corridor were granted prior to 1930 when the Crown lands were under federal ownership and administration whereas the underlying ownership today is provincial. This dichotomy presents interesting questions for lawyers concerning what statute laws and cases must be applied to provide answers to current questions about ownership (Ketchum 1984).

To resume the story of the "twin-tracking" dispute, the Indians concluded that CN Rail was proceeding with its twin-tracking project through the Thompson River without proper regard for their reserve fishing rights. The extensive placement of rip-rap would convert natural river banks into walls of stone, impeding both access to the river and safe landings from boats, and jeopardizing the spawning and rearing success of the salmon. A government/CN Steering Committee and a Technical Working Group were supervising CN design and construction plans from the points of view of the environmental, fisheries, and social concerns of the governments, and an Environmental Assessment Review Panel (EARP) had been established to review these plans. Nevertheless, the Indians, through their Alliance of Tribal Councils, decided that their interests were

428

not being protected. Therefore, they brought suit in the Supreme Court of British Columbia for an injunction restraining CN Rail proceeding with the twin-tracking. The grounds supporting the claim to an injunction were alleged to be interference with aboriginal rights, trespass to reserve rights, including rights of fishing, and interference with traditional use and occupation of the river banks and beds. These allegations raise the following legal questions. They mainly revolve about the question who has the legal right to control the dumping of rock into the river — in effect, who controls the aquatic resources? While the locale in question in the law suit involves merely a 10-mile stretch of the Thompson River, the legal rulings that are made will have significance not just for the remainder of the railway corridor in British Columbia but also for all other situations in Canada where railroads, bodies of water, and Indian reserves are in close proximity.

A. (i) Do the Indians have subsisting aboriginal rights to the aquatic resources?
   (ii) If so, do these Indian rights include a veto over development activities?
   (iii) Were these Indian rights, if any, superior to the rights of Canada, prior to 1930, to grant rights-of-way to CN Rail?
   (iv) Do these Indian rights, if any, override the right of British Columbia to regulate the use of water resources in the Railway Belt since 1930?
B. (i) Do the reserve rights granted to the Indians include the right to fish?
   (ii) If so, is this right to fish a proprietary right in the legal sense?
   (iii) Does this proprietary right to fish include rights to control the use of the banks and beds of the rivers?
   (iv) If so, are these rights to fish superior to the rights of CN Rail pursuant to its right-of-way grants?
C. (i) What are the legal boundaries of the CN Rail rights of way? Do they extend beyond high water mark into the beds of the rivers?
D. (i) If the Indians do not have proprietary rights in the banks and beds of the river and the right to fish, who does?
   (ii) If British Columbia owns the banks and beds of the river and the right to fish, how has its title been established, considering the history of the Railway Belt, the 1930 Natural Resources Transfer Agreements, and the subsequent British Columbia legislation (the *Land Act* and the *Water Act* purport to vest the ownership of water and of all beds of rivers, lakes and streams in the Crown in right of the Province)? Is this title, if any, superior to the title of the Indians, if any?
E. (i) If British Columbia owns the banks and beds of the rivers and the right to fish, is CN Rail violating these rights by dumping rock into the rivers?
   (ii) Does CN Rail need a licence from the province under the *Water Act* which requires a licence to interfere with a water course?
   (iii) Can provincial laws regulate activities of CN Rail considering that CN Rail is a federally incorporated railway subject to regulation by the Canadian Transport Commission?
F. (i) Considering that CN Rail is federally incorporated, that transcontinental railways are subject to the laws of Parliament under s. 92(10)(a) of the *Constitution Act*, and that fish and fishing are also subject to the laws of Parliament under s. 91(12) of the *Constitution Act*, do these federal heads of power preclude provincial laws and regulations from applying to the twin-tracking project?
   (ii) Can these federal heads of power be exercised in a way that overrides Indian aboriginal ownership rights or reserve fishing rights?

What does this inventory of legal issues tell us about who controls aquatic resources in Canada — apart from emphasizing that the answer is bound to be complicated? Let us go back to the three questions: who owns the aquatic resources, what rights does an owner have, and what constraints bind the owner?

Who Owns the Aquatic Resources in Canada?

The earliest claim to control water resources in Canada is the aboriginal rights claim of the Indian people as yet undefined by the courts or the Constitution.

The next claim is the claim of the governments through ownership of all unalienated lands and resources. This claim can be asserted either by the Crown in right of a province or the Crown in right of Canada, depending on the terms of the Constitution and on subsequent transfers of ownership between the governments.[4] Generally, in Canada, the provincial governments own the public lands and resources including water.[5] The exceptions are the northern and offshore lands and waters where federal ownership prevails.[6]

Both levels of government have been free throughout our history to alienate lands and resources to private persons. Colonial policy and the requirements of a developing nation led to a policy of generous grants of lands and other resources for settlement, agriculture, forestry, industry, mining, and petroleum developments. Consequently, the third tier in the ownership ranks is occupied by private parties who have been granted ownership rights by the provincial or federal governments. At common law, these private ownership rights included certain rights to the use of water flowing by the land. In addition, they normally included ownership of the beds of lakes, rivers, and streams. In fact, if a land grant included the banks or bed of a body of water, the owner was entitled to certain rights in the water as a natural incident to this ownership. These were the riparian rights of the common law — the rights to use the water and the duty to allow the water to flow downstream undiminished in quantity and quality save for certain normal and ordinary uses. These common law rules meant that the private owner of the banks of a body of water could control which water developments would be allowed to proceed and which would not, just as a private owner of mineral rights could control whether or not the minerals would be exploited. The only exception were uses of a normal and ordinary nature such as taking water for domestic purposes (Chesman 1984a).

However, there has been a tendency in the western provinces and the northern territories of Canada to consider that renewable natural resources like water and wildlife should be treated differently from nonrenewable natural resources and given special status as resources to be enjoyed by all the people. For example, the *Water Act* in British Columbia reverses the common law riparian rules by declaring that the property in water is and shall be deemed always to have been owned by the Crown and useable only by licence from the government. Also, the British Columbia *Land Act* states that the beds of all lakes, rivers, and streams are conclusively deemed to belong to the Crown subject to limited exceptions in favour of earlier-granted rights and to the right of the government, even today, to make an express grant of the bed to a private person in an appropriate case.

Since development water uses can be licensed and express grants of the bed can be made under these public ownership regimes, the ultimate effect of these statutory policies is not substantially different from the results under the common law riparian rules. That is, in both cases control over the development of water resources can be gained by private parties. The difference is that formerly water development rights had to be acquired from the multitude of private riparian owners whereas today they can be acquired from government.

In the older provinces of Canada, riparians rights have not been abolished in outright fashion by legislation, as they have in western and northern Canada, but the common law rights of riparian owners have nevertheless been substantially abridged. This results from legislation that requires government approval if structures like dams are to be placed in water bodies, if substantial quantities of water are to be diverted, or if polluting substances are to be discharged into water. The difference in some situations will be that the developer will have to get the consent of the riparian owners as well as the permission of government before proceeding with a development project. Even in this case the difference from the water licensing regimes of the western provinces and northern territories is more a matter of form than of substances because the licensing statutes of the latter exempt ordinary domestic uses from the requirement of licensing and give riparian owners the opportunity to intervene in licensing procedures to argue that a development licence should not be granted unless the interests of the riparian owners are protected.

To recapitulate, the tiers of ownership of the aquatic resources are, first, aboriginal rights (as yet undefined by Canadian courts), second, government ownership, either federal or provincial, and third, private ownership rights (riparian rights) or their equivalents under licensing arrangements.

What Rights Does an Owner Have?

We normally think of an owner as having rights to use and control the object of ownership, to receive its profits, and to sell it to someone else. Ownership of water in Canada today carries all of these rights, though modified to reflect the special characteristics of water as a flowing resources in its natural state (Rueggeberg et al. 1984). Thus, water cannot be owned in its natural state in the same way that one can own a book or a pen. That would require an excursion into metaphysics because the water itself is constantly changing as it flows past. However, when water is changed from its natural state, as when spring water is bottled for sale, the bottled water is just as capable of ownership as any other chattel and someone drinking the water without the owner's consent is guilty of theft.

There is one important limitation resulting from the flow characteristic of water in its natural state. This limitation prevails both under common law rules and under licensing statutes. It is that water rights cannot be entirely separated from the riparian lands to which they originally appertain and consequently are not freely marketable by themselves.[7] This restriction stems from the practical fact that a right to use water at location A on a water course cannot be transferred to location B without affecting intermediate and downstream users and calling for the intervention of a regulating authority. Through the licensing process, this authority will impose terms and conditions on the licensed use regarding quantity and quality designed to protect the interests of downstream users.

In summary, owners of water rights can use the resource and control its use by others, whether under the older riparian law or under modern licensing statutes, but there are restrictions on the extent of these rights of use stemming from the flow characteristics of water in its natural state.

The subject of ownership of water is not complete without a reference to fishing rights. At common law the right to fish is enjoyed as part of the ownership of the bed of a body of water. It is an exclusive right of the owner, and, unlike rights to use water, it can be transferred apart from a transfer of the land rights. That is, the right to fish enjoyed by owner of bed A can be transferred as a property right to B who is not a riparian or bed owner at all, but may be an urban sports fishing club that leases fishing rights for the enjoyment of its members. This is the system that exists in England and to some extent in the eastern provinces of Canada.

But in western and northern Canada these common law rights to fish have been virtually repealed pursuant to a policy to make fishing a public right rather than a private one. In the prairie provinces, for example, property in all wildlife is vested in the Crown in right of the province under wildlife legislation that restricts fishing except in accordance with a licence to fish.

A special situation has existed from time immemorial in English common law countries with respect to tidal and coastal waters where there is a public right of fishing to which all ownership rights (if any) are subordinated.[8]

Finally, underground water stands in a special position so far as ownership is concerned. Unlike the riparian law, the common law governing subterranean waters imposed no duty on a land owner to restrain the use of such waters so as to avoid injury to others who might want to use them (McIntyre 1969). Probably this rule reflected the practical difficulties of defining underground aquifers so as to identify and quantify effects caused by their use.

Today, withdrawals from underground aquifers are sometimes controlled through government licensing (Alberta; Ontario),[9] but in general the common law rules still prevail. They have led to the need for zoning regulation in cases where urbanization has resulted in excessive groundwater withdrawals. As well, contamination of groundwater is indirectly regulated by pollution control legislation which, in Canada, operates to restrict discharges into the air and water environments that exceed defined parameters.

All of these ownership rights are subject to restraints of many kinds imposed by federal and provincial statutes. This is the subject to which we turn in answering the third question: "what constraints bind the owner?"

What Constraints Bind the Owner?

Ownership rights and duties are only part of the answer to the question "who controls aquatic resources?" The remaining part is to be found in the laws enacted by Parliament and by the legislatures of the provinces and territories that regulate the exercise of ownership rights. The importance of these restrictive laws is abundantly clear in the case of the water rights legislation. Ohter statutory provisions are equally important. It is first necessary to explain the law-making powers in Canada.

Constitutional Constraints

Canada is a federal state with the power to legislate divided between the federal Parliament and the provincial legislatures.[10] The *Constitution Act* specifies those subjects upon which either the federal Parliament or the provincial legislatures may exclusively make laws. All the provinces are subject to the same rules. Section 91 of the *Constitution Act* is the primary list of those subjects (termed "heads of power") about which the federal Parliament may legislate; s. 92 states the provincial legislatures' heads of power. The two lists are intended to cover all possible subjects. More importantly, the two lists are intended to be exclusive so that neither the federal Parliament nor the provincial legislatures may make laws concerning a matter which does not fall within one of their respective heads of power. When a court finds that a given federal or provincial law concerns a matter that is within a head of power belonging to the other, the court rules that the law is *ultra vires*, which means that the legislature which made the law did not have the power to make it. Once the court declares a law to be *ultra vires*, it has no force or effect.

The courts, however, often have difficulty declaring whether any law is *ultra vires* because the various heads of power in both s. 91 and s. 92 are, by necessity, expressed in very general terms. As a result, the lawyers for the federal and provincial sides of a constitutional dispute may each argue that the disputed law falls within one or more of the broadly worded heads of power given to their side by s. 91 or s. 92.

Provincial Heads of Power

Jurisdiction over aquatic resources is divided between federal and provincial powers. Under s. 109 of the *Constitution Act*, water is owned by the provinces, and under s. 92(5) the provincial legislatures may make laws governing the management and sale of aquatic resources. In addition, three other heads of power in s. 92 allow the provinces to make laws concerning other aspects of aquatic resources. These are s. 92(10), "local works and undertakings"; s. 92(13), "property and civil rights"; and s. 92(16), "matters of purely local and provincial concern." Furthermore, s. 92 A(1)(c) expressly states that the provinces have management powers with respect to hydroelectric power developments.[11] Collectively, these heads of power give the majority of legislative jurisdiction over aquatic resources to the provinces.

Federal Heads of Power

The federal Parliament, however, is allotted three kinds of power which give it constitutional authority to make laws which either directly or indirectly affect aquatic resources. These are the federal "residual" power (Peace, Order, and Good Government or "POGG"), the power to make laws concerning coastal and inland fisheries (s. 91(12)), and the power to make laws concerning navigation and shipping (s. 19(10)).[12] As well, Parliament can pass laws regulating the export of water from Canada.

This division of law-making power respecting aquatic resources raises two central questions. First, given that the provinces own and manage water, how and in what circumstances can the federal parliament make laws affecting water? Second, if the federal laws should conflict with provincial laws, which will prevail in any given situation? Some answers may come from looking at the nature of the three federal heads of power in more detail.

The residual power — POGG

The preamble to s. 91 of the *Constitution Act* gives the federal Parliament the power "to make laws for the peace, order, and good government of Canada..." or "POGG," as lawyers like to call it. This power is termed "residual" as it allows the federal Parliament to legislate on subjects of a national dimension which were not included in either s. 91 or s. 92 heads of power such as radio and aeronautics. The real importance of this power, however, is that it may also be used by the federal Parliament to legislate, under special circumstances, on matters which are constitutionally under exclusive provincial power, such as controlling inflation in costs and prices.

There is a lot of uncertainty about just what these special circumstances are. According to the decisions of the courts in cases which have dealt with this question, there are two possible circumstances where provincial powers may be overridden: where the subject of the law in question is of "national concern" and where the subject constitutes a "national emergency." Defining exactly what these expressions legally mean has been hotly debated in the courts (Chesman 1984b). It is sufficient to say here that some lawyers believe that only if the circumstances constitute a "national emergency — in that the entire country is affected in some immediately damaging way — will the courts allow the federal Parliament to exercise its residual power in a way that substantially affects provincial powers. This was so in the *Anti-Inflation Case* when the Supreme Court of Canada upheld federal wage control laws that invaded provincial legislative powers. Other lawyers contend that in matters of "national concern — that is, the effects are not necessarily nationwide or of "crisis" proportions — the federal Parliament can still act under its POGG power. Some recent cases suggest that the courts are learning towards the latter interpretation.

Fisheries

Using its s. 91(12) power, the federal Parliament passed the *Fisheries Act*, which regulates, among other things, the deposit of contaminants in water. But is that not a provincial concern under a province's water management power? Recent court decisions suggest that the *Fisheries Act* is valid federal law only to the extent that its regulation of water quantity or quality is directly linked to the protection of the fishery.[13] For example, s. 33(3) of the *Fisheries Act* was ruled invalid because it created an offence of merely depositing slash or debris in a water course without requiring that there be harm to fish caused by the slash or debris. Nevertheless, this power still provides Parliament a wide scope to enact laws and regulations affecting water resources.

Navigation

Using its s. 91(10) power, the federal parliament passed the *Navigable Waters Protection Act*. The Act is potentially an important federal instrument in the area of water management, depending on how broadly or narrowly the courts interpret this head of power. For example, s. 6 of the Act allows the federal Minister to order the removal or destruction of any work (bridge, boom, dam, etc.) that, without federal approval, interferes with navigation.

In summary, the provincial legislatures have the constitutional powers to make laws concerning the ownership and management of aquatic resources. The federal Parliament, under POGG and through its powers over fisheries and navigation, may make laws that directly interfere with provincial jurisdiction as well as indirectly affect provincial water management. The result is that now, approval from both levels of government is often needed for activities that affect aquatic resources. For example, if a dam is to be built, provincial approval is necessary because of provincial ownership and management of water while federal approval is also necessary if the dam would affect the fishery or navigation.

Federal Involvement in Interprovincial Disputes

Whatever authority is encompassed under "POGG" is still, as Professor Percy noted, "notoriously controversial" (Percy 1984). This is particularly true in considering whether the federal government should intercede, in the interests of peace, order, and good government, in interprovincial disputes. For instance, acid rain might be agreed to be a "national emergency" needing federal action, since its effects are so pervasive. But can a dispute between British Columbia and Alberta over building a dam on the Peace River, or a disagreement between Quebec and Newfoundland over the use of Churchill Falls to generate electricity, be considered a "national emergency"? Are they even of "national concern" such that the federal government should step in even in some limited way? So far, there are no legal authorities that can give a definite answer. The *Canada Water Act* was the legislative flagship launched in 1970 as the means for asserting federal jurisdiction over transboundary water resources should provincial governments be unwilling to join with federal agencies in cooperative water management schemes, but the Act has never been used by the federal Department of Environment in a way that would test the constitutional authority of Parliament to override provincial objections.[14] The issues and dilemmas that both provincial and federal governments may find themselves in as a result of this jurisdictional ambiguity are at the forefront of water law issues in Canada today.

Applying the Rules

With these rules established, a clearer picture can now be given of the laws and regulations applicable to the twin-tracking illustration.

To justify placing rocks in the bed of the river, CN Rail must gain a property right from the owner of the bed, provided that its existing right-of-way does not already include the bed of the river. Otherwise, CN Rail is legally a trespasser. If the Indians own the bed, CN Rail must acquire this property right from them; alternatively, CN Rail must acquire this right from the province of British Columbia. If it cannot negotiate for these rights successfully, it can turn to certain rights of expropriation given by federal legislation. Parliament is competent to bestow powers of expropriation on CN Rail because the railway is an interprovincial undertaking under s. 92(10)(a) of the *Constitution Act*.

Placing rocks in the river is an interference with a water course that ordinarily must be licensed under the provincial *Water Act*. However, there is an argument that CN Rail, as an interprovincial work or undertaking, is exempt from this provincial licensing requirement. In fact, CN Rail is using this argument as a reason for ignoring the licensing provisions of the *Water Act*.

The Indians may have fishing rights in addition to, or independently of, ownership rights in the bed of the river. The placing of rock in the river may do injury to these fishing rights by interfering with the spawning, rearing, and migration of salmon. It may also interfere with the way in which Indians fish the rivers. CN Rail must acquire from the Indians permission to injure and interfere with these fish and fishing rights, unless CN Rail is prepared to rely on a legal opinion that neither aboriginal rights nor reserve rights give the Indians a legal right to the fish or to fishing.

But if the Indians have no legal rights to fish and fishing, these rights must belong to the government of British Columbia as owner of aquatic resources in the province. British Columbia could demand that CN Rail treat with it to acquire the right to interfere with these resources.

A consideration of rights to the fish and to fishing does not end the matter, for regardless of who owns these rights, interference with them may be authorized by competent legislation enacted to protect the public interest in natural resources. In this case the dumping of rock could amount to the deposit of waste in water contrary to the British Columbia *Waste Management Act* unless covered by a licence. It could also be the deposit of a substance deleterious to fish contrary to s. 33(2) of the federal *Fisheries Act*. In addition, the dumping could be an alteration or disruption of fish habitat contrary to s. 31 of the *Fisheries Act*. The railway might argue its way out of the need for a provincial waste management licence on the same grounds as it claims exemption from the requirement of a water licence. But it cannot escape federal regulation under the *Fisheries Act*. So, clearly, the federal Department of Fisheries and Oceans must be a consenting party to any final outcome that resolves the dispute between CN Rail and the Alliance of Tribal Councils over fish and fishing rights.

Conclusion

The complexity of issues involved in this illustration is typical of water resource issues whether one is dealing with hydroelectric projects like James Bay in Quebe or Kemano in British Columbia, with pulp mill effluent polluting the Ottawa River, or with mine tailings contaminating creeks in the Yukon. Yet, these issues have arisen in the "twin-tracking" case with respect to water entirely within the boundaries of one province. Where water bodies cross provincial boundaries, the issues are even more complex and the parties even more diverse. In the case of the proposed Slave River hydro-electric project, where the waters of the Slave River flow from Alberta into the Slave Lake in the Northwest Territories and Saskatchewan, the parties directly affected are the governments of Alberta, Saskatchewan, and the Northwest Territories and the federal government of Canada as well as the individuals who use these waters in one fashion or another. How rights and responsibilities are to be defined as between upstream and downstream users in the case of such an interjurisdictional water system is a chapter of Canadian law yet unwritten. Nevertheless, the twin-tracking case shows that legitimate concerns cannot be ignored today in water management, whether their basis in law is fully acknowledged or not. Consequently, it is likely that the next major interjurisdictional river development will show that upstrean/downstream concerns cannot be ignored today, no matter how strong an upstream government or user may consider its legal case to be. Members of the public are too sensitized to the importance of fair and equitable water management to permit perceived rights to be ignored. Interventions will force appropriate outcomes — either the parties will reach agreed solutions to interjurisdictional disputes which recognize upstream/downstream rights and obligations or the courts will begin to define these rights and obligations for them. Court decisions today are more likely to apply some principle of equitable apportionment to distribute benefits and costs among the parties than they are to allow upstream jurisdictions a free hand to interfere with flowing water as they wish. A Peace River Bennett Dam experience, where British Columbia allowed the filling of the reservoir behind the dam without regard to damaging downstreams effects in Alberta and the Northwest Territories, will not be tolerated in the future. If agreement is not reached and the courts do not provide a fair balance between upstream and downstream jurisdictions, there will be an irresistible pressure on Parliament and the federal government to dictate the rules for sharing interjurisdictional waters in Canada regardless of the preferences of the individual provinces.

References

CHESMAN, D. 1984a. Memorandum on riparian rights, p. 61-70. In H. I. Ruggeberg and A. R. Thompson [ed.] Water law and policy issues in Canada. Westwater, University of British Columbia, Vancouver, B.C.

    1984b. Constitutional aspects of water law, p. 71-89. In H. I. Ruggeberg and A. R. Thompson [ed.] Water law and policy issues in Canada. Westwater, University of British Columbia, Vancouver, B.C.

DALE, N. 1968. Pollution, property and prices. University of Toronto Press, Downsview, Ont.

KETCHUM, G. C. 1984. Indian rights to water in the prairie provinces, p. 117-148. In H. I. Ruggeberg and A. R. Thompson [ed.] Water law and policy issues in Canada. Westwater, University of British Columbia, Vancouver, B.C.

MCINTYRE, J. M. 1969. The development of oil and gas ownership theory in Canada. U.B.C. Law Rev. 4: 245.

MEEKISON, J. P., R. J. ROMANO, AND W. D. MOULL. 1985. Origins and meanings of section 92A: the 1982 constitutional amendment on resources. Institute of Research on Public Policy, Montreal, Que.

PERCY, D. R. 1984. Federal/Provincial issues, p. 81-91. In H. I. Ruggeberg and A. R. Thompson [ed.] Water law and policy issues in Canada. Westwater, University of British Columbia, Vancouver, B.C.

RUEGGEBERG, H. I., B. SADLER, AND A. R. THOMPSON [ED.] 1984. Water law and policy issues in Canada. Westwater, University of British Columbia, Vancouver, B.C.

Notes

1. 11th Article of Union, confirmed by B.C. statute 47 Vict., c. 14; federal Order in Council dated May 16, 1871, pursuant to s. 146 of Constitution Act, 1867.
2. The legal consequences of the establishment of the Railway Belt are examined in A.G.-B.C. v. A.G.-Can. (1889), 14 Appeal Cases. 295 (P.C.) (The Precious Metals Case).
3. These agreements were given the status of constitutional amendments to the 1867 B.N.A. Act by legislation of the Imperial Parliament and statutes of Canada and of the provinces.
4. In addition to the Natural Resource Transfer Agreements, 1930 (fn. 3), which transferred ownership of unalienated Crown lands from Ottawa to the western provinces, the federal and provincial governments from time to time transfer Crown lands to each other for various specific purposes such as harbours, airports, schools, etc.
5. Constitution Act, s. 109 reserved natural resource rights to the provinces joining confederation.
6. The northern territories of Canada have not yet achieved provincehood. The offshore regions are mainly federal. In these regions Crown lands are federally owned.
7. For many years economists have argued for the recognition of marketable rights to water or to dispose of wastes into water (pollution rights; effluent charges) on the ground that market transactions will more accurately reflect the needs and wishes of society than a system of regulated rights (Dale, 1968).
8. Donnely v. Vroom, [1907] 40 N.S.R. 585.

9. In British Columbia the legislative provision that would have the effect of subjecting ground water to the same licensing regime that applies to surface waters has been enacted but not yet brought into force; *Water Act*, R.S.B.C. 1979, c. 429, s. 3.

10. Of course, the laws of a province can only apply within its boundaries; outside the provinces, only federal law can apply. See *Offshore Minerals Reference* (1968), 65 D.L.R. (2nd) 353 (S.C.C.).

11. Section 92A was added to the constitution as part of the constitutional amendment process culminating in the *Constitution Act, 1982*. For the most part, s. 92A merely declares the natural resource powers which the provinces held under the 1867 constitution (Meekison et al. 1985).

12. There are other federal heads of power that could be significant in special circumstances: e.g., the criminal law power in s. 91(2); the "defence power", which is an expression of POGG in situations of war or threat of war. In addition, s. 92(10)(c) of the *Constitution Act* gives Parliament the power to declare a work "to be for the general advantage of Canada or for the advantage of two or more provinces," in which event federal laws prevail. This declaratory power is politically controversial and has not been used since 1961.

13. *Fowler v. R.*, [1980] 5 W.W.R. 511 (S.C.C.); *Northwest Falling Contractors Ltd. v. R.*, [1981] 1 W.W.R. 681 (S.C.C.)

14. The Act has been used to support cooperative federal/provincial efforts in water basin planning and management.

Statutes

1. *Canada Water Act*, R.S.C. 1970, 1st Supp. c. 5.
2. *Fisheries Act*, R.S.C. 1970, c. F-14, section 33(2).
3. *Indian Act*, R.S.C. 1970, c. I-6.
4. *Land Act*, R.S.B.C. 1979, c. 214, s. 52.
5. *National Parks Act*, R.S.C. 1970, c. N-13.
6. *Navigable Waters Protection Act*, R.S.C. 1970, c. N-19.
7. *Northern Inland Waters Act*, R.S.C. 1970, 1st Supp., c. 28.
8. *Waste Management Act*, S.B.C. 1982, c. 41.
9. *Water Act*, R.S.B.C. 1979, c. 429, s. 2.
10. *Water Resources Act*, R.S.A. 1980, c. W-5.

CHAPTER 16

Water Resources and Native Peoples

Barry Barton

Canadian Institute of Resources Law, University of Calgary, 2500 University Drive NW, Calgary, Alta. T2N 1N4

Introduction

A general review of Canadian waters calls for a consideration of water issues affecting native peoples for the basic reason that native peoples are not the same as other groups in Canadian society. Native persons have in the past experienced strong pressure to abandon their different ways of life and their different values, and accept those of Euro-Canadian society. Native groups have strongly resisted this pressure for assimilation, although they have always recognized the need to adapt and to look to the future. My objective in this chapter is to explain the significance of these differences in relation to water resources. The main themes that I will explore are the different attitudes to water and the environment as a whole in native societies, the vulnerability of traditional lifestyles to disturbance of the natural regime, the legal rights that native peoples can affirm to resist such disturbances, and the implications for water policy and water management. In particular I shall argue the need to recognize that, to a native group, a water issue is more than an economic, environmental, or recreational matter; it is a question of defending its own identity.

I will first examine the significance that is placed on water by native peoples and the ways that they have been disturbed by changes to the natural water regime or to natural water quality. It will be necessary to outline the general framework of legal and constitutional relations between native peoples and federal and provincial governments. It will then be possible to embark upon a more detailed discussion of native rights to water under the two main headings of rights attaching to reserves and rights under aboriginal land claims. The use of negotiated agreements and the role of self-government will also be discussed.

I should stress at an early stage that I am referring to a wide diversity of cultures under the less than perfect label of "native peoples." The term is used here synonymously with "aboriginal peoples" to include all Indian nations, the Metis, and the Inuit. I should add that in dealing with water issues, I do not intend to embark on any detailed discussion of native fishing and hunting rights, which are substantial topics on their own.

Significance of Water to Native Peoples

The traditional aboriginal way of life involves living off the land by hunting, trapping, fishing, and gathering. Most of the wildlife harvested by natives is directly dependent upon the aquatic environment. In many regions, fish are an important part of the diet. On the West Coast, the salmon is not only a dietary staple, but is central to the native economy and culture. Ducks, geese, and other waterfowl are a seasonally important source of food, while semiaquatic mammals such as beaver and muskrat are a source of furs as well as meat. Bigger game may also be water dependent; moose frequent lakes in the summer, while caribou use frozen lakes and rivers for easy travel during their winter movements. Wild rice is harvested in the shallows of numerous lakes of the Canadian Shield. In many regions, rivers and lakes provide the easiest means of transport both in summer and winter. Even at the present day, with all the changes that have flowed from contact with European ways of life, there are many isolated native communities where people for the most part continue to live off the land. There are many other communities where trapping and fishing and the use of "country food" are significant components of a local economy that also includes nontraditional ways of making a living. In summary, native ways of life strongly depend upon the availability of water in its natural state, and have no practical impact on that natural state.

This closeness to the natural environment, and the waters that are part of it, goes far deeper than simple economic dependence. Living off the land is something that native people are proud of; they are proud to be good hunters or good trappers. It gives them a sense of identity, a sense of self-respect, even if it is only a part-time activity, or even if it is only an option available if other work disappears. Closeness to the land is a central component of any native culture and identity. Native groups from all parts of Canada affirm a sense of oneness with the land and the waters. This was expressed to the Inquiry on Federal Water Policy by an Indian spokesman from Alberta, Clifford Freeman:[1]

> For some time now, the aboriginal people of Canada, and other parts of the world have been trying to get one simple message across to the newcomers to our land. This message is 'we are part of our land — our land is part of us.' There cannot be any divisions or separations between us and our land without dire consequences to both.

The Lillooet Tribe of the Fraser River express the belief that[2]

> ...they are an integral part of the rest of Creation, not individuals separate from the rest of Creation which includes the Great Spirit, the spirits, animals, fish, plants, air, sun, moon, stars, water, mountains, land and all existence.

In this way of thinking, it is no poetical affectation to personify the land or a river. The Mackenzie Valley Pipeline Inquiry heard many such expressions. One Dene said (Berger 1977, Vol. 1, p. 94)

> This land fed us all even before the time the white people ever came to the North. To us it is just like a mother that brought her children up. That's how we feel about this country. It is just like a mother to us. That's how serious it is that we think about the land around here.

And another (Berger 1977, Vol. 1, p. 94):

> We love the Mackenzie River, that's our life. It shelters us when it storms and it feeds us when there is hunger. It takes care of its children, the native people.

In essence these feelings and beliefs about the natural environment are religious convictions. The traditional thanksgiving greeting of the Mohawk Nation acknowledges all things and the people as parts of the same creation; the plants, the animals, the thunderclouds that are called the grandfather, with the responsibility to bring water to renew life, the people, Mother Earth, the bodies of water, the birds, and so forth. The thanksgiving for bodies of water is

> We give thanks to the spirit of waters for our strength of well being. The waters of the world have provided to many; they quench thirst, provide food for plant life, and are the source of strength for many medicines we need. Once acknowledged, this too becomes a great power for those who seek its gift, for mankind himself is made from the waters.
>
> Now our minds are one
Agreed

---

[1]Treaty Eight Tribal Association, Inquiry on Federal Water Policy, 1984 ("I.F.W.P.") Submission No. 85.
[2]Lillooet Tribal Council, I.F.W.P. Submission No. 161.

These ways of thinking can be brought into sharper focus by noting how European attitudes to the environment are perceived by native groups. The idea of a wilderness inimical to man is a foreign concept; so too is the idea of man having dominion over the animals and the earth.[3] The Inquiry on Federal Water Policy heard a number of criticisms of European or industrial society's attitudes. The Union of Ontario Indians pointed out[4]

> The maintenance of the environment, especially clean bodies of water, is so essential to the integrity of the community — a fact that we can not over emphasize and a fact that is disregarded so often by the rest of Canadian society.

The Nishnawbe-Aski Nation, of the Ontario Arctic watershed, argued that the root of the water quality issue was the view of water as an expendable resource for industrial activity, necessitating an accommodation of competing uses. This thinking is not acceptable. "The water is part of a system that cannot be separated from the land that it touches, the bed that lies underneath, the habitat it supports, the fish and wildlife it sustains. If you denigrate the quality of the water, you inevitably harm everything."[5] There was the gentle irony of the Mohawk Council of Kahnawake:[6]

> The First Nations of Canada have always understood the importance of water. This resource has sustained a Nation and maintained the individual.
> We are most happy to present our understanding and views on this complex issue. It has taken only 300 years for your society to understand the basic importance of the Mother Earth. We are now happy that the Canadian society has come to recognize and hopefully understand these concerns.

Ernest Benedict (1985) of the Assembly of First Nations described Indian thinking as follows:

> Our wholesome respect for the land and water has not changed. And many of our people still depend, to a large degree, upon the renewable resource harvest. The spiritual affinity with our environment continues, and we still maintain a deep-rooted appreciation for water's lifegiving and cleansing qualities.
> Neither has the non-Indian attitude changed. Water is seen as a resource that must be exploited; and water management is seen as a mere question of supply and demand. This approach to resources has led to all kinds of problems; not just in relation to water but also in relation to fisheries, forestry and other resources.
> Without wanting to appear chauvinistic, we encourage Canada to adopt our wholesome respect for water: teach this philosophy to Canadian children; teach it to resource managers; and inject it into our existing water management systems.

In short, water issues are much more than mere resource allocation or environmental issues as far as native peoples are concerned. The point is forcefully put by Freeman:[7]

> To us environmental issues are human rights issues. The Charter of Rights in the Canadian Constitution guarantees freedom of religion. Our religion, our culture, our way of life demands that we live in harmony with nature. We have a responsibility to guard Mother Earth from abuse just as she has a responsibility to provide us with our livelihood. This relationship has existed since time immemorial and we now strongly resent the disruptions caused by the newcomers who have no intention of remaining in our area permanently and sharing in our relationship with the earth, rather, simply wishing to remove as much as they can, as quickly as they can, and leave us to cope with the ruins. Any degradation of Mother Earth is a degradation of the Indian people and cannot be accepted by us.

---

[3]Treaty Eight Tribal Association, I.F.W.P. Submission No. 85.
[4]I.F.W.P. Submission No. 48.
[5]I.F.W.P. Submission No. 71.
[6]I.F.W.P. Submission No. 150.
[7]*Supra*, note 1.

Adverse Effects

Native peoples have often suffered from the adverse effects of water resource development and from the abuse of water resources. The Kemano Diversion Project is one example, in which British Columbia gave Alcan control over the waters of a large part of the Fraser and Skeena watersheds in north-central British Columbia. A component of the project was the Skins Lake spillway to divert water into the Cheslatta River, and two storage dams lower on the same river to flood Cheslatta and Murray lakes. The Indian people of the Cheslatta River were first approached on 3 April 1952, 3 d before the dam was closed and flooding began. Tremendous pressure was put on them to sell their land and leave their homes at once. In a month they were gone. All the compensation paid to them was required to buy lands to resettle (Day 1985). In their own words, they were never given an opportunity to discuss the merits of the dam:[8]

> They were forced to build a new life in a farming community with which they had little in common. Many were forced to abandon their traditional occupations of hunting, trapping and fishing. A once proud people had for a time lost all dignity and succumbed to despair and alcohol. Whereas no Indians living at Cheslatta had been reliant on social assistance, now Band members have very few other sources of income.
> However, the people are now expressing themselves and getting back on their feet. They are filled with anger and will never allow such shameful treatment of them again.

Apart from this displacement, Indian graveyards were washed away by erosion, the artificial lakes created by the project are still unsafe because the reservoirs were not logged off, and salmon populations, although difficult to ascertain, may have suffered significant periods of decline (Day 1985; see also chapter by Rosenberg et al., this volume).

Another northern British Columbia project, the W.A.C. Bennett Dam on the Peace River, also flooded out native communities when it was completed in 1967,[9] but additionally it affected the Cree, Chipewyan, and Metis people of the Peace–Athabasca Delta, hundreds of miles of downstream. For several years when the dam was being filled, summer water levels in Lake Athabasca were 4 or 5 ft lower than usual (Peace–Athabasca Delta Project Group 1972; Howell 1978). Fur trapping was almost wiped out, and fish and wildlife were severely affected. Neither the federal government nor the Alberta government had been at all vigilant in defending the economy of the delta, even though information about the likely effects of the dam had been available since 1959. Once the damage was done, remedial works were constructed, and research work continues, but the aquatic system has never been fully restored[10] (see also chapter by Rosenberg et al., this volume).

Quebec's James Bay hydroelectric project was announced in 1971 without consulting or even informing the 6650 Cree and 4380 Inuit who were resident in the area. Phase I alone, being completed in 1985, is an enormous undertaking that includes interbasin diversions of water that are by far the largest in Canada (Day 1985; Moss 1985). The Cree and the Inuit relied heavily on hunting and fishing, and as a result the inundations and the arrival of an industrial work force had the gravest implications for them. When they attempted to discuss the project, the Quebec government's position was that the plans were not negotiable and that native peoples had no special rights. The government ignored its obligation to recognize and negotiate native rights, an obligation it had assumed in 1912 on having the territory added to the province.[11] In 1973 the Cree and Inuit sought an injunction to halt construction until their interests were

---

[8]Carrier-Sekani Tribal Council, I.F.W.P. Submission No. 186.
[9]*Ibid.*
[10]Athabasca Chipewyan Indian Band, I.F.W.P. Submission No. 36.
[11]*Quebec Boundaries Extension Act*, S.C. 1912, c.45, s.2; S.Q. 1912, c.7.

recognized. The Superior Court granted the injunction, and although it was immediately overruled on appeal, it had the effect of forcing the Quebec government to agree to negotiate a settlement. Negotiations proceeded under a tight schedule, culminating in the signing of the James Bay and Northern Quebec Agreement in 1975. This Agreement will be considered further below.

These are by no means the only cases in recent times where hydroelectric projects have been planned in utter disregard of the losses to the imposed on native residents. Manitoba Hydro's Churchill River Diversion is another example (see chapter by Rosenberg et al., this volume). These experiences have given native groups good cause to be on their guard against being so treated in the future. Current hydro proposals that disquiet them include Alcan's Kemano Completion Project[12] and the Archipel Project near Montreal.[13]

Pollution of water has also taken its toll in native communities. The most notorious case is the discharge of mercury from a chlorine plant in Dryden into the English-Wabigoon River system. It is disputed how far this pollution affected the health of the Ojibway people of Grassy Narrows and White Dog who consumed fish from the river, but less debatable are the damage to the guiding business and the domestic fishing, the fears caused by years of equivocation by the government agencies about the health hazard, and the resulting social devastation (Ontario 1978; Shkilnyk 1985). Only in 1985, after 8 yr in a mediation process, did the bands win an agreement from the governments and the paper companies to pay compensation. Many other cases, less well publicized but just as serious, have demonstrated the special difficulties that water pollution presents to native peoples (Pearse et al. 1985a, 1985b, p. 48).

Hydroelectricity and industrial pollution are not the only ways that water resources have been affected to the detriment of the native communities. Consumptive uses of water such as irrigation are reducing the amount and quality of water in the Saskatchewan River to the detriment of the trapping and fishing of the Cumberland House Band.[14] The construction of the St.Lawrence Seaway caused a temporary employment boom for Mohawk communities, but damaged their fisheries and marshlands, and continues to cause soil erosion (Lickers 1978). Bands along the Fraser and Thompson rivers in British Columbia are fighting to have a say in the double-tracking of the CN Railway line so that they may protect the salmon runs, their heritage sites, and their reserves (see chapter by Thompson, this volume).

These examples support the proposition that alterations to natural water conditions affect native groups more than other groups in society because of their distinctive closeness to the natural environment. Natives are also more likely to be the ones who suffer the consequences when developers and regulators continue to believe that disturbance of the aquatic environment is acceptable in regions distant from the urban centres and assumed to be empty and wasted wilderness.

Concerns with Water Resources

It would be helpful if the native concern with water resources could be concisely stated, but that is out of the question. In spite of the generalizations that I have ventured here about attitudes to water, native societies vary enormously, and so do the environments within which they exist across the country. Their interests in water are diverse, as the submissions made to the Inquiry on Federal Water Policy demonstrated. Yet certain concerns recurred in those submissions. One of them was that water policy be recognized as an instrument of social policy, that natural waters are a significant requirement for

---

[12]Carrier-Sekani Tribal Council, I.F.W.P. Submission No. 186.
[13]Mohawk Council of Kahnawake, I.F.W.P. Submission No. 150.
[14]Cumberland House Band, I.F.W.P. Submission No. 39.

a distinctive way of life.[15] A second was the need for native peoples to be heard in the planning of water resource projects that affect them. They took pains to point out that they are not anti development, but did insist on being properly informed and consulted.[16] A third was the tangle of jurisdictional responsibilities between the federal and provincial governments and within each government[17] (see chapter by Thompson, this volume), and the fourth was the need to assert or to defend native rights to water, whether they be based on an aboriginal claim or on a treaty.[18]

There is another side to the coin; non-native people also have an interest in the native position. Both Saskatchewan and Alberta are on the recent record as asserting provincial jurisdiction over water and water pollution on Indian reserves,[19] and similar assertions in denial of any special rights to water for native peoples can be found elsewhere.

An Overview of Native Rights to Land and Resources

It becomes apparent that water management issues with respect to native peoples are largely concerned with legal and constitutional rights. This section takes a brief but general survey of native rights to land and resources, without focussing particularly on water. It will provide a background for the more specific discussions in the subsequent sections on native rights to water in relation to reserves and land claims.

Aboriginal Rights

Any discussion of rights to lands and resources as between natives and non-natives must ultimately refer back to the fact that native people occupied and used the lands of North America before European settlement began (Cumming and Mickenberg 1972; Elliott 1985). This prior occupation and use, in accordance with the traditional tenets of organized societies, is the source of aboriginal rights, or aboriginal title, to land. It is simple enough to observe that this conception of aboriginal rights gives native peoples rights to possession and enjoyment of their traditional lands until those rights are extinguished by lawful means. However, the recognition to be given to aboriginal rights in the modern legal system is a question that has become immensely complicated over the years. The Royal Proclamation of 1763[20] is often referred to as a starting point, and did result in legal interests in land for the Indians in those parts of eastern Canada to which it applied.[21] However, for many years it was most uncertain whether the courts would recognize any broader concept of aboriginal title deriving directly from the simple fact of aboriginal occupancy (Elliott 1985, p. 56). In 1973 the Supreme Court of Canada gave that concept strong credibility — although short of final confirmation — in *Calder* v. *Attorney-General of British Columbia*,[22] the Nishga case. (On the substantive issues the judges were evenly divided, and a procedural issue became the turning-point of the case.) Not long afterwards came the interim injunction bringing

---

[15]Gitskan-Wet'suwet'en Council, I.F.W.P. Submission No. 158; Tungavik Federation of Nunavut, I.F.W.P. Submission No. 165.
[16]Nishnawabe-aski Nation, I.F.W.P. Submission No. 71; Fond du Lac Band, I.F.W.P. Submission No. 119.
[17]Nishnawabe-aski Nation, I.F.W.P. Submission No. 71; Lillooet Tribal Council I.F.W.P. Submission No. 161.
[18]For example, British Columbia Aboriginal Peoples' Fisheries Commission, I.F.W.P. Submission No. 114.
[19]Pearse et al. 1985a, p. 48; *Calgary Herald*, 10 August 1984.
[20]R.S.C. 1970, Appendix II, No. 1
[21]*St. Catherines's Milling and Lumber Co* v. *The Queen* (1888), 14 App. Cas. 46.
[22][1973] S.C.R. 313.

the James Bay project to a halt.[23] The reason was a finding that the Cree and Inuit had aboriginal rights that were legally recognizable. The decision was reversed on appeal, but the potential of such litigation was plain for all to see. Two other cases, *Re Paulette*[24] and *Baker Lake*,[25] made similar findings about aboriginal title in different parts of the Northwest Territories, although *Paulette* was reversed on other grounds on appeal.

While the concept of aboriginal title as a common law right has gradually been gaining judicial acceptability, it was recognized and elevated into an entirely new role by the new Constitution in 1982. Section 35 (1) and (2) provides

> (1) The existing aboriginal and treaty rights of the aboriginal peoples of Canada are hereby recognized and affirmed.
> (2) In this act, 'aboriginal peoples of Canada' includes the Indian, Inuit and Metis peoples of Canada.

This section is sure to have far-reaching consequences for the defence of aboriginal title, especially against legislation that purports to interfere with it. However, on a wide range of issues one can do little more than speculate. One issue is how aboriginal title may be extinguished or circumscribed, both before and after 1982. Another is the scope and content of aboriginal rights. Exactly what are the rights that are recognized and affirmed? In particular, do they include water rights? Given the dependence of the aboriginal way of life on the water environment, for hunting and fishing in particular, it seems inevitable that any interference with natural waters traditionally used by native peoples would be an interference with aboriginal rights. In the *Calder* case, Hall J. described aboriginal title as a right to possess lands and a right "to enjoy the fruits of the soil, of the forest and of the rivers and streams".[26] Moreover, a sound legal basis for aboriginal water rights appears to have emerged in the United States (Merrill 1980). Inevitably, the courts will proceed with care in addressing any question of aboriginal title. They are reluctant, especially at the higher levels, to launch into a comprehensive examination of such a complex and weighty matter unless it is demanded by the case before them. Even when they do, they are unwilling to make global decisions on questions outside the facts of the individual case as proved by the parties.

However aboriginal title may develop in the courts, the *Calder* case and the James Bay case had important political consequences when it was realized that aboriginal title claims could make cogent legal challenges to resource development. The federal government had adopted a policy of refusing to recognize aboriginal title, a policy that flew in the face of more than 150 yr of treaty-making. Immediately after the *Calder* decision in 1973, it announced that it would revoke that policy and negotiate with non-treaty Indians and Inuit. These negotiations are now institutionalized in the comprehensive land claims process. Since then three claims have been settled, several more are in the process of negotiation, and a number of others have been accepted for negotiation. I shall return later to consider how water issues are being addressed in this process.

Treaties and Reserves

In much of Canada, aboriginal title has long been extinguished by treaties made between native nations and the Crown. British colonial policy, confirmed by the Royal Proclamation of 1763, recognized the need to obtain from native peoples surrender

---

[23]*Kanatewat* v. *James Bay Development Corporation*, [1974] R.P. 38 (C.S.Q.), rev'd [1975] C.A. 166.

[24](1973) 42 D.L.R. (3d) 8 (N.W.T.S.C.), rev'd on other grounds (1976) 63 D.L.R. (3d) 1 and [1977] 2 S.C.R. 628.

[25]*Hamlet of Baker Lake* v. *Minister of Indian Affairs and Northern Development* (1980), 107 D.L.R. (3d) 513 (F.C.T.D.).

[26][1973] S.C.R. 313 at 422.

of their land rights to the Crown before European settlement proceeded. There were a number of small treaties signed in the Maritimes, Southern Ontario, and Vancouver Island, but the first major treaties were the Robinson–Huron and Robinson–Superior treaties in 1850 (Wildsmith 1985). They established a pattern of surrender of native land rights to the Crown, in exchange for the establishment of reserves, cash payments, annuities, and recognition of the natives' fishing and hunting privileges in the ceded areas. This pattern was followed in the 11 numbered treaties signed between 1871 and 1921 to cover the prairies, western and northern Ontario, and parts of the Northwest Territories (Cumming and Mickenberg 1972; Zlotkin 1985).

Under these treaties, many Indian reserves were established in Ontario and on the prairies. Other reserves were set aside simply by executive action — an Order-in-Council — under public lands legislation. This was the normal procedure in the Maritimes and British Columbia. These procedures and a variety of others were used in southern Ontario and Quebec (Bartlett 1985). Once established, a reserve is governed by the *Indian Act*,[27] which exerts close control over the use and the alienation of the reserve land and its resources by the band in possession of it.

Constitutional Jurisdiction

The final part of this very summary review of native rights to land is the constitutional division of lawmaking powers between the federal government and the provincial governments (Hogg 1985). The starting point is the jurisdiction given to the federal Parliament to legislate for "Indians, and Lands reserved for the Indians".[28] This extends to the Inuit people,[29] and has generally been taken to give the federal government the sole power to accept surrenders of aboriginal title through treaties or other settlements (Bartlett 1986). On Indian reserves, it gives the federal government the sole power to legislate with respect to the land and resources of the reserve.[30] For the provinces' part, they have the ownership and control of the public lands generally.[31] This has left the federal government singularly ill-equipped to carry out its treaty obligations (Bankes 1986):

> During the first three decades of confederation it gradually became clear that while the Dominion could sign treaties and accept a surrender of Indian title, once it had done so it could not dispose of those lands itself or set aside Indian reserves out of those lands. The logic of section 109 of the Constitution Act 1867 was that upon surrender the lands vested in the Crown in right of the province. Furthermore, despite the fact that the province was the main beneficiary of the surrender, it was under no legal or constitutional obligation to transfer land to the Dominion for reserves or contribute to the extinguishment of the Indian title. Similarly, once reserves were surrendered the full beneficial title vested once again in the province.

The result of this state of affairs, which became apparent around the turn of the century, was that the federal government was obliged to enter into agreements with the provinces for the establishment of reserves (Bartlett 1985). The leverage that this gave the provinces enabled them to secure varying measures of provincial control over resources on the reserves, including water resources (Bartlett 1986). The Prairie provinces obtained no such leverage until the Dominion transferred natural resources to them in 1930, and even then their ability to secure powers over reserves were restricted by the terms of the transfers.

---

[27]R.S.C. 1970, c.I-6.
[28]*Constitution Act 1867*, s.91(24).
[29]*Re Eskimos*, [1939] S.C.R. 104.
[30]*Corporation of Surrey v. Peace Arch Enterprises Ltd.* (1970) 74 W.W.R. 380 (B.C.C.A.).
[31]*Constitution Act 1867*, s.109.

Water Rights Attaching to Reserves

Against that background, this section will discuss in more detail native water rights in relation to reserve lands under the *Indian Act*. Native persons resident on reserve land can claim most of the same rights that any other person in the same province can claim with respect to water. They can attempt to sue polluters, or, more realistically, they can put their case forward at public hearings on water resource matters within the province. What is of more interest here are the ways that they can assert rights that are different from others in the province, and that are beyond the control of the provincial government. The question in not a simple one, because there are several different categories of rights to water resources, and there is no such thing as a typical Indian reserve. Each reserve is different in terms of how and when it was created, and what events of legal consequence have affected it since then. In the words of Professor Bartlett (1985, p. 576):

> The ascertainment of rights in Indian reserve lands is a complex task requiring resolution of federal, provincial and Indian claims and entailing recourse to statutes, judicial decisions, the royal prerogative, treaties, and federal-provincial agreements. All must be examined in the context and the history of the particular reserve under consideration.

Using Reserve Land for Water Development

It will pay to recall at the outset that many water development projects need to have the use of reserve lands, whether it be for installations, for an access road or for flooding to create a reservoir. Without authorization, any of these acts is a trespass, for which the trespasser could be sued, or, under the *Indian Act*, prosecuted.[32] The province cannot authorize such acts by simply resorting to its expropriation legislation, because land reserved for Indians is a subject that the constitution gives to Parliament's authority. The consent of the Governor in Council is required by the *Indian Act* before a province can so proceed.[33] Reserve land can only be sold by the band first surrendering the land to the Crown, and even a temporary right of occupation or use requires a permit.[34]

Use of Water

For Indian reserves, rights to take and use water from rivers or lakes within the reserve or forming one boundary of it can be traced to two sources, namely, treaty rights and riparian rights (Bartlett 1980). Although the treaties did not deal with water or water rights separately from the lands that were ceded to the Crown, it is clear that on general principles of judicial interpretation the treaties had the effect of including rights to water in the surrender of all rights, titles, and privileges in the lands concerned (Bartlett 1980). At common law, a grant of land is presumed to include with it a grant of the water rights attaching to it.[35] Further, the descriptions of the territories covered by the treaties generally used lakes and rivers as boundaries, often specifying that the boundary followed the middle line of a river or lake. Treaty Number 1, in Manitoba, is representative (Morris 1880, p. 314):

> The Chippewa and Swampy Cree Tribes of Indians, and all other the Indians inhabiting the district hereinafter described and defined, do hereby cede, release, surrender, and yield up to Her Majesty the Queen, and her successors for ever, all the lands included within the following limits, that it to say: Beginning at the International boundary line

---

[32]R.S.C. 1970, c.I-6, ss.30 & 31.

[33]*Ibid.*, s.35.

[34]*Ibid.*, ss.28, 37, 38.

[35]*Canham* v. *Fisk* (1981), 2 C. & J. 126, 149 E.R. 53; Bartlett (1980) p. 62.

near its junction with the Lake of the Woods, at a point due north from the centre of Roseau Lake, thence to run due north to the centre of Roseau Lake; thence northward to the centre of White Mouth Lake, otherwise called White Mud Lake; thence by the middle of the lake and the middle of the river issuing therefrom, to the mouth thereof in Winnipeg River; thence by the Winnipeg River to its mouth; thence westwardly, including all the islands near the south end of the lake, across the lake to the mouth of the Drunken River; thence westwardly, to a point on Lake Manitoba, half way between Oak Point and the mouth of Swan Creek; thence across Lake Manitoba, on a line due west to its western shore; thence in a straight line to the crossing of the Rapids on the Assiniboine; thence due south to the International boundary line, and thence easterly by the said line to the place of beginning; to have and to hold the same to Her said Majesty the Queen, and her successors for ever.

As for many treaty questions, there is room for serious doubt about how the Indian signatories understood the effect of the treaties on water rights. The main study of the signing of Treaties 8 and 11 concluded that the Indians saw the treaties as friendship pacts and assurances of hunting, trapping, and fishing rights, and never contemplated land surrender or relinquishment of title (Fumoleau 1973, p. 100 and 212).

Although the treaties did not mention water rights expressly, and although the surrender of lands appears generally to have included the surrender of water rights, the treaties can be interpreted to have protected at least one type of water rights. This relates to the promise in the treaties to set aside reserves. In the treaties and the discussions leading up to them, the hope was often expressed that reserves would enable the Indian peoples to abandon their lifestyles of travel in pursuit of fish and game, and instead take up the settled agricultural ways of the European (Morris 1880). It is reasonable to infer that the treaties intended to assure the Indian bands of enough water rights for the economic development of their reserve lands, especially for farming (Bartlett 1980). The Canadian courts have not decided the point, but there is no reason why they should not follow the United States Supreme Court, which reached that conclusion in 1908 in *Winters v. United States*.[36] When the United States government created an Indian reservation, it had to be taken to have intended to deal fairly with the Indians by reserving for them the waters without which their lands would have been useless. The *Winters* doctrine is the cornerstone of Indian water rights in the United States. It ensures that native groups are in a favourable position to negotiate with other water users. The amount of water to which a reserve is entitled is measured by its irrigable acreage, and the date of the priority of the right goes back to the date of the creation of the reserve (Getches et al. 1979). The significance of this interpretation of treaty rights is that it secures larger volumes of water than riparian rights would allow.

(It should also be noted that the treaties promised the native signatories the rights to hunt, trap, and fish throughout the unoccupied lands of the treaty area, and not only on their reserves. These rights are secured from the incursions of provincial hunting laws.[37] There is an argument that without adequate wildlife habitat these rights would be useless, and that therefore, by necessary implication, native persons must possess implied ancillary rights to prevent damage to wildlife habitat — including the aquatic habitat. This argument has not been tested and would have to overcome counterarguments based on provincial rights to control lands and resources.)

Riparian rights are the second source of rights to take and use water. They may simply be described as rights to water that may be enjoyed by any owner or occupier of land by virtue of the fact that the land adjoins a lake or a river, whether or not it includes the bed of the lake or river (La Forest 1973). They are common law rights, which may be varied or abolished by statute, and, as will be seen shortly, there have been many

---

[36]207 U.S. 564, 52 L.Ed. 340 (1908).
[37]*Indian Act*, R.S.C. 1970, c.I-6, s.88; Natural Resource Transfer Agreements, R.S.C. 1970, Appendix II, No. 25, clause 12 (Alta. & Sask.) and clause 13 (Manitoba). See Cumming and Mickenberg (1972, p. 207) and Hogg (1985, p. 560).

such statutes. The civil law provides very similar rights to landowners in Quebec (Lord 1977). Riparian rights include only limited rights to make use of water, but correspondingly they include rights for a downstream riparian owner to have the flow of water and the quality of water to remain substantially unaffected. A classic statement is as follows:[38]

> A riparian proprietor is entitled to have the water of the stream, on the banks of which his property lies, flow down as it has been accustomed to flow down to his property, subject to the ordinary use of the flowing water by upper proprietors, and to such further use, if any, on their part in connection with their property as may be reasonable under the circumstances. Every riparian proprietor is thus entitled to the water of his stream, in its natural flow, without sensible diminution or increase and without sensible alteration in its character or quality.

Riparian rights do not purport to give ownership of the water itself. Their main significance in the present context is that they provide a legal basis for protecting waters from disruption of the natural state of a watercourse upstream. Pollution, water diversion, or water storage are the types of activity that would affect riparian rights and would therefore give rise to a right to sue — in the absence of statutory authorization for that activity.

Indian bands may claim water rights on the basis of riparian rights by reason of their possession of reserve lands (subject to the limitations discussed below).[39] It is unnecessary to ascertain the exact nature of a band's interest in reserve lands for this purpose, or to ascertain whether the underlying title to the reserve is vested in the Crown in right of Canada or in right of the province. It is notable that many of the reserves provided for in the treaties were to be located on the shores or banks of specified bodies of water, or in some cases including a particular lake or bay (Morris 1880, p. 307). In Treaty 5 one reserve was described as follows: (Morris 1880, p. 345):

> ...a reserve commencing at the outlet of Berens River into Lake Winnipeg, and extending along the shores of said lake and up said river and into the interior behind said lake and river, so as to comprehend one hundred and sixty acres for each family of five, a reasonable addition being, however, to be made by Her Majesty to the extent of the said reserve for the inclusion in the tract so reserved of swamps, but reserving the free navigation of the said lake and river, and free access to the shores and waters thereof for Her Majesty and all her subjects.

It is clear that riparian rights were definitely contemplated as accruing to reserves. The proviso for free navigation and access may have been added out of an abundance of caution.

Hence, all other things being equal, Indian bands potentially possess significant rights to the use of water on their reserves by virtue of both treaty rights and riparian rights. Two complicating factors, however, are conflicting legislation and restrictive terms imposed at the time of the creation of the reserve. As to legislation, there is no doubt that clear and validly enacted legislation can abrogate these rights. Less clear, however, is how far a provincial legislature can abrogate riparian or treaty rights that are legally regarded as incidents, or inherent components, of the land of the reserve,[40] not

---

[38] *John Young & Co.* v. *Bankier Distillery Co.*, [1893] A.C. 691 at 698.

[39] *Indian Act*, R.S.C. 1970, c.I-6, ss.20, 31(3); and Bartlett (1980, p. 66-67).

[40] One of the leading cases states (*Chasemore* v. *Richards* (1859) 7 H.L.C. 349 at 382, 11 E.R. 140 at 153) "It has been now settled that the right to the enjoyment of a natural stream of water on the surface, *ex jure naturæ*, belongs to the proprietor of the adjoining lands, as a natural incident to the right to the soil itself, and that he is entitled to the benefit of it, as he is to all the other natural advantages belonging to the land of which he is the owner. He has the right to have it come to him in its natural state, in flow, quantity and quality, and to go from him without obstruction; upon the same principle that he is entitled to the support of his neighbour's soil for his own in its natural state. His right in no way depends upon prescription, or the presumed grant of his neighbour."

depending on ownership of the bed of the waterbody. Any attempt to abrogate such rights would be as much beyond the authority of the province as a confiscation of any other element of the land rights of the reserve. For the province, however, it could be argued that these reserve rights cannot be asserted so as to constrain the provincial jurisdiction over property and civil rights in the province outside the reserve. This argument would leave the province with jurisdiction to take control of all waters and to authorize works such as a dam that would affect water flows past a reserve downstream. In any event some provinces have not gone as far as others in vesting all property in and rights to water in the Crown, thereby leaving private riparian rights in those provinces, including reserve rights, largely undisturbed (Franson and Lucas 1976; Murphy 1977). Speaking generally, Indian bands in Ontario, Quebec, and Atlantic Canada, with the possible exception of Nova Scotia,[41] are likely to enjoy the above-described riparian rights and, in some cases, treaty rights attaching to their reserves.

The case of British Columbia is an example of how native water rights can be affected by legislative restrictions put in place before Indian reserves were established and by the terms of agreements for the transfer from a province to Canada of lands for Indian reserves (Bankes 1986). When British Columbia joined confederation in 1871, no treaties had been made, except for some small portions of Vancouver Island totalling 358 mi^2 in all (Cail 1974; Wildsmith 1985). A few reserves were set apart by the colonial government under a policy that had become very parsimonious under Joseph Trutch (Fisher 1976). The Terms of Union provided for the Dominion to assume responsibility for Indians and Indian lands, continuing the "liberal" policy of the British Columbia Government.[42] Land for reserves was to be agreed upon and conveyed to the Dominion. It took 67 yr, until 1938, to do so. The delay was partly a product of the emerging complexities that the constitution imposed on native title, and partly a product of the reluctance of the provincial government to relinquish any more land than it had to. The problems were approached by joint commissions and by a series of intergovernmental agreements, such as the McKenna–McBride Agreement of 1912.

Indian reserve water rights were one of the issues involved in this process. British Columbia had enacted water rights legislation from an early date. It culminated in the *Water Privileges Act of 1892*[43] which unequivocally declared that all unrecorded water in the Province was vested in the Crown. Common law riparian rights were greatly restricted. Against this background, it was important for Indian reserves to acquire statutory water rights if water was needed for any use over and above domestic use and stock supply. In the British Columbia interior this was a matter of special concern. Reserve Commissioners in the 1870s and 1880s made allocations of water as they located reserves, but they had no legal authority to do so until the Dominion passed an Act for the Railway Belt in 1912.[44] After federal lobbying, the Province followed suit for reserves outside the belt in 1921. Legislation provided for a Board of Investigation which could hear and determine Indian water claims, although the date of priority could be no earlier than the date of the claim.[45] This measure afforded considerable relief (Bankes 1986). The water rights obtained under these special procedures are significant, as it is likely that most reserves in British Columbia do not have riparian rights. However, it must be remembered (certainly the native nations have not forgotten) that for most of the Province, aboriginal rights to land and water have never been surrendered.

---

[41]Water rights in Nova Scotia were vested in the Crown in 1919, and reserves, created at various times, were tranferred by the province to Canada in 1959 (Bartlett 1985, p. 548).
[42]16 May 1871, R.S.C. 1970, Appendix II, No. 10, cl. 13.
[43]S.B.C. 1892, c.47. See Armstrong (1962) and Lucas (1969).
[44]*Railway Belt Water Act*, S.C. 1912, c.47.
[45]*Indian Water Claims Act*, S.B.C. 1921 (2nd Sess.), c.19.

Should it be thought that the special procedures agreed to by British Columbia for native water rights were particularly generous, it is necessary to notice the terms on which the Province was finally persuaded in 1938 to fulfil the 1871 Terms of Union and transfer the reserves to the Dominion. Among other terms designed to protect the Province's interest was a proviso for the repossession of water rights (Bartlett 1985, p. 497):

> PROVIDED also that it shall be lawful for any person duly authorized in that behalf by Us [His Majesty The King in Right of British Columbia], Our heirs and successors, to take and occupy such water privileges, and to have and enjoy such rights of carrying water over, through or under any parts of the hereditaments hereby granted, as may be reasonably required for mining or agricultural purposes in the vicinity of the said hereditaments, paying therefor a reasonable compensation.

Having been approved by the Dominion as a term on which it took title for the Indian bands, such a repossession requires no consent or surrender under the *Indian Act* (Bartlett 1985, p. 499). It is a possibility that must overshadow all reserves in the Province.

In the Prairie provinces, Indian reserve water rights may have been severely curtailed by the *North-west Irrigation Act*.[46] This was a federal Act of 1894, at which time Parliament had unrestricted legislative powers in the region. It was made in response to widespread demands for irrigation in the dry parts of the prairies. To facilitate irrigation it was necessary to abolish riparian rights, which would prevent major irrigation diversions, and instead to introduce governmental regulation of water use (see Percy 1977). The Act deemed all property and rights of use in any river or water body to be vested in the Crown except to the extent that an inconsistent private right of use was established. Persons holding water rights before the commencement of the Act were required to obtain licences for them before 1 July 1896, failing which the rights were forfeited to the Crown. After the passing of the Act, acquisition of water rights by riparian title or Crown grant was barred except in pursuance of an "agreement or undertaking" existing at the time of the passing of the Act. Although the exact effect of the statute has not been judicially determined, and is open to debate, it is clear that the common law rights to the use and flow of water have been effectively abolished (Bartlett 1980, p. 69; Percy 1977).

It is likely that the *North-west Irrigation Act* applied to Indian reserves, confiscating Indian treaty and riparian rights to water. Apparently the Department of Indian Affairs did not apply for water licences to protect reserves (Bartlett 1980). It is arguable, however, that the *North-west Irrigation Act* did not apply to Indian reserves, either because such was not the intention of the Act, or because the Act did not override the land surrender provisions of the *Indian Act*. In any event, the provincial legislation derived from the *North-west Irrigation Act* could not affect reserves set apart after 1930 pursuant to outstanding treaty land entitlements. The provinces cannot unilaterally amend the Natural Resources Transfer Agreements of 1930.[47] Further, there is an anomaly with respect to reserves set aside after 1896 but pursuant to a treaty signed before 1894, and also southern Manitoba, to which the *North-west Irrigation Act* never applied. Subject then to these exceptions, and to any other qualifications that might appear from a detailed study of the position of any one reserve, a prairie Indian band seeking water for other than domestic purposes on its reserve may be obliged to apply for a licence under the provincial legislation (Bartlett 1980).[48]

---

[46]S.C. 1894, c.30, as am. by S.C. 1895, c.33. Generally, see Bartlett (1980).
[47]R.S.C. 1970, Appendix II, No. 25.
[48]The province's hand is strengthened by the Natural Resource Transfer Agreements, *id.*, and an amendment, S.C. 1938, c.36, giving the province the Crown's power under the *North-west Irrigation Act*.

Ownership of River and Lake Beds and Water Power

A different type of right with respect to waters is the right of ownership of the land covered by the waters of a river or lake. It brings with it control of the building of dams, weirs, bridges, wharves, and other such structures. It also brings with it the exclusive right of fishing in the water body,[49] and may often include ownership of underlying mineral resources. However, ownership of the bed, as a water right, may be of limited use to a band if the province has control of the use and flow of the water itself. In these cases, works to develop water resources cannot proceed without agreement between the two.

The rule in Canadian common law is that unless it is stated otherwise, the ownership of the bed of a non-navigable body of water is vested in the owners of the land on either side, up to the middle line of the water body (*ad medium filum aquae*) (La Forest 1973, p. 234). This rule has been modified in favour of Crown ownership by enactments that declare that after their date all grants of land by the Crown will exclude the land forming the beds of waterbodies, subject to express words to the contrary. Such enactments were in place in the Prairie provinces in 1894 in section 5 of the *Northwest Irrigation Act*, in Ontario in 1911, and in Quebec in 1918, to give some examples. Similar results were achieved earlier in Quebec and in parts of New Brunswick by statutory policies which reserved three chains of land in a strip on each side of a river, reserving ownership of the bed along with it (La Forest 1973; Franson and Lucas 1976; Lord 1977). These provisions have given rise to difficult disputes about the sufficiency of express words in a grant to allow ownership of the bed,[50] or about whether a swamp or a slough is indeed a water body.[51]

These complexities also affect the boundaries of an Indian reserve that is bounded by water. There is the additional question of whether or not the setting apart of a reserve for Indians is indeed a "grant" or transfer of land by the Crown. Bartlett (1980) concluded that it is in terms of the *North-west Irrigation Act*. In any event, reserves set out before the effective date of such legislation clearly do include the beds of non-navigable waters as far as the middle line, subject, of course, to any express definition of the water boundaries in the instruments establishing the reserve. For example, when the Eden Valley Reserve Number 216 was set apart in Alberta in 1958, the legal description of the land was given as "saving and excepting thereout and therefrom all lands that lie under the waters of Gravel Lake."[52] Tyler (1982) cited cases in Manitoba where the beds of rivers are expressly excluded from reserves, and on the other hand cases where the beds of rivers and lakes are expressly included.

Works for hydroelectric power generation raise all these questions, but also raise specific questions of their own. The value of hydro sites was very much in the mind of the Ontario government in 1905 when it refused to allow any site suitable for the development of water power exceeding 500 hp to be included within the boundaries of any reserve to be set apart pursuant to the James Bay Treaty, Number 9, which was about to be signed (Bankes 1986; Zlotkin 1985). A similar condition was put on lands for reserves for the Northwest Angle Treaty, Number 3, when those lands were transferred from Ontario to the Dominion in 1915.[53] In 1924, Ontario was able to impose a requirement that no water power in any reserve in the province could be disposed of by Canada without Ontario's consent, and only subject to agreement on the division of revenues from it.[54] This requirement was incorporated by reference in

---

[49]La Forest (1973, p. 235); *The Queen* v. *Robertson* (1882), 6 S.C.R. 52.
[50]For example, *Milk River* v. *McCombs* [1978] 4 W.W.R. 615 (Alta. App. Div.).
[51]For example, *The Queen in Right of Alberta* v. *Very* (1983), 27 Alta. L.R. (2d) 119.
[52]P.C. 1958-1168, 21 August 1958.
[53]S.O. 1915, c.12
[54]S.C. 1924, c.48; Bartlett (1985).

the Natural Resource Transfer Agreements on the prairies in 1930, although the provincial right to revenues on the water powers on pre-1930 reserves was dropped. Meanwhile, in 1919, Parliament had enacted the *Dominion Water Power Act*,[55] declaring that water power rights on Dominion lands were thenceforth vested in the Crown. Reserve lands set apart between 1919 and 1930 therefore did not receive water power rights (Bartlett 1980, p. 74-77).

The obscurity of this picture of reserve water rights becomes even worse when we attempt to analyze the effects that activities on provincial lands may have on water rights in a reserve, and vice versa. A province certainly has jurisdiction to authorize a diversion out of a river, whether or not it harms members of the public; but can it also interfere with the identifiable water rights of an Indian reserve, over which it has no jurisdiction? The question may equally be asked of a diversion by band members. This problem of the unequal distribution of costs and benefits across a jurisdictional boundary is much the same as the case of an interprovincial river. The problem has no simple constitutional answer, but one would hope that each party's rights, however constitutionally valid in themselves, would have to make accomodations to avoid major damage to the equally valid rights of the other (La Forest 1973, p. 322; Percy 1983). Reference is often made to the doctrine of equitable apportionment that the United States Supreme Court applies to interstate water disputes (Johnson 1986). Precisely the same problem affects water pollution, which I have scarcely mentioned in the context of reserves. How far can one party go, whether it is the province or the band, in managing its own water and land activities to allow a certain amount of pollution, even if it affects the right of the other party to water that is undiminished in quality? There is one Supreme Court of Canada case on interprovincial pollution,[56] but it gave very little guidance for the future.

At this point I should repeat that each Indian reserve is different and has its own history. To ascertain the legal situation with respect to any given question relating to water, it may be necessary to look not only at matters of general application such as the *Constitution Acts 1867 to 1982*, the manner in which the province concerned joined confederation, the relevant treaty (if any), the *Indian Act* and all significant water land legislation, both federal and provincial, but also specific items such as survey maps and notes, Orders-in-Council setting the reserve apart, federal-provincial agreements that governed the establishment of the reserve and that may restrict reserve rights in the province's favour, and any band bylaws on water resource matters.

The exceeding complexity of this pattern of enactments and instruments can hardly be laid at the door of the Indian peoples, for they had very little hand in creating it. Much of the complexity is the result of controversies between one level of government and the other about land and resources, deriving ultimately from the unforeseen implications of the *Constitution Act of 1867*. Many of the measures adopted were in breach of treaty promises. Some stipulations purposely reduced the resource base of the reserves, while others were made in neglect of such effects. Nonetheless, it should be clear that Indian reserves often do have water rights, however difficult they are to ascertain. There is no justification in most parts of the country for a claim that the province has full rights to the water resources within Indian reserves.

The Peigan Band and the Province of Alberta

The experience of the Peigan Band illustrates some of the difficulties that native groups can encounter in relation to water resources on reserves.[57] It especially illustrates the

---

[55]S.C. 1919, c.19.
[56]*Interprovincial Co-operatives Ltd.* v. *The Queen in Right of Manitoba*, [1976] I.S.C.R. 477.
[57]Generally, see Peigan Band, I.F.W.P. Submission No. 80; *Calgary Herald*, 10 August 1984; Alberta Native Affairs Secretariat (1984); Environmental Council of Alberta (1979).

importance of a band's right to control use and development of its reserve land, and its right to control entry. The Peigan Band has a reserve in southern Alberta, through which the Oldman River flows. At a point within the reserve, water is diverted from the river for the Lethbridge Northern Irrigation District. The headworks and the canal are operated by the provincial government. In 1976 the Province sought to reconstruct the headworks and the canal in order to expand the system's capacity substantially. When it first approached the Band for further reserve land for this purpose, the Band raised a wider and older set of issues about the Provinces's rights to use reserve land. The Band argued that the Province had no valid title to the right-of-way for the canal across the reserve, or for a small parcel of land in the bed of the Oldman River, on the ground that the grant of rights in 1922 was defective. It sought compensation for the unauthorized use of the land over the years, while also seeking the best possible bargain for the future. The Bank insisted on a linkage between compensation for the old grievance and the giving of any new authorizations for redevelopment. The Province felt that the only relevant issue was some interim authorization to get works under way while the title question was negotiated in the fullness of time.

In May 1978 the Band revoked the government's access permit, cut off access to the headworks, and cut off the flow of water. In a region heavily dependent on irrigation, at the end of May, this was an action that could not be ignored. An interim injuction was obtained against the Band and the blockade was lifted, but negotiations resumed in September 1978. An interim agreement was signed in May 1979 for access over a 2-yr period, but without prejudice to the title questions. In August the Band commenced an action to obtain a judicial determination of the Province's title. Eventually in April 1981 an agreement was signed to settle both the access and title questions. The Band agreed to provide the necessary land, including land in the river bed, to provide rights-of-way, and to discontinue its legal challenges. In return the Province agreed to pay $4 000 000 to settle all past claims, to make an annual payment for the use of the lands of $300 000 per annum, adjusted for inflation, to provide technical support for the development of agriculture and irrigation on the reserve, and to give Band members preference in employment on the construction and maintenance of the project on the reserve. Construction then proceeded over the next 3 yr.

It has been said that this experience illustrates the importance of the negotiating process as a powerful method of dispute resolution (Alberta Native Affairs Secretariat 1984). It certainly does, but it also illustrates that the negotiating process would have been entirely one-sided had the Band not possessed significant legal rights, the rights to control the use of reserve lands and of access across them, and shown a willingness to use those rights to the best effect. The point is proven by the distance that the Province had to move from its first proposal, especially in agreeing to the annual payment of $300 000, which goes far beyond any level of rent that might be payable for the acreage concerned.

Arguably, the same point is also borne out by subsequent developments. A water storage dam above the diversion works on the Oldman River had been under consideration by the Province for some time. One possible site was inside the reserve; another was just upstream. Siting it inside the reserve could bring a large financial benefit to the Band, although the irrigation potential in the reserve was apparently low (Environmental Council of Alberta 1979). The options were studied in detail, but Band members were divided on the on-reserve dam, focussing on the basic question of the ownership of the river. In August 1984 the Province announced that it would go ahead with the off-reserve site upstream. The Band affirms that it has full jurisdiction over Peigan water supplies, and has passed a bylaw to exercise that jurisdiction. Its view is that the Province should acknowledge its rights and be prepared to negotiate for them. However, it is apparent that the Band is less able to influence the construction of a dam outside the reserve, relying perhaps on treaty or riparian rights to an uninterrupted flow of water, than it is able to influence works on the reserve.

The Churchill-Nelson Diversion Project

The Churchill-Nelson Diversion of Manitoba Hydro is another water resource development that has led to confrontation between developers and the native people (Day 1985). This massive hydroelectric project diverts most of the flow of the Churchill River south at Southern Indian Lake into the Burntwood River and thence the Nelson River. It also regulates the outflow of Lake Winnipeg. The ultimate objective is 7000 MW of generation capacity on the Nelson River. The project began in 1966 and the diversion was opened in 1976. The project has had substantial biophysical effects including shore erosion and fishery disruption. These effects were only partially considered as the project was being planned (see chapter by Rosenberg et al., this volume). The disruption of native residents was severe. The project directly affected some 10 500 Indian and Metis people, including three reserves along the diversion route and two on the upper Nelson River. These people were largely engaged in seasonal trapping and hunting. The project has flooded approximately 213 680 ha of land, including 4730 ha on the five reserves. One village had to be relocated.

Manitoba Hydro's initial position was that it was only responsible for facilities directly affected by the project, leaving effects on fishing, hunting, and transportation to the Province. However, the Province did not assume this responsibility. No attention was paid to social impacts, or to the mitigation and compensation of losses. There was no consultation with the native population, nor was adequate information made available. It was not even admitted that there would be flooding of reserve lands. As a result, Manitoba Hydro was forced over the next 15 yr to react on an *ad hoc* basis to the claims made against it.

The native communities formed the Northern Flood Committee in 1974 to begin negotiations with the federal and provincial governments for a proper settlement. Funding to engage expert help was provided by the federal government. Little progress was made initially. One proposal, for all claims to be put in the hands of an arbitrator, was rejected by the communities because they would not have an outsider, even a neutral one, imposing his views on their rights to their lands and resources. However, in February 1976 the parties agreed to appoint a mediator and to negotiate on defined issues, without prejudice to their legal rights. The process of mediated negotiation that followed was not an easy one, but a tentative agreement was reached in August 1977 (Mitchell 1983). It was modified slightly to accomodate Manitoba's desire to preserve its flexibility with respect to matters of policy, and December 1977 it was finally signed by the Northern Flood Committee, Manitoba Hydro, the Manitoba government, and the federal government.

In the Northern Flood Agreement, the parties dealt with many of the project's impacts. Each acre of reserve land lost by flooding and by easements in favour of Manitoba Hydro was to be compensated for by 4 acres of provincial Crown land. Compensation was to be paid for loss of income from trapping and fishing, for the loss of community infrastructure and amenities, and for remedial works. Manitoba Hydro agreed to pay half the costs of restoring domestic water supplies. An economic development fund was set up. Hydro undertook a number of actions to mitigate the effects of the project, including tree-clearing to ensure free navigation. Native rights to wildlife were given further protection, and native participation in policy-making on wildlife, land use, and the environment was secured. Hydro was required to give adequate notice of future plans and changes in operations, and to consult with the communities affected. Yet all these arrangements ran the risk of being outmoded by unforeseen future effects of the project. The parties therefore agreed on the key provision of a permanent arbitrator to settle any claim arising out of the project or out of noncompliance with the Agreement. The arbitrator has broad powers to fashion appropriate remedies, and is to give priority to mitigatory and remedial measures rather than monetary compensation. If Hydro or the governments do not accept the arbitrator's recommendations for mitigatory or remedial measures, the arbitrator may fix damages in lieu.

455

Implementation of the Agreement has proved to be a very protracted affair. In the first years progress was slow, as Canada and Manitoba did not take the necessary administrative steps to charge specific agencies with implementation (Day 1985, p. 58). The pace picked up somewhat when each government set up special offices in 1982. Some payments of compensation have been made and a number of claims have been settled (DIAND 1984a). However, many of the most difficult claims remain unresolved and are likely to remain so for some years yet. Little progress has been made with the transfer of lands to Canada for the reserve lands exchange. In 1981 the arbitrator ordered Canada and Manitoba to carry out studies of mercury contamination resulting from the diversion. This work is still under way (Environment Canada 1984). In March 1985 the Manitoba Court of Appeal dealt with three appeals from decisions of the arbitrator relating to the disclosure of Hydro reports, the core funding of the Northern Flood Committee, and the payment of the costs in arbitrations.[58]

Many of the flaws of the Northern Flood Agreement can be traced to the fact that it was a belated response to a project that was already operating, a project, moreover, that was designed in almost complete disregard for the damage it would cause to the residents of the area (Day 1985, p. 60 and 64). The time available for negotiation was insufficient to plan the improvement of physical, social, and economic conditions for the bands in the way that was hoped. The Agreement and the subsequent experience stand as a catalogue of the impacts that water development projects can have on native people, and as testimony to the difficulties that occur when those impacts are not considered from the outset of the planning of a project.

Land Claims and Water Resources

The next subject, now that reserve lands and treaty rights may be put to one side, is water resource issues in relation to native land claims. This encompasses two aspects, claims that have been or are being negotiated and claims that are based on unextinguished aboriginal title but that are not under any meaningful negotiation. Aboriginal title was described in general terms in an earlier section of this chapter, along with the court cases that brought it into prominence in the 1970s, the recognition of it in the constitution, and the probable inclusion of rights to water within its ambit.

Since 1973, when the federal government changed its policy, aboriginal title claims have been negotiated under what is known as the comprehensive land claims process (DIAND 1981; Morse 1985). Settlements have been concluded for James Bay and northern Quebec in 1975, northeastern Quebec in 1978 (along similar lines), and the western Arctic in 1984. Several other claims are under negotiation, the main ones being those of the Dene and Metis and the Inuit of the Northwest Territories, the Council for Yukon Indians, the Nishga of northern British Columbia, and the Attikamek and Montagnais of Quebec. Under this process, a claim that is lodged by a native group is reviewed by the Office of Native Claims and the Department of Justice. If it passes the review, negotiators are appointed and the negotiations themselves begin. The settlements that have been completed so far have followed a general pattern of a surrender of all claims to aboriginal title in return for ownership of selected land areas, financial compensation, hunting, trapping and fishing rights, social and economic benefits, and participation in local government and environmental management. This process is a great advance on the previous refusal to negotiate, let alone the old policy that made it a criminal offence to raise money to advance a land claim.[59] However, it has deficiencies which have been strongly criticized (Canadian Arctic Resources Committee

---

[58]*Cross Lake Indian Band* v. *Manitoba Hydro-electric Board* , Manitoba C.A., 20 March 1985, Nos. 361/82, 329/83, 29/84, 79/84 and 111/84.
[59]S.C. 1926-7, c.32, s.6.

456

1984, p. 29–86, esp. p. 55; Emond 1984). It suffers from a fundamental contradiction in that the government sees it as a means of clearing up the legal uncertainty that aboriginal title represents, and doing so fairly but finally, while on the other hand the native groups seek reparation for the losses they have already suffered and hope to negotiate the basis for the entire relationship between a native society and the larger non-native society. Further, the negotiating process has become lengthy and frustrating and has resulted in agreements that are extremely involved. These criticisms of the claims policy are being addressed by a federal review that began in July 1985. In spite of the deficiencies in the process, land claim settlements are very significant influences on the use of land and resources in the areas that they cover. Their influence on water resources may be considered in two settlements that have now been concluded, and in one that is still under negotiation.

The James Bay and Northern Quebec Agreement

The events that preceded the James Bay and Northern Quebec Agreement of 1975 have already been described. The Agreement was made between the Cree, the Inuit, Quebec, Canada, the James Bay Energy Corporation, the James Bay Development Corporation, and Hydro-Quebec (Quebec 1976).[60] It is a comprehensive land claim settlement, in that the native groups released all their claims based on aboriginal title to land. They also agreed that the James Bay Project could proceed, subject to certain conditions. In consideration, they obtained exclusive rights to selected parcels of land, cash compensation, specific hunting, trapping and fishing rights, environmental protection guarantees, local and regional governments, and a wide range of socioeconomic rights and benefits. Addressing all these issues, the Agreement fills a book of 450 pages. Its implementation has required dozens of new statutes and regulations (Moss 1985; Bankes 1983). There is no entirely new water regime established for the area of the Agreement, but in three aspects the Agreement does have significant implications for water management.

The first is the way that parcels of land were set aside in different categories for the Cree and Inuit. The land regime for the Cree and the Inuit is not quite the same, but in both cases there is Category I land, which is under exclusively native control, Category II land, over which native persons have exclusive hunting, trapping, and fishing rights, but over which the province has general control and the right to take the land for development purposes, if it provides replacement Category II land or compensation. Category III is basically Crown land. It was agreed that there would be 5288 mi^2 of Category I land, and 60 130 mi^2 of Category II — out of a total of 410 000 mi^2.[61] Specific provisions dealt with water boundaries, with the effect that lakes, rivers, and islands in rivers are included in the surrounding land category.[62] However, all Category I lands were separated from major rivers or other waterbodies by a strip of 200 ft of Category II land running along the high water mark, except for 1 mi in either direction from the centre of a community.[63] This strip allows substantial alteration of river and lake regimes for hydroelectric purposes without impinging directly on Cree or Inuit lands. The strip was stipulated less comprehensively on the Inuit lands, north of 55°N, than to the south, probably because hydroelectric development is less foreseeable there, and because no Category I Inuit land was being transferred to Canada, as was the case for much of the Category I Cree land.

The second significant aspect of the James Bay Agreement is its confirmation of permission for the James Bay Energy Corporation and Hydro-Quebec to proceed with

[60]Generally, see James Bay Crees, I.F.W.P. Submission No. 198; Day (1985); Moss (1985); Bankes (1983).
[61]Quebec (1976), p. xiii, 55, 66, 95, 105.
[62]Ibid., sections 4, 5.1.5, 6.1.2.
[63]Ibid., sections 4, 4 (Annex I), 5.1.5, 6.5, 6 (Schedule 3), 7.1.9.

their development plans.[64] The LaGrande Complex, which was already being built, was to proceed unobstructed. The Cree also agreed not to oppose the next stages of the James Bay Project. A large number of conditions were agreed upon to define the impact of these plans, for instance by specifying the maximum water level of various proposed reservoirs. On the other hand, the Cree and Inuit had to make various accommodations as to the selection and use of their lands. Detailed provisions were made for compensation for and mitigation of the adverse effects of the project on water flows. For instance, new water supply systems were to be built for Fort George and Eastmain, and some reservoir areas were to be cleared of trees. A joint corporation, SOTRAC, was formed to carry out remedial works and was funded by the James Bay Energy Corporation. General provisions authorize the hydro companies to modify or regulate the flow of rivers even if those rivers flow through or by Category I lands, or if there are downstream effects, so long as the water levels are not raised above their previous recorded maximums and compensation is paid for damage to shore facilities.[65]

Thirdly, the Agreement creates environmental and social protection systems that guarantee a special status and involvement for the Cree and Inuit when future development projects are being considered. The systems, which differ somewhat between the Crees and the Inuit, provide for a whole series of bodies to carry out advisory, consultative, administrative, and socioenvironmental impact assessment functions. They include substantial native representation, but they do not operate to give the native peoples a veto over a development except on Cree Category I lands.[66]

The Agreement with the native peoples of James Bay and northern Quebec therefore has a significant effect on the development and management of water resources in the region. It provides substantial protection for the interests of the native peoples in water, by virtue of their rights over Category I and II lands, which include areas within their boundaries that are covered by water bodies. This protection is also afforded by a large degree of participation in the environmental review process and by the many specific limitations and obligations that are undertaken by the James Bay Energy Corporation in building its projects. However, the Category I lands over which the Cree and Inuit exercise a real measure of control are only a minute fraction of the area of the region, and their boundaries are designed to reduce native control over major waterways. The hydro companies secured the right to affect downstream river regimes in almost any way; as long as they do not exceed the highest water level recorded, they can cause a flood or cut flows off at any time of the year. Native participation in making water resource decisions is directed mainly towards mitigation; there is no question about the James Bay Project actually proceeding at the proponents' pace. In all, the James Bay Agreement has not dealt with native claims to lands and resources by dealing with water as a separate resource, except to establish a regime for hydroelectric development.

The Inuvialuit Final Agreement

A more recent land claims settlement is the Inuvialuit Final Agreement in the western Arctic in 1984 (DIAND 1984b).[67] In its general outline it is similar to the James Bay Agreement, but substantially more land was transferred to native ownership, a proportion of it including mineral rights. The Agreement provides that the Inuvialuit shall be granted ownership of the beds of all lakes, rivers, and other water bodies found in Inuvialuit lands, bringing with it important rights in respect of water; but it also provides

---

[64]*Ibid.* section 8.
[65]*Ibid.*, sections 5.5, 7.4.2.
[66]*Ibid.*, sections 22 and 23. As to vetos, see the powers of the Administrator, who for Cree Category I lands is a representative of the Cree Local Government; see sections 22.1.1, 22.4.1, 22.6.15.
[67]*Western Arctic (Inuvialiut) Claims Settlement Act*, S.C. 1983-84, c.24.

that the Crown retains ownership of all waters in the region.[68] This, together with a clause making Inuvialuit lands subject to all laws of general application to private lands, makes it clear that the government continues to regulate all aspects of water allocation and water quality on Inuvialuit lands under the *Northern Inland Waters Act*.[69] In addition, the Agreement authorizes Canada to retain the right to manage and control water bodies for the purposes of fisheries, migratory birds, navigation, flood control, and the like; and title to the bed does not give the Inuvialuit proprietary rights to fishing.[70] Water management for the protection of community water supplies is also under government control.[71] There is established an Environmental Impact Screening and Review Process for all development proposals in the region. The process results in recommendations that are not binding on the government authority responsible, but as in the James Bay Agreement it gives the native people a central role in decision-making.

The Inuvialuit Agreement is therefore and important influence over water resource management in two ways. The Inuvialuit own extensive areas of land, giving them the power to control development in the same way as any other landowner, subject only to a limited government power of expropriation;[72] and they have a role in the environmental impact review process. Notheless, substantial government control remains, even on Inuvialuit lands, in particular under the *Northern Inland Water Act*.

The Inuit of the Eastern Arctic

A different approach to water resources in land claims is seen in the negotiations between the federal government and the Tungavik Federation of Nunavut, representing the Inuit of the eastern Arctic.[73] Agreement in principle was reached in January 1985 on the regulation of water use by a Nunavut Water Board, the powers and responsibilities of which would be drawn from the *Northern Inland Waters Act*. Four of its members would be Inuit, four of them government appointees, and the Chairman would be appointed after consultation. Representation on the Board is also provided for other aboriginal peoples who use an area of Nunavut that is affected by an application.

An innovative feature of the agreed regime is the emphasis that it puts on integrating water management with land use management. Water use applications, along with all other development applications, will be made to the Nunavut Planning Commission and will be checked for conformity with land use plans. If water applications do conform, they will be forwarded to the Nunavut Water Board for decision. The handling of water applications will also be coordinated with impact assessment procedures. Minor water applications will be dealt with more rapidly under an abbreviated procedure.

While this agreement was confined to water use and regulation in Nunavut, another agreement in principle was reached in December 1985 on Inuit property rights in water, a question that relates particularly to lands which are selected for Inuit fee simple title. In 1983 it had been agreed that those lands would not include the beds of freshwater bodies. Yet in 1985 the Inuit successfully negotiated the ownership of the banks of all waters in Inuit lands, and the exclusive rights to use those waters, subject to regulation by the Nunavut Water Board. Even more significantly they obtained the right to

---

[68]DIAND (1984b), sections 7(2), 7(3).

[69]R.S.C. 1970, c.28 (1st Supp.)

[70]DIAND (1984b), sections 7(85) to 7(92).

[71]*Ibid.*, section 7(85).

[72]*Northern Inland Waters Act, supra,* note 69, s.24, permits expropriation for water projects, but this power is probably subject to the special rules, including obtaining the consent of the Governor in Council and the provision of alternative lands, that are set out in section 7(50) of the Agreement, as confirmed by legislation.

[73]Generally, see *Nunavut Newsletter*, Tungavik Federation of Nunavut, Ottawa, 4:1, 4:3 (1985); Tungavik Federation of Nunavut, I.F.W.P. Submission No. 165.

have waters flow through Inuit lands "substantially unaffected in quality and quantity and flow." Where a project elsewhere in the Northwest Territories may substantially affect such quality, quantity, or flow, the proponent must enter into a compensation agreement with the Inuit before a Water Board approval is granted. (If the parties cannot agree, compensation will be determined by the Board or Boards concerned.) The Inuit have thus secured rights akin to riparian rights for their lands and have avoided any argument that the *Northern Inland Waters Act* and the *Dominion Water Power Act* extinguished such rights. This is precisely the same as the issue encountered in ascertaining the water rights that belong to an Indian reserve.

The approach to water resources that has been manifested in the Tungavik Federation of Nunavut's land claim is very different from the approach taken in the James Bay and Inuvialuit settlements, or, indeed, in any other negotiations between a native group and the government. It puts greater value on water as a resource, and proposes a management system that is innovative, not only in the native context, but anywhere.

Other Land Claims

These developments are being watched by other land claims negotiators. The Dene and Metis of the Northwest Territories have not finalized a position on water, but feel the need to have all resource allocations subjected to an environmental planning and review process controlled by the local people. They place special emphasis on the need to obtain control of water quality.[74] Overall, it seems that there will be variations in the treatment of water in different claims, but that progressively more attention will be focussed on it. Native groups are unlikely to settle in the future for roles that secure only mitigative measures, while the main decisions about resource use and project approval are made elsewhere. It seems inevitable that changes will flow from the review of government policy and procedures on land claims, particularly with respect to native views that agreements should not necessarily result in the complete extinguishment of aboriginal title. In any case, land claims agreements will continue to create new institutions and new systems that will be highly significant not only for native persons but also for anybody involved in land or water management (Pearse et al. 1985b, p. 146).

Another aspect of native claims must also be mentioned: the claims that are not under any meaningful negotiation at all. In the provinces, land claims negotiations cannot proceed unless the provincial government, as the proprietor of Crown lands, is prepared to join in negotiations with the federal government and the native organizations. The Quebec government has shown itself to be prepared to do so, both in the James Bay case and in others. The British Columbia government, however, has long denied the validity of all arguments based on aboriginal title, and has steadfastly refused to negotiate the 14 claims that have been lodged.[75] It has joined in negotiations on the claim of the Nishga Tribal Council only on very restricted terms that focus on fisheries and the delivery of specific services (Fisher 1982; DIAND 1984a). Faced with the Province's unyielding stance, British Columbia native groups have seen little choice but to resort to other strategies. One is legal action to prove the existence of aboriginal title, such as the proceedings commenced by the Gitskan-Wet'suwet'en Tribal Council 1984 for a declaration that the provincial government has no claim over its traditional territories.[76] In the recent past, the British Columbia Court of Appeal agreed with the Nuu-Chah-Nulth that there was sufficient merit in arguments in favour of aboriginal title to justify the suspension of logging on Meares Island until the case is heard in full.[77]

---

[74]Dene Nation, I.F.W.P. Submission No. 93.
[75]*Calgary Herald*, 18 June 1985.
[76]Gitskan-Wet'suwet'en Tribal Council, I.F.W.P. Submission No. 158.
[77]*MacMillan Bloedel Ltd.* v. *Mullin*, [1985] 3 W.W.R. 577; leave to appeal to S.C.C. denied, [1985] 5 W.W.R. lxiv.

A second strategy that was used by the Nuu-Chah-Nulth, and also by the Haida Nation in respect of Lyell Island, is the defence of claimed land by protests, blockades, publicity, and even civil disobedience. These methods marshall public support and bring pressure to bear on the provincial government. In late 1985 the Haida's blockades and political action injected a new sense of urgency into the situation. Assisted by the promptings of the federal government, they finally persuaded the British Columbia government to meet with them to discuss aboriginal title — something that the province had long refused to do. Whether these meetings will lead to actual land claims negotiations remains to be seen.

Future Outlook

Negotiated Agreements

There are two other likely future developments that deserve some comment. The first is the role that negotiated agreements can be expected to play in relations between the proponents of major projects and the native residents of the area affected. Contractual arrangements, in substitution for government regulation, are particularly useful for dealing with social and environmental impacts and providing for the infrastructure requirements of the project (Thompson 1984; Saunders 1985). The costs imposed by the project, and the measures that should be taken to deal with them, are better determined by direct negotiations between the parties affected than through government intervention. Agreements can be more effective than statutory regulation in gaining compliance with social goals. One such agreement was negotiated between the proponents of a liquified natural gas terminal and the Port Simpson Band Council, on the British Columbia coast. Agreements are also effective in cases of legal or jurisdictional uncertainty, such as on interprovincial rivers (Barton 1986). Native interests in water resources are often bedevilled by legal uncertainty, as we have seen.

One attempt to obtain the resolution of a water resources dispute by means of mediated negotiation is seen in the twin-tracking of the CN Railway line along the Fraser and Thompson Rivers (see chapter by Thompson, this volume). Another is the Manitoba Northern Flood Agreement. In each of these cases, however, the project proponent came reluctantly into negotiations, and only long after its planning (and even its construction works) were well under way. One must hope that development organizations will eventually realize that at some time or another they will have to come to terms with native communities that they disturb. They will show more foresight if they take the initiative themselves and secure an understanding with affected groups before their projects even begin. Native groups now expect to have a say in the early stages of project planning, where their input can be significant. Belated public relations gestures will not satisfy them.

The use of negotiation does have its limitations, however. Native groups have often found that their efforts to enter into useful discussions have been resisted until they make a legal challenge that is cogent enough to make the developers realize that they may be brought to a halt by an injunction unless they find some accomodation. James Bay is a case in point. Legal redress was sought against the Churchill River diversion. The Peigan Band put the Province's title in question to good effect in its legal proceedings. A claim was filed in the CN twin-tracking case. A feature of this general pattern is that the dispute is not finally disposed of by the courts. Usually the litigation is only taken to a preliminary or interim stage, at which the parties then sit down to some useful discussions about settling their differences. Negotiation, therefore, cannot be taken in isolation from other means of resolving disputes.

Moreover, negotiated solutions cannot fulfil unrealistic expectations of sweeping away all points of difference between the parties for all time. Circumstances change, and experience can show that the terms that the parties agree to may be quite unworkable in practice. Further, implementing an agreement can be just as difficult as negotiating it in the first place. The Northern Flood Agreement is a good example. Nonetheless an appropriate agreement can help enormously in laying down the principles and establishing the procedures that are to govern the relations between the parties.

Aboriginal Self-Government

The second issue is the drive towards aboriginal self-government. It is more and more of a force in land claims negotiations and it is strongly expressed in the aspirations for Nunavut, a new territory to be carved out of the Northwest Territories in the Inuit-dominated eastern Arctic. Self-government aims at an entirely new relationship between native peoples and the federal government. The existing relationship is one of close control by DIAND of all aspects of native social services, economic development, land, and resources. Even with some devolution of administrative responsibilities to the band level, many significant decisions are taken for native people by others who have little accountability to them.

Native groups are no longer willing to accept this domination of their affairs by out-siders. They seek a new relationship that will recognize the rights of native peoples to govern their internal affairs, and that will give their order of government a distinct place in the Canadian constitution (House of Commons 1983). The main objective is to have the right of native peoples to self-government entrenched in the constitu-tion, with their jurisdictions defined, lest it be subject to parliamentary amendment. This was the subject of a first ministers' constitutional conference in April 1985, but not even compromise proposals were able to bridge the gap between the native represen-tatives and the less enthusiastic provinces. The other more immediately possible task is to replace the *Indian Act* with a new system for the recognition of Indian First Nations with governments accountable to their own people, new economic and fiscal systems, and full rights to control their own lands, waters, and resources.

The signs of this change is native thinking may already be discerned in water manage-ment. One example is the Mohawk Nation of St. Regis, who make the ironic claim that their reserve on the St. Lawrence River is the most polluted in Canada (Lickers 1978). In 1976, with the emergence of the threat of mercury contamination, they realized that they had to take the initiative themselves in managing their own natural resources, and formed their own St. Regis Environmental Division. It carried out a pro-gram of fish sampling, finding unsafe levels of mercury. The Band Council was able to act by warning its people accordingly. The Division is kept busy with airborne and waterborne pollutants, and damage from the operation of the St. Lawrence Seaway through dredging, erosion, and the flooding of marshlands. With their own Environmen-tal Division, the Mohawk are less dependent on outside organizations to tell them what risks they face and how to deal with them. It also enables them to make themselves heard by taking a direct role in International Joint Commission organizations and other forums. Another example is the Chipewyan Band in the Peace–Athabasca Delta, who have made a modest start on a water management program to ensure that there is enough water in basins in the Delta to enhance muskrat production.[78] In British Columbia's Skeena Valley, the Gitskan-Wet'suwet'en people assert that their traditional political, legal and social systems for managing their resources remain intact and func-tional. They have been carrying out biological research as a basis for a fisheries manage-ment plan, and have been training fishery technicians. They have been frustrated in their attempts to negotiate with the federal and provincial governments to establish a joint fisheries management plan for their area based on principles of local control and consensus decision-making.[79]

---

[78]Athabasca Chipewyan Indian Band, I.F.W.P. Submission No. 36.
[79]Gitskan-Wet'suwet'en Tribal Council, I.F.W.P. Submission No. 158.

Conclusion

In this review I have sought to demonstrate that, in comparison with other Canadians, native peoples have different attitudes to the waters of lakes and rivers, and a different affinity for them. However, the interests of native residents have often been considerably damaged by developments that have affected water flows or water quality. Redress has often been hard to obtain and the benefits conferred by development have often bypassed the native people who have had to pay the costs. The Wemindji community of James Bay, for instance, is having to build its own mini hydro dam, largely out of its own resources, because Hydro-Quebec finds it too expensive to extend its distribution network to the communities.[80] Other communities have had to negotiate hard to have domestic water supplies restored after disruption by the Churchill River diversion.[81]

An unusually complex pattern of laws governs native rights to water in different situations. There is often a sound legal basis for native residents to act to ensure that their concerns are genuinely taken into account when projects are being planned in their area. We should not expect them to have the same concerns about water resources as the larger society. To native peoples, water policy is not just a natural resources development matter, or even an environmental matter. It is a question of social policy in the broad sense, a question of survival for societies that seek to preserve their own cultural identities.

References

ALBERTA NATIVE AFFAIRS SECRETARIAT. 1984. Peigan Band/Government of Alberta: resolution of the Oldman River irrigation system dispute. Edmonton, Alta. 31 p.

ARMSTRONG, W.S. 1962. The British Columbia Water Act: the end of riparian rights. U.B.C. Law Rev. 1: 583-594.

BANKES, N.D. 1983. Resource-leasing options and the settlement of aboriginal claims. Canadian Arctic Resources Committee, Ottawa, Ont. 236 p.

1986. Indian resource rights and constitutional enactments in western Canada, 1871-1930. In L. A. Knafla, [ed.] Law and justice in a new land. Carswell, Calgary, Alta.

BARTLETT, R. H. 1980. Indian water rights on the Prairies. Man. Law J. 11: 59-90.

1985. Reserve lands, p. 467-578. In B. W. Morse, [ed.] Aboriginal peoples and the law. Carleton University, Ottawa, Ont.

1986. Provincial jurisdiction and resource development on Indian reserve lands. In J. O. Saunders, [ed.] Managing natural resources in a federal state. Carswell, Calgary, Alta.

BARTON, B. J. 1986. Co-operative management of interprovincial water resources. In J. O. Saunders, [ed.] Managing natural resources in a federal state. Carswell Calgary, Alta.

BENEDICT, E. 1985. Submission to Canadian waters, the state of the resource. Rawson Academy of Aquatic Science, Conference, 26-29 May 1985, Toronto, Ont. 32 p.

BERGER, T. R. 1977. Northern frontier, northern homeland: the report of the Mackenzie Valley Pipeline Inquiry to the Minister of Indian Affairs and Northern Development, Ottawa, Ont. 2 vols.

CAIL, R.E. 1974. Land, man and the law: the disposal of crown lands in British Columbia, 1871-1913. U.B.C. Press, Vancouver, B.C. 333 p.

CANANDIAN ARCTIC RESOURCES COMMITTEE. 1984. National and regional interests in the north: third national workshop. Ottawa, Ont. 758 p.

CUMMING, P. A., AND N. H. MICKENBERG. [ED.] 1972. Native rights in Canada. 2nd ed. Indian-Eskimo Association of Canada, Toronto, Ont. 352 p.

---

[80]James Bay Crees, I.F.W.P. Submission No. 108.
[81]Northern Flood Agreement, section 6.

DAY, J. C. 1985. Canadian interbasin diversions. Research Paper No. 6 for the Inquiry on Federal Water Policy, Ottawa, Ont. 111 p.

DEPARTMENT OF INDIAN AFFAIRS AND NORTHERN DEVELOPMENT. 1981. In all fairness: a native claims policy. Ottawa, Ont. 30 p.

1984a. Annual report. Ottawa, Ont. 34 p.

1984b. The western arctic claim: the Inuvialuit final agreement. Ottawa, Ont. 115 p.

ELLIOTT, D. W. 1985. Aboriginal title, p. 48-121. In B. W. Morse, [ed.] Aboriginal peoples and the law. Carleton University, Ottawa, Ont.

EMOND, D. P. 1984. Alternative resolution processes for comprehensive native claims. In Current issues in aboriginal and treaty rights. Continuing Legal Education Seminar, Canadian Bar Association, Ontario Branch, Ottawa, Ont.

ENVIRONMENT CANADA. 1984. Annual report. Inland Waters Directorate, Western and Northern Region, Ottawa, Ont.

ENVIRONMENTAL COUNCIL OF ALBERTA. 1979. Management of water resources within the Oldman River Basin: report and recommendations. Edmonton, Alta. 245 p.

FISHER, P. 1982. The federal policy on comprehensive claims. In Indians and the law. Continuing Legal Education Society of British Columbia, Vancouver, B.C.

FISHER, R. 1976. Joseph Trutch and Indian land policy, p. 256-280. In J. Friesen and H. K. Ralston [ed.] Historical essays on British Columbia. Carleton University, Ottawa, Ont.

FRANSON, R. T., AND A. R. LUCAS. 1976. Canadian environmental law. Butterworths, Toronto, Ont. 7 vols.

FUMOLEAU, R. 1973. As long as this land shall last. McClelland and Stewart, Toronto, Ont. 415 p.

GETCHES, D. H., D. M. ROSENFELT, AND C. F. WILKINSON. 1979. Federal Indian law. West, St. Paul. 660 p.

HOGG, P. W. 1985. Constitutional law of Canada. 2nd ed. Carswell, Toronto. Ont. 988 p.

HOUSE OF COMMONS. 1983. Indian self-government in Canada; Report of the Special Committee, Ottawa, Ont. 203 p.

HOWELL, J. E. 1978. The Portage Mountain Hydro-electric Project, p. 21-64. In E. B. Peterson, and J. B. Wright, [ed.] Northern transitions. Vol. I. Canadian Arctic Resources Committee, Ottawa, Ont.

JOHNSON, R. W. 1986. Multi-state management of rivers: the American experience. In J. O. Saunders, [ed.] Managing natural resources in a federal state. Carswell, Calgary, Alta.

LA FOREST, G. V. 1973. Water law in Canada: the Atlantic Provinces. Department of Regional Economic Expansion, Ottawa, Ont. 550 p.

LICKERS, H. 1978. Saint Regis, the shrouded nation. Alternatives 8(1): 33-36.

LORD, G. [ED.] 1977. Le droit québécois de l'eau. Centre de recherche en droit public, Université de Montréal. Québec, Ministère des Ressources Naturelles, 2 vols.

LUCAS, A. R. 1969. Water pollution control law in British Columbia. U.B.C. Law Rev. 4: 56-86.

MERRILL, J. L. 1980. Aboriginal water rights. Nat. Resour. J. 20: 45-70.

MITCHELL, L. 1983. The northern Manitoba Hydro Dispute: a case study. Presented at a seminar, Environmental Mediation in Canada, Ottawa, Ont., 14-15 April 1983.

MORRIS, A. 1880. The treaties of Canada with the Indians. Belfords Clarke, Toronto, Ont. Reprinted 1979, Coles Publishing, Toronto, Ont. 375 p.

MORSE, B. W. 1985. The resolution of land claims, p. 617-683. In B. W. Morse [ed.] Aboriginal peoples and the law, Carleton University, Ottawa, Ont.

MOSS, W. 1985. The implementation of the James Bay and Northern Quebec Agreement, p. 684-694. In B. W. Morse [ed.] Aboriginal peoples and the law, Carleton University, Ottawa, Ont.

MURPHY, M. 1977. A survey of Canadian water appropriation laws. Rev. Juridique (Thémis) 12: 113-135.

ONTARIO. 1978. Royal Commission on the Northern Environment, Interim Report, Toronto, Ont. 41 p.

PEACE-ATHABASCA DELTA PROJECT GROUP. 1972. The Peace Athabasca Delta: a Canadian resource: Summary Report, Ottawa, Ont. 144 p.

PEARSE, P. H., F. BERTRAND, AND J. W. MACLAREN. 1985a. Hearing about water: a synthesis of public hearings of the Inquiry on Federal Water Policy. Ottawa, Ont. 74 p.

    1985b. Currents of change: final report, Inquiry on Federal Water Policy, to the Minister of the Environment. Ottawa, Ont. 222 p.

PERCY, D. R. 1977. Water rights in Alberta. Alta. Law Rev. 15: 142-165.

    1983. New approaches to inter-jurisdictional problems, p. 113-123. In B. Sadler [ed.] Water policy for western Canada: the issues of the eighties. University of Calgary Press, Calgary, Alta.

QUEBEC. 1976. The James Bay and Northern Quebec Agreement. Quebec, Que. 455 p.

SAUNDERS, J. O. 1985. New directions in negotiating resource agreements: the single window, p. 257-270. In Bankes and J. O. Saunders [ed.] Public disposition of natural resources. Canadian Institute of Resources Law, Calgary, Alta.

SHKILNYK, A. M. 1985. A poison stronger than love. Yale University Press, New Haven, CT. 275 p.

THOMPSON, A. R. 1984. Contractual v. regulatory models for major resource development projects. Resour. (Can. Inst. Resour. Law, Calgary): 8: 1-2.

TYLER, K. J. 1982. Indian resource and water rights. Can. Native Law Rep. 4: 1-39.

WILDSMITH, B. H. 1985. Pre-confederation treaties, p. 122-271. In B. W. Morse [ed.] Aboriginal peoples and the law. Carleton University, Ottawa, Ont.

ZLOTKIN, N. K. 1985. Post-confederation treaties, p. 272-407. In B. W. Morse [ed.] Aboriginal peoples and the law, Carleton University, Ottawa, Ont.

CHAPTER 17

Northern Water Management: From Federal Fiefdom to Responsible Decision Making[1]

Terry Fenge

Director of Research, Tungavik Federation of Nunavut, 130 Slater Street, Ottawa, Ont. K1P 6E2

Introduction

In this chapter I argue that authority to manage the North's natural resources and, in particular, its water resources, should be devolved to northern residents as soon as possible. To support this position I suggest that the federal government's approach to land and water management in the Yukon and the Northwest Territories (N.W.T.) is inadequate as well as outdated, and that the territorial governments, territorial water boards, and northern aboriginal groups are willing and better able to decide how, when, where, and by whom northern land and water should be used.

I shall briefly examine the water management system in the North and suggest that the absence of water- and land-use planning, and current overdependence on water- and land-use regulation, reflects the federal government's lack of ideas about how to manage northern natural resources. In particular, current approaches to water management undervalue in situ conservation uses of water that underly the still vibrant hunting, fishing, and trapping economy in the Yukon and N.W.T. It is argued that this situation will change only when northerners own the natural resources they use, and have the authority and institutions to manage them according to northern and particularly aboriginal people's values.

Recent amendments to the Northern Inland Waters Regulations (NIWR) and an unsuccessful attempt to amend the *Northern Inland Waters Act* (NIWA) are outlined to illustrate the federal government's difficulty in dealing with northern water policy and management questions. A short case study of placer mining regulation in the Yukon is presented to illustrate one of the most intractable water-use issues in the North today. It illustrates the apparently insurmountable difficulties that face Ottawa in dealing with a local and regional water-use conflict 3500 miles distant. Water and land management proposals made by the Government of the Northwest Territories (GNWT), the N.W.T. Water Board, and the Tungavik Federation of Nunavut (TFN) are then presented as evidence of the capability and readiness of northern groups to assume more decision-making authority. Finally, this chapter suggests that the federal government should outline what it perceives to be the national interest in northern waters and that, following this, devolution north of water planning and management functions should be accomplished without delay.

Many southern politicians and decision makers seem to believe that the territorial governments, aboriginal groups, and other northern interests are too immature and parochial and too financially dependent upon the south to manage the North's natural resources wisely. These people feel that the federal government, with its experienced personnel, vast financial resources, and breadth of vision, is best equipped to decide how northern natural resources should be used. Some federal agencies want the terri-

---

[1]Paper presented at the Rawson Academy of Aquatic Science National Symposium on Canadian Waters, Toronto, Ont., 26–29 May 1985.

torial governments, aboriginal organizations, and northern groups to provide evidence of their maturity and good sense before entrusting them with authority to make decisions about natural resource use. Northern groups thus are trapped in a dilemma familiar to many unemployed young people — how to get a first job when the employer insists on experienced help. I question this belief. In my view, parochial attitudes toward northern resource management are to be found in Ottawa whereas imaginative and innovative proposals for northern resource management are coming from the territorial governments, water boards, and aboriginal groups in the North.

The Northern Water Management System

The legislation, regulations, and institutions that govern the use of water in the Yukon and N.W.T. have been described and analyzed exhaustively through research conducted for the Inquiry on Federal Water Policy (Rueggeberg and Thompson 1984). The legislation for northern water management is generally applauded as it brings issues of water quality and water allocation and waste disposal under one statute, NIWA. This act is administered in each territory by a water board whose primary responsibility is to review water-use applications and to issue licences for water use and waste discharge (Rueggeberg 1985; Rueggeberg and Thompson 1984). The legislation defines the mandate of the water boards as:

> to provide for the conservation, development and utilization of the water resources of the Yukon Territory and the Northwest Territories in a manner that will provide the optimum benefit therefrom for all Canadians and for the residents of the Yukon Territory and the Northwest Territories in particular.

The Department of Indian Affairs and Northern Development (DIAND), however, has argued that it is responsible for planning and managing northern water resources, and that the water boards should perform only regulatory functions through their licensing responsibilities (Rueggeberg and Thompson 1984). Unfortunately, regulations dealing with water-use priorities and water-quality standards have not been made even though they are allowed under NIWA. Hence, MacLeod's (1977) conclusion is valid today:

> There has been no establishment of priorities of water use in water management areas, nor reservation of lands or water resources for comprehensive planning.

The water boards have been unable to plan for the "conservation, development and utilization" of northern water resources in the absence of regulations dealing with water-quality standards and water-use priorities and in the face of DIAND's insistence that it is the water manager. The federal government, however, is satisfied with a regulatory approach to northern natural resource management that disposes of rights to use land and water in an orderly manner. This attitude appears to be changing, albeit slowly. In 1981 the federal Cabinet approved a Northern Land Use Planning Policy (DIAND 1981) which may alter this approach to resource management (Simmons et al. 1984). Implementation of this policy has been agonizingly slow, due mainly to DIAND's insistence that it is the northern land manager and that the territorial governments and northern aboriginal people's organizations participate in an advisory capacity only (Fenge 1984).

The N.W.T. Water Board for some years has wanted to develop water-use plans and to set water-quality standards and water-use priorities, but DIAND's limited vision of the task at hand and its preoccupation with defending its jurisdiction and role as water manager has prevented this. The irony is that DIAND does not actively manage northern waters, yet it has prevented the water boards from doing so, even though they are established under federal legislation and report directly to the minister of Indian Affairs and Northern Development, and according to NIWA, they are "to provide for the conservation, development and utilization of the water resources of the Yukon and N.W.T. The N.W.T. Water Board did establish guidelines in 1981 for municipal waste water discharge (N.W.T. Water Board 1981) and in 1985 set out, in conjunction with the GNWT, to establish criteria to determine water-quality standards and water-use priorities (Cournoyea 1985[2]; Hubert 1985).

Water Use Regulation

The most visible task of the water boards has been to license water uses. However, Section 26(g) of NIWA enables the Governor in Council to make regulations authorizing the use of water without a licence. Until February 1984, Section 11 of NIWR specified that water uses of less than 50 000 gal/d continued for fewer than 270 d/yr could be authorized without a license. Based on these provisions, a twin-track water-use permit system developed. The water boards issued licenses for relatively major uses of water and the water controllers, DIAND employees in both territories, issued authorizations for relatively minor uses of water. In the N.W.T., where DIAND personnel were in direct contact with the board, administrative arrangements were made to bring all authorizations to the attention of the board with an informal system for redesignating authorizations as license applications should the board feel that was necessary. Hence, in the N.W.T., de facto oversight of all permitted water uses was provided by the board (D. Gamble, Member, N.W.T. Water Board, 14 March 1984, pers. comm.). Nevertheless, while the licensing process was subject to public scrutiny and was conducted by water boards based in the North, the authorization process involved only the applicants and the water controllers. The vast majority of water-use permits soon were issued through the authorization process which, notwithstanding the informal referral arrangements between DIAND's Yellowknife-based water controller and the N.W.T. Water Board, confirmed DIAND as the main agency dealing with northern water-use issues.

The Dene Nation and Metis Association of the N.W.T. felt the water controllers were exercising too great an influence on water management issues and were infringing upon the authority of the water boards. Hence in 1980 they challenged DIAND in the federal court claiming that NIWA gave no power to the minister of Indian Affairs and Northern Development to delegate authority to the water controllers to issue water-use authorizations. In February 1984, Madame Justice Reed agreed with this position and issued a judgement in which she argued that the authorization system established under Section 11 of NIWR amounted to an illegal delegation to the water controllers of the discretionary powers conferred on the boards by NIWA (Thompson and Qurom 1984[3]; Reed 1984). The day after this judgement, an order in council was passed amending NIWR, but without consultation or discussion with the water boards, territorial governments, the plaintiffs, or other northern interests. The amendment of NIWR deleted the requirement that the water controllers authorize water uses of less than 50 000 gal/d and broadened the scope and range of water uses exempt from licensing. DIAND hoped the land-use permit process it administered under the Territorial Land Use Regulations could handle water-use issues and instructed the water controllers to scrutinize land-use permit applications with this in mind (Faulkner 1985[4]). This response to the legal challenge was an attempt to gloss over an issue vital to sensible water management in the North. As a unilateral move by Ottawa, it highlights the problem faced by officials and water users in the North.

---

[2]N. Cournoyea, Minister of Renewable Resources, 16 April 1985, letter to G. Warner, N.W.T. Water Board.

[3]A. R. Thompson and A. Qurom. 1984. Legal and administrative arrangements for Yukon River Basin management. Unpublished.

[4]N. Faulkner, Assistant Deputy Minister of Department of Indian Affairs and Northern Development, February 1985, telex to H. Beaubier, Regional Director General, N.W.T. Region, DIAND.

This change to NIWR means that, in effect, the water boards and others in the North and Ottawa have little means of knowing the many uses of northern waters. The knee jerk regulatory change makes it even more difficult to manage northern waters according to any rational plan. Whatever the excuses offered, and there are many, the amendment was an abrogation of DIAND's duty to regulate and manage the use of northern water resources. Far from supporting a planned approach to northern water management, the amendment was a step backward to a time when northern water was mistakenly treated as a limitless and free resource. In response to the amendment, the Dene Nation and Metis Association of the N.W.T. (Kakfwi and Tourangeau 1984[5]) presented to the minister of Indian Affairs and Northern Development a carefully worded critique of the situation and a plea for a more rational way of doing things:

> While the court case was intended to challenge *all* water-use authorizations, the impetus for the case arose partially from our frustration during the 1981 N.W.T. Water Board hearings on the IPL-Norman Wells [oil pipeline] application, when it became clear that Dene and Metis representatives would not have any opportunity for input into the terms and conditions attached to the 142 stream and river crossings not covered by a water licence.

> We feel that this ruling provided the ideal opportunity for our long-standing concerns to be dealt with in a positive manner.

> We were therefore shocked to learn of the passage of Order-in-Council number PC 1984-407 amending the Regulations to the Northern Inland Waters Act. The amended Section 11.1 is not only completely unsatisfactory, but actually makes the situation worse, by removing any requirement for authorization for a wide range of water uses.

> A new procedure must be drafted for issuing water-use authorizations. The Dene Nation and Metis Association should be involved in this process. We are concerned not only with the question of who will have authority to issue the authorizations, but also who will make the decision about whether an authorization or a water license is required for a specific water use.

> A Water Management Plan for the Mackenzie River Basin is urgently required to provide a framework for decision-making on all water uses concerning the river. Too many major decisions are being made in isolation without any attempt to come to terms with the cumulative impact of major projects in the Mackenzie River and associated resources. Such a plan can provide guidelines to the Water Board for decisions on major water use proposals.

This critique illustrates the main concerns of northern native peoples regarding water-use questions. They point out that for more than a decade, DIAND has demonstrated by its own actions that it is incapable of perceiving the "big picture" and that it is consumed with patching up a very leaky water-use regulatory system. Moreover, they suggest that water-use planning for the Mackenzie basin be used to guide water-use regulation, and that local and regional interests be adequately represented on decision-making bodies. This position is reasonable, rational, and worthy of broad support by southern interests. It is recommended in various Environmental Assessment and Review Process (EARP) reports of proposed nonrenewable resource development in the N.W.T., by the N.W.T. Water Board, and by many academics active in northern policy research (Barton 1984). The TFN has presented a similar critique of northern water and land management in land claim negotiations, but has gone further than either the Dene Nation or Metis Association of the N.W.T. in articulating an alternative resource management system. This alternative system is described later in this chapter.

---

[5]S. Kakfwi (President, Dene Nation) and L. Tourangeau (President Metis Association of the N.W.T.), 13 February 1984, letter to J. Munro, Minister of Indian Affairs and Northern Development.

Amending the Northern Inland Waters Act

In May 1984, DIAND attempted unsuccessfully to amend NIWA. Scheduled amendments were to enable the water boards and water controllers to set site-specific conditions for the deposition of waste in water bodies, to clarify sections of the act dealing with notification of intent to intervene at public hearings, to make not reporting a violation of a water-use licence an offense, to increase fines for violations of the act, and to put the water-use authorization process back on a legal footing (DIAND 1984a). To do this, however, DIAND sought an agreement with the Progressive Conservative and New Democratic parties that a bill would be passed quickly with no debate (Gingras 1984[6]). Both opposition parties were prepared to allow this if they could study the draft bill in advance of its first reading in the House of Commons. In response to this proposal, DIAND gave them only a Cabinet discussion paper that outlined very generally the need for legislative changes to NIWA but did not specify how the water-use authorization process would be conducted in the future.

Some months earlier, the minister of Indian Affairs and Northern Development had agreed with the chairmen of the Yukon and N.W.T. water boards that NIWA would be amended to bring the water controllers under the functional direction of the boards. However, the discussion paper did not commit DIAND to this course of action, and it characterized the proposed amendments as "housekeeping" items only. The opposition parties pressed the minister's parliamentary secretary for a copy of the draft bill, and argued that they could not be expected to approve something that they had not seen. To the embarrassment of DIAND and surprise of the two opposition parties, it was discovered that the legislation that they were asked to approve had not been drafted and that the Department of Justice refused to do so until the opposition parties gave consent to its speedy and undebated passage. The proposed amendments went no further.

The hurried amendment of NIWR and the inept fashion in which proposed amendments to NIWA were handled demonstrate the federal government's poor ability to deal with northern water policy and management questions. The federal government does have huge financial and personnel resources, but in Ottawa where the final judgements are made, it has neither the political will nor vision to put in place a first-class northern water management system. Northern resource policy is not considered a glamorous or urgent topic in Ottawa. It is, therefore, essential that the territorial governments, water boards, and aboriginal organizations step more fully into the water policy, planning, and management spheres.

Later in this chapter it is suggested that the GNWT and TFN are ready to assume northern water and land management responsibilities and that each has a clear idea of how it will exercise those responsibilities. Before discussing how these organizations see their future roles in land and water management, a case study of placer mining regulation in the Yukon is presented. The purpose of this case study is to show, once more, the limits of Ottawa's ability to regulate and manage northern water, but it is also presented as a caveat, for devolution north of authority to manage natural resources will not alone guarantee better management of those resources. For this to happen, northern-based institutions must acknowledge the primacy and overriding importance of conservation and in situ uses of water. In other words these institutions must not propogate only southern values and norms of development. Instead they must search for a northern way that will include the perspective, goals, and values of aboriginal peoples.

---

[6]R. Gingras, M.P. (Abitibi), Parliamentary Secretary to J. Munro (Minister of Indian Affairs and Northern Development), 18 May 1984, letter to J. Fulton, M.P. (Skeena).

A Northern Water Use Conflict:
Regulation of Placer Mining in the Yukon

Placer mining has been an important economic endeavour in the Yukon since the gold-rush era late in the nineteenth century. In 1983 there were 260 placer mining operations employing 750 individuals on a seasonal basis. Gold valued at $50.5 million was produced in that year. Before 1972, placer mining was carried out under the *Yukon Placer Mining Act*, which contains no provision for environmental protection or for comparing the cost and benefits of placer mining with other resource uses (Fox et al. 1983). Placer mining is exempt from the provisions of the Territorial Land Use Regulations that, elsewhere in the North, are the main regulatory tools used to set environmental terms and conditions under which land may be used. After 1972 the water-use authorization process under NIWA was used to regulate placer mining. In 1977, extensive amendments to the *Fisheries Act* were passed. Section 31 of the revised act prohibits "any work or undertaking that results in the harmful alteration, disruption or destruction of fish habitat," and Section 33(2) prohibits "the deposit of a deleterious substance of any type in water frequented by fish." If interpreted literally, these provisions would close down the placer mining industry, which routinely damages fish habitat and deposits sediment in water that is used by fish. Regulations to govern placer mining have not been passed under the *Fisheries Act* or NIWA, and in the absence of such regulations the blanket provisions of the *Fisheries Act* apply.

Following the 1984 judgement that the water-use authorization process is illegal, placer mining has been licensed by the Yukon Water Board. Rueggeberg (1985) noted that this situation is inadequate for the protection of the environment:

> Although water licence terms and conditions (for placer mining) must not vary from the standards of the Fisheries Act and its Regulations, protecting fish habitat to the extent called for under the Fisheries Act is not within the mandate of the Water Board or of NIWA. Therefore, a water license issued under NIWA is not a guarantee to the placer miner that he is exempt from prosecution under the Fisheries Act, even if he complies fully with the terms of his licence.

Officials from the federal Department of Fisheries and Oceans (DFO) have not enforced the *Fisheries Act* in relation to placer mining but the placer miners complain bitterly of the regulatory uncertainties to which they are subject (Rueggeberg 1985).

After years of fruitless effort by federal agencies and the Yukon Water Board to implement environmental regulations acceptable to all parties, the minister of Indian Affairs and Northern Development appointed a Public Review Committee in March 1983 to examine the Yukon Placer Mining Guidelines developed jointly by officials from his department, DFO, the Department of the Environment (DOE), and the Yukon Territorial Government. Following extensive public hearings, this committee presented a report in January 1984 (DIAND 1984b) designed to provide placer miners and public officials with "legal certainty" and to protect existing investment in the industry. The committee stressed as a basic principle that

> the placer mining industry must be brought under a regulatory regime that is consistent with the nature and scope of environmental regulations that are applicable to other industries.

The committee recommended that regulations under the *Fisheries Act* and/or NIWA be established to allow placer miners to carry out "established industry practises" and to make the Yukon Territory Water Board the "one window" for issuance of water-use authorizations or licences. The committee suggested that effluent standards be applied flexibly to streams of differing value for fish and that a stream classification system be used to outline fishery values in areas subject to placer mining. Furthermore, it was recommended that new placer mining operations conform with the proposed guidelines immediately and that existing operations have 12 yr within which to adjust to the proposed guidelines.

In May 1984 the minister of Indian Affairs and Northern Development issued a "policy directive" to the Yukon Water Board on issuance of water-use licences for placer mining (Munro 1984[7]). The policy directive, which was based upon the "main thrust" of the Report of the Yukon Placer Mining Guidelines Public Review Committee, endorsed the proposed stream classification system and the principle of "moving (effluent) standards reflecting best practicable technology" (Munro 1984[7]). Application of effluent standards was delayed, however, to let a government–industry research and development committee examine ways of improving the economic and environment performance of the industry and to focus specifically on criteria with which to measure the impact of placer mining on fish and fish habitat. This committee is to report shortly.

The Yukon placer mining industry is still not regulated in an environmentally sound fashion, and the placer miners still feel they are subject to arbitrary regulation and inspection by federal officials (Ruggeberg 1985). Conflict between the placer miners, other resource users, and officials from federal agencies has grown worse since amendments to the *Fisheries Act* were passed in 1977 and since the number of placer mining operations increased following the dramatic increase in the price of gold in the early 1980s. The placer mining industry continues to resist the imposition of environmental regulations, and federal government agencies, looking for some measure of industry acceptance of regulation, continue to move forward very slowly. The territorial government has played a marginal role in the whole affair. The report of the Placer Mining Guidelines Public Review Committee is a ray of light in a murky atmosphere. Its central principle that the placer mining industry be subject to regulations similar to other extractive industries is now accepted by most groups. Fox (1984) has suggested that mediation between the interested parties should be used to draft placer mining regulations, but this has not occurred yet (Fox 1984).

This very slow progress toward balancing placer mining with environmental protection was recently upset by an extraordinary intervention by Mr. Erik Nielsen, the Yukon Territory Member of Parliament and Deputy Prime Minister. Mr. Nielsen wrote to Mr. David Crombie, minister of Indian Affairs and Northern Development, on 1 March 1985 recommending significant changes to regulation of placer mining:

> 1. That site inspection and review be clearly mandated as the responsibility of the Mining Inspection Branch of the Department of Indian and Northern Affairs.
>
> 2. That the Water Resources Branch staff of the Department of Indian and Northern Affairs and the Department of Environmental officials in Yukon be directed to assist the placer mining industry and to fully cooperate with requirements set out by the Mining Inspection Branch and the regulatory agency, the Yukon Water Board.
>
> 3. That the Department of Fisheries and Oceans redirect their energies and expenditures to the enhancement of fish stocks and the fish industry.
>
> The main requirement embodied in these recommendations is for a change in attitude by these departmental officials. The directions to be given as recommended, therefore, cannot be subtle. A message of cooperation and of working with the placer mining industry is a must.

If implemented, these recommendations would exclude DFO from assisting in placer mining regulation, would downgrade the roles played by DOE and the Water Resources Branch of DIAND, and would concentrate regulatory authority in the Mining Inspection Branch of DIAND and the Yukon Water Board. Such action would directly oppose

---

[7]J. Munro, Minister of Indian Affairs and Northern Development, 14 May 1984, letter to M. Stehelin, Chairman, Yukon Territory Water Board.

the main recommendation of the report of the Placer Mining Guidelines Public Review Committee and the policy directive issued by the minister of Indian Affairs and Northern Development to the Yukon Water Board in May 1984, both of which seek to balance placer mining and environmental protection. Altering the regulatory process as suggested would inevitably threaten fish and wildlife resources in placer mining areas, for the placer mining industry would be regulated by that section of DIAND that exists to serve the industry.

DIAND, DFO, and DOE have yet to respond to Mr. Nielsen's recommendations. The Deputy Prime Minister's Office is very powerful but it remains to be seen whether DFO and DOE, in particular, are prepared to passively accept Mr. Nielsen's suggestions. A magnanimous intervention by the Deputy Prime Minister bringing together parties long at odds to compromise on a solution true to the management principles outlined by the Placer Mining Guidelines Public Review Committee would have been most helpful. As it is the conflict between the placer mining industry and government regulators promises to drag on.

Some Northern Proposals for
Northern Natural Resource Management

Many groups in the Yukon and N.W.T. have long been dissatisfied with the federal government's approach to northern land and water management. Several EARP panels have heard complaints from northerners about the lack of resource-use priorities to guide decision making (FEARO 1981, 1984). Mr. Justice Berger's 1973-77 inquiry into the environmental, social, and economic impacts of a proposed gas pipeline across the northern Yukon and up the Mackenzie Valley drew attention to the lack of resource-use and conservation planning in the North and to the federal government's preoccupation with northern oil and gas exploitation to the detriment of renewable resource development (Berger 1977). It is only in the last few years, however, that northern interests have articulated alternative approaches to manage northern natural resources. These alternative approaches were aired in public hearings held by the Inquiry on Federal Water Policy in autumn 1984 (Inquiry on Federal Water Policy 1985). Aboriginal peoples' groups have used land claims negotiations to press for new northern-based land and water management institutions and decision-making processes. Meanwhile, the territorial governments have bargained with DIAND directly to ensure the 1981 Northern Land Use Planning Policy is implemented by institutions based in the North (Fenge 1984). These policy processes, taken together, indicate clearly what northerners have in mind regarding the management of their natural resources.

Land- and Water-use Planning

The Northern Land Use Planning Policy approved by the federal Cabinet in 1981 was designed to allocate both land and water in the North to would-be users. Northern interests criticized the policy because under it, Ottawa-based institutions would control land-use planning in the North and because the policy focused overwhelmingly on energy development (Fenge 1984). It seemed to many observers that the policy was designed primarily to ease implementation of the National Energy Program in the North (Thompson 1981[8]). A planning implementation strategy produced by DIAND late in 1982 was similarly criticized by virtually every organized interest in the N.W.T. and Yukon knowledgeable about natural resource-use issues (DIAND 1982). Faced with

---

[8] G. Thompson, Executive Director, Inuit Tapirisat of Canada, 15 September 1981, memorandum to the Board of the Inuit Tapirisat of Canada.

uniform and vehement opposition to its proposal, DIAND negotiated directly with the territorial governments to define an acceptable basis for implementing the Land Use Planning Policy. In March 1983, DIAND and the GNWT agreed to such a basis, and in July, following negotiations with the Dene Nation, Metis Association of the N.W.T., and TFN, an all-party agreement on land-use planning was finalized (DIAND and GNWT 1983). The federal minister of Indian Affairs and Northern Development and the GNWT minister of Renewable Resources signed a letter of agreement to this effect in June 1984. Similar negotiations were conducted in the Yukon but a final agreement for this territory has not been signed.

Very similar perspectives were brought to the land-use planning negotiations by the GNWT and aboriginal peoples' groups. These parties insisted that land- and water-use planning be geared to the social, economic, and cultural needs of local residents. To ensure that this would be so, they argued that land-use planning should cover land and water uses that support renewable resource harvesting. The aboriginal peoples' groups, in particular, wanted land-use planning to protect and conserve areas important for wildlife harvesting. They strove in the negotiations to ensure that land-use planning would provide this protection as well as facilitate orderly mineral and hydrocarbon exploration and development. They insisted furthermore that land-use planning should not be an academic goal-setting exercise but that it should affect all decision making for renewable and nonrenewable resource use. Northern groups know well the labyrinthine nature of policy and decision making in Ottawa, and they were concerned that some agencies, such as the Canada Oil and Gas Lands Administration (COGLA), might ignore land-use plans. They wished to make land-use planning binding upon all parties with decision-making roles in northern natural resource use.

Having stressed the broad principles that should guide land-use planning, the northern groups proposed that a Land Use Planning Commission be established in the North to do the actual planning. They argued that the commission should have significant northern representation and be given significant freedom to develop plans through open, participatory processes. The commission, they suggested, should report directly to the federal minister of Indian Affairs and Northern Development and the territorial minister of Renewable Resources, in this way bypassing civil servants who, it was feared, might preempt the planning process. A Policy Advisory Committee with representatives of federal and territorial government agencies and all northern aboriginal peoples' groups was endorsed to coordinate policy advice to the commission and to the politicians.

The GNWT and N.W.T. Water Board briefs to the Inquiry on Federal Water Policy continued the line adopted by the northern interests in the land-use planning negotiations (GNWT 1984). The GNWT brief outlined the need for a "coherent water resources policy" to ensure the long-term economic development of industry and tourism in the North as well as of the native subsistence economy. A water resource policy implemented through effective land-use planning would, it was hoped, replace ad hoc decision making. The GNWT and N.W.T. water boards are now jointly trying to identify water-use priorities and water-quality standards for adoption throughout the N.W.T.

The GNWT recognizes that it has some distance to go before it will have policies, programmes, and personnel sufficient for it to manage northern water, and it therefore proposes a phased transfer of water management responsibilities to the North (GNWT 1984). Enhancing the authority of the N.W.T. Water Board and ensuring that this body reports to both the federal and territorial governments is an immediate goal. Such an arrangement would parallel the reporting relationship of the soon to be established N.W.T. Land Use Planning Commission. Federal authorities would retain a veto over resource-use decisions, while territorial authorities would gain experience in making decisions and in being held accountable for the outcome.

An important feature of the northern land-use planning negotiations and the northern briefs presented to the Inquiry on Federal Water Policy was that northern interests did not claim decision-making roles for themselves alone, but were willing to share authority with federal agencies. What was unacceptable to northern groups was the unilateral fashion in which many federal agencies made northern resource-use decisions (Rees 1984). Northern interests brought to these two policy processes well thought out ideas about how a land and water management system should function in the North. The land-use planning positions put forward by the northern interests were more imaginative, relevant, and, most important, workable than the model promoted by DIAND and supported by other federal agencies (Fenge 1984). No longer are northern interests intellectually second best to DIAND and other federal agencies.

The Tungavik Federation of Nunavut Approach to Northern Resource Management

Land claim negotiations are providing another opportunity for aboriginal peoples' groups to suggest fundamental changes in management and decision making for northern land, water, wildlife, and subsurface resource development. The Inuvialuit Final Agreement, proclaimed into law in July 1984, is the only comprehensive claim that has been settled in the North (DIAND 1984c). It is not a radical settlement, but it awards Inuvialuit 5000 mi^2 of land outright and an additional 30 000 mi^2 without rights to subsurface resources. Inuvialuit are also guaranteed representation on natural resource-use decision making and advisory bodies.

This settlement confirms Crown ownership of all waters in the claim area, but Inuvialuit own the beds of all lakes, rivers, and streams on Inuvialuit land. Rueggeberg (1985) felt that

> ...the Crown, through the Water Board, still has the authority to allocate water rights, but...any rights that require the use of or affect the beds of a water body cannot be exercised without the permission of the appropriate Inuvialuit authority.

The settlement attempts to link together land- and water-use planning and management with environmental impact analysis (EIA) so that nonrenewable resource development projects can serve predefined resource-use goals. However, the Inuvialuit settlement moves only hesitatingly to this "integrated" form of resource management. A more radical and comprehensive approach was proposed in land claim negotiations in 1982 by TFN, the aboriginal organization representing Inuit of the eastern Arctic (TFN 1982). TFN's proposal stresses resource-use planning at a territorial level for a new political entity called Nunavut and a second level of resource and management planning for municipalities and regions. Official plans are called for at both levels. Every proposed nonrenewable resource development project would be evaluated to assess its conformity with the goals and objectives outlined in the official plans. Projects would not proceed to the regulatory stage (formal environmental and social impact assessment) without having first obtained planning conformity certificates from both Nunavut and municipal–regional officials. Water-use projects would be assessed by a Nunavut Water Board whose functions would be similar to the existing N.W.T. Water Board. The proposed Nunavut Water Board would issue water-use licenses with terms and conditions designed to promote achievement of goals and objectives outlined in Nunavut and municipal–regional plans. Project monitoring, inspection, and enforcement would also be conducted by the proposed Nunavut Water Board. Hydrocarbon and mineral development projects would be assessed by a Nunavut Impact Review Board and, if accepted, would receive certificates authorizing development. To ensure that water, land, hydrocarbon, and mineral development projects result in social, cultural, and economic benefits to local residents, each proponent would be required to negotiate a Nunavut and a municipal–regional Impact Benefit Agreement with Inuit authorities. These agreements would specify Inuit employment targets and outline how the proposed project should promote local economic development.

TFN's proposed resource management system has been called by some naive, mechanistic, and overly legal in approach, and this may be true (Bankes 1986).

However, these criticisms do not detract from the proposal's key strength which is to link together land, water, and subsurface resource management into a rational and organic system. The proposal is an example of expansive thinking and vision.

The proposed resource management system fits well with the federal government's policy of "balanced" northern development, first enunciated in 1972, and is a clear improvement over the ad hoc and reactive approach the federal government takes now to northern resource-use matters (DIAND 1972). For a project proponent, the current resource management system is like a game of pinball. The project bounces backwards and forwards between different government agencies, all of whom guard jealously their own regulatory powers and all of whom require their own terms and conditions to be satisfied before the project is given the signal to proceed, which ends the game. This system is fundamentally irrational and inefficient, for the government agencies involved work at cross purposes. TFN's proposal, in contrast, is rational and would hopefully be efficient. If implemented, it should add a sense of common purpose to land- and water-use regulation, for these activities would be designed to serve predefined goals and objectives. Individual projects and uses of land and water would not be evaluated and regulated solely on their merits but in terms of their cumulative impact, both positive and negative. The federal government should have instituted a land- and water-management system similar to TFN's proposal many years ago. Unfortunately it did not do so, and now northerners themselves are demanding the opportunity to do things right in the North.

On Policy, Planning, and Management

The Yukon and N.W.T. are colonial entities whose land, water, and subsurface resources are controlled by and administered from Ottawa. Northern interests want to share decision-making authority with the federal government and have provided much evidence in recent years of their capability to make decisions responsibly. This claim by northerners is made not only in terms of equity and social justice but also rests on principles of good resource management. Northerners seem more likely than Ottawa to search diligently for and to achieve the balance between renewable and nonrenewable resource development and environmental conservation that the federal government has espoused since 1972 as its policy for northern development.

At the Canadian Arctic Resources Committee (CARC) Third National Workshop on People, Resources, and the Environment North of 60° held in 1983, the leaders of both territorial governments criticized federal authorities for refusing to outline clearly the national interest in the North. Richard Nerysoo, leader of the GNWT, stated in his opening remarks (Nerysoo 1983)

> It is easy to articulate the regional interest; however, national interests in the North tend to be, with few exceptions, undefined. An exception is the National Energy Program. There is no national, much less northern, development policy or strategy to guide decisions on resource management. Without this national direction, tradeoffs between the regional and national interests are difficult to make because they lack an overall framework as a guide.

This statement applies well to water resources in the North. Until the federal government articulates clearly its vision of the national interest in the North for the late 1980s and beyond, we will have no basis against which to evaluate proposed water resource developments. Without such a framework, local and regional interests are likely to be shunted aside. This, of course, is what the territorial governments and northern aboriginal peoples' organizations fear. Whether and when the two northern territories become provinces and so attain ownership and control over natural resources is a moot question (Robertson 1985). Yet surely it is now in the national interest to allow northerners to decide how, when, and by whom water and lands in both territories will be used.

While the federal governments may retain ownership of these resources for years to come, authority to control and manage them should be devolved North as soon as possible. This imperative should underlie future federal policy on land and water use in the Yukon and N.W.T.

Acknowledgements

I thank Don Gamble, John Donihee, Irving Fox, Glen Warner, Brian Wilson, Harold Mundie, and Mike Healey for their constructive and helpful comments on an earlier draft of this chapter.

References

BANKES, N. 1986. The place of land-use planning in the TFN claim. CARC, Ottawa, Ont.

BARTON, B. 1984. The prairie provinces water board as a model for the Mackenzie Basin, p. 37-67. In B. Sadler [ed.] Institutional arrangements for water management in the Mackenzie River Basin. Banff Centre School of Management, University of Calgary Press, Calgary, Alta.

BERGER, T. R. 1977. "Northern frontier," "northern homeland." The report of the Mackenzie Valley Pipeline Inquiry. 2 vols. Lorimer and Minister of Supply and Services Canada.

DIAND (DEPARTMENT OF INDIAN AFFAIRS AND NORTHERN DEVELOPMENT). 1972. Canada's north 1970-1980. Ottawa, Ont.

1981. Northern land use planning discussion paper.

1982. Land use planning in northern Canada. Ottawa, Ont.

1984a. Northern Inland Water Act discussion paper.

1984b. Report of the Yukon Placer Mining Guidelines Public Review Committee. Ottawa, Ont. p. vi.

1984c. The Western Arctic Claim: a guide to the Inuvialuit final agreement. Ottawa, Ont.

DIAND AND GNWT (Government of the Northwest Territories). 1983. Land use planning Northwest Territories basis of agreement, p. 185-193. In National and regional interests in the north: third national workshop on people, resources and the environment north of 60°. CARC, Ottawa, Ont. 758 p.

FEARO (FEDERAL ENVIRONMENTAL ASSESSMENT REVIEW OFFICE). 1981. Norman Wells oilfield development and pipeline project. Report of the Environmental Assessment Panel. Minister of Supply and Services Canada, Ottawa, Ont.

1984. Beaufort Sea hydrocarbon production and transportation. Report of the Environmental Assessment Panel. Minister of Supply and Services Canada, Ottawa, Ont.

FENGE, T. 1984. Environmental planning in northern Canada: musical chairs and other games. A paper presented at the Kativik Environment Conference, Kuujjuak, Que.

FOX, J. K. 1984. An assessment of legal and administrative arrangements related to placer mining in the Yukon River Basin. Westwater Research Centre, University of British Columbia, Vancouver, B.C.

FOX, J. K., P. J. EYRE, AND W. MAIR. 1983. Yukon water resources policy and management issues. Westwater Research Centre, University of British Columbia, Vancouver, B.C.

GNWT (GOVERNMENT OF THE NORTHWEST TERRITORIES). 1984. Department of Renewable Resources. Government of the Northwest Territories brief to Federal Inquiry on Water Resources. Yellowknife, N.W.T.

HUBERT, B. 1985. Water use priorities, water quality and water management decision making. Report of a workshop hosted by the Water Board of the Northwest Territories, Yellowknife, N.W.T.

INQUIRY ON FEDERAL WATER POLICY. 1985. Hearing about water, April, Ottawa, Ont.

MACLEOD, W. 1977. Water management in the Canadian north. CARC, Ottawa, Ont.

NERYSOO, R. 1983. In National and regional interests in the north: third national workshop on people, resources, and the environment north of 60°. p. 12. CARC, Ottawa, Ont. 1984.

N.W.T. WATER BOARD. 1981. Guidelines for municipal type waste water discharges in the N.W.T. Outcrop. Yellowknife, N.W.T.

REED, J. 1984. Reasons for judgement between the Dene Nation and Metis Association of the N.W.T. and Her Majesty the Queen. Federal Court of Canada. Trial Division. Court No. T. 3536-81.

REES, W. 1984. Northern land-use planning. In search of a policy, p. 199-227. *In* National and regional interests in the north: third national workshop of people, resources, and the environment north of 60°. CARC, Ottawa, Ont. 758 p.

ROBERTSON, G. 1985. Northern political development within Canadian federalism, p. 123-131. *In* The North. Collected research studies of the Royal Commission on the Economic Union and Development Prospects for Canada. University of Toronto Press, Downsview, Ont.

RUEGGEBERG, H. I. 1985. Northern water issues. Inquiry on Federal Water policy research paper number 12. Westwater Research Centre, University of British Columbia, Vancouver, B.C.

RUEGGEBERG, H. I., AND A. R. THOMPSON. 1984. Water law and policy issues in Canada. Westwater Research Centre, University of British Columbia, Vancouver, B.C.

SIMMONS, N. M., J. DONIHEE, AND H. MONAGHAN. 1984. Planning for land use in the N.W.T., p. 343-364. *In* Northern ecology and resource management. University of Alberta Press, Edmonton, Alta.

TFN (TUNGAVIK FEDERATION OF NUNAVUT). 1982. Land and resource elements of a basis of agreement. Ottawa, Ont.

CHAPTER 18

Research for Water Resources Management: The Rise and Fall of Great Expectations

A. H. J. Dorcey

Westwater Research Centre, The University of British Columbia,
2075 Wesbrook Mall, Vancouver, B.C. V6T 1W5

Introduction

Water resources management in Canada has become progressively more complicated as it has responded to the growing and diversifying demands for use of water. A century ago, or even only half-century ago, management involved little more than issuing licenses to allocate the use of water. Today, it involves an array of licenses, leases, and permits, together with associated regulations, that both directly and indirectly allocate the water resource. Decisions may involve assessments of impacts on the resource and their broader ecosystemic (ecological, economic, and social) consequences. In some situations, the decisions may be in the context of a previously developed management plan for the water and land resources of a region. Supporting the allocation, assessment and planning decisions there may be a wide variety of contributing research programs.

In spite of all these advances, however, there is deep concern throughout the nation about the state of the resource and a general lack of confidence in the existing approaches to management of water resources in Canada.[1] Yet only 20 years ago, on the eve of major innovations in impact assessment, planning, and research, there were great expectations of our ability to manage the resource. This chapter analyses these innovations; in particular, it focusses on the role of research in management of the resource. Reasons for the fall from great expectations are identified and strategies for renewed innovation in research for water resources management are proposed.[2]

The Evolution of Water Resources Management in Canada

Three phases can be identified in the evolution of water resources management in Canada.[3] Each phase is characterized by different goals and approaches.

---

[1] See "Hearing about water: a synthesis of public hearings of the Inquiry on Federal Water Policy" conducted throughout Canada (Inquiry on Federal Water Policy 1985).

[2] In this chapter, the term "management" is used to embrace all activities (including research, allocation, impact assessment, planning, implementation, and monitoring) involved in managing the use of water resources. The term "research" is also used broadly and embraces all activities involved in developing new knowledge and methods. Five investigatory techniques for generating new knowledge are distinguished: inventory, monitoring, desk analysis, experimental management, and experimental research.

[3] The three phases are comparable with those used first by Quinn (1977) and later developed by Tate (1981), but the analysis of them here differs in several important ways. For a broader analysis of the phases in the development of natural resources policy in Canada, see Burton (1972), who examined the responses to the competing pressures for economic growth and conservation.

Up to the Mid-1960s — Developing the Economy

In the years immediately following World War II, water resources management in Canada was focussed on economic development, as it had been since before Confederation.[4] Pioneer attitudes prevailed: water was a free good and the primary rule of access was "first come, first served." Little thought was given to pollutants or downstream consequences of modifying flow regimes. The *British North America Act* had nothing to say on the subject of water quality management and it divided responsibilities between the federal and provincial governments in such a way as to make holistic approaches difficult (see chapter by Thompson, this volume). Water resources management was mostly a matter of recording stream flows and devising strategies for putting them to use in support of economic development. Conflicts in water use emerged soon after Confederation and resulted in legislation during the nineteenth century to supercede common law allocations by provincial allocation mechanisms. It was not until the mid-twentieth century, however, that the conflicts created by economic development became so pervasive and serious that more fundamental questions began to be asked about the adequacy of the mechanisms for resource allocation.

In the immediate postwar years, management was primarily concerned with providing hydroelectricity and water supply for the expansion of industries and communities, protecting communities from flood, and facilitating water transportation. To a large extent, management was preoccupied with large-scale projects such as those on the Churchill Falls, the St. Lawrence River, and the Columbia River. There were also many small projects, such as the water conservation dams under the Prairie Farm Rehabilitation Administration and the irrigation projects for the drylands of British Columbia that had begun to be built during the 1930s, and the flood protection projects of the Ontario Conservation Authorities during the latter part of this period. A multitude of licenses was issued for small-scale removals or diversions of water. But the major expenditures were on a few large projects.

While early hydroelectric power projects tended to be private, as in Quebec until the 1960s, later projects were primarily public, as exemplifed by Ontario Hydro from its outset. The well-recognized economic benefits of multiple purpose projects were often an important reason for major public sector involvement. To capture the benefits of such large-scale projects, large-scale financing was required. The St. Lawrence Seaway, perhaps the most important project of the period, was promoted and justified not only in terms of facilitating water transport of iron ore but also in terms of hydroelectric power. In the final analysis the hydroelectric generation cost more than the navigation improvements (Richardson et al. 1969, cited in Gossage 1985).

The emphasis on economic expansion was reflected in limited project assessments and impact management, although these were slowly coming to the fore. It was not until the 1950s that provision was routinely made for the passage of migratory fish, such as salmon and trout, around dams and other obstructions even though the federal *Fisheries Act* was unequivocal in its requirement that fish not be obstructed. It is somewhat ironic that the construction of the Welland Canal inadvertently provided access for the sea lamprey (*Petromyzon marinus*) to the upper Great Lakes. Prior to the 1950s, professional expertise in fishway design was severely limited in Canada. The Hell's Gate fishways in the Fraser Canyon were designed chiefly by Milo Bell, an American citizen associated with the International Pacific Salmon Fisheries Commission. The acceleration of dam building in the 1950s led to the creation of a corps of fisheries engineers, led by Charles Clay, that provided technical advice across Canada on fish passage problems confronting the then Department of Fisheries and Forestry.

---

[4]For a description of the importance of water to Canadian societies from the original Native peoples, through the early settlers and the expansion of agricultural and industrial communities to the 1960s including the emergence of management approaches, see Gossage (1985).

At the same time, there was a major expansion of limnological and fisheries studies in most parts of Canada. Great emphasis was given to defining Canada's resources so that, by the early 1960s, the regional limnology and fisheries of Canada had been broadly sketched.[5] By the mid-1960s, fisheries biologists, limnologists, and fisheries engineers were involved in virtually every major water resource project (but see chapter by Rosenberg et al., this volume).

A parallel development in technical strength took place in wildlife biology, particularly concerning waterfowl in which both federal and provincial governments had interests and responsibilities. Accompanying both these increases in competence there was a growing cadre of social scientists, particularly at Canadian universities, with wide-ranging interests in regional planning, natural resources management, and the role of natural resources in national development.[6]

The emphasis on economic development was also reflected in project evaluations which tended to be restricted to narrowly defined cost-benefit assessments. It was during this period that the technique of cost-benefit analysis was undergoing major development, particularly in the United States with respect to water resource projects.[7] In 1961, "A guide to benefit-cost analysis" was published in Canada as part of the proceedings of the Resources for Tomorrow Conference (Sewell et al. 1961). The major pupose of the guide was to promote the use of these techniques in the evaluation of all natural resource projects. It was, however, only partially successful, as the principles were not consistently adhered to when they were used and sometimes the techniques were grossly misused.[8] However, as the primary concern was with economic efficiency, costs such as the environmental damages from reservoir development and the less tangible social impacts were generally not considered. Project investigations were largely limited to design questions, such as the prediction of expected flow regimes and the configuration of a least-cost power system for the developer. To aid in such investigations, major developments were made in the United States in the techniques of hydrologic modelling (Maass et al. 1962), techniques that were not widely used in Canada until later (see also chapter by Rosenberg et al., this volume).[9]

---

[5]An excellent review is provided in the chapters on Canada (Frey 1963), western Canada (Northcote and Larkin), Ontario and Quebec (Fry and Legendre), the Atlantic Provinces of Canada (Smith), and the St. Lawrence-Great Lakes (Beeton and Chandler).

[6]For example, see contributors to Resources for Tomorrow Conference (1961) and references in Sewell (1968).

[7]In the postwar years, innovations in policy analysis for water resources management often emerged firstly in the United States and then were adapted to Canadian needs. References to these innovations are footnoted for each of the major Canadian advances that are considered. White (1969) described the development of methodologies for measuring costs and benefits in the design of optimal river basin projects. He traced these developments from the introduction of the project evaluation guidelines in the federal so-called "Green Book" in 1950 through the work by Eckstein (1958), Krutilla and Eckstein (1958), McKean (1958), Tolley and Riggs (1961), and Hirshleifer et al. (1960). He also pointed to the role of the Harvard Water Resources Seminar in refining these evaluation methodologies and training federal personnel to implement them (Maass et al. 1962).

[8]This is the conclusion of Sewell (1968), one of the authors of the guide, when later he assessed its application to water resources projects and cited, for example, Burton (1965). An interesting contrast, however, is provided by Krutilla's assessment of the Columbia River Treaty negotiations in the 1950s and early 1960s (Krutilla 1967, p. 199). As noted above Krutilla had been at the forefront in the United States development of evaluation methods for water resources projects, and he concluded that "Canadian negotiators generally had excellent information on the economic implications of proposals coming up for consideration" and that they outmanoeuvered their United States counterparts in this regard.

[9]Modelling analyses were used in the Columbia River studies in the early 1960s (Krutilla 1967).

During this period, water resources research of more general applicability was limited in size and scope in Canada.[10] Almost half of the research expenditures addressed water cycle issues (46.4%) and another 30% dealt with water quality (Fig. 1).[11] Funds were provided mostly by the senior governments (82.7%). Almost half of the research was performed directly by the federal government (44.4%); the remainder almost equally by provincial government (21.0%), the universities (19.4%), and industry (14.0%) (Fig. 2). More than half the total research expenditure of all organizations was made in Ontario (52.4%, of which 18.6% was in Ottawa).

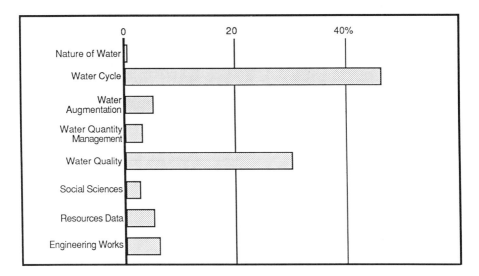

FIG. 1. Distribution of water resources research, 1966. Source: Bruce and Maasland (1968, fig. 3).

Federal research was performed by a wide array of departments and agencies with responsibilities related to water resources including Departments of Agriculture, Energy, Mines and Resources, Fisheries, Forestry and Rural Development, National Health and Welfare, Transport, the Fisheries Research Board, and the National Research Council. Most of the research was oriented to particular departmental responsibilities and was carried out in establishments throughout the country.

Bruce and Maasland's (1968) overall conclusion was that "very few Canadian (research) projects operate at the fringe of world knowledge, or attempt to search for generalized theories useful to water resources development. More specifically, most projects are adaptations to specific Canadian regions of theories, methods and techniques originated elsewhere." They attributed this to the "very low level of support" for all research prior to the mid-1960s, the lack of research and graduate training programs at universities, and the perception that there were no "serious problems" to

<hr>

[10]The Bruce and Maasland study of water resources research in Canada (1968) provided a landmark analysis of the research that had been developed by the mid-1960s. Although the majority of their data was collected for the year 1966–67, activities in this year reflected the fruition of ideas in the early 1960s. The report also contains the results of widespread consultation with those involved in water resources research and management about the value of past programs and priorities for the future.

[11]Bruce and Maasland (1968) provided details of the subtopics included under these categories and the breakdown of total expenditure by federal, provincial, municipal, industry, university, and nonprofessional organizations.

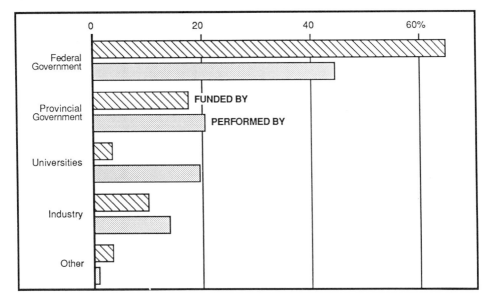

FIG. 2. Source of research funds and sector of performance, 1966. Source: Bruce and Maasland (1968, fig. 2).

address. In an appended report on the contribution of social science research, Sewell (1968) concluded that the expenditures were "extremely small...at less than $280,000 per year, about 3.3% of all research expenditures on water problems in Canada. This sum is minute when compared with investment in water development in Canada, some $750 million a year, or expenditures on physical science aspects of water management, estimated by the Study Group to be about $8,110,300 a year. Many of the problems on the horizon can be traced in part to the lack of attention to the human dimensions of water management."

Although research and management agencies had grown and diversified by the early 1960s, they had only just begun to react to the major pressures for change that were to mount during the next decade. Expenditures on research were relatively small in total, fragmented among a wide variety of organizations and focussed on a limited number of issues. Although there were exceptions to these general characteristics, it was not until the latter part of the 1960s that substantial changes began to take place in the breadth of research and in the relationship between research and management.

Mid-1960s to Mid-1970s — Major Innovations

The decade beginning in 1965 was marked by a dramatic increase in the attention given to the environmental and social consequences of development and a renewed interest in water resource conservation. These changes were foreshadowed in 1961 at the Resources for Tomorrow Conference, which was later described as "a major turning point for natural resource policy and planning in Canada" (Burton 1972).[12] The Canadian Council of Resource Ministers, established as a result of the conference, initially devoted much of its attention to water resources and in 1966 organized a national symposium on Pollution and Our Environment. In the ensuing years, concerns

---

[12]In the United States there had been a growing concern through the 1950s that natural resource limitations could pose serious problems for economic development. It was reflected and magnified by the reports of the Hoover Commissions, the Paley Commission, the President's Water Resources Policy Commission, and the Mid-Century Conference on Resources for the Future (RFF 1977).

expanded to include habitat destruction (e.g. gravel removal, loss of estuarine wetlands) and the health of aquatic ecosystems (e.g. lake eutrophication, biomagnification of toxic pollutants). At the same time, attention was focussed on cultural, social, and economic consequences that had previously been ignored (e.g. displacement of a community by reservoir construction[13]). Cost-benefit techniques were increasingly perceived to be inadequate for comprehensive evaluation. New techniques for environmental and social impact assessment, as well as planning, were introduced.[14]

In 1966, federal responsibilities for water resources management in Canada were consolidated with the establishment of a water sector within the Department of Energy, Mines and Resources (EMR) (Quinn 1977). An Interdepartmental Committee on Water, chaired by EMR, was set up to coordinate all federal agency programs.[15] In the following year, the EMR Minister offered to fund a comprehensive basin planning experiment with the provinces in each of the major regions in Canada. Subsequently, intergovernmental cost-shared agreements were negotiated for the Okanagan, Qu'Appelle, and Saint John basins.[16] This new approach was formalized in 1970 when the *Canada Water Act* was introduced to provide a legislative framework for federal–provincial cooperation in the planning and management of water resources. The attention of the federal government was focused on pollution at that time and, as a result, the Act emphasized water quality management.[17] It provided for regional water quality management agencies to be established sequentially by the federal and provincial governments which would set objectives, standards, schedules, and effluent fees. It thus included the key elements of water quality management strategies then being advocated by researchers (Kneese and Bower 1968).

During the first half of the 1970s, major planning studies were completed throughout Canada (Table 1).[18] They represented an ambitious attempt to apply and further develop techniques for generating information for management of water resources that had been evolving during the preceding two decades.[19] Comprehensive studies were undertaken that analysed the multiple purposes for which the water resources of the basin might be developed, the multiple means by which this might be achieved, and the multiple objectives that might thus be served. Multiple purpose development of water resources had been long practiced and methods for evaluating these options in

---

[13]For example, Wilson's (1973) study of the communities displaced by the Columbia River projects.

[14]To a large extent, these changes echoed events in the United States where, in the mid-1960s, major new programs were introduced for river basin planning and expanded research on water resources management. Two federal acts were particularly important. In 1964 the *Water Resources Research Act* authorized federal support for state water centres and nonfederal research activities. This was followed in 1965 by the *Water Resources Planning Act* that established the federal Water Resources Council and authorized the creation of regional planning commissions.

[15]This grew out of an interdepartmental committee on water that had been operating for some time (Quinn 1985).

[16]Ontario and Quebec turned down the federal offer to finance a joint study of the Ottawa basin (Brule et al. 1981).

[17]Tinney and Van Loon (1972, cited in Quinn 1977) pointed out that the Act was, at the time it was introduced, just one of six dealing with pollution, of which five concerned water pollution.

[18]Progress on these studies was reported in Annual Reports on the *Canada Water Act* and related to the development of water resources management more generally in a series of Canada Water Year Books; the two continuing publications have greatly increased knowledge of the many innovations.

[19]For a perspective on the techniques used in these studies, see "Monograph on comprehensive river basin planning" (Environment Canada 1975). The monograph was written by people who had been involved in the various basin studies. It was produced as one of Canada's contributions to the Inland Water Pollution Project of the NATO Committee on Challenges of a Modern Society, a project that Canada had promoted beginning in 1969.

terms of economic efficiency had been substantially developed. It was a major innovation to expand these methods to include the multiplicity of purposes that might be envisaged for a particular body of water (e.g. municipal and industrial water supply, irrigation, waste disposal, hydroelectric power production, navigation, flood control, fisheries, and recreation). The Okanagan study, for example, considered all of these purposes, except hydroelectric power and navigation, in analysing future scenarios for water use.

The consideration of multiple means was typified by innovations in flood control and waste disposal. Earlier, flood control had been accomplished primarily by structural measures, in particular dams and dykes. Now, attention was given to the appropriate use of flood proofing of structures, taking emergency measures, flood plain zoning,

TABLE 1. Programs and studies completed under the *Canada Water Act*, and status of federal and federal-provincial water management programs. Source: Environment Canada (1984, tables 1 and 2).

A: Programs and studies completed under the Canada Water Act	
Peace-Athabasca Delta Planning	1972
Qu'Appelle River Basin Planning	1972
Saskatchewan-Nelson Basin Planning	1973
Okanagan Basin Planning	1974
Saint John Basin Planning	1975
Lake Winnipeg, Churchill, and Nelson Rivers Planning	1975
Great Lakes Shore Damage Survey	1975
Fraser River Upstream Storage Planning	1976
Churchill River Basin Planning (Sask.-Man.)	1976
Montreal Region Flow Regulation Planning Study	1976
Peace-Athabasca Delta Implementation	1976
Northern Ontario Water Resources Planning	1978
Southeastern New Brunswick Dyking Implementation	1978
St. Lawrence Water Quality Planning Study	1978
Souris Basin Planning	1978
Metropolitan Toronto Flood Control Implementation	1978
Lower Saskatchewan Basin Preplanning	1979
Southwestern Ontario Dyking Implementation	1979
Upper Thames Flood Control Implementation	1979
Yukon Basin Preplanning	1979
Ottawa River Regulation Planning Report	1980
Thompson Basin Preplanning	1981
Great Lakes Shore Damage Survey Implementation	1981
Dykes and Flow Regulation Works — Montreal Region	1981
Mackenzie Basin Planning	1982
Shubenacadie-Stewiacke Basin Planning	1982
Ottawa River Water Quality Report	1982
Okanagan Basin Implementation	1982
Prairie Provinces Water Board's Water Demand Study	1983
North Shore (St. Lawrence) Ecological Inventories	1983
Winter River Preplanning	1983
Qu'Appelle Basin Implementation	1984

TABLE 1. (*Concluded*)

B: *Status of federal and federal-provincial water management programs*

Regulation, apportionment, monitoring, and survey programs

Under negotiation	New during 1983–84	Ongoing during 1983–84
Water Quality Surveys		Water Quantity Surveys Prairie Provinces Water Board Mackenzie River Basin Committee Water Quality Monitoring — Garrison Diversion Lake of the Woods Control Board Ottawa River Regulation Planning Board Ottawa River Water Quality Coordinating Committee

Water management programs

Under negotiation	New during 1983–84	Ongoing during 1983–84
Qu'Appelle Conveyance Regina-Moose Jaw Filtration Plant[a]	Winter Basin Planning Special Recovery Capital Projects in Ontario[a]	Fraser Estuary Planning Wabigoon-English Mercury Contamination Study Waterford River Urban Hydrology (Planning) Study Mercury in Churchill River Diversion System Yukon River Basin Study North Shore (St. Lawrence) Ecological Inventories Study Qu'Appelle Basin Implementation Lower Fraser Valley Flood Control Canada-Ontario Agreement on Great Lakes Water Quality

Flood damage reduction programs

Under negotiation	New during 1983–84	Ongoing during 1983–84
Initial Agreements with Alberta, British Columbia, and Yukon Amending Agreements with Saskatchewan, Ontario, and Nova Scotia Agreement for Flood Control on the Saint-François River in Richmond Amendment to Agreement to Upgrade Ring Dykes, Red River Valley	Mille Isles Control Structure Amending Agreements with Quebec and Newfoundland Quebec City Flood Control Agreement	Flood Damage Reduction Works and Dykes, Montreal Region Initial Agreements with New Brunswick, Newfoundland, Manitoba, and the Northwest Territories Amending Agreements with Ontario, Manitoba, and New Brunswick Upgrading Ring Dykes, Red River Valley

[a]Special funds were made available to Environment Canada for this program.

compensation of victims, and flood insurance.[20] The federal-provincial Flood Damage Reduction Program initiated in 1975 included use of all these means with the exception of state-run insurance (Spargo and Watt 1976, cited in Tate 1981). Similarly, waste disposal was considered within the broader problem of residuals management, wherein options included not just treatment before discharge to the water but also reduced generation of the waste by changing raw materials and production processes, recycling of materials, disposal to the land and atmosphere, measures to increase the assimilative capacity of the receiving environment, such as instream aeration, and effluent charges (Kneese and Bower 1968). This kind of comprehensive approach was applied in the analysis of water quality management in the Lower Fraser River (Dorcey 1976).

At the same time, evaluation was expanded to consider objectives other than national economic efficiency. Presidential approval of the United States Water Resources Council's "Principles and standards for planning water and related land resources" (1973) was the culmination of extensive development and review of the preceding decade. It established two national objectives for evaluating water resources projects, "National Economic Development (NED)" and "Environmental Quality (EQ)", and a system of four accounts for displaying beneficial and adverse effects, NED Account, EQ Account, Regional Development (RD) Account, and Social Well-Being (SWB) Account.[21] These principles were applied in the Canadian studies in the early 1970s. For example, in the Okanagan study three objectives were used in the evaluation of alternative plans: (1) "To increase economic development in the Okanagan basin and adjacent regions, as measured by net regional income"; (2) "To maintain an enhanced environmental quality by management, preservation and improvement of certain natural resource and ecological systems"; and (3) "To enhance social betterment by creating a more equitable distribution of income, employment, population densities and environmental quality between regions of the basin" (Environment Canada 1975). In turn, consideration of such objectives stimulated the development of new methods for measuring both economic and non-economic effects. One focus was on methods for estimating the less tangible values associated with water-based recreations and aesthetics.[22] Another was on methods for assessing environmental impacts.

In 1969 the *National Environmental Policy Act (NEPA)* was introduced in the United States and impact assessment techniques began to proliferate.[23] Environmental assessment procedures were introduced in Canada shortly afterwards. By the mid-1970s, most provinces and the federal government had introduced various impact assessment policies (Mitchell and Turkheim 1977). The techniques were to a large extent developed in the United States and included checklists, flow diagrams, matrices, overlays, and simulation methods (Munn 1975). Assessments were made of past projects that had resulted in environmental problems, such as the W.A.C. Bennett Dam on the Peace River that had major effects on the wildlife and fish in the Peace-Athabasca Delta over 1100 km downstream; of approved but only partially completed projects, such as the James Bay hydro project and exploration drilling projects in the Beaufort Sea; and increasingly, of projects before approvals were given, as in major proposed develop-

---

[20]The merits of considering a mix of structural and nonstructural measures for flood control had been argued in the landmark studies by White (1942) and White et al. (1958); Sewell (1965) later applied these ideas to problems of flood management in the Fraser River.

[21]For a discussion of the United States development of multiobjective water resources planning during this phase, see Haith and Loucks (1976).

[22]See Coomber and Biswas (1973) for a review of these methods in general and Environment Canada (1975, p. 117-137) for discussion of their application in the Okanagan study.

[23]By 1975, 7100 statements had been completed under the United States federal legislation of which 35% dealt with roads, 23% concerned watershed developments, 10% related to parks, forest, and timber, 7% reviewed energy-related projects, 6% studied airports, and 19% covered other topics (Mitchell and Turkheim 1977).

ments such as the Mackenzie pipeline and smaller projects such as municipal subdivisions (Mitchell and Turkheim 1977; see also chapter by Rosenberg et al., this volume). Experience with environmental impact assessments was beginning to direct attention to the need to measure social and cultural impacts that were still inadequately considered by the new techniques for measuring economic and ecological effects (Wolf 1974; Boothroyd 1975; O'Riordan and Sewell 1981; MacLaren and Whitney 1985).

An integral component of the revolution in planning and impact assessment was an unprecedented demand for public involvement in the management of water resources.[24] Public concern about developments that did not take account of peoples' interests, emerging evidence of serious environmental degradation, and an inability to become involved in making decisions resulted in growing demands for public participation in water resource management throughout North America and Europe (Economic Commission for Europe 1971; Environment Canada 1975). The United States federal *NEPA* of 1969 and *Water Pollution Control Act* of 1972 included public participation requirements. In Canada, the Okanagan, Qu'Appelle, and St. John river basin studies (1970–75) conducted under the *Canada Water Act* each included experiments in public involvement (Environment Canada 1975). They all were characterized by much greater efforts than in the past to communicate from the beginning with anyone who might have an interest in the water resource management issues. Communications strategies ranged from newsletters through community hearings to the creation of public and private interest task forces. Comparable experiments were incorporated in the emerging federal and provincial impact assessment processes. The Mackenzie Valley Pipeline Inquiry (Berger 1977), initiated in 1974, elevated public involvement to unimagined heights and promoted a dialogue on the social and economic impacts of resource development on ordinary people, the likes of which had never been seen before (or since) (Sewell 1981).

The advances in planning and impact assessment were also associated with significant progress in the techniques of modelling and thrusts towards more quantitative prediction. The tools of control theory and systems analysis, aided by high-speed computers, began to be widely applied to water resources management problems, and were of great assistance in the increasingly complex analyses that were required when management considerations broadened to include multiple objectives, purposes, and means.[25] Mathematical models that relied on computers because of their analytical intractability became common in hydrology, water quality studies, and in the depiction of ecological systems.[26] At the same time, economic models were refined and integrated with physical–chemical–biological models for both simulation studies and optimization analyses. Particular attention was given to the use of water quality modelling in the Saint John studies (Acres Company Ltd. 1971), and an International Symposium on Modelling Techniques in Water Resources Systems was organized in Ottawa in 1972 (Environment Canada 1972). Workshop approaches were designed for building and utilizing computer models in "adaptive management" strategies: observing, as a project or management program proceeds, the effects of perturbations to natural systems and the manner in which they depart from model prediction so that adaptive corrections may be made along the way (Walters 1975; Holling 1978).

---

[24]See the chapter by Sewell in Environment Canada (1975, p. 73–109) for a review of public involvement techniques and experience with their application in planning for water resources management in Canada and the United States; also see Sadler (1977, 1979) for more recent and broader reviews of public involvement experience.

[25]See Biswas (1976) for an overview of the systems approach to water resources management as it was perceived by the end of this second phase.

[26]See the extensive chapter on the use of mathematical models in the "Monograph on comprehensive river basin planning" (Environment Canada 1975, p. 141–205).

There were great expectations in Canada for expansion of the research needed to evaluate and improve the new trends in management. The Bruce-Maasland study, begun in 1966, reflected this optimism and led to recommendations by the Science Council of Canada that expenditures on water resources research should expand at 20% per year for the next 5 years (Science Council of Canada 1968). In 1966 the federal government decided that two research centres would be established at Burlington, Ont., and at Winnipeg, Man.

The Canada Centre for Inland Waters (CCIW) at Burlington was designed to be a "National centre of expertise in water resources research" (Bruce 1970, p. 187) and James Bruce became its Director in 1970. The Centre brought together in one establishment research activities of EMR, the Fisheries Research Board (FRB), and National Health and Welfare (NHW). With the formation of Environment Canada in 1970 these activities were further consolidated as part of the Inland Waters Branch of the new department. From the beginning, however, CCIW gave major attention to the issues in the Great Lakes and to water pollution in particular: "No other water pollution problem in Canada and indeed in the Industrial Heartland of North America can compare with its magnitude and importance".[27] An initial task of the Centre was to provide information for the International Joint Commission (IJC) on pollution of Lakes Erie and Ontario and the international section of the St. Lawrence River. The resulting research attracted a great deal of attention to eutrophication and the importance of controlling phosphate discharges, which together with the identification of other pollution problems (pesticides, oil spills, and thermal wastes) had a major influence on the research programs of the Centre in the first half of the 1970s.

The Freshwater Institute (FWI) in Winnipeg was established to develop the freshwater fisheries research activities of the FRB.[28] The emerging problem of eutrophication in the lower Great lakes led to a major focus on eutrophication and limnology (Johnstone 1977; Scott 1978). Early research by the Institute on experimental lakes was central to the development of understanding of the role of phosphates in eutrophication.[29] In the first half of the 1970s the research expanded into the north and to other problems resulting from economic and land use development, including acidification of lakes, heavy metal toxicity, impoundments, and the impacts on the fisheries of developments such as the Mackenzie Valley pipelines.

Also, by 1971, six research centres, each focussing on different aspects of aquatic resources management, had been established at universities across Canada with development grants from the National Advisory Committee on Water Resources Research of the Inland Waters Directorate, in the newly created Department of the Environment (Table 2). During 1971-72, $841 335 was allocated in development grants and an additional $864 490 to some 130 individual researchers at 32 universities.[30] Close ties to the universities were also stressed in the design of CCIW and FWI.

---

[27]Treasury Board submission for CCIW, quoted in the Planning Report (Public Works Canada 1968).

[28]Fisheries research was being developed by national organizations long before anything comparable for water resources. The Board of Management (1898) was the forerunner of the Biological Board of Canada (1912) which became the Fisheries Research Board of Canada (1937) (Hachey 1965). The early emphasis was on marine fisheries; the marine biological stations were established at St. Andrews, N.B., and Nanaimo, B.C., in 1908. Freshwater fisheries research began at Winnipeg in 1944 but was temporarily moved to London, Ont., between 1957 and 1966.

[29]The results of the remarkable long-term program of research associated with the Experimental Lakes Area are reported in three special issues of the Journal of the Fisheries Research Board of Canada: Vol. 28, No. 2 (1971); Vol. 30, No. 10 (1973); and Vol. 37, No. 3 (1980). For a discussion of the results of the early research and its application in lake management, see Vallentyne (1974).

[30]More than 90% of this funding went to natural science studies; 15 researchers received $85 250 for social science investigations.

TABLE 2. NACWRR development grants, 1971–72. Source: Canada National Advisory Committee Water Resources Research (1972).

Amount ($)	Granted to	Purpose
90 000	University of British Columbia, Water Resources Centre	Studies of water quality management in the Vancouver and Lower Mainland area
130 000	University of Toronto	Research in water quality management in conjunction with its Environmental Sciences and Engineering Program
182 160	University of Manitoba, Agassiz Centre for Water Studies	Water studies with special emphasis on social science investigations of water management
217 980	Université Laval, Centre de Recherches sur l'Eau	Program centered on water management problems of the St. Lawrence River Basin
123 000	University of Saskatchewan, Division of Hydrology	Hydrologic studies in prairie and northern environments
98 195	McMaster University, Department of Chemical Engineering	Research into waste treatment processes, with particular reference to the identification of potentially hazardous pollutants

Realizing that it would be impossible to provide adequate research facilities to each of the university groups active in water resources research, space was planned for up to 70 university professors and their students from the nine major universities within an hour's drive of CCIW (Bruce 1970). FWI was built on the campus of the University of Manitoba and designed with the idea that senior staff would hold adjunct professorships (Johnstone 1977).

With the development of planning studies and impact assessment processes, research began to be related more closely to the needs of management. For example, in the Okanagan River Basin study, researchers from agencies of both the federal and provincial governments and universities undertook component investigations (Anonymous 1983; Stockner and Northcote 1974). The impact assessment for the expansion of Vancouver International Airport, one of the first projects to be assessed through the federal Environmental Assessment and Review Process (EARP), resulted in government and university researchers beginning a series of investigations into the functioning of the estuarine ecosystem (Dorcey 1981; Mitchell and Gardner 1983).

Overall, during the decade beginning in 1966, funding of research for water resources management in Canada expanded substantially.[31] The increase in funding reflected the emergence of the new priorities for water and environmental quality management; to a great extent they were associated with efforts to deal with water pollution problems in the Great Lakes but also, more generally, with federal–provincial collaboration in watershed management studies. The Lower Lakes study was just getting underway in 1966 and represented a major expansion in effort by both the federal government and the Province of Ontario. The research programs were given added support in 1972 by the signing of the Canada–United States Agreement on Great Lakes Water Quality. The Agreement called for two major studies to be undertaken through the auspices of the IJC. The first was a water quality study of Lakes Superior and Huron, which

---

[31]Although there was not another major estimate of research funding until 1979 (Lefeuvre 1984), discussion in the Lefeuvre report of the trends in funding provides a basis for making some specific comments about the decade 1966–75.

was similar to the Lower Lakes studies conducted between 1966 and 1969. The second was to estimate the pollution loads to the Great Lakes from land use activities.

Thus the second phase in the postwar evolution of water resources management in Canada was one of major innovation and great expectations. In quick succession, ambitious attempts were launched to develop and implement new techniques of planning and impact assessment that took into account not only multiple uses of the water resource but also the multiple means that might be employed and the multiple objectives that might thus be accomplished. These attempts generated an array of new questions that stimulated new research, in particular to address environmental quality questions. There was a pervasive optimism that the management problems could be dealt with and the research questions answered if only there were the political will. Carried along by this optimism, the federal and most provincial governments created environment ministries and experimented with public participation in the governance of water resources. For their part, universities established new environmental studies programs and departments, as well as broadened existing programs within such disciplines as engineering, biology, economics, law, and geography, to undertake the emerging interdisciplinary research challenges and to train a new genre of resource managers.

The trends in water resource management were associated with more general changes in public attitude during the late 1960s and early 1970s. The perception that all resources were limited, that the past record indicated shameful wastefulness, and that materialism had gotten out of hand led to the movement from a "consumer society" to a "conserver society" (Science Council of Canada 1977). Notions of stewardship of resources, more responsible use of pesticides and herbicides, recycling of materials, and particularly a clamour for public participation were a large part of the context in which water resource management advanced in the second phase.

Mid-1970s to Mid-1980s — Retrenchment and Adaptation

In the second half of the 1970s, the weaknesses in the Canadian economy and increasing disenchantment with the results of planning and impact assessment led to questioning of the innovations that had been introduced into water resources management.[32] Up to and including 1975-76, federal funding was provided on the basis of individual projects. Thereafter, the Treasury Board established a ceiling on *Canada Water Act* expenditures of about $18 million per year (Environment Canada 1984). With subsequent reductions it was about $12 million by 1979, of which $5 million was committed to river basin planning and implementation (Table 3). The cutbacks have

TABLE 3. Environment Canada expenditures by program. Source: Environment Canada (1984).

River Basin Planning			River Basin Plan Implementation	
Person-years	$000	Year	Person-years	$000
31.0	2115	1977-78	14.5	1249
31.5	2318	1978-79	12.1	3474
32.0	1812	1979-80	11.3	3234
21.9	1835	1980-81	8.3	2382

NOTE: Information was not readily available for 1975-77. Also, on the basis of a more accurate reporting system established in 1981, it now appears that the above estimates are too high. The 1981-82 data are likely to show person-years of approximately 14 for RBP and 6 for RBPI and funding for both programs at about $2 million.

---

[32]Two publications provide a perspective on these events. Brule et al. (1981) reported the results of an evaluation of the river basin planning and implementation programs of Environment Canada. A book edited by Mitchell and Gardner (1983) contains the papers from a national symposium on Canadian experiences in river basin management, held in 1981.

continued in recent years, and no new intergovernmental basin investigations have been negotiated since 1981 (Pearse et al. 1985). With the onset of the recession in the 1980s, provincial funding was also cut back. For example, in British Columbia, where major new resource planning initiatives were still underway at the end of the 1970s, the onset of the recession brought restraint measures which reduced the personnel in the Ministry of Environment by 30%. Cutbacks in government funding when combined with the effects of high rates of inflation resulted in a major reduction in the resources for planning and implementation.

The cutbacks in joint federal–provincial projects were in part a result of some significant doubts about the productivity of earlier innovations (Brule et al. 1981). Many of the planning studies were perceived to be too long and costly, with almost every study requiring extension of the agreement to complete the program. The studies were criticized for overemphasizing inventory programs, not focussing on key issues, and producing recommendations that were too numerous and vague. The experiments with public involvement were criticized for delaying the process, overemphasizing the interests of the active publics, and usurping the role of elected officials. There were commonly delays of up to 2 years in implementing recommendations, which frustrated the heightened expectations that had been created by public involvement and sometimes meant that the recommendations were overtaken by events. On balance, the negative perceptions seemed to drown out the positive.

One response was to make the studies less comprehensive and revert to narrower or single-issue investigations, such as those dealing with mercury contamination (English–Wabigoon rivers) and urban runoff effects (Waterford River) (Table 1). Another more progressive response was to search for ways to keep the studies and programs comprehensive while completing them more quickly and cheaply. In the case of the *Canada Water Act* studies, this included the use of preplanning agreements to give more careful consideration to the design of studies before they were undertaken (e.g. Lower Saskatchewan, Thompson, and Yukon basins). Some provinces began to develop their own planning programs. In British Columbia a strategic planning program was launched in the late 1970s to develop regional, usually watershed, plans (O'Riordan 1981). The objective was to undertake multipurpose, multiobjective, and multimeans analyses but complete them more quickly and cheaply by relying on data already available. Alberta (Primus 1981), Ontario (Pope 1981), New Brunswick (Cardy 1981), and Saskatchewan (Pearse et al. 1985) also adopted regional and strategic approaches.

There was equal, if not greater, dissatisfaction with impact assessment processes. Project proponents increasingly questioned their costs and benefits (Economic Council of Canada 1979, 1981), environmental interests were critical of their effectiveness (Rees 1981), and analysts savagely critiqued them (Rosenberg et al. 1981). The lack of "before and after" studies, which allow experience to guide future assessments, was a particular point of criticism (Larkin 1984).[33] One response by government was to organize impact assessment processes so that the steps and expectations were clearer and to introduce procedures for fast-tracking and priorizing the amount of attention they received (e.g. British Columbia Guidelines for Mine Development) (Fig. 3). Another response was to refine the methodologies of impact assessment to improve the practice of science in public policy contexts and to give greater attention to scoping before and monitoring after assessments are made (e.g. Holling 1978; Beanlands and Duinker 1983). In addition, increased attention was given to the assessment and implementation of mitigation and compensation opportunities. Finally, at least conceptually, impact

---

[33]Notable exceptions were studies by Day (e.g. Peet and Day 1980), Hecky and Newbury (e.g. Hecky et al. 1984), and studies presented at the conference on Follow-up/Audit of Environmental Assessment Results, held at the Banff Centre, October 13–16, 1985.

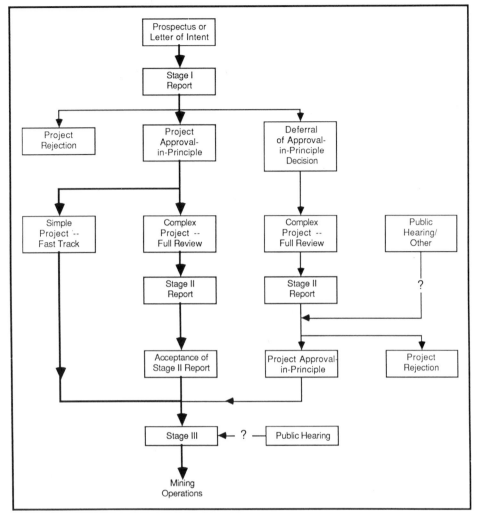

FIG. 3. Current procedures for obtaining mine development approvals. Source: Adapted from a poster for the British Columbia Ministry of Energy, Mines and Petroleum Resources (1984), showing the revised Mine Development Review Process.

assessment and planning processes began to be integrated (e.g. Dorcey and Hall 1981; Cornford et al. 1985; O'Riordan 1985).

A response by some analysts has been to question more fundamentally the design of institutional arrangements for all these aspects of water resources management. There has been continuing interest in describing the continual innovation in institutional arrangements for resource allocation, impact assessment, planning, and implementation (Mitchell 1975; Mitchell and Gardner 1983). Substantial attention, particularly by lawyers, has been given to the ways in which arrangements were developed for coordinating the roles of the federal and provincial governments as each tested the limits of its jurisdiction in these new approaches to resource management (Thompson 1981; Rueggeberg and Thompson 1984). However, various social scientists have become increasingly interested in the analysis of institutional arrangements. Detailed studies have been undertaken of "who" is involved and "how" in numerous aspects of water resources management.[34] Each discipline has promoted its own perspective on reform. Thus, for example, economists have emphasized the need to increase economic efficiency by making greater use of market allocation mechanisms through the establishment of marketable rights and charging for uses of the resource,[35] and political scientists have focussed on the need to redesign decision-making arrangements so that they more closely reflect the ideals of democratic governance to which Canadians aspire (Swainson 1976; Fox et al. 1983). In recent years, however, there have begun to be attempts to integrate the analyses and reforms suggested from various disciplinary perspectives (Dorcey 1986a).

Through the period since the mid-1960s, the demands for information continued to increase dramatically but funding for research did not expand as had been proposed. The development of strategic planning (e.g. O'Riordan 1981) and the attempts to improve impact assessments (e.g. Beanlands and Duinker 1983) pointed to the need for better information about aquatic resource systems. Research demands also increased as major new questions involving toxicity, acid rain, and climatic change came to the fore. The research community has, thus, not only had to respond to growing demands to become more involved in management but also to address new problems.

All of this occurred over a period when research funding declined in real terms.[36] Between 1979 and 1983 there was a 25% reduction in the funding of research in terms of 1979 prices (Table 4) and it is believed that this trend had already begun by the mid-1970s (Lefeuvre 1984). The shift in research expenditures from water quantity to water quality issues that began during the second phase was by 1979 clearly evident (Table 5).[37] Social science research continued to be no more than 5% of the total. Funding was increasingly dominated by the senior governments (Fig. 4). The approximate doubling of the proportion funded by provincial governments in 1979 reflects the expanded involvement of the provincial governments and declines in federal funding during the mid-1970s (Brule et al. 1981).[38] By 1983, however, the provincial proportion had dropped back to just over 10% and the proportion of research carried out by the federal government had expanded to slightly more than half. The proportion carried out by universities in the 1980s increased by 25% but they provided little of their own funding. The proportion carried out by industry shrank by more than 30%. It appears that the research carried out by all research organizations has become even more concentrated in Ontario than it was before; 58.4% of the projects were carried

---

[34]For example, see Sproule-Jones (1980) for a case study of pollution control in the Fraser Estuary, Munton's (1980) analysis of the politics of Canadian and American water pollution control in the Great Lakes during the 1960s and 1970s, Salter's (1985) investigation of the assessment of the pesticide Captan, and Dorcey and Martin's (1985, 1986) studies of the determinants of agreements reached by scientists involved in two instances of marine disposal of mine tailings.

[35]For example, see the seminal study of property rights by Dales (1968), the recent recommendations for greater use of pricing by Pearse et al. (1985), and the assessment of the difficulties of implementing them in practice by Dewees (1980).

[36]In 1983 a third major study was made of research funding in Canada by Mitchell and McBean (1985) (see also McBean and Mitchell 1985), under the sponsorship of the Science Council of Canada and the Inquiry on Federal Water Policy (Pearse et al. 1985). The Mitchell-McBean study, together with the Lefeuvre (1984) study, provides a basis for describing the trends in research funding during the third phase, 1976-85.

[37]Because of inconsistencies in the responses to the questionnaires, it is more appropriate to aggregate categories 500 and 900. A breakdown of these expenditures by subtopics and research groups that can be compared with the earlier data for 1966 is available in McBean and Mitchell (1985).

[38]The magnitude of this change may be overestimated because of incomplete coverage of federal expenditures.

TABLE 4. Research funds for water resources, over time. Source: McBean and Mitchell (1985, table 2).

Year	Research Funds in given year ($)	Research Funds in 1983 dollars[a]
1966	8 389 000	29 640 000
1979	52 000 000	75 400 000
1983	55 820 000	55 820 000

[a]Consumer price index used to update funding levels.

TABLE 5. Percentage of effort within different catagories of research (all values as percentage of total; na = category not included in 1966 study). Source: McBean and Mitchell (1985, table 8).

Category	1966	1979	1983
Nature of water	0.5	0.5	0.3
Water cycle	47.0	24.0	27.6
Supply augmentation and conservation	5.0	1.0	3.3
Quantity management and control	4.0	4.5	2.2
Quality management and protection	30.0	40.5	52.7
Economic, social, and institutional aspects	3.0	2.0	5.0
Resources data	5.0	4.5	4.7
Engineering works	5.5	7.0	3.4
Environmental management and protection	na	16.0	0.8

out in Ontario and an additional 12.2%, higher than any other province, was carried out in Ottawa.[39]

The Water Resources Research Support Program of Environment Canada that had provided the development grants to establish research centres at universities in the early 1970s was drastically cut back and eroded by inflation during the ensuing decade. From a budget of $1.41 million in 1972-73, it was cut to $1.14 million in 1973-74, to $1.0 million in 1974-75, and to $0.25 million in 1979-80. Until the major cut in 1979 these funds were distributed among 20 to 24 universities, but by 1983 this had declined to 13. Of the six centres developed in the early 1970s, four have continued to operate but only by diversifying their activities and finding other sources of funding.[40].

To illustrate the changes in funding at the major federal government research centres, Mitchell and McBean reported trends in the eighties at the National Water Research Institute (NWRI).[41] Between 1981 and 1984 the regular staff declined by 4.6%, a decline partially countered by hiring on temporary funds. Capital, operation, and maintenance funds fluctuated with the initiation of new projects, but there was a small

[39]Mitchell and McBean (1985) warned that these estimates have to be interpreted with particular care because they were unable to allocate all expenditures to the location at which the project was carried out. No data on the location of expenditures were reported by the Lefeuvre study. The consolidation of the National Hydrology Research Institute in a new building in Saskatoon in 1986 will reduce some of the concentration in Ontario. However, it will also decrease the regionalization of some other research units; for example, the National Water Research Institute unit on the West Coast will be relocated to Saskatoon.

[40]University of British Columbia's Westwater Research Centre, Université Laval's Centre de Recherches sur l'Eau, McMaster University's Department of Chemical Engineering, and University of Saskatchewan's Division of Hydrology.

[41]NWRI is part of the Inland Waters Directorate (IWD) and is headquartered at Burlington as a unit of CCIW. In 1983 it spent 15.3% of the estimated total expenditures on water resources research in Canada; its parent organization, IWD, spent almost 40% of the Canadian total.

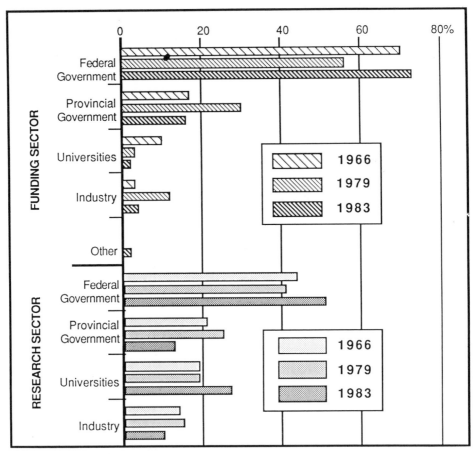

FIG. 4. Changes with time of percent of total funding, by source and expenditure, 1966-83. Source: McBean and Mitchell (1985, tables 3 and 4).

decline in the total resources available, with a decrease in the proportion going to research.[42]

These declines in funding for in-house research were to a small extent offset by the federal government agencies increasingly contracting out their research. In 1983 the Department of Supply and Services let water research contracts worth almost $3 million, 88% of which went to the private sector and the remainder to the universities (Mitchell and McBean 1985).

Thus, in the third phase there has been a major questioning of the great expectations for water resources management and research that arose in the second phase. Disappointment with the initial results of the ambitious innovations in planning, impact assessment, and research led to faltering support for the new approaches, particularly in the federal government. The fall in expectations was hastened by the decline in funding during the late 1970s and first half of the 1980s. In a climate of retrenchment and doubt, adaptations to the lessons from earlier experience and to the new realities of restraint have been slow to emerge but also often have gone unrecognized.

The principles of water resources management have changed radically in the years since World War II. Values have shifted. Increasingly sophisticated techniques of evaluation have proliferated. The task of management has become more complicated. Many

[42]Although comparable data were not available for other organizations, Mitchell and McBean concluded that this was a general trend.

new and ambitious experiments in management were tried. In retrospect, difficulties and failures should have been expected. The expectations, particularly for research, were unrealistically great, and the fall from them all the harder because of cutbacks in funding. In this situation the Inquiry into Federal Water Policy initiated in 1984 was most timely. The final report (Pearse et al. 1985) vigorously reasserts the basic principles of water resources management that were largely developed in the second phase and makes 55 specific recommendations for improving federal water policy. In the remainder of this chapter, the focus is on ways to improve the relationship between research and water resources management, including consideration of the Inquiry's recommendations for doing this.[43]

Innovation in Research for Water Resources Management

The innovative principles and techniques of water resources management which were largely developed in the second phase were appropriate. It was in putting them into practice that difficulties arose. In the turbulence of innovation and of other events during phase three, the appropriate relationship between research and management has become particularly confused and problematic.

Clearly there are major questions that need to be addressed, including: (a) How should research activities be related to the larger context of management? (b) How much should be invested in research? (c) What should be the priorities for research to meet management needs? (d) What should be the balance of research across the basic-applied spectrum? (e) What should be the distribution of effort between natural and social sciences?

Bruce and Maasland (1968) found it relatively easy to suggest answers to these questions because of the great expectations for new initiatives and additional funds to support them. By the time Lefeuvre (1984) and Mitchell and McBean (1985) revisited them, the answers appeared more complex and difficult. However, the experience and expectation of declining real resources for research and management have not yet catalysed a vigorous analysis of such questions.

The Inquiry on Federal Water Policy recognized that "the appropriate criterion for judging the adequacy of effort is whether research needs are being met" but was forced to conclude that it was a task which they were "unable to undertake within the scope of this inquiry" (Pearse et al. 1985, p. 116). The Inquiry estimated that in 1985–86 the federal government alone would spend $60 million on research, as part of $373 million in water-related programs that contribute to the generation of between $8 billion and $26 billion each year from the use of water resources in Canada. The implication of all these reviews of research is that nobody can, nor yet knows how to, provide very good answers if you ask: Is $60 million too much or too little? Could the $60 million be spent in more productive ways? What is most troubling about these occasional one-shot assessments is that they continue to reveal how poorly the relevant questions are being addressed in the ongoing routine of research and management. For these reasons, I argue that a strategy of experimentation that is explicitly designed to refine the statement of the questions and to test potential ways of answering them should be adopted.

A Strategy of Experimentation in Linking Research and Management

If we focus on the development of processes for setting research priorities, then the other important considerations will be revealed. Dorcey and Hall (1981) suggested a preliminary set of questions that should be considered in such a process[44]: (a) How

---

[43]The recommendations to the Inquiry on research policy by Mitchell and McBean (1985) are also considered.
[44]They also demonstrated how they could be answered in an application to the Fraser Estuary.

well can management decisions be made with presently available information? (b) How much better could decisions be with different improvements in the information available? (c) How might information be improved through different types of research? (d) What is the likelihood of improving information by different types of research? (e) What are the costs of different types of research? (f) How long will it take to improve information by different types of research?

Obviously, to address such questions it is necessary to have both managers and researchers involved in the process. Since management issues vary greatly in type and priority across the regions of Canada, the process has to involve managers along with relevant researchers in each of the regions.

The final report of the Inquiry on Federal Water Policy, "Currents of change," recognized some of these requirements but stopped short of dealing with crucial and fundamental aspects. The report recommended that the federal Minister of the Environment appoint a Canadian Water Resources Research Council, with members drawn from all key interests, to review and make recommendations annually on research funding and priorities; the Interdepartmental Committee on Water be assigned responsibility for coordinating research within the federal government; considerations be given to recognition of the primary geographical rather than functional focus of the mandate of existing federal government research institutes (e.g. the National Water Research Centre might become Great Lakes Research Centre, and the National Hydrology Research Institute might become the Prairie and Northern Rivers Research Centre); an external advisory board, with members drawn from all key interests, be established for each federal water research institute, to assess research programs and plans and report annually to the relevant minister and the proposed Research Council; and support be given to development of centres of excellence at regional universities with each specializing in a different specific and relevant field of water research.

If implemented, each of these recommendations would help in the forming of more links between researchers and managers, but I doubt that significant improvements in the setting of research priorities for management will occur unless there are also more fundamental changes: the currents of change must run much deeper and more strongly than recognized in the Inquiry's report. During the last 20 years, many groups and processes have brought together a mix of interests and expertises in the management of water resources, just as was proposed in "Currents of Change." But it has seldom been possible to make them work well.

Structuring the Links

The recommendations in "Currents of change" are a move in the right direction but do not go as far as would be required to create the structural changes needed to encourage researchers and managers to work more closely together. Improvements in water resources management depend on researchers and managers becoming more intimately involved in each other's work. This implies much greater emphasis on structuring their interactions in the regions where management takes place. How best to structure is unclear and will vary between the regions of the country. Therefore a strategy of experimentation should be adopted that includes experiments in linking managers and researchers within selected regions and at the national level.[45]

The diverse regions of Canada pose different opportunities and problems for research and management. Regional differences in management priorities were clearly brought out in the Inquiry on Federal Water Policy: "In a way, water policy is typically Canadian

---

[45]Although there have been various approaches in the past, they have not been designed with the objective of evaluating them and there has been little analysis of them. The literature on program evaluation stresses the importance of designing programs from the outset so that they can be evaluated, because without this it has proved to be extremely difficult to assess them post hoc (e.g. Rossi et al. 1979).

insofar as the predominant problems are regionally different. In the far north it is the problem of protecting the delicate ecosystem; on the Prairies, it is looming scarcity; in the Great Lakes - St. Lawrence basin and Atlantic Canada, it is pollution, with a special concern for ground water in the Maritimes. On the Pacific slope, the most urgent issue is the increasingly conflicting demands on water systems" (Pearse 1984, p. 3).

The regions also vary greatly in their research capabilities. At one extreme is the rich diversity and abundance within government organizations, universities, and colleges and the private sector in the Great Lakes-St. Lawrence region. At the other extreme is the paucity of capability in the north, which is primarily served by a variety of organizations from the south. In the other regions, there are substantial capabilities but generally they are not as diverse or as abundant as in the Great Lakes-St. Lawrence region.

In designing experiments to test how the researchers might be more effectively linked to managers, it will be necessary to recognize not only the particular opportunities and problems of the regions but also some general requirements. Within any region it will be necessary to link research into management processes in its basins or sub-basins. Unless this happens, researchers will not get close enough to management decision-making. As indicated earlier, strategic planning and management processes have begun to be developed within many of the regions of Canada, but this has generally not been accompanied by processes for involving researchers in the decision-making.

The case study in the Fraser Estuary (Dorcey and Hall 1981) suggested how an experiment might be structured to link research and management relevant to the particular characteristics of its situation. There is a cornucopia of research capability in the region but it is not marshalled to meet management needs, in spite of critical requirements for information. In close proximity there are three major federal government research institutes, three large universities, several colleges, and an unusually large number of private research organizations, all of which potentially could be involved in generating the information required. As there was no structure that encourged the researchers to interact amongst themselves or with managers concerned with the Fraser, it was suggested that a Fraser Estuarine Research Council (FERC) should be established. The essential characteristics of the proposal were as follows.

(1) The purpose of FERC would be to take the lead in
(a) stimulating and coordinating the design, conduct, and assessment of investigations undertaken by people in government and nongovernment organizations and
(b) facilitating the bargaining between these organizations to determine research priorities.

(2) The Council would be established by joint action of the federal and provincial governments, as an independent body with a fixed life, initially perhaps 10 years.

(3) It would have a secretariat, including a director and a small staff, that would carry out the work of the council on a day-to-day basis.

(4) A budget would be provided to FERC by a Federal-Provincial-Industry Agreement; a minimum of 90% of the budget would be for allocation to people in government and nongovernment organizations to undertake priority investigations.

(5) The Council would consist of highly regarded individuals, drawn in equal proportions from government and nongovernment organizations, and appointed by funding sources.

(6) The Fraser Estuary Management System (FEMS) has been evolving slowly through the last 5 years. Today, FEMS involves a Management Committee, consisting of an Executive of five lead agencies and 27 other members at large, a small secretariat, and an Information System (IS).[46] FERC, through its directors and staff, would be the

---

[46]For a more detailed analysis of what is required in FEMS, see Dorcey and Hall (1981) and Dorcey (1981). The present system is outlined by O'Riordan and Wiebe (1984) and evaluated in Dorcey (1986a).

critical link between FEMS and the community of researchers. The directors of FERC and FEMS would be ex-officio members of each other's executive bodies in order to facilitate interaction.

(7) The funding of FERC and FEMS would be conditional on prior public review and written comment on the proposed budgets and programs of each, in order to create an incentive for interaction.

(8) The information system required by FEMS and FERC would be operated jointly.

(9) The director of FERC would arrange with one or more of the relevant professional organizations to conduct an annual review of the investigations' program.[47] He would also participate in the public involvement process of FEMS, coordinating the selective involvement of individual investigators.

In other basins, different experiments would likely be appropriate to the research capabilities and the management systems that are in place or could reasonably be developed. For example, the Okanagan, the Mackenzie, the Churchill–Nelson, the Great Lakes, the St. Lawrence, and the St. John would each have different requirements, and would present different problems and opportunities in responding to them. Note that the experiments might be conducted at different scales: while there is reason to focus on just the estuary of the Fraser, it may be more appropriate to focus on larger areas in other situations, such as the Mackenzie basin or one or more of the Great Lakes.

If a series of such basin experiments was conducted, it would be possible to use what was learned from them to design and experiment with possibilities for region-wide structures. At this point, while the need for region-wide linking of management and research can be recognized, it is difficult to suggest what might be cost-effective experiments. The need for region-wide structures for planning and management has become more and more evident with the growing number of interbasin issues and will become increasingly important (e.g. allocation of developments such as hydroelectric power, fisheries enhancement, forestry, and agriculture between basins; interbasin transfers; and regional effects of climatic change and acid precipitation). Thus in a second stage of experimentation, it might be desirable to test some form of Pacific Region Water Resources Management System and a related Pacific Regional Water Research Council as one of a series of such experiments in the various regions of Canada.

The fragmented and poorly designed structures for setting research priorities in Canada have become increasingly inadequate. The first stage of experiments alone has the potential to improve this situation greatly by developing techniques and processes for setting priorities together with substantive results in a selected number of key basins. The second stage could produce further improvement by refining the techniques and methods as well as generating perspectives on priorities within key regions, which would provide a basis for determining national priorities and strategies for their pursuit in appropriate regions. The ultimate form of the structure sketched here will depend on the costs and benefits found in the experiments. I expect, however, that it will not be necessary to develop a monolithic structure, but rather that it will be more productive to design structures specific to each region and that only a small number of basins would merit the degree of attention suggested for the Fraser Estuary.

Allocating Research and Management Resources

Whoever allocates the resources for research and management clearly has a major impact on incentives. It is for this reason that the Fraser example provides for a pool of funds that can be allocated in the sub-basin and a mechanism for doing this that involves the relevant interests and expertises. This should be a characteristic in any

---

[47]Potential professional organizations would include the Pacific Estuarine Research Society, the Association of Professional Biologists, the B.C. Water and Waste Association, the B.C. Association of Professional Economists, the Planning Institute of British Columbia, etc.

selected sub-basin or region. If researchers can obtain funding by working more closely with managers, then in an era of shortage of funds, it can be expected that at least some researchers will shift their attention towards management information needs. At the same time, if managers see an opportunity to obtain additional resources, they too will be induced to consider how they can interest researchers in their particular information needs. This has been happening during the last decade but only in a limited and fragmented way.

For interactions between researchers and managers to happen reasonably efficiently, there need to be arrangements for encouraging and facilitating them. Hence the suggestion of FERC in the Fraser situation. It would need a small staff to carry this out. In this way it should be possible to orchestrate the interactions between researchers and managers so that they happen efficiently and do not unnecessarily divert their attention from their primary responsibilities. The staff should not only have knowledge and experience of relevant research fields but also management experience. They cannot be just administrators. They need to have a continuing interest in both camps. Such professionals are relatively scarce. It would therefore be necessary to create new career opportunities that enable them to move between these positions and either management or research positions. They could become the nucleus for building in the research community a stronger capability for management-oriented research and in the management community a stronger capability for integrating research into management.

Without funds to allocate, such organizations will not have any significant influence, nor would some of the problems with present funding allocation be dealt with. In general, the distribution of research funding in Canada is controlled by the dominant role of the federal government and its relatively centralized allocation processes. Where the allocation of research funds is controlled in the region, it is for the most part determined by researchers, and basin and sub-basin managers have only fragmented influence. Without any rigorous process for evaluating research priorities, they are set primarily by crisis issues of the moment (e.g. acid rain or climate change) and/or disciplinary priorities (e.g. the next step forward in understanding a benthic ecosystem). This is not to say the topics are unimportant but rather that it cannot be assumed they are the only and most important ones. At the same time, managers are scrambling to generate new information as best they are able. Sometimes this means the information is not generated. Other times information is generated but it is poorer in quality than it might have been because of the lack of expertise and resources. It has become common practice to conduct such investigations by secondments of agency personnel. The relatively small number of managers and researchers who try to generate the information required in such situations tend to be spread so thinly that they have difficulty doing justice to any of their responsibilities. The major problem is that nobody ever really assesses what resources are required to do an adequate job, or how limited resources should be allocated among competing priorities, and as a result, in spite of heroic efforts by all concerned, neither the researchers nor the managers are pleased by what is produced. It is for these reasons that organizations such as FERC are required. They focus attention on the research needs for management of the resources in a basin. By giving them a budget, they not only have the means to influence the practices of researchers and managers but also to focus attention on the scarcity of research resources and hence on the need to set priorities. If this kind of perspective were being developed in an ongoing fashion in selected critical sub-basins of the regions and in each of the key regions of the country, then it would be possible to improve greatly the national priority setting process.

Dorcey and Hall (1981) suggested that funding should be supplied under a federal–provincial–industry agreement in order to involve all three parties in determining the amount required and its allocation. In the last decade, industry has been making increasingly large expenditures for impact assessment, monitoring, and management, as well as for associated research. So far there has been relatively little effort made to integrate these activities into overall priorities; they tend to be locked into project-by-project considerations.[48] The case study in the Fraser suggested that some of the industry expenditures could be reallocated to more productive investigation and FERC could provide the mechanism for doing this. At the same time, FERC could be a local link to national organizations outside of the federal line departments, such as CCREM, NSERC, and SSHRC, that are involved in setting priorities and allocating funding for research that could address regional needs. Eventually, if they are created, FERC could be the essential link to the proposed Canadian Water Research Council and the Pacific Regional Water Research Council.

Only if these various changes are made will it be possible to give credible answers to the question of the adequacy of the resources that Canada and the provinces put into research for water resources management. If Canadians cannot even be given a reasonable account of how much they are presently spending to generate the information needed by managers, let alone the merits of alternative investigation priorities, it is unlikely that there will be adequate support for appropriate levels of funding. Organizations such as FERC should not only generate this information but also must take special steps to foster a more informed public and be seen to be responsive. By working with the relevant professional organizations and other interests through the public involvement program of the management system, this can be done effectively and efficiently.

Developing Interaction Skills

Clearly the success of the innovations I am suggesting will depend on the productivity of the interactions between researchers, managers, and others who are involved. Productivity will depend on not only the structure of the processes but also on the interaction skills of the participants, in particular their ability to communicate effectively, challenge constructively, and bargain successfully.[49]

Research and management success increasingly hinges on effective communication between individual researchers, managers, users, politicians, and the general public.[50] Each therefore has to know how to transmit and receive information in ways appropriate to the differing attitudes, perceptions, needs, and comprehension of the other. There are several reasons why communication has become particularly difficult. First, there is a need to transmit and receive information among an ever expanding group of people. Second, the wide variety of sciences involved and their highly specialized terminologies generate problems in sharing understanding both with nonscientists and with different specialists. Third, it has become increasingly necessary to convey understanding of complex natural, socioeconomic, and institutional systems. Fourth, growing recognition of the uncertain knowledge of how these systems function has compounded the difficulty. Finally, the consequent pervasive and pressing needs for technical and political judgement have greatly challenged communication skills.

A large proportion of communications goes beyond ensuring faithful understanding to contesting the assertions of others. When doubt and disagreements arise, it becomes essential to challenge constructively to determine the assumptions, data, and logic upon which the assertions are based.[51] Unfortunately the factors that are making com-

---

[48]There are some notable exceptions (e.g. the Environmental Studies Revolving Fund for energy projects, and the federal-provincial-industry study of forestry conflicts on the West Coast).
[49]See Dorcey (1986a, 1986c) for an elaboration of the argument in this section and detailed examples of the relevance to different contexts in the management of water resources in Canada.
[50]See Stanton (1982) for a review of basic principles and skills in communications and Arnold et al. (1983) for their application in the general field of planning, an application that is particularly relevant to the present context; both are applied to water resources management in Dorcey (1986a).
[51]See Mason and Mitroff (1981) for a discussion of the growing need in decision analyses to challenge the assumptions of others and methods for doing this; their concepts and methods are applied to water resources management in Dorcey (1986a).

munications more difficult also increasingly create scope for doubt and inhibit clarification of the basis for disagreements.

Since resources are scarce and there are competing interests, trade-offs will be inevitable. This increasingly applies not only to the water resource itself but also to management resources, of which research resources are just one specific component. Whenever possible, it is desirable to avoid making trade-offs, but when they are unavoidable, it is preferable to minimize the total losses. Bargaining has the potential to provide a most appropriate process for making such decisions in water resources management.[52] Clarifying the interests of the parties involved in the bargaining, identifying options for meeting these interests, evaluating the options in terms of the criteria important to the participants, and reaching an agreement on an option that will endure are all becoming much more difficult for the same reasons as communications and challenging.

Observation of water resources management reveals that many participants have major weaknesses in these skills and that these seriously undermine research and management productivity (Dorcey 1986a, 1986c; Dorcey and Martin 1986). Communications break down frequently and in numerous ways; for example the water chemist responds to the economist's simple query by sending his latest journal article, the fisheries biologist fails to check that she has understood correctly the implications of the benthic ecologist's remarks, the company does not summarize the results of its monitoring program in terms that address the interests of the local community, the planners provide the minister with a complex plan that does not make evident how it will respond to his constituents' concerns, the researcher does not notify the native Indian band of the plans to sample in the streams near their reserve, and so on. When doubt and disagreement lead to challenging, it is often done in a negative and adversarial manner. The objective is to destroy the argument of the other party. This is evident not only in the courts and other quasi-judicial processes, such as inquiries, where it is expected, but also in many other management situations, for example the comments by researchers on each others work in internal correspondence, closed meetings, and public conferences and reviews by regulators of a draft study report. As a result of such weaknesses in communication and in challenging, bargaining may not even begin or may be long drawn out. Often, in addition, there are weaknesses in bargaining skills that result in it breaking down or reaching agreements which do not last long, for example the regulatory agency, company, and research scientists are unable to reach any agreement on a monitoring and special studies program because of differences over several small components and the second year of funding is withheld for a research program when problems arise that the researchers knew could happen but had been afraid to mention less they jeopardize the initial funds. The major impact of these weaknesses on the productivity of research and management is further confirmed by observation of the beneficial effects in institutions where just one participant is skilled in conducting and promoting interactions, for example resolution of a long disagreement between the regulatory agency and the waste discharger as a result of the arrival of a new regional pollution control officer, and the breakdown in a planning exercise, involving many different research and management agencies, when a skilled task force director left.

Conclusion

Expectations of water resources management and research rose to such great heights in the heady days of the 1960s that managers and researchers had a long way to fall

---

[52]Bargaining can be defined as a process whereby two or more parties attempt to settle what each shall give and take, or perform and receive, in a transaction between them (Rubin and Brown 1975). Fisher and Ury (1981) provided a set of principles for successful bargaining that are applied to water resources management in Dorcey (1986a).

if they should falter. In retrospect, it was somewhat naive not to expect that such major innovations would inevitably encounter surprising difficulties. When they did falter, the fall in expectations was made even steeper by the onset of a series of ever-worsening crises in the Canadian economy. To date, it has proved to be extremely difficult to arrest the decline, to establish more realistic expectations, to recognize the successful adaptations, and to revitalize innovation.

From the three attempts over the last two decades to assess the contribution of research to management of Canada's water resources, it has become clear that we have great difficulty in measuring how much is being spent on research, let alone estimating its payoff in improved management. Because there are growing demands for research to meet management needs and declining resources for research, I have argued that the innovations for the late 1980s must include a strategy with two major thrusts: (1) experimentation with processes for setting research priorities in the regions and (2) development of the interaction skills of researchers and managers. Although not explored here, these two thrusts suggest important questions about their implications for each other; for example: How should the structure of institutional arrangements for setting research priorities be further adapted to facilitate productive interactions? How should job performance criteria be revised to encourage and reward the development of interaction skills?[53] Simple as the proposals are, they will have far-reaching consequences, some of which will undoubtedly be surprising. For example, in teaching short courses and graduate programs in water resources management, I have found that the basic interaction skills can be readily taught to students from widely different backgrounds. In doing this, it has become evident to me that the development of these interaction skills stimulates new perspectives on technical skills. An illustration of the far-reaching effects that can be expected is provided by the researcher specializing in benthic ecology who becomes more productively involved with others in the management of the resource. Over time the researcher becomes more skilled in communicating his knowledge to the other scientists and nonspecialists; in turn, this leads him to adjust the variety of ways he reports the results of his research and to spend more time discussing its design and conduct, which in turn leads him to change his research priorities and methods, which in turn leads him to give new advice to his local university on the technical skills required by the next generation of benthic ecologists. Cumulatively, across the various disciplines and over time, I expect that such adaptations could induce major and fundamental changes in our perspectives on what constitutes appropriate technical skill. Thus, if the innovations in skill development and priority setting are boldly implemented, more fundamental changes will be catalysed and there will be good reason, once again, to have great expectations of research and water resources management in Canada.

Acknowledgements

I am deeply grateful to P. A. Larkin for encouraging me to write this chapter and for providing detailed comments and suggestions on the early ideas. Several people reviewed a first draft and offered many valuable suggestions for revisions and additions: I. K. Fox, B. Mitchell, T. G. Northcote, J. O'Riordan, W. R. S. Sewell, and M. H. Sproule-Jones. Comments were also provided by P. H. Pearse, F. Quinn, and J. R. Vallentyne. B. A. Jaffray assisted me in reviewing the literature and J. M. Olynyk prepared the final bibliography, tables, and figures. I am most appreciative of the patient understanding and thoughtful advice provided by M. C. Healey throughout the writing and editing of the manuscript. As might be expected, reviewers did not always agree with each other nor I with them. The responsibility for the content of the final manuscript is entirely my own.

---

[53]These more fundamental implications for innovation in water resources management are examined in Dorcey (1986a, 1986b, 1986c).

References

ACRES COMPANY LTD., H. G. 1971. Water quality management methodology and its application to the Saint John River. Report prepared for the Environment Canada Policy Planning Directorate. Available from Environment Canada, Ottawa, Ont.

ANONYMOUS. 1973. Findings and recommendations of the consultative board under the Canada–British Columbia Okanagan basin agreement. Department of the Environment, Ottawa, Ont.

ARNOLD, D. S., C. BECKER, AND E. K. KELLER. 1983. Effective communications: getting the message across. International City Managers Association, Washington, DC.

BEANLANDS, G. E., AND P. N. DUINKER. 1983. An ecological framework for environmental impact assessment in Canada. Dalhousie University Institute for Resource and Environmental Studies, Halifax, N.S., and the Federal Environmental Assessment Review Office, Ottawa, Ont.

BERGER, T. R. 1977. Northern frontier, northern homeland — the report of the Mackenzie Valley pipeline inquiry. 2 vols. Supply and Services Canada, Ottawa, Ont.

BISWAS, A. K. [ED.] 1976. Systems approach to water management. McGraw-Hill, New York, NY.

BOOTHROYD, P. 1975. Review of the state of the art of social impact research in Canada. Ministry of State for Urban Affairs, Ottawa, Ont.

BRUCE, J. P. 1970. Water pollution control and the role of the Canada Centre for Inland Waters. Can. Geogr. J. LXXX(6): 182-193.

BRUCE, J. P., AND D. E. L. MAASLAND. 1968. Water resources research in Canada. Science Council Secretariat, Special Study No. 5, Ottawa, Ont.

BRULE, B., F. QUINN, J. WIEBE, AND B. MITCHELL. 1981. An evaluation of the river basin planning and implementation programs, Inland Waters Directorate, Environmental Conservation Service. Planning and Evaluation Directorate, Corporate Planning Group, Environment Canada, Ottawa, Ont.

BURTON, I. 1965. Investment choices in public resource development. In A. Rotstein [ED.] The prospect of change. McGraw-Hill, Toronto, Ont.

BURTON, T. L. 1972. Natural resource policy in Canada: issues and perspectives. McClelland and Stewart Ltd., Toronto, Ont.

CANADA NATIONAL ADVISORY COMMITTEE ON WATER RESOURCES RESEARCH. 1972. Water resources research grants awarded for water resources research, 1971-72. NACWRR, Ottawa, Ont.

CARDY, W. F. G. 1981. River basins and water management in New Brunswick. Can. Water Resour. J. 6(4): 66-79.

COOMBER, N. H., AND A. K. BISWAS. 1973. Evaluation of environmental intangibles. Genera Press, New York, NY.

CORNFORD, A., J. O'RIORDAN, AND B. SADLER. 1985. Planning, assessment and implementation: a strategy for integration. In B. Sadler [ed.] Environmental protection and resource development: a strategy of convergence. University of Calgary Press, Calgary, Alta.

DALES, J. 1968. Pollution, property and prices. University of Toronto Press, Downsview, Ont.

DEWEES, D. 1980. Evaluation of policies for regulating environmental pollution. Westwater Research Centre, University of British Columbia, Vancouver, B.C. Unpubl. rep.

DORCEY, A. H. J. [ED.] 1976. The uncertain future of the Lower Fraser. Westwater Research Centre, University of British Columbia, Vancouver, B.C.

DORCEY, A. H. J. 1981. The uncertain quest for a management strategy in the Fraser estuary. Can. Water Resour. J. 6(4): 95-118.

　　　1986a. Bargaining in the governance of Pacific coastal resources: research and reform. Westwater Research Centre, University of British Columbia, Vancouver, B.C.

　　　1986b. Techniques for joint management of natural resources: getting to yes. In J. O. Saunders [ed.] Managing natural resources in a federal state. The Carswell Company Ltd., Toronto, Ont.

　　　1986c. The myth of interagency cooperation in water resources management. Presented at 39th annual conference, Canadian Water Resources Association, Montebello, Quebec, 27-30 May, 1986. (To be published by Can. Water Resour. J.)

DORCEY, A. H. J., AND K. J. HALL. 1981. Setting ecological research priorities for management: the art of the impossible in the Fraser estuary. Westwater Research Centre, University of British Columbia, Vancouver, B.C.

DORCEY, A. H. J., AND B. R. MARTIN. 1985. Reaching agreement in impact management: a case study of the Utah and Amax mines. In B. Sadler [ed.] Audit and evaluation in environmental impact assessment. University of Calgary Press, Calgary, Alta.

    1986. Science and scientists in impact management. Westwater Research Centre, University of British Columbia, Vancouver, B.C. (In press)

ECKSTEIN, O. 1958. Water-resource development: the economics of project evaluation. Harvard University Press, Cambridge.

ECONOMIC COMMISSION FOR EUROPE (UNITED NATIONS). 1971. Proceedings of ECE Symposium on problems relating to the environment. ECE, New York, NY.

ECONOMIC COUNCIL OF CANADA. 1979. Responsible regulation: an interim report by the Economic Council of Canada. Supply and Services Canada, Ottawa, Ont.

    1981. Reforming regulation. Supply and Services Canada, Ottawa, Ont.

ENVIRONMENT CANADA. 1972. International symposium on modelling techniques in water resources systems. Environment Canada, Ottawa, Ont.

    1975. Monograph on comprehensive river basin planning. Information Canada, Ottawa, Ont.

    1984. The Canada Water Act Annual Report, 1983-84. Supply and Services Canada, Ottawa, Ont.

FISHER, R., AND W. URY. 1981. Getting to yes: negotiating without giving in. Houghton Mifflin, Boston, MA.

FOX, I. K., P. J. EYRE, AND W. MAIR. 1983. Yukon water resources management: policy and institutional issues. Westwater Research Centre, University of British Columbia, Vancouver, B.C.

FREY, D. G. 1963. Limnology in North America. University of Wisconsin Press, Madison, WI.

GOSSAGE, P. 1985. Water in canadian history: an overview. Inquiry on Federal Water Policy, Res. Pap. No. 11, Ottawa, Ont.

HACHEY, H. B. 1965. History of the Fisheries Research Board of Canada. Fish. Res. Board Can. MS Rep. 843.

HAITH, D. A., AND D. P. LOUCKS. 1976. Multiobjective water resources planning. In A. K. Biswas [ed.] Systems approach to water management. McGraw-Hill, New York, NY.

HECKY, R. E., R. W. NEWBURY, R. A. BODALY, K. PATALAS, AND D. M. ROSENBERG. 1984. Environmental impact prediction and assessment: the Southern Indian Lake experience. Can. J. Fish. Aquat. Sci. 41: 720-732.

HIRSHLEIFER, J., J. C. DEHAVEN, AND J. E. MILLIMAN. 1960. Water supply. University of Chicago Press, Chicago, IL.

HOLLING, C. S. [ED.] 1978. Adaptive environmental assessment and management. International Institute for Applied Systems Analysis, international series on applied systems analysis No. 3. John Wiley & Sons, Toronto, Ont.

INQUIRY OF FEDERAL WATER POLICY. 1985. Hearing about water: a synthesis of public hearings of the Inquiry on Federal Water Policy. Environment Canada, Ont.

JOHNSTONE, K. 1977. The aquatic explorers: a history of the Fisheries Research Board of Canada. University of Toronto Press, Downsview, Ont.

KNEESE, A. V., AND B. T. BOWER. 1968. Managing water quality: economics, technology, institutions. Johns Hopkins Press for Resources for the Future, Baltimore, MD.

KRUTILLA, J. 1967. The Columbia River Treaty: the economics of an international river basin development. Johns Hopkins Press for Resources for the Future, Baltimore, MD.

KRUTILLA, J., AND O. ECKSTEIN. 1958. Multiple purpose river development: studies in applied systems analysis. Johns Hopkins Press for Resources for the Future, Baltimore, MD.

LARKIN, P. A. 1984. A commentary on environmental impact assessment for large projects affecting lakes and streams. Can. J. Fish. Aquat. Sci. 41: 1121-1127.

LEFEUVRE, A. R. 1984. Water resources research in Canada in the late 1970s. Technical Workshop Series No. 4. Inland Waters Directorate, Environment Canada, Ottawa, Ont.

MAASS, A., M. M. HUFSCHMIDT, R. DORFMAN, H. A. THOMAS JR., S. A. MARGLIN, AND G. M. FAIR. 1962. Design of water-resource systems. Harvard University Press, Cambridge.

MacLAREN, V. W., AND J. B. WHITNEY. 1985. New directions in environmental impact assessment in Canada. Methuen, Toronto, Ont.

MASON, R. O., AND I. I. MITROFF. 1981. Challenging strategic planning assumptions: theory, cases and techniques. John Wiley & Sons, New York, NY.

McBEAN, E. A., AND B. MITCHELL. 1985. Water resources research in Canada: funding levels. Can. Water Resour. J. 10(2): 56-66.

McKEAN, R. 1958. Efficiency in government through systems analysis, with emphasis on water resources development. John Wiley & Sons, New York, NY.

MITCHELL, B. 1975. Institutional arrangements for water management. Univ. Waterloo Dep. of Geogr. Publ. Ser. No. 5.

MITCHELL, B., AND J. S. GARDNER. 1983. River basin management: Canadian experiences. Univ. Waterloo Dep. Geogr. Publ. Ser. No. 20.

MITCHELL, B., AND E. McBEAN. 1985. Water resources research in Canada: issues and opportunities. Inquiry on Federal Water Policy, Res. Pap. No. 16, Ottawa, Ont.

MITCHELL, B., AND R. TURKHEIM. 1977. Environmental impact assessment: principles, practices and Canadian experiences, p. 47-66. In R. R. Krueger and B. Mitchell [ed.] Managing Canada's renewable resources. Methuen, Toronto, Ont.

MUNN, R. E. [ED.] 1975. Environmental impact assessment: principles and procedures. SCOPE Rep. No. 5. John Wiley & Sons, Toronto, Ont.

MUNTON, D. 1980. Great Lakes water quality: a study in environmental politics and diplomacy. In O. P. Dwivedi [ed.] Resources and the environment: policy perspectives. McClelland and Stewart, Toronto, Ont.

O'RIORDAN, J. 1981. New strategies for water resources planning in British Columbia. Can. Water Resour. J. 6(4): 13-43.

1985. Environmental planning, project assessment and regulations: forging the links in the chain. Paper presented at the Workshop on Environmental Impact Procedures in Australia, Canada and New Zealand. (To be published by the New Zealand Commission for Environment)

O'RIORDAN, J., AND J. WIEBE. 1984. An implementation strategy for the Fraser River Estuary Management Program. Prepared by Fraser River Estuary Management Review Committee, Vancouver, B.C.

O'RIORDAN, T., AND W. R. D. SEWELL [ED.] 1981. Project appraisal and policy review. John Wiley & Sons, Toronto, Ont.

PEARSE, P. H. 1984. Of home and the river. Can. Water Resour. J. 9(2): 1-6.

PEARSE, P. H., F. BERTRAND, AND J. W. MacLAREN. 1985. Currents of change: final report of the Inquiry on Federal Water Policy. Cat. No. En 37-71/1985/1E, Environment Canada, Ottawa, Ont.

PEET, S. E., AND J. C. DAY. 1980. The Long Lake diversion: an environmental evaluation. Can. Water Resour. J. 5(3): 34-48.

POPE, A. W. 1981. Remarks of the Honourable Alan W. Pope. Can. Water Resour. J. 6(4): 7-12.

PRIMUS, C. 1981. River Basin planning in Alberta: current status — future issues. Can. Water Resour. J. 6(4): 44-50.

PUBLIC WORKS CANADA. 1968. Canada Centre Inland Wates — Planning Report 1968. Prepared by the Department of Public Works, Canada for the Department of Energy, Mines and Resources, Canada.

QUINN, F. 1977. Notes for a national water policy, p. 226-238. In R. R. Krueger and B. Mitchell [ed.] Managing Canada's renewable resources. Methuen, Toronto, Ont.

1985. The evolution of Federal Water Policy. Can. Water Resour. J. 10(4): 21-33.

REES, W. E. 1981. Environmental assessment and the planning process in Canada, p. 3-39. In S. D. Clark [ed.] Environmental assessment in Australia and Canada. Westwater Research Centre, University of British Columbia, Vancouver, B.C.

RESOURCES FOR TOMORROW CONFERENCE. 1961. Conference background papers. Queen's Printer, Ottawa, Ont.

RFF. 1977. Resources for the future: the first 25 years. Johns Hopkins University Press, Baltimore, MD.

RICHARDSON, R. E., W. G. ROOKE, AND G. H. NcNEVIN. 1969. Developing water resources: the St. Lawrence Seaway and the Columbia/Peace power projects. Ryerson/MacLean-Hunter, Toronto, Ont.

ROSENBERG, D. M., V. H. RESH, S. S. BALLING, M. A. BARNABY, J. N. COLLINS, D. V. DURBIN, T. S. FLYNN, D. D. HART, G. A. LAMBERT, E. P. McELRAVY, J. R. WOOD, T.E. BLANK, D. M. SCHULTZ, D. L. MARRIN, AND D. G. PRICE. 1981. Recent trends in environmental impact assessment. Can. J. Fish. Aquat. Sci. 38: 591-624.

ROSSI, P. H., H. E. FREEMAN, AND S. R. WRIGHT. 1979. Evaluation: a systematic approach. Sage Publications, Beverly Hills, CA.

RUBIN, J. Z., AND B. R. BROWN. 1975. The social psychology of bargaining and negotiation. Acadamic Press, New York, NY.

RUEGGEBERG, H. I., AND A. R. THOMPSON. 1984. Water law and policy issues in Canada. Westwater Research Centre, University of British Columbia, Vancouver, B.C.

SADLER, B. 1977. Involvement and environment. Environmental Council of Alberta, Edmonton, Alta.

   1979. Public participation in environmental decision-making: strategies for change. Environmental Council of Alberta, Edmonton, Alta.

SALTER, L. 1985. Observations on the politics of assessment: the Captan case. Can Public Policy XI(1): 64-76.

SCIENCE COUNCIL OF CANADA. 1968. A major program of water resources research in Canada. Sc. Counc. Can. Rep. No. 3.

   1977. Canada as a conserver society: resource uncertainties and the need for new technologies. Sci. Coun. Can. Rep. No. 27.

SCOTT, D. P. 1978. Biennial report for 1974-76 Western Region, Fisheries and Marine Service. Fish. Mar. Serv. Tech. Rep. 813.

SEWELL, W. R. D. 1965. Water management and floods in the Fraser River basin. Department of Geography, University of Chicago, Chicago IL.

   1968. Special report on the contribution of social science research to water resources management in Canada, p. 111-169. In J. P. Bruce and D. E. L. Maasland [ed.] Water resources in Canada. Science Secretariat, Special Study No. 5, Ottawa, Ont.

   1981. How Canada responded: the Berger Inquiry In T. O'Riordan and W. R. Derrick Sewell [ed.] Project appraisal and policy review. John Wiley & Sons, Toronto, Ont.

SEWELL, W. R. D., J. DAVIS, A. D. SCOTT, AND D. W. ROSS. 1961. Guide to benefit-cost analysis. Queen's Printer, Ottawa, Ont.

SPARGO, R. A., AND W. E. WATT. 1976. The Canadian flood damage reduction program. In Canadian background papers for the World Water Conference, Inland Waters Directorate, Environment Canada, Ottawa, Ont.

SPROULE-JONES, M. 1980. The real world of pollution control. Westwater Research Centre, University of British Columbia, Vancouver, B.C.

STANTON, N. 1982. What do you mean, Communication? Pan Books, London.

STOCKNER, J. G., AND T. G. NORTHCOTE. 1974. Recent limnological studies of Okanagan basin lakes and their contribution to comprehensive water resource planning. J. Fish. Res. Board Can. 31: 955-976.

SWAINSON, N. 1976. Managing the water environment. University of British Columbia Press in association with the Westwater Research Centre, University of British Columbia, Vancouver, B.C.

TATE, D. 1981. River basin development in Canada, p. 151-179. In B. Mitchell and W. R. D. Sewell [ed.] Canadian resource policies: problems and prospects. Methuen, Toronto, Ont.

THOMPSON, A. R. 1981. Environmental regulation in Canada: an assessment of the regulatory process. Westwater Reseach Centre, University of British Columbia, Vancouver, B.C.

TINNEY, E. R., AND R. J. VAN LOON. 1972. Canadian Federal Water Policy. In Water management. Organization for Economic Cooperation and Development, Paris.

TOLLEY, G. S., AND F. E. RIGGS [ED.] 1961. Symposium on the economics of watershed planning. Iowa State University Press, Ames, IA.

UNITED STATES WATER RESOURCES COUNCIL. 1973. Principles and standards for planning water and related land resources. Water Resources Council, Washington, DC.

VALLENTYNE, J. R. 1974. The algal bowl: lakes and man. Fish. Res. Board Can. Misc. Publ. 22.

WALTERS, C. 1975. An interdisciplinary approach to development of watershed simulation models. J. Fish. Res. Board Can. 32: 177-195.

510

WHITE, G. F. 1942. Human adjustment to floods: a geographical approach to the flood problem in the United States. Univ. Chicago Dep. of Geogr. Res. Pap. No. 29.

    1969. Strategies of American water management. University of Michigan Press, Ann Arbor, MI.

WHITE, G. F., W. C. CALEF, J. W. HUDSON, H. M. MAYOR, J. R. SHEAFFER, AND D. J. VOLK. 1958. Changes in urban occupancy of flood plains in the United States. University of Chicago Press, Chicago, IL.

WILSON, J. W. 1973. People in the way: the human aspects of the Columbia River project. University of Toronto Press, Downsview, Ont.

WOLF, C. P. 1974. Social impact assessment: the state of the art, p. 1-44. In C. P. Wolf [ed.] Social impact assessment, Section 2 of Part 1. In D. C. Carson [ed.] Man-environment interactions. Hutchinson and Ross Inc., Stroudsberg, Pennsylvania, Dowden.

CHAPTER 19

Managing Human Uses and Abuses of Aquatic Resources in the Canadian Ecosystem

J. R. Vallentyne

Department of Fisheries and Oceans, Great Lakes Laboratory for Fisheries and Aquatic Science, Canada Centre for Inland Waters, P.O. Box 5050, Burlington, Ont. L7R 4A6

and A. L. Hamilton

International Joint Commission, Canadian Section, 100 Metcalfe St., 18th floor, Ottawa, Ont. K1P 5M1

Introduction

During the 1960s there was a widespread recognition among industrialized nations that a more integrated management framework was needed to deal with problems of air and water pollution. The concept of environment provided an appropriate basis for that integrated framework; however, environment seemed to mean everything, and a department of everything was politically unthinkable.

The political solution in Canada, as in most nations, was to establish federal and provincial departments of the environment, restricting the meaning of environment to "green" environment. Social and economic aspects of environmental problems were relegated to other departments of government. This raised the political profile of "green" environmental concerns, but failed to grapple with the need for better integration of social, economic, and environmental issues. In consequence, unresolved problems began to emerge in new forms in the 1970s, for example, unemployment, pollution from toxic chemicals in old dump sites, and global disruptions to local supplies and prices of petroleum.

Even as governments were reorganizing to integrate "green" environmental concerns, the need for a broader ecosystem perspective was emerging. By this we mean a perspective that views social, economic, and environmental issues within the context of nature and that relates political systems to larger ecological systems that contain them, rather than as interacting entities among themselves.

The essential difference between environmental and ecosystem perspectives depends on how we see ourselves in relation to our environment. Looking outwards, our environment is perceived to be external. The view is like that of a camera taking a picture of its environment; the camera never appears in the picture. The ecosystem view is like that of a second camera taking a picture of the first camera taking a picture of its environment. The second camera sees the first camera and its environment as an integral system. Alternation of the two views gives rise to an ecosystem perspective.

An ecosystem perspective is implicit in most religions and stems from firm philosophic, evolutionary, and scientific foundations (Vernadsky 1926; Odum 1971; Great Lakes Research Advisory Board 1978). Politically, it is also noteworthy that the 1968 UNESCO Biosphere Conference (Bourlière and Batisse 1978), which set the global stage for implementation of ecosystem management, preceded the United Nations Conference on the Human Environment (Stockholm, 1972). Based on these and other reasons — notably pollution and other negative forms of environmental feedback — we believe that human societies are now in the early stages of a transition from a concept of *environmental management in a political context* to a concept of *political management in an ecosystem context.* In other words, political systems are being forced to take account of ecological realities. This chapter examines the extent to which an ecosystem perspective has penetrated the "mind" and operations of the Canadian federal government with respect to managing human uses and abuses of aquatic resources.

Pearse et al. (1985), Dorcey (this volume), and Thompson (this volume) have summarized Canadian policy and legislation related to human uses and abuses of aquatic resources. Our purpose is to evaluate influences bearing on future policies and legislation in terms of the extent to which those influences reflect an ecosystem approach to management. Three recent reports have been particularly relevant to this undertaking: the final report of the Inquiry on Federal Water Policy (Pearse et al. 1985), popularly known as the "Pearse Inquiry," the final report of the Royal Commission on the Economic Union and Development Prospects for Canada (1985), popularly known as the "Macdonald Commission," and the Ministerial Task Force report "Improved program delivery: environment" (Study Team 1986), which formed part of the process popularly referred to as the "Nielsen Review."

Our presumptions, openly stated, are that we do not manage aquatic resources, only our uses and abuses of those resources, that we must take greater ecosystemic and Biospheric account of our relationships with those resources, and that Canadians are in the process of developing a political framework for managing human uses and abuses of the Canadian ecosystem.

Since we are discussing a new framework for relating people and natural resources, clarity is essential to avoid misinterpretation. For this reason, we start with a summary of certain concepts essential to understanding the meaning of our use of the term *ecosystem approach.*

Setting the Ground

The Biosphere

The Biosphere is the living and life-supporting system in the outer part of the Earth.[1] Its boundaries are essentially fixed. We depend on The Biosphere for supplies of energy, fresh air, potable water, uncontaminated food, and the recycling of wastes. Recognition of this dependency establishes the causal order inherent in ecological doctrine: that the health of the political system is dependent on the health of the economic system, which is in turn dependent on the health of the natural resource system. It follows directly from this that the greater the stresses imposed by human populations on natural resource systems the greater the extent to which those stresses will spill over into social and economic systems, and into adjacent ecosystems, ultimately extending to The Biosphere as a whole. Radioactive elements — whether from nuclear bombs or reactor malfunctions such as those at Three Mile Island (United States) or Chernobyl (U.S.S.R.) — eventually spread throughout the atmosphere, hydrosphere, and food webs. Political and economic repercussions then ensue. Global climatic change from the accumulation of carbon dioxide in the atmosphere is another example. If continued, it is likely to be accompanied by major ecological, economic, and political disruptions from rising sea level, changes in agricultural patterns, and shifts in the distribution of species.

The Ecosystem Concept

Ecosystem is used here to mean a subdivision of The Biosphere with boundaries arbitrarily defined according to particular purposes in hand. At one end of the spectrum, we use *personal ecosystem* to refer to a person and that person's environment,

---

[1]Following the editorial practice of the International Society for Environmental Education and the journal *Environmental Conservation*, The Biosphere and Earth are capitalized as place names.

jointly. At the other end of the spectrum is our shared planetary ecosystem, The Biosphere. Between these extremes, forests, lakes, oceans, farms, and even cities can be viewed as ecosystems. We also view nations as ecosystems, i.e. as politically defined subdivisions of The Biosphere. *Canada as an ecosystem encompasses the interacting components of sunshine, air, water, soil, plants, and animals, including man, within the geographic boundaries of Canada.*

The use of political boundaries to define ecosystems may seem strange to ecologists and politicians alike; nevertheless we believe it is essential to effective management of the human uses and abuses of natural resources. The entire Biosphere has changed to such an extent that it is unrealistic to imagine that ecological systems can be analyzed independently of human influences. Indeed, a strong case can be made for viewing nations as ecosystems in the sense that they are functionally held together by systems of industrial production, transport, communication, agriculture, law, and politics that are inextricably linked to other (nonhuman) systems of nature. Skeptics need only consider the drastic changes that would take place in the Canadian ecosystem if for some reason humans were to disappear. The compelling point, however, is that only governments have the resources to undertake systemic analyses of the flows of energy, materials, and information in large-scale systems.

The Egosystem and the Ecosystem

We begin with a holistic perception of nature as a series of systems within systems at various levels of integration ranging from atoms to galaxies and beyond. In this heirarchy no one viewpoint can be said, scientifically, to be better than another. However, with human interests in mind, two important viewpoints may be contrasted. One is from the human system looking outward to a larger system at a higher level of integration, and the other from the larger system looking inward at the human component. These viewpoints may be described as egocentric and ecocentric, respectively, and the systems from which they originate as egosystems and ecosystems, respectively. The egosystem can be a person, corporation, government, or voluntary membership organization. The ecosystem can be any larger form of organization that integrates actions of the egosystem with other systems of nature.

A visual model may help to give reality to these abstract concepts. Consider some boys throwing stones (egosystems) into a pond. The falling stones give rise to concentric waves (analagous to outward moving spheres of influence) that interact with one another and, ultimately, with the shoreline of the pond (the ecosystem). Reflections from these interfaces give rise to new patterns, which arise as products of interactions of the waves with each other and with the boundaries of the pond. This model may represent the interactions of people in a home, provinces in the Canadian ecosystem, or nations in relation to The Biosphere as a whole.

In terms of the issues raised here, egocentric views typically begin with the assertion of political rights to the exploitation of natural resources, e.g. the rights of a province in relation to the rest of Canada, or the "sovereign" rights of Canada in relation to the rest of the world. The assumptions underlying egosystemic behavior are "me first," that nature is under the egosystem's command, and that egosystems have an inherent right to unlimited exploitation of natural resources.

In contrast, ecocentric views take account of constraints imposed by ecological, geographical, political, or other boundaries (e.g., those of the Canadian ecosystem in relation to the provinces, or those of The Biosphere in relation to Canada). Resources may be developed by the part (the egosystem), but from the point of view of the whole (the ecosystem), they are for the benefit of the whole. The assumptions underlying ecosystemic behavior are ecosystem first, that, ultimately, people are under nature's command, and that we share natural resources with other forms of life. In other words, we can manage the human uses of natural resources but not the ecosystems from which those resources stem.

These viewpoints are not as incompatible as the above characterizations might suggest; nor is one right and the other wrong. Generally, ecosystem behavior only comes into play when ecosystemic causes of stress are known; otherwise, egosystem behavior prevails. The question is not so much who or what is in command, but the need for egosystems to take better account of ecosystemic influences on their behavior. Francis Bacon's aphorism "Nature to be commanded must be obeyed" neatly captures this circular causal relationship.

Ecosystem Approach

Ecosystem approach is used here to mean an integrative approach to management that relates people to ecosystems that contain them rather than to resources or environments with which they interact. Historically, approaches to the management of

TABLE 1. Comparison of four approaches to resolving man-made ecosystem problems.

Problem	Approach			
	Egosystemic	Piecemeal	Environmental	Ecosystemic
Transmission of disease	Causes unknown	Conduits pills	Curative	Preventive, rehabilitative
Organic waste	Hold your nose	Discharge downstream	Reduce BOD	Energy recovery
Eutrophication	Mysterious causes	Discharge downstream	Phosphorus removal	Nutrient recycling
Acid rain	Unaware	Not yet a problem	Taller smokestacks	Recycle sulfur
Energy shortages	Hunt a scapegoat	Increase supply	Expand grid	Inverted rate schedules
Toxic chemicals	Unaware	Not yet a problem	Discharge permits	Design with nature
Greenhouse effect	Unaware	Not yet a problem	Sceptical analysis	Carbon recycling
Pests	Run for your life	Broad spectrum insecticides	Selective degradable poisons	Integrated pest management
Traffic congestion	More roads	More superhighways	Staggered hours	Public transport, decentralize
Demotechnic growth	Unaware	Measure it	Zoned development	Conserver society
Attitude to nature	Indifferent	Dominate	Cost/benefit	Respect
View of future	Egocentric	Linear, predictable	Wary	Emergent, evolving

human affairs have developed reactively in response to ecosystemic stresses. The shift has been from egocentric (unaware) to piecemeal (one by one) to environmental ("green") and, most recently, to ecosystemic approaches. Table 1 lists examples of typical actions taken under each of these management styles.

Criteria for an Ecosystem Approach

The criteria we use to determine whether a set of measures constitutes an ecosystem approach have been drawn from the Great Lakes Research Advisory Board (1978), Lee et al. (1982), and Christie et al. (1986) and are as follows.

(1) *Integration of knowledge (synthesis)*. Integrated knowledge is required to anticipate (predict) the ecosystemic consequences of human actions and inactions, to assess (monitor and evaluate) those consequences, and to adapt (change in a positive direction) the attitude, perceptions, and behavior of egosystems so as to take better account of ecosystemic realities. Integrated knowledge refers both to the development of new knowledge through research and to the dissemination of knowledge through educational systems.

(2) *A holistic perspective*. By this we mean a perspective that relates systems at different levels of integration (e.g. people in relation to regional or national ecosystems, nations in relation to The Biosphere). This can be contrasted with piecemeal or environmental perspectives that attempt to interrelate parts within a given level of integration (e.g. fish in relation to water quality, people in relation to natural resources, or nations in relation to each other).

(3) *Ecological actions*. By this we mean actions that take connectivity and recycling into account, including exchanges with neighboring ecosystems. Recycling may come about through human intervention as in sewage treatment plants or industrial recycling of metals or, alternatively, through other parts of the ecosystem as in the microbial transformation of human wastes in soil. The principles of conservation fall under this heading.

(4) *Anticipatory actions*. Because phenomena tend to recur in nature, the consequences of human actions can often be foreseen. Anticipatory actions are those which promote human and environmental health by forestalling conditions that might lead to undesirable and irreversible changes, e.g. pollution or the extinction of species. On the other hand, new domains of ignorance ("loose ends") are created by human innovations. When the uncertainty of prediction is such that ecosystemic consequences cannot be foreseen, models may be helpful to improve understanding; otherwise, prudence combined with monitoring becomes the rule. An ecosystem approach is founded on the principle that prevention is preferable to cure; account is taken of the need for safety factors and lead times.[2]

(5) *Ethical actions*. When the consequences of human actions knowably impinge on other living organisms, directly or indirectly through shared environments, actions are based on respect for nature and a belief in the inherent good of other forms of life and the habitats that sustain them. An ecosystem ethic is essentially a renaming of the "land" ethic of Leopold (1966).

On first sight, these criteria seem to call for radical changes in behavior; and, in fact, they do. Yet, on reflection, what is called for is only an extension in time and space of older moral norms (Taylor 1981). Transformed into an ecosystem approach, the

---

[2]Following the aphorism that an ounce of prevention is worth a pound of cure, JRV views prevention as a central pillar of ecosystem philosophy. ALH, on the other hand, is not convinced that the future is sufficiently knowable to prevent the appearance of troubles. Since responses to environmental injury can simultaneously be viewed as reactive in terms of the past and preventive in terms of the future, these views are not irreconcilable.

command on the Temple of Apollo at Delphi "Know Thyself" becomes "Know Your Ecosystem" (Criterion 1). Self-interest becomes enlightened self-interest based on an understanding of environmental backlashes (Criteria 2,3,4). The Golden Rule "Do unto others as you would have others do unto you" becomes the Golden Ecosystem Rule: "Do unto the ecosystems you share with others as you would have others do unto the ecosystems they share with you" (Criterion 5).

With this as background we are now in a position to examine the reports of the Pearse Inquiry, the Macdonald Commission, and the Nielsen Review for their perceptions of the need for an ecosystem approach to managing the human uses and abuses of aquatic resources.

The Inquiry on Federal Water Policy

The Inquiry on Federal Water Policy came into being in January 1984. Its terms of reference were broad. They called for recommendations and specific strategies on emerging water issues in support of "...conservation, development and utilization of water resources so as to ensure the enhancement of the health, well-being and prosperity of the people of Canada including continued regional economic growth and the quality of the Canadian environment..."

The final report, issued in September 1985, is a well-written, superbly illustrated, and generally excellent summary of the supplies and human uses of Canadian water resources and of the Canadian legal administrative and policy instruments through which the human uses of water resources are managed. It presents a balanced analysis of the strengths and weaknesses of federal policies and programs. The principles stated in Chapter 1 are ecologically sound: the watershed as the natural unit of management; the interdependence of land and water; the continuity of the hydrologic cycle; a focus on health and productivity of water systems; water development projects to take systemic account of economic and environmental risks and uncertainties; intrinsic value of natural flows to be considered in major or irreversible alterations; user pays principle; and public participation in decision-making.

We are less enthusiastic about the solutions put forward in the report. They suggest that all that is needed is to adjust the dials of the management system rather than to redesign the circuits. Of the 54 recommendations, we see only 5 as specifically addressing the need for an ecosystem approach. These are as follows.

"7.2 The *Environmental Contaminants Act* should be amended substantially to place the onus on producers importers and users of toxic substances...to seek approval and registration of those substances before they are marketed or used. (Criteria 1,5).

"10.1 The federal government should adopt integrated watershed management as a principle of federal water policy. (Criteria 2,3).

"10.3 Comprehensive and consistent criteria should be adopted for evaluating water development projects in which the federal government participates. (Criteria 1,2,3).

"10.4 Projects involving transfers of water from one basin to another should be considered only with great caution and only when alternative means of serving the purpose are infeasible. Federal involvement in such projects should be based on comprehensive evaluation of proposals including careful analysis of economic, social and environmental impacts in both donor and receptor basins. (Criteria 1,3).

"10.9 Administration of the deleterious substances section (Sec. 33(2)) of the *Fisheries Act* should be based not only on national baseline standards for industrial effluents, but also, and more importantly, on site-specific controls consistent with the integrated resource management requirements of particular watersheds." (Criterion 3).

The remaining recommendations seem designed to reinforce a system of piecemeal management of water in a political context. This was not accidental. "In formulating our recommendations," the authors wrote, "we have tried to be pragmatic in building on existing arrangements and capabilities. We have not proposed an abrupt wholesale

518

reform but rather a shift in direction to focus on new and emerging issues." This is inconsistent with the authors' own assessment of the need for an integrated policy and management framework.

In our opinion, most of the recommendations of the Inquiry of Federal Water Policy are too conservative. They will not lead to important and necessary changes in the ways we address water issues. The recommendations as a whole reflect too great a willingness to accept current dogmas, inadequate accounting systems, and insufficiently integrated policy directions. The effect of the recommendations, if implemented, would be to reinforce the present unecological way of doing business. In short, "Currents of Change" is neither a blueprint nor a strategy for change.

Report of the Macdonald Commission

In the press release, dated November 5, 1982, announcing the appointment of the Honorable Donald S. Macdonald as chairman of the Royal Commission on the Economic Union and Development Prospects for Canada, the Prime Minister described the terms of reference for the Commission as "perhaps the most important and far-reaching that have ever been assigned to any Commission in our history."

The general charge to the Commission was "...to inquire into and report upon the long-term economic potential, prospects and challenges facing the Canadian federation and its respective regions, as well as the implications that such prospects and challenges have for Canada's economic and governmental institutions and for the management of Canada's economic affairs."

In the final report, released in May 1985, five assumptions were stated: interdependence among nations is likely to increase in the future; the eradication of poverty in developing nations is "not on the horizon"; the general increase in human well-being is at risk; the limits to growth "are not physical, but political, social and institutional"; and broad, long-term goals are needed.

These assumptions are realistic and represent a healthy balance of egosystem and ecosystem views, although it must be recognized that The Biosphere does impose physical limits to growth. On the other hand, of the total of 208 recommendations listed at the end of the report, we can only see 6 that clearly reflect an ecosystem approach. Because of their bearing on later discussion, we quote these at length.

From Vol. I, p. 209-210: "Combining humanitarian and pragmatic interests, Canada should:

"Seek to broaden Canadian and world understanding of the meaning of interdependence and the threats and opportunities which confront civilization. This approach includes giving higher priority to issues relating to the natural environment and especially to the implications for global well-being of the continuing population explosion. (Criteria 2,5).

"Vigorously support reform of the multilateral system represented by the United Nations and its specialized agencies to bring the institutional machinery into line with the substantive problems and opportunities of the future. (Criterion 4).

From Vol. I, p. 209-210: "In particular, the Commissioners recommend that:

"In the decades ahead, Canada's policy makers integrate environmental decisions with those of economic development. This policy will be essential, for there is, in this Commission's view, no ultimate trade-off between economic development and the preservation and enhancement of a healthy environment and a sustainable resource base. (Criteria 1,5).

"Canadian economic policies be developed increasingly in a global context. This process requires a fuller recognition of the long-term and structural changes evolving, particularly in the areas of trade, technology and the role of governments, in the struggle for competitiveness. The incentives for work effort and productive contribution should be enhanced in a more flexible market environment. (Criterion 2).

"Canadian social support mechanisms and programs be designed more efficiently to accomplish the feasible task of removing the blight of poverty within Canada and to establish our national economy on a flexible, but secure, social infrastructure (Criteria 1,5).

From Vol. II, p. 530: "For the environment, we recognize the growing challenge and the need to integrate decisions related to environment and economic development. We recommend a series of measures to correct the incentives which are aimed at protecting the environment; and, in general, we propose strengthening the regulatory framework." (Criteria 1,3).

To evaluate the Report solely on the basis of its individual recommendations would be a gross injustice. In Part IV (Natural Resources and Environment), for example, some general principles are stated to set the stage for specific recommendations. Because of their bearing on this paper, we quote them at length.

"We conclude that in Canada, with all its rich resource heritage, there is no conflict, in the long-term between the stewardship, preservation and enhancement of the natural resource base and growth prospects for the traditional resource industries. Consequently, we perceive a vital need to integrate environmental decisions and decisions relating to economic development, and our proposals for action in each of the particular resource sectors reflect this perception. Thus we recommend a study of the loss of prime farm land to non-agricultural uses and emphasize our concern about the problem of soil deterioration and soil salinity. We support the infusion of large sums of both public and private monies into reforestation and silviculture, and we recommend that the duration of leasing agreements between governments and forest companies be increased in order to provide an incentive for long-term management of forest tracts. Finally, in recognition of the fact that natural resources belong to the Canadian people and must be passed on to future generations, we believe that private developers should continue to pay governments a royalty based on production for oil and gas and minerals.

"In many other places in this Report, we call for less government intervention; in the area of environmental regulation, however we are obliged to call for more. Over the long-term, the task of environmental regulation promises to be immense. We shall have to deal with growth in the number and size of projects that may adversely affect the environment, with an increasing number of pollutants and hazards, with the irreversible, and sometimes unquantifiable, effects of a growing range of industrial substances and processes, and with the emerging international aspects of our environmental responsibility. Consequently, we recommend that governments increase their spending to provide the analytical resources needed to support the long-term regulatory task. We further recommend that federal environmental processes be put on a statutory basis, and that federal and provincial review processes be brought into greater harmony."

The above assumptions, principles, and recommendations show that there was no doubt in the minds of the Commisssioners about the unity in the long term of environmental and economic considerations. Furthermore, while economic growth was viewed as a key means of increasing the welfare of all members of society, the Commissioners did not advocate the pursuit of growth at all costs. It is also clear from other sections of the Report that the Commissioners were alert to the need for integration of social concerns and economic considerations.

By far the most interesting conclusions and recommendations in respect to the questions posed here appeared at the end of Chapter 13 on "The Environment, Society and the Economy". We quote them in full as they were given at the end of the chapter rather than in the condensed form in which they appeared in the General Summary of Conclusions and Recommendations. Even a cursory reading shows the close correspondence between the Commission's assessment of needs in respect to environment, society, and the economy with what we call an ecosystem approach. The recommendations appeared under four headings (Vol. II, p. 526-528).

"Taking Preventive Measures

"Commissioners recommend greater use of a preventive approach to environmental decision-making, an approach that reflects and reinforces the growth in public support for policies that contribute to the regeneration of ecological systems. The concern for environmental values should be incorporated into a variety of decision-making processes, such as those that affect the work-place, the regulation and approval of large-scale projects, and the introduction of new products into the market-place.

"Greater consideration should be given to the development of a combined social and economic accounting system that covers not only the conventional economic indicators, but also such matters as soil depletion, forest degeneration, the costs of restoring a damaged environment, and the effects of economic activity on health.

"Formalizing the Environmental Framework

"...We therefore recommend that:

"Efforts be made to establish, on a sustained basis, the institutional arrangements through which environmental decisions are made.

"Government give greater emphasis to the scientific and analytical capacity of their environmental departments and increase the resources available for the enforcement of environmental policy.

"A national body with a core of independent scientific expertise be created to identify hazards that are, or are likely to become, seriously injurious. It would be the responsibility of such a body, styled "The Environmental Council of Canada", to provide information and advice about hazards that are of high national or regional priority; those, for example, that involve major water systems, significant industrial groups, and the actions of federal and provincial crown corporations.

"In recognition of the important role that research and development play in support both of the environmental regulatory function and of self-monitoring by concerned private sector businesses and associations, funding should be made available to permit research to be undertaken on a continuing basis.

"Environmental Review and Assessment

"This Commission recommends that:

"Project-approval hearings be co-ordinated or consolidated as a remedy for the excessive "regulatory lag" that results from multiple hearings requirements. Major projects, in particular, almost always require the approval of more than one government and, frequently, more than one agency in the same government. Consequently, we urge that efforts be made to harmonize requirements when multiple hearings are unavoidable and to develop common federal-provincial-municipal/review procedures whenever possible.

"The federal environmental-assessment process be placed on a statutory basis, and that threshold sizes be established for complusory project assessment so that for smaller projects, assessment would not be compulsory.

"Hearings procedures give greater attention to the effective analysis of technical and scientific controversy not only to improve our understanding of complex scientific questions, but also to expose more clearly the underlying economic and political dimensions of what sometimes are cast as purely scientific decisions.

"Visibility and Participation

"Commissioners recommend that:

"Measures be implemented to ensure a sustained public monitoring of environmental progress involving government-industry negotiations on environmental performance standards. Visibility and accountability should be increased.

"Increased public funding be made available to environmental groups to enable them to provide a more continuous presence in hearings and in monitoring activity."

These recommendations show that the Commission not only grasped the full sense of an ecosystem approach but translated the essentials into necessary political actions. This is particularly significant because none of the Commissioners had any professional training or experience in ecology; nor did the Commission avail itself through contracts of ecosystem expertise as it did on economic expertise. We are quite frankly impressed. Our only criticisms of Chapter 13 are minor: Chapter 13 would have been better placed at the beginning of the Report, and an "Ecosystem Council of Canada," rather than an Environmental Council of Canada, would have been more in line with the Commissioner's views on the need for closer integration of economic and environmental interests.

On the other hand, while the Report goes a long way toward identifying profitable paths into the future, except for Chapter 13, it fails to achieve the necessary level of synthesis. In some parts, social, economic, and environmental factors seem to be viewed as additive rather interactive and systemic. As with the Pearse Inquiry, more attention could have been given to redesigning circuits and less to adjusting dials. The Report does not go far enough in terms of its own perception of the need for better integration of environmental and economic decisions. The nature of the schism is most clearly expressed in Vol. II, p. 508-509, where the rationale for better integration of economic and environmental values is expressed in dollars.

We examined the supplementary statements of individual Commissioners at the end of the Report, looking for one that would give the sense of an ecosystem approach. The closest approximation was the supplementary statement of Commissioner John R. Messer which, significantly, began: "Economic man is an imperfect fiction created by economists. He does not exist as a whole person in real life."

Our conclusion is that the Commissioners, at least those involved in drafting the Report, and Chapter 13 in particular, actually did perceive the essence of an ecosystem approach to managing human uses of natural resources. The seeds of an ecosystem approach had evidently fallen onto the field of the Commission, germinated and thrust out roots and shoots; but another season would have to pass before the shoots would burst into flower.

The Nielsen Review

The Study Team (1986) report on Environment to the Ministerial Task Force on Program Review was charged to examine the effectiveness and relevance of programs under the federal Minister of the Environment in relation to present and future needs. Unlike the Pearse Inquiry and Macdonald Commission, government policies were accepted as givens. The Study Team was directed by a senior private sector chairman assisted by a senior official from the Department of the Environment and five private sector members. The focus was on four major concerns: beneficiairies/impact, efficiency, gaps, and fiscal projections.

The Study Team did not address the need for an ecosystem approach; however, two recommendations under the heading "Programs for Sustaining a Healthful Environment" are pertinent in that connection. The first emphasized a need to "confirm the mandate of those programs concerned with sustaining a healthful environment in Canada and with this confirmation provide an indication of how they should be administratively achieved." The second proposed the creation of a "Ministerial Task Force of Cabinet Ministers from the four core departments (Environment, Health and Welfare, Agriculture, and Fisheries and Oceans) concerned with environmental issues." These recommendations, if implemented, would go a long way toward improving and giving explicit recognition to an ecosystem approach.

In the area of water management research, the Study Team recommended that "A Water Management Research Committee be established in conjunction with a framework of water management priorities established by CCREM (Canadian Council of Resource and Environment Ministers), that would be responsible for recommending research activities for water management research." The rationale for establishing a Water Management Research Committee was the need for integration of fragmented responsibilities for water issues in the federal government.

The Great Lakes Basin Ecosystem

In this section we attempt to combine in a positive and constructive way elements of the reports of the Pearse Inquiry, the Macdonald Commission, and the Nielsen Review with the philosophy and criteria of an ecosystem approach. The Great Lakes Basin provides a convenient focus for the discussion because of the general consensus (Christie et al. 1986; Caldwell 1987) on the part of scientists, governments, and citizen groups in the basin on the need for an ecosystem approach to improving Great Lakes water quality. Pertinent background can be found in Great Lakes Research Advisory Board (1978), the Great Lakes Water Quality Agreement (1978), Lee et al. (1982), Christie et al. (1986), and the review of the National Research Council of the United States and the Royal Society of Canada (1986) on the Great Lakes Water Quality Agreement.

The Great Lakes Basin can be viewed as a simplified model of the world — reduced in area (754 000 km^2) and political complexity (11 governments: 8 state, 1 provincial, and 2 federal), yet large enough to be of interest in showing how cooperative efforts on shared interests can jointly benefit adjacent political jurisdictions. The Great Lakes contain nearly one fifth of all freshwater at the Earth's surface. Except for Lake Michigan, which lies wholly within the United States, the international boundary between Canada and the United States approximately bisects the waters of the Great Lakes. The waters are thus "boundary waters" rather than international waters.

Human population in the Great Lakes Basin at the beginning of the nineteenth century is estimated to have been about 300 000 with a d-index[3] of less than 2 (Vallentyne 1981). By 1986, the population had risen to approximately 38 million with a d-index in the range of 80-90. In other words, during the past two centuries there has been more than a 5000-fold increase in demotechnic metabolism. Comparable changes were taking place in other parts of Canada and the United States.

In 1909 near the inflection point of this phase of rapid demotechnic growth, the governments of Canada and the United States signed a Boundary Waters Treaty. The purpose of the Treaty was to provide a mechanism to avoid conflict in instances where actions in respect to water on one side of the border affected, or were likely to affect, health or property on the other side of the border. The Treaty contained provision for an International Joint Commission composed of six persons, three appointed by Canada and three by the United States, to advise the governments jointly on shared interests. The Commissioners are thus not appointed to represent their governments, but to advise the governments in common interest. The powers assigned to the Commission were *regulatory* with respect to levels and flows, *investigatory*, including surveillance and coordination, *advisory* with respect to references from the two governments, and, under special conditions, not yet brought into play, *decision-making*.

We regard the Boundary Waters Treaty as an example of the first and most essential institutional step in managing the human uses of aquatic resources of politically shared ecosystems: it acknowledged the transboundary influence of drainage basins and created

---

[3]*Demotechnic* (*demos*, population; *techne*, technology) refers to the combined physiological and technological metabolism of a human population. It is a simplification of an earlier term, *demophoric* (*demos*, population; *phora*, technological production), first used by Vallentyne and Tracy (1972). The demotechnic index or d-index is the ratio of technological energy consumption (as coal, oil, natural gas, and hydroelectric and nuclear power) to physiological energy consumption (as food). For further explanation, consult Vallentyne (1978).

an institutional framework to resolve conflicts that could otherwise jeopardize peaceful relations between adjacent nations. The Treaty is, however, dated in several respects. For example, it contains a general clause stating that boundary waters and waters flowing across the boundary shall not be polluted, but no specific provisions for controlling water pollution. The use of aquatic ecosystems for fisheries and recreation is not mentioned, nor are other transboundary influences such as air flows, species invasions, or transport of hazardous materials. Nevertheless, it is a landmark piece of legislation in terms of providing administrative arrangements for managing the human uses of shared resources.

In 1964, the governments of Canada and the United States delivered a reference to the International Joint Commisssion to examine the extent, causes, and measures for the control of pollution in the lower Great Lakes and their connecting channels. After detailed investigations over a period of 5 yr, the Commission recommended that a continuing mechanism be established to coordinate international programs to control water pollution in the Great Lakes. This led to the signing of the first Great Lakes Water Quality Agreement in 1972. The International Joint Commission was empowered to oversee implementation of the Agreement.

The primary focus of the Water Quality Agreement of 1972 was eutrophication and the treatment of municipal and industrial wastes. Vollenweider's (1968) ecosystemic model of the rules of phosphorus and nitrogen in eutrophication became the rationale for control. Even at the time of signing, however, new problems associated with toxic chemicals and acid rain were emerging as major concerns. Significantly, neither could be controlled by water management agencies alone.

The impetus for a change in the context of the Agreement from *water* to *water in an ecosystem* came from a special report of the Great Lakes Research Advisory Board (1978) to the International Joint Commission entitled "The Ecosystem Approach." This called for a radical philosophic shift from a system-external-to-man approach (e.g. water quality management) to a man-in-the-system approach (management of human activities in the Great Lakes Basin ecosystem). The benefits of adopting such an approach were illustrated in terms of billions of dollars that could be saved by reducing the excessive use of road salt for deicing, and benefits to human health from improved screening of toxic industrial chemicals with respect to "designing with nature" (see also Table 1). The need to reshape human attitudes, perceptions, and behavior in line with an ecosystem perspective was regarded as essential.

The presentation of "The Ecosystem Approach" to the International Joint Commission in July 1980 coincided with two other significant events, both of which shifted the focus of attention from water to land. These were presentation to the Commission of the final report of the Pollution from Land Use Activities Reference Group (PLUARG) and the declaration by the New York State Commissioner of Health of the site of the Love Canal in Niagara Falls, NY, as a hazard to human health. Before the year ended, a new Great Lakes Water Quality Agreement was signed incorporating elements of an ecosystem approach.

In the words of the Macdonald Commission (Vol. II, p. 520) the Great Lakes Water Quality Agreement of 1978 is " a milestone document because it was one of the first international statements that promoted the integration of human activity with the realities of the biota, land, water and air phenomena of the environment." Caldwell (1987) commented: "Viewed retrospectively from the future, this binational commitment, reiterated and reinforced, may be seen as not less significant than the Boundary Waters Treaty of 1909 upon which it has been based...a giant step in concept and principle."

The point we wish to emphasize is not that a full-fledged ecosystem approach is in operation in the Great Lakes Basin but that there is a general consensus in the basin on the necessity for an ecosystem approach to improving water quality, and that elements of such an approach are in early stages of development. Significant progress includes a basin-wide phosphorus control program based on bilaterally agreed loading reductions, acceptance of the necessity for an ecosystem perspective by governments and citizen groups, common reference to the ecosystem framework in discussions of environmental issues, increasing governmental focus on preventive strategies to reducing environmental harm, the recent Great Lakes Charter and Great Lakes Toxic Substances Control Agreement on the part of the Council of Great Lakes Governors, and the integration of public interest groups through organizations such as Great Lakes United and The Center for the Great Lakes (with offices in the United States and Canada).

In 1983, a workshop was organized to determine the extent to which an ecosystem approach might be compatible with the joint interests of people, industries, environmental groups, and governments in the Great Lakes Basin (Christie et al. 1986). Three main obstacles were identified: lack of a holistic perspective, predominance of egosystem thinking, and lack of preventive approaches to environmental disruptions. "Enlightened self-interest" was identified as the most promising strategy. Thirty-three initiatives were proposed to accelerate implementation of an ecosystem approach.

Canada as an Ecosystem

In his book *The Poverty of Power*, which appeared at the height of the energy crisis, Commoner (1976) argued that the environment-energy-economic dilemma facing the United States was based on a "tangled knot" of poorly understood problems involving "complex interactions among three basic systems...that, together with the social or political order, govern all human activity." He identified the three systems as the ecosystem, the production system, and the economic system.

Commoner used *ecosystem* to mean the natural resource system, external to man, that supports human life and activity. This is a more restricted and less holistic use than the one we have adopted. By *production system* he meant the man-made network of agricultural and industrial processes that convert natural resources into goods and services, the real wealth that sustains human society. By *economic system* he meant the system that transforms the wealth created by the production system into earnings, profit, credit, savings investment, and taxes and governs how that wealth is distributed, and what is done with it.

While it is clear that the economic system is dependent on the wealth yielded by the production system, and the production system on the resources provided by the natural resource system, Commoner observed that the governing influence has flowed in exactly the opposite direction and that "the relationships among the great systems on which society depends are upside down." The bottom line of his book was that the current system is unsustainable. We concur with this conclusion.

In our view, this structural inversion is leading to an alarming "ecosystem deficit" in terms of the quality productivity of Canadian land, air, water, and human resources. We are mortgaging the future — puting our descendents in the position of having to forego opportunities that we have enjoyed. In the long term, the social and environmental parts of the ecosystem deficit are likely to be a much greater threat to our descendents than the current economic deficit. This should not be interpreted as a lack of concern about the economic deficit, but rather, alarm over the acceptance and even indifference to long-term and even permanent ecosystem deficits. The Macdonald Commission also shared this view.

Some noteworthy Canadian examples of ecosystem deficits (including several cited by the Macdonald Commission) are dryland salinization of the prairies due to summer fallowing practices, the paving over of class 1 agricultural land and attendant rise in urban runoff, threats to human health of unacceptable concentrations of long-lived toxic industrial chemicals in rain and mother's milk and the runoff of fertilizers from farmer's fields leading to rising levels of nitrates in groundwaters and surface waters. Contamination of fish in Canadian waters can even be linked to the use of pesticides such as mirex and toxaphene in the southern United States and Central America.

Why are these ecosystem deficits occurring? In one sense they can be viewed as water management problems; in another sense they can be viewed as farm management problems and defined in terms of farming practices. Yet, if one looks more deeply, it is clear that many aspects of these deficits are inevitable consequences of an economic system that forces farmers to mine their land to meet short-term demands of the market. Dryland salinization and toxic chemicals in fish are not unrelated issues; both arise from an economic system that shapes agricultural policies and practices.

The single, most essential point we wish to stress is that while environmental issues may not find favor in a dollar-conscious society, in the long-term the deck is stacked in favor of the natural resource system.[4] To continue to ignore this reality will bring costly consequences to Canadians and the Canadian ecosystem. In short, there is a need for a complete reformulation of political objectives and strategies in the game that we are playing with nature; if not, the Canadian people are certain to lose.

This reformulation must simultaneously proceed from top/down and bottom/up. Starting from The Biosphere it becomes an overriding principle that national ecosystems are to be managed in long-term Biospheric interest. Vernadsky (1945) described this as "the reconstruction of The Biosphere in the interests of freely thinking humanity at a single totality." It is only on such a perspective that the various international treaties, agreements, and conventions on such topics as endangered species, the ozone layer, ocean dumping, and so on make any sense. The Biosphere is the commons. New rules and procedures need to be developed to protect it from national short-term interests.

Starting from the bottom/up, it must be recognized that all systems are programmed "me first" and also, that systems and their operating environments are interdependent in terms of survival. These are just as true of industries and their markets as of people and natural resources, or fish and aquatic habitats. Based on the power of reflective thought, the optimal human strategy is enlightened self-interest through environmental education (Christie et al. 1986).

Our greatest disappointment with the final reports of the Pearse Inquiry and Macdonald Commission is that the authors did not challenge outdated traditions and practices stemming from present "upside-down" views. Their own reports provided ample evidence of the need to view Canada as an ecosystem. They grappled with many of the essential problems, but failed to realize that the ecosystemic domain within which they were working transcended the economic and political domains. Consequently, their conclusions and recommendations tended to focus on symptoms rather than causes and on refining the status quo rather than on strategies to reduce stresses on the Canadian ecosystem.

The very idea that any jurisdiction or agency manages water is a barrier to developing holistic approaches to addressing water issues. In the Canadian context it has encouraged a continuing belief in man's "dominion over all," and it has helped to foster a bureaucratic division of responsibilities that limits cooperation between agencies, jurisdictions, and disciplines. It has also led to inward-looking programs, jealously guarded on the basis of legislative mandates and bureaucratically defined needs. These tend to focus unduly on short-term, tactical considerations that are consistent with preserving the status quo, and bear little relation to major issues of the present or future.

Constitutional, political, bureaucratic, scientific, and technical dimensions of "water management" all point to the fact that governments do not manage water resources.

---

[4]This was also the main point of the submission of the Rawson Academy for Aquatic Science (1984) to the Pearse Inquiry, of Environment Canada's (1984) submission, "Sustainable Development," to the Macdonald Commission, and of the booklet "Mandate for Change: Key Issues, Strategies and Workplan" prepared by the World Commission on Environment and Development (1985).

Governments have a responsibility to improve the human uses of shared resources so that their policies and actions are consistent with restoring and maintaining the health of the shared natural resource systems on which all life depends.

Science in Support of Sound Water Policy

Most current and emerging water issues are complex ecosystem issues. They arise from human uses and abuses of air, land, and biota in addition to water. They are commonly interjurisdictional and often international. In most cases they are not amenable to resolution by the piecemeal approaches of the past. The question is: to what extent is the ecosystem approach now developing in the Great Lakes basin applicable on a wider scale to similar problems in Canada?

Unfortunately, the Canadian policy framework for environmental sciences in general, and freshwater science in particular, has not encouraged the development of a sound support base for an ecosystem approach. The problem, in our opinion, primarily stems from a failure on the part of politicians and bureaucrats at all levels of government to recognize that the context has changed from water to ecosystem, and from nation to Biosphere. There has also been a failure of environmentalists to recognize the political and economic nature of their struggle (Bailey 1978). The result has been a costly and continuing succession of social, economic, and environmental ills, including allergies such as the "twentieth century syndrome", toxic chemical and other industrial forms of pollution, overexploitation of renewable resources (fisheries, forests, topsoil), and social disruptions in proximity to waste dumps.

Some senior officials within federal agencies have been remarkably successful at selling their science policy ideas, perhaps because they were consistent with the current political perceptions of how the economy works. These examples of successful salesmanship have not always been in the best interest of Canadian aquatic science, which provides the ecological knowledged base for effective management. For example, in 1969, Simon Reisman, then Secretary of Treasury Board, made the following statement to the Senate Science Committee, chaired by Senator Maurice Lamontagne (Hayes 1973):

"I would like to say a few words about science. In the eyes of the Board, science is not regarded as a thing in itself but as a means to an end. Scientific projects are not examined on their merits but as components of programs...This kind of thinking has enabled the Treasury Board to identify selected areas of research and development that justified priority treatment."

In essense, Mr. Reisman was telling parliamentarians: don't bet on the horse; bet on the track. At the time of his testimony, Planning-Programming-Budgeting (PPB) was in vogue; but, as Hayes (1973) pointed out, there was a fallacy in the assumption that PPB could be directly transferred from simple operational tasks, such as road building, to complex creative endeavors such as science. Today's acronyms are different, but little else has changed; nor can change be expected until more attention is payed to the horse and less attention to the track.

Another important aspect of Canadian science policy has been the federal government's stance on "make or buy." This policy effectively moved resources from in-house federal research programs to the consultant industry via the Department of Supply and Services. The consultant industry has been able to reorganize data and sell the resulting syntheses as proprietary information in a manner that meets contractual requirements. One would be hard pressed, however, to demonstrate that this has contributed significantly to the stock of knowledge about Canadian freshwater ecosystems. An alarming outcome of the "make or buy" policy coupled with the policy of no growth in the federal public service has been reduced recruitment of new scientific personnel in federal research institutes. The result has been that few institutes now have many research scientists under 40 yr of age. Another is the counterproductive tendency for scientists in universities to set themselves up as consultants, diverting their efforts from knowledge-generating activities to information-marketing activities.

We are not able to assess whether the aspects of science policy discussed above are applicable to the technology side of "science and technology," nor can we assess whether the Treasury Board model is likely to help Canadians to produce smaller and better microchips than Japanese or Americans. However, we cannot conceive that the "bet on the track mentality" can encourage the integrated science needed to improve and restore health to the Canadian ecosystem. The present state of affairs is simply not conducive to generating strategic advice on matters of long-term importance to Canada.

The community involved in developing ecosystem science has a responsibility to (a) recaste science and scholarship to reflect the importance of integrative, transdisciplinary studies, (b) help to identify and define emerging ecosystem issues, (c) review and assess the consequences of past egosystemic activities, (d) help to anticipate the consequences of present trends and planned activities, and (e) recommend appropriate programs and measures to prevent or ameliorate undesirable results of these trends.

It is increasingly coming to be recognized that threats to national and global security can stem from ecological as well as military and economic causes. Global climatic change from the "greenhouse effect" and cancer-producing chemicals that seep in and out of Canada without clearance by Customs or Health and Welfare officials are pertinent examples. These threats are more complex and less manageable than military issues. To discover and understand them demands a high degree of scientific knowledge and synthetic skills.

Institutional Opportunities

Major policy and structural changes are required if the scientific establishment is to contribute effectively to a sustainable egosystem–ecosystem relationship. Managing the human uses and abuses of freshwater ecosystems is but one facet of this larger challenge. Nevertheless, freshwater ecosystems can be prime indicators of stresses on both local and global life-support systems as early warning signals. This is highlighted by the manner in which the acid rain issue evolved from effects on small headwater lakes and streams.

The institutional challenge is to develop more effective mechanisms for involving scientific expertise in support of public policy (Royal Commission on the Economic Union and Development Prospects for Canada 1985). These mechanisms must be designed to have a sustained impact on the tangled web of interactive policies, programs, actions, and inactions that lead to allocative conflicts over rights to use and abuse the Canadian ecosystem.

The freshwater heritage that we pass to future generations of Canadians will reflect our ability to recognize water as (a) a *strategic issue* more fundamental to the health and integrity of the nation than the railroad at the time of confederation (b), a *horizontal issue* that cuts across territorial barriers imposed by legislative mandates and bureaucratic compartments, and (c) an *intergrative, cross-media issue* linked to life-styles and land-use practices by virtue of the transport of materials across air–water–land interfaces.

Within the existing constitutional and legislative framework it is clear that the federal government must play a leading watchdog role in safeguarding the health of the Canadian ecosystem. This has much more to do with "peace, order, and good government" than with the federal government's limited mandates for resource management and environmental protection. It also has a great deal to do with our cultural identity and the way Canadians, as a people, relate to the natural world. Canadians have a right to expect their federal government to ensure that short-term self-interest does not impair the health of the Canadian ecosystem — now, and for future generations.

Both internal and external advice is needed by governments; and it is desirable that the internal and external roles be clearly separated (McTaggart-Cowan 1974). Internally, the federal institution in the best position to provide leadership in ecosystem management is Environment Canada. It has a sufficient mandate, if it chooses, to adopt a more proactive leadership role with respect to its "horizontal" responsibilities within government. It also has a broad public support base and the internal flexibility to address these responsibilities.

By highlighting its horizontal responsibilities, Environment Canada could contribute significantly to shaping and influencing policies, programs, and actions that impinge on the quality, productivity, and utility of the land, air, water, and living resources on which the health of the Canadian economy depends. An excellent start in developing a framework to define the problems was made in Canada's State of the Environment Report (Bird and Rapport 1986) with its focus on ecosystem stress analysis.

Yet, Environment Canada has never really crystallized into a unified institution. We attribute this to a failure to articulate and exercise a horizontal mandate and a failure to describe a clear policy with regard to the role of research within that mandate. In the past, Environment Canada has made a deliberate decision to regionalize and defend its research activities on the basis of legislative mandates and operational priorities. Such an approach must, by definition, lead to fragmented, piecemeal research. At the same time, it impedes transdisciplinary research on linkages between ecosystem compartments.

The decision within the Inland Waters Directorate (IWD) of Environment Canada to justify its research on the basis of internal support to IWD "water management" responsibilities is reminiscent of a decision made within the Fisheries and Marine Service in the late 1970s when that Service was a part of the Department of Fisheries and the Environment. That decision — to integrate its research and management functions — created an internally justifiable unit that soon separated as the Department of Fisheries and Oceans. Probably, an internally justifiable IWD could fit just as well in the Department of Energy, Mines and Resources or the Department of Agriculture. However, if Environment Canada is to exercise a horizontal influence at the policy level, the water research function must remain with Environment Canada.

There are two fundamentally different strategies that Environment Canada might follow to enhance its influence over issues pertaining to natural resources and the environment. One is to seek greater legislative and bureaucratic authority to impose its view on other departments and sectors of society. The other is to develop the knowledge, expertise, and networks to infuse ecosystem thinking into all levels of policy-making and resource management. The first is a "power over" strategy; the second is an "influence with" strategy. While both strategies have their place, it is clear that the "influence with" strategy must be the primary mode for exerting horizontal influence at the policy-making and resource management levels.

From the foregoing it should be clear that the institutional framework within which federal water research is conducted needs to be changed. The changes should reflect a conscious decision to focus the research effort not on water per se, but on water in an ecosystem context. In other words, it is not the water management mandate of IWD that needs to be addressed; rather, it is the national need for improved management of human uses and abuses of water in the Canadian ecosystem.

In addressing this need it is important that the institutional framework be equipped with mechanisms to increase (a) the strategic role of research in support of sound policy and resource management, (b) the support for cross-media research on ecosystem issues of national significance, (c) the support of transdisciplinary research on ecosystem issues of national significance, and (d) implementation of an ecosystem approach to managing the human uses and abuses of natural resources.

These criteria are consistent with the need to increase the horizontal influence of Environment Canada on its external clients. For the Department to meet the criteria, there would have to be a reversal of the earlier decision to rationalize and defend its research activities on the basis of its own rather limited mandates for resource management and environmental protection. Were this to happen, a department Science Service, headed by an Assistant Deputy Minister, could be a productive approach. Alternatively, there could be an Ecosystem Research Board of Canada headed by a Chairman who reported directly to the Minister of the Environment, to the Ministerial Task Force proposed by the Nielsen Review (Study Team 1986), or to the Prime Minister. Regardless of the reporting relationship, it is important that there be a strong mandate for long-term strategic research on current, emerging, and anticipated ecosystem issues. There should be a responsibility to (a) serve national interests rather than the interests of specific departments in accordance with their own legislative mandates and program elements, (b) help Canadians gain a greater appreciation of the need to safeguard the health of the Canadian ecosystem and global Biosphere, (c) develop and implement an intergrated research program designed to influence major decisions on federal policies and programs affecting the health of the Canadian ecosystem, (d) develop and maintain close working relationships with the university and industrial communities, and (e) foster excellence, relevance, and openness.

There is also a need for an institutional mechanism at a high level of government to interface with global problems. For this, we foresee the desirability of an *Ecosystem Intelligence Service* to provide comprehensive policy advice to the highest corporate levels of government on the state of the Canadian ecosystem and Biosphere. This is necessary if the federal government is to grapple effectively with emerging continental and global issues such as energy, toxic chemicals, the greenhouse effect, acid rain, deforestation, and ethical issues. Such a Service could have as its primary mandate the provision of policy advice on the state of the Canadian ecosystem, and an ongoing responsibility to evaluate ecosystem implications of a wide range of policies and programs under consideration by the federal government.

Finally, we wish to stress the important, informal role played by external (nongovernmental) organizations in the formulation of policy at both national and global levels. In a penetrating analysis of the science–policy interface, Caldwell (1984) stated "The importance of NGOs (nongovernmental organizations) in international environmental policy-making cannot be overemphasized. NGOs have been essential to environmental policy-making from its beginning, both within and among nations, and they have been the instigators of numerous treaties and international cooperative arangements. Because NGOs form extra-governmental networks among as well as within nations, they are not constrained by the characteristic inhibitions of diplomatic protocol and bureaucratic procedure. In both the forming and execution of international policy they may act more rapidly and directly, and with less risk to national sensitivities than can the official intergovernmental agencies."

Caldwell's observations were made in the context of environmental decision-making; however, they pertain with even greater force in the context of ecosystemic decision-making. In the complexities of the modern world, perhaps only "senates" of nongovernmental organizations can speak effectively for the long-term interests of people in respect to national and global ecosystems. Environment Canada's public participation policy with respect to nongovernmental organizations is a first and good step in that direction. A comparable development at the international level is needed.

Concluding Remarks

The task of managing a sustainable egosystem–ecosystem relationship is formidable. Nevertheless, it is a national and global challenge that must be taken up. Demotechnic stresses on natural resource systems are likely to intensify in the future. Vigilance will be required to prevent abuse to the ecosystems that sustain these resources. The interjurisdictional nature of the necessary solution adds to the complexity of the issue and the greatness of the challenge.

The many local, regional, national, and international initiatives that have arisen in the past two decades attest to a growing perception of the need for a healthy egosystem-ecosystem relationship. These include the increasing number of organizations concerned with environmental health, the vastly improved reporting of environmental issues by the media, national incentives for energy conservation, local depots for recycling paper, glass, and metals, a rise of interest in physical fitness, and an evolving planetary consciousness.

At the international level, new organizations such as the United Nations Environment Program and World Commission on Environment and Development have appeared on the scene. The latter's "Mandate for Change: Key Issues, Strategy and Work Plan" (World Commission on Environment and Development 1985) is a particularly pertinent document for its clear focus on the necessity of integrating economic and environmental operations. Similarly, the "World Conservation Strategy: Living Resource Conservation for Sustainable Development" (International Union for the Conservation of Nature and Natural Resources 1980) has had wide international exposure. Canada hosted major events on behalf of the World Commission on Environment and Development and the International Union for the Conservation of Nature and Natural Resources in 1986 and will probably continue to play key roles in facilitating their further development.

At a bilateral level the recent joint report of the special envoys on acid rain indicates official recognition that the governments of Canada and the United States agree that "acid rain is a serious problem." Close examination shows that the Ontario Environmental Assessment Act is, in fact, an Ecosystem Assessment Act in that it addresses social, economic, and environmental issues. The report of the Conservation Council of Ontario (1986), "Towards a Conservation Strategy for Ontario: An Assessment of Conservation and Development in Ontario," is an excellent document for its ecological focus. The initiatives of the Council of Great Lakes Governors referred to earlier are significant in terms of basin-wide interest. Finally, the title of the review of the National Research Council of the United States and the Royal Society of Canada (1986) is significant: "The Great Lakes Water Quality Agreement: An Evolving Instrument for Ecosystem Management."

Canadians have played leading roles in developing the concept of an ecosystem approach to problem-solving. In implementing the concept, politicians, civil servants, scientists, industrialists, the media, and leaders of citizen groups all have important roles to play. The opportunities we leave our descendents will hinge on the extent to which we are able to maintain the vital productive capacity of the Canadian ecosystem.

NOTE ADDED IN PRESS: After this chapter was completed, the Canadian Wildlife Federation (Beamish et al. 1986) completed a comprehensive report on freshwater fisheries in Canada. An additional publication containing recommendations arising from phase I of the report is scheduled for publication early in 1988.

References

BAILEY, C. 1978. Environmentalism, the left and the Conserver Society. Conserv. Soc. Notes 1(2): 22-28.

BEAMISH, F. W. H., P. J. HEALEY, AND D. GRIGGS. 1986. Freshwater fisheries in Canada: report on phase I of a national examination. Canadian Wildlife Federation, 1673 Carling Ave., Ottawa, Ont. K2A 3Z1. xxxiv + 155 p.

BIRD, P. M., AND D. J. RAPPORT. 1986. The State of the Environment Report for Canada. Supply and Services Canada, Ottawa, Ont. Cat. No. EN21-54/1986E. 276 p.

BOURLIÈRE, F., AND M. BATISSE. 1978. Ten years after the Biosphere Conference: from concept to action. Nat. Resour. 14(3): 14-17.

CALDWELL, L. K. 1984. The President as Convenor of Interests: policy development for the twenty-first century. School of Public and Environmental Affairs, Indiana University, Bloomington, IN. 25 p.

1987. Implementating an ecological approach to basin-wide management. *In* Management for the Great Lakes: implementing international agreements". State University of New York Press, Buffalo, NY. (In press)

CHRISTIE, W. J., M. BECKER, J. W. COWDEN, AND J. R. VALLENTYNE. 1986. Managing the Great Lakes Basin as a home. J. Great Lakes Res. 12(1): 2–17.

COMMONER, B. 1976. The poverty of power: energy and the economic crisis. Bantam Books Inc., New York, NY. 298 p.

CONSERVATION COUNCIL OF ONTARIO. 1986. Towards a conservation strategy for Ontario: an assessment of conservation and development in Ontario. ISBNO-199856-05-5. 167 p. (Available from the Conservation Council of Ontario, Suite 202, 74 Victoria Street, Toronto, Ont.)

ENVIRONMENT CANADA. 1984. Submission to the Royal Commission on the Economic Union and Development Prospects for Canada. 18 p. (Unpubl.)

GREAT LAKES RESEARCH ADVISORY BOARD. 1978. The ecosystem approach: scope and implications of an ecosystem approach to transboundary problems in the Great Lakes Basin. Great Lakes Regional Office, International Joint Commission, 100 Ouellette Avenue, Windsor, Ont. ix + 47 p.

GREAT LAKES WATER QUALITY AGREEMENT. 1978. Great Lakes Regional Office, International Joint Commission, 100 Ouellette Ave., Windsor, Ont. 52 p.

HAYES, F. R. 1973. The chaining of Prometheus: evolution of a power structure for Canadian science. University of Toronto Press, Downsview, Ont., and Buffalo, NY. xix + 217 p.

INTERNATIONAL UNION FOR THE CONSERVATION OF NATURE AND NATURAL RESOURCES. 1980. World conservation strategy: living resource conservation for sustainable development. IUCN-UNEP-WWF. ISBN 2-88032-101-8. (Available from the International Union for the Conservation of Nature and Natural Resources, CH-1196 Gland, Switzerland).

LEE, B. J., H. A. REGIER, AND D. J. RAPPORT. 1982. Ten ecosystem approaches to the planning and management of the Great Lakes. J. Great Lakes Res. 8(3): 505–519.

LEOPOLD, A. 1966. A Sound County almanac, with essays on conservation from Round River. Sierra Club/Ballantine, New York, NY. 295 p.

MCTAGGART-COWAN, P. D. 1974. What type of Board are you? J. Fish. Res. Board Can. 31: 1281–1282.

NATIONAL RESEARCH COUNCIL OF THE UNITED STATES AND THE ROYAL SOCIETY OF CANADA. 1986. The Great Lakes Water Quality Agreement: an evolving instrument for ecosystem management. National Academy Press, Washington, DC. xix + 224 p.

ODUM, E. P. 1971. Fundamentals of ecology. W. B. Saunders, Philadelphia, PA. xix + 574 p.

PEARSE, P.H., F. BERTRAND, AND J. W. MACLAREN. 1985. Currents of change. Final report, Inquiry on Federal Water Policy. Environment Canada, Ottawa, Ont. 222 p.

RAWSON ACADEMY FOR AQUATIC SCIENCE. 1984. Canadian waters: some strategic considerations affecting the future of the resource. Submission to the Inquiry on Federal Water Policy. 19 p. (Unpubl., available from Rawson Academy of Aquatic Science, No. 200, 601–17th Ave, S.W., Calgary Alta. T2S 0B3

ROYAL COMMISSION ON THE ECONOMIC UNION AND DEVELOPMENT PROSPECTS FOR CANADA. 1985. Can. Gov. Publ. Centre, Supply and Services Canada, Cat. No. Z1-1983/1-1E. Vol. I, xxii + 385 p; Vol. II, 827 p; Vol. III, 699 p.

STUDY TEAM. 1986. Report to the Task Force on Program Review. Improved program delivery: environment. Supply and Services Canada, Ottawa, Ont. Cat. No. CP32-50/14-1985E. 256 p.

TAYLOR, P. W. 1981. The ethics of respect for nature. Environ. Ethics 3: 197–218.

VALLENTYNE. J. R. 1978. Today is yesterday's tomorrow. Verh. Int. Ver. Limnol. 21: 1–12.
          1981. The ecosystem approach to planning, research and management in the Great Lakes Basin. Verh. Int. Ver. Limnol. 21: 1749–1752.

VALLENTYNE, J. R., AND H. L. TRACY. 1972. New term introduced at the First Conference on the Environmental Future. Biol. Conserv. 4(5): 371–372.

VERNADSKY, V. I. 1926. Biosfera (in Russian). Leningrad. 146 p. La Biosphère (in French, 1929). Alcan, Paris. 232 p.
          1945. The biosphere and the noosphere. Am. Sci. 33(1): 1–12.

VOLLENWEIDER, R. A. 1968. Scientific fundamentals of the eutrophication of lakes and flowing waters, with particular reference to nitrogen and phosphorus as factors in eutrophication.

OECD,DAS/CSI/68.27, 1-159. (Available from the Organization for Economic Cooperation and Development, Paris.)

WORLD COMMISSION ON ENVIRONMENT AND DEVELOPMENT. 1985. Mandate for change: key issues, strategies and work plan. 43 p. (Available from the World Commission on Environment and Development, Palais Wilson, CH-1201, Geneva, Switzerland.)